Springer Series in Astrophysics and Cosmology

The series covers all areas of astrophysics and cosmology, including theory, observations, and instrumentation. It publishes monographs and edited volumes. All books are authored or edited by leading experts in the field and are primarily intended for researchers and graduate students.

Ali Akil · Cosimo Bambi
Editors

The Black Hole Information Paradox

A Fifty-Year Journey

Springer

Editors
Ali Akil
Department of Computer Science
University of Hong Kong
Hong Kong SAR, China

Cosimo Bambi
Department of Physics
Fudan University
Shanghai, China

ISSN 2731-734X ISSN 2731-7358 (electronic)
Springer Series in Astrophysics and Cosmology
ISBN 978-981-96-6169-5 ISBN 978-981-96-6170-1 (eBook)
https://doi.org/10.1007/978-981-96-6170-1

This Springer imprint is published by the registered company Springer Nature Singapore Pte Ltd.
The registered company address is: 152 Beach Road, #21-01/04 Gateway East, Singapore 189721, Singapore

Preface

Fifty years have passed since Stephen Hawking published his seminal papers, first showing that black holes emit radiation, then showing that this radiation, being thermal, may lead to a paradox, a problem, the loss of information encoded in the matter that fell into the black hole. In the early years after the paper was published, the reaction to the computations and approximations therein was generally highly skeptical. Later, several alternative derivations were drawn and they mostly reached the same conclusions. Then a plethora of approaches to resolving the paradox was triggered. A tremendous number of pathways were explored, which contributed to the understanding of both gravity and quantum theory, as both are heavily involved in black hole physics. Sometimes this even helped understand phenomena that were not so directly connected to the paradox.

In general, the literature on the black hole information paradox is huge, and the background it requires is enormous, thus probably there is no way one author or perhaps even five could cover it all. This may explain why after 50 years of research on the topic, there is not yet a comprehensive book covering more than one approach to the paradox. This led us to contact many of the world's experts in every direction and approach that we know of, and ask for their contribution to the book that will cover most of the different perspectives. We are very grateful (and delightfully surprised) that most of them kindly agreed and generously contributed a chapter. We regret that some important approaches are not covered by this book (e.g., Soft Charges, Black-To-White hole tunneling, and a few more), but we can proudly say that we have managed to cover the great majority of the chapters that we wanted.

In our opinion, the importance of having such a book cannot be overstated. This makes it possible for a curious reader to open one table of contents and find most of the resolution attempts all at once. Moreover, this helps preventing the temptation to stick to one direction and live inside it. This also facilitates comparative studies and makes open-mindedness one step easier. We tried to include every possible topic that could be written, and emphasize pedagogical writing for as much as the space and time allow. Some chapters are review chapters, covering a wide existing literature under one approach, whereas some others are original contributions, published for the first time. We hope that the reader will learn and benefit from the book. We hope that this book, if it does not retrieve information from stellar black holes, it at least retrieves the information that is lost in the black hole of the impenetrable literature. We hope as well that this can trigger a discussion that is more informed until eventually we can all say confidently that the paradox or information loss problem is resolved.

Last, we would like to thank all authors who enthusiastically agreed to contribute to this volume. Their dedication and hard work have been pivotal in making this project successful. We value their expertise and are grateful for their generous contribution of knowledge and time.

Shanghai, China
March 2025

Ali Akil
Cosimo Bambi

Contents

Chapter 1
Deriving the Paradox: Original Derivation of Hawking Radiation

Mariano Cadoni, Salvatore Mignemi, Mirko Pitzalis, and Andrea P. Sanna

Abstract We revisit Hawking's original derivation of the evaporation process in a non-stationary spacetime, presenting it in a clear and pedagogical manner, with a focus on the spherical collapse of a star into a black hole. Our analysis highlights the underlying assumptions in the calculations, clarifying their physical significance, potential implications, and the limitations of this approach.

1.1 Introduction

Hawking's original derivation of black-hole thermal emission is almost fifty years old [1]. To date it stands as a cornerstone of theoretical physics, combining physical intuition with mathematical rigour, each complementing the other as the derivation grows increasingly complex.

Hawking's paper not only firmly established black-hole thermodynamics on solid physical grounds, elevating it from a mere analogy, but also paved the way for the "golden age" of quantum field theory (QFT) in curved spacetime. It sparked groundbreaking ideas and developments among the physics community, even beyond black-hole physics, ranging from the Fulling-Davies-Unruh effect [2–4], cosmological particle creation [5–9], quantum information theory up to black-hole analogues

---

M. Cadoni (✉) · M. Pitzalis · A. P. Sanna
Dipartimento di Fisica, Università di Cagliari, Cittadella Universitaria, Monserrato, Italy
e-mail: mariano.cadoni@ca.infn.it

M. Pitzalis
e-mail: mirko.pitzalis@ca.infn.it

M. Cadoni · S. Mignemi · M. Pitzalis · A. P. Sanna (✉)
I.N.F.N, Sezione di Cagliari, Cittadella Universitaria, Monserrato, Italy
e-mail: asanna@dsf.unica.it

S. Mignemi
e-mail: smignemi@unica.it

S. Mignemi
Dipartimento di Matematica, Università di Cagliari, Cagliary, Italy

A. Akil and C. Bambi (eds.), *The Black Hole Information Paradox*, Springer Series in Astrophysics and Cosmology, https://doi.org/10.1007/978-981-96-6170-1_1

in condensed matter physics [10–16]. Hawking radiation also lies at the root of one of the most challenging open problems in fundamental theoretical physics: the black-hole information problem [10–12, 17–19].

The literature now offers several derivations of the Hawking effect, covered here in Chap. 2. Each of them employs different setups for the background geometry, ranging from collapsing bodies to eternal black holes. Some of them also differ in their characterization of the quantum effects at play (like, for instance, conformal anomalies). Additionally, derivations that address the information problem and concepts like complementarity [20–24], such as the Firewall proposal [25], typically use a set of quantum-field modes that are purely ingoing at the horizon, whereas the original Hawking derivation employs ingoing modes at null past infinity.

Here we present a revisited version of the original Hawking derivation. We closely follow the line of reasoning of the original derivation, modifying it only in some minor technical aspects, aiming at enhancing its pedagogical clarity and simplifying certain details that are not essential for the understanding of the whole subject.

In the following sections, we will first discuss the physical framework (Sect. 1.2), followed by an introduction to the basics of QFT in curved spacetime (Sect. 1.3). Finally, we will present the derivation of the Hawking effect (Sect. 1.4).

1.2 Hawking Radiation: Physical Framework

The strongest argument in favor of black-hole thermal emission is provided by black-hole thermodynamics. In the stationary uncharged case, the mass M, the horizon area A and the angular momentum J satisfy the differential relation [26]

$$\mathrm{d}M = \frac{\mathscr{K}}{8\pi}\mathrm{d}A + \Omega \mathrm{d}J, \tag{1.1}$$

where $\mathscr{K}$ is the surface gravity at the event horizon and Ω the black-hole angular velocity. This expression is analogous to the first law of thermodynamics if one identifies A as the black-hole entropy (as first conjectured by Bekenstein [27, 28]) and $\mathscr{K}/2\pi$ as the temperature.[1] Furthermore, as argued by Jacobson in [29, 30], this is not merely an analogy: a black hole must genuinely behave as a thermodynamic system. If this is true, consistency with the basic principles of thermodynamics and statistical mechanics requires a black hole to be capable of both emitting and absorbing particles. The primary conceptual challenge in this reasoning, as opposed to a standard black body, lies in the lack of a microscopic description of black holes. Hawking's striking physical intuition was that, being gravity linked to curvature, particle creation by a black hole can be understood as a consequence of the dynamics of quantum fields in a curved spacetime. Qualitatively, particle creation in a region of spacetime with curvature of the order of R can be understood as resulting from

[1] Throughout the chapter, we will adopt units in which $G = \hbar = c = k_{\mathrm{B}} = 1$.

the uncertainty in the particle number operator $a^\dagger a$ for modes with wavelengths larger than $\sim R^{-1/2}$. This arises because defining positive frequency modes requires a local inertial coordinate system, which can only be set up within a region of size $\sim R^{-1/2}$. The uncertainty in the particle number operator leads to an uncertainty in the local energy density of order R^2. This uncertainty, therefore, can be thought of as corresponding to the local energy density of particles created by the gravitational field. As long as $R^{-1/2}$ remains much larger than the Planck length ℓ_P, Einstein's equations imply that the corresponding uncertainty in the spacetime curvature due to this energy density uncertainty can be safely neglected. Consequently, the semi-classical description provides an ideal framework for discussing particle creation by macroscopic black holes. In this framework, gravity is treated classically, while the matter fields are quantized on the curved black-hole background. Following the same reasoning, the gravitational background of a large black hole can be treated as fixed, allowing backreaction effects on the geometry to be neglected as a first approximation.

A consequence of particle emission is the decrease in the black-hole mass and in the area of the event horizon.[2] This process can be understood as arising from a negative energy flux across the event horizon, balancing the positive energy flux emitted to infinity. Heuristically, the negative energy flux is due to the tunneling of negative energy particles created from the continuous spontaneous creation of virtual particle-antiparticle pairs around the black hole. These particles emerge from a classically forbidden region, tunnel through the event horizon and enter the black hole, where the time-translation Killing vector becomes spacelike. This allows negative energy particles to exist as real particles with a timelike momentum vector having negative energy relative to infinity, as measured by the time translation Killing vector. The positive energy partner of the virtual pair can escape to infinity, where it contributes to the thermal emission. The probability of the negative energy particle tunneling through the event horizon is governed by the surface gravity $\mathcal{K}$, as this quantity measures how fast the Killing vector becomes spacelike. Virtual particle pairs created with wavelength λ separate temporarily to a distance $\sim\lambda$. For λ of the order of the black-hole mass, strong tidal forces prevent re-annihilation, allowing the black hole to radiate quanta with wavelengths of order of M. The thermal emission carries away energy, causing the black hole to lose mass.

However, in most cases, particle creation is locally negligible. The black-hole temperature is roughly $T \sim 10^{-6}$ $(M_\odot/M)$ K, meaning that for solar-mass black holes, it is much lower than that of the Cosmic Microwave Background. As a result, these black holes absorb more radiation than they emit. It may nonetheless lead to significant global effects when considered over time scales comparable to the lifetime of the Universe. For smaller black holes, for instance, such as those potentially

[2] Black-hole evaporation represents a violation of the area theorem [31], which implies that the horizon area can never decrease in any dynamical process. It is important to stress, however, that the theorem rests on the validity of the weak energy condition, which is violated during quantum thermal emission. This violation can be understood as a consequence of the indeterminacy of particle number and energy density in a curved spacetime mentioned above.

formed in the early Universe [32, 33], the Hawking process can drive their complete evaporation.

Another remarkable aspect of the Hawking effect, which can be inferred from general principles, is the universality of the radiation flux. Surprisingly, it is independent of the details of the gravitational collapse and the form of the quantum matter fields. This property can be logically deduced from the no-hair theorems [34–36]. This indicates that the key factor is the event-horizon structure. This universality, however, holds strictly only when greybody factors are neglected, as we will assume in the following sections.

This universality can also be leveraged to simplify calculations by considering the simplest setup: massless scalar quantum fields in the simplest spherically-symmetric collapsing geometry, the Vaidya spacetime. To describe the semiclassical dynamics of quantum fields, it is necessary to define a complete set of purely ingoing and outgoing modes, each containing only positive frequencies relative to the time coordinate used by the corresponding observer. There are several methods to achieve this. In the next section, we will adopt the approach originally used by Hawking [1]. We define:

(a) a complete set of purely ingoing modes ϕ^{in}, containing only positive frequencies with respect to the affine parameter on past null infinity $\mathscr{I}^-$. These are completely determined by their Cauchy data on $\mathscr{I}^-$;
(b) a complete set of modes constructed from purely outgoing ϕ^{out} and internal modes ϕ^{int}. The former contains only positive frequencies relative to the affine parameter at future null infinity $\mathscr{I}^+$ and has zero Cauchy data on $\mathscr{I}^-$. The latter has Cauchy data on the event horizon, but zero data on both $\mathscr{I}^-$ and $\mathscr{I}^+$.

As long as one is concerned with calculating the outgoing flux of Hawking particles at late times, the ϕ^{int} modes seem to be completely irrelevant. Particle creation is a global process not localized on the event horizon. The key to understanding this lies in the relationship between the "in" and "out" modes, governed by a Bogoliubov transformation. This transformation underpins the entire framework: from Hawking radiation to the information paradox.

Furthermore, as long as the flux of Hawking particles at $\mathscr{I}^+$ is considered, a different set of modes can be used, i.e., ϕ^{out} and ϕ_h^{in} instead of ϕ^{in} and ϕ^{int}. The modes ϕ_h^{in} have positive frequency relative to a free-falling observer near the event horizon. This setup is commonly adopted in recent formulations addressing the information problem, like those leading to the Firewall hypothesis [25]. This approach will not be employed in the derivation contained in the following sections.

Hawking's derivation is far from being devoid of conceptual issues, as it is known to be plagued by the so-called transplanckian problem [37–40]. As detailed in Sect. 1.4, Hawking's results involve integrating outgoing modes through the collapsing body. This is unphysical due to blue-shift effects. Nevertheless, Hawking recognized that the essential element of his derivation was the horizon structure, implying that the results should be independent of ultraviolet (UV) gravitational physics.

In the modern framework based on the modes ϕ^{out} and ϕ_h^{in}, this independence from UV physics is rephrased in terms of the adiabatic principle. By considering a free-falling observer near the horizon and foliating the near-horizon geometry with smooth slices, it becomes clear that the geometry evolves slowly compared to the blue-shifted frequency of the Hawking modes at $\mathscr{I}^+$. Consequently, according to the adiabatic principle, these modes must remain in their ground state relative to the infalling observer frame. Even though, to date, a complete description of UV gravitational physics remains elusive, as long as the adiabatic and equivalence principles hold, the extreme blue-shift remains irrelevant, and all relevant information is encoded in the event-horizon structure.

In the following sections, we will go through the derivation of the Hawking effect using the setup elucidated in this section. For the technical computation, we will mainly follow Ref. [41].

1.3 Quantization in Curved Spacetime

In a curved spacetime, unlike in the flat Minkowski one, Poincaré symmetry is no longer preserved. As a consequence, one cannot globally define field modes with positive frequency across the entire spacetime, because in curved spacetime they lose their invariant meaning, as also the notion of vacuum or particle number. Two local free-falling observers at two different points in spacetime will each employ distinct timelike coordinates, leading to two different notions of positive frequency modes. Thus, the standard quantization procedure must be extended to accommodate curved geometries. The minimal generalization is as follows

- We consider the dynamics of matter fields in a fixed curved gravitational background, focusing specifically on massless scalar fields ϕ. At the classical level, ϕ satisfies the standard wave equation, with the Minkowski metric replaced by a classical spacetime metric $g_{\mu\nu}$, solution of Einstein's equations. Therefore, the Klein-Gordon (KG) equation is generalized as usual by simply replacing partial derivatives with covariant derivatives

$$\Box\phi = g^{\mu\nu}\nabla_\mu\nabla_\nu\phi = 0; \tag{1.2}$$

- Assuming the spacetime to be globally hyperbolic ensures the existence of an initial data Cauchy hypersurface Σ, on which the KG inner product can be generalized and computed as follows

$$(\phi_1, \phi_2) = -\mathrm{i}\int_\Sigma \mathrm{d}\Sigma^\mu \left(\phi_1\partial_\mu\phi_2^* - \phi_2^*\partial_\mu\phi_1\right), \tag{1.3}$$

  where $\mathrm{d}\Sigma^\mu = \mathrm{d}\Sigma\, n^\mu$, with $\mathrm{d}\Sigma$ representing the spacelike volume element, while n^μ the future-directed unit normal vector to Σ.

  A key property of this inner product is that it is *independent* of the choice of Σ. To see this, consider two hypersurfaces Σ and Σ', enclosing a four-dimensional volume S. The difference between the inner product computed on Σ and on Σ' can be expressed as a volume integral over S, thanks to Gauss' law. Using the KG Eq. (1.2), one finds $(\phi_1, \phi_2)_\Sigma - (\phi_1, \phi_2)_{\Sigma'} = 0$, confirming the assertion.

In flat spacetime, the standard quantization procedure involves decomposing the field ϕ, solution of the KG equation $\eta^{\mu\nu}\partial_\mu\partial_\nu\phi = 0$, into positive and negative frequency modes

$$\phi = \sum_i \left(\phi_i \, a_i + \phi_i^* \, a_i^\dagger\right), \tag{1.4}$$

where the $\{\phi_i\}$ form a complete orthonormal set of solutions to the KG equation. Here, a_i and $a_i^\dagger$ represent the annihilation and creation operators, respectively, for particles in the i-th state. The vacuum state is uniquely defined as the state for which $a_i|0\rangle = 0$ for all i.

In curved spacetime, however, the splitting between positive and negative frequency modes is not unique, leading to the absence of a single, well-defined vacuum state. Different choices of positive frequency modes result in different vacuum states and hence a different Fock space. Indeed, since generically there will not be any timelike Killing vector field, one will not in general be able to find solutions to the wave equation that separate into time-dependent and space-dependent factors, and correspondingly will not be able to classify modes into positive- or negative-frequency ones. Nevertheless, in stationary spacetime, we can still select a natural set of positive frequency solutions. In this case, a timelike Killing vector field ξ^μ exists, leaving the metric invariant under infinitesimal transformations expressed as $\delta_\xi g_{\alpha\beta} = 0$. This allows for a global definition of positive frequency modes as those satisfying

$$\xi^\mu \nabla_\mu \phi_j = -\mathrm{i}\omega_j \phi_j, \qquad \omega_j > 0. \tag{1.5}$$

In more general settings, where the spacetime is non-stationary (for instance during gravitational collapse) or when the timelike Killing vector field is not globally defined (e.g., in the presence of an event horizon), it is no longer possible to define globally a set of positive frequency modes. Even in a general case, one can still have a natural choice of positive frequency modes for those spacetimes possessing asymptotic stationary regions in the past and in the future, commonly referred to as "in" and "out" regions, respectively. There, one can construct orthonormal sets of modes, which are exact solutions of the wave equation, with positive frequency with respect to the inertial time in the past $\left\{\phi_i^{in}\right\}$ and in the future $\left\{\phi_i^{out}\right\}$.

A key consequence of not having a unique, globally-defined vacuum state is that, if quantum fields start in the vacuum state as observed by an inertial observer in the "in" region $|in\rangle$, the changing, non-stationary gravitational field causes these fields to no longer appear in the vacuum state for an inertial observer in the "out" region at later times. This phenomenon can be precisely quantified by resorting to the Bogoliubov transformations.

1.3.1 Bogoliubov Transformations

Let us expand the scalar field ϕ in the basis of the positive frequency solutions $\{\phi_i^{in}\}$ in the "in" stationary region

$$\phi = \sum_i \left[a_i^{in} \phi_i^{in} + a_i^{\dagger in} \phi_i^{in*} \right], \tag{1.6}$$

while the same expansion can be performed in the "out" stationary region, in terms of the basis $\{\phi_i^{out}\}$

$$\phi = \sum_i \left[a_i^{out} \phi_i^{out} + a_i^{\dagger out} \phi_i^{out*} \right]. \tag{1.7}$$

As discussed in Sect. 1.2, the "out" region represents a Cauchy hypersurface only if we include the event horizon. In other words, a complete set of modes in the "out" region is represented by purely outgoing modes ϕ^{out} and modes in the interior of the black hole ϕ^{int}. However, we will neglect the latter, as they do not alter particle production in the "out" region.

The modes satisfy the relations, both valid for "in" and "out" modes (we omit the superscripts)

$$\left(\phi_i, \phi_j\right) = -(\phi_i^*, \phi_j^*) = \delta_{ij}, \qquad (\phi_i, \phi_j^*) = 0, \tag{1.8}$$

while the creation and annihilation operators satisfy the commutation relations (again we drop the superscripts "in" and "out")

$$\left[a_i, a_j^\dagger\right] = \delta_{ij}, \qquad \left[a_i, a_j\right] = \left[a_i^\dagger, a_j^\dagger\right] = 0. \tag{1.9}$$

Since both sets of modes are complete, it is possible to expand one set in the basis of the other. However, it is not guaranteed that positive frequency solutions of one set can be expanded solely in terms of the pure positive frequency solutions of the other. As a result, the expansion will generally involve a mixture of both positive and negative modes, with the expansion coefficients given by the Bogoliubov matrices α and β, namely

$$\phi_j^{out} = \sum_i \left(\alpha_{ji}\, \phi_i^{in} + \beta_{ji}\, \phi_i^{in*}\right). \tag{1.10}$$

Using the orthonormal relations (1.8), we can invert the above to write

$$\alpha_{ij} = \left(\phi_i^{out}, \phi_j^{in}\right), \qquad \beta_{ij} = -\left(\phi_i^{out}, \phi_j^{in*}\right). \tag{1.11}$$

Equation (1.8) can also be used to write the following relations between the Bogoliubov coefficients, which will be essential in deriving the spectrum of Hawking radiation

$$\sum_k \left(\alpha_{ik}\alpha^*_{jk} - \beta_{ik}\beta^*_{jk}\right) = \delta_{ij};$$
$$\sum_k \left(\alpha_{ik}\beta_{jk} - \beta_{ik}\alpha_{jk}\right) = 0. \tag{1.12}$$

We can also invert Eq. (1.10) and write

$$\phi_i^{in} = \sum_j \left(\alpha^*_{ji}\,\phi_j^{out} - \beta_{ji}\,\phi_j^{out*}\right). \tag{1.13}$$

Using the previous relations, together with the fact that, from Eqs. (1.6) and (1.7), we also have $a_i^{in} = (\phi,\ \phi_i^{in})$ and $a_i^{out} = (\phi,\ \phi_i^{out})$, respectively, we can expand one of the two sets of creation and annihilation operators in terms of the other

$$a_i^{in} = \sum_j \left(\alpha_{ji}\, a_j^{out} + \beta^*_{ji}\, a_j^{\dagger out}\right);$$
$$a_i^{out} = \sum_j \left(\alpha^*_{ij}\, a_j^{in} - \beta^*_{ij}\, a_j^{\dagger in}\right). \tag{1.14}$$

As always, the "in" and "out" vacuum states are defined as $a_i^{in}|in\rangle = 0$ and $a_i^{out}|out\rangle = 0$ for all i, respectively. We can exploit this to compute the expectation value of the "out" particle number operator for the i-th mode

$$N_i^{out} \equiv a_i^{\dagger out} a_i^{out}, \tag{1.15}$$

in the state $|in\rangle$

$$\begin{aligned}\langle in|N_i^{out}|in\rangle &= \langle in|a_i^{\dagger out} a_i^{out}|in\rangle \\ &= \langle in|\sum_j \left(-\beta_{ij}a_j^{in}\right)\sum_k \left(-\beta^*_{ik}\, a_k^{\dagger in}\right)|in\rangle \\ &= \sum_j |\beta_{ij}|^2 .\end{aligned} \tag{1.16}$$

Thus, if the β_{ij} coefficients are non-zero, the particle content of the $|in\rangle$ vacuum state, with respect to the "out" Fock space, is non-trivial. Thus, what appears as an empty vacuum from one perspective is seen as teeming with particles from another. This exactly corresponds to particle creation in the "out" region from the vacuum in the "in" region. Conversely, if all β_{ij} vanish, we have, from Eq. (1.12) that $\sum_k \alpha_{ik}\alpha^*_{jk} = \delta_{ij}$, meaning that the positive frequency mode bases $\{u_i^{in}\}$ and $\{u_i^{out}\}$ are related by a unitary transformation. In this case, the definition of the vacuum remains unchanged, i.e., $|in\rangle = |out\rangle$.
The main result is that the β_{ij} coefficients are the essential ingredient to determine the number of particles created by the gravitational field during the gravitational

collapse and detected by an observer in the asymptotic region. As we will see, the Bogoliubov coefficients have a simple and *universal* asymptotic form, depending only on the surface gravity of the newly formed black hole.

1.4 Hawking Radiation from a Collapsing Body

1.4.1 The Vaidya Spacetime

To understand the origin of particle creation, it is essential to consider not only the quasi-stationary final state of the black hole, but also the time-dependent formation phase. The Vaidya spacetime provides the simplest solution of Einstein's equations describing a spherically-symmetric collapse leading to black-hole formation. In Eddington-Finkelstein coordinates, the metric is given by

$$\mathrm{d}s^2 = -\left(1 - \frac{2M(v)}{r}\right)\mathrm{d}v^2 + 2\mathrm{d}v\mathrm{d}r + r^2\mathrm{d}\Omega^2, \quad \mathrm{d}\Omega^2 = \mathrm{d}\theta^2 + \sin^2\theta\mathrm{d}\varphi^2, \tag{1.17}$$

where v is the advanced time. When $M(v) = \text{constant} \equiv M \neq 0$, we recover the static Schwarzschild spacetime, whereas when $M(v) = 0$ the metric reduces to the flat Minkowski spacetime. Instead, if the mass is solely a function of v, the stress-energy tensor sourcing the solution takes the form

$$T_{vv} = \frac{L(v)}{4\pi r^2}, \qquad \frac{\mathrm{d}M}{\mathrm{d}v} = L(v), \tag{1.18}$$

describing a purely ingoing radial flux of radiation and representing the simplest case of gravitational collapse. In more realistic scenarios, there would also be an outgoing flux as well as non-spherical deformations. We assume, additionally, that the influx radiation is turned on at some finite advanced time v_i and turned off at v_f. In this way, we can divide the spacetime into three regions

1. For $v < v_i$, $M(v) = 0$ and we have flat Minkowski spacetime;
2. For $v_i < v < v_f$, $M(v)$ grows and we have the collapse region;
3. For $v > v_f$, $M(v) = M$ and we have the final Schwarzschild black hole.

The original analysis considers the complete gravitational collapse of a spherical distribution of matter. As noted by Hawking, however, only the regions outside the collapsing matter in the asymptotic future are relevant. While the exact form of the Bogoliubov coefficients will depend somewhat on the details of the gravitational collapse, their asymptotic form is completely determined by the surface gravity of the black hole. Instead, the modes of any quantum field propagating through the interior of the collapsing body will be severely disrupted; thus, even though some particle creation will depend on the details of the collapse, these particles will disperse. In conclusion, only the first and last regions above are important and, at late retarded

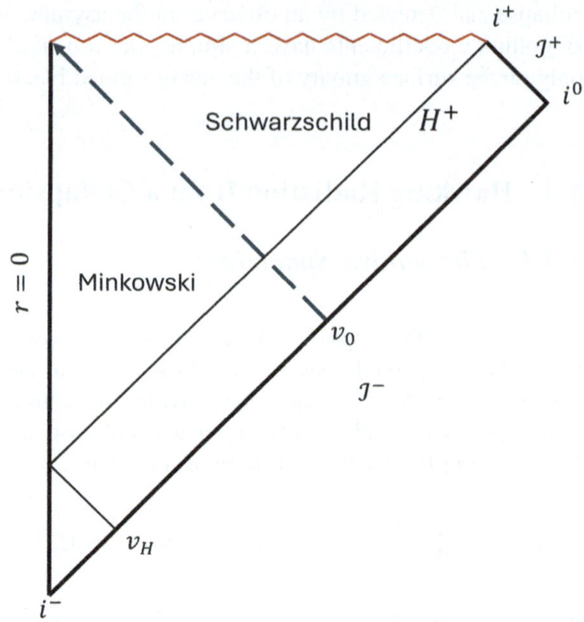

Fig. 1.1 Penrose diagram of the Vaidya spacetime. A Schwarzschild solution at late times is jointed with a Minkowski one at early times along the shock-wave at advanced time $v = v_0$ of the ingoing flux of matter (blue dashed arrow)

times at $\mathscr{I}^+$, there will be a steady flux of particles fully determined by the asymptotic form of the Bogoliubov coefficients. The discussion above, therefore, allows one to make a further simplifying assumption: the region 2 can be freely narrowed down to an idealized single null surface. This is equivalent to treating the ingoing flux as a shock-wave at some $v = v_0$, namely

$$L(v) = M\delta(v - v_0), \qquad M(v) = M\theta(v - v_0). \tag{1.19}$$

Thus, the full solution is obtained by jointing the Minkowski and Schwarzschild spacetimes across the shock-wave. The Minkowski region corresponds to the "in" region, while the Schwarzschild one represents the "out" region (see Fig. 1.1). Introducing a shockwave that connects a Schwarzschild spacetime to Minkowski spacetime does not alter the original analysis, but instead provides a clearer definition of the initial conditions.

1.4.2 Massless Fields and Wave Equations

We consider the simplest case of matter fields propagating in the Vaidya spacetime: a massless scalar field, obeying the KG Eq. (1.2). Given that the background geometry is spherically symmetric, the mode solution of the wave Eq. (1.2) can be decomposed into time, radial and angular components, with the latter expressed by spherical harmonics $\mathscr{Y}_{lm}(\theta, \varphi)$

$$\phi(t, r, \theta, \varphi) = \sum_{l,m} \frac{\phi_l(t, r)}{r} \mathscr{Y}_{lm}(\theta, \varphi). \tag{1.20}$$

Neglecting the angular part, the problem reduces to a two-dimensional one in t and r coordinates. Since we have two jointed regions, we will have two corresponding equations

- In the "in" Minkowski region $v < v_0$, we have

$$\left[-\partial_t^2 + \partial_r^2 - \frac{l(l+1)}{r^2}\right] \phi_l(t, r) = 0; \tag{1.21}$$

- In the "out" Schwarzschild region $v > v_0$, we instead have

$$\left[-\partial_t^2 + \partial_{r_*}^2 - \mathscr{V}_l(r)\right] \phi_l(t, r) = 0, \tag{1.22}$$

  where r_* is the tortoise coordinate

$$r_* = \int \frac{\mathrm{d}r}{1 - \frac{2M}{r}} = r + 2M \ln\left|\frac{r}{2M} - 1\right|, \tag{1.23}$$

  while $\mathscr{V}_l(r)$ is the effective potential

$$\mathscr{V}_l(r) \equiv \left(1 - \frac{2M}{r}\right)\left[\frac{l(l+1)}{r^2} + \frac{2M}{r^3}\right]. \tag{1.24}$$

Note that the potential vanishes both at the event horizon $r = 2M$ ($r_* \to -\infty$) and at spatial infinity $r \to \infty$ ($r_* \to +\infty$). Since the relevant physics occurs near the horizon, we introduce a further simplification and neglect the potential everywhere. This approximation is justified within the geometric optics approximation and is particularly well-suited for the "s-wave" component, i.e., the one with $l = 0$, as this mode is the least affected by the potential.

Thus, within this approximation, we deal with free KG equations in both regions, the only difference being the form of the radial coordinate. We are, therefore, left with the following two equations

- For $v < v_0$

$$\left(-\partial_t^2 + \partial_r^2\right) \phi(t, r) = 0, \tag{1.25}$$

  with the boundary condition $\phi(t, r = 0) = 0$ to ensure regularity (notice that we dropped the subscript l in ϕ_l since we are considering only the s-wave sector).
- For $v > v_0$

$$\left(-\partial_t^2 + \partial_{r_*}^2\right) \phi(t, r) = 0. \tag{1.26}$$

We will allow only positive frequency modes. Thus, we assume the following harmonic dependence on time

$$\phi(t,r) = e^{-\mathrm{i}\omega t}\phi(r), \tag{1.27}$$

and the KG equations become

$$\begin{aligned} \frac{\mathrm{d}^2\phi}{\mathrm{d}r^2} + \omega^2\phi(r) = 0, \qquad &\text{“in” region} \\ \frac{\mathrm{d}^2\phi}{\mathrm{d}r_*^2} + \omega^2\phi(r) = 0, \qquad &\text{“out” region} \end{aligned} \tag{1.28}$$

The solutions of these equations are better expressed in terms of null coordinates $u = t - r$ and $v = t + r$. Since we have two different spacetimes across the shockwave, we will have different transformations

- In the Minwkoski “in” region, we define $u_{in} = t_{in} - r_{in}$ and $v = t_{in} + r_{in}$ and the metric becomes

$$\mathrm{d}s_{in}^2 = -\mathrm{d}u_{in}\mathrm{d}v + r_{in}^2\mathrm{d}\Omega^2. \tag{1.29}$$

- In the Schwarzschild “out” region, we instead define $u_{out} = t_{out} - r^*_{out}$ and $v = t_{out} + r^*_{out}$, and the metric reads

$$\mathrm{d}s_{out}^2 = -\left(1 - \frac{2M}{r_{out}}\right)\mathrm{d}u_{out}dv + r_{out}^2\mathrm{d}\Omega^2. \tag{1.30}$$

The solutions of the wave equations will consist of a combination of ingoing and outgoing waves. In both cases, the ingoing waves will have a time dependence of the form $\sim e^{-\mathrm{i}\omega v}$. The outgoing waves, however, will behave as $\sim e^{-\mathrm{i}\omega u_{in}}$ in the Minkowski region, and as $\sim e^{-\mathrm{i}\omega u_{out}}$ in the Schwarzschild one.

1.4.3 Ingoing and Outgoing Modes

We first define the “in” Fock space associated with the natural time parameter at $\mathscr{I}^-$. The complete set of positive frequency modes are

$$\phi_\omega^{in} = \frac{1}{4\pi\sqrt{\omega}}\frac{e^{-\mathrm{i}\omega v}}{r}. \tag{1.31}$$

These modes are completely determined by their Cauchy data on $\mathscr{I}^-$.

It is straightforward to show that they satisfy the normalization condition

$$(\phi_\omega^{in}, \phi_{\omega'}^{in}) = -\mathrm{i} \int_{\mathscr{I}^-} \mathrm{d}v \, \mathrm{d}\Omega^2 \, r^2 \left(\phi_\omega^{in} \partial_v \phi_{\omega'}^{in*} - \phi_{\omega'}^{in*} \partial_v \phi_\omega^{in}\right) = \delta(\omega - \omega'). \tag{1.32}$$

Then, we can construct the Fock space at $\mathscr{I}^+$, associated, instead, to the natural time parameter at $\mathscr{I}^+$, whose complete set of positive frequency modes is

$$\phi_\omega^{out} = \frac{1}{4\pi\sqrt{\omega}} \frac{e^{-\mathrm{i}\omega u_{out}}}{r}. \tag{1.33}$$

These modes have zero Cauchy data on $\mathscr{I}^-$ and satisfy the normalization condition

$$(\phi_\omega^{out}, \phi_{\omega'}^{out}) = -\mathrm{i} \int_{\mathscr{I}^+} \mathrm{d}u_{out} \, \mathrm{d}\Omega^2 \, r^2 \left(\phi_\omega^{out} \partial_{u_{out}} \phi_{\omega'}^{out*} - \phi_{\omega'}^{out*} \partial_{u_{out}} \phi_\omega^{out}\right) = \delta(\omega - \omega'). \tag{1.34}$$

$\mathscr{I}^+$ is not a proper Cauchy surface in the "out" region, but we must add also the future horizon H^+. In other words, the modes ϕ_ω^{out} do not form a complete space. To construct the complete "out" Fock space, we must add the modes that cross the future horizon ϕ_ω^{int}. As first noted in Sect. 1.2, however, it turns out that the final expression describing particle production at future infinity is insensitive to the specific expression of these modes.

In Sect. 1.3.1, we have seen that the most important ingredient to describe particle production is the Bogoliubov coefficient β, given by Eq. (1.11). In this case, we have

$$\beta_{\omega\omega'} = -\left(\phi_\omega^{out}, \phi_{\omega'}^{in*}\right) = \mathrm{i} \int_{\mathscr{I}^-} \mathrm{d}v \, \mathrm{d}\Omega^2 \, r^2 \left(\phi_\omega^{out} \partial_v \phi_{\omega'}^{in} - \phi_{\omega'}^{in} \partial_v \phi_\omega^{out}\right). \tag{1.35}$$

Since the scalar product is independent of the choice of the Cauchy surface, for mathematical and physical convenience we choose to evaluate it at $\mathscr{I}^-$. However, to do so, we must compute the behavior of the outgoing modes ϕ_ω^{out} at $\mathscr{I}^-$. This requires tracing the modes backward in time, which introduces certain difficulties and leads to the "transplanckian problem" discussed in Sect. 1.2.

The form of the Bogoliubov coefficients in the region $v > v_0$ at $\mathscr{I}^+$ is always given by Eq. (1.33). To determine their form in the Minkowski region $v < v_0$, we can impose two conditions:

1. Matching along the shock-wave $v = v_0$. Just before this, the form of the "out" modes in the u_{in} coordinates reads

$$\phi_\omega^{out} = \frac{1}{4\pi\sqrt{\omega}} \frac{e^{-\mathrm{i}\omega u_{out}(u_{in})}}{r}. \tag{1.36}$$

We have to determine $u_{out}(u_{in})$ by matching the two metrics along the shock-wave. Requiring continuity across $v = v_0$ implies $r(v_0, u_{in}) = r(v_0, u_{out})$, where, using the relations between (t, r) and (u, v)

$$
\begin{aligned}
r(v_0, u_{in}) &= \frac{v_0 - u_{in}}{2}; \\
r(v_0, u_{out}) &= \frac{v_0 - u_{out}}{2} - 2M \ln\left[\frac{r(v_0, u_{out})}{2M} - 1\right].
\end{aligned}
\tag{1.37}
$$

Equating the above expressions gives

$$
u_{out} = u_{in} - 4M \ln \frac{|v_0 - 4M - u_{in}|}{4M}. \tag{1.38}
$$

2. The regularity condition at $r = 0$ forces this form of the modes in the Minkowski part of the spacetime

$$
\phi_\omega^{out} = \frac{1}{4\pi\sqrt{\omega}}\left[\frac{e^{-\mathrm{i}\omega u_{out}(u_{in})}}{r} - \frac{e^{-\mathrm{i}\omega u_{out}(v)}}{r}\theta(v_{\mathrm{H}} - v)\right], \tag{1.39}
$$

where $v_{\mathrm{H}} \equiv v_0 - 4M$ is the location of the null ray that will form the horizon at $u_{out} \to \infty$.

We analyze the behavior of ϕ_ω^{out} in the two limiting regions. At early times, i.e., $v \to -\infty$, from Eq. (1.38), we have $u_{out}(v) \simeq v$. At $\mathscr{I}^-$, therefore, we have

$$
\phi_\omega^{out} \simeq -\frac{1}{4\pi\sqrt{\omega}}\frac{e^{-\mathrm{i}\omega v}}{r}. \tag{1.40}
$$

This is still a positive-frequency mode with respect to the affine parameter ν at $\mathscr{I}^-$. Substituting this into $\beta_{\omega\omega'}$ in (1.35) yields zero. Consequently, at early times and asymptotically, no particle emission reaches $\mathscr{I}^+$.

As u_{out} increases, instead, at late times $u_{out} \to +\infty$ (corresponding to $v \to v_{\mathrm{H}}$), using Eq. (1.38) we have

$$
\phi_\omega^{out} \simeq -\frac{1}{4\pi\sqrt{\omega}}\frac{e^{-\mathrm{i}\omega\left(v_{\mathrm{H}} - 4M \ln \frac{v_{\mathrm{H}} - v}{4M}\right)}}{r}\theta(v_{\mathrm{H}} - v). \tag{1.41}
$$

The infinite blueshift in the exponent and the existence of the critical time v_{H}, typical of processes involving black hole formation, will be responsible for the presence of the outgoing flux at $\mathscr{I}^+$. The equation above is a superposition of positive and negative frequency modes with respect to the inertial time at $\mathscr{I}^-$, which is at the basis of the result of Hawking.

We can now provide a better intuitive argument to justify the existence of Hawking radiation. Consider an asymptotic observer in the initial region sending a light ray with very large frequency before the advanced time ν_H. This light ray propagates through the Minkowski region without modifying its frequency ω. Once it enters the

Schwarzschild region, crossing the dynamical line $\nu = \nu_0$, the geometry will alter the amplitude and the frequency of the wave. The effective frequency of the outgoing modes measured by the inertial observer at $\mathscr{I}^+$ will suffer an exponentially increasing redshift

$$\omega \sim \omega_r e^{-u_{out}/4M} \tag{1.42}$$

with an e-folding time of $\kappa^{-1} = 4M$ which is the surface gravity of the Schwarzschild black hole. As a consequence of this gravitational redshift, the energy of the outgoing wave is much smaller than that of the incoming wave, without having increased the black-hole internal energy. To reestablish energy conservation, an additional emission of particles, carrying the necessary energy, is required. This should correspond to a process of stimulated emission, which in turn suggests the existence of a mechanism of spontaneous emission, corresponding to the Hawking process.

Thus, the non-stationary character of the metric is essential for the Hawking effect. Moreover, the presence of the event horizon is also crucial. Without it, the redshift is not divergent and the energy deficit in the process would be very small.

1.4.4 Wave Packets and Computation of the Bogoliubov Coefficients

The quantity of interest is the expectation value of the number of particles with a given frequency emitted at $\mathscr{I}^+$, which, according to Eq. (1.16), is given by

$$\langle in|N_\omega^{out}|in\rangle = \int_0^\infty \mathrm{d}\omega' \, |\beta_{\omega\omega'}|^2. \tag{1.43}$$

We assume that the quantum state of the matter fields is the natural vacuum $|in\rangle$ at $\mathscr{I}^-$. Considering the transplackian problem mentioned earlier, this assumption is quite strong. Although the matter field is in its natural vacuum at $\mathscr{I}^-$, the extreme blue-shift it experiences near the horizon could, in principle, bring it into a highly excited state. The validity of the adiabatic principle prevents this from happening (see Sect. 1.2).

The expression (1.43) gives the average number of created particles at $\mathscr{I}^+$ in the frequency range $(\omega, \omega + \mathrm{d}\omega)$. Since $|\beta_{\omega\omega'}|$ behaves like $(\omega')^{-1/2}$ for large ω', this integral has a logarithmic divergence. The collapsing ball produces a steady flux of radiation at $\mathscr{I}^+$, so the infinite total number of created particles corresponds to a finite steady rate of emission continuing for an infinite time. This is due to the fact that using states with a definite frequency implies an absolute uncertainty in time, meaning that the above expression provides the number of particles with frequency ω emitted at any time. Here, instead, we are more interested in the mean number of particles produced at late retarded time $u_{out} \to +\infty$, when the black hole has settled down to a stationary configuration. Therefore, for our purposes, it is more convenient to replace the plane-wave type modes with wave packets. The complete orthonormal set of wave packet modes at $\mathscr{I}^+$ is the following

$$\phi_{jn}^{out} = \frac{1}{\sqrt{\varepsilon}} \int_{j\varepsilon}^{(j+1)\varepsilon} \mathrm{d}\omega \, e^{2\pi \mathrm{i}\omega \frac{n}{\varepsilon}} \, \phi_{\omega}^{out}, \tag{1.44}$$

where n and $j \geq 0$ are integers, while ε is a parameter controlling the width of the wave packets. The latter are peaked around $u_{out} = \frac{2\pi n}{\varepsilon}$, with width $2\pi/\varepsilon$. Taking $\varepsilon \gg 1$ ensures that the modes are narrowly centered around $\omega \simeq \omega_j = j\varepsilon$. Therefore, $\langle N_{jn}^{out} \rangle$ counts the particles with frequencies in the range $(\omega_j - \varepsilon, \omega_j + \varepsilon)$, emitted over a time interval $2\pi/\varepsilon$ at time $u_{out} = 2\pi n/\varepsilon$. The late-time modes are characterized by a very large quantum number n, and these are the modes we must propagate backwards into the past and project onto $\mathscr{I}^-$.

To determine the number of particles created, we need to evaluate the coefficients

$$\beta_{jn,\omega'} = -\left(\phi_{jn}^{out}, \phi_{\omega}^{in*}\right) = \mathrm{i} \int_{\mathscr{I}^-} \mathrm{d}v \, \mathrm{d}\Omega^2 \, r^2 \left[\phi_{jn}^{out} \partial_v \phi_{\omega'}^{in} - \phi_{\omega'}^{in} \partial_v \phi_{jn}^{out}\right]. \tag{1.45}$$

Since ϕ_{jn}^{out} vanishes at $v \to \pm\infty$, we can perform a partial integration and discard the boundary term, yielding

$$\beta_{jn,\omega'} = 2\mathrm{i} \int_{\mathscr{I}^-} \mathrm{d}v \, \mathrm{d}\Omega^2 \, r^2 \phi_{jn}^{out} \partial_v \phi_{\omega'}^{in}. \tag{1.46}$$

Since we are primarily interested in particle production at late time, we will evaluate the above expression as $n \to \infty$. At large n, the wave packets at $\mathscr{I}^+$, when traced backwards in time, become localized within an infinitesimally small region around v_{H} at $\mathscr{I}^-$. Therefore, in the limit $n \to \infty$, we can use the expression given in Eq. (1.41). Substituting the latter into u_{jn}^{out}, we obtain the following expression for $\beta_{jn,\omega'}$

$$\begin{aligned}
\beta_{jn,\omega'} &= 2\mathrm{i} \int_{\mathscr{I}^-} \mathrm{d}v \, \mathrm{d}\Omega^2 r^2 \Bigg\{ \frac{1}{\sqrt{\varepsilon}} \int_{j\varepsilon}^{(j+1)\varepsilon} \mathrm{d}\omega \, e^{2\pi \mathrm{i}\omega n/\varepsilon} \times \\
&\times \left[-\frac{1}{4\pi\sqrt{\omega}} \frac{1}{r} e^{-\mathrm{i}\omega\left(v_{\mathrm{H}} - 4M \ln \frac{v_{\mathrm{H}} - v}{4M}\right)} \theta(v_{\mathrm{H}} - v) \right] \partial_v \left(\frac{1}{4\pi\sqrt{\omega'}} \frac{e^{-\mathrm{i}\omega' v}}{r} \right) \Bigg\} \\
&= -\frac{1}{2\pi\sqrt{\varepsilon}} \int_{-\infty}^{v_{\mathrm{H}}} \mathrm{d}v \int_{j\varepsilon}^{(j+1)\varepsilon} \mathrm{d}\omega \, e^{2\pi \mathrm{i}\omega n/\varepsilon} \sqrt{\frac{\omega'}{\omega}} \, e^{-\mathrm{i}\omega\left(v_{\mathrm{H}} - 4M \ln \frac{v_{\mathrm{H}} - v}{4M}\right) - \mathrm{i}\omega' v},
\end{aligned} \tag{1.47}$$

where in the second step, we have applied the Heaviside theta function on the extrema of the integral over v and performed the integral over the angular coordinates. In the following, also the other Bogoliubov coefficient α will be useful. It reads

$$\begin{aligned}\alpha_{jn,\omega'} &= (\phi^{out}_{jn}, \phi^{in}_{\omega'}) \\ &= -2\mathrm{i}\int_{\mathscr{I}^-} \mathrm{d}v\,\mathrm{d}\Omega^2\, r^2 \phi^{out}_{jn}\partial_v \phi^{in*}_{\omega'} \\ &= -\frac{1}{2\pi\sqrt{\varepsilon}}\int_{-\infty}^{v_{\mathrm{H}}} \mathrm{d}v \int_{j\varepsilon}^{(j+1)\varepsilon} \mathrm{d}\omega\, e^{2\pi \mathrm{i}\omega n/\varepsilon}\sqrt{\frac{\omega'}{\omega}}\, e^{-\mathrm{i}\omega\left(v_{\mathrm{H}}-4M\ln\frac{v_{\mathrm{H}}-v}{4M}\right)+\mathrm{i}\omega' v}.\end{aligned} \tag{1.48}$$

We now introduce the variable $x \equiv v_{\mathrm{H}} - v$. Equation (1.47) becomes

$$\beta_{jn,\omega'} = -\frac{e^{-\mathrm{i}\omega' v_{\mathrm{H}}}}{2\pi\sqrt{\varepsilon}}\int_0^\infty \mathrm{d}x \int_{j\varepsilon}^{(j+1)\varepsilon} \mathrm{d}\omega\, e^{2\pi\mathrm{i}\omega n/\varepsilon}\sqrt{\frac{\omega'}{\omega}}\, e^{-\mathrm{i}\omega(v_{\mathrm{H}}-4M\ln\frac{x}{4M})+\mathrm{i}\omega' x}. \tag{1.49}$$

The integral over ω can be readily evaluated, considering that it varies over a small interval. Since $\varepsilon \ll 1$ and we are close to $v = v_{\mathrm{H}}$ (so to $x \sim 0$), the dominant contributions to the integral come from the exponential term $e^{2\pi\mathrm{i}\omega n/\varepsilon+\mathrm{i}\omega 4M\ln x/4M}$, and thus we have

$$\beta_{jn,\omega'} = -\frac{e^{-\mathrm{i}(\omega_j+\omega')v_{\mathrm{H}}}}{\pi\sqrt{\varepsilon}}\sqrt{\frac{\omega'}{\omega_j}}\int_0^\infty \mathrm{d}x\, e^{\mathrm{i}\omega' x}\,\frac{\sin(\varepsilon L(x)/2)}{L(x)}\, e^{\mathrm{i}L(x)\omega_j}, \tag{1.50}$$

where $\omega_j = j\varepsilon \simeq (j+\frac{1}{2})\varepsilon$, while

$$L(x) \equiv \frac{2\pi n}{\varepsilon} + 4M\ln\frac{x}{4M}. \tag{1.51}$$

With the same procedure and computations, Eq. (1.48) yields

$$\alpha_{jn,\omega'} = -\frac{e^{-\mathrm{i}(\omega_j-\omega')v_{\mathrm{H}}}}{\pi\sqrt{\varepsilon}}\sqrt{\frac{\omega'}{\omega_j}}\int_0^\infty \mathrm{d}x\, e^{-\mathrm{i}\omega' x}\, e^{\mathrm{i}L(x)\omega_j}\,\frac{\sin(\varepsilon L(x)/2)}{L(x)}. \tag{1.52}$$

Both Bogoliubov coefficients, thus, depend on an integral of the form

$$I(\omega') \equiv \int_0^\infty \mathrm{d}x\, e^{-\mathrm{i}\omega' x} e^{\mathrm{i}L(x)\omega_j}\,\frac{\sin(\varepsilon L(x)/2)}{L(x)}. \tag{1.53}$$

From the definitions in Sect. 1.3.1, we know that $\omega' > 0$ for the α coefficients, while $\omega' < 0$ applies to the β ones. We now proceed to derive an important relation between the two coefficients.

The integral (1.53) possesses a branch point at $x = 0$, arising from the logarithm in the definition of $L(x)$. Let us assume the branch cut to lie in the negative real semi-axis. The integrand remains analytic in the whole complex plane, except along the real negative axis.

For $\omega' < 0$, we can rotate the contour of integration to the negative imaginary axis and then set $x \equiv -\mathrm{i}y$. Therefore, $\ln(x/4M) = \ln(-\mathrm{i}y/4M) = -\mathrm{i}\frac{\pi}{2} + \ln(y/4M)$, and Eq. (1.53) yields

$$\begin{aligned} I(\omega' > 0) &= -\mathrm{i}\int_0^\infty \mathrm{d}y\, e^{-\omega' y}\frac{\sin\left(\varepsilon L_y/2\right)}{L_y}\, e^{\mathrm{i}L_y\omega_j} \\ L_y &\equiv \frac{2\pi n}{\varepsilon} + 4M\left(-\frac{\mathrm{i}\pi}{2} + \ln\frac{y}{4M}\right). \end{aligned} \tag{1.54}$$

Similarly, if $\omega' < 0$, we can rotate the contour to positive imaginary axis and define $x \equiv \mathrm{i}z$, from which we have

$$\begin{aligned} I(\omega' < 0) &= \mathrm{i}\int_0^\infty \mathrm{d}z\, e^{\omega' z}\,\frac{\sin\left(\varepsilon L_z/2\right)}{L_z}\, e^{\mathrm{i}L_z\omega_j} \\ L_z &\equiv \frac{2\pi n}{\varepsilon} + 4M\left(\frac{\mathrm{i}\pi}{2} + \ln\frac{z}{4M}\right). \end{aligned} \tag{1.55}$$

Combining the two, we have thus

$$\begin{aligned} I(\omega' > 0) &= -\mathrm{i}e^{2\pi M\omega_j}\, e^{2\pi \mathrm{i}n\omega_j/\varepsilon}\int_0^\infty \mathrm{d}y\, e^{-\omega' y}\,\frac{\sin\left(\varepsilon L_y/2\right)}{L_y}\, e^{4M\mathrm{i}\omega_j\ln(y/4M)}; \\ I(\omega' < 0) &= \mathrm{i}e^{-2\pi M\omega_j}\, e^{2\pi \mathrm{i}n\omega_j/\varepsilon}\int_0^\infty \mathrm{d}z\, e^{\omega' z}\,\frac{\sin\left(\varepsilon L_z/2\right)}{L_z}\, e^{4M\mathrm{i}\omega_j\ln(z/4M)}. \end{aligned} \tag{1.56}$$

For narrow wave packets ($\varepsilon \ll 1$), centered around ω_j, and at late times, i.e., $n \to \infty$, we have $\varepsilon L_y \simeq \varepsilon L_z$, and this yields the above relation between the integrals

$$I(\omega' > 0) = -e^{4\pi M\omega_j}\, I(\omega' < 0). \tag{1.57}$$

Using Eqs. (1.50) and (1.52), this entails a relation between the Bogoliubov coefficients

$$|\alpha_{jn,\omega'}| = e^{4\pi M\omega_j}\, |\beta_{jn,\omega'}|. \tag{1.58}$$

As we shall see, this is essential in determining the spectrum of the emitted particles.

1.4.5 Planck Spectrum

We can now derive the spectrum of particles detected at future infinity. The continuum version of the first equation in Eq. (1.12) gives

$$\int_0^\infty \mathrm{d}\omega'\,\left(\alpha_{jn,\omega'}\alpha^*_{j'n',\omega'} - \beta_{jn,\omega'}\beta^*_{j'n',\omega'}\right) = \delta_{jj'}\delta_{nn'}, \tag{1.59}$$

which, for $j = j'$ and $n = n'$ becomes

$$\int_0^\infty d\omega \left(|\alpha_{jn,\omega'}|^2 - |\beta_{jn,\omega'}|^2\right) = 1. \tag{1.60}$$

Using Eq. (1.58), we also have $|\alpha_{jn,\omega'}|^2 = e^{8\pi M\omega_j}|\beta_{jn,\omega'}|^2$, which gives

$$\int_0^\infty d\omega' \, |\beta_{jn,\omega'}|^2 = \frac{1}{e^{8\pi M\omega_j} - 1}. \tag{1.61}$$

The left hand side is exactly Eq. (1.43), namely the average number of particles detected at future infinity at late times. The right hand side coincides with a Planck distribution at a temperature

$$T_{\mathrm{H}} = \frac{1}{8\pi M}, \tag{1.62}$$

which is the Hawking temperature.

References

1. S.W. Hawking, Particle creation by black holes. Commun. Math. Phys. **43**, 199–220 (1975) [erratum: Commun. Math. Phys. **46**, 206 (1976)]
2. S.A. Fulling, Nonuniqueness of canonical field quantization in Riemannian space-time. Phys. Rev. D **7**, 2850–2862 (1973)
3. P.C.W. Davies, Scalar particle production in Schwarzschild and Rindler metrics. J. Phys. A **8**, 609–616 (1975)
4. W.G. Unruh, Notes on black hole evaporation. Phys. Rev. D **14**, 870 (1976)
5. L. Parker, Quantized fields and particle creation in expanding universes. 1. Phys. Rev. **183**, 1057–1068 (1969)
6. L. Parker, Quantized fields and particle creation in expanding universes. 2. Phys. Rev. D **3**, 2546–2546 (1971) (erratum)
7. L. Parker, Thermal radiation produced by the expansion of the Universe. Nature **261**, 20–23 (1976)
8. Y.B. Zeldovich, Particle production in cosmology. Pisma Zh. Eksp. Teor. Fiz. **12**, 443–447 (1970)
9. Y.B. Zeldovich, A.A. Starobinsky, Particle production and vacuum polarization in an anisotropic gravitational field. Sov. Phys. JETP **34**, 1159–1166 (1972); Zh. Eksp. Teor. Fiz. **61**, 2161–2175 (1971)
10. D.N. Page, Information in black hole radiation. Phys. Rev. Lett. **71**, 3743–3746 (1993)
11. G. Penington, Entanglement wedge reconstruction and the information paradox. JHEP **09**, 002 (2020)
12. A. Almheiri, T. Hartman, J. Maldacena, E. Shaghoulian, A. Tajdini, The entropy of Hawking radiation. Rev. Mod. Phys. **93**(3), 035002 (2021)
13. S. Raju, Lessons from the information paradox. Phys. Rept. **943**, 1–80 (2022)
14. C. Barcelo, S. Liberati, M. Visser, Analog gravity from Bose-Einstein condensates. Class. Quant. Grav. **18**, 1137 (2001)
15. S. Weinfurtner, E.W. Tedford, M.C.J. Penrice, W.G. Unruh, G.A. Lawrence, Measurement of stimulated Hawking emission in an analogue system. Phys. Rev. Lett. **106**, 021302 (2011)
16. J. Drori, Y. Rosenberg, D. Bermudez, Y. Silberberg, U. Leonhardt, Observation of stimulated Hawking radiation in an optical analogue. Phys. Rev. Lett. **122**(1), 010404 (2019)

17. S.W. Hawking, Breakdown of predictability in gravitational collapse. Phys. Rev. D **14**, 2460–2473 (1976)
18. G. Penington, S.H. Shenker, D. Stanford, Z. Yang, Replica wormholes and the black hole interior. JHEP **03**, 205 (2022)
19. A. Almheiri, T. Hartman, J. Maldacena, E. Shaghoulian, A. Tajdini, Replica wormholes and the entropy of Hawking radiation. JHEP **05**, 013 (2020)
20. G. 't Hooft, On the quantum structure of a black hole. Nucl. Phys. B **256**, 727–745 (1985)
21. G. 't Hooft, The black hole interpretation of string theory. Nucl. Phys. B **335**, 138–154 (1990)
22. L. Susskind, L. Thorlacius, J. Uglum, The stretched horizon and black hole complementarity. Phys. Rev. D **48**, 3743–3761 (1993)
23. D.A. Lowe, J. Polchinski, L. Susskind, L. Thorlacius, J. Uglum, Black hole complementarity versus locality. Phys. Rev. D **52**, 6997–7010 (1995)
24. Y. Kiem, H.L. Verlinde, E.P. Verlinde, Black hole horizons and complementarity. Phys. Rev. D **52**, 7053–7065 (1995)
25. A. Almheiri, D. Marolf, J. Polchinski, J. Sully, Black holes: complementarity or firewalls? JHEP **02**, 062 (2013)
26. J.M. Bardeen, B. Carter, S.W. Hawking, The four laws of black hole mechanics. Commun. Math. Phys. **31**, 161–170 (1973)
27. J.D. Bekenstein, Black holes and the second law. Lett. Nuovo Cim. **4**, 737–740 (1972)
28. J.D. Bekenstein, Black holes and entropy. Phys. Rev. D **7**, 2333–2346 (1973)
29. T.A. Jacobson, Introduction to black hole microscopy. arXiv:hep-th/9510026
30. T. Jacobson, Introductory lectures on black hole thermodynamics. Institute for Theoretical Physics University of Utrecht (1996)
31. S.W. Hawking, Gravitational radiation from colliding black holes. Phys. Rev. Lett. **26**, 1344–1346 (1971)
32. S. Hawking, Gravitationally collapsed objects of very low mass. Mon. Not. Roy. Astron. Soc. **152**, 75 (1971)
33. A. Escrivà, F. Kuhnel, Y. Tada, Primordial black holes. arXiv:2211.05767
34. W. Israel, Event horizons in static vacuum space-times. Phys. Rev. **164**, 1776–1779 (1967)
35. W. Israel, Event horizons in static electrovac space-times. Commun. Math. Phys. **8**, 245–260 (1968)
36. B. Carter, Axisymmetric black hole has only two degrees of freedom. Phys. Rev. Lett. **26**, 331–333 (1971)
37. T. Jacobson, Black hole evaporation and ultrashort distances. Phys. Rev. D **44**, 1731–1739 (1991)
38. T. Jacobson, Black hole radiation in the presence of a short distance cutoff. Phys. Rev. D **48**, 728–741 (1993)
39. S. Massar, R. Parentani, From vacuum fluctuations to radiation. 2. Black holes. Phys. Rev. D **54**, 7444–7458 (1996)
40. T. Jacobson, On the origin of the outgoing black hole modes. Phys. Rev. D **53**, 7082–7088 (1996)
41. A. Fabbri, J. Navarro-Salas, *Modeling Black Hole Evaporation* (World Scientific, 2005)

Chapter 2
Alternative Derivations of Hawking Radiation

Chon Man Sou

Abstract Since the original derivation of Hawking radiation, there have been lots of alternative approaches to show the same fact that black holes emit particles as hot bodies with a temperature. These alternative methods generally rely on different conditions and physical quantities to manifest the radiation, providing various points of view of this effect in the intersection of gravity and quantum theory. This chapter presents some alternative derivations of Hawking radiation in the literature, including the tunneling, anomaly and Green's function methods. From these methods, various features of the black hole system can be seen, such as the gravitational and trace anomalies of the $(1 + 1)$-dimensional effective theory and the analytical continuation of the complexified spacetime.

2.1 Introduction

The original derivation by Hawking [28, 29] compares the wave modes of quantum field before and after a gravitational collapse of forming a black hole from a star, and the existence of Hawking radiation is reflected by the expectation value of the particle number in each mode, which agrees with thermal radiation with a specific temperature. It was soon after understood that such a state with non-zero particle number exactly corresponds to a thermal state [47, 62], and the gravitational collapse can also be replaced by the usual Schwarzschild spacetime to obtain the same result [58].[1] Thus, the conditions required to derive the Hawking radiation are certainly not unique as the original one. Remarkably, the necessary step of this type of derivations of Hawking radiation is the comparison of different vacua, quantified by the Bogoliubov transformation, which has already become a standard technique for

[1] On the other hand, if we stay with the gravitational collapse, some minimal conditions to derive the radiation have been studied in [4].

C. M. Sou (✉)
Department of Physics, Tsinghua University, Beijing, China
e-mail: cmsou@mail.tsinghua.edu.cn

A. Akil and C. Bambi (eds.), *The Black Hole Information Paradox*, Springer Series in Astrophysics and Cosmology, https://doi.org/10.1007/978-981-96-6170-1_2

studying particle production in curved spacetime [10, 19, 43], with various applications such as cosmology [20]. Not only can we reconsider the conditions to derive the radiation, but it is also reasonable to consider alternative methods.

Without using the standard technique of quantum field theory in curved spacetime with the Bogoliubov transformation, there are still alternative derivations of Hawking radiation utilizing various features of black holes, such as the tunneling of particles across the horizon determined by energy conservation [46], the anomalies of the effective theory near the horizon highlighting the role of energy-momentum tensor [14, 49] and the emission of particles from the thermal feature of their propagator [23]. For the significance of deriving the same result with different methods, it may be a way to see how various properties and points of view of black holes can consistently exist and imply the same effect, because some of them may be related indirectly. In the literature, there is a review including a brief introduction of many alternative derivations [12]. Here we choose three types of alternative methods to present and discuss: tunneling, anomaly and Green's function, which are extensively studied in the literature.

This chapter is organized as follows. In Sect. 2.2, the tunneling method is introduced, including the Parikh-Wilczek null geodesic method and the Hamilton-Jacobi method. In Sect. 2.3, the anomaly method is introduced, starting by deriving the $(1+1)$-dimensional effective theory near the horizon, and the Hawking radiation is derived with the gravitational and trace anomalies in this theory based on some corresponding conditions. In Sect. 2.4, we first introduce analytical properties for general thermal Green's functions, and the Hartle-Hawking path integral derivation based on the analyticity of Green function will be demonstrated. Section 2.5 is the discussion and conclusion.

Here are some conventions used in this chapter. We set the unit to be $G = c = k_B = \hbar = 1$, and the metric signature is $(-+++)$. Unless stated explicitly, we denote $\varepsilon \to 0^+$ as an infinitesimally small positive number, which will be used in many places in the derivations.

2.2 The Tunneling Method

Compared to the original Hawking's derivation using quantum field theory in curved spacetime [28, 29], the tunneling method emphasizes an intuitive picture of emitting the Hawking radiation: a virtual particle pair created near (inside or outside) the black-hole horizon can materialize if one of them tunnels to another side of the horizon, in which the positive-energy particle in the pair radiates away, whereas the negative-energy particle goes inward and lead to the decrease of black hole's mass. To quantify such a tunneling process, we only need the classical action of the particle in order to calculate the tunneling probability, and this avoids the ambiguity of defining particle states in curved spacetime, showing the advantage and simplicity of this method.

The well-known derivation by Parikh and Wilczek is shown in [44], known as the null geodesic method in the tunneling literature [60], and some additional discussions by Parikh are in [45, 46]. Besides the derivation by Parikh and Wilczek, the idea of understanding the Hawking radiation as a tunneling process is also seen in [15] by Damour and Ruffini, and [55] by Srinivasan and Padmanabhan, in which the latter has been further developed into the Hamilton-Jacobi method of the tunneling [3, 42, 53, 54]. For a more detailed introduction of the tunneling method, there is also a review with various applications to stationary and dynamical black holes, as well as cosmological horizons [60].

2.2.1 The Null Geodesic Method

We start with the Schwarzschild black hole with the metric

$$ds^2 = -\left(1 - \frac{2M}{r}\right) dt^2 + \left(1 - \frac{2M}{r}\right)^{-1} dr^2 + r^2 d\Omega^2, \tag{2.1}$$

which is well-known to have a coordinate singularity at the horizon $r = 2M$, problematic for evaluating the across-horizon tunneling. To avoid the coordinate singularity, the Painlevé-Gullstrand coordinates [26, 48] are introduced by considering a radially free-falling observer with vanishing velocity at infinity, whose geodesic follows the normalized four vector [36]

$$u^\mu = \left(\frac{1}{1 - \frac{2M}{r}}, -\sqrt{\frac{2M}{r}}, 0, 0\right), \tag{2.2}$$

which is fixed by the normalization

$$g_{\mu\nu} u^\mu u^\nu = -1, \tag{2.3}$$

and the constant of motion (the ratio of energy to rest mass)

$$- g_{\mu\nu} \xi^\mu u^\nu = 1, \tag{2.4}$$

where $\xi^\mu = (\partial_t)^\mu$ is the timelike Killing vector. We then define a new time t_p such that its constant-time surface is orthogonal to the observer's geodesic, i.e. $\partial_\mu t_p = -u_\mu$, and its total differentiation is

$$dt_p = \frac{\partial t_p}{\partial t} dt + \frac{\partial t_p}{\partial r} dr \tag{2.5}$$

$$= dt + \left(1 - \frac{2M}{r}\right)^{-1} \sqrt{\frac{2M}{r}} dr, \tag{2.6}$$

which gives

$$t_p = t + 2\sqrt{2Mr} + 2M \log\left(\frac{\sqrt{r} - \sqrt{2M}}{\sqrt{r} + \sqrt{2M}}\right). \tag{2.7}$$

With the new coordinates, the line element becomes regular across the horizon

$$ds^2 = -\left(1 - \frac{2M}{r}\right) dt_p^2 + 2\sqrt{\frac{2M}{r}} dt_p dr + dr^2 + r^2 d\Omega^2, \tag{2.8}$$

which also has features of time independence and flat Euclidean space on constant-time surfaces.

For the null geodesic method, the energy conservation of the full system of particles and black hole has to be considered because the tunneling "path" evaluated in this method is defined in the back-reaction sense. As the black hole's mass fluctuates when the particles are emitted, for a spherically symmetric thin shell (s-wave) of particles,[2] it moves along the geometry (2.8) with $M \to M - \omega$

$$ds^2 = -\left[1 - \frac{2(M-\omega)}{r}\right] dt_p^2 + 2\sqrt{\frac{2(M-\omega)}{r}} dt_p dr + dr^2 + r^2 d\Omega^2, \tag{2.9}$$

which agrees with the Birkhoff's theorem. At the first glance, such a decrease of black hole's mass seems to be a trivial consequence of the energy conservation, but it is necessary to determine the barrier of quantum tunneling [45, 46]. For the usual tunneling process encountered in quantum mechanics, the length of the barrier is determined by classical turning points where the kinetic energy vanishes. If the same point of view is applied to the case of black hole, we would expect that the length of the barrier is infinitesimally small because classical outgoing null geodesics exist just outside the horizon.[3] However, with the knowledge that the black hole's mass fluctuates as $M \to M - \omega$, the barrier of quantum tunneling can be understood as the forbidden region from the initial horizon $r = 2M$ to the final horizon $r = 2(M - \omega)$. In other words, there is no preexisting barrier, but it depends on the particle itself [46].

As a particle near the horizon is infinitely blue-shifted, the wavelength of its wave packet is small enough to consider the geometric optics limit, implying the WKB approximation for the field solution [45]

$$\phi(t_p, r) \approx e^{iS(t_p, r)}, \tag{2.10}$$

[2] Besides making the calculation simpler, the case with spherical symmetry circumvent the quantization of gravitons when back reaction is considered [45].

[3] The geodesic equation of outgoing radiation in the geometry (2.8) is $\frac{dr}{dt_p} = 1 - \sqrt{\frac{2M}{r}}$, which is positive just outside the horizon.

where $S(t_p, r)$ is the classical action of particle determined by the Hamilton-Jacobi equation. Classically, the particle follows the trajectory determined by initial conditions, and the action is real. Quantum mechanically, it is possible for the particle to pass through some classically forbidden regions, known as the tunneling, and the tunneling probability is evaluated by the imaginary part of the action

$$\Gamma = e^{-2\mathrm{Im}\,S}. \tag{2.11}$$

For a particle pair created just inside the horizon, the positive-energy particle travels from the initial position $r_i = 2M$ to the final position $r_f = 2(M - \omega)$, and the action is

$$S = \int_{2M}^{2(M-\omega)} p_r\, dr, \tag{2.12}$$

where p_r is the conjugate momentum of r. Without using the explicit solution of p_r, the integral can be evaluated with the Hamilton's equations

$$\begin{aligned} \frac{dr}{dt_p} &= \dot{r} \\ &= \frac{\partial H}{\partial p_r} \\ &= \frac{dH}{dp_r}\bigg|_r, \end{aligned} \tag{2.13}$$

and the imaginary part of (2.12) becomes

$$\begin{aligned} \mathrm{Im} S &= \mathrm{Im} \int_{2M}^{2(M-\omega)} \int_0^{p_r} dp_r'\, dr \\ &= \mathrm{Im} \int_{2M}^{2(M-\omega)} \int_M^{M-\omega} \frac{dH}{\dot{r}} dr. \end{aligned} \tag{2.14}$$

With the geodesic equation

$$\dot{r} = 1 - \sqrt{\frac{2(M-\omega)}{r}}, \tag{2.15}$$

we can explicitly express $\mathrm{Im} S$ as a double integral

$$\mathrm{Im} S = \mathrm{Im} \int_{2M}^{2(M-\omega)} \int_M^{M-\omega} \frac{dH}{1 - \sqrt{\frac{2H}{r}}} dr. \tag{2.16}$$

There is a pole when $r = 2H$, and the correct way to obtain the imaginary part is to deform the integration contour by the $i\varepsilon$ prescription: $H \to H + i\varepsilon$ or equivalently $r \to r - i\varepsilon$. With the Sokhotski-Plemelj theorem, the integrand becomes

$$\frac{1}{1-\sqrt{\frac{2H}{r}}-i\varepsilon} = i\pi\delta\left(1-\sqrt{\frac{2H}{r}}\right) + \text{p.v.}\left(\frac{1}{1-\sqrt{\frac{2H}{r}}}\right), \tag{2.17}$$

where p.v. is the Cauchy principal value.[4] Only the delta function $\delta\left(1-\sqrt{\frac{2H}{r}}\right)$ contributes to the imaginary part, and (2.16) can be easily evaluated:

$$\begin{aligned} \text{Im}S &= \pi \int_{2M}^{2(M-\omega)} \int_{M}^{M-\omega} \delta\left(1-\sqrt{\frac{2H}{r}}\right) dH dr \\ &= -\pi \int_{2M}^{2(M-\omega)} r\, dr \\ &= 4\pi\left(M - \frac{\omega}{2}\right)\omega. \end{aligned} \tag{2.18}$$

Similarly, the calculation can be done for a particle pair created just outside the horizon in which the negative-energy particle tunnels into the black hole, and we change the null geodesic equation to the correct direction and with the self-gravitational correction [37]

$$\dot{r} = -1 + \sqrt{\frac{2(M+\omega')}{r}}, \tag{2.19}$$

where ω' ranges from 0 to $-\omega$. With the $i\varepsilon$ prescription with $\varepsilon \to -\varepsilon$, the imaginary part of the action is again the same:

$$\begin{aligned} \text{Im}S &= -\text{Im} \int_{2M}^{2(M-\omega)} \int_{M}^{M-\omega} \frac{dH}{1-\sqrt{\frac{2H}{r}}+i\varepsilon} dr \\ &= 4\pi\left(M - \frac{\omega}{2}\right)\omega. \end{aligned} \tag{2.20}$$

With the imaginary part of action (2.18) and (2.20), the tunneling probability (2.11) is

$$\begin{aligned} \Gamma &= e^{-8\pi\left(M-\frac{\omega}{2}\right)\omega} \\ &= e^{\Delta S_{\text{B-H}}}, \end{aligned} \tag{2.21}$$

[4] For an integrand $f(x)$ has a singularity at b, the Cauchy principal value is p.v. $\int_a^c f(x)dx = \lim_{\varepsilon->0^+}\left(\int_a^{b-\varepsilon} f(x)dx + \int_{b+\varepsilon}^c f(x)dx\right)$.

which is related to the change of the Bekenstein-Hawking entropy. For the leading order with $\omega \ll M$, (2.21) reduces to the Boltzmann factor, and the Hawking temperature is recovered

$$T_H = \frac{1}{8\pi M}. \tag{2.22}$$

For the next order $O(\omega^2)$ in (2.21), it originates from the energy conservation that emitting radiation reduces the black hole's mass, and hence increasing its temperature.

As a final remark of the null geodesic method, there are two conditions required during the derivation which should have a better justification:

- Back-reaction by energy conservation in (2.9) is necessary to define the tunneling path, but there are other classically forbidden paths defined in a way not relying on the back-reaction.
- The sign of $i\varepsilon$ prescription in (2.17) should be justified with the analytical properties of S since flipping the sign $\varepsilon \to -\varepsilon$ leads to minus the imaginary part.

These two aspects will be addressed with the Hamiltonian-Jacobi method and analytical continuation in Sect. 2.2.2.

2.2.2 *The Hamiltonian-Jacobi Method*

This method directly applies the WKB approximation (2.10) to the massless Klein-Gordon equation

$$-\frac{1}{\sqrt{-g}}\partial_\mu\left(\sqrt{-g}g^{\mu\nu}\partial_\nu\phi\right) = 0, \tag{2.23}$$

and thus the action $S(x^\mu)$ satisfies the relativistic Hamiltonian-Jacobi equation [3]

$$g^{\mu\nu}\partial_\mu S\partial_\nu S = 0. \tag{2.24}$$

This method has been widely applied to generalize the tunneling process to complex paths by analytical continuation of coordinates [42, 53–55]. Firstly, we follow [60] to repeat the calculation for a null geodesic from just inside the horizon to just outside, but without considering back-reaction, and we will see that the result agrees with the previous calculation. The Hamilton-Jacobi Eq. (2.24) in the Painlevé-Gullstrand coordinates (2.8) is

$$\begin{aligned} & -\left(\partial_{t_p} S\right)^2 + 2\sqrt{\frac{2M}{r}}\partial_{t_p} S \partial_r S + \left(1 - \frac{2M}{r}\right)(\partial_r S)^2 \\ & = -\omega^2 - 2\omega\sqrt{\frac{2M}{r}}\partial_r S + \left(1 - \frac{2M}{r}\right)(\partial_r S)^2 \\ & = 0, \end{aligned} \tag{2.25}$$

where the constant energy $\omega = -\partial_{t_p} S$, and the angular part is neglected, so we solve for $\partial_r S$ with the correct sign for outgoing

$$\partial_r S = \frac{\omega}{1 - \sqrt{\frac{2M}{r}}}. \tag{2.26}$$

For a null geodesic near the horizon

$$\begin{aligned} ds^2 & = 2dt_p dr + dr^2 \\ & = 0, \end{aligned} \tag{2.27}$$

we have $dt_p = -\frac{dr}{2}$, and thus the imaginary part of the action with (2.26) is

$$\begin{aligned} \text{Im} S & = \text{Im} \int \left(\partial_r S dr + \partial_{t_p} S dt_p\right) \\ & = \text{Im} \int_{2M-\varepsilon}^{2M+\varepsilon} \frac{\omega}{1 - \sqrt{\frac{2M}{r}} - i\varepsilon} dr \\ & = 4\pi M\omega, \end{aligned} \tag{2.28}$$

where the real temporal part does not contribute any imaginary value, and the $i\varepsilon$ prescription $r \to r - i\varepsilon$ is applied. Compared to (2.20), this agrees with the null geodesic method when the back-reaction term is neglected, suggesting that we can also have a tunneling process without considering any back-reaction, in which the path is the classically-forbidden outgoing null geodesic. In fact, the path in the integral (2.28) can be extended to $r_1 < r < r_2$ where $r_1 < 2M$ and $r_2 > 2M$ are inside and outside the horizon respectively because the contributions in $[r_1, 2M - \varepsilon]$ and $[2M + \varepsilon, r_2]$ are real, so general null geodesics from inside to outside the horizons give the same tunneling probability.

On the other hand, the sign of $i\varepsilon$ prescription can be justified if the tunneling path is analytically continued to a complex path. With the Kruskal extension, a Schwarzschild black hole can be separated into four regions, including outside the horizon (region I), inside the horizon (region II) and inside the horizon and with time reversal (region III), as shown in Fig. 2.1. As demonstrated in [61], instead of directly tunneling from region II to I, there exists a complex path tunneling from region II to III and then following the classically allowed path from region III to I, described by the complex rotation of the Kruskal coordinates with $t \to t - 4Mi\theta$:

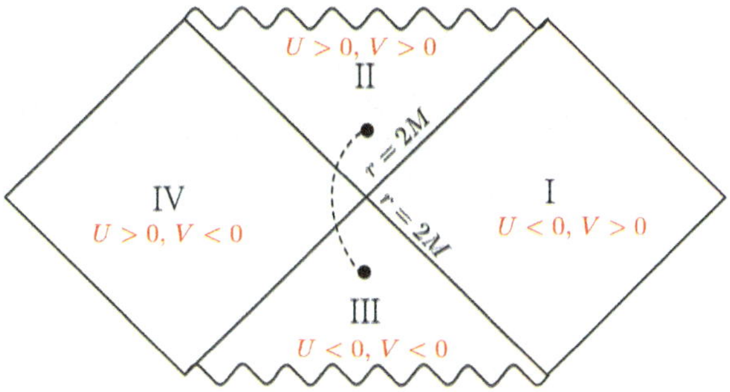

Fig. 2.1 A Schwarzschild black hole with the Kruskal extension. The dashed path indicates the complex path (2.29) from the region II to III

$$\begin{aligned}
\tilde{U} &= e^{i\theta}\left(1-\frac{r}{2M}\right)^{\frac{1}{2}} e^{\frac{r-t}{4M}} \\
&= e^{i\theta} U \\
\tilde{V} &= e^{-i\theta}\left(1-\frac{r}{2M}\right)^{\frac{1}{2}} e^{\frac{r+t}{4M}} \\
&= e^{-i\theta} V,
\end{aligned} \tag{2.29}$$

where $\theta = 0$ and $\theta = \pi$ correspond to the region II and III respectively. With the Kruskal coordinates, the action integral $\int \partial_\mu S dx^\mu$ is expressed from $\int \left(\partial_r S dr + \partial_{t_p} S dt_p\right)$ to $\int \left(\partial_{\tilde{U}} S d\tilde{U} + \partial_{\tilde{V}} S d\tilde{V}\right)$, and the imaginary part of the action is

$$\begin{aligned}
\text{Im} S &= \text{Im} \int_{II \to III} \left(\partial_{\tilde{U}} S d\tilde{U} + \partial_{\tilde{V}} S d\tilde{V}\right) \\
&= \text{Im}\left[i \int_0^\pi d\theta \; (U \partial_U S - V \partial_V S)\right] \\
&= -4M \int_0^\pi d\theta \, \partial_t S|_r \\
&= -4M \int_0^\pi d\theta \, \partial_{t_p} S \\
&= 4\pi M \omega.
\end{aligned} \tag{2.30}$$

It is also easy to check that the following path from region III to I, described by the time-reversed (equivalently ingoing) one from (2.25)

$$\partial_r S = -\frac{\omega}{1+\sqrt{\frac{2M}{r}}}, \tag{2.31}$$

is classically allowed since there is no any singularity at $r = 2M$. Therefore, the imaginary part contributed by the complex path (2.30), obtained by the imaginary change of the Schwarzschild time t, agrees with (2.20) (ignore the back-reaction term) and (2.28), suggesting that the $i\varepsilon$ prescription makes sense.[5] Such an analytical continuation to the region III was also applied in the Hartle-Hawking path-integral derivation of Hawking radiation [27], in which the analytical properties of Green function is studied, suggesting that the choice of complex path (2.29) is reasonable. We will discuss this derivation in detailed in Sect. 2.4.2, and the same tunneling probability is obtained.

2.3 The Anomaly Method

The relationship between anomaly and Hawking radiation has been noticed by Christensen and Fulling in 1977, soon after Hawking's discovery [14], but the result relies on the trace anomaly of the energy-momentum tensor for a field with conformal invariance in $1+1$ dimension, limiting the applicability of this method in real cases and general spacetimes. A more feasible anomaly method was shown much later by Robinson and Wilczek in 2005 [49], in which the anomaly in general covariance, the gravitational anomaly, was applied to derive the Hawking radiation. Soon after the original paper by Robinson and Wilczek, the gravitational-anomaly method was generalized to Reissner-Nordström [30] and Kerr [31, 41] black holes.

The detailed derivation of black hole effective field theory is demonstrated in Robinson's thesis [50]. The recipe is as follows:[6]

Near the horizon, i.e. $0 \leq r - r_h \leq \varepsilon$

- A scalar field can be described by $(1+1)$-dimensional theory in the $r - t$ section with the partial wave decomposition.
- Ingoing modes near the horizon are discarded because they cannot affect the effective theory outside the horizon, leading to a $(1+1)$-dimensional chiral scalar field.
- Such a chiral scalar field has the gravitational anomaly in its energy-momentum tensor, and the full quantum theory must admit a flux to compensate this. This flux will be the Hawking radiation.

---

[5] Certainly, a better justification should connect the complex continuations between t and r, and the demonstration here is useful to determine the sign of the $i\varepsilon$ prescription.

[6] For the original paper by Robinson and Wilczek [49], the outgoing modes near the horizon are discarded, and see the remark at the end of Sect. 2.3.3 for a detailed discussion.

2.3.1 *The Effective* $(1+1)$*-Dimensional Massless Theory Near the Horizon*

We consider a scalar field which generally has mass and interaction terms in its action

$$\begin{aligned} S &= -\frac{1}{2}\int d^4x\sqrt{-g}\left(g^{\mu\nu}\partial_\mu\phi\partial_\nu\phi + \sum_{n=2}^{\infty}\lambda_n\phi^n\right) \\ &= \frac{1}{2}\int d^4x\sqrt{-g}\left[\frac{1}{\sqrt{-g}}\phi\partial_\mu\left(\sqrt{-g}g^{\mu\nu}\partial_\nu\phi\right) - \sum_{n=2}^{\infty}\lambda_n\phi^n\right] \\ &= \frac{1}{2}\int d^4x\sqrt{-g}\left(\phi g^{\mu\nu}\nabla_\mu\nabla_\nu\phi - \sum_{n=2}^{\infty}\lambda_n\phi^2\right), \end{aligned} \tag{2.32}$$

where we did an integration by parts in the second line, ∇_μ is the covariant derivative, and $\lambda_2 = m^2$ is the mass term. For the simplest case, the metric is static and spherically symmetric

$$ds^2 = -f(r)dt^2 + \frac{1}{f(r)}dr^2 + r^2d\Omega^2, \tag{2.33}$$

so the Schwarzschild black hole (2.1) corresponds to $f(r) = 1 - \frac{2M}{r}$, and the horizon is thus determined by $f(r_H) = 0$ as well as the surface gravity

$$\kappa = \frac{1}{2}f'(r_H), \tag{2.34}$$

is finite. For making the following calculation simple, the tortoise coordinates are introduced

$$\frac{dr}{dr_*} = f(r), \tag{2.35}$$

$$ds^2 = f(r)(-dt^2 + dr_*^2) + r^2d\Omega^2, \tag{2.36}$$

and the variable r_* near the horizon is

$$\begin{aligned} r_* &\approx \frac{1}{f'(r_H)}\log\left(\frac{r-r_H}{r_0-r_H}\right) \\ &= \frac{1}{2\kappa}\log\left(\frac{r-r_H}{r_0-r_H}\right), \end{aligned} \tag{2.37}$$

where r_0 is an unimportant integration constant, and therefore $r_* \to -\infty$ when $r \to r_H$. The action (2.32) in this coordinate system is

$$S=\frac{1}{2}\int dtdr_*d^2\Omega\left[\phi\left(-r^2\partial_t^2+r^2\partial_{r_*}^2+2rf\partial_{r_*}-fL^2-fm^2r^2\right)\phi\right.$$
$$\left.-\sum_{n=3}^{\infty}\lambda_n fr^2\phi^n\right],\tag{2.38}$$

where

$$L^2=-\frac{1}{\sin\theta}\partial_\theta\left(\sin\theta\partial_\theta\right)-\frac{1}{\sin^2\theta}\partial_\varphi^2,\tag{2.39}$$

is the orbital angular momentum.

In the static and spherically symmetric metric, the field has a mode expansion

$$\phi=\frac{1}{r}\sum_{l,m}\varphi_{lm}(t,r_*)Y_{lm}(\Omega),\tag{2.40}$$

where $Y_{lm}(\Omega)$ is the spherical harmonics, and the action (2.38) can be further simplified by integrating over the angles

$$S=\frac{1}{2}\int dtdr_*\left[\sum_{l,m'}\varphi_{lm'}\left(-\partial_t^2+\partial_{r_*}^2-\frac{ff'}{r}-f\frac{l(l+1)}{r^2}-fm^2\right)\varphi_{lm'}\right.$$
$$\left.-\sum_{n=3}^{\infty}\frac{\lambda_n f}{r^{n-2}}\sum_{l_1,\ldots,l_n,m_1,\ldots,m_n}{}^nC_{\{\mathbf{l}\}}^{\{\mathbf{m}\}}\varphi_{l_1m_1}\ldots\varphi_{l_nm_n}\right],\tag{2.41}$$

where the $l(l+1)$ term is obtained from (2.39)

$$L^2Y_{lm}=l(l+1)Y_{lm},\tag{2.42}$$

and

$$\int d^2\Omega\, Y_{lm}(\Omega)Y_{l'm'}(\Omega)=\delta_{l,l'}\delta_{m,m'}\tag{2.43}$$

$${}^nC_{\{\mathbf{l}\}}^{\{\mathbf{m}\}}=\int d^2\Omega\, Y_{l_1m_1}(\Omega)\ldots Y_{l_nm_n},\tag{2.44}$$

are the integrals of the product of n spherical harmonics. With (2.37), the function $f(r)$ near the horizon is exponentially suppressed in $r_*\to-\infty$

$$\begin{aligned}f(r_*)&\approx f'(r_H)(r-r_H)\\&\approx 2\kappa(r_0-r_H)e^{2\kappa r_*},\end{aligned}\tag{2.45}$$

and thus the terms with $f(r)$ in the action (2.41) can be neglected, leading to an effectively free $(1+1)$-dimensional massless field near the horizon in the $r-t$ section

$$S \approx \frac{1}{2} \int dt dr_* \sum_{lm} \varphi_{lm} \left(-\partial_t^2 + \partial_{r_*}^2\right) \varphi_{lm}, \tag{2.46}$$

where we can simply neglect the subscript lm.

The equation of motion from (2.46) is just the usual wave equation

$$\begin{aligned}\left(\partial_t^2 - \partial_{r_*}^2\right) \varphi &= \partial_u \partial_v \varphi \\ &= 0,\end{aligned} \tag{2.47}$$

where the variables

$$\begin{aligned} u &= t - r_* \\ v &= t + r_*. \end{aligned} \tag{2.48}$$

It is clear that the field near the horizon can have two independent sets of solutions: the outgoing and ingoing modes with the forms

$$\begin{aligned}\varphi(u) &= \frac{e^{-i\omega u}}{\sqrt{4\pi\omega}} \\ \varphi(v) &= \frac{e^{-i\omega v}}{\sqrt{4\pi\omega}},\end{aligned} \tag{2.49}$$

respectively, which play roles like right- and left-handed fields.

2.3.2 A Simple Derivation of the Trace and (Covariant) Gravitational Anomalies of the Effective Theory

We follow [11] to give a simple derivation of the trace anomaly of the effective theory (2.64), and a more detailed discussion can be found in [17, 18]. For the $(1+1)$-dimensional metric (2.70), we can rewrite it with null coordinates

$$ds^2 = -C(U, V) dU dV, \tag{2.50}$$

where for example $U = u = t - r_*$ and $V = v = t + r_*$ corresponds to $C = f$, and we can further rescale U and V to get the form in (2.50). The components of the energy-momentum tensor of the field is

$$T_{UU} = \partial_U \varphi \partial_U \varphi, \; T_{VV} = \partial_V \varphi \partial_V \varphi, \tag{2.51}$$

and classically we will get $T_{UV} = 0$ which agrees with the vanishing trace.

As a first attempt, the quantum expectation value of T_{UU} can be evaluated as follows

$$\begin{aligned}\langle T_{UU}\rangle &= \lim_{U\to U'} \partial_U \partial_{U'} \langle \varphi(U)\varphi(U')\rangle \\ &= -\frac{1}{4\pi} \lim_{U\to U'} \partial_U \partial_{U'} \log \left|U - U'\right| \\ &= -\frac{1}{4\pi} \lim_{U\to U'} \frac{1}{(U-U')^2},\end{aligned} \tag{2.52}$$

where the two-point function is calculated with the mode functions (2.49)

$$\begin{aligned}\langle \varphi(U)\varphi(U')\rangle &= \frac{1}{4\pi} \lim_{L\to\infty} \int_{L^{-1}}^{+\infty} \frac{d\omega}{\omega} e^{-i\omega(U-U')} \\ &= -\frac{\log\left|U-U'\right|}{4\pi} + \lim_{L\to\infty} \frac{1}{4\pi}\left(-\gamma_E - \frac{i\pi}{2} + \log L\right),\end{aligned} \tag{2.53}$$

in which the second constant term cannot contribute after taking derivatives. The evaluation of $\langle T_{VV}\rangle$ can be done similarly by replacing U to V. To resolve the divergence in (2.52) when $U \to U'$, we have to renormalize it by subtracting the one by local inertial modes, in which the inertial coordinates $(\hat{U}, \hat{V})$ are defined by

$$\begin{aligned}\hat{U} - \hat{U}_0 &= \int_{U_0}^{U} \frac{C(U', V_0)}{C(U_0, V_0)} dU' \\ \hat{V} - \hat{V}_0 &= \int_{V_0}^{V} \frac{C(U_0, V')}{C(U_0, V_0)} dV',\end{aligned} \tag{2.54}$$

which can be interpreted as the affine parameters of light-like geodesics. In these coordinates, the inertial modes are

$$\begin{aligned}\tilde{\varphi}(\hat{U}) &= \frac{e^{-i\omega\hat{U}}}{\sqrt{4\pi\omega}} \\ \tilde{\varphi}(\hat{V}) &= \frac{e^{-i\omega\hat{V}}}{\sqrt{4\pi\omega}},\end{aligned} \tag{2.55}$$

so the renormalized energy-momentum tensor is

$$\begin{aligned}\langle T_{UU}\rangle_{\text{ren}} &= -\frac{1}{4\pi} \lim_{U\to U'} \partial_U \partial_{U'} \left(\log|U-U'| - \log|\hat{U}(U) - \hat{U}(U')|\right) \\ &= -\frac{1}{12\pi}\sqrt{\partial_U \hat{U}}\, \partial_U^2 \left(\frac{1}{\sqrt{\partial_U \hat{U}}}\right)\end{aligned}$$

$$= -\frac{1}{12\pi} C^{\frac{1}{2}} \partial_U^2 C^{-\frac{1}{2}}, \tag{2.56}$$

and similarly we have

$$\langle T_{VV} \rangle_{\text{ren}} = -\frac{1}{12\pi} C^{\frac{1}{2}} \partial_V^2 C^{-\frac{1}{2}}. \tag{2.57}$$

The T_{UV} component is obtained by requiring the conservation

$$\begin{aligned} \nabla_\nu \langle T_U^\nu \rangle_{\text{ren}} &= C^{-1} \partial_V \langle T_{UU} \rangle_{\text{ren}} + \partial_U \left(C^{-1} \langle T_{UV} \rangle_{\text{ren}} \right) \\ &= 0, \end{aligned} \tag{2.58}$$

which gives the T_{UV} component

$$\langle T_{UV} \rangle_{\text{ren}} = -\frac{1}{24\pi} \partial_U \partial_V \log C, \tag{2.59}$$

From here onwards, we can neglect $\langle \ldots \rangle_{\text{ren}}$ for simplicity, and $T_{\mu\nu}$ should be understood as the quantum expectation value. Combining (2.56), (2.57) and (2.59), the energy-momentum tensor can be collectively expressed as [17]

$$T_{\mu\nu} = \frac{R}{48\pi} g_{\mu\nu} + \theta_{\mu\nu}, \tag{2.60}$$

where the Ricci scalar R is

$$R = 4C^{-1} \partial_U \partial_V \log C, \tag{2.61}$$

and the components of $\theta_{\mu\nu}$ are

$$\begin{aligned} \theta_{UU} &= -\frac{1}{12\pi} C^{\frac{1}{2}} \partial_U^2 C^{-\frac{1}{2}} \\ \theta_{VV} &= -\frac{1}{12\pi} C^{\frac{1}{2}} \partial_V^2 C^{-\frac{1}{2}} \\ \theta_{UV} &= \theta_{VU} = 0. \end{aligned} \tag{2.62}$$

From the conservation of $T_{\mu\nu}$, we also have

$$\nabla^\mu \theta_{\mu\nu} = -\frac{1}{48\pi} \partial_\nu R. \tag{2.63}$$

From (2.60), it is clear that the energy-momentum tensor has a trace anomaly

$$T^{\alpha}_{\alpha} = \frac{1}{24\pi} R. \tag{2.64}$$

Next we follow a simple argument in [6, 21] to show the (covariant) gravitational anomaly for the chiral theory obtained by neglecting ingoing modes. As U and V are null variables representing outgoing (right-handed) and ingoing (left-handed) modes, (2.62) implies that we can separate $\theta_{\mu\nu}$ into two parts

$$\begin{aligned} \theta_{\mu\nu} &= \theta^{(R)}_{\mu\nu} + \theta^{(L)}_{\mu\nu} \\ \theta^{(R)}_{UU} &= \theta_{UU} \\ \theta^{(L)}_{VV} &= \theta_{VV} \\ \theta^{(R)}_{VV} &= \theta^{(R)}_{UV} = \theta^{(L)}_{UU} = \theta^{(L)}_{UV} = 0. \end{aligned} \tag{2.65}$$

With this separation, the energy-momentum tensor is also separated as $T_{\mu\nu} = T^{(R)}_{\mu\nu} + T^{(L)}_{\mu\nu}$ where

$$T^{(R/L)}_{\mu\nu} = \frac{R}{96\pi} g_{\mu\nu} + \theta^{(R/L)}_{\mu\nu}, \tag{2.66}$$

and for the chiral theory with only the outgoing (right-handed) modes, $T^{(R)}_{\mu\nu}$ is not conserved:

$$\begin{aligned} \nabla^{\mu} T^{(R)}_{\mu\nu} &= \frac{\partial_{\nu} R}{96\pi} + \nabla^{\mu}\theta^{(R)}_{\mu\nu} \\ &= \frac{\partial_{\nu} R}{96\pi} + \delta^{U}_{\nu} \nabla^{\mu}\theta_{\mu U} \\ &= \begin{cases} -\frac{\partial_U R}{96\pi} & \nu = U \\ \frac{\partial_V R}{96\pi} & \nu = V \end{cases}, \end{aligned} \tag{2.67}$$

where (2.63) is used, and this can collectively be rewritten as

$$\nabla^{\mu} T^{(R)}_{\mu\nu} = \frac{1}{96\pi} \varepsilon_{\mu\nu} \partial^{\mu} R, \tag{2.68}$$

where $\varepsilon_{\mu\nu}$ is the Levi-Civita tensor. This is the covariant gravitational anomaly in this chiral theory.

2.3.3 *Hawking Radiation by Resolving the Gravitational Anomaly*

Since the ingoing modes are not relevant to the effective theory outside the horizon, they should be discarded, leading to a $(1+1)$-dimensional chiral scalar theory. Generally, such a chiral theory is known to have a (consistent) gravitational anomaly in the energy-momentum tensor, obtained from loop diagrams [2, 9]

$$\begin{aligned}\nabla_\mu T_{(\chi)}{}^\mu_\nu &= \frac{1}{\sqrt{-g_{(2)}}}\partial_\mu N^\mu_\nu \\ &\equiv A_\nu,\end{aligned} \tag{2.69}$$

where $g_{(2)\mu\nu}$ is the $(1+1)$-dimensional metric of the $r-t$ section of (2.33)

$$ds^2 = -f(r)dt^2 + \frac{1}{f(r)}dr^2, \tag{2.70}$$

and

$$N^\mu_\nu = \eta\frac{1}{96\pi}\varepsilon^{\beta\mu}\partial_\alpha\Gamma^\alpha_{\nu\beta}, \tag{2.71}$$

in which the convention of the Levi-Civita tensor is $\varepsilon^{01} = -\varepsilon^{10} = 1$. Here we denote η as a sign discrepancy appeared in the literature, such as $\eta = 1$ used in [30–32], whereas $\eta = -1$ used in [41], and we will show that this affects the sign of T^r_t for the flux of Hawking radiation.

On the other hand, there is also a covariant version of the gravitational anomaly, previously derived in (2.68) from a relatively simple set up[7]

$$\begin{aligned}\nabla_\mu \tilde{T}_{(\chi)}{}^\mu_\nu &= -\eta\frac{1}{96\pi}\varepsilon_{\mu\nu}\partial^\mu R \\ &\equiv \tilde{A}_\nu,\end{aligned} \tag{2.72}$$

where $\tilde{T}_{(\chi)}{}^\mu_\nu$ is obtained by adding the Bardeen-Zumino polynomial to $T_{(\chi)}{}^\mu_\nu$ [8], and R is the Ricci scalar of the metric (2.70),[8] and $\varepsilon_{01} = -\varepsilon_{10} = -1$ by lowering indices. In the literature such as [30], the consistent version (2.69) is commonly used to fix the form of the flux of Hawking radiation, and the covariant version (2.72) is used to fix a boundary condition at the horizon in the final step. We choose to follow this derivation. On the other hand, directly using the covariant version also leads to the correct result [5].

---

[7] In [30], there is also a $\sqrt{-g_{(2)}}$ in the denominator (2.72), but this does not affect the result since it is equal to 1.

[8] The Bardeen-Zumino polynomial is basically determined by the right-hand sides of (2.69) and (2.72), so we do not write it down explicitly.

In the $r-t$ section, the non-vanishing components of the Christoffel symbol $\Gamma^{\alpha}_{\nu\beta}$ for the metric (2.33) are

$$\Gamma^{t}_{tr} = \Gamma^{t}_{rt} = \frac{f'}{2f}, \tag{2.73}$$

$$\Gamma^{r}_{tt} = \frac{ff'}{2}, \; \Gamma^{r}_{rr} = -\frac{f'}{2f}, \tag{2.74}$$

and thus the explicit components of N^{μ}_{ν} in the $r-t$ section can thus be calculated with (2.71)

$$\begin{aligned} N^{r}_{t} &= \eta \frac{1}{192\pi}\left(f'^{2} + f'' f\right) \\ N^{t}_{r} &= -\eta \frac{1}{192\pi f^{2}}\left(f'^{2} - f'' f\right) \\ N^{t}_{t} &= 0 \\ N^{r}_{r} &= 0, \end{aligned} \tag{2.75}$$

where the trick to simplify the calculation is that all the components of $\Gamma^{\alpha}_{\nu\beta}$ are t-independent, so only the spatial derivatives $\partial_r \Gamma^{r}_{\nu\beta}$ in (2.71) contribute, which involve the two components in (2.74). With $\Gamma^{\alpha}_{\nu\beta}$, the A_ν in (2.69) is obtained easily

$$\begin{aligned} A_t &= \eta \frac{1}{192\pi}\left(f'^{2} + f'' f\right)' \\ A_r &= 0. \end{aligned} \tag{2.76}$$

Similarly, with the Ricci scalar R, we can determine $\tilde{A}_\mu$ in (2.72)

$$\begin{aligned} R &= -f'' \\ \tilde{A}_t &= \eta \frac{1}{192\pi}\left(2ff'' - f'^{2}\right)' \\ \tilde{A}_r &= 0. \end{aligned} \tag{2.77}$$

For the static metric, $T_{(X)}{}^{\mu}_{\nu}$ is time-independent, and (2.69) reduced to analytically solvable ordinary differential equations

$$\begin{aligned} \partial_r T_{(X)}{}^{r}_{t} &= A_t(r) \\ \partial_r T_{(X)}{}^{r}_{r} + \frac{f'}{2f}\left(2T_{(X)}{}^{r}_{r} - T_{(X)}{}^{\alpha}_{\alpha}\right) &= 0, \end{aligned} \tag{2.78}$$

where $T_{(X)}{}^{\alpha}_{\alpha} = \partial_r T_{(X)}{}^{t}_{t} + \partial_r T_{(X)}{}^{r}_{r}$ is the trace of energy-momentum tensor. The solutions of the differential equations fix all the four components of $T_{(X)}{}^{\mu}_{\nu}$

$$
\begin{aligned}
T_{(\chi)}{}^r_t &= -K_\chi + \int_{r_H}^r A_t(r')dr' \\
&= -f^2 T_{(\chi)}{}^t_r \\
T_{(\chi)}{}^r_r &= \frac{K_\chi + Q_\chi}{f} + \frac{1}{2f}\int_{r_H}^r T_{(\chi)}{}^\alpha_\alpha(r') f'(r')dr' \\
&= T_{(\chi)}{}^\alpha_\alpha - T_{(\chi)}{}^t_t,
\end{aligned}
\tag{2.79}
$$

where K_χ and Q_χ are integration constants.

Now we take a closer look to the energy-momentum tensor of the chiral theory with neglecting ingoing modes near the horizon. Besides the anomalous part in $r_H \le r \le r_H + \varepsilon$, the energy-momentum tensor conserves as usual outside the near-horizon region ($r > r_H + \varepsilon$), so T^μ_ν is separated into two parts

$$
\begin{aligned}
T^\mu_\nu &= \begin{cases} T_{(\chi)}{}^\mu_\nu, & r_H \le r \le r_H + \varepsilon \\ T_{(o)}{}^\mu_\nu, & r > r_H + \varepsilon \end{cases} \\
&= T_{(\chi)}{}^\mu_\nu H + T_{(o)}{}^\mu_\nu \Theta_+,
\end{aligned}
\tag{2.80}
$$

where the Heaviside step functions $\Theta_+ = \Theta(r - r_H - \varepsilon)$, and $H = 1 - \Theta_+$. The divergences of the two parts of T^μ_ν are

$$
\begin{aligned}
\nabla_\mu \left(T_{(\chi)}{}^\mu_\nu H\right) &= \nabla_\mu T_{(\chi)}{}^\mu_\nu H + T_{(\chi)}{}^\mu_\nu \partial_\mu H \\
&= \frac{1}{\sqrt{-g_{(2)}}}\partial_\mu N^\mu_\nu H + T_{(\chi)}{}^\mu_\nu \partial_\mu H \\
&= \frac{1}{\sqrt{-g_{(2)}}}\partial_\mu \left(N^\mu_\nu H\right) + \left(T_{(\chi)}{}^\mu_\nu - \frac{1}{\sqrt{-g_{(2)}}} N^\mu_\nu\right)\partial_\mu H \\
&= \partial_\mu \left(N^\mu_\nu H\right) - \left(T_{(\chi)}{}^\mu_\nu - N^\mu_\nu\right)\partial_\mu \Theta_+,
\end{aligned}
\tag{2.81}
$$

where (2.69) and $\sqrt{-g_{(2)}} = 1$ are applied, and

$$
\begin{aligned}
\nabla_\mu \left(T_{(o)}{}^\mu_\nu \Theta_+\right) &= T_{(o)}{}^\mu_\nu \partial_\mu \Theta_+ + \nabla_\mu T_{(o)}{}^\mu_\nu \Theta_+ \\
&= T_{(o)}{}^\mu_\nu \partial_\mu \Theta_+,
\end{aligned}
\tag{2.82}
$$

where the conservation outside the near-horizon region is applied

$$
\nabla_\mu T_{(o)}{}^\mu_\nu = 0.
\tag{2.83}
$$

The conservation in $r > r_H + \varepsilon$ also fixes the form of $T_{(o)}{}^\mu_\nu$, which is just the homogeneous version of (2.79)

$$
\begin{aligned}
T_{(o)}{}^r_t &= -K_o \\
&= -f^2 T_{(o)}{}^t_r
\end{aligned}
$$

$$\begin{aligned} T_{(o)}{}^{r}_{r} &= \frac{K_o + Q_o}{f} + \frac{1}{2f}\int_{r_H}^{r} T_{(o)}{}^{\alpha}_{\alpha}(r')f'(r')dr' \\ &= T_{(o)}{}^{\alpha}_{\alpha} - T_{(o)}{}^{t}_{t}. \end{aligned} \tag{2.84}$$

To see how the gravitational anomaly should be compensated by some fluxes, we start with the effective action W obtained by path integrating out the scalar field

$$W(g_{\mu\nu}) = -i \log\left(\int \mathscr{D}\varphi e^{iS(\varphi, g_{\mu\nu})}\right). \tag{2.85}$$

Since the fundamental theory is generally covariant, the variation of effective action W under coordinate transformation $x^{\mu} \to x^{\mu} - \lambda^{\mu}$

$$\begin{aligned} -\delta W &= \int d^2x \sqrt{-g_{(2)}}\lambda^{\nu}\nabla_{\mu}T^{\mu}_{\nu} \\ &= 0, \end{aligned} \tag{2.86}$$

implying that the anomaly in the near-horizon region (2.69) has to be cancelled by some fluxes, and we will show that this gives the flux Hawking radiation.

With the divergences of energy-momentum tensors (2.81) and (2.82), the variation of the chiral effective action with neglecting the incoming modes is

$$\begin{aligned} -\delta W_{(\chi)+(o)} = \int d^2x \Big\{ &\lambda^t \left[\partial_r\left(N^r_t H\right) + (T_{(o)}{}^{r}_{t} - T_{(\chi)}{}^{r}_{t} + N^r_t)\partial_r\Theta_+\right] \\ &+ \lambda^r \left[\left(T_{(o)}{}^{r}_{r} - T_{(\chi)}{}^{r}_{r}\right)\partial_r\Theta_+\right]\Big\}. \end{aligned} \tag{2.87}$$

It is noteworthy that the first total-derivative term $\partial_r\left(N^r_t H\right)$ is thought to be cancelled by the quantum effect of the incoming modes in the literature [30, 32, 41], represented by [31]

$$T_{(L)}{}^{\mu}_{\nu} = -N^{\mu}_{\nu}, \tag{2.88}$$

from the corresponding gravitational anomaly with flipping the sign for incoming modes

$$\nabla_{\mu}T_{(L)}{}^{\mu}_{\nu} = -\frac{1}{\sqrt{-g_{(2)}}}\partial_{\mu}N^{\mu}_{\nu}. \tag{2.89}$$

By adding this incoming part, the divergence (2.81) becomes

$$
\begin{aligned}
\nabla_\mu \left(T_{(\chi)}{}^\mu_\nu H + T_{(L)}{}^\mu_\nu H\right) &= \nabla_\mu \left(T_{(\chi)}{}^\mu_\nu + T_{(L)}{}^\mu_\nu\right) H + \left(T_{(\chi)}{}^\mu_\nu + T_{(L)}{}^\mu_\nu\right) \partial_\mu H \\
&= -\left(T_{(\chi)}{}^\mu_\nu - N^\mu_\nu\right) \partial_\mu \Theta_+,
\end{aligned} \tag{2.90}
$$

so the total derivative in (2.87) is canceled. With (2.79), (2.84) and (2.87), the variation of the effective action is evaluated explicitly

$$
\begin{aligned}
-\delta W &= \int d^2x \left[\lambda^t \left(-K_o + K_\chi + N^r_t\right) + \frac{\lambda^r}{f} \left(K_o + Q_o - K_\chi - Q_\chi\right)\right] \delta(r - r_H - \varepsilon) \\
&= 0,
\end{aligned} \tag{2.91}
$$

for arbitrary λ^μ, implying that their coefficients have to be zero, so we get two conditions

$$
\begin{aligned}
-K_o + K_\chi + N^r_t|_{r_H} &= 0 \\
K_o + Q_o - K_\chi - Q_\chi &= 0,
\end{aligned} \tag{2.92}
$$

which satisfy the relations

$$
\begin{aligned}
K_o &= K_\chi + N^r_t|_{r_H} \\
Q_o &= Q_\chi - N^r_t|_{r_H}.
\end{aligned} \tag{2.93}
$$

Such a shift of the energy-momentum tensor can be understood as a flux

$$
\begin{aligned}
\Phi &= \mathrm{sign}(\eta) N^r_t|_{r_H} \\
&= \frac{\kappa^2}{48\pi}.
\end{aligned} \tag{2.94}
$$

To finally fix the flux (K_o), we need to impose a vanishing condition for the covariant energy-momentum tensor at the horizon $\tilde{T}_{(\chi)}{}^\mu_\nu|_{r_H} = 0$, and with (2.76), (2.77) and (2.79) it is

$$
\begin{aligned}
\tilde{T}_{(\chi)}{}^r_t &= T_{(\chi)}{}^r_t + \int_{r_0}^r \tilde{A}_t(r') - A_t(r') dr' \\
&= -K_\chi + \int_{r_H}^r A_t(r') dr' + \eta \frac{f f'' - 2f'^2}{192\pi}
\end{aligned} \tag{2.95}
$$

which r_0 is determined by $f'(r_0) = f''(r_0) = 0$, and this gives $K_\chi = -\eta \frac{\kappa^2}{24\pi}$. Inserting this to (2.93), we have

$$
\begin{aligned}
T_{(o)}{}^r_t &= \eta \frac{\kappa^2}{48\pi} \\
&= \eta \Phi.
\end{aligned} \tag{2.96}
$$

For an outgoing flux of real scalar field, we have $T_{(o)}{}^{r}_{t} = g_{tt} T_{(o)}{}^{rt} = g_{tt} \left(\partial_t \phi(t - r_*)\right)^2 < 0$, suggesting that $\eta = -1$ which is also consistent to the sign derived by the trace anomaly [14] shown in Sect. 2.3.4, and the size of the flux outside the horizon is indeed Φ. By comparing to the flux of thermal radiation with temperature T in $(1 + 1)$ dimension:

$$\begin{aligned} \Phi_{\mathrm{Th}} &= \langle \rho v^r \rangle_{\mathrm{Th}} \\ &= \int_0^{+\infty} \frac{dk}{2\pi} \frac{k}{e^{\frac{k}{T}} - 1} \\ &= \frac{\pi T^2}{12}, \end{aligned} \tag{2.97}$$

where $\langle \ldots \rangle_{\mathrm{Th}}$ means taking a thermal average for the energy flux ρv^r, we obtain the correct Hawking temperature $T_H = \frac{\kappa}{2\pi}$. It is noteworthy that this method only proves the flux as a thermal radiation, and more directly related thermal features such as spectrum cannot be shown explicitly [16]. However, the gravitational-anomaly method can derive the thermal distribution in some cases such as higher-spin currents with a charged black hole [33–35].

Remark on the original derivation by Robinson and Wilczek

There are a few differences between the original derivation in [49] (see also [16, 59]) and the one presented in this chapter, which is based on several subsequent papers [30, 31, 41]:

- **The discarded modes are outgoing modes instead of the ingoing ones**
  As the Boulware vacuum causes a diverging energy-momentum for a freely falling observer at the horizon, they chose to exclude outgoing ("horizon-skimming") modes attributing such a divergence, which also leads to a $(1 + 1)$-dimensional chiral theory. However, most of the subsequent papers adopt the exclusion of ingoing modes in [30], and this difference was also discussed in the appendix of [32] by Iso, Morita and Umetsu, as well as an essay by Das, Robinson and Vagenas [16].
  As explained in [32], both the outgoing and ingoing choices give the same answer in the Schwarzschild black hole, since the factors like f'^2 and f'' in (2.75) are unchanged. However, for the cases of charged or rotating black holes, excluding ingoing modes gives the correct Hawking radiation, whereas another choice cannot, so we adopt the former in this book.
- **Consider also the region inside the horizon and the justification of the total-derivative term**
  Robinson and Wilczek [49] also include the region inside the horizon, and T^{μ}_{ν} (2.80) is now separated into three parts

$$T^{\mu}_{\nu} = \begin{cases} T_{(o)}{}^{\mu}_{\nu}, & r > r_H + \varepsilon \\ T_{(X)}{}^{\mu}_{\nu}, & |r - r_H| \leq \varepsilon \\ T_{(i)}{}^{\mu}_{\nu}, & r < r_H - \varepsilon \end{cases}$$
$$= T_{(X)}{}^{\mu}_{\nu} H + T_{(o)}{}^{\mu}_{\nu} \Theta_+ + T_{(i)}{}^{\mu}_{\nu} \Theta_-, \tag{2.98}$$

where the Heaviside step functions $\Theta_\pm = \Theta(\pm r \mp r_H - \varepsilon)$, and the boxcar (top-hat) function $H = 1 - \Theta_+ - \Theta_-$. Correspondingly, the divergences (2.81) and (2.82) now become

$$\nabla_\mu \left(T_{(X)}{}^{\mu}_{\nu} H\right) = \nabla_\mu T_{(X)}{}^{\mu}_{\nu} H + T_{(X)}{}^{\mu}_{\nu} \partial_\mu H$$
$$= \partial_\mu \left(N^{\mu}_{\nu} H\right) - \left(T_{(X)}{}^{\mu}_{\nu} - N^{\mu}_{\nu}\right) \partial_\mu \left(\Theta_+ + \Theta_-\right), \tag{2.99}$$

and

$$\nabla_\mu \left(T_{(o)}{}^{\mu}_{\nu} \Theta_+ + T_{(i)}{}^{\mu}_{\nu} \Theta_-\right) = T_{(o)}{}^{\mu}_{\nu} \partial_\mu \Theta_+ + T_{(i)}{}^{\mu}_{\nu} \partial_\mu \Theta_-, \tag{2.100}$$

respectively. The energy-momentum tensor in the region $|r - r_H| > \varepsilon$ is conserved $\nabla_\mu T_{(o/i)}{}^{\mu}_{\nu} = 0$, and this fixes the form of $T_{(o/i)}{}^{\mu}_{\nu}$:

$$T_{(o/i)}{}^{r}_{t} = -K_{o/i}$$
$$= -f^2 T_{(o/i)}{}^{t}_{r}$$
$$T_{(o/i)}{}^{r}_{r} = \frac{K_{o/i} + Q_{o/i}}{f} + \frac{1}{2f} \int_{r_H}^{r} T_{(o/i)}{}^{\alpha}_{\alpha}(r') f'(r') dr'$$
$$= T_{(o/i)}{}^{\alpha}_{\alpha} - T_{(o/i)}{}^{t}_{t}. \tag{2.101}$$

With these modifications, the variation of effective action (2.91) now includes more terms involving $T_{(i)}{}^{\mu}_{\nu}$ and Θ_-

$$\delta W$$
$$= \int d^2x \Big\{ \lambda^t \left[\partial_r \left(N^r_t H\right) + (T_{(o)}{}^{r}_{t} - T_{(X)}{}^{r}_{t} + N^r_t) \partial_r \Theta_+ + (T_{(i)}{}^{r}_{t} - T_{(X)}{}^{r}_{t} + N^r_t) \partial_r \Theta_-\right]$$
$$+ \lambda^r \left[\left(T_{(o)}{}^{r}_{r} - T_{(X)}{}^{r}_{r}\right) \partial_r \Theta_+ + \left(T_{(i)}{}^{r}_{r} - T_{(X)}{}^{r}_{r}\right) \partial_r \Theta_-\right] \Big\}. \tag{2.102}$$

The total-derivative term $\partial_r \left(N^r_t H\right)$ in (2.102) is assumed to be vanishing when $\varepsilon \to 0$ [49, 50]. But to the knowledge of the chapter's author, the explicit justification of this is unclear:

$$\partial_r \left(N^r_t H\right) \approx 2\varepsilon N^r_t \delta'(r - r_H) + \mathcal{O}(\varepsilon^2), \tag{2.103}$$

as this also contributes a term with $\varepsilon\delta'(r - r_H)$ which will be shown relevant to determine the flux. On the other hand, the same total-derivative term in (2.87) is argued to be cancelled by the quantum effect of the incoming modes [30].

- **More conditions to fix the flux**
  To fix the flux from (2.102), we need more conditions to determine K_i and Q_i in (2.101), requiring the derivative of Heaviside step functions to be expanded to $\mathcal{O}(\varepsilon)$

$$\partial_\mu \Theta_\pm \approx \delta^r_\mu \left(\pm\delta(r - r_H) - \varepsilon\delta'(r - r_H)\right) + \mathcal{O}(\varepsilon^2), \tag{2.104}$$

and thus (2.102) (if the total-derivative term is neglected) becomes

$$\begin{aligned}\delta W \approx & \int d^2x \Big\{\lambda^t \left[(K_o - K_i)\delta(r - r_H) - \varepsilon\left(K_o + K_i - 2K_\chi - 2N^r_t\right)\delta'(r - r_H)\right] \\ & - \lambda^r \Big[\frac{K_o + Q_o + K_i + Q_i - 2K_\chi - 2Q_\chi}{f}\delta(r - r_H) \\ & - \varepsilon\frac{K_o + Q_o - K_i - Q_i}{f}\delta'(r - r_H)\Big]\Big\} + \mathcal{O}(\varepsilon^2),\end{aligned} \tag{2.105}$$

By requiring $\delta W = 0$ for arbitrary λ^μ, we require all the coefficients of $\delta(r - r_H)$ and $\delta'(r - r_H)$ to be zero, giving four equations

$$\begin{aligned} K_o - K_i &= 0 \\ K_o + K_i - 2K_\chi - 2N^r_t|_{r_H} &= 0 \\ K_o + Q_o + K_i + Q_i - 2K_\chi - 2Q_\chi &= 0 \\ K_o + Q_o - K_i - Q_i &= 0, \end{aligned} \tag{2.106}$$

which satisfy the relations

$$\begin{aligned} K_o &= K_i = K_\chi + N^r_t|_{r_H} \\ Q_o &= Q_i = Q_\chi - N^r_t|_{r_H}. \end{aligned} \tag{2.107}$$

This gives the same condition to determine the Hawking radiation (2.93), but more boundary condition, such as the vanishing $\tilde{T}^\mu_\nu$ at the horizon, is still needed to determine the value of K_o.

It is noteworthy that the derivation with gravitational anomaly only requires the information near the horizon, such as the $(1 + 1)$-dimensional approximation (2.46) and N^μ_ν (2.94) at the horizon. We will see in Sect. 2.3.4 that the old derivation by trace anomaly requires more assumptions.

2.3.4 Hawking Radiation from the (1 + 1)-Dimensional Trace Anomaly

Before ending the section of anomaly method, we introduce an old derivation of Hawking radiation from the (1 + 1)-dimensional trace anomaly [14], and we will see some similarities and differences compared to gravitational-anomaly method in Sect. 2.3.3.

Classically, the energy-momentum tensor of a (1 + 1)-dimensional massless scalar field

$$T_{\nu}^{\mu} = \partial^{\mu}\varphi\partial_{\nu}\varphi - \frac{\delta_{\nu}^{\mu}}{2}\partial^{\alpha}\phi\partial_{\alpha}\varphi, \tag{2.108}$$

which clearly has zero trace $T_{\alpha}^{\alpha} = 0$. However, when consider a (renormalized) quantum expectation value of the trace, this can acquire a non-zero value related to the background geometry [17], as shown in (2.64). With this trace anomaly for the metric (2.70), Christensen and Fulling [14] showed that the flux of the Hawking radiation is determined by the conservation

$$\nabla_{\mu}T_{\nu}^{\mu} = 0, \tag{2.109}$$

which has been solved in (2.84) when we consider the outside region in the gravitational-anomaly method, and the solution depends on two constants K and Q, as well as a function T_{α}^{α}. These three unknown constants and function are fixed now as follows, and there are some conditions different to the gravitational-anomaly method.

- The trace anomaly (2.64) fixes T_{α}^{α} and thus the integral term in (2.84)

$$T_{\alpha}^{\alpha} = -\frac{f''}{24\pi}$$
$$\frac{1}{2f}\int_{r_H}^{r} T_{\alpha}^{\alpha}(r')f'(r')dr' = -\frac{f'^2 - (2\kappa)^2}{96\pi f}. \tag{2.110}$$

It is noteworthy that the gravitational-anomaly method does not rely on T_{α}^{α} to determine the integration constants in (2.93), which is replaced by the gravitational anomaly in $\nabla_{\mu}T_{\nu}^{\mu}$.

- The T_{uu} vanishes at least quadratically at the horizon

$$|T_{uu}f^{-2}||_{r_H} < \infty, \tag{2.111}$$

for no singularity for a freely falling observer, implying that

$$\frac{1}{4}\left(-fT_t^t + fT_r^r - 2T_t^r\right)\Big|_{r_H} = K + \frac{Q}{2}$$
$$= 0, \tag{2.112}$$

where (2.84) is applied, and this gives one condition of K and Q (the definitions of K and Q in [14] are different to ours: $K_{\text{there}} = -K$ and $Q_{\text{there}} = -2K - Q$). For the gravitational-anomaly method, this is replaced by the vanishing boundary condition for the covariant energy-momentum tensor at the horizon $\tilde{T}_t^r = 0$.

- At spatial infinity, the energy density is equal to the flux $T_t^t = T_t^r$ because for the outgoing modes of scalar field

$$T_t^t = T_t^r = -\left(\partial_t \phi(t-r)\right)^2, \tag{2.113}$$

which gives another condition of K and Q:

$$-K - Q - \frac{\kappa^2}{24\pi} = -K. \tag{2.114}$$

For the gravitational-anomaly method, it is not required to have such a matching at the infinity, which is replaced by a matching between the chiral and non-chiral parts at $r = r_H + \varepsilon$ (near the horizon) in order to cancel the anomaly for the full theory.

These conditions imply the same flux as the one by gravitational anomaly (2.96), that

$$T_t^r = -K$$
$$= -\frac{\kappa^2}{48\pi}, \tag{2.115}$$

and again this is consistent to the previous choice that $\eta = -1$ in (2.71).

The drawback of the trace anomaly is restricted to the case of massless scalar in the whole spacetime (not just the approximation near the horizon), and therefore this seems to be a special feature. This in fact motivates the derivation of gravitational anomaly, introduced in the previous Sect. 2.3.3, which is applicable in more general cases. On the other hand, by comparing the consistent answers by the two anomaly methods, one can also justify some boundary conditions essential to derive the results, such as the vanishing covariant energy-momentum tensor at the horizon used in the gravitational anomaly [41].

2.4 The Green's Function Method

Soon after Hawking's original derivation, the thermal feature of the Green's function for particles propagating in the black hole's spacetime was noticed [23, 24], and its mathematical properties were used in the well-known Hartle-Hawking path-integral derivation [27]. The thermal character of Green's function has also been generalized to the cases of particle production in cosmology [13, 25] and accelerated frame [56, 57], suggesting the universality of this method in various flat or curved spacetimes.

2.4.1 *Thermal Green's Function*

The thermal feature of Hawking radiation can be seen in the periodicity of Green's function when analytically continued to complex coordinates, consistent to the behavior of a thermal Greens's function [23, 24]. We leave the detailed analysis of the analytical properties of Hawking radiation in Sect. 2.4.2, and here we first discuss the thermal case.

For a system in a thermal state described by the following density matrix

$$\rho = \frac{e^{-\beta H}}{\mathrm{Tr}\left(e^{-\beta H}\right)}, \tag{2.116}$$

where $\beta = \frac{1}{T}$ is the inverse of temperature, and H is the system's Hamiltonian, and the expectation value of operators is defined by

$$\begin{aligned} \langle \ldots \rangle_\beta &= \mathrm{Tr}(\rho \ldots) \\ &= \frac{\mathrm{Tr}\left(e^{-\beta H} \ldots\right)}{\mathrm{Tr}\left(e^{-\beta H}\right)}. \end{aligned} \tag{2.117}$$

In particular, for a scalar field in the Heisenberg picture

$$\phi(t, \mathbf{x}) = e^{iHt}\phi(0, \mathbf{x})e^{-iHt}, \tag{2.118}$$

the following two Green functions

$$\begin{aligned} G_T^+(t, \mathbf{x}; t', \mathbf{x}') &= i\frac{\mathrm{Tr}\left(e^{-\beta H}\phi(t, \mathbf{x})\phi(t', \mathbf{x}')\right)}{\mathrm{Tr}\left(e^{-\beta H}\right)} \\ &= i\frac{\mathrm{Tr}\left(e^{-\beta H}\phi(t, \mathbf{x})e^{\beta H}e^{-\beta H}\phi(t', \mathbf{x}')\right)}{\mathrm{Tr}\left(e^{-\beta H}\right)} \\ &= i\frac{\mathrm{Tr}\left(\phi(t + i\beta, \mathbf{x})e^{-\beta H}\phi(t', \mathbf{x}')\right)}{\mathrm{Tr}\left(e^{-\beta H}\right)} \end{aligned}$$

$$
\begin{aligned}
&= i\frac{\mathrm{Tr}\left(e^{-\beta H}\phi(t',\mathbf{x}')\phi(t+i\beta,\mathbf{x})\right)}{\mathrm{Tr}\left(e^{-\beta H}\right)} \\
&= G_T^-(t+i\beta,\mathbf{x};t',\mathbf{x}'),
\end{aligned} \tag{2.119}
$$

and similarly

$$
\begin{aligned}
G_T^-(t,\mathbf{x};t',\mathbf{x}') &= i\frac{\mathrm{Tr}\left(e^{-\beta H}\phi(t',\mathbf{x}')\phi(t,\mathbf{x})\right)}{\mathrm{Tr}\left(e^{-\beta H}\right)} \\
&= i\frac{\mathrm{Tr}\left(e^{-\beta H}e^{\beta H}\phi(t,\mathbf{x})e^{-\beta H}\phi(t',\mathbf{x}')\right)}{\mathrm{Tr}\left(e^{-\beta H}\right)} \\
&= i\frac{\mathrm{Tr}\left(e^{-\beta H}\phi(t-i\beta,\mathbf{x})\phi(t',\mathbf{x}')\right)}{\mathrm{Tr}\left(e^{-\beta H}\right)} \\
&= G_T^+(t-i\beta,\mathbf{x};t',\mathbf{x}'),
\end{aligned} \tag{2.120}
$$

give the Kubo-Martin-Schwinger (KMS) condition for a system in thermal equilibrium [38, 39]. From (2.119) and (2.120), the thermal Green function

$$
G_T(t,\mathbf{x};t',\mathbf{x}') = i\frac{\mathrm{Tr}\left(e^{-\beta H}T\phi(t,\mathbf{x})\phi(t',\mathbf{x}')\right)}{\mathrm{Tr}\left(e^{-\beta H}\right)}, \tag{2.121}
$$

has to be defined in the complex t-plane, and there is a subtlety in defining the time-ordering T in a complex way [22]:

$$
T\phi(t,\mathbf{x})\phi(t',\mathbf{x}') = \begin{cases}
\phi(t,\mathbf{x})\phi(t',\mathbf{x}') & \mathrm{Re}(t) > \mathrm{Re}(t'),\ \mathrm{Im}(t) = \mathrm{Im}(t') \\
\phi(t',\mathbf{x}')\phi(t,\mathbf{x}) & \mathrm{Re}(t) < \mathrm{Re}(t'),\ \mathrm{Im}(t) = \mathrm{Im}(t') \\
\phi(t,\mathbf{x})\phi(t',\mathbf{x}') & \mathrm{Im}(t) < \mathrm{Im}(t') \\
\phi(t',\mathbf{x}')\phi(t,\mathbf{x}) & \mathrm{Im}(t) > \mathrm{Im}(t')
\end{cases} \tag{2.122}
$$

where $|\mathrm{Im}(t'-t')| < \beta$. Such a definition puts smaller imaginary time to the left is consistent with the case to put larger real time to the left if $\mathrm{Im}(t) = \mathrm{Im}(t')$, by the analytic continuation shown in Fig. 2.2. For $\mathrm{Im}(t-t') = -\varepsilon$ (making $\mathrm{Im}(t-t'+i\beta) < \beta$), the thermal Green's function with the time ordering (2.122) is

$$
\begin{aligned}
G_T(t+i\beta,\mathbf{x};t',\mathbf{x}') &= G_T^-(t+i\beta,\mathbf{x};t',\mathbf{x}') \\
&= G_T^+(t,\mathbf{x};t',\mathbf{x}') \\
&= G_T(t,\mathbf{x};t',\mathbf{x}'),
\end{aligned} \tag{2.123}
$$

where (2.119) is applied. Similarly, for $\mathrm{Im}(t-t') = \varepsilon$ (making $\mathrm{Im}(t-t'-i\beta) > -\beta$),

$$
\begin{aligned}
G_T(t-i\beta,\mathbf{x};t',\mathbf{x}') &= G_T^+(t-i\beta,\mathbf{x};t',\mathbf{x}') \\
&= G_T^-(t,\mathbf{x};t',\mathbf{x}')
\end{aligned}
$$

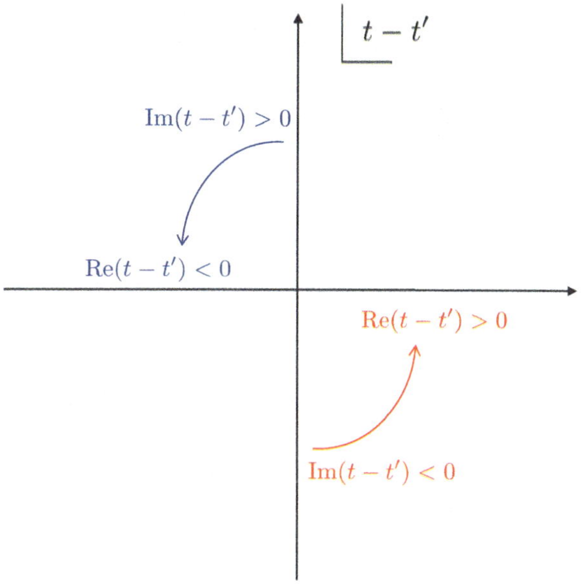

Fig. 2.2 The time ordering defined in (2.122) agrees with the analytic continuation

$$= G_T(t, \mathbf{x}; t', \mathbf{x}'), \tag{2.124}$$

where (2.120) is applied. Therefore, the thermal Green's function is periodic in the imaginary time with the period of inverse temperature $i\beta = i\frac{1}{T}$, and such a characteristic for a thermal system can also be seen in a system with a black hole.

2.4.2 The Path-Integral Approach

We follow the path-integral approach by Hartle and Hawking [27], in which the Hawking radiation is manifested by comparing the amplitudes of emitting or absorbing a particle by a black hole. The feature of this approach is to obtain the propagator in the field-theory sense from the path integral of point particles via the Schwinger's proper time formalism [51, 52], providing a more intuitive derivation compared to the direct approach with Green's function.

For a point particle, we choose its action as

$$S[x] = \frac{1}{4}\int_0^W g_{\mu\nu}\frac{dx^\mu}{dw}\frac{dx^\nu}{dw}dw, \tag{2.125}$$

where w and W are parameter times, which play the roles of the Schwinger's proper time in the curved spacetime. Such a quadratic form of action, which is different to the usual action by proper time, has advantages in calculating path integrals. The amplitude for a particle from x' at $w = 0$ to x at $w = W$ is a path integral, obtained by integrals over spacetime in infinitely many small time intervals with size $\varepsilon = \frac{W}{N+1}$:

$$\begin{aligned} & F(W, x, x') \\ &= \int \mathscr{D}x\, e^{iS[x]} \\ &= \lim_{N\to\infty} \int \frac{d^4x_N}{A}\sqrt{-g(x_N)} \ldots \int \frac{d^4x_1}{A}\sqrt{-g(x_1)} \exp\left(i \sum_{i=0}^{N} S\left(\varepsilon, x_{i+1}, x_i\right)\right), \end{aligned} \tag{2.126}$$

where $x_0 = x'$ and $x_{N+1} = x$ correspond to the two end points, and the normalization is

$$A = \int d^4x\sqrt{-g(x)}e^{iS(\varepsilon,x,x')}. \tag{2.127}$$

It is expected that the path integral (2.126) is oscillating rapidly, and the standard procedure is to analytically continue the time variable to imaginary value such that the integral is well-defined. For a Schwarzschild black hole in the Kruskal coordinates

$$ds^2 = \frac{32M^3e^{-\frac{r}{2M}}}{r}\left(-dT^2 + dX^2\right) + r^2 d\Omega^2, \tag{2.128}$$

where the T and X satisfy

$$\begin{aligned} T &= \begin{cases} \left(\frac{r}{2M} - 1\right)^{\frac{1}{2}} e^{\frac{r}{4M}} \sinh\left(\frac{t}{4M}\right) & r > 2M \\ \left(1 - \frac{r}{2M}\right)^{\frac{1}{2}} e^{\frac{r}{4M}} \cosh\left(\frac{t}{4M}\right) & r < 2M \end{cases} \\ X &= \begin{cases} \left(\frac{r}{2M} - 1\right)^{\frac{1}{2}} e^{\frac{r}{4M}} \cosh\left(\frac{t}{4M}\right) & r > 2M \\ \left(1 - \frac{r}{2M}\right)^{\frac{1}{2}} e^{\frac{r}{4M}} \sinh\left(\frac{t}{4M}\right) & r < 2M \end{cases} \\ -T^2 + X^2 &= \left(\frac{r}{2M} - 1\right) e^{\frac{r}{2M}}, \end{aligned} \tag{2.129}$$

the analytic continuation is done by letting $T = i\zeta$, and the parameter times w and W are continued to negative imaginary values $-i\omega$ and $-i\Omega$ respectively. The corresponding metric $\gamma_{\mu\nu}$ is now defined by

$$ds^2 = \frac{32M^3e^{-\frac{r}{2M}}}{r}\left(d\zeta^2 + dX^2\right) + r^2 d\Omega^2, \tag{2.130}$$

and another useful expression in the following calculation is the local Riemann normal coordinates around a point x with a small deviation z

$$\gamma_{\mu\nu} \approx \delta_{\mu\nu} - \frac{1}{3} R_{\mu\alpha\nu\beta} z^\alpha z^\beta + \mathscr{O}(z^3). \tag{2.131}$$

With these, the normalization factor A (2.127) evaluated from x to $x + z$ within a small interval ε is analytically continued to

$$\begin{aligned} A = & \int d^4 z \sqrt{\gamma(x+z)} e^{-S(\varepsilon, x, x+z)} \\ \approx & \int d^4 z \left(1 - \frac{1}{6} R_{\mu\nu} z^\mu z^\nu + \mathscr{O}(z^3)\right) e^{-\frac{1}{4\varepsilon}\delta_{\mu\nu} z^\mu z^\nu} \\ \approx & \; 16\pi^2 \varepsilon^2 + \mathscr{O}(\varepsilon^3). \end{aligned} \tag{2.132}$$

Instead of directly evaluating the path-integral (2.126), solving the partial differential equation satisfied by it is an easier way, obtained by the change of the (analytically continued) amplitude $F(\Omega + \varepsilon, x, x')$ with respect to the parameter time

$$\begin{aligned} F(\Omega + \varepsilon, x, x') = & \int \frac{d^4 y}{A} \sqrt{\gamma(y)} e^{-S(\varepsilon, x, y)} F(\Omega, y, x') \\ = & \int \frac{d^4 z}{A} \sqrt{\gamma(x+z)} e^{-S(\varepsilon, x, x+z)} F(\Omega, x+z, x') \\ \approx & \int \frac{d^4 z}{A} e^{-\frac{1}{4\varepsilon}\delta_{\mu\nu} z^\mu z^\nu} \left[\left(1 - \frac{1}{6} R_{\mu\nu} z^\mu z^\nu\right)\left(1 + z^\alpha \partial_\alpha + \frac{1}{2} z^\alpha z^\beta \partial_\alpha \partial_\beta\right)\right. \\ + & \left. \mathscr{O}(z^3)\right] F(\Omega, x, x') \\ = & \; F(\Omega, x, x') + \varepsilon \delta^{\mu\nu} \left(\partial_\mu \partial_\nu - \frac{R_{\mu\nu}}{3}\right) F(\Omega, x, x'), \end{aligned} \tag{2.133}$$

where $\mathscr{O}(z^3)$ terms are neglected because they contribute higher orders in ε after integrating over z. (2.133) obtained with the local normal coordinates can be generalized to the covariant form by replacing $\delta^{\mu\nu} \to \gamma^{\mu\nu}$ and $\partial_\mu \partial_\nu \to \nabla_\mu \nabla_\nu$, and thus we have the partial differential equation

$$\begin{aligned} \frac{\partial F}{\partial \Omega} & = \left(\gamma^{\mu\nu} \nabla_\mu \nabla_\nu - \frac{R}{3}\right) F \\ & = \gamma^{\mu\nu} \nabla_\mu \nabla_\nu F, \end{aligned} \tag{2.134}$$

as $R = 0$ for the Schwarzschild black hole. By analytically continuing back to the original metric (2.128), we have

$$i \frac{\partial F}{\partial W} = -g^{\mu\nu} \nabla_\mu \nabla_\nu F, \tag{2.135}$$

which has the form of the Schrödinger equation and satisfies the initial condition

$$F(0, x, x') = \frac{1}{\sqrt{-g}} \delta^4(x - x'). \tag{2.136}$$

With a small W, $F(W, x, x')$ has a WKB ansatz

$$\begin{aligned} F(W, x, x') &\approx N(W, x, x') e^{i S(W, x, x')} \\ &= N(W, x, x') e^{i \frac{s(x, x')}{4W}}, \end{aligned} \tag{2.137}$$

where $s(x, x')$ is the square of distance between x and x'. For flat space, one can easily check that

$$N(W) = \frac{i}{(4\pi W)^2}, \tag{2.138}$$

gives an exact solution of (2.135). For general curved spacetime, initial condition (2.136) implies that the solution for small W is dominated by the neighborhood of x, which can be approximated by local flatness, so it is expected that

$$N(W, x, x') \approx \frac{D(x, x')}{W^2}, \tag{2.139}$$

generally, and a more detailed discussion can be found in [40]. We follow the Schwinger's proper time formalism to obtain the Feynman propagator between the two points x and x' by a weighted sum of the amplitude over all possible W

$$G_F(x, x') = \lim_{\varepsilon \to 0^+} i \int_0^{+\infty} dW \, e^{-im^2 W - \frac{\varepsilon}{W}} F(W, x, x'), \tag{2.140}$$

where m is the particle's mass, and we introduce a small suppression factor $e^{-\frac{\varepsilon}{W}}$ to regulate the integral for avoiding the divergence at $W \to 0$ due to (2.139). With (2.135), one can easily check that

$$\begin{aligned} \left(g^{\mu\nu} \nabla_\mu \nabla_\nu - m^2\right) G_F &= i \int_0^{+\infty} dW \, e^{-im^2 W} \left(g^{\mu\nu} \nabla_\mu \nabla_\nu - m^2\right) F(W, x, x') \\ &= \int_0^{+\infty} dW \, e^{-im^2 W} \left(\frac{\partial}{\partial W} - im^2\right) F \\ &= \int_0^{+\infty} dW \frac{\partial}{\partial W} \left(e^{-im^2 W} F\right) \\ &= -F(0, x, x') \\ &= -\frac{1}{\sqrt{-g}} \delta^4(x - x'), \end{aligned} \tag{2.141}$$

where the initial condition (2.136) and boundary condition $\lim_{W\to+\infty} F(W, x, x') = 0$ are applied.[9] Therefore, the propagator $G_F(x, x')$ defined in (2.140) is indeed the Green's function of the Klein-Gordon equation.

To derive Hawking radiation, the relationship between the amplitudes of emitting and absorbing a particle by a black hole is needed, and this is determined by the analytical properties of the propagator $G_F(x, x')$ as follows:

- $G_F(x, x')$ is periodic in $\mathrm{Im}(t)$ with a period $8\pi M$, which can be easily seen from the periodicity of X and T in the Kruskal coordinates (2.129), consistent to the thermal Green's function discussed in Sect. 2.4.1 with temperature $T_H = \frac{1}{8\pi M}$.
- For x a point with fixed (r, θ, φ) inside the horizon and x' a fixed point outside the black hole, $G_F(x, x')$ is analytical in a strip in the complex t plane with $-4\pi M \le \mathrm{Im}(t) \le 0$. By deforming the integration contour for calculating the amplitude of emitting a particle to $\mathrm{Im}(t) = -4\pi M$, the amplitude of absorbing is obtained.

Now we demonstrate the second point, starting by proving the analyticity. From (2.140), we know that the singular points of $G_F(x, x')$ are contributed by the small-W part of the integrand, where the regularization with ε is important, and we have seen that this part has the WKB form with summing over all geodesics connecting x and x'

$$F(W, x, x') \approx \sum_c \frac{D_c(x, x')}{W^2} e^{i\frac{s_c(x,x')}{4W}}. \tag{2.142}$$

With this, we can calculate $G_F(x, x')$ with $m = 0$ by separating the integral by a singular part $[0, W_0]$ (W_0 is small) and an analytical part $[W_0, +\infty]$ denoted by $G_{F,0}$:

$$G_F(x, x') \approx -i \sum_c \frac{e^{i\frac{s_c(x,x')}{4W_0}}}{s_c(x, x') + i\varepsilon} D_c(x, x') + G_{F,0}, \tag{2.143}$$

so the singular points are located at $s_c(x, x') + i\varepsilon = 0$. To see the locations of these points for null geodesics from x' to x, it is convenient to define the null variables

$$\begin{aligned} U &= T - X \\ V &= T + X \\ ds^2 &= -32\frac{M^3}{r} e^{-\frac{r}{2M}} dU dV + r^2 d\Omega^2, \end{aligned} \tag{2.144}$$

---

[9] It relies on the fact that the parabolic Eq. (2.135) spreads the initial localized data (2.136) at late time (large W) as $F(W, x, x') \approx \mathcal{O}(W^{-2})$.

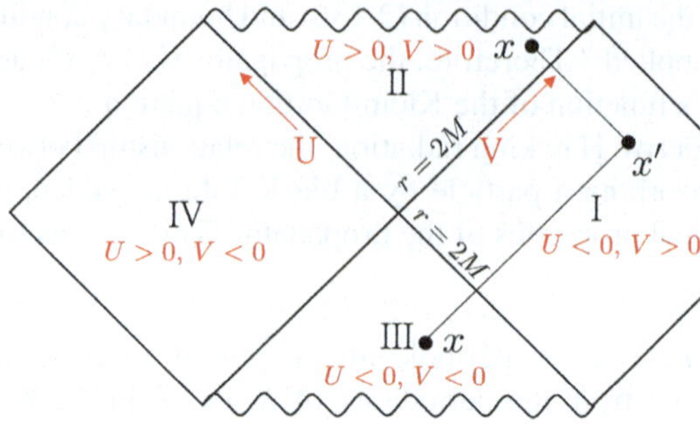

Fig. 2.3 The Penrose diagram for a Schwarzschild black hole. We denote x' as an observer in region I, and x is the point of emitting (region II) or absorbing (region III) particles

and four regions are defined according to the signs of the (U, V) pair, according to the Penrose diagram in Fig. 2.3. Therefore, a null geodesic with $s(x, x') = 0$ connecting regions I and II is $V = V_0$. Now with the $i\varepsilon$ prescription, the singular points are determined by[10]

$$\begin{aligned} s(x, x') &\approx \frac{\partial s}{\partial V}(V - V_0) + \mathscr{O}(\varepsilon^2) \\ &= -i\varepsilon, \end{aligned} \tag{2.145}$$

where for the simplest case with $d\Omega^2 = 0$ in (2.144)[11]

$$\begin{aligned} \frac{\partial s}{\partial V} &= -\int_{U(x')}^{U(x)} \frac{32M^3}{r} e^{-\frac{r}{2M}} dU \\ &< 0, \end{aligned} \tag{2.146}$$

where we use the fact that $U(x) - U(x') > 0$ since

$$U = \begin{cases} -\left(\frac{r}{2M} - 1\right)^{\frac{1}{2}} e^{\frac{r-t}{4M}} < 0 & \text{region I } (r > 2M) \\ \left(1 - \frac{r}{2M}\right)^{\frac{1}{2}} e^{\frac{r-t}{4M}} > 0 & \text{region II } (r < 2M) \\ \left(1 - \frac{r}{2M}\right)^{\frac{1}{2}} e^{\frac{r-t}{4M}} < 0 & \text{region III } (r < 2M), \end{cases} \tag{2.147}$$

and therefore we have $\mathrm{Im}(V - V_0) > 0$ from (2.145). From the relationship between V and (r, t)

$$V = \begin{cases} \left(\frac{r}{2M} - 1\right)^{\frac{1}{2}} e^{\frac{r+t}{4M}} > 0 & \text{region I } (r > 2M) \\ \left(1 - \frac{r}{2M}\right)^{\frac{1}{2}} e^{\frac{r+t}{4M}} > 0 & \text{region II } (r < 2M) \\ \left(1 - \frac{r}{2M}\right)^{\frac{1}{2}} e^{\frac{r+t}{4M}} < 0 & \text{region III } (r < 2M), \end{cases} \tag{2.148}$$

[10] Do not be confused with the notation that $s(x, x')$ is the square of distance.

[11] For more general cases with $\Delta\varphi \neq 0$ is considered in [27].

we know that the singular points correspond to $\mathrm{Im}(t) = \varepsilon$ in the region II. On the other hand, for a point x in the region III with $\mathrm{Im}(t) = -4\pi M$, the corresponding null geodesic with $s(x, x') = 0$ is $U = U_0$. Similar to (2.145), the singular points in U is determined by

$$\begin{aligned} s(x, x') &\approx \frac{\partial s}{\partial U}(U - U_0) + \mathcal{O}(\varepsilon^2) \\ &= -i\varepsilon, \end{aligned} \tag{2.149}$$

and

$$\begin{aligned} \frac{\partial s}{\partial U} &= -\int_{V(x')}^{V(x)} \frac{32M^3}{r} e^{-\frac{r}{2M}} dV \\ &> 0, \end{aligned} \tag{2.150}$$

where $V(x) - V(x') < 0$ from (2.148), so we have $\mathrm{Im}(U - U_0) < 0$. From (2.147), we know that the corresponding singular points in the region III has $\mathrm{Im}(t) = -4\pi M - \varepsilon$. Therefore, the two cases with the regions II and III prove the fact that $G_F(x, x')$ is analytical in a strip with $-4\pi M \leq \mathrm{Im}(t) \leq 0$.

With the analyticity of G_F, we can calculate the amplitude of emitting and absorbing particles from the black hole. As a demonstration of the relationship between the amplitude and the propagator, we pick a field ϕ to calculate. Consider initial and final states, $|f_i\rangle$ and $|g_f\rangle$ respectively [24]

$$\begin{aligned} |f_i\rangle &= \int d\Sigma^{\mu} f(x) \overleftrightarrow{\partial_{\mu}} \phi(x)|0_-\rangle \\ |g_f\rangle &= \int d\Sigma'^{\nu'} g(x') \overleftrightarrow{\partial_{\nu'}} \phi(x')|0_+\rangle, \end{aligned} \tag{2.151}$$

where $|0_{\pm}\rangle$, Σ and Σ' are the vacua and hypersurfaces at initial and final times, and $f \overleftrightarrow{\partial_{\mu}} g = f(\partial_{\mu} g) - (\partial_{\mu} f)g$. Therefore, the amplitude from the initial to final states corresponds to a detector at x' receiving a particle from x

$$\begin{aligned} i\frac{\langle g_f|f_i\rangle}{\langle 0_+|0_-\rangle} &= \int\int \frac{d\Sigma^{\mu} d\Sigma'^{\nu'}}{\langle 0_+|0_-\rangle} \langle 0_+| \Big[g^* \partial_{\nu'} \left(f\phi(x')\partial_{\mu}\phi(x) - \phi(x')\phi(x)\partial_{\mu} f \right) \\ &\quad - \left(f\phi(x')\partial_{\mu}\phi(x) - \phi(x')\phi(x)\partial_{\mu} f \right) \partial_{\nu'} g^* \Big] |0_-\rangle \\ &= \int\int d\Sigma^{\mu} d\Sigma'^{\nu'}\, g^*_{\nu'} \left(f\partial_{\mu} G_F - G_F \partial_{\mu} f \right) - \left(f\partial_{\mu} G_F - G_F \partial_{\mu} f \right) \partial_{\nu'} g^* \\ &= \int\int d\Sigma^{\mu} d\Sigma'^{\nu'}\, g^* \overleftrightarrow{\partial_{\nu'}} \left(f \overleftrightarrow{\partial_{\mu}} G_F \right) \\ &= -\int\int d\Sigma^{\mu} d\Sigma'^{\nu'}\, g^*(x') \overleftrightarrow{\partial_{\nu'}} \left(G_F(x', x) \overleftrightarrow{\partial_{\mu}} f(x) \right), \end{aligned} \tag{2.152}$$

where we used

$$G_F(x', x) = i\frac{\langle 0_+|T\phi(x')\phi(x)|0_-\rangle}{\langle 0_+|0_-\rangle}. \tag{2.153}$$

(2.152) implies that we can determine the amplitude of emission by the propagator that obtained previously in (2.140). Due to the time translational symmetry of the Schwarzschild black hole, we expect that there exists modes with $g(x') \propto e^{-iEt'}$ and $f(x) \propto e^{-iEt}$, together with the form of the propagator

$$G_F(x', x) = G_F(0, \mathbf{r}'; t - t', \mathbf{r}) . \tag{2.154}$$

By choosing the hypersurface with constant r in region II, the amplitude (2.152) is proportional to

$$\begin{aligned} \mathscr{E}_E(\mathbf{r}', \mathbf{r}) &= \int_{-\infty}^{+\infty} dt\, e^{-iEt} G_F(0, \mathbf{r}'; t, \mathbf{r}) \\ &= e^{-4\pi ME} \int_{-\infty}^{+\infty} dt\, e^{-iEt} G_F(0, \mathbf{r}'; t - 4i\pi M, \mathbf{r}) \\ &= e^{-4\pi ME} \tilde{\mathscr{E}}_E(\mathbf{r}', \mathbf{r}), \end{aligned} \tag{2.155}$$

where we deform the contour to $\mathrm{Im}(t) = -4\pi M$ in the second line, and $\tilde{\mathscr{E}}_E(\mathbf{r}', \mathbf{r})$ corresponds to the contribution to the amplitude from a point x in the region III to x' in the region I. As the region III is the time reversal of the region II, we can also interpret $\tilde{\mathscr{E}}_E(\mathbf{r}', \mathbf{r})$ contributing the amplitude from x' in the region I to the reflected point of x in the region II (going into the black hole), obtained by changing back to $\mathrm{Im}(t) = 0$. After taking the absolute square of the amplitude, from (2.155) we know that the probabilities of emitting and absorbing particles are related to

$$\begin{aligned} P_{\text{emit}} &= e^{-8\pi ME} P_{\text{absorb}} \\ &= e^{-\frac{E}{T_H}} P_{\text{absorb}}, \end{aligned} \tag{2.156}$$

giving the Boltzmann factor with the Hawking temperature $T_H = \frac{1}{8\pi M}$.

2.5 Discussion and Conclusion

In this chapter, we have presented three types of derivations of Hawking radiation: the tunneling, anomaly and Green's function methods. Different setups in each type have also been compared, such as the null geodesic and Hamilton-Jacobi in the tunneling type, and the gravitational and trace anomalies in the anomalous type. The relationship between different types of methods can be seen as follows:

- **Tunneling versus Green's function (path integral)**
  It is easy to see that the tunneling method is the WKB approximation of the Hartle-Hawking path-integral approach [60], and the physical quantity to calculate is essentially the same: the probability of emitting particles in (2.21) and (2.156). Some authors also consider that the path-integral approach is also as semi-classical as the tunneling method in the sense that the wave modes of the quantum field are not required, and one of the differences is that the former requires the Kruskal extension in (2.155), whereas in the latter it is not necessary [54, 55] (note that a complex path with the Kruskal extension can also be used, as demonstrated in Fig. 2.1). Another similarity (and also difference) of the two methods is the use of analytical continuation: deforming the path to evaluate the singular integral in the tunneling method (2.17) and analytically continuing the Green function to obtain the ratio of emitting to absorbing probability of radiation in (2.155).
- **Tunneling versus gravitational anomaly**
  Clearly, the physical quantities calculated in these two methods are different: tunneling focuses on probability, whereas anomaly relies on the flux of radiation to deduce the Hawking temperature. However, it has been suggested that the two methods can be connected via chirality [6, 7]. The idea is to argue that the Kruskal variables (T, X) inside and outside the black holes (2.129) are related via

$$
\begin{aligned}
t_{\text{in}} &= t_{\text{out}} - 2i\pi M \\
r_{*,\text{in}} &= r_{*,\text{out}} + 2i\pi M \\
u_{\text{in}} &= u_{\text{out}} - 4i\pi M \\
v_{\text{in}} &= v_{\text{out}},
\end{aligned}
\tag{2.157}
$$

where the relationship $e^{\frac{r_*}{4M}} = \left(\frac{r}{2M} - 1\right)^{\frac{1}{2}} e^{\frac{r}{4M}}$ is used. Therefore, the probability for a left-handed (ingoing) mode moves inward is

$$
\begin{aligned}
P_L &= \left|e^{-i\omega v_{\text{in}}}\right|^2 \\
&= \left|e^{-i\omega v_{\text{out}}}\right|^2 \\
&= 1,
\end{aligned}
\tag{2.158}
$$

whereas for a right-handed (outgoing) mode tunneling to outside is

$$
\begin{aligned}
P_R &= \left|e^{-i\omega u_{\text{in}}}\right|^2 \\
&= \left|e^{-i\omega(u_{\text{out}} - 4i\pi M)}\right|^2 \\
&= e^{-\frac{\omega}{T_H}},
\end{aligned}
\tag{2.159}
$$

so the tunneling probability can be derived from the correct chiral mode. However, the connection between this probability and the tunneling calculation with classical action in Sect. 2.2 is not manifest with matching the variables in (2.157). On the other hand, there are some situations, such as the Rindler spacetime for an accelerating observer, that the tunneling and anomaly methods give different results [1], suggesting that the relationship between the two methods is still subtle.

From the various derivations of Hawking radiation, many features of black holes reflect their rich physical and mathematical structures, and we should expect that there are still many unexplored aspects. The Hawking radiation is thus a useful effect to verify the consistency of different points of view on black holes, a subtle intersection between gravity and quantum physics.

Acknowledgements CM Sou thanks the helpful discussion with Ali Akil and Wei-Xiang Feng, and the support from the Shuimu Tsinghua scholar program, NSFC under Grant No. 12275146, the National Key R&D Program of China (2021YFC2203100), and the Dushi Program of Tsinghua University.

References

1. V. Akhmedova, T. Pilling, A. de Gill, D. Singleton, Comments on anomaly versus WKB/tunneling methods for calculating Unruh radiation. Phys. Lett. B **673**, 227–231 (2009). https://doi.org/10.1016/j.physletb.2009.02.022. arXiv:0808.3413
2. L. Alvarez-Gaume, E. Witten, Gravitational anomalies. Nucl. Phys. B **234**, 269 (1984). https://doi.org/10.1016/0550-3213(84)90066-X
3. M. Angheben, M. Nadalini, L. Vanzo, S. Zerbini, Hawking radiation as tunneling for extremal and rotating black holes. JHEP **05**, 014 (2005). https://doi.org/10.1088/1126-6708/2005/05/014. arXiv:hep-th/0503081
4. C. Barcelo, S. Liberati, S. Sonego, M. Visser, Minimal conditions for the existence of a Hawking-like flux. Phys. Rev. D **83**, 041501 (2011). https://doi.org/10.1103/PhysRevD.83.041501. arXiv:1011.5593
5. R. Banerjee, S. Kulkarni, Hawking radiation and covariant anomalies. Phys. Rev. D **77**, 024018 (2008). https://doi.org/10.1103/PhysRevD.77.024018. arXiv:0707.2449
6. R. Banerjee, B.R. Majhi, Connecting anomaly and tunneling methods for Hawking effect through chirality. Phys. Rev. D **79**, 064024 (2009). https://doi.org/10.1103/PhysRevD.79.064024. arXiv:0812.0497
7. R. Banerjee and B.R. Majhi, Hawking black body spectrum from tunneling mechanism. Phys. Lett. B **675**, 243–245 (2009). https://doi.org/10.1016/j.physletb.2009.04.005. arXiv:0903.0250
8. W.A. Bardeen, B. Zumino, Consistent and covariant anomalies in gauge and gravitational theories. Nucl. Phys. B **244**, 421–453 (1984). https://doi.org/10.1016/0550-3213(84)90322-5
9. R.A. Bertlmann, E. Kohlprath, Two-dimensional gravitational anomalies, Schwinger terms and dispersion relations. Ann. Phys. **288**, 137–163 (2001). https://doi.org/10.1006/aphy.2000.6110. arXiv:hep-th/0011067
10. N.D. Birrell, P.C.W. Davies, *Quantum Fields in Curved Space* (Cambridge University Press, 1984). ISBN 978-0-521-27858-4. https://doi.org/10.1017/CBO9780511622632

11. R. Brout, S. Massar, R. Parentani, P. Spindel, A primer for black hole quantum physics. Phys. Rept. **260**, 329–454 (1995). https://doi.org/10.1016/0370-1573(95)00008-5. arXiv:0710.4345
12. S. Carlip, Black hole thermodynamics. Int. J. Mod. Phys. D **23**, 1430023 (2014). https://doi.org/10.1142/S0218271814300237. arXiv:1410.1486
13. D.M. Chitre, J.B. Hartle, Path integral quantization and cosmological particle production: an example. Phys. Rev. D **16**, 251–260 (1977). https://doi.org/10.1103/PhysRevD.16.251
14. S.M. Christensen, S.A. Fulling, Trace anomalies and the Hawking effect. Phys. Rev. D **15**, 2088–2104 (1977). https://doi.org/10.1103/PhysRevD.15.2088
15. T. Damour, R. Ruffini, Black Hole evaporation in the Klein-Sauter-Heisenberg-Euler formalism. Phys. Rev. D **14**, 332–334 (1976). https://doi.org/10.1103/PhysRevD.14.332
16. S. Das, S.P. Robinson, E.C. Vagenas, Gravitational anomalies: a recipe for Hawking radiation. Int. J. Mod. Phys. D **17**, 533–539 (2008). https://doi.org/10.1142/S0218271808012218. arXiv:0705.2233
17. P.C.W. Davies, S.A. Fulling, W.G. Unruh, Energy momentum tensor near an evaporating black hole. Phys. Rev. D **13**, 2720–2723 (1976). https://doi.org/10.1103/PhysRevD.13.2720
18. P.C.W. Davies, S.A. Fulling, Quantum vacuum energy in two dimensional space-times. Proc. Roy. Soc. Lond. A **354**, 59 (1977). https://doi.org/10.1098/rspa.1977.0056
19. L.H. Ford, Quantum field theory in curved space-time. arXiv:gr-qc/9707062
20. L.H. Ford, Cosmological particle production: a review. Rept. Prog. Phys. **84**(11), 116901 (2021). https://doi.org/10.1088/1361-6633/ac1b23. arXiv:2112.02444
21. S.A. Fulling, The gravitational anomaly: an elementary, coordinate space approach. Gen. Rel. Grav. **18**, 609–615 (1986). https://doi.org/10.1007/BF00769929
22. S.A. Fulling, S.N.M. Ruijsenaars, Temperature, periodicity and horizons. Phys. Rept. **152**, 135–176 (1987). https://doi.org/10.1016/0370-1573(87)90136-0
23. G.W. Gibbons, M.J. Perry, Black holes in thermal equilibrium. Phys. Rev. Lett. **36**, 985 (1976). https://doi.org/10.1103/PhysRevLett.36.985
24. G.W. Gibbons, M.J. Perry, Black holes and thermal Green's functions. Proc. Roy. Soc. Lond. A **358**, 467–494 (1978). https://doi.org/10.1098/rspa.1978.0022
25. G.W. Gibbons, S.W. Hawking, Cosmological event horizons, thermodynamics, and particle creation. Phys. Rev. D **15**, 2738–2751 (1977). https://doi.org/10.1103/PhysRevD.15.2738
26. A. Gullstrand, Allgemeine Lösung des statischen Einkörperproblems in der Einsteinschen Gravitationstheorie (Almqvist & Wiksell, 1922)
27. J.B. Hartle, S.W. Hawking, Path integral derivation of black hole radiance. Phys. Rev. D **13**, 2188–2203 (1976). https://doi.org/10.1103/PhysRevD.13.2188
28. S.W. Hawking, Black hole explosions. Nature **248**, 30–31 (1974). https://doi.org/10.1038/248030a0
29. S.W. Hawking, Particle creation by black holes. Commun. Math. Phys. **43**, 199–220 (1975) [erratum: Commun. Math. Phys. **46**, 206 (1976)]. https://doi.org/10.1007/BF02345020
30. S. Iso, H. Umetsu, F. Wilczek, Hawking radiation from charged black holes via gauge and gravitational anomalies. Phys. Rev. Lett. **96**, 151302 (2006). https://doi.org/10.1103/PhysRevLett.96.151302. arXiv:hep-th/0602146
31. S. Iso, H. Umetsu, F. Wilczek, Anomalies, Hawking radiations and regularity in rotating black holes. Phys. Rev. D **74**, 044017 (2006). https://doi.org/10.1103/PhysRevD.74.044017. arXiv:hep-th/0606018
32. S. Iso, T. Morita, H. Umetsu, Quantum anomalies at horizon and Hawking radiations in Myers-Perry black holes. JHEP **04**, 068 (2007). https://doi.org/10.1088/1126-6708/2007/04/068. arXiv:hep-th/0612286
33. S. Iso, T. Morita, H. Umetsu, Fluxes of higher-spin currents and Hawking radiations from charged black holes. Phys. Rev. D **76**, 064015 (2007). https://doi.org/10.1103/PhysRevD.76.064015 [arXiv:0705.3494
34. S. Iso, T. Morita, H. Umetsu, Higher-spin currents and thermal flux from hawking radiation. Phys. Rev. D **75**, 124004 (2007). https://doi.org/10.1103/PhysRevD.75.124004. arXiv:hep-th/0701272

35. S. Iso, Hawking radiation, gravitational anomaly and conformal symmetry: the origin of universality. Int. J. Mod. Phys. A **23**, 2082–2090 (2008). https://doi.org/10.1142/S0217751X08040627. arXiv:0804.0652
36. Y. Kanai, M. Siino, A. Hosoya, Gravitational collapse in Painleve-Gullstrand coordinates. Prog. Theor. Phys. **125**, 1053–1065 (2011). https://doi.org/10.1143/PTP.125.1053. arXiv:1008.0470
37. P. Kraus, F. Wilczek, Selfinteraction correction to black hole radiance. Nucl. Phys. B **433**, 403–420 (1995). https://doi.org/10.1016/0550-3213(94)00411-7. arXiv:gr-qc/9408003
38. R. Kubo, Statistical mechanical theory of irreversible processes. 1. General theory and simple applications in magnetic and conduction problems. J. Phys. Soc. Jap. **12**, 570–586 (1957). https://doi.org/10.1143/JPSJ.12.570
39. P.C. Martin, J.S. Schwinger, Theory of many particle systems. 1. Phys. Rev. **115**, 1342–1373 (1959). https://doi.org/10.1103/PhysRev.115.1342
40. C. Morette, On the definition and approximation of Feynman's path integrals. Phys. Rev. **81**, 848–852 (1951). https://doi.org/10.1103/PhysRev.81.848
41. K. Murata, J. Soda, Hawking radiation from rotating black holes and gravitational anomalies. Phys. Rev. D **74**, 044018 (2006). https://doi.org/10.1103/PhysRevD.74.044018. arXiv:hep-th/0606069
42. T. Padmanabhan, Entropy of horizons, complex paths and quantum tunneling. Mod. Phys. Lett. A **19**, 2637–2643 (2004). https://doi.org/10.1142/S0217732304015257. arXiv:gr-qc/0405072
43. L.E. Parker, D. Toms, *Quantum Field Theory in Curved Spacetime: Quantized Field and Gravity* (Cambridge University Press, 2009). ISBN 978-0-521-87787-9, 978-0-511-60155-2. https://doi.org/10.1017/CBO9780511813924
44. M.K. Parikh, F. Wilczek, Hawking radiation as tunneling. Phys. Rev. Lett. **85**, 5042–5045 (2000). https://doi.org/10.1103/PhysRevLett.85.5042. arXiv:hep-th/9907001
45. M.K. Parikh, Energy conservation and Hawking radiation. https://doi.org/10.1142/9789812704030_0155. arXiv:hep-th/0402166
46. M.K. Parikh, A secret tunnel through the horizon. Int. J. Mod. Phys. D **13**, 2351–2354 (2004). https://doi.org/10.1142/S0218271804006498. arXiv:hep-th/0405160
47. L. Parker, Probability distribution of particles created by a black hole. Phys. Rev. D **12**, 1519–1525 (1975). https://doi.org/10.1103/PhysRevD.12.1519
48. P. Painlevé, La mécanique classique et la théorie de la relativité. C. R. Acad. Sci. (Paris) **173**, 677–680 (1921)
49. S.P. Robinson, F. Wilczek, A relationship between Hawking radiation and gravitational anomalies. Phys. Rev. Lett. **95**, 011303 (2005). https://doi.org/10.1103/PhysRevLett.95.011303. arXiv:gr-qc/0502074
50. S.P. Robinson, Two quantum effects in the theory of gravitation
51. J.S. Schwinger, On gauge invariance and vacuum polarization. Phys. Rev. **82**, 664–679 (1951). https://doi.org/10.1103/PhysRev.82.664
52. M.D. Schwartz, *Quantum Field Theory and the Standard Model* (Cambridge University Press, 2014). ISBN 978-1-107-03473-0
53. S. Shankaranarayanan, K. Srinivasan, T. Padmanabhan, Method of complex paths and general covariance of Hawking radiation. Mod. Phys. Lett. A **16**, 571–578 (2001). https://doi.org/10.1142/S0217732301003632. arXiv:gr-qc/0007022
54. S. Shankaranarayanan, T. Padmanabhan, K. Srinivasan, Hawking radiation in different coordinate settings: complex paths approach. Class. Quant. Grav. **19**, 2671–2688 (2002). https://doi.org/10.1088/0264-9381/19/10/310. arXiv:gr-qc/0010042
55. K. Srinivasan, T. Padmanabhan, Particle production and complex path analysis. Phys. Rev. D **60**, 024007 (1999). https://doi.org/10.1103/PhysRevD.60.024007. arXiv:gr-qc/9812028
56. W. Troost, H. Van Dam, Thermal effects for an accelerating observer. Phys. Lett. B **71**, 149–152 (1977). https://doi.org/10.1016/0370-2693(77)90764-X
57. W. Troost, H. van Dam, Thermal propagators and accelerated frames of reference. Nucl. Phys. B **152**, 442–460 (1979). https://doi.org/10.1016/0550-3213(79)90091-9
58. W.G. Unruh, Notes on black hole evaporation. Phys. Rev. D **14**, 870 (1976). https://doi.org/10.1103/PhysRevD.14.870

59. E.C. Vagenas, S. Das, Gravitational anomalies, Hawking radiation, and spherically symmetric black holes. JHEP **10**, 025 (2006). https://doi.org/10.1088/1126-6708/2006/10/025. arXiv:hep-th/0606077
60. L. Vanzo, G. Acquaviva, R. Di Criscienzo, Tunnelling methods and Hawking's radiation: achievements and prospects. Class. Quant. Grav. **28**, 183001 (2011). https://doi.org/10.1088/0264-9381/28/18/183001. arxiv:1106.4153
61. L. Vanzo, On tunneling across horizons. EPL **95**(2), 20001 (2011). https://doi.org/10.1209/0295-5075/95/20001. arXiv:1104.1569
62. R.M. Wald, On particle creation by black holes. Commun. Math. Phys. **45**, 9–34 (1975). https://doi.org/10.1007/BF01609863

Chapter 3
Is the Information Loss Problem a Paradox?

Luca Buoninfante and Francesco Di Filippo

Abstract The aim of this chapter is twofold. First, we introduce the information loss problem; second, we provide a critical assessment by thoroughly inspecting the assumptions underlying its formulations. In particular, we argue that if we work in the regime of validity of semiclassical gravity and do not add additional assumptions that are not necessary for the Hawking calculation, the answer to the question in the title is NO. In other words, the black hole evaporation is certainly unitary as predicted by quantum field theory in curved spacetime. However, if additional assumptions are added, such as a universal area bound on the entropy, contradictions may arise even in regimes where we would expect the semiclassical approximation to be valid. We show that a contradiction indeed arises, but not between the laws of semiclassical general relativity and quantum mechanics, but rather between the former and the additional (holographic) requirement of the area limit, according to which an exterior observer describes a black hole as a quantum system whose entropy is bounded by its area.

3.1 Introduction

Fifty years ago Hawking wrote his seminal papers on the Hawking radiation [13, 14] and for the first time raised the question of whether information is lost due to black hole evaporation, thus giving rise to the notoriously known "information loss paradox" [9, 15]. The usual argument is that semiclassical gravity and quantum mechanics are incompatible because the black hole evaporation violates unitarity

L. Buoninfante (✉)
High Energy Physics Department, Institute for Mathematics, Astrophysics, and Particle Physics, Radboud University, Nijmegen, The Netherlands
e-mail: luca.buoninfante@ru.nl

F. Di Filippo (✉)
Institute of Theoretical Physics, Faculty of Mathematics and Physics, Charles University, Prague, Czech Republic
e-mail: francesco.difilippo@matfyz.cuni.cz

Institut fur Theoretische Physik, Goethe Universität, Frankfurt am Main, Germany

A. Akil and C. Bambi (eds.), *The Black Hole Information Paradox*, Springer Series in Astrophysics and Cosmology, https://doi.org/10.1007/978-981-96-6170-1_3

by evolving an initial pure state into a final mixed state. Despite its simplicity, this statement hides several assumptions and can in fact be proven logically incorrect if we work within the validity regime of the semiclassical approximation and do not introduce additional assumptions that are unnecessary for the standard calculation of quantum field theory in curved spacetime.

The aim of this article is to review different formulations of the information loss "paradox" and then critically inspect the various assumptions that are usually made. We will present two different formulations of the problem and especially focus on the modern version which is due to Page [16, 17]. We refer to the latter formulation as the entropy problem. This states that a logical contradiction can arise even for macroscopic objects, i.e. for time scales much smaller than the lifetime of the black hole, where we would expect the semiclassical approximation to still be valid. The usual claim is that the unitarity of quantum mechanics clashes with the laws of semiclassical general relativity even in low-energy regimes. However, we disagree with this reasoning and will show that the laws of quantum mechanics and general relativity are certainly compatible, at least as long as the semiclassical approximation does not breakdown. In particular, this is the case if we consider time scales much shorter than the duration of black hole evaporation. What we will carefully argue instead is that a contradiction may arise only when further and generally unjustified assumptions are made. In particular, if the *area limit* [3]—according to which a black hole can be described as a quantum object whose entropy is always bounded by its area—is imposed as one of the starting assumptions, then a contradiction is found. However, the latter is not between semiclassical gravity and quantum mechanics, but between semiclassical gravity and the additional requirement of the area limit [9]. This represents a radically different formulation of the black hole information loss problem.

The work is organized as follows.

Section 3.2: Before discussing the information loss problem, we find it useful to briefly review some key elements of the physical mechanism underlying the Hawking radiation and clarify some aspects that are sometimes misunderstood or misinterpreted. In particular, we will explain how to define vacuum states at early and late times for an evaporating black hole, why a time-dependent geometry is needed to have particle creation, why the standard heuristic picture of particle-antiparticle pair production is unphysical and incorrect, and what is the role of the horizon.

Section 3.3: Two different formulations of the information loss problem will be presented: the unitarity problem and the entropy problem. In both scenarios, the underlining assumptions will be critically inspected and it will be clarified that there is no incompatibility between quantum mechanics and semiclassical gravity. In particular, we will explain that, within its regime of validity, semiclassical gravity predicts a unitary evolution of quantum states during the black hole evaporation. The computation from quantum field theory in curved spacetime will imply that the number of degrees of freedom contained within a black hole cannot be limited by its area.

Section 3.4: We will make several important remarks on the information loss problem by emphasizing that contradictions may arise only if the area limit is taken as a starting assumption. We will argue that the combination of established theories, such as quantum mechanics and semiclassical general relativity in their regimes of validity, does not require and is in fact incompatible with the holographic principle. We will also discuss the open question of if, how and when the information retrieval occurs.

Section 3.5: The key points of our discussion will be concisely summarized and the take-home message of this work will be sharply conveyed with the hope that less misunderstanding will arise in the future about the information loss problem.

Conventions. Throughout the article we will work with the mostly plus convention for the metric signature $(- + ++)$ and adopt the Natural units ($\hbar = 1 = c$).

3.2 Particle Creation

Before delving into the information loss problem, we believe it is useful to recall the main features of particle creation because its physical mechanism is sometimes misunderstood and misinterpreted. First of all, we should distinguish at least two different types of physical scenarios that could give rise to particle creation: (i) particle creation due to relative observations made by observers in two different frames; (ii) particle creation due to some time-dependent process that takes an initial empty vacuum state into a non-empty final state. An example of the first type is the Unruh effect [22] according to which an inertial observer sees an empty vacuum, while an accelerating observer sees a thermal bath. The second type of scenario for particle creation is not frame-dependent but is due to some dynamical process that is able to convert an initial empty vacuum state into a non-empty excited state. This can happen either thanks to some time-dependent potential [10] (e.g. a scattering process or the Schwinger effect) or to some dynamical background geometry (e.g. an expanding universe [19, 26] or a collapsing black hole [13, 14]).

Both types of particle creation phenomena are characterized by a flux of energy that is produced in the form of radiation. Energy conservation must still be respected in some way, thus the energy produced must come from somewhere. For example, in the case of the Unruh effect, the energy carried by the radiation that is observed in the non-inertial frame comes from the energy that is needed to keep the observer accelerating, e.g. the work done by the engine to accelerate. Furthermore, the time-dependency of a potential can convert potential energy into emitted radiation. An expanding cosmological background and a collapsing star can give rise to some net redshift effects on the propagating waves whose final states are characterized by larger amplitudes, i.e. by larger occupation numbers. In the latter case, the energy of the emitted radiation comes from a variation over time of the gravitational field perceived by the propagating wave, due to the expansion of the universe or the contraction of a collapsing star.

In the following, we will analyze the main aspects of the mechanism of particle creation due to the collapse of a star into a black hole. After introducing some elements of quantum field theory in curved spacetime and defining the concept of vacuum, we will explain in more detail the physical reason why particles are created in this dynamical geometry, thus it will also become clear why no particle creation can occur in a static (eternal) black hole spacetime. Furthermore, we will show that the semiclassical computation of Hawking radiation is consistent with a unitary quantum mechanical evolution once all the outgoing Hawking modes are correctly taken into account.

3.2.1 Definition of the Vacuum State

To understand the phenomenon of particle creation, we need to properly define the concept of the vacuum state. For simplicity, we only analyze a free massless scalar field whose dynamical equation is

$$g^{\mu\nu}\nabla_\mu\nabla_\nu\phi = 0. \tag{3.1}$$

In flat spacetime, we can decompose the field into positive and negative energy modes, i.e.[1]

$$\phi = \sum_i \left(a_i f_i + a_i^\dagger f_i^*\right), \tag{3.2}$$

where a_i and $a_i^\dagger$ are the so-called annihilation and creation operators, while the mode-functions are given by

$$f_i \propto e^{-i\omega_{\mathbf{k}_i} t \pm \mathbf{k}_i\cdot\mathbf{x}}, \qquad |\mathbf{k}_i| = \omega_{\mathbf{k}_i} \geq 0. \tag{3.3}$$

The vacuum state is defined as the state annihilated by the operators a_i, i.e.

$$a_i\,|0\rangle = 0. \tag{3.4}$$

The problem with this definition, is that it requires a choice for the time and spatial coordinates, that is, in general it is not unique. However, in flat spacetime it can be proven that the vacuum state is Poincaré invariant: every inertial observer would agree on the definition of the vacuum state. This does not mean that there is a unique vacuum choice, even in flat spacetime. In fact, according to the Unruh effect, an accelerated observer measures a thermal bath, while an inertial observer sees the same state as an empty vacuum.

[1] It is worth to mention that the summation over i is a short-hand notation for a more general case of an integral over the angular components of $\mathbf{k}_i$ and its modulus $|\mathbf{k}_i| = \omega_{\mathbf{k}_i}$. Below we will also use the notation a_ω and f_ω to indicate the modes with frequency ω.

In a generic spacetime, the situation is more involved. In general, there is no notion of preferred observer. Hence, there is no preferred choice of the vacuum state. The problem is alleviated for stationary spacetimes with an everywhere defined killing vector ξ. In this case, the killing vector provides a preferred notion of time and thus a way to split the positive and negative frequency modes, i.e. we have Eq. 3.2 where

$$\xi^{\mu} \nabla_{\mu} f_i = -i\omega_i f_i \,, \qquad \omega_i \geq 0. \tag{3.5}$$

The existence of a preferred vacuum choice is also the reason why particle creation cannot happen in a stationary spacetime. In particular, in an eternal black hole spacetime no particle is created.[2]

While this is only possible for stationary spacetimes, the physically relevant scenarios correspond to backgrounds that are approximately stationary at early times and, after a non-trivial time-dependent dynamical transient, return to being approximately stationary at sufficiently late times. Therefore, we can identify *two* timelike vector fields ξ_{in} and ξ_{out} that are killing vectors in the two asymptotic regions and write two different mode decompositions of the field ϕ in two different bases [6, 12]:

$$\phi = \sum_i \left(a_i^{in} f_i^{in} + a_i^{in\dagger} f_i^{in*} \right), \qquad \phi = \sum_i \left(a_i^{out} f_i^{out} + a_i^{out\dagger} f_i^{out*} \right). \tag{3.6}$$

We can then define two vacuum states, one for the in region in the asymptotic past and one for the out region in the asymptotic future:

$$a_i^{in} \left| in \right\rangle = 0 \,, \qquad a_i^{out} \left| out \right\rangle = 0. \tag{3.7}$$

These two vacua only agree if the two killing vectors are equal (e.g. for stationary spacetimes). Generically, they differ and can be shown to be related by the general relation [12]

$$\left| in \right\rangle = \left\langle out | in \right\rangle e^{\frac{1}{2} \sum_{ij} V_{ij} a_i^{out\dagger} a_j^{out\dagger}} \left| out \right\rangle , \tag{3.8}$$

where V_{ij} are functions of the Bogoliubov coefficients that relate the annihilation and creation operators between two asymptotic regions; however, their explicit expressions are not important for our discussion.

As we will see later, a relation between $\left| in \right\rangle$ and $\left| out \right\rangle$ that is not compatible with 3.8 could be a sign of unitarity violation or an indication that we have done something wrong in the way we have treated the outgoing radiation, for example by leaving out some of the outgoing modes. The latter, in fact, is the mistake that

[2] It is often claimed that Hawking radiation is present even in an eternal black hole spacetime. However, while energy fluxes may indeed be present, they are not due to particle creation. What happens in this case is that one can define the so-called Hartle-Hawking state according to which both vacua on the past null infinity and the future null infinity are not empty, i.e. there is a pre-existing energy flux flowing from past to future and the black hole is in thermal equilibrium with these fluxes (see e.g. [5] for a pedagogical discussion).

is sometimes committed and that leads to an apparent contradiction between the unitarity of quantum mechanics and the validity of semiclassical general relativity.

3.2.2 Time Dependent Geometries

The origin of the Hawking radiation is to be found in the dynamical nature of the spacetime, given by a contracting star that eventually collapses into a black hole. Let us give an intuitive physical explanation of why particles are created in this situation. Consider a scalar wave that propagates and enters the star, passes through it and escapes back to infinity. In a static configuration, the wave would reach the future asymptotic region unmodified as any effect during the propagation towards the central object would be exactly compensated by opposite effects when the wave propagates away from it. However, if the star is collapsing, this is no longer true. It is possible to show that the wave would emerge with a net redshift and a net amplification [8, 12]. This physically means that the state of the outgoing wave is characterized by a larger occupation number as compared to the initial incident one. Therefore, an asymptotic observer on $\mathscr{I}^+$ would detect a flux of energy, i.e. Hawking radiation. We now provide some more details on the spectrum of the radiation.

The Penrose-Carter diagram of a spherically symmetric black hole formed by gravitational collapse is depicted in Fig. 3.1. If we start with a vacuum state on $\mathscr{I}^-$ which is empty, the same state on $\mathscr{I}^+$ will have a non-trivial matter content. The seminal Hawking computation tells us that the expectation value of the particle number operator on $\mathscr{I}^+$ is [13, 14]

$$\langle in|\, N_\omega^{out}\, |in\rangle = \frac{\Gamma_{l,\omega}}{e^{8\pi GM\omega} - 1}, \tag{3.9}$$

where we have introduced the number operator $N_\omega^{out} = a_\omega^{out\dagger} a_\omega^{out}$. The expression in Eq. 3.9 is the same distribution of the thermal spectrum of radiation for bosons. This implies that an asymptotic observer on $\mathscr{I}^+$ sees a flux of radiation that follows a Planckian distribution with temperature inversely proportional to the mass of the black hole, i.e.

$$T_H = \frac{1}{8\pi GM}. \tag{3.10}$$

Furthermore, the quantity $\Gamma_{l,\omega}$ is the so-called gray-body factor that physically represents the transmission coefficient of the wave through the potential that is peaked around the radius of the photosphere; $l \in \{0, 1, 2, \dots\}$ is the angular momentum number associated with the spherical harmonic decomposition of the scalar field. This means that a fraction $1 - \Gamma_{l,\omega}$ is backscattered by the potential, falls back into the black hole and does not reach the asymptotic region at infinity.

To confirm that the radiation is truly thermal, one has to check that different quanta with frequencies ω and ω' on $\mathscr{I}^+$ are uncorrelated. A direct computation shows that

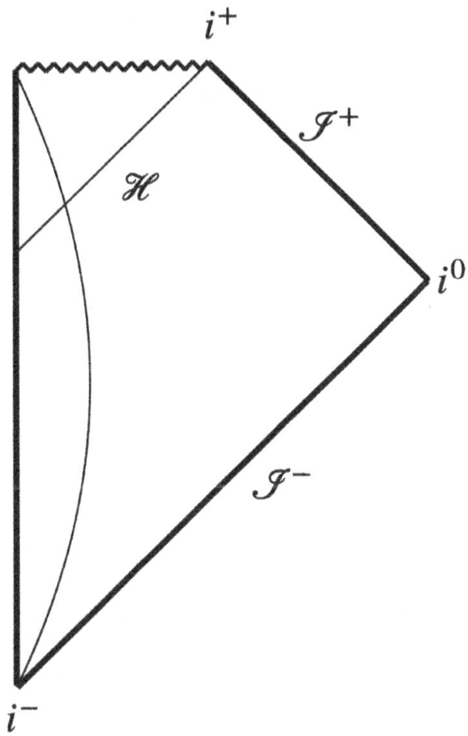

Fig. 3.1 Penrose-Carter diagram of a black hole formed via gravitational collapse

this is the case [25] (see also [12] for a pedagogical analysis). In particular, we have

$$\langle in|\, N^{out}_{\omega} N^{out}_{\omega'}\, |in\rangle = \langle in|\, N^{out}_{\omega}\, |in\rangle \, \langle in|\, N^{out}_{\omega'}\, |in\rangle \qquad \text{for} \quad \omega \neq \omega', \tag{3.11}$$

which indeed means that there is no correlation between the modes on $\mathscr{I}^+$. Therefore, starting from a pure state on $\mathscr{I}^-$ we obtain a mixed state on $\mathscr{I}^+$, in contradiction with 3.8 which would instead tell us that the in-state is pure also for observers in the out region as it can be written as a linear combination of pure states. Given that unitary operators must evolve pure states into pure states, we might be temped to conclude that we have reached a paradox. However, this conclusion would be too hasty. As we will explain below, this apparent contradiction is caused by the fact that we have left out some of the outgoing modes, in particular those that do not reach $\mathscr{I}^+$ but that instead cross the event horizon and enter the black hole [25].

It is worth mentioning that Hawking's original computation neglects the backreaction, i.e. the effect of the emitted radiation on the spacetime geometry. Approximating the black hole as a infinite reservoir[3] is definitely a good approximation for a certain interval of time, but eventually the radiated energy would become comparable to the

[3] The black hole can be approximated as an infinite reservoir by taking the double limit $M \to \infty$, $G \to 0$, with GM kept fixed. In this way, the black hole mass is infinite and the right-hand side

mass of the black hole so that the assumption of a fixed background geometry (i.e. no backreaction) would not hold anymore [1, 4, 11].

To clarify the language we use in this chapter, let us emphasize that with the expression "semiclassical approximation" we refer to a low-energy regime in which general relativity can be considered a reliable effective field theory, that is, the characteristic energy scales are lower than the effective-field-theory cutoff which we take to be equal to the Planck mass, in particular the black hole mass has to be larger than the cutoff. This means that matter falls into a black hole according to the laws of general relativity, for example at the horizon scale nothing special happens due to the equivalence principle. Furthermore, time scales of the order of half the black-hole lifetime can still be described within the semiclassical approach, the backreaction of Hawking radiation on the black hole geometry does not have to be neglected and the radiation does not have to be completely thermal. Indeed, backreaction and non-thermal corrections can be consistently taken into account in the adiabatic approximation $\dot{T}_H/T_H^2 \ll 1 \Leftrightarrow 1/GM^2 \ll 1$ and within the same framework of quantum field theory in curved spacetime [4]. In our definition of semiclassical approximation, quantum-gravitational effects can be present as long as the black-hole mass is bigger than the Planck mass, for example we can have gravitons interacting in an evaporating black-hole spacetime.

Before concluding this section, we want to clarify some additional aspects regarding the physical mechanism of Hawking radiation.

3.2.3 Heuristic Picture and Its Limitations

The heuristic argument that is often invoked for understanding Hawking evaporation is quite simple and is worth recalling it here to explain why it is physically incorrect.

It is sometimes said that in quantum mechanics the vacuum is not empty, but it is populated by virtual particle-antiparticle fluctuations. When a pair is created near a black hole horizon, it is possible that the two particles are separated by tidal forces and become real. The positive energy particle escapes to infinity, while the negative energy one enters the horizon and is destroyed at the singularity. For an asymptotic observer, this process is equivalent to a flux of positive energy leaving the black hole. This heuristic explanation of the origin of Hawking radiation is widely spread due to its simplicity. However, while it has some merit in partly capturing the physics at play, it falls short in several regards and can lead to confusions.

One of the main limitations of this heuristic argument is that it makes no mention to the time-dependence of the spacetime and it seems to be applicable to static geometries as well. However, we have seen in Sect. 3.2.1 that particle creation only happens when the spacetime is time-dependent as only in that case the two vacuum states defined in the asymptotic regions do not agree. In fact, even if the early- and

---

of the Einstein's equation vanishes (i.e. backreaction is neglected) but, at the same time, the black hole radius $2GM$ and its evaporation rate $\sim 1/(GM)^2$ are finite.

late-time geometries are stationary, the root of Hawking radiation lies in the time-dependence of the geometry that is necessary to dynamically form the black hole. Furthermore, while the heuristic picture requires that the ingoing and outgoing particles are equally correlated throughout the entire evaporation process, the rigorous semiclassical gravity result shows that the outgoing modes are mostly correlated only with the early-time ingoing modes.

From the discussion above, it should also be clear that the cause of Hawking radiation is not to be found in the presence of a horizon. In fact, it is possible to have particle production even without the formation of a horizon. However, if the horizon does not form, the geometry eventually relaxes into a static configuration and the radiation stops. The formation of the horizaon makes the evaporation continue forever (or at least for a very long but finite time if we take into account the backreaction according to which the mass has to decrease). Finally, the presence of a horizon leads to universal radiation, whose the thermal distribution does not depend on the details of the gravitational collapse [14] (see also [5, 12] for pedagogical explanations).

In the remaining of this chapter, we will clarify that information loss problems may arise but only if some extra ingredients and assumptions are added beyond the framework of semiclassical general relativity.

3.3 Standard Formulations of the Information Loss Problem

In the previous section, we have seen that a pure state on $\mathscr{I}^-$ does not evolve into a pure state on $\mathscr{I}^+$. Given that a unitary evolution must evolve pure states into pure states, we might be temped to conclude that this result, derived within semiclassical gravity, is paradoxical. However, this is not the case. In fact, contrary to $\mathscr{I}^-$, the null region $\mathscr{I}^+$ is not a Cauchy hypersurface.[4] Therefore, we do not need to recover a pure state on $\mathscr{I}^+$ simply because some modes can be lost inside the event horizon $\mathscr{H}$. A unitary evolution operator must evolve a pure state on $\mathscr{I}^-$ into a pure state on $\mathscr{I}^+ \cup \mathscr{H}$. This is indeed the case. The framework of semiclassical gravity, in particular Eq. 3.8, tells us that the *in* state is a pure state even when expressed in terms of particle states in the *out* region, if all the outgoing modes $a_i^{out\dagger}$ are correctly taken into account [25]. Hence, unitarity is not violated and there is no logical contradiction.

What still needs to be understood is where the correlations are. We have discussed that the radiation on $\mathscr{I}^+$ is uncorrelated. Therefore, the only possibility is that modes on $\mathscr{I}^+$ are correlated with modes that enter the horizon. There are subtleties to verify this possibility as the geometry in some regions of the event horizon is time-dependent and thus it is not easy to provide a well-defined notion of particles. However, there

[4] Note that this is an abuse of nomenclature. A Cauchy hypersurface must be acausal, while $\mathscr{I}^-$ and $\mathscr{I}^+$ are null and, hence, neither is a Cauchy hypersurface. What we mean here, is that $\mathscr{I}^-$ intercepts all the modes, while $\mathscr{I}^+$ does not.

are reasonable ways to define the modes that enter the horizon and, indeed, explicit computations show that the late-time modes on $\mathscr{I}^+$ are maximally entangled with the modes that enter the horizon at early times [25]. If these correlations are properly taken into account, then the relation in Eq. 3.8 can be shown to be true, which also means that an initial pure state $|in\rangle$ evolves into a final pure state $|out\rangle$.

So far so good, but where is the paradox if there is a way to understand the correlations between ingoing and outgoing quanta and still be consistent with unitarity? The short answer is that there is no paradox. In fact, logical contradictions can arise only if additional assumptions are introduced. The standard formulations of the information loss problem require some extra ingredients that go beyond what is necessary for the semiclassical computation of the Hawking radiation. In the reminder of this section we will carefully explain what are the ingredients that are added to the semiclassical description and that could lead to contradictions; we will closely follow the discussion in Ref. [9].

3.3.1 Complete Evaporation and Unitarity Problem

The original information loss problem is usually stated as a contradiction between the unitarity of the evolution operator in quantum mechanics and the validity of the semiclassical approximation until the end of black hole evaporation. However, this incompatibility is true only if the black hole fully evaporates leaving behind an empty regular spacetime plus radiation. More precisely, what is usually referred to as "unitarity problem" can be formulated as the impossibility of coexistence of the following three assumptions:

A1: The evolution operator in quantum mechanics is unitarity, in particular pure states evolve into pure states;
A2: Semiclassical general relativity is valid during the entire process of black hole evaporation, in particular the emitted radiation on $\mathscr{I}^+$ is in a thermal state;
A3: The black hole evaporates completely, leaving behind a regular empty spacetime, plus radiation.

If the black hole fully evaporates according to semiclassical gravity and leaves a regular empty spacetime (see the first diagram in Fig. 3.2), then the final state of the process is mixed: the radiation is in a mixed state and nothing is left to purify it. This then implies that unitarity is violated because an initial pure state would evolve into a mixed state.[5] However, we should emphasize that if any of the three assumptions is given up, then no problem or paradox arises. For example, if A1 is not required, then the evolution operator is allowed to be non-unitary. If A2 is dropped, then we cannot

[5] It is not important that the initial state is pure. Indeed, even if the initial state is mixed, the unitarity problem would still arise because such a mixed state would be evolved into another mixed state that is not the one predicted by a unitary evolution.

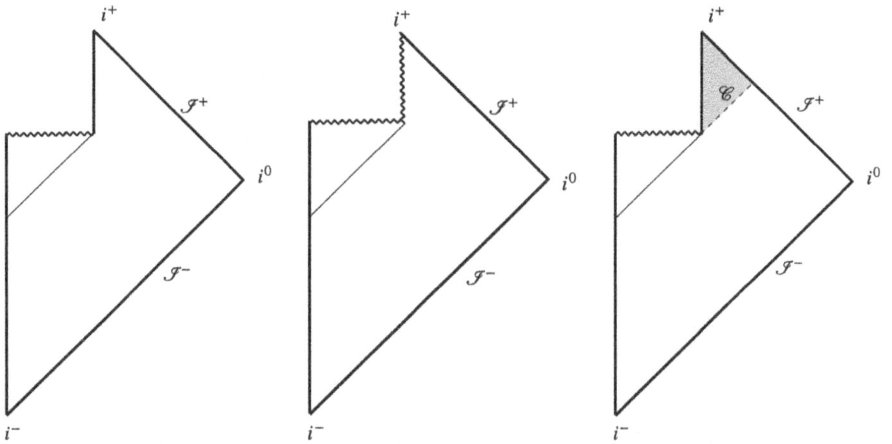

Fig. 3.2 Penrose-Carter diagrams of different possibilities for the endpoint of the evaporation. The left diagram corresponds to the case in which the black hole completely evaporates and leaves an empty regular spacetime plus radiation. The middle diagram represents a scenario in which the endpoint of black hole evaporation corresponds to a naked singularity plus radiation. The right diagram emphasizes the fact that the shaded gray region is in the causal future of the singularity, thus the dashed line $\mathscr{C}$ represents a Cauchy horizon

trust semiclassical gravity until the endpoint of the evaporation and the final state of the Hawking radiation does not have to be mixed. If A3 is not assumed, it may happen that the black hole evaporation leaves a naked singularity (like in the middle diagram in Fig. 3.2) or even some remnant that could be entangled with the outgoing radiation and eventually purify the final joint state. Therefore, this original formulation of the information loss problem only arises if A1, A2 and A3 are simultaneously assumed. It is generally believed that A2 cannot be true because quantum-gravity effects are expected to become relevant close to the endpoint of black hole evaporation.

3.3.2 *Entropy Problem*

The previous formulation of the information loss problem is mainly related to the physics of the endpoint of the evaporation. However, there exists a stronger formulation according to which a paradox could arise at time scales much shorter than the black hole lifetime. This is originally due to Page [16, 17] and is sometimes called "entropy problem". It is formulated as an incompatibility among the following three assumptions:

B1: The evolution operator in quantum mechanics is unitarity, in particular pure states evolve into pure states;

B2: Semiclassical general relativity is a valid low-energy effective field theory as long as the black mass is much larger than the Planck mass;

B3: As seen from an outside observer, a black hole can be described as a quantum object whose entropy is bounded by $A/4G$, where A is the horizon area.[6]

As compared to the previous formulation, we have A1 = B1, but the second assumption is now weaker, i.e. B2 $\subset$ A2. This means that the semiclassical approach can be considered reliable only for black holes whose mass is much larger than the Planck mass, and which are therefore still in an evaporation stage far from the endpoint. The new important ingredient is the assumption B3 which is telling us that the total thermodynamic entropy of the black hole[7] $S_{\rm th}$ is bounded by the area, i.e. $S_{\rm th} \leq A/4G$. We now explain why B1, B2, B3 cannot be simultaneously compatible.

Let us call $\mathbb{H}_{\rm bh}$ and $\mathbb{H}_{\rm rad}$ the Hilbert spaces of the black hole and the emitted radiation, respectively. If the initial state before the formation of the black hole was pure, then B1 implies that the joint state of black hole plus radiation must also be pure, that is, the state $|\psi\rangle \in \mathbb{H}_{\rm bh} \otimes \mathbb{H}_{\rm rad}$ must have zero von Neumann entropy. Tracing over the degrees of freedom of the black hole, we obtain the density matrix for the subsystem of radiation whose von Neumann entropy $S_{\rm rad}$ coincides with that of the black hole subsystem which we call $S_{\rm bh}$. During the evaporation the area decreases. On the other hand, the entropy $S_{\rm rad}$ of the emitted radiation (and similarly $S_{\rm bh}$) will continue to increase according to the semiclassical computation which can be trusted as long as B2 is true. Therefore, there exists a time scale[8]—denoted by $t_{\rm Page}$ and known as Page time—after which the von Neumann entropy of the radiation (and thus of the black hole subsystem) exceeds $A/4G$; see Fig. 3.3. Consequently, B3 implies that $S_{\rm rad}$ becomes larger than the total thermodynamic entropy of the black hole. In formula we have

$$t > t_{\rm Page} \quad \Rightarrow \quad S_{\rm rad} = S_{\rm bh} > \frac{A}{4G} \geq S_{\rm th}. \tag{3.12}$$

From the the last inequality, it follows that for time scales $t > t_{\rm Page}$ the joint state $|\psi\rangle \in \mathbb{H}_{\rm bh} \otimes \mathbb{H}_{\rm rad}$ becomes mixed, thus contradicting the hypothesis B1 and giving rise to a paradox.

As a way to solve this puzzle and preserve unitarity, it has been argued that the von Neumann entropy of Hawking radiation must start decreasing at time scales of the order of $t \sim t_{\rm Page}$ and tend to zero at the end of evaporation [2, 3, 16–18]. In this way the von Neumann entropy of radiation would be purified consistently with the requirement of unitarity in B1 and would follow the so-called Page curve shown in Fig. 3.3. However, this type of resolution clearly requires giving up B2 because

[6] For the purposes of the discussion it is irrelevant what type of horizon we are referring to when talking about the area limit as in the standard picture the area of the event horizon is very close to the area of the trapping horizon.

[7] Here, by the expression "total thermodynamic entropy" we mean the thermodynamic entropy of the black hole taking into account the degrees of freedom of both the geometry and the matter inside the black hole, i.e. the horizon and the bulk degrees of freedom, respectively.

[8] It is worth to mention that different notions of time are possible. Here we are considering a time function whose vector field is orthogonal to the Cauchy foliation.

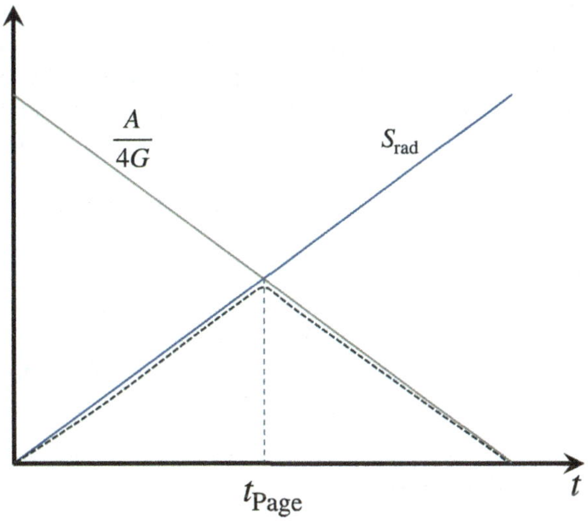

Fig. 3.3 Behavior of various entropies. The gray line is the area $(A/4G)$ of the black hole that decreases in time; the blue line corresponds to the entanglement entropy of the outgoing radiation (S_{rad}) which increases according to the prediction of semiclassical gravity; the dashed black line is the entropy of the outgoing radiation that follows the Page curve if the assumption B3 on the area limit is required as an additional hypothesis

new physics beyond the standard semiclassical gravity result is needed at time scales much shorter than the black hole lifetime in order to obtain the Page curve.

It is worth mentioning that some attempts have been made to obtain a behavior compatible with the Page curve by simply taking into account the backreaction of the emitted radiation. However, it seems that this may not be sufficient [15].

3.3.3 Violation of Semiclassical Gravity

We have inspected the standard formulations of the information loss problem and noticed that a logical contradiction only arises once extra assumptions are added. In particular, the first formulation of the problem needs additional requirements on the endpoint of the evaporation. Therefore, it is not particularly worrisome since it is reasonable to believe that we cannot trust the semiclassical approximation in that regime where quantum-gravity effects should not be negligible.

The second formulation of the problem is usually considered more problematic as the validity of semiclassical gravity is only assumed in regimes that are still far from the endpoint of black hole evaporation. However, in this case a limit on the numbers of degrees of freedom that can be contained inside a black hole is conjectured. This corresponds to an additional requirement that is not needed for the standard semiclassical computation of Hawking radiation and actually cannot even be obtained by a self-consistent computation within the model. Interestingly, this second formulation violates the spirit of semiclassical gravity from the very beginning. In fact, the consistent way to deal with the evolution of a quantum state in curved spacetime would be to choose a Cauchy foliation and consider the evolution of the state from one spacelike hypersurface to the other. In doing so, we are forced to consider Cauchy hypersurfaces that enter the event horizon. By correctly taking into account the outgoing modes on the region of the hypersurface inside the black hole we can make sure that unitarity is satisfied. Indeed, semiclassical gravity predicts that a pure state evolves into a pure state, but it does so in a way that violates the entropy limit as shown in Eq. 3.12.

There is no local observation that can detect the presence of an event horizon, and the equivalence principle implies that nothing special should happen there. However, the requirement B3 implicitly assumes that the number of degrees of freedom on the regions of the Cauchy hypersurfaces that are inside the black hole is limited by the area of the horizon. This means that new physics beyond semiclassical general relativity is needed to explain the entropy limit on $S_{\rm bh}$ and $S_{\rm rad}$ [9, 20, 21].

3.4 Discussion

The information loss problem is often formulated as the incompatibility between the assumption of unitary evolution in quantum mechanics, in particular the fact the pure states evolves into pure states, and the validity of semiclassical gravity which predicts that the entanglement entropy of the Hawking radiation can become larger than the area limit. However, after a careful analysis of the physical assumptions and logical implications, we have shown that in fact unitarity and semiclassical gravity are compatible, i.e. B1 and B2 can coexist. A contradiction arises only when the area limit on the total black hole entropy (B3) is required.

We will now make some important observations that also reflect our view on the information loss problem and on the fact that the area limit cannot be satisfied in the standard framework of semiclassical gravity, but requires the introduction of new physics.

Area limit for eternal geometries. The first point worth discussing is that there are several arguments in favor of the area limit on the number of degrees of freedom contained in a black hole [3]. However, to our knowledge, all these arguments are made for stationary eternal geometries. The causal structure is very different from that of a dynamical black hole. It would therefore be essential to quantify the evidence for the area limit in dynamical spacetimes before claiming that B3 can be a justified assumption.

Gravitational entropy in dynamical configurations. In relation to the previous point, a major difficulty lies in the fact that it is still not known how to give a definition of gravitational entropy in dynamical configurations. Let us consider a star that collapses and forms a black hole. It is widely accepted that the star will have an entropy which scales with its volume. On the other hand, it is equally widely accepted that once the black hole is formed the entropy will scale as its area. Perhaps, understanding the transition from a volume law to an area law could shed light on understanding B3 and thus whether there is an information loss problem. However, we believe that the existence of an area limit on the total black hole entropy will still need some new physics beyond semiclassical gravity.

Decoupling between horizon and bulk degrees of freedom. An intriguing possibility for the transition from a volume law to an area law could simply be related to the fact that the bulk degrees of freedom decouple and become inaccessible from the outside. Therefore, external observers can describe thermodynamical properties of the black hole simply by looking at the horizon degrees of freedom whose number is limited by $A/4G$. In this picture, the degrees of freedom of the bulk decouple but do not disappear, in particular they still contribute to the von Neumann entropy of the black hole. However, even in this case the total entropy of the black hole would still not be limited by the area and the hypothesis B3 would still not be justified.

Role of the singularity. Within the context of semiclassical gravity, a spacetime singularity forms. The theory cannot predict the fate of anything that reaches the singularity. In particular, information falling into the singularity will be destroyed and leave the manifold from the point of view of classical or semiclassical observers. This suggests that to fully understand the problem of information loss in evaporating black holes we must be able to provide a description of the singularity and predict the fate of observers who fall into it. On the other hand, solving the information loss problem by keeping the area limit and assuming a Page curve-like behavior for the entanglement entropy of Hawking radiation would imply that the information can be retrieved without understanding the fate of the singularity.

Event or trapping horizon. Related to the previous point, we should distinguish between event horizons and long-lived dynamical horizons. If we form a true event horizon, anything that crosses it will be lost from the point of view of classical asymptotic observers [23, 24]. On the other hand, if the horizon is a long-lived trapping horizon, the information should eventually leak out; see Fig. 3.4 for an illustration of these two possibilities. However, event horizons are teleological in nature [24] and a classical observer cannot detect their presence in finite time. With reference to Fig. 3.4, the situation is slightly different as observer on $\mathscr{I}^+$ can detect the presence of an event horizon if they reach the Cauchy horizon. However, this is only possible at the end of the evaporation which is way after the Page time. On the other hand, retrieving the information in a way that is consistent with the area limit (i.e. with the Page curve) would require knowing that an event horizon is absent, but the semiclassical gravity approach is not able to provide this knowledge in a regime still very far from the endpoint of evaporation. Note that there are studies on particle creation in the presence of dynamical horizons (while an event horizon is not required) [1]: in this case one can show that the information cannot leak out of the trapped region within the regime of validity of semiclassical gravity. Therefore, an information retrieval that is consistent with the area limit seems to require new physics beyond the semiclassical approach already in the low-energy regimes far from the endpoint of evaporation.

Holography versus semiclassical gravity. The existence of an area limit on the entropy of a system is usually advocated by the holographic principle [7]. A

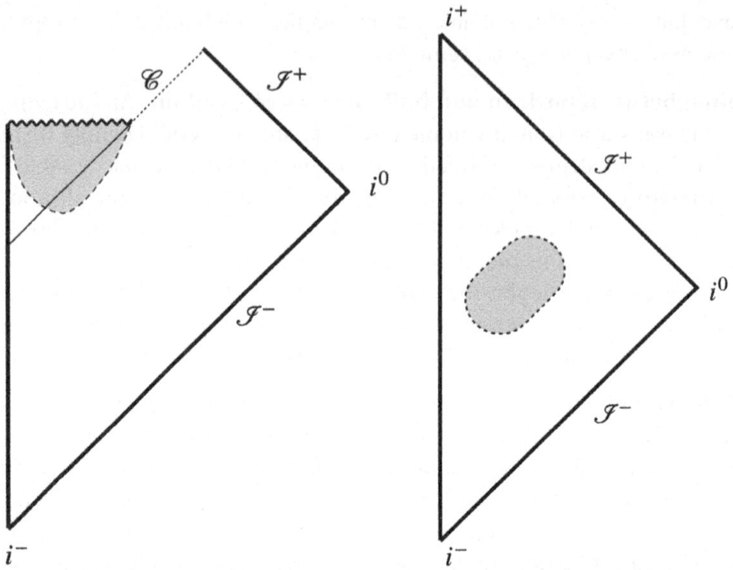

Fig. 3.4 Causal structures of two evaporating black holes. On the left, a singular black hole evaporates and leaves a spacetime with a Cauchy horizon. We do not draw the spacetime beyond the Cauchy horizon because points in that region would lie in the causal future of the singularity. The spacetime has both an event horizon (black solid null line) and a trapping horizon (dashed line enclosing the shaded region). On the right, a regular black hole with no singularity evaporates completely and leaves a regular spacetime. There is no event horizon. The trapping horizon (dashed line enclosing the shaded region) closes on itself. In the left spacetime, we expect information to be lost into the singularity or hidden behind the Cauchy horizon. In the right spacetime, we expect the information to be completely recovered at the end of evaporation

claim that is often made is that holography should be a fundamental property of a consistent theory of quantum gravity, and thus several researchers consider this to be strong support for the area limit. However, conjecturing holography is the same as requiring the additional assumption B3 in the second formulation of the information loss problem discussed in the previous section. We therefore know that holography will require new physics beyond semiclassical gravity already in the low-energy regime, that is, on time scales much shorter than the lifetime of the black hole.

3.5 Take-Home Message

We are now in the position to answer the question in the title of this chapter "Is the information loss problem a paradox?". A paradox is a logical impossibility such as a self-contradictory statement. In this regard, the information loss problem is *not* a paradox. The popular formulation of the problem often expresses it as an incompatibility between the low-energy description of semiclassical gravity and the unitary evolution of quantum states. Given that the equations of motion of semiclassical gravity are built to lead to a unitary evolution, such an incompatibility would imply a paradoxical conclusion. However, we have seen that the formulation of the problem requires extra assumptions that cannot be derived within the semiclassical model. Therefore, the answer to the question is NO, there is no paradox.

The information loss problem is not an incompatibility between quantum mechanics and semiclassical general relativity, but it is an incompatibility between the validity of semiclassical general relativity and the assumption of the area limit on the total entropy of a black hole. This is not a paradox, but the community should investigate why these two well-motivated hypotheses are incompatible. Once all the outgoing modes of the radiation are properly taken into account both on $\mathscr{I}^+$ and on the event horizon, then one can show that indeed an initial pure state is evolved into a final pure state. The mistake that is sometimes made is to neglect the outgoing modes that cross the event horizon; doing this would obviously give rise to an inconsistent quantum evolution which violates unitarity. Hence, the growth of the entanglement entropy of the Hawking radiation is consistent with unitarity but does not respect the area limit.

Although there is no paradox, there are still open questions about black hole evaporation. These are whether, how, and when information retrieval occurs. We stressed that the presence or absence of the singularity is crucial for this question. If the singularity is there, and thus an event horizon is present, then all the information which falls into the black hole will be crashed into the singularity or hidden behind the Cauchy horizon, that is, in both cases it will be lost for an outside observer. This is not a paradox or a violation of unitarity, but only a consequence of the fact that part of the information has leaked out of the region of spacetime observable by an external observer. If instead the singularity is not there and the event horizon is replaced by a trapping horizon, then all the information can be retrieved. This means that quantum-gravity effects that aim to solve the singularity problem might also shed light on the question of information retrieval in black hole evaporation.

On the other hand, if one postulates an area limit on the total entropy of the black hole, then information retrieval occurs automatically thanks to the Page curve-like behavior of the entanglement entropy of Hawking radiation, which decreases and vanishes at the end of evaporation. However, as we carefully explained above, requiring the area limit means assuming new physics beyond semiclassical gravity at time scales that are much shorter than the black hole lifetime. In particular, this means that the validity of semiclassical general relativity as a low-energy effective field theory and the holographic principle are incompatible.

Acknowledgements L. B. acknowledges financial support from the European Union's Horizon 2020 research and innovation programme under the Marie Sklodowska-Curie Actions (grant agreement ID: 101106345-NLQG). F. D. F. acknowledges financial support from the PRIMUS/-23/SCI/005 and UNCE24/SCI/016 grants by Charles University, and the GAR 23-07457S grant from the Czech Science Foundation.

References

1. I. Agullo, P.C. Cabrera, B. Elizaga Navascués, Phys. Rev. D **110**(8), 085002 (2024). arXiv:2407.03031
2. A. Almheiri, R. Mahajan, J. Maldacena, Y. Zhao, JHEP **03**, 149 (2020). arXiv:1908.10996
3. A. Almheiri, T. Hartman, J. Maldacena, E. Shaghoulian, A. Tajdini, Rev. Mod. Phys. **93**(3), 035002 (2021). arXiv:2006.06872
4. C. Barcelo, S. Liberati, S. Sonego, M. Visser, Phys. Rev. D **83**, 041501 (2011). arXiv:1011.5593
5. I. Basile, L. Buoninfante, F. Di Filippo, B. Knorr, A. Platania, A. Tokareva, *Lectures in Quantum Gravity*. arXiv:2412.08690
6. N.D. Birrell, P.C.W. Davies, *Quantum Fields in Curved Space* (Cambridge University Press, 1982)
7. R. Bousso, Rev. Mod. Phys. **74**, 825–874 (2002). arXiv:hep-th/0203101
8. R. Brout, S. Massar, R. Parentani, P. Spindel, Phys. Rept. **260**, 329–454 (1995). arXiv:0710.4345
9. L. Buoninfante, F. Di Filippo, S. Mukohyama, JHEP **10**, 081 (2021). arXiv:2107.05662
10. G.V. Dunne, T. Hall, Phys. Rev. D **58**, 105022 (1998)
11. G. Dvali, Fortsch. Phys. **64**, 106–108 (2016). arXiv:1509.04645
12. A. Fabbri, J. Navarro-Salas, *Modeling Black Hole Evaporation* (Imperial College Press, 2005)
13. S.W. Hawking, Nature **248**, 30–31 (1974)
14. S. W. Hawking, Commun. Math. Phys. **43**, 199–220 (1975) [erratum: Commun. Math. Phys. **46**, 206 (1976)]
15. S.D. Mathur, Class. Quant. Grav. **26**, 224001 (2009). arXiv:0909.1038
16. D.N. Page, Phys. Rev. Lett. **71**, 3743–3746 (1993). arXiv:hep-th/9306083
17. D.N. Page. arXiv:hep-th/9305040
18. G. Penington, S.H. Shenker, D. Stanford, Z. Yang, JHEP **03**, 205 (2022). arXiv:1911.11977
19. L. Parker, Phys. Rev. Lett. **21**, 562–564 (1968)
20. C. Rovelli, Black holes have more states than those giving the Bekenstein-Hawking entropy: a simple argument. arXiv:1710.00218
21. C. Rovelli, Entropy **21**(9), 839 (2019). arXiv:1902.03631
22. W.G. Unruh, Phys. Rev. D **14**, 870 (1976)
23. W.G. Unruh, R.M. Wald, Rept. Prog. Phys. **80**(9), 092002 (2017). arXiv:1703.02140
24. M. Visser, JHEP **07**, 009 (2015). arXiv:1409.7754
25. R.M. Wald, Commun. Math. Phys. **45**, 9–34 (1975)
26. Y.B. Zeldovich, A.A. Starobinsky, Zh. Eksp, Teor. Fiz. **61**, 2161–2175 (1971)

Chapter 4
Alternative Theory for the Quantum Black Hole and the Temperature of Its Quantum Radiation

Gerardus 't Hooft

Abstract A mechanism is found that explains how matter falling into the future event horizon of a black hole leaves information there, which it sends to the past event horizon, and there it determines how particles are emitted. This way information can be conserved. The mechanism is a calculable gravitational effect. We also show how it is avoided that the "hidden region" of the black hole gets involved, which has been standing out as a problem ever since Hawking found that particles must be emitted. The most striking consequence of our mechanism is that the radiation temperature is not what Hawking calculated, but twice that value. This is a direct consequence of the fact that what is usually handled as the hidden region, is not there.

4.1 Summary of the Techniques Used in This Chapter

The Schwarzschild solution of Einstein's equations [1] is a smooth, non quantum mechanical exact solution, describing the periphery of a black hole, where Standard Model physics should still be applicable. We may be tempted to start with describing the situation with 'no particles present', and call that the vacuum state. But this definition turns out not to be sufficiently precise. When an object such as an elementary particle falls into a black hole, it gradually becomes invisible, and similarly, the particles emitted by a black hole, will be difficult to follow far backwards in time. Yet we show that with only a few natural assumptions, one can set up equations that can be solved. We find that black holes produce an almost perfect thermal distribution of out-going particles.

However, there is an important point where many colleagues do not agree with the author of this chapter. What is the *temperature* of this radiation? We explain in this chapter how one can argue that the temperature will be twice as high as mostly thought, by Stephen Hawking and his followers. This factor 2 emerges, when we continue the Schwarzschild solution analytically, as far as we can. There, one discovers an entirely new universe, as if there were two universes. One is ours, full

G. 't Hooft (✉)
Utrecht, The Netherlands
e-mail: g.thooft@uu.nl

A. Akil and C. Bambi (eds.), *The Black Hole Information Paradox*, Springer Series in Astrophysics and Cosmology, https://doi.org/10.1007/978-981-96-6170-1_4

of particles, while the other is not well understood. We insist to analyse this feature as accurately as we can, using equations that must make sense.

The temperature of the radiation emitted by a black hole can be calculated in several ways. The best way is to first calculate the entropy of a black hole, by meticulously counting the number of independent quantum states. This has to be done with care. At every moment in time we only count the particle states that are visible in principle, so that these are the building blocks of all black hole states. A particle is visible if an observer far away can use his operators to create or annihilate that particle. We must count the orthonormal states, as it is done in ordinary physics, not the entangled states, since parts of these are often invisible. Going to different values of the time parameter, this set of visible particles is continuously transformed into different sets, but we have no problems with that. All this will be further explained.

By identifying the operators that create or annihilate elementary particles near a black hole, one can follow the evolution of the system as a function of time. The operators far from the horizon are related to the operators close to the horizon, by Bogoljubov transformations. This implies that an observer near the crossing point of future and past event horizon, sees particles differently from what the far-away observer sees. We calculate how these particles are related using the Bogoljubov transformation.

And then we apply a time translation in the future direction, for the distant observer. This transformation reduces the visibility of the in-going particles, but the out-particles become better visible. This transformation is called a 'firewall transformation'. The most central process in our formalism uses the fact that removing an in-particle causes a shift in the wave functions of all out-particles. This shift has been calculated, it is basically the Shapiro effect for radiation grazing past a heavy object, due to the gravitational field. At early times, this shift may be of the order of the Planck scale, but it increases exponentially in time. The mathematics is basically very easy. The position of the emerging out-particle is identified with the momentum of the disappearing in-particle. And so it happens that, while an in-particle is annihilated, an out-particle is formed, in such a way that the out-wave function is simply a Fourier transform of the in-wave function.

This is how, as time passes, the set of visible particles continuously changes, exactly as one should have expected just by staring at the physics.

But then there is one thing that still has to be worked out: *Which of the two universes carries the in-particles, and in which of the two universes does the outgoing particle go?* Remember that this universe actually contains two universes; it is too big. Again there is a natural answer: *the 'hidden' universe is an exact copy of the visible universe.*

We explain why this is so.

A consequence of this is that the wave functions of all in-particles must be *even* functions of their momenta p,

$$\psi^{\text{in}}(p) = \psi^{\text{in}}(-p) \, .$$

The Fourier transform of this gives an out-wave function that is even as well:

$$\psi^{\text{out}}(u) = \psi^{\text{out}}(-u) .$$

This implies that knowing ψ in one universe is all we need to obtain the Fourier transformed wave function in the other universe. They both remain even. The mathematics of 'even Fourier transforms', or Fourier transforms on half-lines, is easy. Instead of the usual expression our momentum space in-wave functions $\psi(p)$ for positive momenta p, is transformed by the Fourier operator F^+ as follows:

$$F^+\psi(u) = \sqrt{\tfrac{2}{\pi}} \int_0^\infty \mathrm{d}p \, \cos(p\,u)\, \psi(p) .$$

This function is its own inverse. This, we claim, leads to the solution to the information loss problem. Technical questions will remain.

To see what the temperature of the black hole radiation is, we consider the time dependence of the visible particle states. One finds that the Bogoljubov transformation gives the time dependence of the states in the two universes, which we call region I and region II, in terms of the Hamiltonian,

$$H^{\text{tot}} = H_I - H_{II} .$$

This operator is exactly conserved, because, locally, it is the generator of the Lorentz transformations in the $r,\ t$ direction, and because of exact Lorentz invariance its value starts out as zero, so that $H^{\text{tot}} = 0$ always. The reason for that is that the black hole is switched on in the Schwarzschild metric, where the two universes are identical (This is not so in Rindler space, where one starts out with non-Lorentz invariant states, and this is why, in Rindler space, the treatment must be different). We see that the wave function used by distant observers will be

$$|\Omega\rangle = \sum_\psi |\psi\rangle\, \langle\psi| ,$$

where we use Dirac's bra-ket notation just to indicate that the states $\langle\psi|$ evolve with minus the Hamiltonian used for the states $|\psi\rangle$.

This is an exact density matrix ρ that evolves with the evolution law

$$\dot{\rho} = i[\rho, H],$$

From the Bogoljubov transformation, we find the density matrix ϱ from the state $|\Omega\rangle$ as follows :

$$\langle n_{II}|\, \varrho\, |n_I\rangle = \langle n_I,\ n_{II}|\Omega\rangle = C^{st} e^{-\pi n_I\, \omega}\, \delta_{n_I, n_{II}}$$

where n is the number of particles with energy ω, and C^{st} is a normalisation factor. This gives a density matrix $e^{-\pi H}$ if $H = n\,\omega$ and $\beta = 1/T = \pi$:

$$\rho = e^{-\pi H} .$$

H is the Hamiltonian in black hole units. According to thermodynamics, the thermal density matrix is $e^{-\beta H}$, where $\beta = 1/T$, so this gives the temperature $T = 1/\pi$.

Hawking did the calculation in a slightly different way, comparing the black hole with Rindler space. The local vacuum state, $|\Omega\rangle$, may contain n_I particles in region I and n_{II} particles in region II. He calculated the probability that there are n particles visible in region I:

$$P(n_I = n) = C^{st} |\sum_{n_{II}} \langle n_I = n,\ n_{II}|\Omega\rangle|^2 = C^{st} |e^{-\pi n\omega}|^2 = C^{st} e^{-2\pi H} .$$

Therefore, he found a temperature $1/\beta = 1/2\pi$. This disagreement is important. Forcing the number of particles in region II to be equal to the number of particles in I was the primary reason for this difference, but it also is the reason why the Fourier transforms mentioned above are unitary in the subspace I only. Therefore, if we want a theory where the quantum information is preserved, we should accept this new value for the temperature.

The value for Rindler space stays what it always has been. Our theory simply states that black holes have nothing to do with Rindler universes.

The next section contains 11 subsections where we repeat the arguments and calculations a bit more precisely, but still without too technical derivations.

4.2 Introduction

4.2.1 The Meaning of Quantum Logic

When this book was written, Quantum Mechanics was approximately one century old, and during all this time this theory was presented as a marvel of 20th century analysis, requiring an entirely different way of thinking not only about space, time and matter, but also about *reality*. This torment of logical thinking was almost thought to be over now; the non existence of local reality was accepted and a lesson was learned: 19th century logic was over, whereas surprises would continue to come.

This left the author of this chapter as a solitary bastion, being one of the last representatives of 19th century logic. He still maintains that classical logic is not over, but our understanding of the equations is still wanting. In the 1960s, our attitude turned out to pay off in a big way, when new pieces of insight told us that elementary particles can be tamed with equations from *quantum field theory*. We learned to ignore warnings from science philosophers who claimed that special relativity is incompatible with quantum mechanics and locality. What we did learn was that we should rely on the results of experimental scientists, whose measurements indicated that fields are important.

Imposing internal logic there, led to light-cone commutation relations and dispersion equations. The procedure of *renormalisation* was found to be an essential ingredient of these theories, and the scheme arrived at was the *Standard Model of the elementary particles*. Thanks are due to numerous experimental physicists who made relentless efforts to figure out how to measure almost everything there was to know about elementary particles. It was concluded that quantum field theories are just fine in spite of what those philosophers said. For ordinary elementary particles, the right equations were discovered, and how to use these was almost completely understood. These particles obey quantum mechanics as well as special relativity—*and* locality.

My claim [2, 3] is that the world can be described by local non-quantum equations, and classical logic will be fine. According to this theory, the reason why quantum computers will seem to do calculations of unbelievable complexity and rapidity, is that they will be able to harvest classical vacuum fluctuations that will go all the way to the Planck scale. Yet unlike standard quantum mechanics, these fluctuations may represent only one reality.

This book is not about the interpretation of quantum mechanics. It is about including *general relativity* in our quantum models. And now the new problem is that experimental information about this question is very difficult to produce. Our present strategy is to apply quantum mechanics to black holes. Our hope is that, just requiring that the answers should be sensible, could yield new information on how to proceed. But, as one will see, the math is quite demanding.

This chapter is organised as follows. The remainder of this Sect. 4.2 contains most of our arguments. first, we show how Hawking's result can be understood as a consequence of Bogoljubov transformations acting on the region surrounding the horizon. Our first attempt leads to a picture of particles radiating away from a black hole, which we call 'scenario 1'. This scenario assumes that black hole radiation has precisely the temperature as calculated by Hawking [4] . But then we show that the Bogoljubov transformation can be reconciled with preservation of quantum information, only if we use 'scenario 2' for the interpretation. Scenario 2 however would imply that the temperature of the radiation is twice as high as Hawking's value.

We introduce the various techniques that are indispensable to know about, using a sequence of subsections. During this entire first section, Sects. 4.2.1–4.2.11, we shall avoid most of the equations. Therefore, this part of the chapter is not too technical. In the Sects. 4.3–4.12 we display the calculations backing our result.

4.2.2 Quantum Mechanics in Black Holes

Black holes occur at all scales in physics, and therefore it is conceivable that their role in uniting general relativity with quantum mechanics may well turn out to be crucial. We use the quantum language here the way we were taught to, as it is described in our elementary text books; just don't leave reality out of sight. In particular, resist the temptation to make our cover story about quantum mechanics look even more fantastic than it does now. Don't let observers run through wormholes and the like.

Our first attempts must be to bring our understanding of the gravitational force to the same level as our understanding of elementary particles.

The primary reason why the basic synthesis of gravity with quantum mechanics is so difficult to understand is, that direct experimental data are so much more difficult to acquire. As could be expected, a majority of researchers in this field concluded that black hole theories would have to be non-local [5]. I do not believe them, and there are no non-local phenomena to be found in the treatment proposed in this chapter, but we are not there yet; let us continue.

Throughout this chapter, our basic goal is to arrive at unitary evolution equations. To get an impression, the reader is invited to glance through the last pages (such as Sects. 4.11 and 4.12), just to see in what direction we are going. We do not succeed completely, but we think we are on our way.

4.2.3 Schwarzschild Coordinates

We introduce the Schwarzschild coordinates in Sect. 4.3. At first sight, they do not reveal much about the numerous excited states a black hole can be in, but we can easily formulate these excited states in terms of particles that may be frolicking around near the horizon. The case we are primarily interested in is the black hole that is large and heavy compared to the Planck scale. One then finds that the particles that give more structure to the horizons, will be much lighter than the Planck scale. This allows us to treat the Schwarzschild metric as a sturdy background, and the equations for the light and tiny particles can be derived from the Standard Model, generalised to handle the case of weakly interacting particles, in a moderately perturbed background metric.

4.2.4 The No Hair Theorem

Note that we ignore the 'no hair theorem'. This theorem is one of the errors one can commit by not understanding the laws already known. Dust falling into a black hole has its time reverse coming out; together they form lots of hair. Time reversal symmetry applies to all of the known laws of nature at small scales, so why forgetting that?

4.2.5 The Kruskal–Szekeres Coordinates

The weakly perturbed black hole can be analytically extended to form a maximal analytic structure, using the Kruskal–Szekeres coordinate system, to be introduced in Sect. 4.4. This description of the same space-time for which the Schwarzschild coordinates had been introduced, is nevertheless quite different: it connects the universe surrounding the black hole to another universe, whose role will at first be mysterious. It is necessary first to include this 'spectator universe' in our descriptions, to leave its true role in the physics of black holes for later. Only near the end, in Sect. 4.12, the building blocks of our theory will merge into a solid picture. It has to do with conical singularities.

4.2.6 Vacuum States

The vacuum state is defined to be the state from which no particles can be removed. That is, the annihilation operators a, to be derived from the fields and their equations, just give us the null state: $a|\psi\rangle = 0$. But wait, one can always consider the removal of particles that were involved in 'dressing' the black hole, and indeed, technically, this means that genuine black holes cannot be surrounded by a real vacuum. Or in other words, there is only one vacuum state, which is the state where we removed the black hole itself.

But we can reformulate empty states in the immediate vicinity of a black hole. These are good approximations for describing a black hole with no other particles in their vicinity, and their role can be derived accurately: there are different kinds of such vacuum states! In Sects. 4.5 and 4.6, it is explained why this definition of vacuum states depends on who is watching. The observer outside the black hole or someone very close to the horizon? It depends on the time coordinate used by the observer.

4.2.7 Thermal Radiation Emitted by a Black Hole: Scenario 1

In Sects. 4.5 and 4.6, we shall handle the transformation from one time coordinate to another, and explain the resulting Bogoljubov transformations, which do not keep the vacuum states invariant.

Next, in Sect. 4.7, we describe how Hawking found that a distant observer sees a black hole that emits radiation. This radiation is purely thermal, and he finds a temperature

$$T_H \stackrel{?}{=} 1/8\pi kGM \; . \qquad (4.2.1)$$

In this section, we derive how our formalism can yield this result, including this value for the temperature, by assuming a prescription that we call 'scenario 1'.

4.2.8 The Thermal Density Matrix: Scenario 2

The above calculation was borrowed from a similar calculation in *Rindler space* [6]. Yet there is something odd with this result (hence the question mark in Eq. (4.2.1)). It was derived from the wave function of the vacuum state $|\Omega\rangle$ for the local observers. In terms of the particles in regions I and II, this state is found to be

$$\Psi = C^{st} \sum_{n_{II} \geq 0} |n_I\rangle\, e^{-n_I \pi \omega}\, |n_{II} = n_I\rangle \tag{4.2.2}$$

(multiplied over all positive values of the variable ω), and in view of the restriction $n_{II} = n_I$, there is hardly any summation. However, if it were really a thermal expression valid for two independent universes, then all states with arbitrary values for n_I and n_{II} should be included. The fact that, in the expressions we get $n_I = n_{II}$, implies that region II is just an exact copy of region I. That is not quite the situation in Rindler space. In Rindler space, one really has two different universes connected naturally at one point (more precisely: a sheet, if we add the transverse coordinates).

The expression obtained from the Bogoljubov transformation is actually an amplitude, to be squared to obtain the probabilities of the various states in region I.

This is also the reason why, in Rindler space, one can never recover information that disappeared across the horizon. If, in contrast, we insist to describe a black hole in such a way that information is returned, we have to identify the two universes as being a single one.

4.2.9 The Quantum Clone

This is why we advocate to use not scenario 1 for a black hole, but scenario 2, which states that the wave function we find by employing the Bogoljubov transformation, actually plays the role of a *density matrix* for the black hole:

$$(4.1.2) \;\rightarrow\; \varrho = e^{-\beta H}\,, \tag{4.2.3}$$

for a real thermal density matrix, but this differs in various ways from scenario 1:

First, a density matrix naturally only consists of bras multiplied with the kets, describing exactly the same physical state, which explains our restriction $n_I = n_{II}$.

Second, the bra state $\langle\psi|$ runs backwards in time. This is why it is actually controlled by *minus* the Hamiltonian for space I. Consequently, this density matrix would apply only for one universe. Being a density matrix, it should not be squared to obtain the probabilities—it is a square already.

But this implies that, in scenario 2, the exponent being $e^{-\beta E}$, gives a β whose value is half of what Hawking found; so the temperature is now twice his.

Thus one finds a temperature twice the values given by Eq. (4.2.1):

$$T \stackrel{!}{=} 1/4\pi kGM\ . \tag{4.2.4}$$

In Sect. 4.9 we discuss this difference. The question that must be considered concerns the interpretation of the universe in region II of the KS metric. Do we have indeed an other universe there? We claim to have convincing arguments that region II is not another, 'alien' universe. It is a copy of our own universe. We explain this in Sect. 4.9, and add more clues in Sect. 4.11. For black holes, scenario 2 is indeed more natural than scenario 1.

4.2.10 The Shapiro Effect

A crucial ingredient in our treatment is the requirement that a radiating black hole should behave in agreement with the fundamental laws of quantum mechanics: the Schrödinger equation can be inverted in time, so that, consequently quantum information should not get lost. A beautiful mechanism for the retrieval of black hole information is obtained by considering the gravitational forces produced by the Shapiro effect. This is the gravitational pull by which particles going into a black hole affect the particles coming out. The emerging particles are shifted along the horizon where they originate. How this effect can repair information loss is discussed in Sect. 4.12.

4.2.11 Discussion

This work is not finished. There are several weak points in our approach. We were forced to use approximations and simplifications. For instance the Shapiro effect that we use to see information be returned to us in the form of Hawking radiation, only partly does its job: only *geometric features* generate the gravitational forces that send particles back out of the black hole. It remains to be seen whether among the 'in particles', quantum numbers not related to energy and momentum can be restored to re-emerge with the 'out particles'.

And then, consider the crossing points of future and past horizons. The author has neither time nor space to do further calculations on this singular point; it might well turn out to play a pivotal role in future formalisms. Let us just observe that, identifying regions I and II, implies a division of space-time by an isotropy operator $\mathbb{Z}(2)$. It actually acts as a mirror, the ideal instrument to send information back to where it came from, but also it induces a mild *conical singularity* This singularity actually has some of the characteristics of a membrane. One can imagine membranes of this sort to affect the Standard Model interactions, but how this goes is not understood at present.

These are not drawbacks that would be likely to invalidate our ideas; rather, they pose new challenges, to be addressed.

In any case, our procedure works best if the two universes are handled as being exact *quantum clones* of one another. That doubles the temperature of the radiation. And we resist the temptation to return to the subject by which we began writing this chapter: It may well be that, eventually, the interpretation of quantum mechanics will have to enter this discussion.

The Sects. 4.3–4.12, contain many of the details and calculations that explain our findings in a systematic formulation.

4.3 Schwarzschild Coordinates and Horizons

4.3.1 The Region $r \to 2GM$

Shortly after Albert Einstein wrote his famous equations for the gravitational force, well-known astronomer Karl Schwarzschild [1] found an exact solution for the spherically symmetric case, in the absence of matter:

$$\mathrm{d}s^2 = \Big(\frac{2GM}{r} - 1\Big)\mathrm{d}t^2 + \Big(1 - \frac{2GM}{r}\Big)^{-1}\mathrm{d}r^2 + r^2\mathrm{d}\Omega^2\,; \quad \begin{cases} \Omega \equiv (\theta, \varphi)\,, \\ \mathrm{d}\Omega \equiv (\mathrm{d}\theta,\ \sin\theta\,\mathrm{d}\varphi)\,, \end{cases} \tag{4.3.1}$$

where $\mathrm{d}s$, the distance between two infinitesimally close space-time points x and $x + \mathrm{d}x$, is defined to be real in the space-like direction and imaginary when time-like. This metric describes a black hole of mass M that remains unchanged both in the future and in the past.

Its space-time geometry is sketched in Fig. 4.1a. The angular coordinates Ω are suppressed here; the dashed vertical line is the *horizon*, corresponding to the point $r = 2GM$. We see that the coefficient in front of $\mathrm{d}t^2$ vanishes there, while the coefficient in front of $\mathrm{d}r^2$ generates a singularity. This singularity can easily be removed by replacing the radial coordinate r with ρ:

$$\rho = \sqrt{r - 2GM}\,, \quad \text{so that} \quad r = 2GM + \rho^2, \quad \text{and} \quad \mathrm{d}r = 2\rho\mathrm{d}\rho\,. \tag{4.3.2}$$

In terms of these coordinates, the metric (4.3.1) can be written in a compact form:

$$\mathrm{d}s^2 = \frac{-\rho^2\,\mathrm{d}t^2}{r^2} + 4r\,\mathrm{d}\rho^2 + r^2\,\mathrm{d}\Omega^2\,. \tag{4.3.3}$$

Note that Eq. (4.3.2) implies that ρ is imaginary when $r < 2GM$. This will make the region $r < 2GM$ unphysical to some extent—but not quite: it is better to say that the radial space coordinate becomes time-like and the time coordinate becomes space-like there, see the blue shaded regions in Figs. 4.1 and 4.2. Later we will see

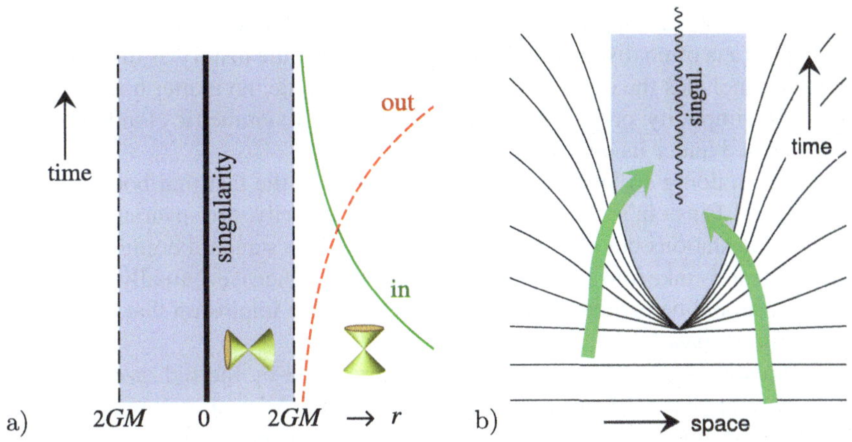

Fig. 4.1 **a** The stationary black hole, in the Schwarzschild coordinates, $(r,\ t)$, showing the geodesics of a particle going straight in (green) and a particle going straight out (red, dashed). The dashed vertical line is the horizon; the cones indicate the orientation of the local light cones. The angles $(\theta,\ \varphi)$ are not shown. The central singularity cannot be crossed. **b** An artist's impression of spacetime for a black hole being formed. The black curves are constant t lines, approaching the horizon. Thick green arrows represent imploding matter. The wavy line is the central singularity

more precisely what this means, and indeed the region $r < 2GM$ will only play a secondary role.

Note on the other hand, that this metric naturally continues to *negative* values of ρ, where, as we shall see in the next section (or by carefully examining Eq. (4.3.3)), the physics is exactly the same as for the positive values. Nothing becomes complex or imaginary there. However, clearly, this seems to make space-time twice as big as what it should be. It looks as if black holes are some kind of worm holes towards another universe, but it is a big mistake to think that. We return to this feature when the regions I and II of the Kruskal–Szekeres coordinates are discussed, see Sect. 4.4 of this chapter.

4.3.2 *The Time Coordinate t*

Allow us to make a somewhat extensive digression. We cannot proceed without observing that the black hole's past seems to be represented incorrectly in the metric as represented in Fig. 4.1a. This is often brought forward as an objection against using the metric as represented in Eq. (4.3.1). And it is true that all matter that has entered into the black hole during its entire life in the past, should still be visible at the horizon, the vertical dotted line. The collection of all in-going particles (in-particles for short) as represented by all green solid curves below the curve shown, must have an effect on the horizon. In fact, they change the entire metric into the one sketched in

Fig. 4.1b. Therefore, the approach starting with the eternal (that is, time-independent) metric Fig. 4.1a is often dismissed. Authors then include the history as they prefer to see it, which includes the event that created the black hole, no matter how long ago that was. For simplicity, one often imagines a spherically symmetric cloud of matter, which collapsed under its own weight.

However, in doing so, one would completely ignore the fact that both quantum mechanics and Einstein's equations for General Relativity, are invariant not only under time translations but also under time-reversal. The standard arguments in the literature [5] rarely take into account that the black hole, when it eventually evaporates completely, should be handled as the time-reverse of the implosion that created the black hole in the past.[1]

The evaporating particles (out-particles), represented by the red dashed line in Fig. 4.1a, should be treated just like the in-particles are. All evaporating particles in the future cause similar deformations of the metric near the horizon. In this entire chapter, we shall treat the features of the collapse and of the evaporation in a time-symmetric way.

4.3.3 The Heat Bath

The way to handle the situation systematically is now as follows:
we assemble all possible states of a black hole by first considering how these states should behave when we assume the black hole to be stationary. This is realised if the black hole is situated in a heat bath formed by particles that have the same temperature as the Hawking particles generated by the black hole. Then, consider the region near the center where all in- and out-going particles meet. Let us assume that an observer there sees no particles. In that case, the situation should be completely stationary. As we shall derive however, if a local observer sees no particles, not only does a distant observer in the future see Hawking particles emerging from the hole, but by time reversal invariance, an observer in the distant past should also see particles entering the black hole—that is, the particles in the heat bath. In conclusion, the distant observers might not agree with the local observer concerning the particle density, but they both agree that the situation can be looked upon not only as entirely stationary in time, but also symmetric under time reversal.[2]

---

[1] Hawking [4, 7, 8] did take into account that the black hole may evaporate completely, but the space-time metric he produced was asymmetric under time reversal. This was because he took into account the fact that the *entropy* of any configuration will be allowed only to increase, and that would be asymmetric under time reversal. However entropy refers to the probability density of states, not the states themselves, which should display invariance or covariance under time reversal, or CPT symmetry to be precise, as we know from elementary particle physics.

[2] Though stationary, this black hole will not be absolutely stable; this is not a problem in practice, as long as we restrict ourselves to describing relatively short stretches of time.

4.3.4 Small Excitations

To consider more complete sets of black hole solutions, we subsequently add or remove particles by applying creation and/or annihilation operators defined by the ubiquitous elementary particle fields. This then allows us to consider black holes with time-dependent features. One may always consider entangled particles later; they will not affect the primary equations affecting the original Schwarzschild solution.

This is the way we go in this chapter. Our point is that both the distant past and the distant future will be irrelevant as soon as one can describe the evolution of the entire set of *all states that may be involved* and that we may encounter in any given short stretch of time (see the green region in Fig. 4.2a). Every element of this set evolves into an other element of this set. The set should be defined in such a way that neither the states describing the distant past, nor the ones in the distant future, have a direct effect on the evolution law in the present. Think of an electron. The equations from which one can derive its behavior at present, involving the entire Hilbert space, do not require the knowledge whether the electron emerged long ago by pair creation, or beta radiation or whatever. Neither do we need to know how it will end its life.

Important remark alluded to earlier: at times earlier or later than the green region, the particle excitations eventually will also invade the blue regions but we shall have to ensure that all important events that will give us local physics will be at the visible sides of the black hole only (or possibly we'll enter with infinitesimal steps in the forbidden domains). Thus, there is no need to worry about particles moving from the green zone to the forbidden region close to the horizon; they are taken care of by the mechanism alluded to earlier and to be discussed later in this Chapter.

We only put particles where they can directly be seen, or at least their interactions must play a role *at the present*. This means that, for the time being, we *don't* put particles anywhere in the blue regions of Fig. 4.1a and b, or Fig. 4.2a and b. What happens in those regions is postponed to later concern (Sect. 4.4 and onwards). This is the first and not at all the last remark concerning locality, an important concept in this approach. So far, we have merely added small particles in the green region, at some safe distance from the horizon; how these evolve in earlier or later domains remains to be calculated.

The state we choose for a black hole, at any given time, which may include entanglements of several states, depends on the past. But in general, the physical laws here are well captured by the Schrödinger equation active in the present, just as we have it for all particles in the Standard Model, where any further reminders of the past are unnecessary. This simple observation already departs from claims by some other observers, which we ignore until forced otherwise (and this does not seem to happen anywhere in the description we shall arrive at, in this chapter).

Eventually, of course, we should formulate the entire set of equations to see what really happens. This is our aim. Thus we proceed using the metric of Fig. 4.1a. It is used as a starting point: neither the distant past not the distant future can affect the e.o.m. for the behaviour of the black hole at present. But, as we shall see, the complete set of all excited states of a black hole, having light particles flying around everywhere in the green region, needs to be defined with care. More about this in Sect. 4.10 of this chapter; we shall indeed find that one can define other excitations, but, surprisingly perhaps, these can be transformed to elements of the set we already have.

End of digression—although this needs not be the end of the discussion of time reversal symmetry. We are aware of the fact that nature does not *have to* be symmetric under any symmetry transformation of the form of CPT. We find that our first attempts to formulate a synthesis of general coordinate invariance and quantum mechanics must share the known symmetries of both theories. It is entirely acceptable to study other options, but we leave that to other investigators.

4.4 The Kruskal–Szekeres Coordinate Frame

4.4.1 Using Short—But Not Too Short—Time Intervals

The conclusion of the previous subsection is that we should first try to use the metric of a completely stationary black hole, and consider particles added to that (or removed, see later), in order to get the most generic states. Starting from that, we plan to construct the Schrödinger equation,[3] which must be accurate during the entire evolution process of the black hole. First we must make sure that the equations work accurately during a sufficiently short time interval.

During such an interval, the black hole mass will stay almost constant, but one can be more precise, by allowing the mass to make small, perturbative fluctuations.[4] Since the mass is observable, standard quantum mechanical techniques will suffice to understand how the mass will slowly evolve during longer time intervals. This will in practice be exactly what we need to know.

Of course, an important question will be how to knit successive time intervals together; we'll come to that. Having that, the next task will be to see how transitions to lighter or heavier black hole states can take place, and this should enable us to handle the initial and final states of the black hole correctly after all.

Space-time will look as in Fig. 4.2a. Here, the green part is the region that we allow to be filled with particles of the Standard Model (or beyond). All particles must be sufficiently light to allow us to neglect their gravitational fields, *in as far as they stay in the green region*. However, their geodesics will continue, and either in the far past, or in the far future, or both, they will approach a horizon too tightly to be called a particle at all. As seen by the local observer, heir energy and momentum will increase exponentially as we get further away from the green region. It so happens that we have

[3] In the not so distant future, but not in this chapter, unfortunately.

[4] It is easy to imagine where fluctuations in the mass can come from: particles will be flying in and out. The mass/energy of the entire system is constant, but the mass of the black hole itself depends on the extent to which we regard particles very close to its horizon, to be integral parts of the black hole, or alternatively, whether a particle may be considered to be separated too far for that.

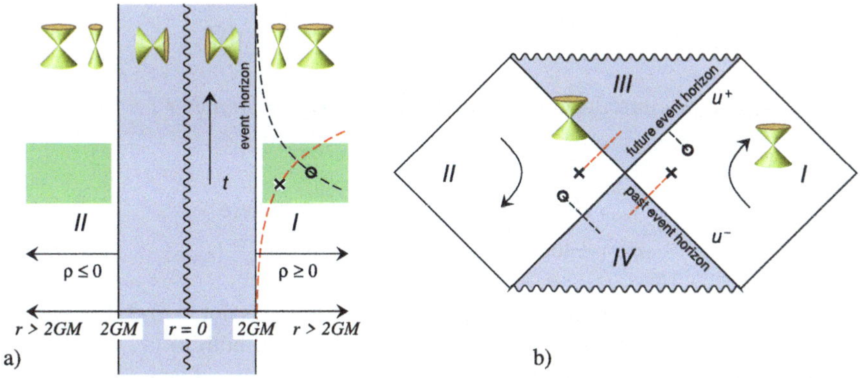

Fig. 4.2 **a** Schwarzschild metric showing coordinates $(\rho,\ t)$. Solid angle Ω is suppressed. Green is the time slot where we choose the in- and out-going particles (marked O and X). The yellow cones show the orientation of the local light cones in the various domains. **b** Kruskal–Szekeres [9, 10] coordinates in the form of a Penrose diagram. The coordinates $u^{\pm}$ are defined in Eq. (4.4.4). Region I represents the outside region in (**a**), region II will first be treated as if the physical events taking place there are independent of the ones in region I. Only later we shall consider the option of quantum clones. Arrows in (**b**) show how local time proceeds. In both figures, the yellow cones show the orientation of the local light cones

identified an *other way* to represent such an 'ex particle' as an elementary component of the black hole state. This procedure will remove exactly all particles that we do not want to regard as particles anymore. This is further explained in Sect. 4.12.2.

4.4.2 The Coordinates x and y

We have now come to the point of considering other coordinate frames for the same metric, that are not singular on the horizons (while the original Schwarzschild coordinates do generate singularities there). Nevertheless, we need to keep track of time translation invariance. The ideal coordinate system for this purpose[5] turns out to be the Kruskal–Szekeres coordinates [9, 10], where r and t are replaced by x and y. These are defined by

$$x\,y = (\frac{r}{2GM} - 1)e^{r/2GM}\,,$$
$$y/x = e^{t/2GM}\,. \tag{4.4.1}$$

One derives that in he Kruskal–Szekeres frame, the metric (4.3.1) reads

[5] There are other choices that can be useful but they will not be considered here.

$$ds^2 = \frac{32(GM)^3}{r} e^{-r/2GM} dx\,dy + r^2 d\Omega^2 . \tag{4.4.2}$$

For later use, we introduce a new dimensionless time parameter τ,

$$\begin{aligned} &\tau = t/4GM, \quad \text{so that} \\ &y = e^{q+\tau}, \quad x = e^{q-\tau}, \quad \text{where} \\ &q = \tfrac{1}{2}\log\left(r/2GM - 1\right) + r/4GM . \end{aligned} \tag{4.4.3}$$

An important feature of this re-parametrisation is that it is not 1-to-1. Every point (r, t, Ω) in the Schwarzschild frame, Fig. 4.2a, is mapped both to (x, y, Ω) and to $(-x, -y, \Omega)$. Indeed this repetition is fundamentally *exact*, a feature that is often ignored. Does any event in region I have a *quantum clone* in region II? Does an observer have a quantum clone? We shall return to this (Sects. 4.5 and 4.9 of this chapter).

4.4.3 Penrose Diagrams

After replacing the light cone coordinates x and y by coordinates $u^\mp$ that cover a finite domain, one gets a picture of space and time in the form of the diagram as sketched in Fig. 4.2b. A good choice is

$$x = C \tan u^-, \quad y = C \tan u^+, \quad \text{with} \quad C = \frac{\sqrt{e/2}}{2GM}, \quad |u^\pm| < \tfrac{1}{2}\pi . \tag{4.4.4}$$

The picture of the universe that we thus obtain, has the useful property that local light cones everywhere are oriented in the same direction (see the cones in Fig. 4.2b). Such diagrams are called *Penrose diagrams* [11, 12].

One good thing about these coordinates is that they are conformal, which here means that the coefficients in front of the terms with dx^2 and dy^2 both vanish. A consequence of this is that, at given, fixed angles Ω, the lines with x constant and the lines with y constant, are light rays. If we rotate the (x, y) frame by 45°, as in Fig. 4.2b, we see that time-like curves all have a slope of less than 45° with respect to a vertical line, while information cannot travel along the lines with slopes larger than 45°. The transformation to the coordinates $u^\pm$ shown in Eq. (4.4.4) was chosen to turn the diagram shown in Fig. 4.2b into a compact shape. The boundaries of the physically important regions are formed by the lines $|u^\pm| = \frac{1}{2}\pi$ and $u^\pm = 0$. The local light cones are oriented as is shown in the figure.

The only singularities in the coordinates $(x,\ y)$ (or $u^{\mp}$) are at the edges. The wiggly line is a true natural singularity, but that is entirely in the domain where $r \to 0$, so that ρ, see Eq. (4.3.2), is imaginary. One also notes[6] that this natural singularity is always beyond the infinite future (in region III) or before the infinite past (in region IV).

One can now see that there are two different horizons. The solid line where $u^- = 0$ is the *future event horizon* and the line $u^+ = 0$ is the past event horizon.

In particular, the metric is completely smooth at the origin of the diagram. There is, however a very peculiar, somewhat worrying aspect of this particular Penrose diagram: it seems to consist of *two universes*, connected at the origin. Only space-like curves rigorously connect these two universes. What does the second universe stand for? Many authors have different feelings about what it means for a black hole to involve two universes. How do we understand this physically? One popular view is what here will be called the 'science fiction' view: a traveller who, somehow, manages to move faster than light during a short lapse of time (*eigen* time to be precise) should be able to travel from one universe to the other. Unfortunately, the present author does not believe in such science fiction stories (one can't travel faster than light), so this option will quickly be dismissed.

We shall frequently refer to the different regions in the Kruskal–Szekeres (KS) coordinate frame [9, 10], which we label I, II, III, and IV, as shown in the diagrams.[7] And indeed, fortunately, there are more rigorous procedures to figure out how to deal with the physics of what we call region II, as we shall see.

Let us emphasise again: region II appears to be an exact copy of region I. In Fig. 4.3a, we see how other researchers often describe the time evolution in the Kruskal - Szekeres coordinates. they take this diagram as an ordinary manifold where region I describes our entire universe, while region II, an exact image of region I, somehow stands for something else, *inside* the black hole. To us, region II cannot be anything else than a copy of our own universe. It only makes sense to assume that it is an exact copy of our universe, a quantum clone, to be precise, but we shall come to that, see Sects. 4.9 and 4.11.

4.4.4 Cauchy Surfaces

The time evolution of physical states (states with particles flying around in regions I–IV) is to be described by drawing different *Cauchy surfaces*, 3-dimensional sub-

[6] This suffices to ensure that these singularities are also physically insignificant. The future event horizon itself is already at the infinite future for any observer in region I. We would like to keep things that way, but remarkably, we may observe later that infinite future and infinite past might become interchangeable.

[7] Other authors favour the interchange $II \leftrightarrow III$, so that their labels are ordered anticlockwise. We prefer the order used here, as regions III and IV often play not such a big role in our derivations. As yet, particles in regions III or IV seem to be insignificant—but please hold on.

spaces of the space-time manifold where we describe all possible states at a given moment of time.

In flat space-time we can consider all planes where $t = Cst$ to form a useful set of Cauchy surfaces enumerated by this constant.

For a black hole described by the KS metric we use the $\tau = 0$ plane (see Eq. (4.4.3)) that separates the *future* from the *past.* This is our primary Cauchy surface. As time τ proceeds, many researchers draw the Cauchy surfaces as in Fig. 4.3a. This would be the picture drawn by a local observer. However, as time τ proceeds, time in region II, as defined by a distant observer, runs backwards. Therefore, one should draw the Cauchy surfaces as in Fig. 4.3b. These are the surfaces $\tau = Cst$. Indeed, he most remarkable feature of this metric is that in one part of spacetime, region I, time τ runs forward while, equally swiftly, time in the opposite part, region II, *runs backwards*, see the arrows in this Figure.

The novelty of this situation is that, the point at the center stays fixed to the center at all times as seen by the distant observer. It so happens that the blue line in Fig. 4.3b is much more suitable than Fig. 4.3a for the calculations to be done next. It represents a valid set of Cauchy surfaces, since the data at $\tau = 0$ completely fix the state the black hole is in; at every other fixed value of the τ variable these data can also be fixed.

In regions III and IV, the direction of the time flow is ambiguous; both these regions may be thought of as representing time, either after the infinite future, or before the infinite past. The data in these two regions are not needed for understanding observed phenomena.

Consequences of this situation often seem to be ignored, while, actually, they are essential. Some authors complained that our choice for the Cauchy surfaces looks too complicated. That's a matter of taste; our essential point is that time translations for the time coordinate, $\tau = t/4GM$ in Eq. (4.4.3), keep this set of Cauchy surfaces unchanged, a must if we wish to understand time translation invariance (or covariance).

Let us again make an attempt to describe the complete set of states for a black hole, taking into account the fact that we have the two regions, I and II. We introduce them

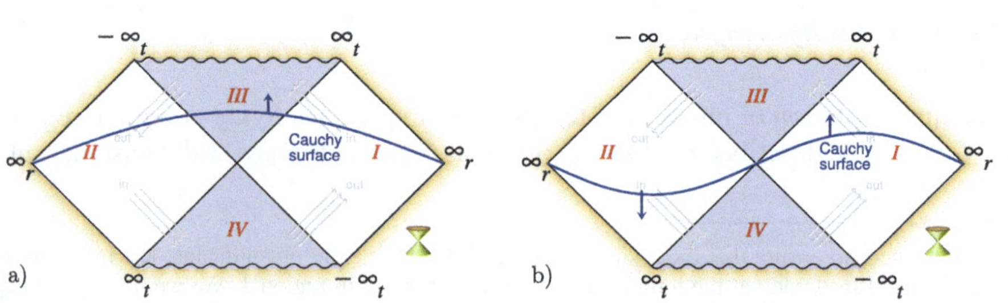

Fig. 4.3 **a** Cauchy surface (blue curve) as often adopted in the literature. This choice is not compatible with time translation covariance. **b** Cauchy surfaces of this shape form a set that is completely invariant under time translations and therefore our favoured choice

by first considering the entire black hole region, and, just like most other authors, we consider pumping in particles, one by one, anywhere in the blank regions of Fig. 4.2b. For the time being, we allow any kind of particles, both gravitons and Standard Model particles.

As already stated, this will be an important restriction that we may have to reconsider later, but for the time being it is the first thing to try.[8]

This is done often in physics: particles whose interactions are sufficiently weak are represented by fields that only interact linearly. It will be an inevitable simplification, as yet.

We now define particles in terms of creation and annihilation operators of Standard Model fields, adapted to this curved space-time background. The metric (4.3.1) describes the eternal black hole. See Fig. 4.2a. Only one horizon is visible there, but in terms of the Kruskal–Szekeres (KS) coordinates, we see that there are two horizons, see Fig. 4.2b.

It is important to realise how different observers will describe what they see. Local observers will see nothing special when a particle crosses a horizon. But for a distant observer a horizon is a boundary that cannot be crossed. This observer will see the particle slowing down, become infinitely heavy, and stop completely. It will also become more and more difficult to see the particle. Nevertheless, something changes in the black hole whose horizon it is. We shall see in Sect. 4.10 what happens: due to the Hawking effect, particles come out of the black hole, and these can be affected sufficiently to preserve all information carried in by the in-particles.

The out-particles must have been invisible at early times, but they turn into visible objects later. Naturally, we must understand how to relate those particles that have entered at early times (the early in-particles), to the Hawking particles (out-particles) that emerge at much later times. Can we *restore unitarity* by transforming in-particles into out-particles and back? In principle, indeed we can! In fact, as will be shown, the mapping is dictated primarily by known laws of physics. And these are local![9] See Sect. 4.10.

Thus, on the future event horizon we expect a boundary condition that maps all physical data of the particles crossing that horizon, onto particles that, with an appropriate time lapse,[10] emerge from the past event horizon.

It was found to be crucial that all added particles are excitations due to the action of Standard Model fields, with the restriction that:

> a small number $\epsilon > 0$ exists (to be specified later), such that these operators are all situated where they are visible to the outside world, at $r > 2GM_{\rm BH} + \epsilon$, inside a sufficiently large interval of the time t. All particles found at earlier or later times, or the particles that may have crossed a horizon ($r = 2GM_{\rm BH}$), can be included by subjecting the given ones to the

---

[8] Indeed, a suggestion made recently is that *heavy objects*, such as rocks, planets, neutron stars, or even other black holes, may enter into a black hole. It could be that the final states then not only contain familiar clouds of Hawking particles, in the form of gases of Standard Model particles, but also heavier objects, including microscopic black holes. This is an exciting idea that needs to be followed up.

[9] But there still are important parts that will require further clarification.

[10] This time lapse is precisely dictated by the equations.

evolution equations dictated by the Standard Model, taking in mind the horizon boundary conditions (which in this chapter will be derived only briefly near the end).

We leave for later consideration how to fill the remaining spaces with particles that cannot be seen from the outside. In particular, we must postulate what should be done with particles allowed in region II. As stated before, region II is exactly as large as the entire universe, normally represented by region I only.

Since we define the particle content of the black hole by limiting ourselves to a small time interval, there is no place for the imploding particles that produced the black hole in the (distant) past, or for the particles involved in the final puff of the black hole. These must be derived at a later stage by solving Schrödinger equations. In line with this, as was elaborated in Sect. 4.3, we only need to consider, as our background, *eternal black holes*, as described by the metric (4.3.1).

4.5 The Transformation from Global to Local Observers

4.5.1 The Central Region

We continue with the description of the Kruskal–Szekeres metric, Fig. 4.2, but now we concentrate on a small part of the central region, small enough to allow us to ignore local curvature. Near the origin, where $r \to 2GM$, or $\rho \to 0$ (see Eq. (4.3.2)), this coordinate frame then obtains the metric (4.4.2)–(4.4.4), or

$$\mathrm{d}s^2 \to 2\mathrm{d}u^-\mathrm{d}u^+ + (2GM)^2\mathrm{d}\Omega^2 ; \tag{4.5.1}$$

the angles Ω are defined as in Eq. (4.3.1).

Any given point (r, t, Ω) in the Schwarzschild metric does not correspond to a single point (u^-u^+, Ω), but to two points: (u^-, u^+, Ω) and $(-u^-, -u^+, \Omega)$, since, in Eq. (4.4.1), the sign of x cancels out against the sign of y, see Eqs. (4.4.1)–(4.4.4). No other ambiguities in the mapping are seen to occur, see Fig. 4.2.

For large black holes ($2GM \gg 1$), the direct effects of single particles on the Schwarzschild metric are assumed to be negligible. These effects take place at time scales much longer than the time scales at which the changes take place that we wish to study, particularly since the particles involved are much lighter than the Plank energy. Also, we expect that there will be perturbation expansions that can yield sufficient accuracies to allow us, eventually, to study long time scale effects as well. This is why the expressions (4.4.4) and (4.5.1) are accurate starting points for calculations on macroscopic black holes.[11]

[11] In many recent publications the idea is circulating that particle entanglement features [13] will invalidate such a view categorically. We suspect that this is due to miscommunication; the entropy of the particles that might be expected to emerge from region II does play a role, but is interpreted quite differently here. In any case, we decide first to investigate the more conventional, and to my

At a small region near both horizons we limit ourselves to the longitudinal space-time coordinates $(Z,\ T)$. We may approximate $\tan u^{\pm} \to u^{\pm}$, and write Eq. (4.5.1) as

$$ds^2 = dZ^2 - dT^2 = 2du^+du^-, \tag{4.5.2}$$
$$Z = X^3 = \tfrac{1}{\sqrt{2}}(u^+ + u^-), \quad T = \tfrac{1}{\sqrt{2}}(u^+ - u^-), \tag{4.5.3}$$

(the transverse coordinates Ω and angular momenta $(\ell,\ m)$, may be added later). Note that, the metric (4.5.2), (4.5.3) describes flat space-time; this means that we are performing the flat-spacetime approximation in the immediate vicinity of the central point.

We rescale the light cone coordinates ρ and τ by

$$\sqrt{\frac{y}{x}} = \sqrt{\frac{u^+}{u^-}} = e^{t/4GM} = e^{\tau}\ , \quad \text{and} \quad Z^2 - T^2 = \varrho^2\ , \tag{4.5.4}$$

or

$$Z = \varrho \cosh \tau, \qquad T = \varrho \sinh \tau, \tag{4.5.5}$$

where $\varrho = 2\rho\sqrt{r}$ (see the definition of the coordinate ρ in Eq. (4.3.2)), and τ is a rescaled version of the time parameter t for the black hole, as experienced by a distant observer.

4.5.2 A Scalar Field in the Central Region

We now need to formulate precisely the transformations of fields in the locally flat region near both horizons, to the fields in region I and II of the KS coordinates. From a classical viewpoint, these fields are the same, but in quantum mechanics, their structures in terms of creation and annihilation operators are different. An observer using the coordinate frame $(\varrho,\ \tau)$, sees particles that are not the same as the ones seen in the frame $(Z,\ T)$. The relation is given by Eq. (4.5.4). The derivation is not always done correctly. Here is how we derive it (Since Z is the third space coordinate, we write $\partial/\partial Z = \partial_3$; furthermore, the transverse component of a 4-vector k is indicated as $\tilde{k}$).

In the coordinates (4.5.2), consider a real scalar field $\phi(Z,\ T)$ obeying

$$\frac{\partial^2}{\partial T^2}\phi = (\partial^{\,2} - m^2)\phi = (\partial_3^2 - \mu^2)\phi\ , \qquad \mu = \sqrt{m^2 + \tilde{k}^2}\ , \tag{4.5.6}$$

and Fourier transform it in the space direction, defining the function $k^0(k_3) = +\sqrt{k_3^2 + \mu^2}$:

---

mind more natural, approaches, focussing on elementary basis elements of Hilbert space, rather than superpositions.

$$\phi(Z,T)=\int\frac{\mathrm{d}k_3}{\sqrt{4\pi k^0}}\left(e^{ik_3Z-ik^0T}a(k_3)+e^{-ik_3Z+ik^0T}a^\dagger(k_3)\right)\,; \tag{4.5.7}$$

$$\dot\phi(Z,T)=\int\frac{-ik^0\mathrm{d}k_3}{\sqrt{4\pi k^0(k_3)}}\left(e^{ik_3Z-ik^0T}a(k_3)-e^{-ik_3Z+ik^0T}a^\dagger(k_3)\right)\,. \tag{4.5.8}$$

Here and further on, we use the notation $\partial_T\phi\equiv\dot\phi\equiv\partial\phi/\partial T$.

At time $T=0$, the inverse of this transformation is

$$a(k_3)=\frac{1}{\sqrt{4\pi\,k^0(\mathbf{k})}}\int\mathrm{d}Z\,\Big(k^0\phi(Z)+i\dot\phi(Z)\Big)e^{-ik_3\cdot Z}\,. \tag{4.5.9}$$

Here, $\dot\phi=\partial\phi/\partial t$, and $k^0(\mathbf{k})=+\sqrt{k_3^2+\mu^2}$. The operator $a^\dagger$ is obtained by replacing $i\leftrightarrow -i$ in Eq. (4.5.9). Notice that these expressions are the standard ones for a scalar, quantised field ϕ in a flat region of space-time. The functions a and $a^\dagger$ are the standard annihilation and creation operators for particles with momentum k_3 in the Z direction. The transverse components μ of the momenta will be negligible as yet, as these are not involved in the transformations that will be considered.

At time $T=0$, we have the usual commutation relations:

$$[\phi(Z),\,\dot\phi(Z')]=i\delta(Z-Z'),\quad\text{while}\quad[\phi(Z),\,\phi(Z')]=[\dot\phi(Z),\,\dot\phi(Z')]=0, \tag{4.5.10}$$

from which one derives

$$[a(k_3),\,a(k_3')]=0\,,\quad[a(k_3),\,a^\dagger(k_3')]=\delta(k_3-k_3')\,; \tag{4.5.11}$$

(the normalisation factors in Eqs. (4.5.7), (4.5.8) and (4.5.9) were chosen so as to arrive at the usual commutation relations for a and $a^\dagger$). These commutation relations ensure that, in this locally flat coordinate frame, $a^\dagger(k_3)$ and $a(k_3)$ create and annihilate an amount of energy $k^0(k_3)$ as usual. Thus, particle states will be regarded as being produced from the vacuum by applying $a^\dagger$ to vacuum states, in the usual manner: $H=k^0a^\dagger a$; $k^0=\sqrt{k_3^2+\mu^2}$. The general expression for the field ϕ is such that A corresponds to annihilation and $A^\dagger$ to creation:

$$\phi(Z,\,T)=A(Z,T)+A^\dagger(Z,T), \tag{4.5.12}$$

with

$$A(Z,T)=\int\frac{\mathrm{d}k_3}{\sqrt{4\pi k^0}}\,a(k_3)e^{ik_3Z-ik^0\,T}; \tag{4.5.13}$$

$$\dot A(Z,T)=-i\int\mathrm{d}k_3\sqrt{\frac{k^0}{4\pi}}a(k_3)e^{ik_3Z-ik^0\,T}\,. \tag{4.5.14}$$

These equations show the creation and annihilation operators that can be used to generate any state with particles in the central region, as seen by a local observer. There will be upper and lower limits for the momenta k_3. The particles can be in region I, where $Z > 0$ or in region II, where $Z < 0$. So far, so good.

4.6 The Bogoljubov Transformation

4.6.1 *Processing Different Annihilation and Creation Operators*

Our point is now that in a locally flat space, the distinction between created and annihilated particles is fixed as soon as it is agreed upon what the time variable is. We need to replace the local time parameter T by the parameter τ as it occurs in the metric (4.5.4). This is a transformation on the time direction, which does not keep energy invariant. As we shall see, if one observer annihilates a particle, this may be regarded as a creation by an other observer.

In the following, we make use of the momentum parameters as used above:

$$k^{\pm} = \tfrac{1}{\sqrt{2}}(k_3 \pm k^0), \quad k_3 = \tfrac{1}{\sqrt{2}}(k^+ + k^-)\,, \quad k^0 = \tfrac{1}{\sqrt{2}}(k^+ - k^-)\,,$$

$$\text{obeying} \quad 2k^+k^- = {k_3}^2 - {k^0}^2 = -\mu^2, \quad \mu = \sqrt{m^2 + \tilde{k}^2}\,. \tag{4.6.1}$$

The light cone coordinates $u^{\pm}$ are defined in Eq. (4.5.3).

4.6.2 *Relating Operator Fields a to a_2*

At first sight, our problem seems to be easy. τ is conjugate to $\ln(u^+)$, or $-\ln u^-$, since an advance $\delta\tau$ in the τ variable corresponds to multiplying $u^{\pm}$ by $e^{\pm\delta\tau}$. Define[12]

$$a_2(\omega) = \int_0^\infty \frac{\mathrm{d}k^+}{k^+}\sqrt{\frac{k^0}{2\pi}}\, a(k_3)\, e^{-i\omega \ln\left(\frac{k^+\sqrt{2}}{\mu}\right)}, \tag{4.6.2}$$

$$\text{with inverse:} \quad a(k_3) = \frac{1}{\sqrt{2\pi k_0}} \int_{-\infty}^{\infty} \mathrm{d}\omega\, a_2(\omega) e^{i\omega \ln\left(\frac{k^+\sqrt{2}}{\mu}\right)}. \tag{4.6.3}$$

The importance of this expression, the pre-factors, and in particular the factor $\sqrt{k^0}$, can be understood as follows:

---

[12] Our notation follows the notation of Ref. [14] , except that we skipped the use of the operator a_1 there.

Note that, in Eqs. (4.6.2) and (4.6.3), the integrals are one-dimensional, since we are keeping $\tilde{k}$ or μ fixed (these do not participate in the transformations). Taking k^3 as the only independent integration variable, and using Eq. (4.6.1), we can write

$$\mathrm{d}k^+ = \frac{\mathrm{d}k^3}{\sqrt{2}}\left(1 + \frac{k^3}{\sqrt{\mu^2 + k_3^2}}\right) = \frac{k^+\,\mathrm{d}k^3}{k^0}, \quad \text{so that} \quad \frac{\mathrm{d}k^+}{k^+} = \frac{\mathrm{d}k_3}{k^0}\,. \tag{4.6.4}$$

This allows us to derive from the definitions (4.6.2) and (4.6.3):

$$[a_2(\omega), a_2^\dagger(\omega')] = \delta(\omega - \omega') \tag{4.6.5}$$

Of course, the pre-factors in Eqs. (4.6.2) and (4.6.3) were chosen such that (4.6.5) is valid.

4.6.3 Relating A to a_2

Now let $A(Z, T)$ be the part of the field $\phi(Z, T)$ that is analytically proportional to the annihilation operators a, see Eq. (4.5.12). Then let us plug Eqs. (4.6.3) into (4.5.13), using (4.6.4). Here is the detailed calculation[13]:

$$\begin{aligned}
A(Z,T) &= \int \frac{\mathrm{d}k_3}{\sqrt{4\pi k^0}}\, a(k_3) e^{+ik_3Z - ik^0T} = \\
&= \int_{-\infty}^{\infty} \frac{\mathrm{d}k_3}{2\pi k^0\sqrt{2}} \int_{-\infty}^{\infty} \mathrm{d}\omega\, a_2(\omega) e^{i\omega\,\ln\left(\frac{k^+\sqrt{2}}{\mu}\right) + ik_3Z - ik^0T} = \\
&= \frac{1}{2\pi\sqrt{2}} \int_{-\infty}^{\infty} \mathrm{d}\omega \int_0^{\infty} \frac{\mathrm{d}k^+}{k^+}\, a_2(\omega)\, e^{i\omega\ln\left(\frac{k^+\sqrt{2}}{\mu}\right) + i\frac{Z}{\sqrt{2}}\left(k^+ - \frac{\mu^2}{2k^+}\right) - i\frac{T}{\sqrt{2}}\left(k^+ + \frac{\mu^2}{2k^+}\right)} = \\
&= \frac{1}{2\pi\sqrt{2}} \int_{-\infty}^{\infty} \mathrm{d}\omega \int_0^{\infty} \frac{\mathrm{d}k^+}{k^+}\, a_2(\omega)\, e^{i\omega\ln\left(\frac{k^+\sqrt{2}}{\mu}\right) + i\left(k^+e^{-\tau} - \frac{\mu^2}{2k^+}e^{\tau}\right)\frac{\varrho}{\sqrt{2}}} = \\
&= \frac{1}{2\pi\sqrt{2}} \int_{-\infty}^{\infty} \mathrm{d}\omega \int_0^{\infty} \frac{\mathrm{d}k^+}{k^+}\, a_2(\omega)\, e^{i\omega\ln\left(\frac{k^+\sqrt{2}}{\mu}\right) + i\,\frac{1}{2}\mu\varrho - i\,\omega\,\tau}\,.
\end{aligned} \tag{4.6.6}$$

Equation (4.6.1) were used to write

$$k_3 = \frac{k^+}{\sqrt{2}}\left(k^+ - \frac{\mu^2}{2k^+}\right); \quad \text{and} \quad k_0 = \frac{k^+}{\sqrt{2}}\left(k^+ + \frac{\mu^2}{2k^+}\right); \tag{4.6.7}$$

[13] Of course I do not expect that many readers will follow all these details; we wrote down the calculation to show that the Bogoljubov [15] transformation (4.6.15)–(4.6.19) can be rigorously derived.

Substituting $\frac{\sqrt{2}}{\mu}\,k^+ = s\,,\quad \alpha = \frac{1}{2}\mu\varrho,$ we define a function F:

$$F(\omega, \alpha) = \int_0^{\infty} \frac{\mathrm{d}s}{s} s^{i\omega}\, e^{i\alpha(s - 1/s)}\,, \tag{4.6.8}$$

to rewrite Eq. (4.6.6) as

$$A(Z, T) = \frac{1}{2\pi\sqrt{2}} \int_{-\infty}^{\infty} \mathrm{d}\omega\, a_2(\omega)\, F(\omega, \tfrac{1}{2}\mu\varrho) e^{-i\omega\tau}\,. \tag{4.6.9}$$

We see clearly that, in terms of the annihilation operator $a_2(\omega)$, the τ parameter acts as time, and ω represents the energies being annihilated.

4.6.4 Properties of the Function F

Properties of function F are easy to derive:

$$F^*(\omega, \alpha) = F(-\omega, -\alpha)\,, \tag{4.6.10}$$

$$F(-\omega, \alpha) = e^{-\pi\omega} F^*(\omega, \alpha) \quad \text{if} \quad \alpha > 0, \tag{4.6.11}$$

$$F(-\omega, \alpha) = e^{+\pi\omega} F^*(\omega, \alpha) \quad \text{if} \quad \alpha < 0\,. \tag{4.6.12}$$

The last two equations follow from replacing $s \leftrightarrow s\, e^{\lambda i}$. In Eq. (4.6.11), the rotation $\lambda = 0$ to $\lambda = \pi$ makes the integral convergent only if $\alpha > 0$. To prove Eq. (4.6.12), we rotate the s variable in the opposite direction.

Equation (4.6.9) is crucial for understanding the Hawking radiation phenomenon. The $\omega\tau$ combination found there implies that the operator $a_2(\omega)$ annihilates an amount ω from the energy in the field A. This is the energy of the light cone variables in region I of the Penrose diagram, see Fig. 4.2.

4.6.5 How a_2 and $a_2^\dagger$ Mix to form a_I and a_{II}

However, now comes the crucial point. In Eq. (4.6.3), the integration goes from $\omega = -\infty$ to $+\infty$, so that ω can be negative. An operator that annihilates a negative amount of energy, is actually a creation operator, and should be handled as such. We see that the integrals in Eq. (4.6.6) have to be split in two parts, depending on the sign of ω.

In short: in region I, a_2 is an annihilation operator if $\omega > 0$ and a creation operator if $\omega < 0$. In region II it is the other way around. However, also another step is needed:

An observer in region I can only observe the fields in region I. These are the fields where the coordinate $\varrho > 0$. Thus, in region I we have the annihilation operators that have $k^0 > 0$ and $\omega > 0$. Etc. Therefore, we split the integrals (4.6.2) and (4.6.3) in accordance with the sign of ω, and the sign of ϱ. Consider the fields in region I and take $\varrho > 0$. Using the properties (4.6.10)–(4.6.12), we rewrite (4.6.9) as

$$\phi(\varrho, \tau) = A(\varrho, \tau) + A^\dagger(\varrho, \tau) = \tag{4.6.13}$$

$$= \int \frac{dk_3}{2\pi\sqrt{2}} \int_0^\infty d\omega F(\omega, \tfrac{1}{2}\mu\varrho)\Big(a_2(\omega) + e^{-\pi\omega} a_2^\dagger(-\omega)\Big) + h.c. \tag{4.6.14}$$

Here, the terms inside the large brackets are to be associated with $A(\varrho, \tau)$ and their hermitian conjugates to $A^\dagger(\varrho, \tau)$. This assures that all terms for $A(\varrho, \tau)$ annihilate positive amounts of energy, while $A^\dagger(\varrho, \tau)$ creates positive energy.

Note that in Eqs. (4.6.13) and (4.6.14), $\varrho > 0$. The expressions inside the large brackets always enter in this combination, as long as we are dealing exclusively with the operators defined in region I.

We see that, if we restrict ourselves to the fields in region I only, we find that they do generate pure annihilation operators, but only in this combination: the operator $a_2^\dagger(-\omega)$ is to be added with the reduction factor $e^{-\pi\omega}$, which originates in Eqs. (4.6.10)–(4.6.12). This mixture of creation and annihilation operators is characteristic for a *Bogoljubov transformation* [15]. Take $\omega > 0$. We define $a_I(\omega)$ in region I and and $a_{II}(\omega)$ in region II:

$$a_I(\omega) = \mathcal{N}\big(a_2(\omega) + e^{-\pi\,\omega} a_2^\dagger(-\omega)\big)\,. \tag{4.6.15}$$

$$a_{II}(\omega) = \mathcal{N}\big(a_2(-\omega) + e^{-\pi\,\omega} a_2^\dagger(\omega)\big)\,. \tag{4.6.16}$$

Here, $\mathcal{N} = 1/\sqrt{1 - e^{-2\pi\omega}}$ is a normalisation factor, to ensure that

$$[a_I(\omega),\, a_I^\dagger(\omega')] = [a_{II}(\omega),\, a_{II}^\dagger(\omega')] = \delta(\omega - \omega')\,. \tag{4.6.17}$$

The inverse of this set of equations are: (again when $\omega > 0$):

$$a_2(\omega) = \mathcal{N}\big(a_I(\omega) - e^{-\pi\,\omega} a_{II}^\dagger(\omega)\big)\,, \tag{4.6.18}$$

$$a_2(-\omega) = \mathcal{N}\big(a_{II}(\omega) - e^{-\pi\,\omega} a_I^\dagger(\omega)\big)\,. \tag{4.6.19}$$

For a local observer to whom the region at the origin of the Penrose diagram seems flat, all operators $a_2(\omega)$ are superpositions of the annihilation operator $a(k_3)$, for all ω, see Eq. (4.6.6). On the other hand, for an observer who only has access to region I, only the operators depending on $\phi(\varrho, \tau)$ with $\varrho > 0$ are visible. These are the operator combinations $a_I(\omega)$ with $\omega > 0$.

What are the physical consequences to be deduced from these transformation rules?

4.7 Hawking's Original Result in This Formalism

4.7.1 The Vacuum State $|\Omega\rangle$ for the Local Observer

All states with or without particles in them, can be obtained by allowing creation operators, such as $a^\dagger(k_3)$, $a_2^\dagger(\omega)$, $A_I^\dagger(\varrho,\tau)$ or $a_{II}^\dagger$ to act on a vacuum state. Let us start from the vacuum state $|\Omega\rangle$ as experienced by an observer near the center of the Penrose diagram. For this observer, the annihilation operators $a(k_3)$ and $a_2(\pm\omega)$ all take the value zero, as there is nothing to annihilate:

$$a(k_3)|\Omega\rangle = 0\,; \quad \text{therefore} \quad a_2(\pm\omega)|\Omega\rangle = 0\,. \tag{4.7.1}$$

This enables us to use Eqs. (4.6.18) and (4.6.19) to derive all elements $\langle n_I,\, n_{II}|\Omega\rangle$ up to an overall constant:

We find that

$$a_I(\omega)|\Omega\rangle = e^{-\pi\omega} a_{II}^\dagger|\Omega\rangle, \quad \text{and} \quad a_{II}(\omega)|\Omega\rangle = e^{-\pi\omega} a_I^\dagger(\omega)|\Omega\rangle\,. \tag{4.7.2}$$

Since all annihilation operators $a(\omega)$ act on a state with n ω-objects as

$$a(\omega)|n\rangle = \sqrt{n}\,|n-1\rangle\,, \quad \text{or} \quad \langle n-1|a(\omega)|\Omega\rangle = \sqrt{n}\langle n|\Omega\rangle, \tag{4.7.3}$$

we derive for all basis elements where $n_I = n_{II} = n \geq 0$:

$$\langle n+1,\, n+1\,|\Omega\rangle = e^{-\pi\omega}\langle n,\, n\,|\Omega\rangle\,. \tag{4.7.4}$$

In this case, the factors $\sqrt{n(\omega)}$ cancel out. They do not cancel out if we apply Eq. (4.7.2) to states where $n_{II} \neq n_I$. In that case, Eq. (4.6.15) gives

$$\langle n+1, m|\Omega\rangle = \sqrt{\tfrac{m}{n+1}}\,\langle n, m - 1|\Omega\rangle e^{-\pi\omega}\,; \tag{4.7.5}$$

using this recursively would lead to states with negative n_I or n_{II}, but for $n < 0$ we cannot use Eq. (4.7.5). Thus, we arrive at

$$\langle n_I,\, n_{II}|\Omega\rangle = \langle 0,\, 0|\Omega\rangle e^{-n_I\,\pi\omega}\,\delta_{n_I,\,n_{II}}, \tag{4.7.6}$$

where $\langle 0,\, 0|\Omega\rangle = \sqrt{1 - e^{-2\pi\omega}}$, since $|\Omega\rangle$ must be normalised: $\langle\Omega|\Omega\rangle = 1$.

The Hamiltonian that measures the energy associated to advances in the time-coordinate τ, can be written as (note the sign of ω):

$$H_\tau = \int_{-\infty}^{\infty} \mathrm{d}\omega\,\omega a_2^\dagger(\omega)a_2(\omega) - C(\omega) \quad = \int_0^\infty \omega\mathrm{d}\omega\Big(a_I^\dagger(\omega)a_I(\omega) - a_{II}^\dagger(\omega)a_{II}(\omega)\Big)\,. \tag{4.7.7}$$

This identity follows directly from Eqs. (4.6.18) and (4.6.19). $C(\omega)$ is a constant that tunes the energy of the vacuum to zero, through normal ordering. From here on, take $\omega > 0$, and we define the signs of the energy operators H_I and H_{II} both to be positive.

We then see that region II contributes negatively to what seems to be the total energy. This happens because the time evolution as a function of τ, in region II, goes in the opposite direction, see Fig. 4.3b.

4.7.2 The Generator of Local Lorentz Transformations

The Hamiltonian for the distant observer can be seen to be the generator of Lorentz transformations in the radial direction for the local observer. This means that, if L is the local Lorentz generator, H_I is the Hamiltonian for the distant observer in region I, and H_{II} is the (positive) Hamiltonian in region II, whereas

$$L = H_\tau = H_I - H_{II} \,. \tag{4.7.8}$$

The operators $a_2(\pm\omega)$ give zero because these are linear functions of $a(k_3)$, see Eqs. (4.5.13) and (4.6.3). But the operators $a_I(\omega)$ and $a_{II}(\omega)$ do not vanish, since they receive admixtures of $a^\dagger(k_3)$ (or equivalently, $a_2^\dagger(\omega)$), see Eqs. (4.6.15)–(4.6.19)). It is important that the operator $a_I(\omega)$ only annihilates particles in region I, and $a_{II}(\omega)$ annihilates only particles in region II.

The operator L is the local generator of the Lorentz transformations and as such more basic than the operator $H_I + H_{II}$.

Now consider a given value for ω. If $a_I(\omega)$ and $a_{II}(\omega)$ do not vanish, this must mean that the observer in regions I or II do observe particles. let us calculate what they see. Given ω, write a state containing n_1 particles in region I and n_2 particles in II, as $|n_1,\, n_2\rangle$, or $|n_1\rangle_I\, |n_2\rangle_{II}$.

$|\Omega\rangle$ is defined by (4.7.1) and (4.7.6), which is Lorentz invariant. Therefore, one may expect $|\Omega\rangle$ itself to be Lorentz invariant. Consequently, for all ω, we must have:

$$L|\Omega\rangle = (H_I - H_{II})|\Omega\rangle = 0\,; \tag{4.7.9}$$

$$\langle n_1,\, n_2|L|\Omega\rangle = (n_1 - n_2)\langle n_1, n_2|\Omega\rangle = 0\,; \tag{4.7.10}$$

hence, $\langle n_1, n_2|\Omega\rangle = 0$ unless $n_1 = n_2$. This is what we derived in Eq. (4.7.6). We see here that it is to be interpreted by observing that the state $|\Omega\rangle$ must be Lorentz invariant.

4.7.3 The Emergence of Hawking Radiation, Scenario 1

This derivation leads to an important physical result: *A distant observer experiences the apparent emergence of particles, both in region I and in region II.* These particles are neatly described by a wave function that puts equal numbers of particles in region I and region II. The state $|\Omega\rangle$, which can also be considered for an entirely flat space-time, is called the Unruh state [16].

Adding a bit of physical intuition, one now continues to ask the question: suppose we have an observer in region I, who has no access to region II. According to the local observer, the state $|\Omega\rangle$ is the only time-translation invariant one. What would the observer at infinity see? The answer was given immediately. In nearly all treatises about quantum black holes, we find an argumentation, that, in our notation, amounts to *scenario 1*:

The wave function for n_1 particles in region I and n_2 particles in region II is

$$\psi(n_1, n_2) = C^{st}\, e^{-n_1\omega\pi}\,\delta_{n_1,n_2}\,. \tag{4.7.11}$$

This is a quantum state, which needs to be normalised to one[14]:

$$\sum_{n_1,n_2} |C^{st}|^2\, e^{-2\pi n_1\omega}\delta_{n_1,n_2} = |C^{st}|^2 \sum_{n=0}^{\infty} e^{-2\pi\omega n} = 1\,. \tag{4.7.12}$$

Now, since the sum here is $(1-e^{-2\pi\omega})^{-1}$, we find

$$|C^{st}|^2 = 1 - e^{-2\pi\omega}\,. \tag{4.7.13}$$

In region I one cannot observe the particles in region II. Therefore, the probability to observe n_I particles in region I is the absolute square of this wave function, *summed over all unobserved states $|n_{II}\rangle$ in region II*. But in the state $|\Omega\rangle$, the number n_2 can take only one value: $n_2 = n_1$. Therefore, the probability that, given the wave function $|\Omega\rangle$, we see n particles in region I is

$$|C^{st}|^2 e^{-2\pi n\omega} = (1-e^{-2\pi\omega})\, e^{-2\pi n\omega} \tag{4.7.14}$$

This is a typical thermodynamical expression, and, since the energy of an n particle state is $n\omega$, the temperature is given by

$$\beta = 2\pi\,, \quad \text{or, we have a temperature} \quad T = 1/(2\pi)\,, \tag{4.7.15}$$

in the units that fit with the given black hole mass. Replacing the time unit τ by $t/4GM$, Eq. (4.5.4), we get the Hawking temperature,

---

[14] We repeated the calculation of expression (4.7.6) here, because it is essential to understand where it comes from.

$$T_H \stackrel{?}{=} 1/8\pi G M \,. \tag{4.7.16}$$

A peculiar phenomenon is that, if one *would* have access to both regions, one would find that the particles in region I and region II are precisely correlated: if, with the appropriate thermal suppression factor, a heavy particle or state would have been observed in I, then, for certain, a heavy particle would be guaranteed to emerge also in region II, while the expected additional thermal suppression factor there, would be totally absent.

And yet, we put a question mark in Eq. (4.7.16). Not everything is right. We calculated that the number of particles emerging in region II always equals the number emerging in region I. How do the particles in universe II know what the number n is in region I?

4.8 Thermal Density Matrix and Radiation Temperature

4.8.1 Scenario 2

The calculation of the state of the particles emerging from the black hole, Eq. (4.7.6), seems fine, technically. But it is the quantum state (4.7.11) that we find here, in 'scenario 1' , that leads to concern.[15]

The particles in region II obey field equations exactly identical to the ones in region I. Their numbers are always equal to the numbers in region I: $n_{II} = n_I$. Why are they in a 'hidden' universe?

This author claims that they are not hidden. A perfect scenario can be arranged, which puts everything in a frame that should look familiar. This we call scenario 2: *thermodynamics in a single universe*. Look at the evolution equations. The Hamiltonian H_{II} controlling region II is mathematically perfectly equal to the Hamiltonian H_I controlling region I, except for one thing: according to the laws of the time evolution, where the variable τ acts as time, the Hamiltonian H_{II} should be *minus* H_I. We already saw, in Eq. (4.7.9), that Lorentz invariance connects H_I with $-H_{II}$. A time translation in τ sends all wave functions in region II to the minus direction. In short, the wave function for the entire system evolves with the total Hamiltonian $H_I - H_{II}$.

But this is nothing new. This is how the density matrix, $\varrho(\tau) = |\psi_\tau\rangle\,\langle\psi_\tau|$ evolves. $|\psi_\tau\rangle = e^{-iH\tau}|\psi_0\rangle$ is the ket state, and $\langle\psi_\tau| = \langle\psi_0|e^{iH\tau}$ is the bra state. This seems to make sense. Indeed, Hawking [8] has been exploring exactly this alley in his earlier investigations concerning the information loss problem. In this scenario, nothing about the wave function gets lost; all we have to do is figure out how the black hole evolution looks from this vantage point.

[15] The observation that different approaches yield different Hawking temperatures was first made in Ref. [17].

Indeed, there is a consequence of this that was pointed out by this author long ago [17]; it appears to have enjoyed little attention. If the density matrix ϱ is directly described by the product of states in region I and states in region II, we have, in the stationary case,

$$|\Omega\rangle = \varrho = \sum C^{st}\,|\psi\rangle_{I\ II}\langle\psi|\,, \tag{4.8.1}$$

where C^{st} is a normalisation factor. Indeed, the state $|\Omega\rangle$, Eq. (4.7.11) has exactly this form:

$$|\Omega\rangle = C^{st}\sum_{n\geq 0}|n\rangle e^{-n\pi\omega}\langle n|\,. \tag{4.8.2}$$

However, if we now calculate the temperature, using the Boltzmann-Maxwell distribution, we get

$$\mathrm{Tr}\,\big(C^{st}\sum_n|n\rangle e^{-n\pi\omega}\langle n|\big) = C^{st}\sum_n\langle n|e^{-n\pi\omega}|n\rangle = \sum_n\langle n|e^{-\beta n\omega}|n\rangle\,. \tag{4.8.3}$$

It follows that the inverse temperature β in this scenario, scenario 2, is exactly half the value (4.7.15) obtained in scenario 1. If now we derive the Hawking temperature, we get

$$T = 1/4\pi GM, \tag{4.8.4}$$

This must look confusing to the reader. Expression (4.8.2) looks exactly like Eq. (4.7.11). Whence this factor 2 in the temperature?

4.9 Region II as Quantum Clone of Region I

4.9.1 The Difference Between the Two Scenarios

We started with the same expressions for the Bogoljubov-transformed vacuum state $|\Omega\rangle$ both in Sects. 4.7 (scenario 1), and in 4.8 (scenario 2). Yet the temperature we computed differs by a factor 2. Whence this factor? One answer is that the wave function had to be squared in scenario 1, while we used it linearly in scenario 2. This lead to the term $e^{-2\pi n\omega}$ in scenario 1, while we had $e^{-\pi n\omega}$ in scenario 2, and showing this was the reason for doing the calculation as accurately as possible in Sects. 4.7 and 4.8.

But where did this difference arise? And which of the two calculations, if any, is correct?

The difference comes about in our use of the inner products. In Sect. 4.7, we handled the states $|n\rangle$ in region II, (or $\langle n|$) as all being orthogonal to all states $|n\rangle$.

In Sect. 4.8, we use the inner products $\langle n|n'\rangle$ as non vanishing if $n' = n$. Of course: the inner product $\langle n|n'\rangle = \delta_{n,n'}$ when used in the density matrix.

The conclusion we do arrive at is that region II is filled with particles exactly the same way as region I, apart from the fact that region II evolves not exactly the same way as region I, but as the *hermitian conjugate* of I.

Since probabilities are now obtained directly from a density matrix, rather than a wave function, one might even suspect that this result agrees well with a probabilistic interpretation of quantum mechanics in general. This also does not prove who is right, but we have not yet used all our ammunition. There is something else to worry about: does the quantum evolution preserve information? Do our methods, as used so-far, give us clues for other approaches?

4.9.2 Rindler Space

Rindler space [6] is a description of ordinary flat space time using a parameter space where time is replaced by the generator τ for Lorentz transformations in the z-direction. Here, the physics is well understood, and we emphasise that scenario 1 applies to Rindler space: the temperature detected by an observer who is accelerated with respect to the horizons of this space-time, has always been calculated correctly.

Rindler space hides two entire universes behind its two horizons, and, in this case we understand precisely how to do the calculation: all ket states $|\psi\rangle$ in region I are to be considered orthogonal to all bra states $\langle\psi|$ in region II, so that, in particular, the inner products $\langle\psi|\psi\rangle$ vanish, unlike the way they are used in Eq. (4.8.3). The two universes represented in Rindler space may be entirely different, so they will in general not be identical twins. One can have a solar system in one of them and a totally different alien world in the other. The observer has to decide in which of the two universes to sit when performing measurements.

The two regions exhibited in scenario 2 are identical twins—they are equal in the quantum mechanical sense: quantum clones of one another. This situation is illustrated in Fig. 4.4. Every visible space-time point in the Schwarzschild metric, marked $r,\ t$, corresponds to a point in region I, marked $x,\ y$, and it also corresponds to a point in region II marked $-x,\ -y$. Even the observer is represented in these three frames. This is why we call these three frames quantum clones.

We end this section by anticipating how we will be able to decide what model for region II to use for describing a physical black hole. The extended description of the black hole metric, the Kruskal–Szekeres coordinate system (Sect. 4.4, Fig. 4.2b and 4.3), shows two regions, and as yet, we did not have any use for the particles in region II of this metric. We claim that black holes must be fundamentally different from Rindler space; Rindler space can only be described by scenario 1, while black holes can only be described by scenario 2, if we may assume that they preserve quantum information.

Fig. 4.4 Region II as an exact quantum clone of region I. Not only all particles are the same ones in the two regions, also the observer is copied. In this scenario, it is better to return eventually to the Schwarzschild metric (r, t), of which both region I (x, y) and region II $(-x, -y)$ are obviously quantum copies

4.9.3 *Clones*

What we have so-far is that, if there are n particles in region I, there will also be n particles in region II. This comes in addition to the observation that, if we consider the two regions I and II in *any* black hole solution without matter, we always find these two regions to be identical. Actually, we are only able to admit exactly as many matter particles in region II as we have in region I. This could be regarded as a law of Nature: *any modification* in region I of a black hole, is to be accompanied by an *identical* modification in region II.

Or: there may be *no way* to bring about even the slightest difference between region I and region II. As we already stated, it implies that region I and region II are *quantum clones* of one another. This means that *even the outside observer* is copied in region II as soon as he or she pops up in region I; see Fig. 4.4. A striking consequence of this law will be that region II will contain no information whatsoever that is not already available in region I. Actually, there is a simpler way to phrase this law: Regions I and II also carry a unique copy in the original Schwarzschild coordinate frame. Since the Schwarzschild coordinates do not allow different copies of themselves, there are no other copies. Note that we do not have such a law in Rindler space.

The importance of this carbon-copy law is now two-fold: one, it may well enable us to understand how information can be preserved in an evolving black hole (information has no way to hide since all information in region II is already visible in region I, and in the Schwarzschild frame. This may well enable us to dismiss one or both of the scenarios sketched in Sects. 4.7 and 4.8.

Second, since all contents of region II are now obviously observable in region I, one may already argue that information sent into a black hole has nowhere to hide. We shall see how this point materialises in the next sections.

4.9.4 A Conical Singularity

Perhaps the situation can be clarified a bit more if we use a different language. Identifying one region of a metrical space with another region, is tantamount to dividing this space-time by $\mathbb{Z}(2)$, and the physical effect this has can be described in terms of a conical singularity. Here, the conical singularity is placed at the center of the Kruskal–Szekeres metric. Since the time coordinate is involved in an essential manner, one can easily imagine how this modification of the metric also generates a factor 2 in the temperature. Careful inspection indicates that the unit of time used far from the center differs by a factor 2 from the unit used initially, and, as temperature has the dimension of energy, or the inverse of the time unit, the temperature is modified by a factor 2.

Conical singularities are physically observable, so what we have here is a modification of the local laws of physics! it may be worth-while to investigate how the existence of such singularities may affect the Standard Model. This has not yet been done, as far as we know.

4.10 The Shapiro Effect Linking in- and Out-Particles

4.10.1 How In-Particles Affect Out-Particles and Vice Versa, on the Horizon

One of the main topics in this book, is the question how matter going in ('in-particles'), may be expected to transfer the information they carry, onto particles going out ('out-particles'). Now as long as these particles stay away from any of the horizons, we may assume that a field theory can be used such as the Standard Model, modified so that it can be applied on a modestly curved background metric.

At first sight, one might expect that the effects that in-particles have on out-particles, is far too weak to transmit *all* information from one to the other. Certainly a graviton hopping over should be ineffective.

But then one underestimates this force; note that gravitation scales, to become strong at tiny distances. More to the point, the effect has been calculated. It is found that the curvature of the background metric will cease to be modest when visited by particles *on* one of the horizons. Unfamiliar, but computable, phenomena take place when an in- particle crossing a horizon also crosses the orbit of a particle moving out in a geodesic parallel to that horizon. What happens can be entirely calculated

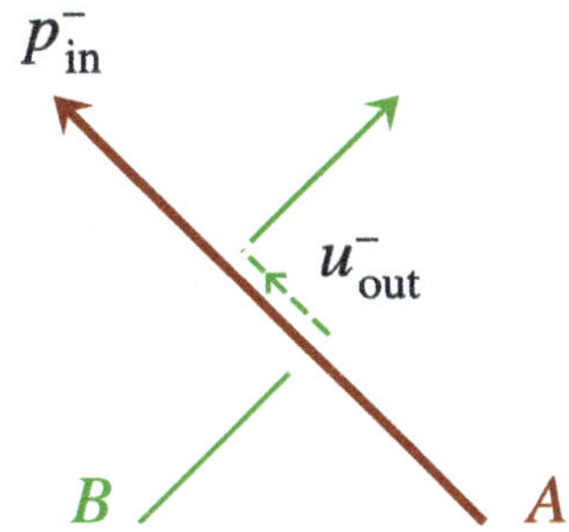

Fig. 4.5 A particle with high momentum (moving to the left) crosses the orbit of a low momentum particle

using standard, non-quantum mechanical equations: Einstein's field equations. All we need are the momenta and the positions of these particles that meet one another. The calculation is completely clean, and can be done with relative ease. The way to discover this is displayed in detail in Ref. [14]. Here, we briefly summarise the main features to enable the reader to follow the developments on this topic.

We use the feature described in Refs. [18, 19] . It is illustrated in Fig. 4.5. With almost the speed of light, a high-energy particle A is crossing the orbit of a light test particle, B. The particle B then experiences a time lapse. The magnitude of the time lapse was calculated by Shapiro [20] , deriving how radio and optical radiation is slowed down when signals are grazing past the Sun. The Sun does to the radio signals what particle A does to particle B. Since the gravitational field is controlled by field equations, the effect on the spectator particle B is described by a Green function.

4.10.2 The Green Function on the Horizon

Let particle A (later to be chosen to be the in-particle) have 4-momentum p_A, which dominates in the light cone direction. The gravitational field of particle A is here Lorentz contracted so strongly that the effect is limited to a light-like plane, and the spectator particle B is dragged along by a distance u_B in the 4-direction of the motion p_A of A. The magnitude of the drag is given by

$$u_B(\tilde{x}) = \int \mathrm{d}^2\tilde{y}\, f(\tilde{x} - \tilde{y}) p_A(\tilde{x}), \tag{4.10.1}$$

The Green function f lives in the two dimensional plane orthogonal to p_A and p_B.

If we Lorentz transform the entire scenery such a way that particle A will almost stand still, we find that not B but A is being dragged along by exactly the same mechanism. The phenomenon was described by Aichelburg and Sexl [18], and Bonner [21]. Applying this to particles near a black hole horizon we see the same phenomenon, where now the horizon plays the same role as the flat plane in the set-up by Aichelburg and Sexl, but now this space is a 2-sphere. On this 2-sphere, the Green function f takes a form given by the equation

$$(1 - \Delta_\Omega) f(\Omega, \Omega') = 8\pi G \delta^2(\Omega, \Omega'), \tag{4.10.2}$$

so that

$$u^{\text{out}}(\Omega) = \int \mathrm{d}^2\Omega' \, f(\Omega, \Omega') p^{\text{in}}(\Omega'), \tag{4.10.3}$$

and conversely,

$$u^{\text{in}}(\Omega) = -\int \mathrm{d}^2\Omega' \, f(\Omega, \Omega') p^{\text{out}}(\Omega') \,. \tag{4.10.4}$$

We used a notation where the momentum of the in-particle is denoted as a distribution, usually of the form $p^{\text{in}}(\Omega) = p_1 \delta^2(\Omega, \Omega_1)$. The derivation of equation (4.10.2) is actually a rather tedious one. It was derived in Ref. [14].

4.10.3 *Information Retrieval*

Using this phenomenon, we shall describe the process of information retrieval for black holes in the remainder of this chapter. The in-going matter is assumed to be described by a distribution of momenta, generated by many particles going in, but not so many that we need a second quantised procedure.[16] Consider N particles going in, enumerated by $k = 1, \cdots, N$. The in-momentum distribution is then given by the sum

$$p^{\text{in}}(\Omega) = \sum_{k=1}^{N} \delta^2(\Omega - \Omega_k) p_k \,. \tag{4.10.5}$$

It must be noted that this is not exactly the form in which the Standard Model particles arrive at the horizon; in this expression, they may be viewed as second-quantised particles only w.r.t. the 2-sphere of the horizon, while, as a function of the radial coordinate $\varrho = r - 2GM$, we only see the single operator p^{in}.

We anticipate that this mismatch can be handled, but, we emphasise that the presentation that we advertise now, will be unitary, in a way to be explained, which is why we here briefly summarise how this is done.

The momentum distributions $p^{\text{in}}(\Omega)$ and the coordinate shift $u^{\text{out}}(\Omega)$ are both expanded in spherical harmonics:

---

[16] One expects in the order of one Hawking particle to be emitted in a period of time in the order of $1/GM_{BH}$.

$$p^{\rm in}(\Omega) = \sum_{\ell,m} p^{\rm in}_{\ell,m} Y_{\ell,m}(\Omega) \; ; \tag{4.10.6}$$

$$u^{\rm out}(\Omega) = \sum_{ell,m} u^{\rm out}_{\ell,m} Y_{\ell,m}(\Omega) \; . \tag{4.10.7}$$

The point is that these particles do not represent the entire set of all particles ever absorbed by the black hole, but only those particles that are seen in a given time interval, representing a stretch of time roughly of the order of GM in natural units, so they are particles that describe only small deviations in a Schwarzschild metric that, by itself, is perfectly time translation and rotation invariant.

We claim that the disturbances caused by much earlier in-going and much later out-going particles will be represented by the particles that are to be described in the other time sectors, and the equations we derive will show how this happens.[17]

Now consider Eqs. (4.10.2) and (4.10.3). In the spherical wave expansion, these take the form

$$u^{-\,\rm out}_{\ell m} = \frac{8\pi G}{\ell^2+\ell+1} p^{-\,\rm in}_{\ell m} , \tag{4.10.8}$$

where the subscript minus (–) stands for the minus-direction in the local Lorentz frame close to the horizon.

The Green function diagonalises when using the partial wave expansion, and hence $u^{-\,\rm out}_{\ell m}$ has become detached from the other partial wave components (ℓ', m'), and, consequently, our system is simplified enormously by this procedure.[18] This is in fact the same procedure as the one that enables the calculations of the electron wave functions for the hydrogen atom. The classical photon field in the atom is the analogue of the classical graviton field in our equations here.

But there is something more. Eq. (4.10.8) is a simple operator expression that can be seen to act on quantum states $|p^{-\,\rm in}_{\ell m}\rangle$ and $|u^{-\,\rm out}_{\ell m}\rangle$, which we may subject to Fourier transformations to get $|u^{+\,\rm in}_{\ell m}\rangle$ and $|p^{+\rm out}_{\ell m}\rangle$. This gives exactly Eq. (4.10.4), with the minus sign.

Indeed, there is a neat way to employ quantum commutation rules to see that the information carried by the in-particles takes the form of a Hilbert space that returns exactly the same information to the out-particles.

---

[17] The Page time [13], often introduced in this context, is a far too long time stretch to be of use here, as explained in a footnote in Sect. 4.5 of this chapter.

[18] Our approach was criticised by claiming that transverse components of the gravitational fields should invalidate the approximations. But, while the transverse components of the momenta are basically time-independent, the radial components increase exponentially with time. This is why we consider these as being dominant. Only at large ℓ values, we may expect stronger transverse perturbations, which at first glance might require a cut-off in angular momentum space, but this might be avoidable. We expect that large ℓ values will not dominate, but the side effects at large ℓ still have to be sorted out.

Note however, that this statement only holds for the two regions, I and II, combined. How to separate the information for the two different regions, is an important question, to be meticulously addressed in the next sections.

Consider the shifts in the coordinates u^- caused by the in-particles with momenta $p_k^{\rm in}$. These momenta increase exponentially in time, which is why the shifts (4.10.8) they cause are essential. The total of the in-going momentum is a distribution $p^-(\Omega)$ where the effects of all in-particles are added together. In turn, they shift all out-particles, at different values of Ω, by the amount $u^-(\Omega)$. Since the effects of early in-particles on late out-particles diverge exponentially with time difference, we only take into account the particles inside the time zone indicated by the green region in Fig. 4.2a. How to handle the other particles will also be addressed.

We have to assume that all in-particles and all out-particles arrive or leave at different values for the angles Ω (which, in fact, is a good approximation; for large black holes, the odds against two particles entering or leaving at exactly the same values of ℓ and m is negligible). Then we see that Eq. (4.10.8) in fact suffices to relate all out-particle information to the information from the in-particles. The relation is in the form of a Fourier transformation. Fourier transformations preserve the norm of a function, and therefore, they are unitary.

4.11 More About the Clones

4.11.1 Out-Particles are the Fourier Transforms of the in Particles

In this section we shall derive further support for the factor 2 that was first encountered in Sect. 4.8, the fact that the temperature of Hawking radiation must be given by Eq. (4.8.4) rather than (4.7.16). One thing that one might conclude from the previous section is, that we did not yet have everything completely right.

Only the *geometric* properties of the in-particles are neatly transmitted by the *geometric* properties of the out-particles, but how non-geometric properties are to be handled is not yet understood. We leave this here for future investigations.

However, if, by way of simplification, we limit ourselves to scalar, massless particles entering and leaving the black hole, much more can be understood.

To do this analysis right for the scalar particles, we first need to achieve optimal understanding of the geometric parts (momenta $p^\pm_{\ell,m}$ and positions $u^\pm_{\ell,m}$). By investigating the gravitational effects of in-particles onto out-particles, we find that the out-*positions* are determined by the in-*momenta*, and *vice versa*. Quantum mechanics clearly states that momenta are the Fourier transforms of the positions, and *vice versa*. The Fourier transformation operator is unitary since the norm of a function is preserved. This is the first step towards understanding why the prescriptions we are providing, relating out-particles to in-particles, will preserve information.

Judicious attempts to map second-quantised Standard Model particles onto the radial first-quantised objects will produce novel insights that might, one day, enable us to find out what the primary source is of the complex structures of the Standard Model itself. In fact, the two transverse dimensions stay second-quantised.[19] Thus we see here the holographic principle at work.

As a first step, we arrive at a description of all states of a black hole in terms of states in a Hilbert space, where the spectrum of in-going particles is exactly as large as the spectrum of out-going particles, *except for one very important point,* that we shall consider now.

Our point is that, as was observed above, the operation to get the out-Hilbert space in terms of the in-Hilbert space is the mathematical procedure of the Fourier transformation. The Fourier transformation does not distinguish the positive halves from the negative halves both in position space, $u^{\pm} > 0$ and momentum space, $p^{\mp} > 0$. Therefore, limiting ourselves only to region I of the black hole might not be correct. Can we resolve this problem?

4.11.2 Keeping the Even in- and the Even Out-Wave Functions Separate

Surprisingly, there is a beautiful solution. It is closely connected to our earlier consideration that the states on the Kruskal–Szekeres world frame must be regarded as elements of the density matrix. More precisely, we address the entire region II as the space of bra states. The only states we shall accept on the entire KS coordinate frame is the set of 'cloned twins':

$$|\psi_I, \psi_{II}\rangle \;\rightarrow\; |\psi\rangle\langle\psi|, \tag{4.11.1}$$

where $|\psi\rangle$ only consist of wave functions defined on region I, and the states $\langle\psi|$ are the same states, in bra notation, exclusively defined on region II. This is not a trivial formalism at all, while many researchers are taking their refuge in much more entangled space-time metric frames. We claim that the approaches using more complex metric structures will perhaps be needed to describe further details at a later stage, but we first must understand and use the Ansatz that region II is a cloned version of region I, as this is totally essential for recovering unitarity. The situation differs from Rindler space [6] in the sense that also the observer is formally copied, see Fig. 4.4. This restriction may be easier to accept by noting that the original Schwarzschild metric, (4.3.1), already automatically exhibits every state only once, and *there is only one observer in the Schwarzschild coordinate system.* See Fig. 4.4.

[19] The functions $u_{\ell,m}$ and $p_{\ell,m}$ are the spherical harmonic transformations of the operators $u(\theta,\varphi)$ and $p(\theta,\varphi)$ and these are fields; this is why we talk of second quantisation in the transverse coordinates.

The way this new picture affects our equations is that now we have a map of the in-going wave function defined exclusively on the positive half of the entire future horizon (while the part where $u^+ < 0$ is fixed as an exact copy of $u^+ > 0$, and we demand in turn also the past horizon to be similar: the part where $u^- > 0$ will be constructed as an exact copy of $u^- < 0$ (see also Fig. 4.2b).

Equation (4.10.8) relates the *entire* future horizon to the entire past horizon. How does this change if the two halves of both horizons are restricted to being copies of the other halves? This happens to be easy. We write

$$\psi_{\text{in}}(p) = \psi_{\text{in}}(-p), \quad \psi_{\text{out}}(u) = \psi_{\text{out}}(-u)\ ; \tag{4.11.2}$$

This is not in conflict with the demand that the functions in u^- space are Fourier transforms of the ones in p^+ space: the Fourier transform of an even function is an even function itself. Therefore, we may assume Eq. (4.10.8) to hold for positive u^{out} and p^{in} separately from the same equations for negative u^{out} and p^{in}.

4.11.3 Replacing the Fourier Transform by an Even Function

We are now in a position to formulate precisely the boundary condition that maps in-going energy-momentum that enters through the positive half of the future horizon, onto the states that return from half the past horizon, while all of this happens primarily in region I; it is exactly copied in region II.

The wave functions in p^- space are the Fourier transforms of the states in u^+ space, even if we only take half of the available space, that is, on the Cauchy surfaces spanned by the u^+ axis we keep only the wave functions that are *even* in u^+. Then we find that also their Fourier transforms are even functions of p^-. Proof:

$$\psi^{\text{out}}(u) = \psi^{\text{out}}(-u) = \frac{1}{\sqrt{2\pi}}\int_{-\infty}^{\infty} \mathrm{d}p\, e^{i\,p\,u}\,\psi^{\text{in}}(p) = \frac{1}{\sqrt{2\pi}}\int_0^{\infty}(e^{ipu}+e^{-ipu})\psi^{\text{in}}(p) = \sqrt{\frac{2}{\pi}}\int_0^{\infty}\cos(pu)\,\psi^{\text{in}}(p), \tag{4.11.3}$$

which is its own inverse:

$$\psi^{\text{in}}(p) = \sqrt{\frac{2}{\pi}}\int_0^{\infty}\cos(pu)\,\psi^{\text{out}}(u), \tag{4.11.4}$$

Exactly the same events happen in region II.

This implies that the subsets of all even wave functions are transformed onto each other. Or: if the in going states form a set of quantum clones, then the out going states also form quantum clones automatically. In other words, even in these subsets the Fourier transform is a unitary transformation. The Fourier transform is then effectively replaced by the transformation (4.11.4).

4.12 Discussion. Information and Temperature

4.12.1 Comparing the Horizon with a Brick Wall or Mirror

All of the above, taken together, gives us an interesting glimpse of a theory where all information is passed on from in-going particles to out-going ones, *via* unitary transmission operators, (4.11.3), (4.11.4). These unitary operators turn the horizons into effective mirrors. Note that, this turns our model into something very much resembling the 'brick wall model', introduced long ago [22]; see Fig. 4.6.

It is instructive to describe the sequences of operations that take proper care of the time dependence. At a given time τ_0, we describe all in-particles as wave functions $|\psi^{\rm in}_{\ell m}(u^+)\rangle$, on the future event horizon u^+, where $u^+ > 0$. As time proceeds, $u^+_{\rm in}$ (and with it, $p^{+\,\rm out}$), will decrease exponentially, so that the wave functions will collapse towards the origin. Their (modified) Fourier transforms, $\psi^{\rm in}_{\ell m}(p)$, see Eq. (4.11.4), will expand exponentially in time, since the product (pu) in this transformation then stays constant. Now Eq. (4.10.8) will relate $p^{\rm in}$ to $u^{\rm out}$. Clearly, $u^{\rm out}$ will also expand. These are the out-particles, coming from the past horizon, doing their best, exponentially, to leave the scenery. Clearly, they contain all information that went in; nothing is lost—apart from the non-geometrical information, as explained.

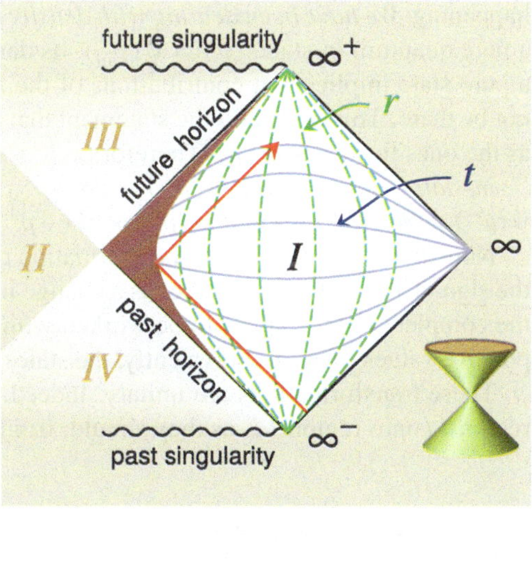

Fig. 4.6 The brick wall model bounces information carrying particles back out

4.12.2 The Transition to Different Time Slices

We apply this mechanism for a short time period. Now consider the next short time period. The particles that have been created before, will rapidly leave. They might still feel the forces from the previous period, but these are now too weak to have any effect. This implies that their spot in (ℓ, m) space is open for the next round. New particles will arrive and the same action will be repeated.

Classically, all these wave functions will also continue into regions III and IV, but we are not interested in that, since they will be invisible from the outside world.

There are also particles that may have entered from far away to merge with the future event horizon. As soon as they come too close to the horizon, they may become effectively invisible. In that case, we apply the inverse of our new transformation, to turn them into particles that leave. This is what we called the 'firewall transformation' in Ref. [23].

Note that this, finally, justifies what we postulated in the beginning, which is that we may omit particles that are either too close to the horizon or too far away. This set continually changes as we go from one time window to the next, keeping all our particles in the range where they are allowed (by obeying the constraint that their energies and momenta are negligible *in this time frame*).

4.12.3 Pure States and Mixed States

Hawking [8], Giddings [5], and many others, conclude rather prematurely that black holes must turn pure quantum states into mixed states. Here, we see nothing of the sort happening. *We have been dealing with density matrices from the beginning.* Consider a pure quantum in-state, $|\psi\rangle = \psi(p^-_{\text{in}})$. Its density matrix is $\varrho_{\text{in}} = |p^-_{\text{in}}\rangle\langle p^-_{\text{in}}|$. Being a pure state implies that contributions of the form $|\psi_1\rangle\langle\psi_2|$, with $\psi_2 \neq \psi_1$, should not be there. This is exactly the statement that the states for positive p^-_{in} are the same as the ones for negative p^-_{in}. Therefore:

in-states of the form $\psi(p^-) = \psi(-p^-)$ *yield out-states of the same form:* $\psi(u^-) = \psi(-u^-)$, *so that* $\psi(p^+) = \psi(-p^+)$.

Note that, according to the transformation rules (4.11.3) and (4.11.4), a change in the sign of p has the same effect as a change in the sign of u. We must conclude that the complete Hilbert space to be worked with consists of the states $\psi(u)$ with only positive values of u, or equivalently, the states $\psi(p)$ with only the positive values of p. These transformations are unitary. Indeed, they map region I into region I and region II into region II, as they should, from a physical point of view.

4.12.4 Temperature

This implies that the density matrices are spanned by the Hilbert space vectors $|\psi\rangle\,\langle\psi|$, so that, indeed, the states $|\Omega\rangle$ are density matrix elements. Their amplitudes for states with n particles are $C^{st}\,e^{-n\pi\omega}$, as they should be when the temperature is $1/\pi$, not $1/2\pi$, and consequently the Hawking temperature for a black hole with mass M must be

$$T = 4\pi GM \tag{4.12.1}$$

rather than Eq. (4.7.16).

There is an interesting argument explaining the relation between temperature and information. The behavior of a field theory at a given temperature $1/\beta$, can be understood in terms of the functional integral for a statistical (thermal) system that extends for a stretch of time in the Euclidean 4-direction, with a length β. We could continue the functional integral for another amount of β in Euclidean time, making the universe twice as big. This would make the total entropy of this field twice as big, since entropy simply counts the number of possible states (logarithmically). The total energy GM, on the other hand, is fixed in relation with the black hole mass. Since

$$T\mathrm{d}S = \mathrm{d}M, \tag{4.12.2}$$

the temperature T must increase by a factor 2 if entropy S shrinks by a factor 2.

Note that identifying region II with region I implies that, in the thermal functional integral, the variables in region II are fixed as soon as those in region I are, so that these cannot be included in the calculation of the total entropy. This argument can be elaborated more, but we merely mention it to show how the physical features of our theory hang together.

Strauss and Whiting [24] report that they tried to comprehend what we are suggesting, and continued by testing the idea in the case of spherically symmetric dust clouds. But it should have been clear that the entire Hilbert space that we investigated, is dominated by non spherically symmetric states, and information retrieval is only conceivable if one considers all components of Hilbert space that can be involved (remember the heat bath in Sect. 4.3.3). It should have been obvious that black holes with spherically symmetric dust clouds added, will emit Hawking radiation that not only consists of such clouds.

Remarks such as those made by Strauss and Whiting will perhaps be useful if one is interested in rather drastic toy models, but it is relatively easy to consider the effects of all angular momenta $(\ell, m) \neq (0, 0)$ combined, while only one r or one p is allowed.

4.12.5 The Conical Singularity

Finally, there is the question whether the identification of regions I and II could generate a special type of singularity on the horizons. The answer is yes, at the points $u^{\pm} \to 0$, this identification can be removed by the transformation $(x, y) \leftrightarrow (x^2, y^2)$, since the operator $\mathbb{Z}(2)$ mentioned earlier then becomes the identity operator, but a mild singularity at $x = y = 0$ then does result. In the equations (4.4.1), we see that the time t goes to $2t$, Elaborating this theme a bit further, we see that the temperature, which has dimension of an inverse time, is affected by a factor two when taking this conical singularity into account.

We thank N. Gaddam, U. Gürsoy and N. Groenenboom for useful discussions and remarks. Also N. Strauss and B. Whiting, for writing up their considerations and findings [24], even if we do not always agree. It is the kind of discussions that may lead us into considering the next steps.

References

1. K. Schwarzschild, Über das Gravitationsfeld eines Massenpunktes nach der Einsteinschen Theorie. Sitzungsberichte der Königlich Preussischen Akademie der Wissenschaften **7**, 189–196 (1916). Bibcode:1916AbhKP1916..189S. Translation: S. Antoci, A. Loinger, On the gravitational field of a mass point according to Einstein's theory (1999). arXiv:physics/9905030
2. G. 't Hooft, Explicit construction of Local Hidden Variables for any quantum theory up to any desired accuracy, in *Quantum Mechanics and Fundamentality: Naturalizing Quantum Theory between Scientific Realism and Ontological Indeterminacy*, Chap. 13. (Springer Nature, Valia Allori ed.). arxiv:2103.04335 [quant-ph]
3. G. 't Hooft, The hidden ontological variable in quantum harmonic oscillators. To be publ in *Frontiers in Quantum Science and Technology*, London. arXiv:2407.18153
4. S.W. Hawking, Particle creation by black holes. Comm. Math. Phys. **43**, 199 (1975); Phys. Rev. D **14**, 2460 (1976)
5. S. Giddings, Black Holes and Quantum Predictability. arXiv:hep-th/9306041
6. W. Rindler, *Essential Relativity* (Van Nostrand Reinhold Co, New York, 1969). ISBN 978-0-387-90201-2. https://doi.org/10.1007/978-1-4757-1135-6
7. S.W. Hawking, The Information Paradox for Black Holes, talk given on 28 August 2015 at 'Hawking Radiation', a conference held at KTH Royal Institute of Technology, Stockholm. ArXiv: 1509.01147 [hep-th]. Accessed 3 Sep 2015
8. S.W. Hawking, The unpredictability of quantum gravity. Comm. Math. Phys. **87**, 395 (1982)
9. M.D. Kruskal, Maximal extension of Schwarzschild metric. Phys. Rev. **119**, 1743 (1960)
10. G. Szekeres, On the singularities of a Riemannian manifold, Publ. Math. Debrecen **7**, 285 (1960) (now available at General Relativity and Gravitation **34** (2002) 2001)
11. R. Penrose, Asymptotic proprierties of fields and space-times. Phys. Rev. Lett. **10**(2), 66-68 (1963). Bibcode:1963PhRvL..10...66P. https://doi.org/10.1103/PhysRevLett.10.66
12. S. W. Hawking, G.F.R. Ellis, *The Large Scale Structure of Space-time*. Cambridge Monographs on Mathematical Physics, 21st edn. (Cambridge University Press, Cambridge). ISBN 978-0-521-09906-6. See Chapter 5
13. D.N. Page, Information in Black Hole Radiation. Phys. Rev. Lett. **71**, 3743 (1993). (arXiv:hep-th/9306083)
14. G. 't Hooft, The scattering matrix approach for the quantum black hole: an overview. J. Mod. Phys. A **11**, 4623–4688 (1996). arXiv:gr-qc/9607022

15. N.N. Bogoljubov, On a new method in the theory of superconductivity. Nuovo Cim **7**, 794–805 (1958). https://doi.org/10.1007/BF02745585
16. W.G. Unruh, Notes on black hole evaporation. Phys. Rev. D **14**, 870 (1976); W.G. Unruh, Phys. Rev. D **14**, 870–92 (1976)
17. G. 't Hooft, An Ambiguity of the equivalence principle and Hawking's temperature. J. Geom. Phys. **1**, 45–52 (1984)
18. P.C. Aichelburg, R.U. Sexl, On the Gravitational field of a massless particle. Gen. Rel. Grav. **2**, 303 (1971)
19. T. Dray, G. 't Hooft, The gravitational shock wave of a massless particle. Nucl. Phys. B **253**, 173 (1985)
20. I. Shapiro, Fourth test of general relativity. Phys. Rev. Lett. **13**(26), 789–791. Bibcode:1964PhRvL..13..789S. https://doi.org/10.1103/PhysRevLett.13.789
21. W.B. Bonnor, The gravitational field of light. Commun. Math. Phys. **13**, 163 (1969)
22. G. 't Hooft, On the quantum structure of a black hole. Nucl. Phys. B **256**, 727–745 (1985)
23. G. 't Hooft, The black hole firewall transformation and realism in quantum mechanics. Universe **7**, 298 (2021). https://doi.org/10.3390/universe7080298. arXiv:2106.11152v2
24. N.A. Strauss, B.F. Whiting, A classical firewall transformation as a canonical transformation. Class. Quantum Grav. **41**, 035001 (2024). https://doi.org/10.1088/1361-6382/ad170f

Chapter 5
Paradox No More: How Stimulated Emission of Radiation Preserves Information Absorbed by Black Holes

Christoph Adami

Abstract Black holes have been implicated in two paradoxes that involve apparently non-unitary dynamics. According to Hawking's theory, information that is absorbed by a black hole is destroyed, and the originally pure quantum state prior to the formation of the black hole is converted to a mixed state upon complete evaporation. Here I address one of the two, namely the apparent loss of (classical) information when it crosses the event horizon. I show that this paradox is due to a mistake in Hawking's original derivation: he ignored the contribution of the *stimulated* emission of radiation that according to Einstein's theory of blackbody radiance must accompany the spontaneous emission (the Hawking radiation). Resurrecting the contribution of stimulated emission makes it possible to calculate the (positive) classical information transmission capacity of black holes, which implies that information is fully recoverable from the radiation outside the black hole horizon. I also briefly discuss the second paradox.

5.1 Historical Background

In September of 1973, Stephen Hawking traveled to Moscow to meet the eminent Soviet physicist Yakov Zeldovitch and his student, Alexei Starobinskii. At the time, Hawking had established a classical "area law" for black holes [20, 21] that implied that the surface area of black holes can never decrease, but was irritated by Jacob Bekenstein's claim that this area law was the gravitational equivalent of the second law of thermodynamics, by identifying the surface area of the black hole with an entropy [10, 13]. Hawking disagreed with Bekenstein's identification [24] because it would imply that black holes have a temperature. Zeldovitch and Starobinskii, however, convinced Hawking that Bekenstein was right. They had been working on a phenomenon called "superradiance": the amplification of incoming radiation by

C. Adami (✉)
Department of Physics and Astronomy, Michigan State University, East Lansing, MI, USA
e-mail: adami@msu.edu

A. Akil and C. Bambi (eds.), *The Black Hole Information Paradox*, Springer Series in Astrophysics and Cosmology, https://doi.org/10.1007/978-981-96-6170-1_5

bulk matter. Zeldovitch had shown that a rotating cylinder would emit such radiation [50], while Starobinskii applied the same thinking to rotating black holes [37]. Zeldovich and Starobinskii suggested to Hawking that black holes would emit radiation not only when impacted by incoming radiation, but *also* when stimulated by vacuum fluctuations. After returning from Moscow, Hawking embarked on the calculation that we all now know [22], showing that black holes indeed emit radiation spontaneously (as if stimulated by the vacuum).

The phenomenon that Zeldovitch and Starobinskii described is a classical phenomenon: even water waves scattering off of a rotating flow can create superradiance [43]. It is a process that leads to the *amplification* of incoming waves, and requires incoming waves of *negative* effective frequency (or, in the case of superradiance in atoms, a population inversion).[1] There is, however, a *quantum* version of superradiance that does not require negative frequencies: this is Einstein's *stimulated emission of radiation* [18]. The effective frequency of radiation emanating from a black hole in curved-space quantum field theory is

$$\omega = \omega_0 - m\Omega - \varepsilon\Phi. \tag{5.1}$$

Here, ω_0 is the frequency of the mode, m is the black hole's orbital angular momentum quantum number, ε the charge, and Ω and Φ are the rotational frequency and the electric potential of the black hole, respectively. For non-rotating uncharged black holes $\omega \geq 0$, and such black holes cannot superradiate in the classical form (which requires $\omega < 0$). As a consequence Hawking, apparently unaware of "quantum superradiance", ignored stimulated emission completely. As we will see, this is the mistake that gave rise to the information paradox.

It turns out that Schwarzschild black holes (that are uncharged and do not rotate) *can* emit stimulated radiation: the energy is provided by the black hole just as the energy in Hawking radiation is provided by the black hole. Intuitively, just like vacuum fluctuations can create particle–anti-particle pairs out of the vacuum which, if the process takes place close to the horizon, leads to Hawking radiation (spontaneous emission of radiation), an incoming particle can stimulate the creation of a particle–anti-particle pair with quantum numbers aligned with the incoming particle. In modern language, an incoming quantum state will stimulate the emission of a *clone-anti-clone* pair at the horizon [35]. This does not violate the quantum no-cloning theorem [17, 49] because (as argued by Mandel shortly after the publication of the no-cloning theorem [29]) the concommittant spontaneous emission (here, Hawking radiation) will always lead to *approximate* cloning, which is not forbidden.

It was Bekenstein himself who first pointed out Hawking's mistake, in a paper where he treated black holes as ordinary black bodies. Using purely statistical arguments, Bekenstein together with his student Amnon Meisels showed that spontaneous

[1] Radiation can also be amplified when a charged particle moves through a dielectric medium at a speed larger than the speed of light in the medium. This form of superradiance, which obtains its energy from the medium, is known as the Čerenkov effect, see for example [12].

radiation without concommittant stimulated emission is inconsistent, and using Einstein's A- and B-coefficient formalism was able to derive the probability distribution of stimulated radiation simply by insisting on probability conservation [11]. A detailed calculation of the response of Schwarzschild black holes to incoming quanta using Hawking's formulation of semi-classical curved-space quantum field theory [3, 9, 32] reproduced the Bekenstein-Meisels result precisely.

But does the stimulated emission effect really save information from destruction in black holes? Not everyone was convinced. Schiffer, for example, argued that spontaneous emission might still "overpower" the stimulated process, leading to a loss of information overall [33]. Here I show using a rigorous application of quantum information theory that the information transmission capacity of black holes is strictly positive, which implies that information can be *perfectly* reconstructed from the radiation outside of the event horizon of a black hole after information has been absorbed by the black hole. This information is *not* encoded in the Hawking radiation, however, but instead can be extracted from the stimulated radiation only, that is, from the almost perfect *clones* (copies) of the incoming states [4].

In the following sections, I discuss the probability distributions of outgoing spontaneous and stimulated emission, outline the calculation of the capacity of quantum black holes to transmit classical information, and discuss its relevance in understanding microscopic time-reversal invariance in black hole dynamics. I close with a brief discussion of the remaining paradox: whether or not the evaporation of a black hole that was formed from a pure state will return to a pure state.

5.2 Probability Distributions

The probability distribution of outgoing quanta—both spontaneous and stimulated—can be obtained via a variety of methods. I discuss the simplest of methods first (maximum entropy analysis), and briefly discuss how the same distributions are obtained in curved-space quantum field theory, before calculating the channel capacity.

5.2.1 Spontaneous Emission

The probability distribution $p(n)$ of quanta in any of n outgoing modes emitted by a black hole via spontaneous emission (the distribution of Hawking radiation), can be derived using only Hawking's result that the mean number of outgoing quanta is [22]

$$\langle n \rangle = \frac{\Gamma}{e^x - 1}. \tag{5.2}$$

Here, Γ is the absorption coefficient of the black hole (the "grey-body factor", so that $1-\Gamma$ is the black hole's reflectivity), and $x=\omega/T_{\rm BH}$ where ω is the effective frequency shown in (5.1) and $T_{\rm BH}$ is the black hole temperature.

You can derive the result (for a single massless scalar bosonic mode) [14]

$$p(n)=(1-e^{-\lambda})e^{-n\lambda} \tag{5.3}$$

simply by demanding that the Shannon entropy of the outgoing radiation

$$S=-\sum_n p(n)\log p(n) \tag{5.4}$$

is maximal, with the constraints $\sum_n p(n)=1$ and $\sum_n np(n)=\frac{\Gamma}{e^x-1}$ implemented via Lagrange multipliers. In Eq. (5.3), the Lagrange multiplier λ is related to black hole parameters via

$$e^{-\lambda}=\frac{\Gamma}{e^x-1+\Gamma}. \tag{5.5}$$

Note that distribution (5.3) (which was not given by Hawking) was also derived independently using full-fledged curved-space quantum field theory by Wald [46]. In the following, we will restrict ourselves to non-rotating and uncharged (Schwarzschild) black holes. Let us also introduce convenient parameters[2] α, β, and γ with the definitions

$$e^x=\frac{\alpha^2}{\beta^2}, \tag{5.6}$$

$$\gamma^2=1-\Gamma, \tag{5.7}$$

with the normalization condition

$$\alpha^2-\beta^2+\gamma^2=1. \tag{5.8}$$

In terms of these parameters

$$\langle n\rangle=\frac{\Gamma}{e^x-1}=\beta^2 \tag{5.9}$$

and

$$e^{-\lambda}=\frac{\beta^2}{1+\beta^2}, \tag{5.10}$$

[2] These parameters originate from the curved-space quantum field-theoretic treatment we will discuss later, but serve here to create particularly simple expressions.

so that the probability distribution of Hawking radiation can be written as

$$p(n) = \frac{1}{1-\beta^2}\left(\frac{\beta^2}{1+\beta^2}\right)^n . \tag{5.11}$$

Introducing further

$$z = \frac{\beta^2}{1+\beta^2} \tag{5.12}$$

allows us to write

$$p(n) = (1-z)z^n, \tag{5.13}$$

which makes the correct normalization of the distribution apparent, owing to the geometric series expansion.

5.2.2 *Stimulated Emission*

Bekenstein and Meisels showed that this distribution $p(n)$ cannot consistently describe a black hole that is in equilibrium with its surrounding radiation. Specifically, they showed that if a black hole was immersed in a heat bath with temperature T with distribution $p_\star(n) = (1-e^{-y})e^{-ny}$ where $y = \omega/T$, then the outgoing radiation

$$p_o(n) = \sum_{m=0}^{\infty} p(n|m)p_\star(m) \tag{5.14}$$

does not satisfy the detailed balance condition

$$e^{-xm}p(n|m) = e^{-xn}p(m|n). \tag{5.15}$$

if $p(n|m)$, the probability that n particles are outgoing if m particles are impinging on the black hole, is due *only to scattering* with a probability $1-\Gamma_0$. We can see this by writing out the probability distribution $p(n|m)$ due to scattering: the binomial

$$p(n|m) = \binom{m}{n}\Gamma_0^{m-n}(1-\Gamma_0)^n. \tag{5.16}$$

It is easily checked that Eq. (5.16) does not respect the detailed balance condition. As a consequence, such a black hole could not be in equilibrium with its surrounding radiation.

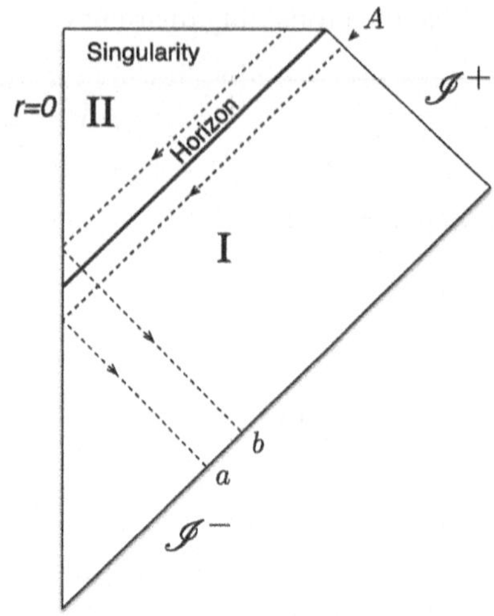

Fig. 5.1 Penrose diagram showing the early-time modes a and b, propagated backwards in time from future infinity $\mathscr{I}^+$ towards past infinity $\mathscr{I}^-$ (just as the black hole formed) and traveling just outside and just inside of the event horizon. The mode at future infinity is annihilated by A. In this depiction, modes a and b are shown separated for clarity, but they are just inside and just outside the horizon and coincide at $\mathscr{I}^-$

Stimulated emission can be studied by introducing incoming particles at the exact moment of the formation of the black hole (see Fig. 5.1). Note that in this diagram, the modes a (particles propagating just outside the horizon) and b (propagating just inside the horizon) are followed backwards towards $\mathscr{I}^-$ from $\mathscr{I}^+$, even though we would like to describe the time-reversed situation, where a mode a is scattered towards $\mathscr{I}^+$. In his description, Hawking eliminated positive frequency modes at $\mathscr{I}^-$ after following them backwards from $\mathscr{I}^+$, preventing him from studying stimulated emission. Here, we will study their fate in forward time. This is achieved by the unitary transformation (here and before $\hbar$ is set to 1)

$$|\Psi\rangle_{\text{out}} = e^{-iH}|m\rangle_a|0\rangle_b, \tag{5.17}$$

where H implements the Bogoliubov transformation relating in-modes to out-modes (for simplicity we treat only a single mode here, see [2, 3] for expressions including a sum over modes)

$$H = ig(a^\dagger b^\dagger - ab) + \text{h.c.}. \tag{5.18}$$

In (5.18), g is a coupling strength that is related to the black hole surface gravity and sets the temperature of the black hole, and we added the hermitian conjugate in order to be able to treat anti-particles as well. This is not strictly necessary, but is convenient to encode logical bits, as we will see later.[3]

Hamiltonian (5.18) has the same form as the so-called "squeezing Hamiltonian" in quantum optics that describes optical parametric amplification [28]. Indeed, the black hole is a quantum amplifier as well.

[3] As a consequence, Hamiltonian (5.18) has an eigenspace that is a product of particle- and anti-particle space, so the in-vacuum in (5.17) should strictly be written $|m, 0\rangle_a|0, 0\rangle_b$.

We can calculate the density matrix of outgoing radiation in region I (outside of the black hole horizon, see Fig. 5.1) using the formalism of curved-space quantum field theory in the case of full absorption ($\Gamma = 1$) using (5.17) to calculate $\rho_{\rm out} = |\Psi\rangle_{\rm out}\langle\Psi|$ and tracing over the horizon modes [3]. The resulting outgoing density matrix $\rho(m)$ factorizes into particle and anti-particle matrices

$$\rho(m) = \rho_m \otimes \rho_0. \tag{5.19}$$

The matrix ρ_0 is just the standard density matrix of Hawking radiation since no anti-particles are sent into the black hole

$$\rho_0 = \sum_{n=0}^{\infty} p(n)|n\rangle\langle n| \tag{5.20}$$

with $p(n)$ given by Eq. (5.13), while [3]

$$\rho_m = (1-z)^{1+m} \sum_{n=0}^{\infty} z^n \binom{n+m}{n} |n+m\rangle\langle n+m| \tag{5.21}$$

with $z = \frac{\beta^2}{1+\beta^2}$ as before.

The matrix ρ_m can be rewritten as

$$\rho_m = \sum_n p(n|m)|n\rangle\langle n|, \tag{5.22}$$

with

$$p(n|m) = (1-z)^{1+m} z^n \binom{n+m}{n}. \tag{5.23}$$

This is the generalization of the spontaneous emission formula to allow for m incoming particles, and reduces to $p(n)$ for $m = 0$. We can use those distributions to calculate the mean number of particles outside the horizon. The number of particles due to spontaneous emission (Hawking radiation) is

$$\langle n\rangle_{\rm spont} = \sum_{n=0}^{\infty} np(n) = \beta^2, \tag{5.24}$$

which is the celebrated Planck distribution as

$$\beta^2 = \frac{e^{-\omega_k/T_{\rm BH}}}{1 - e^{-\omega_k/t_{\rm BH}}}. \tag{5.25}$$

The total number of particles emitted in response to m incoming particles is

$$\langle n \rangle_{\text{tot}} = \sum_{n=0}^{\infty} np(n|m) = \beta^2(m+1). \tag{5.26}$$

Equation (5.26) confirms that beyond the Hawking radiation ("+1"), the region outside the horizon has $m\beta^2$ particles that were stimulated by the m incoming particles, with the same spectrum as Hawking radiation. However, while the spectrum is the same, the stimulated particles carry the imprint of the information that disappeared behind the horizon. This should not be surprising to those who realize that stimulated emission is essentially a quantum cloning process [27].

Figure 5.2 shows a sketch of the particle numbers in front and behind the horizon, as calculated using the unitary transformation (5.17). For that figure, we have reinstated the "grey-body" factor Γ_0, which is the probability for a single quantum to be absorbed at the horizon. This modifies the number of particles in region I to

$$\langle n \rangle_{\text{tot}} = (1 - \Gamma_0)m + \beta^2(m+1) \tag{5.27}$$

with

$$\beta^2 = \frac{\Gamma}{e^x - 1} = \frac{\Gamma_0(1 - e^{-x})}{e^x - 1}. \tag{5.28}$$

The term $(1 - \Gamma_0)m$ refers to the number of particles reflected at the black hole horizon. Reflection at the horizon is natural for particles with non-zero angular momentum with respect to the black hole. Only s-waves do not scatter. We will now see whether the stimulated particles outside the horizon are sufficient to reconstruct any classical information that was attached to the particles absorbed at the horizon.

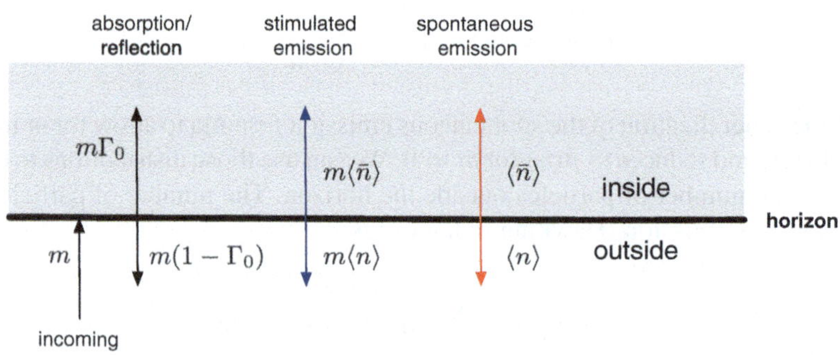

Fig. 5.2 Outgoing particles (arrows pointing down) for m incident particles on the horizon. In this sketch, $\langle n \rangle$ ($\langle \bar{n} \rangle$) refer to the mean number of particles (anti-particles) given by Eq. (5.2) that are generated in front and behind the horizon (to conserve particle number)

5.3 The Holevo Capacity of Black Holes

The Holevo capacity of a channel [25] is the capacity to transmit classical information over a quantum channel. A positive capacity implies that classical information can be sent over this channel with *arbitrary* accuracy, that is, with zero information loss. Just as in the classical theory [34], the capacity represents the rate of perfect information transfer when information enters the channel at one bit per second. Thus, a channel with a capacity of 0.1 s^{-1}, for example, makes it possible to transmit information with perfect accuracy at a rate of 0.1 s^{-1}. The loss in speed is always due to the error-correcting embedding that is necessary in order to recover the information with perfect accuracy. Let us calculate the Holevo capacity for black holes [3].

5.3.1 Early-Time Capacity

We will first calculate the capacity of the black hole channel when information is entering the black hole just after its formation. This is the situation usually represented by the Penrose diagram shown in Fig. 5.1, but it is admittedly an awkward channel: nobody would want to transmit information by timing a packet of particles so that it is absorbed just as the black hole is formed. The following calculation is, in fact, just an instructive exercise that will help us understand the more general calculation that comes later, when we send information into an already-formed black hole.

We imagine that a sender X prepares classical bits using a dual-rail encoding, sending m particles into the black hole signifying a logical '1', and instead m anti-particles when signaling a '0'. I displayed the total density matrix outside the horizon when sending a logical '1' earlier as Eq. (5.19): it has simulated emission (which comprises spontaneous radiation) in the particle sector, and only spontaneous emission in the anti-particle sector.[4]

The distribution $p(n|m)$ shown in Eq. (5.23) suffices to calculate the Holevo capacity χ of the channel. Technically, the Holevo capacity is the maximum Shannon information shared between preparer X (who sends '1' with probability p and '0' with probability $1-p$) and the area outside the horizon, labeled I in Fig. 5.1: $\chi = \max H(X;\mathrm{I})$. It is thus the generalization of the Shannon capacity, but for quantum channels. The information that X has about region I then takes the simple form [3]

$$\chi = \max_p \chi(p) = \max_p \Big[S(\rho) - \Big(pS[\rho(m)] + (1-p)S[\rho(\bar{m})] \Big) \Big]. \tag{5.29}$$

In this expression, ρ refers to the density matrix of the sender X: a probabilistic mixture of the density matrix for a logical '1' $\rho(m)$, and for the logical '0', $\rho(\bar{m})$ using the mixing probability p

[4] There are other ways to encode information in this channel, even if there were no anti-particles: you could send m particles to signify a '1', and zero particles ("silence") to signify a '0'. This will result in a much lower capacity, however, as error correction is significantly more difficult.

$$\rho = p\rho(m) + (1-p)\rho(\bar{m}). \tag{5.30}$$

Further, $S(\rho)$ is the von Neumann entropy

$$S(\rho) = -\mathrm{Tr}\,\rho\log\rho. \tag{5.31}$$

The von Neumann entropy reverts to the Shannon entropy for diagonal (that is, classical) density matrices.

Let us proceed to calculate χ, but restrict ourselves to sending a single particle or anti-particle to keep the calculation tractable. Setting $m=1$, we need to calculate the entropies $S(\rho)$, $S(\rho(0))$, and $S(\rho(1))$. Now, because $S[\rho(m)] = S[\rho(\bar{m})]$ it turns out that

$$\chi(p) = S(\rho) - S[\rho(1)]. \tag{5.32}$$

To calculate $S[\rho(1)]$ we need (recall Eq. (5.19))

$$S(\rho_0) = -\sum_{m=0}^{\infty} p(m|0)\log p(m|0) = -\log(1-z) - \frac{z}{1-z}\log z, \tag{5.33}$$

as well as

$$S(\rho_1) = -\sum_{m=0}^{\infty} p(m|1)\log p(m|1) = -2\log(1-z) - \frac{2z}{1-z}\log z - (1-z)^2\Delta, \tag{5.34}$$

where

$$\Delta = \sum_{m=0}^{\infty} z^m(m+1)\log(m+1). \tag{5.35}$$

Equation (5.35) has a closed-form expression in terms of derivatives of the polylogarithm function $\mathrm{Li}(s,z) = \sum_{n=0}^{\infty} z^n/n^s$

$$\Delta = -\frac{1}{z}\frac{\partial}{\partial s}\mathrm{Li}(-1,z) = -\frac{\partial}{\partial z}\frac{\partial}{\partial s}\mathrm{Li}(0,z). \tag{5.36}$$

using the notation $\frac{\partial}{\partial s}\mathrm{Li}(0,z) = \frac{\partial}{\partial s}\mathrm{Li}(s,z)|_{s=0}$. Gathering (5.33) and (5.34) we obtain

$$S[\rho(1)] = S(\rho_0) + S(\rho_1) = -3\log(1-z) - \frac{3z}{1-z}\log z - (1-z)^2\Delta. \tag{5.37}$$

Let's further calculate $S(\rho) = S[p\rho(1) + (1-p)\rho(\bar{1})] \equiv S(p)$. We find

$$S(p) = -(1-z)^3 \sum_{mm'=0}^{\infty} z^{m+m'} \Big\{ [p(m+1) + (1-p)(m'+1)] \\ \times \log\Big((1-z)^3 z^{m+m'} [p(m+1) + (1-p)(m'+1)]\Big) \Big\} \tag{5.38}$$

It is immediately clear that $S(p) = S(1-p)$ as the replacement $p \leftrightarrow (1-p)$ just relabels m and m'. Then

$$\max_p \chi(p) = \chi(1/2). \tag{5.39}$$

But let's keep p arbitrary for the moment. We can calculate

$$S(p) = -3\log(1-z) - \frac{3z}{1-z}\log z + (1-z)^3 \Lambda(p), \tag{5.40}$$

where

$$\Lambda(p) = -\sum_{mm'} (1 + m' + p(m-m')) \log(1 + m' + p(m-m')) z^{m+m'}. \tag{5.41}$$

Thus, the term $-3\log(1-z) - \frac{3z}{1-z}\log z$ cancels from $\chi(p)$, and we have

$$\chi(p) = S(\rho) - S[\rho(1)] = (1-z)^3 \Lambda(p) + (1-z)^2 \Delta. \tag{5.42}$$

We also note that $\chi(0) = \chi(1) = 0$.

To obtain the capacity $\chi(1/2)$, we need to calculate $\Lambda(1/2)$. By changing the summation in (5.38), we can see that

$$\Lambda(1/2) = -\sum_{u=0}^{\infty}\sum_{v=0}^{u} z^u \frac{1}{2}(u+2)\log[\frac{1}{2}(u+2)] \tag{5.43}$$

$$= -\sum_{u=0}^{\infty} z^u \frac{1}{2}(u+1)(u+2)\log[\frac{1}{2}(u+2)] \tag{5.44}$$

$$= \frac{1}{2}\sum_{u=0}^{\infty} z^u (u+1)(u+2) - \frac{1}{2}\sum_{u=0}^{\infty} u(u+1)\log(u+1) z^{u-1} \tag{5.45}$$

where we just shifted the sum in the 2nd term. Thus,

$$\Lambda(1/2) = \frac{1}{(1-z)^3} - \frac{1}{2}\frac{1}{z}\sum_{u=0}^{\infty} z^u u(u+1)\log(u+1). \tag{5.46}$$

and therefore (renaming $u \to m$)

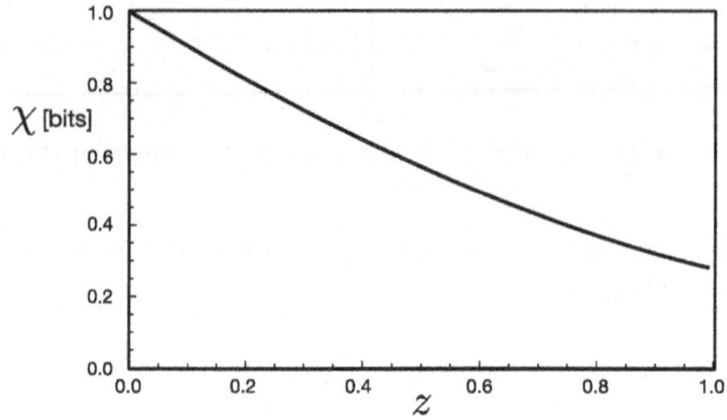

Fig. 5.3 Early-time capacity χ for a binary non-reflecting ($\Gamma = 1$) black hole channel as a function of the parameter $z = e^{-\omega/T_{\rm BH}}$. The limit $z = 0$ corresponds to a black hole with infinite mass (vanishing surface gravity), while $z \to 1$ as the mass of the black hole tends to zero and the Hawking temperature diverges

$$\chi(1/2) = 1 - \frac{1}{2}(1-z)^3\frac{1}{z}\sum_{m=0}^{\infty} z^m m(m+1)\log(m+1) + (1-z)^2\Delta \quad (5.47)$$

Now,

$$\frac{1}{z}\sum_{m=0}^{\infty} z^m m(m+1)\log(m+1) = -\frac{\partial^2}{\partial z^2}\frac{\partial}{\partial s}\text{Li}(0,z) \quad (5.48)$$

so

$$\chi(1/2) = 1 + \frac{1}{2}(1-z)^3\frac{\partial^2}{\partial z^2}\frac{\partial}{\partial s}\text{Li}(0,z) - (1-z)^2\frac{\partial}{\partial z}\frac{\partial}{\partial s}\text{Li}(0,z) \quad (5.49)$$

$$= 1 - \frac{1}{2}(1-z)^2\frac{\partial^2}{\partial z^2}(z-1)\frac{\partial}{\partial s}\text{Li}(0,z). \quad (5.50)$$

This is the "early-time capacity" of the black hole, and its functional dependence on z is shown in Fig. 5.3. We note that the capacity is everywhere positive, which implies that no information is lost within the black hole ever: it is possible to retrieve any information sent into the black hole from the radiation outside the horizon with arbitrary accuracy.

5.3.2 *Late-Time Capacity*

While a calculation of the early-time capacity is instructive, we also need to acknowledge that it is inconsistent. Indeed, the conditional probability distribution $p(n|m)$ from Eq. (5.23) that went into the calculation of the early-time capacity does not obey the detailed balance condition (5.15). This is easily seen by noting that z in (5.23) is just $z = e^{-x}$ (because we had set $\Gamma = 1$), and then[5]

$$e^{-mx}(1-e^{-x})^{m+1}e^{-nx} \neq e^{-nx}(1-e^{-x})^{n+1}e^{-mx} \tag{5.51}$$

The reason for this failure is simple: the use of a black hole's absorptivity $\Gamma = 1$ is unphysical as this would imply that no radiation is emitted back via stimulated emission. Indeed, Bekenstein and Meisels found that in order for detailed balance to be respected, the effective absorptivity of the black hole must be

$$\Gamma = \Gamma_0(1-e^{-x}), \tag{5.52}$$

where Γ_0 is the "naked" absorptivity of the black hole, that is, the absorptivity determined solely from scattering (see Eq. (5.16) and Fig. 5.2), so that Γ is strictly smaller than 1.

The correct way to describe scattering with probability $1 - \Gamma_0$ is to introduce particles sent into the black hole long after formation of the black hole, instead of creating the ludicrous construction where particles in mode a are sent into the black hole precisely at the moment where it had just formed (Fig. 5.1). Sorkin [36] showed that late-time particles can be consistently described by introducing particles in mode c whose creation and annihilation operators commute with the early-time modes a and b because the c-modes are blue-shifted with respect to the early-time modes.[6] The early-time Bogoliubov transformation that links the outgoing annihilation operator A_k at future infinity (see Fig. 5.1) to the operators a_k and b_{-k} at past infinity is (here written for a single mode k)

$$A_k = \alpha_k a_k - \beta_k b_{-k}^\dagger \tag{5.53}$$

with the normalization condition $\alpha_k^2 - \beta_k^2 = 1$ (and implemented by the Hamiltonian (5.18) via $A_k = e^{-iH} a_k e^{iH}$) is then extended to

$$A_k = \alpha_k a_k - \beta_k b_{-k}^\dagger + \gamma_k c_k, \tag{5.54}$$

corresponding to the modes shown in the Penrose diagram in Fig. 5.4. The resolution of the annihilation operator A_k into three modes rather than two is motivated by the same procedure that Hawking used: following the field at future infinity backwards in time, the field is split into a component that is scattered at the horizon (mode c_k), while the remainder continues backwards just inside and just outside the horizon, and splits into the positive- and negative-frequency modes a_k and b_{-k} after reflection at the origin. In the following, I will omit the mode label k as we are going to be

[5] There is no need to include the binomial coefficient $\binom{n+m}{n}$ here because it is symmetric in n and m.

[6] Note that we have changed notatiom from that of Sorkin's (switching a and c) in order to be consistent with previous notation.

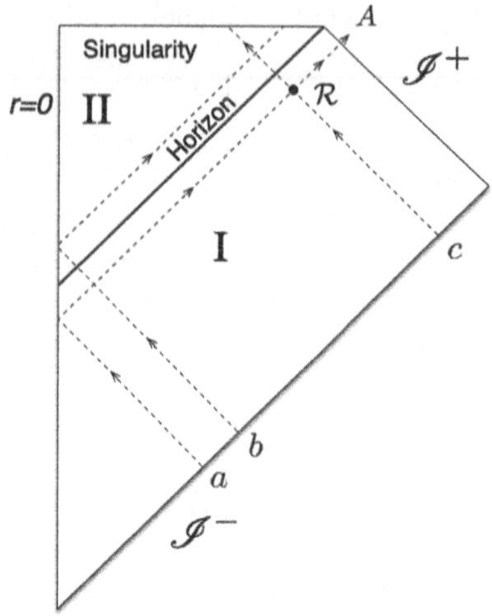

Fig. 5.4 Penrose diagram showing early-time modes (a and b) and late-time modes c. Late-time modes are scattered at the horizon with probability $\mathcal{R} = 1 - \alpha = 1 - \Gamma_0$ (the black hole reflectivity). A perfectly absorbing black hole has $\mathcal{R} = 0$

concerned with a mode with fixed k, but keep in mind that in general the Hamiltonians will imply sums over all modes (Fig. 5.4).

As the three modes a, b, and c are orthogonal, the late-time normalization condition is $\alpha^2 - \beta^2 + \gamma^2 = 1$. We introduced this notation earlier (see Eq. (5.8)) without explanation why such a parameterization allowed us to succinctly rewrite the probability distribution of Hawking quanta in a general fashion, for arbitrary absorption coefficient. We now understand the reason why: it is the only way to consistently describe scattering at the black hole horizon. Introducing a scattering rate $1 - \Gamma$ when following particles backwards from future infinity towards past infinity (as Hawking did) is inconsistent in more ways than one: when reversing time, we also need to flip the absorptivity of the black hole (more on time reversal later).

In this parameterization, the coefficient α^2 does not refer to the effective absorptivity of the black hole, but rather to the absorption coefficient due to scattering only, that is, $\alpha^2 = \Gamma_0$. As before, β^2 controls the rate of spontaneous emission, but the effective absorption coefficient is $\Gamma = 1 - \gamma^2$, so that the normalization condition $\alpha^2 - \beta^2 + \gamma^2 = 1$ is exactly equivalent to Bekenstein and Meisel's relation (5.52), which, in turn, is just the familiar relationship between the effective absorptivity of matter κ and the opacity κ_0 for a medium in thermodynamic equilibrium according to the theory of radiative transport (see, e.g., [16, p. 123]).

A perfectly absorbing black hole then has $\alpha^2 = 1$ (not $\gamma^2 = 0$). In that case, $\beta^2 = \gamma^2$, which implies that the strength of spontaneous emission of late-time particles is the same as that of early-time particles, modulo the difference in red shift, which we have ignored here. In fact, this allows us to see that the late-time formalism resolves the transplanckian paradox (see adjacent box).

Resolving the Transplanckian Paradox

We recall that t Hooft [41] and later Jacobson [26] pointed out that when the initial modes a and b are transported along the horizon to late times, they acquire an infinite redshift (see also the discussion in [19]). Besides leading to conceptual problems associated with a breakdown of the semi-classical approximation, the infinite redshift of early-time modes also contradicts the notion that radiation emanating from an object with the diameter $R = 2M_{\rm BH}$ (where $M_{\rm BH}$ is the black hole mass) should instead have a frequency of $\omega \sim \frac{1}{M_{\rm BH}}$, an essentially infinite discrepancy for large black holes. A popular "resolution" of this paradox was the so-called "black hole complementarity" principle [40, 42]. Black hole complementarity posited that the horizon in reality consists of an "exterior" and an "interior" horizon that are complementary in the sense that states on the surfaces of these two horizons are orthogonal (the Hilbert space factorizes into a product of interior and exterior spaces). We can now see that such a construction (which incidentally renders the energy-momentum tensor infinite at the horizon, that is, it is not renormalizable) is completely unnecessary as it is instead the Hilbert space of early- and late-time modes that factorizes. Because late-time particles do not accumulate a redshift, both the spontaneous and stimulated radiation associated with them is not infinitely redshifted, and their frequency instead is commensurate with the size of the black hole.

The direct calculation of $p(n|m)$ in curved-space quantum field theory is complicated and we will not repeat it here (it was performed independently using three different methods in [3, 9, 32]). The most intuitive calculation writes the unitary transformation from past-infinity vacuum to future-infinity vacuum

$$|0\rangle_{\rm out} = e^{-iH}|0\rangle_{\rm in} \tag{5.55}$$

in terms of a Hamiltonian H that can be written as a sum of two terms: one that is equivalent to an active optical element, the so-called "squeezing Hamiltonian" (OPA stands for "optical parametric amplification") with coupling strength g

$$H_{\rm OPA} = ig(a^\dagger b^\dagger - ab) + \text{h.c.} \tag{5.56}$$

and the Hamiltonian of a passive optical element, the "beam splitter"

$$H_S = ig'(a^\dagger c - ac^\dagger) + \text{h.c.} \tag{5.57}$$

that scatters late incoming modes c into horizon modes a. In (5.57), the scattering strength g' is bounded from below by the coupling strength g and sets the black hole reflectivity via [3]

$$\alpha^2 = \cos^2(\sqrt{g'^2 - g^2}). \tag{5.58}$$

Using the total Hamiltonian $H = H_{\text{OPA}} + H_S$, it is not difficult to show that

$$A = e^{-iH} a e^{iH} = \alpha a - \beta b + \gamma c, \tag{5.59}$$

with α defined in (5.58), and

$$\beta = g \frac{\sin(\sqrt{g'^2 - g^2})}{\sqrt{g'^2 - g^2}}, \tag{5.60}$$

$$\gamma = -g' \frac{\sin(\sqrt{g'^2 - g^2})}{\sqrt{g'^2 - g^2}}. \tag{5.61}$$

It is easy to check that $\alpha^2 - \beta^2 + \gamma^2 = 1$ holds.

Using the full Hamiltonian $H = H_{\text{OPA}} + H_S$ with m incoming late-time particles to calculate $\rho_{\text{out}} = |\Psi\rangle_{\text{out}}\langle\Psi|$ where

$$|\Psi\rangle_{\text{out}} = e^{-iH} \frac{(c^\dagger)^m}{\sqrt{m!}} |0\rangle_{\text{in}} \tag{5.62}$$

and $|0\rangle_{\text{in}} = |0,0\rangle_a |0,0\rangle_b |0,0\rangle_c$ gives rise to the late-time probability distribution $p(n|m)$ for arbitrary m in terms of α, β, and γ [2, 3]

$$p(n|m) = (1-z)^{m+1} z^n (\alpha^2)^m \sum_{k=0}^{\min(n,m)} (-1)^k \binom{m}{k} \binom{m+n-k}{n-k} (1 - \frac{\gamma^2}{\alpha^2\beta^2})^k. \tag{5.63}$$

We can quickly check a few limits. For a perfectly black (fully absorbing) hole, $\alpha^2 = 1$ which makes the term in brackets $1 - \frac{\gamma^2}{\alpha^2\beta^2} = 0$. As a consequence, only the term with $k = 0$ in the sum (5.63) survives, leading to

$$p(n|m) = (1-z)^{m+1} z^n \binom{m+n}{n}. \tag{5.64}$$

This distribution looks exactly like the early-time distribution (5.23), but it is not: here, $z = \frac{\beta^2}{1+\beta^2}$ and it is *not*[7] equal to e^{-x}. Rather, we have

$$e^{-x} = \frac{z}{1-z}, \tag{5.65}$$

[7] In Eq. (5.23), $\gamma^2 = 0$ and therefore $\beta^2 + 1 = \alpha^2$.

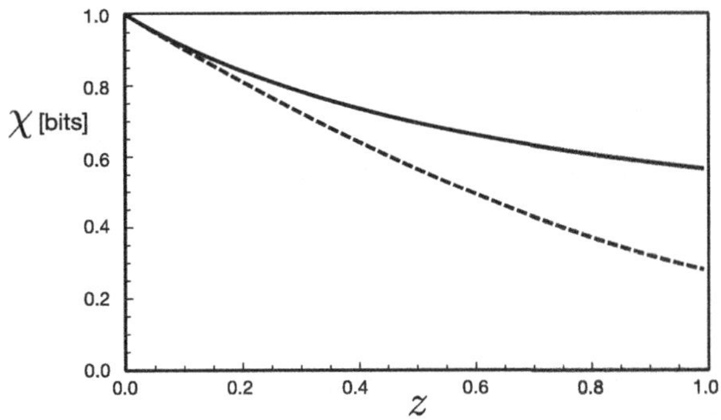

Fig. 5.5 Capacity χ for a binary non-reflecting black ($\Gamma_0 = 1$) hole channel as a function of the parameter $z = e^{-\omega/T_{bh}}$. The solid line represents the late-time capacity, while the dashed line is the capacity for early-time modes. Note that since $g \approx z$, we can see this plot also as depicting the dependence of the information transmission capacity on the mode coupling strength in the black-hole Hamiltonian

which we can use to show that (5.64) indeed satisfies detailed balance. Using (5.64) to calculate the capacity of a fully absorbing black hole gives us the solid curve in Fig. 5.5. The late-time capacity is strictly larger than the early-time capacity (shown as a dashed line in Fig. 5.5 for comparison), because the late-time capacity can be obtained from the early-time capacity simply by a replacement $z \to \frac{z}{1+z}$ [3].

We can also use Eq. (5.63) to derive the probability distribution of outgoing radiation for a white hole ($\alpha^2 = 0$). In that case, it is the *last* term in the sum (5.63) that survives (all other terms vanish on account of $(\alpha^2)^m$ in Eq. (5.63)), giving rise to the white hole distribution

$$p(n|m) = (1-z)^{m+1} z^{n-m} \binom{n}{m}. \tag{5.66}$$

We can use this distribution to calculate the number of particles outside the white hole horizon as

$$\langle n \rangle = \sum_{n=0}^{\infty} n p(n|m) = \frac{m+z}{1-z} = m(1+\beta^2) + \beta^2. \tag{5.67}$$

This is the result (5.27) with $\Gamma_0 = 0$.

5.4 Time-Reversal Invariance

In the previous discussion, I briefly alluded to the problem of time reversal in the presence of black holes. Indeed, it is possible to see the entire black hole information paradox as one that breaks probability conservation due to a failure of time-reversal

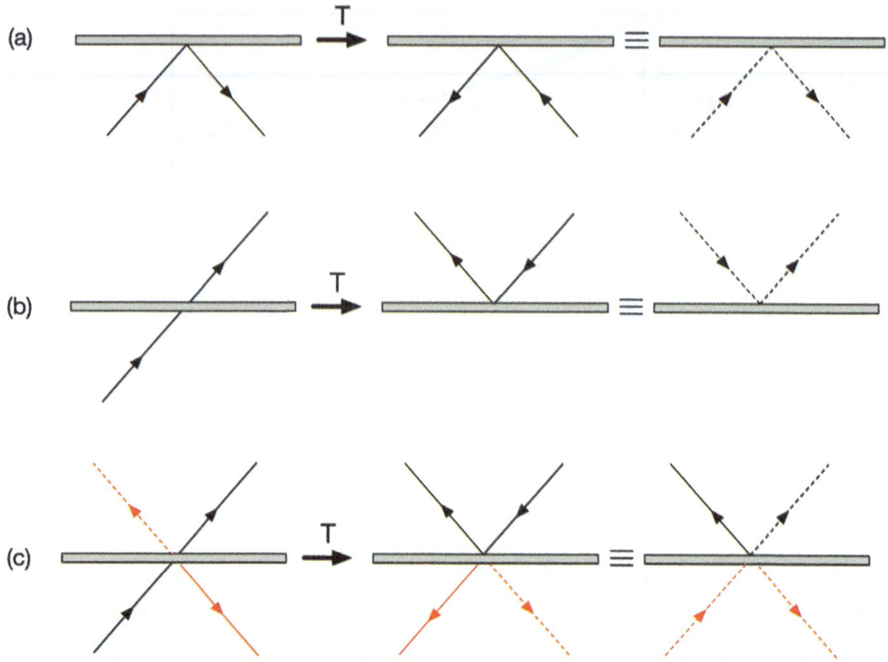

Fig. 5.6 Sketch of particle and anti-particle trajectories in the presence of white- and black hole horizons. Incoming particles are in black, while stimulated particles/anti-particles are in red. Solid lines represent particle trajectories, while dashed lines illustrate anti-particles. **a** Classical trajectory "reflected" at a mirror (i.e., falling back under the influence of gravity) on the left, its time-reversed trajectory (middle), and the equivalent trajectory in which particles moving forward in time are replaced by anti-particles moving backwards in time (right). **b** Classical trajectory of a particle absorbed by a black hole horizon (left), its time-reversed trajectory (middle), and the CP-transformed process (right). **c** Quantum trajectories including stimulated emission effects. The time-reversed process is equal to the CP-transformed process where particles are replaced by anti-particles moving backwards in time

invariance [47]. Newtonian gravity is, of course, invariant under time-reversal at the micro-level: every particle trajectory in a gravitational field, when time-reversed, gives rise to another plausible particle trajectory. One way to illustrate this is as in Fig. 5.6a (left panel), where (to simplify the discussion) the effect of a gravitational field "pulling down" a particle is shown as if a perfect mirror reflected the particle (instead of showing a parabolic trajectory). When reversing time, the trajectory simply reverses (middle panel). Using charge-conjugation symmetry (replacing a particle moving forward in time by an anti-particle moving backwards in time) gives us the right panel. Comparing the first and the last panel, we see that time reversal is equivalent to CP symmetry, which is the content of the celebrated CPT theorem [38].

It is easy to see that classical black holes violate this theorem. If instead of falling back a particle instead encounters a perfectly absorbing black hole horizon (Fig. 5.6b, left panel) reversing the arrow of time (middle panel) does *not* generate the time-reversed picture of the left panel, since a particle from inside of the black hole cannot escape to the outside. Instead, it is "reflected" at the horizon. Indeed, viewed from inside a black hole must act like the perfect mirror in Fig. 5.6a, as a particle cannot escape past the horizon. In other words, viewed from the inside a black hole must look like a white hole. Replacing particles by anti-particles moving backwards in time produces the right panel, which evidently is *not* the anti-particle version of the left panel, thus breaking CPT invariance.

Simply adding Hawking radiation to this picture does not restore CPT invariance. However, including stimulated emission of radiation does. We can see in Fig. 5.6c the stimulated particle/anti-particle pair in red (full absorbtion, i.e., $\Gamma_0 = 1$ is shown here) that must accompany the absorption process. Time-reversing this trajectory produces the middle panel, where the particle from inside of the black hole indeed reflects at the horizon, but it *also* stimulates a particle/anti-particle pair outside of the horizon (shown in red). In fact, this is the white-hole stimulated emission process described earlier via Eq. (5.66). Re-interpreting particles moving backwards in time as anti-particles moving forwards in time produces the right panel in Fig. 5.6c, which indeed is the same process as in the left panel, only with particles replaced with anti-particles. Thus, stimulated emission of pairs restores CPT invariance. I have not shown spontaneous emission of pairs (Hawking radiation) in these diagrams as those particles have no influence on CPT invariance: they only serve to safeguard the no-cloning theorem.

5.5 Discussion

That something was not right with classical black holes was clear early on (as they manifestly violate time-reversal invariance). Witten pointed out [47] that not only classical but also quantum black holes violate time-reversal invariance (along with detailed balance), and this is perhaps the most succinct way of characterizing the "information paradox" discussed in this work. It is now clear that this problem is entirely due to a mistake of Hawking. Made aware of superradiance by Zeldovitch and Starobinskii, he realized (as Zeldovitch and Starobinskii had suggested to him) that black holes must also emit *spontaneously*. Unaware that there is a quantum phenomenon of superradiance that does not require negative effective frequencies, Hawking simply put to zero amplitudes at past infinity (after propagating modes backward from future infinity) because he was treating Schwarzschild black holes that cannot superradiate, unaware of quantum superradiance (stimulated emission).

We might ask why Hawking chose to follow modes backwards in time, rather than forward in time as we do here. I believe the answer is that propagating early-time modes forward is awkward: if a mode is released before the formation of the black hole, then it could not be "reflected though the center" [22, p. 208]. But if it was released after the formation of the hole, it would have to be scattered at the horizon, not by the center. Indeed, our calculation of the early-time capacity using particles

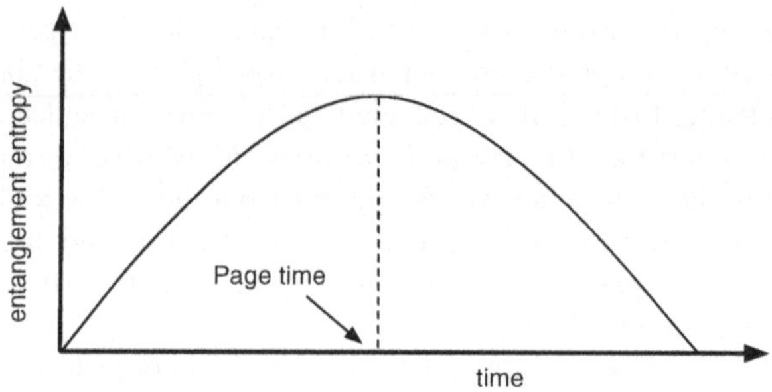

Fig. 5.7 The Page curve examines the relationship between the entanglement entropy of Hawking radiation and time (strictly speaking, the "size" of the subsystem as determined by the number of particles in the subsystem). In a unitary picture of black hole formation and evaporation, the entanglement entropy should be rising up to the Page time, and then fall again until ultimately reaching zero

released with precise timing towards the soon-to-be-forming horizon mirrors that awkwardness: the resulting radiation does not conform to detailed balance and therefore is not physical. The solution to this conundrum was found by Sorkin [36], who introduced a "late-time" Hilbert space that decouples from the in-vacuum Hilbert space that defines the forming black hole. The resulting equation for the probability distribution of outgoing radiation, using conventional curved-space quantum field theory methods, is fully consistent with that obtained by purely statistical (maximum entropy) arguments by Bekenstein and Meisels [11]. It is fair to say that this problem is now solved for all time.

However, all might still not be well with black holes. A second paradox (which is different from the "information paradox") could still be plaguing black hole physics. This problem concerns the unitarity of the evaporation process of the black hole (as pointed out early on by Hawking [23]), and the apparent "lack of predictability" is also related to a failure of time-reversal invariance. However, this problem is not solved by stimulated emission of radiation, as the loss of prediction is not due to losing information that is absorbed by a black hole, but the predictability of spacetime itself. The problem with a unitary time-evolution of a black hole from formation to evaporation is best exemplified by the "Page curve" [31], which we can interpret as a curve showing the entropy of the black hole as a function of time (shown in Fig. 5.7). In a unitary picture of black hole dynamics, a black hole that started out as a quantum mechanical pure state (with vanishing entanglement entropy) should end up as a pure state again, as opposed to a mixed state with positive entropy. It is difficult to calculate the entanglement entropy of a black hole as a function of time because a calculation of the black hole S-matrix (which tracks dynamics over time) requires a Hamiltonian that describes the interaction between the black hole degrees of freedom and Hawking radiation. However, curved-space quantum field theory does not yield such an interaction Hamiltonian: the standard theory (even including the scattering term (5.16)) is a free-field theory.

One way out of this conumdrum is to take the quantum optics analogy seriously, and look towards interaction Hamiltonians whose limit is the free field Hamiltonian (5.18). Indeed, the squeezing Hamiltonian is the limit of a more general Hamiltonian that takes into account the degrees of freedom of the pump (typically a laser) that powers the creation of entangled pairs (called "signal" and "idler" modes). In the limit (5.18) this pump is usually assumed to be so intense that the average number of photons in this mode is constant: the pump is said to be *inexhaustible*, meaning that the source can produce entangled pairs of particles endlessly. Realistic pumps, however, are not "undepletable": the production of pairs can "back-react" on the pump leading to its exhaustion. The quantum Hamiltonian that couples the pairs to the pump degrees of freedom is well-known [5, 44, 45, 48], and accurately describes realistic experimental situations. Not surpringly, such "trilinear" Hamiltonians have been used to model the interaction between black hole modes (standing in for the pump modes, since in this picture the black hole is the quantum amplifier) and the horizon modes a and b [6, 30] . The "backreaction" of the Hawking radiation on the black hole itself is thus modeled precisely in the manner in which the production of signal/idler pairs depletes the pump, until the mean number of photons in the pump is zero (the black hole has disappeared).

Using such a trilinear interaction in a Monte Carlo path integral approach to the S-matrix, it is possible to study formation and decay of a black hole with a fixed mass (fixed number of pump modes), giving rise to Page curves [8, 15]. However, in that work the b modes created behind the horizon were assumed not to interact with the black hole modes anymore, while this is in theory possible. An extended analysis that allows internal b modes to create additional pairs shows that lifting this approximation does not destroy the Page curves [7].

Granted, the trilinear interaction that models the unitary time evolution of black holes does not follow from a fundamental theory of quantum gravity. However, it can be argued that *any* consistent unitary theory that obeys energy conservation rules must have a trilinear interaction term in the low-energy limit (as argued, for example, by Strominger [39]). At the same time, it is remarkable that the process that leads to the ultimate decoupling of black hole degrees of freedom from the radiation degrees of freedom turns out to be the analogue of a well-known quantum information protocol, namely the fully-quantum Slepian-Wolf (FQSW) protocol [1] (see the discussion in [2]). This too is unlikely to be a coincidence.

We thus come to the conclusion that black holes are interesting astrophysical objects that, rather than upending the laws of physics, have taught us much about the most fundamental laws of physics, such as time-reversal invariance and the no-cloning theorem. And perhaps they have more surprises left in them after all.

Acknowledgements I am indebted to my collaborators in black hole physics: Greg Ver Steeg and Kamil Bradler.

References

1. A. Abeyesinghe, I. Devetak, P. Hayden, A. Winter, The mother of all protocols: restructuring quantum information's family tree. Proc. R. Soc. London A **465**, 2537–2563 (2009)
2. C. Adami, Stimulated emission and the black hole information problem. Ann. Phys. **468**, 169739 (2024)
3. C. Adami, G. Ver Steeg, Classical information transmission capacity of quantum black holes. Class. Quantum Gravity **31**, 075015 (2014)
4. C. Adami, G. Ver Steeg, Black holes are almost optimal quantum cloners. J. Phys. A **48**, 23FT01 (2015)
5. G.P. Agrawal, C.L. Mehta, Dynamics of parametric processes with a trilinear Hamiltonian. J. Phys. A **7**, 607 (1974)
6. P.M. Alsing, Parametric down conversion with a depleted pump as a model for classical information transmission capacity of quantum black holes. Class. Quantum Gravity **32**, 075010 (2015)
7. P.M. Alsing, Black hole waterfall: a unitary phenomenological model for black hole evaporation with Page curve (2025). arXiv:2501.00948
8. P.M. Alsing, M.L. Fanto, A discrete analogue for black hole evaporation using approximate analytical solutions of a one-shot decoupling trilinear Hamiltonian. Class. Quantum Gravity **33**, 015005 (2016)
9. J. Audretsch, R. Müller, Amplification of black-hole Hawking radiation by stimulated emission. Phys. Rev. D **45**, 513 (1992)
10. J.D. Bekenstein, Black holes and the second law. Lett. Nuovo Cimento **4**, 737–740 (1972)
11. J.D. Bekenstein, A. Meisels, Einstein A and B coefficients for a black hole. Phys. Rev. D **13**, 2775 (1977)
12. J.D. Bekenstein, M. Schiffer, The many faces of superradiance. Phys. Rev. D **59**, 064014 (1998)
13. J.D. Bekenstein, Black holes and entropy. Phys. Rev. D **7**, 2333–2346 (1973)
14. J.D. Bekenstein, Statistical black-hole thermodynamics. Phys. Rev. D **12**, 3077–3085 (1975)
15. K. Brádler, C. Adami, One-shot decoupling and Page curves from a dynamical model for black hole evaporation. Phys. Rev. Lett. **116**, 101301 (2016)
16. J.P. Cox, R.T. Giuli. *Principles of Stellar Structure, volume 1: physical Principles* (Gordon and Breach, 1968)
17. D. Dieks, Communication by EPR devices. Phys. Lett. A **92**, 271 (1982)
18. A. Einstein, Zur Quantentheorie der Strahlung. Physik. Zeitschr. **18**, 121–128 (1917)
19. A. Fabbri, J. Navarro-Salas, *Modeling Black Hole Evaporation* (Imperial College Press, London, 2005)
20. S.W. Hawking, Gravitational radiation from colliding black holes. Phys. Rev. Lett. **26**, 1344–1346 (1971)
21. S.W. Hawking, Black holes in general relativity. Commun. Math. Phys. **25**, 152–166 (1972)
22. S.W. Hawking, Particle creation by black holes. Commun. Math. Phys. **43**, 199–220 (1975)
23. S.W. Hawking, Breakdown of predictability in gravitational collapse. Phys. Rev. D **14**, 2460 (1976)
24. S.W. Hawking, *A Brief History of Time* (Bantam, New York, N.Y., 1988)
25. A.S. Holevo, The capacity of quantum channel with general signal states. IEEE Trans. Inform. Theor. **44**, 269–273 (1998)
26. T. Jacobsen, Black-hole evaporation and ultrashort distance. Phys. Rev. D **44**, 1731–1739 (1991)
27. A. Lamas-Linares, C. Simon, J.C. Howell, D. Bouwmeester, Experimental quantum cloning of single photons. Science **296**, 712–714 (2002)
28. U. Leonhardt, *Essential Quantum Optics: from Quantum Measurements to Black Holes* (Cambridge University Press, 2010)
29. L. Mandel, Is a photon amplifier always polarization dependent? Nature **304**, 188 (1983)
30. P.D. Nation, M.P. Blencowe, The trilinear Hamiltonian: a zero-dimensional model of Hawking radiation from a quantized source. New J. Phys. **12**, 095013 (2010)

31. D.N. Page, Information in black hole radiation. Phys. Rev. Lett. **71**, 3743 (1993)
32. P. Panangaden, R.M. Wald, Probability distribution of radiation from a black hole in the presence of incoming radiation. Phys. Rev. D **16**, 929–932 (1977)
33. M. Schiffer, Is it possible to recover information from the black-hole radiation? Phys. Rev. D **48**, 1652 (1993)
34. C.E. Shannon, A mathematical theory of communication. Bell Syst. Tech. J. **27**(379–423), 623–656 (1948)
35. C. Simon, G. Weihs, A. Zeilinger, Optimal quantum cloning via stimulated emission. Phys. Rev. Lett. **84**, 2993–2996 (2000)
36. R.D. Sorkin, A simplified derivation of stimulated emission by black holes. Class. Quantum Gravity **5**, L149–L155 (1987)
37. A.A. Starobinskiĭ, Amplification of waves during reflection from a rotating "black hole". Sov. Phys. JETP **37**, 28 (1973)
38. R.F. Streater, A.S. Wightman, *PCT, Spin and Statistics, and All That* (W.A. Benjamin Inc, New York, NY, 1964)
39. A. Strominger, Les Houches lectures on black holes, in editors, *Fluctuating Geometries in Statistical Mechanics and Field Theory*, eds. by F. David, J. Zinn-Justin, P. Ginsparg (Elsevier, Amsterdam, 1996), pp. 699–760. arXiv:hep-th/9410187v2
40. L. Susskind, L. Thorlacius, J. Uglum, The stretched horizon and black hole complementarity. Phys. Rev. D **48**, 3743–3761 (1993)
41. G. t Hooft, On the quantum structure of a black hole. Nucl. Phys. B **256**, 727–745 (1985)
42. G. t Hooft, The black hole interpretation of string theory. Nucl. Phys. B **335**, 138–154 (1990)
43. T. Torres, S. Patrick, A. Coutant, M. Richartz, E.W. Tedford, S. Weinfurtner, Rotational superradiant scattering in a vortex flow. Nat. Phys. **13**, 833–838 (2017)
44. J. Tucker, D.F. Walls, Quantum theory of parametric frequency conversion. Ann. Phys. N. Y. **52**, 1–15 (1969)
45. J. Tucker, D.F. Walls, Quantum theory of the traveling-wave frequency converter. Phys. Rev. **178**, 2036 (1969)
46. R.M. Wald, On particle creation by black holes. Commun. Math. Phys. **45**, 9–34 (1975)
47. E. Witten, Quantum mechanics of black holes. Science **337**, 538–40 (2012)
48. D.F. Walls, R. Barakat, Quantum-mechanical amplification and frequency conversion with a trilinear Hamiltonian. Phys. Rev. A **1**, 446 (1970)
49. W.K. Wootters, W.H. Zurek, A single quantum cannot be cloned. Nature **299**, 802 (1982)
50. Y.B. Zeldovitch, Amplification of cylindrical electromagnetic waves reflected from a rotating body. Sov. Phys. JETP **35**, 1085–1087 (1972)

Chapter 6
There is no Black Hole Information Paradox

Carlo Rovelli and Francesca Vidotto

Abstract The alleged black hole information paradox is the consequence of an arbitrary and unjustified hypothesis about nature.

In this chapter, which follows closely [1, 2] and especially [3], we explain why the alleged black hole information paradox is the consequence of a unjustified hypothesis about nature.

The black hole information problem, as formulated by Don Page in 1993 [4], regards the physics of the spacetime region *before* the quantum gravity onset. Figure 6.1 is the Carter-Penrose diagram of a spherically symmetric spacetime geometry around a collapsing star, on which there is an evolving quantum field ϕ. The geometry takes into account the back-reaction of the Hawking radiation of the field. The collapsed matter distribution generates a trapped region: the black hole. The boundary of this region is the (trapping) horizon. In the limit in which we disregard the back-reaction of the Hawking radiation, the horizon is null, but taking back-reaction into account, it is time-like. The notion of *event* horizon is not defined, because no assumption is made about the distant future, which depends on quantum gravity and is not relevant. Considerations on this region alone are sufficient for Page's argument for the information problem.

C. Rovelli (✉)
Aix-Marseille University, Université de Toulon, CPT-CNRS , Marseille, France
e-mail: rovelli.carlo@gmail.com

Perimeter Institute, Waterloo, Ontario, Canada

Santa Fe Institute, Santa Fe, New Mexico, USA

C. Rovelli · F. Vidotto
Philosophy Department and Rotman Institute, Western University, London, Ontario, Canada

F. Vidotto
Physics and Astronomy Department, Western University, London, Ontario, Canada

Instituto de Estructura de la Materia, IEM-CSIC, Madrid, Spain

A. Akil and C. Bambi (eds.), *The Black Hole Information Paradox*, Springer Series in Astrophysics and Cosmology, https://doi.org/10.1007/978-981-96-6170-1_6

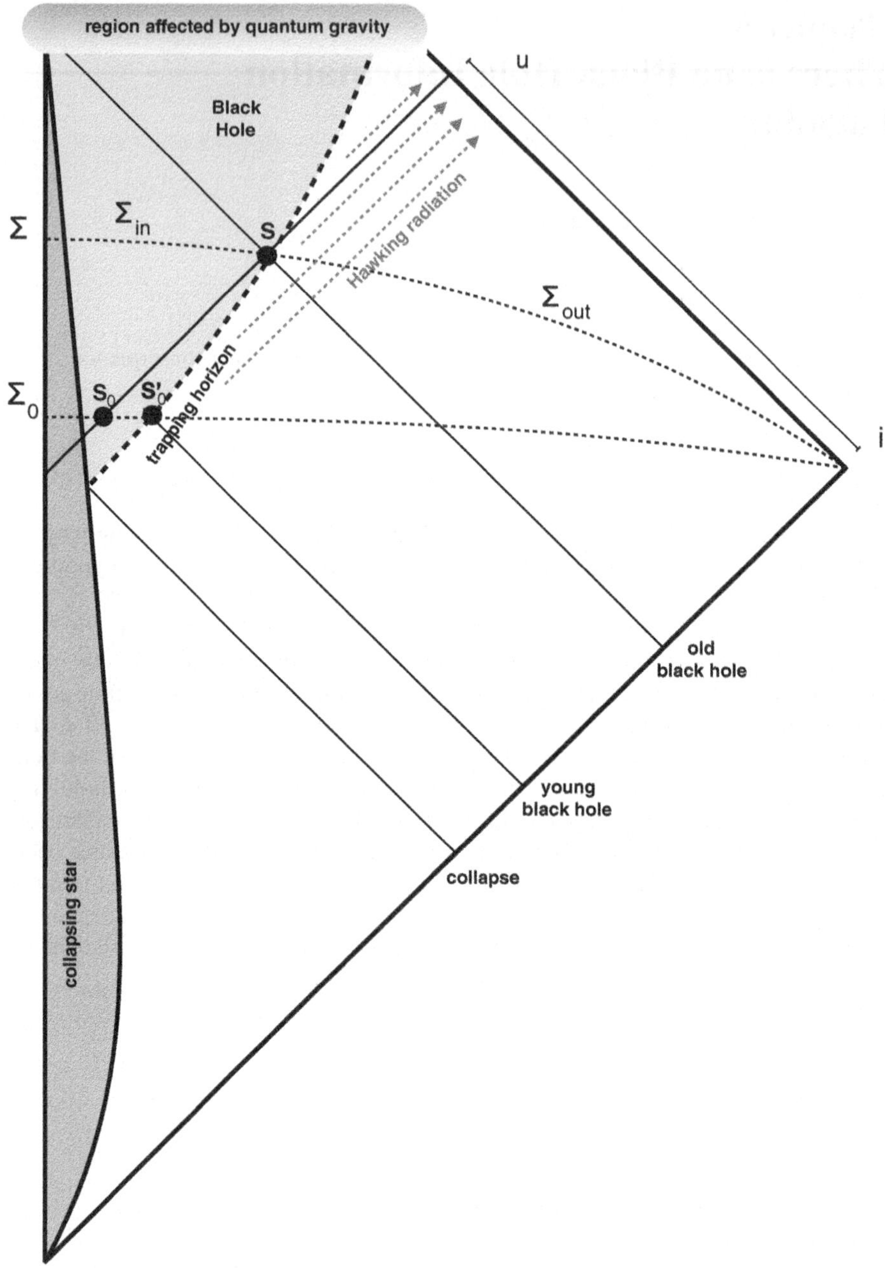

Fig. 6.1 The portion of spacetime relevant for the information 'paradox'

Here is Page's key observation. Consider a sphere S on the horizon at retarded time u (see Fig. 6.1). The region of future null infinity preceding the time u receives the Hawking radiation emitted until the horizon has reached S. Consider the case where the black hole is "old" at S, namely the area A of S is much smaller than the the initial horizon area A_0 at S_0. The Hawking radiation arriving at future infinity around u is in a mixed state. If the initial state of the field was pure and evolution is unitary, this radiation must be correlated with something else: with what?

There are two reasonable possibilities:

(a) it is correlated with degrees of freedom inside the horizon;
(b) it is correlated with degrees of freedom outside the horizon. In particular, late Hawking quanta may be correlated with early Hawking quanta.

A part of the theoretical community has got convinced that the correct answer must be (b) because (a) is ruled out by the following argument by Don Page, based on a specific assumption. Let N be the number of states of a black hole with which external degrees of freedom can be entangled at some given time and let

$$N_{BH} = e^{A/4}. \tag{6.1}$$

where A is the area of the horizon at that time. Consider the following

$$\text{Assumption:} \quad N \sim N_{BH}. \tag{6.2}$$

Then, necessarily (a) is ruled out, because for sufficiently late times A is small and therefore N is insufficient for purifying the Hawking radiation.

But is the assumption justified?

The thermodynamic interaction between a black hole and its surroundings is well described by treating the black hole as a system with $N_{BH} = e^{A/4}$ (orthogonal) states, where A is the horizon area. But a system can have many more degrees of freedom than those determining its thermodynamic interactions. For instance, this is the case when some degrees of freedom are physically constrained from acting back on the exterior, as is precisely the case for a black hole. It is only for ergodic systems that the von Neumann entangled entropy (which purifies the external state) must be equal or smaller than the thermodynamic entropy, and certainly a black hole is not ergodic. So, there is definitely no compelling reason based on known physics for assuming (6.2).

Some theoretical hypotheses, like AdS-CFT or some String formulations elevate (6.2) into a postulate. Juan Maldacena has called this assumption a "central dogma", and, like all dogmas, it can well be false. As many dogmas, this one as well leads to bizarre consequences, like firewalls on the horizon [5, 6].

The possibility that $N \gg N_{BH}$ is supported by the fact that according to classical General Relativity the interaction between a black hole and its surroundings is entirely determined by what happens in the vicinity of the horizon, not by the interior. The number N_{BH} is likely to count only horizon states that (before the transition) can be

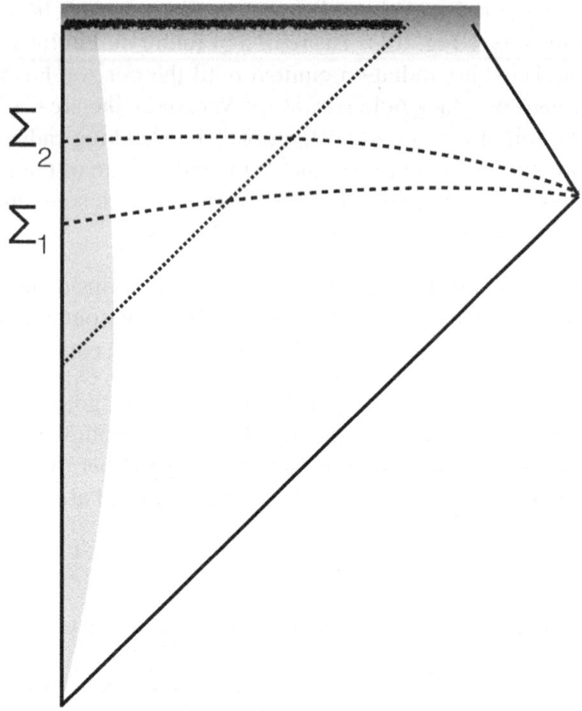

Fig. 6.2 The (lowest part) of the conformal diagram of a gravitational collapse. The clear grey region is the object, the dotted line is the horizon, the thick upper line is the singularity, the dark upper region is where quantum gravity effects may become relevant (this region plays no role in this paper). The two Cauchy surfaces used in the paper are the dashed lines

distinguishable from the exterior. These are "surface" states. On the other hand, N counts also states that can be distinguished by local observables *inside* the horizon. These do not contribute to the thermodynamical entropy $S = A/4$, but can contribute to the von Neumann entropy, which can therefore remain high even when A shrinks.

These generic considerations can be sharpened. The fact that a blackhole can have more states than N_{BH}, in fact, follows from elementary considerations of causality.

To show this, consider a gravitationally collapsed object and let Σ_1 be a Cauchy surface that crosses the horizon but does not hit the singularity, see Fig. 6.2. Let Σ_2 be a later similar Cauchy surface and $i = 1, 2$. Let A_i be the area of the intersection of Σ_i with the horizon. Assume that no positive energy falls into the horizon during the interval between the two surfaces. Let quantum fields live on this geometry, back-reacting on it [7]. Finally, let Σ_i^{in} be the (open) portions of Σ_i inside the horizon.

Care is required in specifying what is meant here by 'horizon', since there are several such notions (event horizon, trapping horizon, apparent horizon, dynamical horizon...) which in this context may give tiny differences in location. For precision, by 'horizon' we mean here the event horizon, if this exists. If it doesn't, we mean the boundary of the past of a late-time spacelike region lying outside the black hole (say outside the trapping region). With this definition, the horizon is light-like.

Because of the back-reaction of the Hawking radiation, the area of the horizon shrinks and therefore

$$A_2 < A_1. \tag{6.3}$$

Now consider the evolution of the quantum fields from Σ_1 to Σ_2. We are in a region far away from the singularity and therefore (assuming the black hole is large) from high curvature. Therefore we expect conventional quantum field theory to hold here, without strange quantum gravity effects, at least up to high energy scales. Since the horizon is light-like, Σ_1^{in} is in the causal past of Σ_2^{in}. This implies that any local observable on Σ_1^{in} is fully determined by observables on Σ_2^{in}. That is, if $\mathscr{A}_i$ is the local algebra of observables on Σ_i^{in} then $\mathscr{A}_1$ is a subalgebra of $\mathscr{A}_2$:

$$\mathscr{A}_1 \subset \mathscr{A}_2. \tag{6.4}$$

Therefore any state on $\mathscr{A}_2$ is also a state on $\mathscr{A}_1$ and if two such states can be distinguished by observables in $\mathscr{A}_1$ they certainly can be distinguished by observables in $\mathscr{A}_2$ as the former are included in the latter. Therefore the states that can be distinguished by $\mathscr{A}_1$—which is to say: on Σ_1^{in}—can also be distinguished by $\mathscr{A}_2$—which is to say: on Σ_2^{in}. Therefore the distinguishable states on Σ_1^{in} are a subset of those in Σ_2^{in}. How many are they? Either there is an infinite number of them, or a finite number due to some high-energy (say Planckian) cut-off. If there is an infinite number of them, then immediately the number of states distinguishable from inside the black hole is larger that N_{NB}, which is finite. If there is a finite number of them, then the number N_2 of distinguishable states on Σ_2^{in} must be equal or larger than the number N_2 of states distinguishable on Σ_1^{in}, because the second is a subset of the first. That is

$$N_2 \geq N_1. \tag{6.5}$$

Comparing Eqs. (6.3) and (6.5) shows immediately that it is impossible that $N_i = e^{A_i/4}$, as the exponential is a monotonic function.

This follows from only assuming the validity of quantum field theory in regions of low curvature. In other words, the "dogma" (6.2) necessarily violates known physics in a region where we have no reasons to assume that it should fail. In our opinion, this is not good scientific method.

The conclusion is that the number of states distinguishable from the interior of the black hole must be different from the number $N_{BH} = e^{A/4}$ of the states contributing to the Bekenstein-Hawking entropy. Since the second is shrinking to zero with the evaporation, the first must overcome the second at some point. Therefore in the interior of a black hole there are more possible states than $e^{A/4}$.

The physical interpretation of the conclusion is simple: the thermal behavior of the black hole described by the Bekenstein-Hawking entropy $S = A/4$ is determined by the physics of the vicinity of the horizon.

A vivid manifestation of the fact that in classical General Relativity the effect of a black hole on its surroundings is independent of the black hole interior is in the numerical simulations of black hole merging and radiation emission by oscillating black holes: in writing the numerical code, it is routine to cut away a region inside the (trapping) horizon: it is irrelevant for whatever happens outside! This is true in classical General Relativity, and there is no compelling reason to suppose it to fail if quantum fields are around.

The idea of interpreting of S_{BH} as determined by the number of states of near-surface degrees of freedom, and not interior ones is of course not a new idea. It has a long history [8–14]. See in particular [15, 16] in support of this idea from two different research camps, loops and strings.

Importantly, this conclusion is *not* in contrast with the various arguments leading to identify Bekenstein-Hawking entropy with a counting of states. To the opposite, evidence from it comes from the membrane paradigm [17] and from Loop Quantum Gravity [15, 18–20], which both show explicitly that the relevant states are surface states, but also from the string theory counting [21, 22], because the counting is in a context where the relevant state space is identified with the scattering state space, which could be blind to interior observables. For a classic discussion on different viewpoints about these alternatives, see [23].

Notice that in popular accounts (for a recent one, see for instance [24]), three assumptions are erroneously said to be proven incompatible: (i) unitarity of the quantum evolution, (ii) equivalence principle (absence of firewalls), and (iii) quantum field theory on curved spacetimes. This is wrong. It is only the further assumption of the dogma, that leads to problems.

In conclusion, if there are more states available in a black hole than $e^{A/4}$, then the Page argument for the information loss paradox fails. States purifying the Hawking radiation are inside the hole even when the horizon shrinks. The information can later escape from the emission of a long living remnant left at the end of the evaporation, as detailed for instance in [3].

Acknowledgements This work was made possible through the support of the ID# 62312 grant from the John Templeton Foundation, as part of the project "The Quantum Information Structure of Spacetime" (QISS). CR acknowledges support from the Perimeter Institute for Theoretical Physics through its distinguished research chair program. FV acknowledges support from the Perimeter Institute for Theoretical Physics through its affiliation program. Research at Perimeter Institute is supported by the Government of Canada through Industry Canada and by the Province of Ontario through the Ministry of Economic Development and Innovation. Research in FV's research group at Western University is supported by the Canada Research Chairs Program and by the Natural Science and Engineering Council of Canada (NSERC) through the Discovery Grant "Loop Quantum Gravity: from Computation to Phenomenology." We acknowledge the Anishinaabek, Haudenosaunee, Lūnaapéewak, Attawandaron, and neutral peoples, on whose traditional lands Western University and the Perimeter Institute are located.

References

1. C. Rovelli, Black holes have more states than those giving the Bekenstein-Hawking entropy: a simple argument. arXiv:1710.00218
2. C. Rovelli, The subtle unphysical hypothesis of the firewall theorem. Entropy **21**(9) (2019). arXiv:1902.03631

3. C. Rovelli, F. Vidotto, Planck stars, White Holes, Remnants and Planck-mass quasi-particles. The quantum gravity phase in black holes' evolution and its manifestations. arXiv:2407.09584
4. D.N. Page, Information in black hole radiation. Phys. Rev. Lett. **71**, 3743–3746 (1993). arXiv:hep-th/9306083
5. A. Almheiri, D. Marolf, J. Polchinski, J. Sully, Black holes: complementarity or firewalls? J. High Energy Phys. **1302**, 62 (2013). arXiv:1207.3123
6. D. Marolf, The black hole information problem: past, present, and future. Rep. Progr. Phys. **80**, 092001 (2017). arXiv:1703.02143
7. R.M. Wald, *Quantum Field Theory in Curved Space-Time and Black Hole Thermodynamics* (Chicago Lectures in Physics University Press, Chicago, USA, 1994)
8. J.W. York Jr., Dynamical origin of black-hole radiance. Phys. Rev. D **28**(12), 2929–2945 (1983)
9. W.H. Zurek, K. Thorne, Statistical mechanical origin of the entropy of a rotating, charged black hole. Phys. Rev. Lett. **54**, 2171 (1885)
10. A. Wheeler, *A Journey into Gravity and Spacetime* (Freeman, New York, 1990)
11. G. T'Hooft, The black hole interpretation of string theory. Nucl. Phys. B **335**, 138 (1990)
12. L. Susskind, L. Thorlacius, R. Uglum, The stretched horizon and black hole complementarity. Phys. Rev. D **54**, 2171 (1985)
13. V. Frolov, I. Novikov, Dynamical origin of the entropy of a black hole. Phys. Rev. D **48**, 4545 (1993)
14. F. Larsen, F. Wilczek, Internal structure of black holes. Phys. Lett. **375**, 37 (1996)
15. C. Rovelli, Loop quantum gravity and black hole physics. Helv. Phys. Acta **69**, 482 (1996). arXiv:gr-qc/9608032
16. A. Strominger, Black hole entropy from near horizon microstates. JHEP **9802**, 9 (1998). arXiv:hep-th/9712251
17. K. Thorne, R. Price, D. Macdonald, *Black Holes: the Membrane Paradigm* (Yale University Press, New Haven, 1986)
18. C. Rovelli, Black hole entropy from loop quantum gravity. Phys. Rev. Lett. **77**(16), 3288–3291 (1996). arXiv:gr-qc/9603063
19. A. Ashtekar, J. Baez, A. Corichi, K. Krasnov, Quantum geometry and black hole entropy. Phys. Rev. Lett. **80**, 904–907 (1998). arXiv:gr-qc/9710007
20. A. Perez, Black holes in loop quantum gravity (2017)
21. A. Strominger, C. Vafa, Microscopic origin of the Bekenstein-Hawking entropy. Phys. Lett. **B379**(1–4), 99–104 (1996). arXiv:hep-th/9601029
22. G.T. Horowitz, A. Strominger, Counting states of near-extremal black holes. Phys. Rev. Lett. **77**, 2368–2371 (1996)
23. T. Jacobson, D. Marolf, C. Rovelli, Black hole entropy: inside or out? (2005). arXiv:hep-th/0501103
24. C. Wuthrich, Quantum gravity from general relativity, in *Routledge Companion to the Philosophy of Physics* (2019)

Chapter 7
UV Effects and Short-Lived Hawking Radiation: Alternative Resolution of Information Paradox

Pei-Ming Ho, Hikaru Kawai, and Wei-Hsiang Shao

Abstract This chapter suggests an alternative solution to the black-hole information paradox by proposing that Hawking radiation ceases around the scrambling time due to trans-Planckian effects inherent in string theory. We consider two toy models in the literature that incorporate stringy effects. The first model utilizes the generalized uncertainty principle, which introduces a minimal length. The second model is inspired by string field theory, where interactions are exponentially suppressed in the UV limit. Both models indicate an early termination of Hawking radiation around the scrambling time, resulting in negligible evaporated energy and a predominantly classical black hole.

7.1 Introduction

The black-hole information paradox [1, 2], arising from Hawking's prediction of black-hole evaporation [3, 4], has been a central challenge in theoretical physics [5]. Regarding the fate of the information carried by the collapsing matter (e.g. the wave functions of all particles in the matter), there are three logical possibilities:

P.-M. Ho (✉) · H. Kawai · W.-H. Shao
Department of Physics and Center for Theoretical Physics, National Taiwan University, Taipei, Taiwan
e-mail: pmho@phys.ntu.edu.tw

H. Kawai
e-mail: hikarukawai@phys.ntu.edu.tw

P.-M. Ho · H. Kawai
Physics Division, National Center for Theoretical Sciences, Taipei, Taiwan

A. Akil and C. Bambi (eds.), *The Black Hole Information Paradox*, Springer Series in Astrophysics and Cosmology, https://doi.org/10.1007/978-981-96-6170-1_7

1. Information is lost from our universe,[1] and the radiation carries little or no information about the collapsing matter.
2. Information is preserved in a small remnant [9], while the radiation carries little or no information.
3. Information is preserved in the radiation, which carries all or almost all information about the collapsing matter and no information is lost.

However, a fourth logical possibility has often been overlooked.

4. *Hawking radiation is turned off at an early stage so that only a tiny fraction of the initial mass is evaporated. Almost all information about the collapsing matter remains trapped within the black hole.*

In this scenario [10–14], the black hole essentially remains classical. Given the negligible quantum effects on macroscopic objects in quantum mechanics, this resolution can be considered the most conservative approach to the information paradox.

The prevailing assumption in the literature has been that Hawking radiation persists until the black hole evaporates to a small fraction of its initial mass. The information paradox is commonly associated with the *Page time* [15, 16] when roughly half of the black hole's mass has evaporated. It has been argued that resolving the paradox requires $\mathcal{O}(1)$ corrections to Hawking radiation [5, 17], indicating nonperturbative quantum gravity effects.[2] If such corrections arise before the Page time, they could significantly modify Hawking radiation to achieve the desired information transfer. On the other hand, while such corrections are yet unknown, they may as well turn off Hawking radiation, and resolve the information paradox as the fourth logical possibility outlined above.

Regardless, as the information paradox is centered around the properties of Hawking radiation, a careful examination of its derivation is essential. While the motivation for this work stems from the information paradox, our primary focus is directed toward achieving a precise description of Hawking radiation, particularly its modifications due to UV physics.

Numerous studies have supported the robustness of Hawking radiation against various hypothetical UV modifications (see, for instance, Refs. [18–30]). These findings provide strong support for the persistence of Hawking radiation beyond the Page time. Nevertheless, it has also been shown that Hawking radiation undergoes $\mathcal{O}(1)$ corrections due to trans-Planckian physics as early as $\mathcal{O}(M \log(M))$ from the onset of radiation for a large black hole with mass M [11–14, 31–33].[3] This time scale $\mathcal{O}(M \log(M))$ was proposed in Ref. [34] to quantify the time it takes for the collapsing matter to "mix" its information content across the horizon. In this chapter,

[1] This includes proposals in which baby universes branch off during the process and preserve the information outside our universe [6, 7]. For example, the black hole might develop a "neck" resembling Wheeler's bag of gold [8]. As the black hole radiates, the neck narrows and eventually pinches off, forming a baby universe.

[2] At low energies, quantum gravity effects can normally be expressed as $1/M_p$ corrections, where M_p is the Planck mass. We need $\mathcal{O}(1)$ effects in this sense.

[3] The possibility that Hawking's derivation may fail after the scrambling time was pointed out in Ref. [35].

we will use the term *scrambling time* to refer to this time scale without assuming its original definition in our discussion. For a large black hole, the scrambling time is significantly shorter than the Page time $\sim \mathcal{O}(M^3)$. As we will demonstrate in this chapter, there are scenarios in which Hawking radiation ceases around the scrambling time, with the radiated energy constituting a negligible fraction of the black hole's mass. This renders Hawking radiation a minor quantum effect that can be effectively ignored.

It is important to note that the series of works [11–14, 31–33] reviewed here did not try to deduce the necessary features of quantum gravity from the assumption that the information paradox has to be resolved. Instead, they focused on identifying overlooked issues in the derivation of Hawking radiation for improvement. The following key issues were uncovered:

1. The background geometry is frequently assumed to be static or stationary, ignoring the collapsing matter. This is only justified in the free field theory as in Hawking's original calculation [4].
2. While renormalizable interactions have been considered [36–40],[4] non-renormalizable interactions have been largely ignored, except that their potential importance was mentioned in Ref. [41].
3. The claimed robustness of Hawking radiation in the literature typically refers only to the Hawking temperature, not the radiation's magnitude. The time dependence of the magnitude of Hawking radiation has been overlooked in most studies, with a few exceptions [10–14, 42–44].

When non-renormalizable interactions between the collapsing matter and the radiation field are considered, the time dependence of Hawking radiation can be significantly affected. Recent studies [11, 31–33] have shown that certain higher-derivative non-renormalizable interactions can lead to exponentially growing contributions to particle creation at large distances, causing the perturbation theory to break down around the scrambling time. This indicates that the low-energy effective theory's prediction of Hawking radiation is unreliable beyond the scrambling time. The origin of this phenomenon lies in the trans-Planckian center-of-mass energy between the collapsing matter and the outgoing Hawking particle after the scrambling time.

As UV physics becomes relevant to Hawking radiation at late times, the information of the collapsing matter could potentially be transmitted into radiation. However, two concrete UV models [13, 14] motivated by string theory and another model [10] motivated by loop quantum gravity have shown that Hawking radiation is terminated around the scrambling time. In another model [12] of modified UV dispersion relations, Hawking radiation can be turned off at any time after the scrambling time, depending on the details of UV physics.

[4] References [36, 37, 39] showed that the corrections to Hawking radiation can be treated perturbatively, but Ref. [38] proposed that ultra-energetic quanta are produced in the gravitational collapse region, while Ref. [40] reported a large one-loop effect on Hawking radiation.

This chapter explores a resolution to the information paradox based on the early termination of Hawking radiation, arguing that trans-Planckian effects inherent in UV-complete theories of quantum gravity lead to its suppression near the scrambling time. Section 7.2 reviews the traditional model of black-hole evaporation, highlighting assumptions often overlooked in the derivation of Hawking radiation. Section 7.3 addresses prevalent arguments against the relevance of UV physics in resolving the information paradox, alongside the difficulties associated with information transfer through Hawking radiation. In Sect. 7.4, the scattering amplitudes between the collapsing matter and the radiation field due to higher-derivative non-renormalizable interactions are calculated, demonstrating their exponential growth over time and the breakdown of the low-energy effective theory around the scrambling time. Section 7.5 explores two UV models of the radiation field that lead to the termination of Hawking radiation around the scrambling time. The first model incorporates the generalized uncertainty principle, while the second mimics the exponential suppression of UV interactions in string field theory. Comments on their implications and comparisons with other proposals addressing the information paradox are provided in Sect. 7.6. Finally, conclusions are presented in Sect. 7.7.

Throughout the chapter, we adopt the following unit system:

$$c = 1, \qquad \hbar = 1, \qquad k_B = 1, \qquad G_N = \ell_p^2 = M_p^{-2}, \tag{7.1.1}$$

where k_B is the Boltzmann constant, and G_N is the Newton constant.

7.2 Traditional Model

This section briefly reviews the formation and evaporation of black holes, drawing upon standard references [45, 46]. This scenario is called a *traditional black hole* in Ref. [5]. In this chapter, we refer to this framework as the *traditional model*. We shall accept all assumptions in the traditional model, except for the properties of Hawking radiation, which will be determined through explicit calculation.

7.2.1 Overview

For simplicity, consider the gravitational collapse of a uniform spherical dust shell with a total mass M and thickness $d \ll a$ (and $d \gg \ell_p$). When the shell enters the horizon at the *Schwarzschild radius*

$$a \equiv 2G_N M, \tag{7.2.1}$$

the average density ρ of the collapsing shell is

$$\rho \simeq \frac{1}{8\pi G_N a d} \propto \frac{1}{a} \tag{7.2.2}$$

since its mass is given by $M \simeq \rho V$, with $V \simeq 4\pi a^2 d$ being its volume. As a result, for sufficiently large black holes (i.e., large a), the density $\rho \propto 1/a$ becomes very small at the moment of horizon crossing. In this chapter, we focus on such a large black hole formed by the collapse of a dilute dust shell, where ρ is small enough for the spacetime curvature to be small everywhere.[5] The traditional model is constructed in such a way that all high-energy events are avoided as much as possible.

We further assume that interactions among the dust particles are negligible so that all dust particles undergo free fall. According to the equivalence principle, particles crossing the horizon will not experience any unusual effects; instead, they will feel like they are floating in empty space. Furthermore, it is assumed in the traditional model that the dust interacts very weakly with the radiation field ϕ, so that the radiation field can stay in an approximate vacuum state.

According to Hawking's calculation [4] (which will be reviewed in Sect. 7.2.3), distant observers start to detect outgoing radiation at the same retarded time u when the radius $R(u)$ of the matter shell is close to the horizon radius (see Fig. 7.1), i.e., when

$$R(u) - a \ll a. \tag{7.2.3}$$

The emitted radiation exhibits the spectrum of thermal radiation characterized by the Hawking temperature

$$T_H = \frac{1}{4\pi a}. \tag{7.2.4}$$

The energy flux in the radiation is thus $\sim\mathcal{O}(1/a^4)$, and the radiated power is $\sim\mathcal{O}(1/a^2)$. This is an extremely slow process. For a black hole of a solar mass, it takes $\sim 10^{67}$ years to evaporate.

As Hawking radiation carries energy away from the black hole, the Schwarzschild radius $a(t)$ gradually decreases over time. Low-energy effective theories apply to the black hole only if the Schwarzschild radius $a(t)$ is much larger than the Planck length ℓ_p:

$$a(t) \equiv 2G_N M(t) \gg \ell_p. \tag{7.2.5}$$

(For a black hole of a solar mass, the Schwarzschild radius is $a_\odot \simeq 3\,\text{km} \simeq 2 \times 10^{38}\ell_p$). We shall always assume the large black hole limit (7.2.5).

In the traditional model, Hawking radiation is derived from the low-energy effective theory in a perturbative expansion in powers of $\ell_p^2/a^2(t)$, or equivalently $1/(a(t)M_p)^2$. The low-energy effective theory includes the quantum field theory for the radiation field propagating on the black-hole background determined by the semi-classical gravity theory. For a sufficiently large black hole, the curvature near

[5] For example, we can assume that ρ is much smaller than the density of air on Earth, allowing the electromagnetic field to be approximately in the vacuum state.

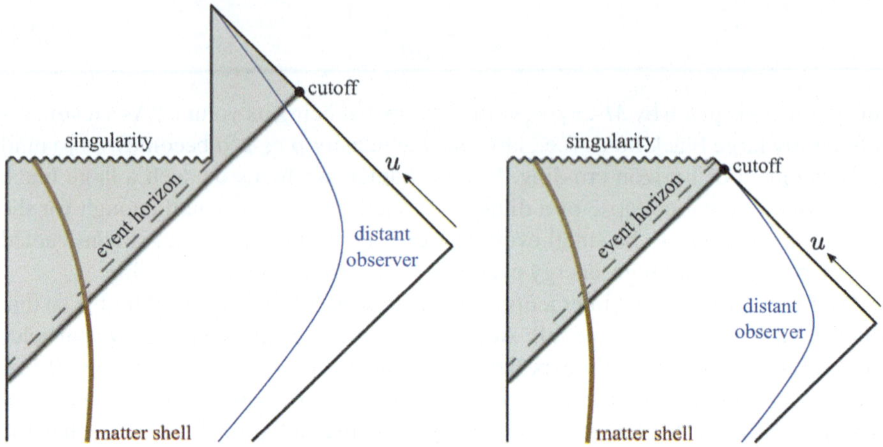

(a) evaporating black hole (b) classical black hole

Fig. 7.1 Panel (**a**) shows the Penrose diagram of a black hole in the traditional model and panel (**b**) that of a classical black hole. The shaded areas refer to the regions beyond a certain cutoff retarded time. In this chapter, we will only be concerned with Hawking radiation before this cutoff time, which will later be identified as the scrambling time. Causally, the region above the cutoff time cannot affect the Hawking radiation below this line. Therefore, the difference between (**a**) and (**b**), as well as the event horizon and the singularity, are all irrelevant in our discussion

its horizon is even smaller than that around the surface of the Earth. There is consequently no reason why low-energy effective theories such as the Standard Model of particle physics and Einstein's theory of general relativity cannot adequately describe the physics near the black-hole horizon and away from the black hole.

In the following, we introduce the geometry of a Schwarzschild black hole formed by the gravitational collapse of a spherical thin shell in Sect. 7.2.2, and review the derivation of Hawking radiation in the traditional model in Sect. 7.2.3.

7.2.2 Black-Hole Geometry

For explicit calculations, we only focus on spherically symmetric configurations for simplicity, but we believe that the underlying physics of Hawking radiation can be extended to a generic black-hole formation process.

The Schwarzschild spacetime is the unique static solution to Einstein's equations for a black hole with spherical symmetry. For the dynamic process of black-hole formation, we now discuss how the Schwarzschild geometry should be modified due to the collapsing matter and the back-reaction of Hawking radiation. We will also describe the geometric properties that will be needed for the calculation of Hawking radiation.

7.2.2.1 Schwarzschild Solution

The Schwarzschild solution of Einstein's equations in vacuum describes a static, spherically symmetric classical black hole:

$$ds^2 = -\left(1 - \frac{a}{r}\right) dt^2 + \frac{dr^2}{(1 - a/r)} + r^2 \left(d\theta^2 + \sin^2\theta \, d\varphi^2\right), \tag{7.2.6}$$

where a is the Schwarzschild radius (7.2.1), and t is the Schwarzschild time. We refer to r as the areal radius because it directly determines the area of the spatial 2-sphere to be $4\pi r^2$.

The line element (7.2.6) can be rewritten as

$$ds^2 = -\left(1 - \frac{a}{r}\right) du \, dv + r^2 \left(d\theta^2 + \sin^2\theta \, d\varphi^2\right) \tag{7.2.7}$$

in terms of the Eddington retarded time u and advanced time v:

$$u \equiv t - r_*, \tag{7.2.8}$$

$$v \equiv t + r_*, \tag{7.2.9}$$

where r_* is the tortoise coordinate defined as

$$r_* \equiv r - a + a \log(r/a - 1). \tag{7.2.10}$$

In terms of the Kruskal light-cone coordinates

$$U \equiv -2a \, e^{-u/2a}, \tag{7.2.11}$$

$$V \equiv 2a \, e^{v/2a}, \tag{7.2.12}$$

the Schwarzschild metric is

$$ds^2 = -\frac{a}{r} e^{-\frac{r-a}{a}} \, dU dV + r^2 \left(d\theta^2 + \sin^2\theta \, d\varphi^2\right). \tag{7.2.13}$$

This metric is regular and smooth at the horizon.

The light-cone coordinates u and v align with the Minkowski light-cone coordinates in the asymptotically flat region ($r \to \infty$), whereas U and V are more naturally employed by freely falling observers. Notably, the retarded coordinates are related by an exponential blue-shift factor

$$\frac{du}{dU} = e^{u/2a} \tag{7.2.14}$$

that goes to infinity at the horizon ($U = 0$ or equivalently $u \to \infty$). This exponential relation (7.2.14) between the retarded times for freely falling observers and distant observers is the single most crucial equation for Hawking radiation.

Since massless outgoing particles move along constant-u (or equivalently, constant-U) curves, it is convenient to express their wave functions in terms of the retarded light-cone coordinates u or U. Consider the wave function of a Hawking particle detected at a large distance ($r \gg a$) between the retarded times u_1 and u_2 ($u_2 > u_1$). In terms of the Kruskal coordinate U, the wave function lies within a range of U given by

$$\Delta U \equiv U_2 - U_1 = 2a\left(e^{-u_1/2a} - e^{-u_2/2a}\right) < 2a\, e^{-u_1/2a}. \tag{7.2.15}$$

To establish a reference time scale, we define the *scrambling time* as

$$u_{scr} \equiv 2a \log\left(a^2/\ell_p^2\right). \tag{7.2.16}$$

For a large black hole ($a \gg \ell_p$), u_{scr} is much shorter than the Page time $\mathcal{O}(a^3/\ell_p^2)$. For instance, a black hole of a solar mass has $u_{scr} \simeq 3\,ms$ and a Page time $\sim 10^{67}$ years. Any wave function detected after half the scrambling time (i.e. $u_1 > u_{scr}/2$) lies within two Planck lengths from the horizon ($U = 0$) in terms of U:

$$\Delta U < |U_1| < 2\ell_p. \tag{7.2.17}$$

The wave function of a Hawking particle detected after the scrambling time u_{scr} is limited to an extremely tiny domain of $\Delta U < 10^{-38}\ell_p$ for a black hole of a solar mass.

7.2.2.2 Near-Horizon Geometry

In the near-horizon region defined by

$$0 < r - a \ll a, \tag{7.2.18}$$

the Schwarzschild metric (7.2.13) can be approximated as

$$ds^2 \simeq -dUdV + r^2(U, V)\left(d\theta^2 + \sin^2\theta\, d\varphi^2\right). \tag{7.2.19}$$

The areal radius $r(U, V)$ is implicitly defined by the relation

$$UV = -4a^2 e^{r_*(r)/a}, \tag{7.2.20}$$

where $r_*(r)$ is the tortoise coordinate (7.2.10).

It is important to note that this analysis focuses on dynamical black holes formed from gravitational collapse, rather than eternal black holes. The Schwarzschild metric applies only to the region outside the collapsing matter. For simplicity, we consider a thin shell collapsing at the speed of light. Generalizing this to finite speeds is straightforward, provided that the instantaneous velocity at the moment of horizon crossing is not fine-tuned to approach zero.

Without loss of generality, we can choose the coordinate system such that the shell crosses the horizon at $v = 0$, or equivalently, $V = 2a$. (The Schwarzschild metric (7.2.6) is invariant under a shift of t, which leads to a shift of v).

Inside the collapsing shell, the metric is Minkowskian and has the same form as Eq. (7.2.19), except that the areal radius $r(U, V)$ is now given by

$$r(U, V) = R(U, V) \equiv \frac{V - U}{2} \tag{7.2.21}$$

up to an additive constant, which can be set to zero by shifting U and V inside the shell.

Across the thin shell at $V = 2a$, the Minkowski light-cone coordinate U inside the shell can be matched with the Kruskal light-cone coordinate U outside the shell in the near-horizon region where $r - a \ll a$. This allows us to use the same symbols U and V throughout, interpreting Eq. (7.2.14) as the blue-shift factor between the Minkowski patches in the infinite past ($V \to -\infty$) and the infinite future ($V \to \infty$) for a massless outgoing wave packet along a fixed U trajectory.

7.2.2.3 Energy-Momentum Tensor and Back-Reaction

The classical Schwarzschild solution assumes a vanishing energy-momentum tensor outside the collapsing shell. A more accurate description requires the *semi-classical Einstein equation*:

$$G_{\mu\nu} \equiv R_{\mu\nu} - \frac{1}{2} g_{\mu\nu} R = 8\pi G_N \langle T_{\mu\nu} \rangle, \tag{7.2.22}$$

where $\langle T_{\mu\nu} \rangle$ is the expectation value of the energy-momentum tensor in the low-energy effective theory for the radiation field ϕ, including the energy flux of the Hawking radiation. The semi-classical Einstein equation (7.2.22) determines how the background geometry is modified by the quantum energy-momentum tensor.

With regard to the information paradox, we assume that the initial state of the radiation field is the Minkowski vacuum in the infinite past. For a matter shell collapsing at the speed of light, a Minkowski vacuum is expected everywhere inside the collapsing shell. Hence, $\langle T_{\mu\nu} \rangle = 0$ inside the shell.

The initial state also determines the quantum state outside the shell. In the traditional model, it is well approximated by the Unruh vacuum [47][6] (see Ref. [48] for a detailed discussion on the validity of this approximation). The Unruh vacuum is defined as the state with zero outgoing energy flux on the shell and zero ingoing energy flux in the infinite past. It is characterized by a finite $\langle T_{\mu\nu}\rangle$ for freely falling observers, as required by the equivalence principle.

In the traditional model, the expectation value $\langle T_{\mu\nu}\rangle$ of the Unruh vacuum is assumed to satisfy the bounds

$$\langle T_{UU}\rangle \sim \left(\frac{du}{dU}\right)^2 \langle T_{uu}\rangle \sim \mathcal{O}(1/a^4), \tag{7.2.23}$$

$$\langle T_{UV}\rangle \sim \left(\frac{du}{dU}\right)^{-1} \langle T_{uv}\rangle \sim \mathcal{O}(1/a^4), \tag{7.2.24}$$

$$\langle T_{VV}\rangle \sim \langle T_{vv}\rangle \sim \mathcal{O}(1/a^4), \tag{7.2.25}$$

$$\langle T_{\theta\theta}\rangle \sim \mathcal{O}(1). \tag{7.2.26}$$

Dictated by the equivalence principle, these bounds have found a lot of supportive evidence [49–57]. As the quantum effect $\langle T_{\mu\nu}\rangle$ is small, one can treat the back-reaction problem perturbatively. A perturbative calculation for a dynamical black hole including the collapsing matter can be found in Ref. [58].

As the back-reaction to the spacetime geometry is very small, one can define an effective time-dependent Schwarzschild radius $a(u)$ such that Eq. (7.2.14) remains approximately correct. The value of $a(u)$ changes very slowly as

$$\left|\frac{da(u)}{du}\right| \sim \mathcal{O}\left(\frac{\ell_p^2}{a^2(u)}\right). \tag{7.2.27}$$

This allows us to treat a as a constant within a time scale much shorter than the Page time:

$$\Delta u \ll \mathcal{O}\left(\frac{a^3(u)}{\ell_p^2}\right). \tag{7.2.28}$$

For the Schwarzschild metric (7.2.7), Eq. (7.2.23) implies

$$\langle T_{uu}\rangle = 0 \quad \text{at} \quad U = 0 \tag{7.2.29}$$

because

$$\frac{du}{dU} \sim \frac{1}{r-a} \tag{7.2.30}$$

[6] For black holes, two other vacua are often considered [46]. The Boulware vacuum, representing an eternal black hole without radiation, has a diverging energy-momentum tensor at the horizon in the freely falling frame. The Hartle-Hawking vacuum is a time-independent state that assumes a balanced ingoing and outgoing energy flux at the Hawking temperature.

in the limit $r \to a$ on the future horizon. From the perspective of a distant observer, Eq. (7.2.29) indicates that the outgoing energy flux $\langle T_{uu} \rangle$ is zero at the future horizon and gradually increases to account for the energy in Hawking radiation at $r - a \gtrsim \mathcal{O}(a)$.

In view of energy conservation, the outgoing energy flux in Hawking radiation is balanced by a negative ingoing energy flux entering the horizon[7] to decrease the black hole's energy.[8] In this sense, the collapsing matter does not lose energy directly to Hawking radiation. Its energy is merely canceled by more and more negative energy inside the black hole. Since the transfer of information is usually associated with the transfer of energy, this suggests that the information in the collapsing matter is not transmitted to the Hawking radiation.

The physical picture behind this energy flow involves a Hawking particle and its partner created in a curved background without directly interacting with the collapsing matter. The Hawking particle carries positive energy away to large distances, while its partner, which carries negative energy, falls into the black hole, effectively reducing its mass.

To illustrate this, we can draw an analogy with the Schwinger effect by substituting the gravitational field and masses with an electric field and charges. Imagine a large positively charged plate that generates a strong electric field, and outgoing positive charges and ingoing negative charges are created in pairs. The background electric field diminishes over time due to the back-reaction resulting from the ingoing negative charges. However, this does not imply that the charges on the plate will gradually disappear, as the ingoing charges are not necessarily the antiparticles of those on the plate. Furthermore, the outgoing charges are not expected to carry any information about the charges on the plate.

7.2.2.4 Apparent Horizon

The time-dependent Schwarzschild radius $a(t)$ does not define the event horizon, which is an outgoing light-like surface extrapolated backward in time from the singularity at $r = 0$. In spacetime configurations with spherical symmetry, the *trapping horizon* is defined by the condition

$$\frac{\partial}{\partial V} r(U, V) = 0 \tag{7.2.31}$$

[7] For observers staying on top of the trapping horizon, the power of the ingoing energy flux is $\mathcal{E} \simeq -1/(2\ell_p^2 a^2)$ [59].

[8] Although the negative ingoing energy flux violates the null energy condition, it is accepted as a quantum effect. If there is no negative ingoing energy flux, and we simply extrapolate the energy flux of Hawking radiation from large distances back to the near-horizon region, the energy flux becomes very large around the horizon due to the blue-shift factor (7.2.14). The KMY model [60–64] is based on this energy-momentum tensor. Similar to the firewall proposal [17, 65], the information paradox is alleviated in the KMY model by a Planckian-scale energy-momentum tenor around the horizon.

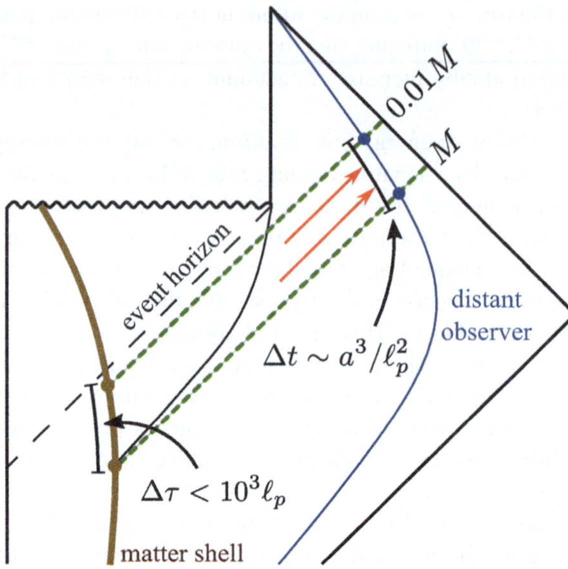

Fig. 7.2 In terms of the Eddington retarded time u, which agrees with the Schwarzschild time t at a fixed radius r, the time for a black hole to evaporate to 1% of its initial mass M is $\mathcal{O}(a^3/\ell_p^2)$. However, in terms of the proper time of a comoving observer, this process occurs over a brief time scale $\sim 1000\ell_p$ [69]

on the areal radius r. Due to the negative ingoing energy flux, the outer trapping horizon is time-like [66–68]. Given the areal radius $r(U, V)$, Eq. (7.2.31) allows us to solve for the trapping horizon $r_{trap}(u)$ as a function of u. The Schwarzschild radius $a(u)$ approximates the (outer) trapping horizon with a difference of $\mathcal{O}(\ell_p^2/a^2)$ [69].

Interestingly, although the time for a black hole to evaporate to a small fraction $1/n$ of its initial mass (assuming $n \ll (a/\ell_p)^{2/3}$) is very long ($\sim \mathcal{O}(a^3/\ell_p^2)$) for distant observers, from the viewpoint of comoving observers in free fall with the collapsing matter, the proper time available to causally deliver information into the Hawking radiation is merely $\mathcal{O}(n^{3/2}\ell_p)$ [69]. For instance, the time it takes for the black hole to evaporate to $1/100$ of its initial mass is, albeit longer than the Page time for a distant observer, as short a time scale as $\sim 1000\ell_p$ for a comoving observer. If one expects 99% of the information contained within the collapsing matter to be transmitted to the Hawking radiation, it has to occur within $1000\ell_p \sim 10^{-40}s$! (see Fig. 7.2).

Therefore, if the information of the collapsing matter has to be transferred to Hawking radiation, there must be nonlocal or acausal physics at work.[9]

[9] Nonetheless, there have been proposals [70, 71] suggesting that the maximal entanglement between Hawking particles and their partners could provide a causal mechanism for information retrieval. This approach relies on postselected quantum teleportation, which requires a special final-state boundary condition to be imposed at the black-hole singularity.

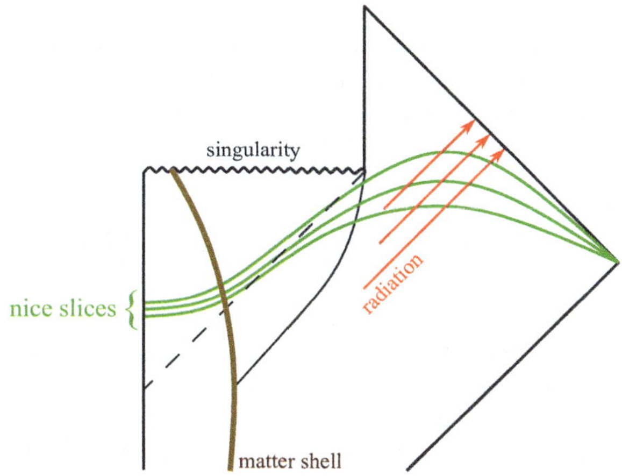

Fig. 7.3 Illustration of nice slices that intersect the infalling matter and both early and late Hawking radiation

7.2.2.5 Nice Slices

Away from the singularity, the spacetime geometry of a large black hole is smooth. It is possible to construct time slices in the Schwarzschild space (see Fig. 7.3) such that the extrinsic and intrinsic curvatures remain small ($\sim\mathcal{O}(1/a^2)$) until the black hole becomes microscopic. These nice slices can extend from the initial time when the radiation field is in the Minkowski vacuum to a late time well past the Page time, so they capture not only the matter falling into the black hole but also the Hawking particles at large distances and their partners inside the black hole [5].

The adiabatic theorem can be applied to the Hamiltonian evolution on these nice slices, concluding that only low-energy states of energies $\sim\mathcal{O}(1/a)$ can be excited from the Minkowski vacuum.

An example of such nice slices can be found in Refs. [72, 73]. In the Kruskal coordinates (U, V), the Schwarzschild black hole has an event horizon at $U = 0$ and a curvature singularity at $UV = 4a^2$. These slices are constructed by joining two space-like segments smoothly at the surface $V = e^{t_{ns}/a}\, U$:

$$\begin{aligned} UV &= a^2 \qquad \text{for} \quad V < e^{t_{ns}/a}\, U, \\ e^{t_{ns}/2a}\, U + e^{-t_{ns}/2a}\, V &= 2a \qquad \text{for} \quad V > e^{t_{ns}/a}\, U. \end{aligned} \tag{7.2.32}$$

In terms of the coordinates (u, v), Eq. (7.2.32) implies that

$$-\, e^{(t_{ns}-t)/2a} + e^{(t-t_{ns})/2a} = e^{-r_*/2a} \simeq 0 \qquad \text{as} \quad r \to \infty\,, \tag{7.2.33}$$

indicating that these slices are asymptotic to surfaces of constant Schwarzschild time, that is, $t = t_{ns}$ in the limit $r \to \infty$.

Combining the idea of nice slices with the equivalence principle and the adiabatic theorem, the "nice-slice" argument [72, 73] (which will be summarized in

Sect. 7.3.1.2) establishes that freely falling observers experience an "uneventful" horizon, with the fields being in an approximate vacuum state. As such a state cannot carry significant information about the collapsing matter, Hawking radiation originating from it is also limited in its information content.

7.2.3 Derivation of Hawking Radiation

The distinction between positive and negative frequency modes in a quantum field theory defines the vacuum state. Due to the blue-shift factor (7.2.14), the Minkowski vacuum in the infinite past (where U is used to define frequencies) differs from the Minkowski vacuum in the asymptotically-flat region (where u is used). Hawking radiation is just the interpretation of the Minkowski vacuum in the infinite past by distant observers. The exponential relation (7.2.11) between the Kruskal retarded time U and the Eddington retarded time u implies that the Minkowski vacuum defined inside the collapsing shell (in terms of U) appears as an excited state with Hawking particles when viewed from large distances (in terms of u).

A derivation of this phenomenon is briefly reviewed in the following discussion, while detailed and elaborated derivations are presented in Chap. 1 "Deriving the Paradox: Original Derivation of Hawking Radiation" and Chap. 2 "Alternative derivations of Hawking Radiation" of this book.

7.2.3.1 Wave Equation

In principle, Hawking radiation includes all fields. However, since the Hawking temperature (7.2.4) is very low for a large black hole, the dominant radiation field is the electromagnetic field, as it is massless and allows for a free-field approximation. For simplicity, we consider below a massless scalar field ϕ whose Hawking radiation is qualitatively the same as that of the electromagnetic field.

For a static, spherically symmetric background with the line element

$$ds^2 = -f(r)\,du\,dv + r^2(d\theta^2 + \sin^2\theta\,d\varphi^2), \tag{7.2.34}$$

the Laplace equation $\nabla^2\phi = 0$ leads to

$$\partial_u\partial_v\psi - \frac{\partial_u\partial_v r}{r}\psi = 0 \tag{7.2.35}$$

for an s-wave solution

$$\phi(u, v) \equiv \frac{\psi(u, v)}{r}. \tag{7.2.36}$$

Equation (7.2.35) would be identical to the 2D wave equation for a massless scalar if the second term $(\partial_u \partial_v r)/r$ can be neglected.

Using the Schwarzschild metric (7.2.6) where $f(r) = 1 - a/r$, we find

$$\frac{\partial_u \partial_v r}{r} = -\frac{1}{4}\frac{a}{r^3}\left(1 - \frac{a}{r}\right). \tag{7.2.37}$$

This term is small both in the near-horizon region ($r - a \ll a$) and at large distances ($r \gg a$). In the intermediate region, it behaves as an effective potential barrier that leads to reflection and transmission. However, its effect will not be important for an order-of-magnitude estimate of Hawking radiation. Thus, we can approximate the s-wave solution as given by Eq. (7.2.36) with

$$\psi(u, v) = \psi_{in}(v) + \psi_{out}(u). \tag{7.2.38}$$

For the purpose of deriving Hawking radiation, only the outgoing sector $\psi_{out}(u)$ of the field is needed.

7.2.3.2 Uncertainty Relations

Before presenting the derivation of Hawking radiation, let us first explain the underlying physics using the uncertainty relation.

Due to the exponential relation (7.2.11), the conjugate momenta of u and U satisfy the commutation relation [33]

$$[P_u\,,\, P_U] = [-i\partial_u\,,\, -ie^{u/2a}\partial_u] = -\frac{i}{2a}P_U, \tag{7.2.39}$$

which leads to the uncertainty relation

$$\Delta P_u\, \Delta P_U \geq \frac{1}{4a}\langle P_U \rangle. \tag{7.2.40}$$

For a typical Hawking particle, we expect $P_u \sim \mathcal{O}(1/a)$ and $\Delta P_u \ll P_u$. Equation (7.2.40) then implies

$$\Delta P_U \geq \frac{1}{4a\,\Delta P_u}\langle P_U \rangle \gg \langle P_U \rangle. \tag{7.2.41}$$

This indicates that the deviation of P_U is significantly larger than its central value $\langle P_U \rangle$, suggesting that both positive and negative-P_U modes are present.

Consequently, a Hawking particle composed of purely positive-P_u states turns out to be a superposition of both positive and negative-P_U states. The mixing of positive and negative energy modes for a given outgoing wave with $P_u > 0$ implies a non-trivial Bogoliubov transformation between the creation and annihilation operator pairs defined in the two coordinate systems, as will be shown explicitly below.

7.2.3.3 Bogoliubov Transformation

Now we derive the Bogoliubov transformation between the ladder operators $(a_\Omega, a_\Omega^\dagger)$ for freely falling observers and those $(b_\omega, b_\omega^\dagger)$ for distant observers. The derivation here has the advantage that it does not rely on the existence of the event horizon.

Assuming a free-field approximation for the propagation of the quantum state, Hawking particles can be detected in the near-horizon region as wave functions with positive frequencies with respect to u (and thus a mixture of positive and negative frequencies with respect to U), similar to their detection at large distances. Theoretically, the detection of a quantum state ψ can be equivalently carried out at an earlier time for the state Ψ if they are related via a unitary time evolution.

For a large black hole, the spacetime curvature is small, and the general s-wave solution for the massless scalar field ϕ in the near-horizon region is approximately

$$\phi(x) \simeq \int_0^\infty \frac{d\omega}{4\pi\sqrt{\omega}\, r}\left(b_\omega\, e^{-i\omega u} + b_\omega^\dagger\, e^{i\omega u} + \tilde{b}_\omega\, e^{-i\omega v} + \tilde{b}_\omega^\dagger\, e^{i\omega v}\right), \tag{7.2.42}$$

or equivalently,

$$\phi(x) \simeq \int_0^\infty \frac{d\Omega}{4\pi\sqrt{\Omega}\, r}\left(a_\Omega\, e^{-i\Omega U} + a_\Omega^\dagger\, e^{i\Omega U} + \tilde{a}_\Omega\, e^{-i\Omega V} + \tilde{a}_\Omega^\dagger\, e^{i\Omega V}\right), \tag{7.2.43}$$

depending on which coordinate system is used in the Fourier expansion. We use ω and Ω as symbols for the frequencies defined in terms of u and U, respectively. In quantum mechanics, they are identified with the conjugate momenta P_u and P_U.

Upon quantization, we have the canonical commutation relations

$$[b_\omega\,, b_{\omega'}^\dagger] = \delta(\omega - \omega'), \qquad [b_\omega\,, b_{\omega'}] = 0 = [b_\omega^\dagger\,, b_{\omega'}^\dagger]\,, \tag{7.2.44}$$

$$[a_\Omega\,, a_{\Omega'}^\dagger] = \delta(\Omega - \Omega')\,, \qquad [a_\Omega\,, a_{\Omega'}] = 0 = [a_\Omega^\dagger\,, a_{\Omega'}^\dagger], \tag{7.2.45}$$

and similarly for the ladder operators associated to the ingoing modes. The Unruh vacuum is defined by

$$a_\Omega|0\rangle = 0 \qquad \forall\, \Omega > 0. \tag{7.2.46}$$

That is, there are no outgoing particles in the near-horizon region for freely falling observers.

Identifying the field ϕ in Eqs. (7.2.42) and (7.2.43) in the near-horizon region where the areal radius r is approximately a constant, one finds the Bogoliubov transformation

$$b_\omega = \int_0^\infty d\Omega \left(\alpha_{\omega\Omega}\, a_\Omega + \beta_{\omega\Omega}\, a_\Omega^\dagger\right), \qquad b_\omega^\dagger = \int_0^\infty d\Omega \left(\alpha_{\omega\Omega}^*\, a_\Omega^\dagger + \beta_{\omega\Omega}^*\, a_\Omega\right), \tag{7.2.47}$$

where the Bogoliubov coefficients are [4]

$$\alpha_{\omega\Omega} \equiv \frac{1}{2\pi}\sqrt{\frac{\omega}{\Omega}} \int_{-\infty}^{\infty} du\, e^{i\omega u - i\Omega U(u)} \simeq \frac{a}{\pi}\sqrt{\frac{\omega}{\Omega}}\, (2a\Omega)^{i2a\omega}\, e^{\pi a\omega}\, \Gamma(-i2a\omega)\,, \tag{7.2.48}$$

$$\beta_{\omega\Omega} \equiv \frac{1}{2\pi}\sqrt{\frac{\omega}{\Omega}} \int_{-\infty}^{\infty} du\, e^{i\omega u + i\Omega U(u)} \simeq \frac{a}{\pi}\sqrt{\frac{\omega}{\Omega}}\, (2a\Omega)^{i2a\omega}\, e^{-\pi a\omega}\, \Gamma(-i2a\omega)\,, \tag{7.2.49}$$

with $U(u)$ given in Eq. (7.2.11).

7.2.3.4 Hawking Radiation in Low-Energy Effective Theory

To discuss the time dependence of Hawking radiation, consider the detection of a Hawking particle in a given wave packet

$$\psi(u) \equiv \int_0^\infty \frac{d\omega}{\sqrt{\omega}}\, f_{\omega_0}(\omega)\, e^{-i\omega(u-u_0)}\,, \tag{7.2.50}$$

which is centered around the retarded time $u = u_0$. The annihilation operator for this particle state is

$$b_\psi \equiv i \int du\, \psi^*(x)\, \overset{\leftrightarrow}{\partial}_u\, (r\phi(x)) = \int_0^\infty d\omega\, f_{\omega_0}^*(\omega)\, e^{-i\omega u_0}\, b_\omega\,, \tag{7.2.51}$$

and the corresponding creation operator is

$$b_\psi^\dagger \equiv \int_0^\infty d\omega\, f_{\omega_0}(\omega)\, e^{i\omega u_0}\, b_\omega^\dagger\,. \tag{7.2.52}$$

The profile function $f_{\omega_0}(\omega)$ is normalized by

$$\int_0^\infty d\omega\, |f_{\omega_0}(\omega)|^2 = 1\,, \tag{7.2.53}$$

so that

$$[b_\psi\,, b_\psi^\dagger] = 1\,. \tag{7.2.54}$$

To confirm the thermal spectrum of Hawking radiation, we need the profile function $f_{\omega_0}(\omega)$ to have a narrow width $\Delta\omega \ll \omega_0 \sim 1/a$, where ω_0 is the central frequency of the wave packet. As a result, we can approximate

$$\int_0^\infty d\omega\, f_{\omega_0}(\omega)\, h(\omega) \simeq \int_0^\infty d\omega\, f_{\omega_0}(\omega)\, h(\omega_0) \tag{7.2.55}$$

for a generic smooth function $h(\omega)$ with a width $\gg \Delta\omega$.

The number operator for particles in the state ψ is

$$n_\psi \equiv b_\psi^\dagger b_\psi. \tag{7.2.56}$$

Using Eqs. (7.2.47) and (7.2.52), the expectation value of n_ψ in the Unruh vacuum can be obtained as

$$\begin{aligned}\langle 0|n_\psi|0\rangle &= \int_0^\infty d\omega \int_0^\infty d\omega'\, f_{\omega_0}(\omega) f^*_{\omega_0}(\omega') e^{i(\omega-\omega')u_0} \int_0^\infty d\Omega\, \beta^*_{\omega\Omega}\, \beta_{\omega'\Omega} \\ &\simeq \int_0^\infty d\omega \int_0^\infty d\omega'\, f_{\omega_0}(\omega) f^*_{\omega_0}(\omega')\, e^{i(\omega-\omega')u_0} \frac{a^2}{\pi^2}\, \omega_0\, e^{-2\pi a\omega_0}\, |\Gamma(i2a\omega_0)|^2 \int_0^\infty \frac{d\Omega}{\Omega} (2a\Omega)^{-i2a(\omega-\omega')} \\ &\simeq \frac{1}{e^{4\pi a\omega_0}-1} \int_0^\infty d\omega\, \left|f_{\omega_0}(\omega)\right|^2 = \frac{1}{e^{4\pi a\omega_0}-1},\end{aligned} \tag{7.2.57}$$

which is precisely the Planck distribution at the Hawking temperature (7.2.4). We have thus completed the derivation of Hawking radiation in the low-energy effective theory.

7.2.3.5 UV Cutoff

We show in this section that the Hawking radiation is turned off around the scrambling time u_{scr} (7.2.16) if there is a UV cutoff at the Planck scale [10, 11].

Truncating the lightcone momentum at $\Lambda_\Omega = \ell_p^{-1}$ leads to a modification of Eq. (7.2.57) as

$$\begin{aligned}\langle 0|n_\psi|0\rangle &\simeq \int_0^\infty d\omega \int_0^\infty d\omega'\, f_{\omega_0}(\omega) f^*_{\omega_0}(\omega')\, e^{i(\omega-\omega')u_0} \frac{a^2}{\pi^2}\, \omega_0\, e^{-2\pi a\omega_0}\, |\Gamma(i2a\omega_0)|^2 \int_0^{\ell_p^{-1}} \frac{d\Omega}{\Omega} (2a\Omega)^{-i2a(\omega-\omega')} \\ &\simeq \frac{1}{2\pi} \frac{\omega_0}{e^{4\pi a\omega_0}-1} \int_{-\infty}^{u_{scr}/2} du\, \left|\psi(u)\right|^2.\end{aligned} \tag{7.2.58}$$

Note that the scrambling time u_{scr} (7.2.16) appears in the upper bound of the u-integration as a consequence of applying the change of variable $\Omega \mapsto u = 2a \log(a\Omega)$ to the Ω-integration.

We observe that the number expectation value of Hawking particles in the state ψ is characterized by a Planck distribution at the same Hawking temperature (7.2.4) but with an amplitude that depends on u_0 through the wave function $\psi(u)$ (7.2.50). For a wave function $\psi(u)$ that is localized in a region $|u - u_0| \lesssim \mathcal{O}(a)$ centered around $u = u_0$, the u-integral in Eq. (7.2.58) essentially vanishes if $u_{scr}/2$ is much smaller than u_0. Therefore, there is a large suppression of the magnitude of Hawking radiation after the scrambling time:[10]

$$\langle 0|n_\psi|0\rangle \to 0 \quad \text{for} \quad u_0 - u_{scr}/2 \gg a\,. \tag{7.2.59}$$

From this result, it becomes clear that Hawking radiation after the scrambling time stems from trans-Planckian modes with respect to the Kruskal retarded time U. Different UV models may in principle predict different behaviors of Hawking radiation at late times. This is a manifestation of the trans-Planckian problem of Hawking radiation [74, 75].

In theories with an absolute minimal length scale (such as a lattice theory), Lorentz symmetry only emerges in the low-energy limit, and the minimal length implies a UV cutoff on both the frequency and momentum of quantum modes. The derivation here implies that Hawking radiation ceases around the scrambling time in all such theories.

The discussion here can be generalized to a class of theories that admit modified canonical commutation relations of the form [11]

$$[a_\Omega\,, a^\dagger_{\Omega'}] = g(\Omega)\,\delta(\Omega - \Omega')\,, \tag{7.2.60}$$

$$[a_\Omega\,, a_{\Omega'}] = [a^\dagger_\Omega\,, a^\dagger_{\Omega'}] = 0\,, \tag{7.2.61}$$

where $g(\Omega)$ characterizes the deformation in a given theory. The model with a UV cutoff $\Omega \leq \Lambda_\Omega$ considered above corresponds to the specific case where $g(\Omega)$ is given by the step function $\Theta(\Lambda_\Omega - \Omega)$.

In fact, as long as one assumes the Unruh vacuum (7.2.46) as well as the Bogoliubov transformation (7.2.47)–(7.2.49), it can be shown (see Appendix C in Ref. [11]) that

$$\langle 0|n_\psi|0\rangle \simeq \frac{1}{2\pi}\frac{\omega_0}{e^{4\pi a\omega_0} - 1}\langle 0|[b_\psi\,, b^\dagger_\psi]|0\rangle\,. \tag{7.2.62}$$

In other words, the standard Hawking radiation persists as long as a particle with the wave function $\psi(u)$ can be defined, meaning there exists a corresponding pair of creation and annihilation operators that satisfy the canonical commutation relation (7.2.54). Conversely, if Hawking radiation is absent, it is simply because no such particles can be defined.

This finding demonstrates the robustness of Hawking radiation: Hawking particles are not detected only if such particles cannot be defined. If a black hole ceases to emit Hawking radiation after the scrambling time, it would imply that a distant observer should also be unable to create a particle with frequency $\sim\mathcal{O}(1/a)$ in the same direction.

It is worth noting that although previous studies [18, 19, 21, 23, 24, 26] on Hawking radiation using subluminal UV dispersion relations effectively introduced

---

[10] For a derivation of this outcome in 2D conformal field theory, see Appendix B in Ref. [11].

a UV cutoff on the frequency in the freely falling frame, these studies did not report significant deviations from conventional Hawking radiation. The key distinction is that the modified dispersion relations considered in these analyses did not exclude the presence of degrees of freedom with trans-Planckian momenta.

7.3 Information Paradox

The information paradox has different meanings at different times for different people. A simple way to phrase the question is this—"*Is the evaporation of a black hole analogous to burning a piece of coal?*" In principle, a low-energy effective theory can describe the transfer of information from a piece of coal to radiation with some residual information remaining in the ashes (the remnant) without violating unitarity, provided that the temperature stays well below the theory's cutoff scale but high enough to tear down the structure of the coal and turn it into radiation or ashes. However, this picture does not apply to the traditional model of black holes. The key difference is the absence of a high enough temperature to tear down the collapsing matter into radiation and ashes, in contrast to the hot burning process of coal.

But what is wrong with keeping the information inside the black hole? A notable problem is the contradiction with the Bekenstein-Hawking entropy [76, 77]

$$S_{\text{BH}} = \frac{\text{horizon area}}{4G_N}, \tag{7.3.1}$$

which is believed to give the thermodynamic entropy of a black hole. As the black hole radiates, the entanglement entropy of the black hole increases over time. On the other hand, the Bekenstein-Hawking entropy (as a bound on the total entropy of the black hole) decreases with time. There is a conflict at the Page time (see Fig. 7.4) when the entanglement entropy exceeds the Bekenstein-Hawking entropy.

The Bekenstein-Hawking entropy and the holographic principle [78–81] are supported by extensive evidence (e.g. [80–84]). It leads to a sharp conflict with Hawking radiation in the low-energy effective theory after the Page time. An explanation of the paradox including recent developments can be found in other chapters in this book as well as Ref. [85].

7.3.1 Is UV Physics Relevant?

There would be no immediate paradox if we did not believe in the low-energy effective-theoretic derivation of Hawking radiation. Here we recall some of the arguments for the folklore that Hawking radiation can be reliably calculated in a low-energy effective theory. We will also explain the reasons why these arguments are in fact invalid.

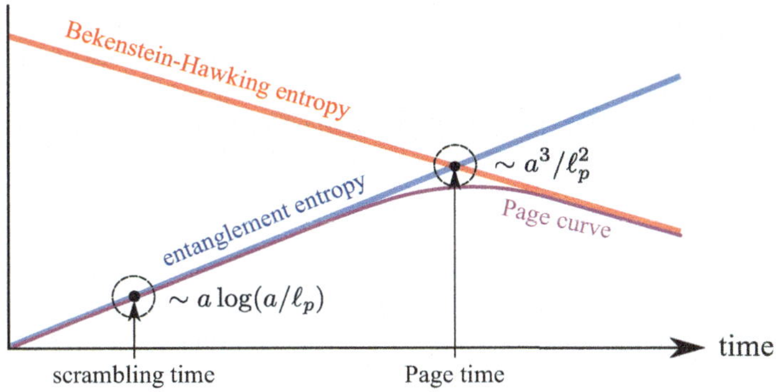

Fig. 7.4 Since the Unruh vacuum is a pure state, Hawking particles outside the horizon are entangled with their partners inside the horizon. Hence, the entanglement entropy of the black hole (blue curve) increases over time. On the other hand, the Bekenstein-Hawking entropy (red curve) decreases as the black hole evaporates. There is a conflict at the Page time when the entanglement entropy of the black hole becomes larger than its thermodynamic entropy. If Hawking radiation is analogous to the radiation from burning a piece of coal, the entropy of the black hole would follow the Page curve [15, 16]

7.3.1.1 Local Lorentz Symmetry

As the Hawking temperature is given by Eq. (7.2.4), a typical Hawking particle is at a low frequency $\omega \sim \mathcal{O}(1/a)$ for a distant observer. But for Hawking particles detected at late times, this frequency can be extremely large from the viewpoint of a freely falling observer due to the infinite blue shift near the horizon.[11] Naively, this suggests that a UV theory might be required to describe the origins of Hawking radiation, particularly the radiation quanta that are emitted at late times.

However, frequency is not a local Lorentz invariant. Any frequency can be made arbitrarily high or low through a local Lorentz boost. This is why we can still use the Standard Model to describe ultra-high-energy cosmic rays ($\sim 10^6\ TeV$) when they collide with particles on Earth. A UV theory is needed only when Lorentz invariants—such as the center-of-mass energy of a collision—reach trans-Planckian scales. As a result, this so-called "trans-Planckian problem" of Hawking radiation is not always taken seriously.

Existence of trans-Planckian Lorentz-invariants

Nevertheless, if one considers the scattering process between an outgoing Hawking particle and a dust particle in the collapsing matter (or the back-reacted geometry due to the quantum energy-momentum tensor—see Sect. 7.4.2.2), the center-of-mass

[11] Even though Hawking radiation is said to emerge at a place away from the horizon where $r - a \sim \mathcal{O}(a)$, theoretically, unitarity allows us to consider the inverse time evolution of a Hawking particle backward in time to a point very close to the horizon.

energy can be extremely large. When the center-of-mass energy exceeds the Planck energy, the effective theory is expected to break down on predictions about Hawking radiation [11, 31–33].

More explicitly, a particle in the infalling matter must have an ingoing momentum

$$P_V \gtrsim \mathcal{O}(1/a) \tag{7.3.2}$$

to fall into the black hole, since it cannot fit inside the horizon if its wavelength is much longer than the Schwarzschild radius. At the same time, a Hawking particle with the characteristic frequency $\sim \mathcal{O}(1/a)$ has an outgoing momentum

$$P_U \sim \mathcal{O}\left(\frac{1}{a}\, e^{u/2a}\right) \tag{7.3.3}$$

in terms of the Kruskal coordinate U due to the blue shift (7.2.14). Thus, the center-of-mass energy between the infalling matter and a Hawking particle emitted around the retarded time u_0 can be estimated as

$$s^2 \sim P_V P_U \sim \frac{1}{a^2}\, e^{u_0/2a}\,. \tag{7.3.4}$$

This Lorentz-invariant quantity becomes trans-Planckian when

$$u_0 \geq u_{scr}\,, \tag{7.3.5}$$

where the scrambling time u_{scr} is given in Eq. (7.2.16). We will see in Sect. 7.4 that this is precisely the condition for low-energy effective theories to break down regarding their predictions of Hawking radiation.

7.3.1.2 Nice-Slice Argument

The *nice-slice argument* [72, 73] is often quoted as the reason why UV physics should be irrelevant to Hawking radiation. The essence of the argument is that, since all intrinsic and extrinsic curvatures are of order $\mathcal{O}(1/a^2)$ or less, in the Hamiltonian formalism, the adiabatic theorem ensures that only degrees of freedom with low energies $\sim\mathcal{O}(1/a)$ can be excited from the ground state. Therefore, assuming that the initial state is the Minkowski vacuum, throughout the spacetime region where nice-slices can be defined, there would be no high-energy excitations (see Sect. 7.2.2.5 for an explicit construction of nice slices in the Schwarzschild spacetime).

The use of low-energy effective theories is thus justified to describe the black-hole geometry. For instance, if one adds higher-derivative corrections to the Einstein-Hilbert action, it will only lead to perturbative corrections to the energy-momentum tensor. Hence, $\mathcal{O}(1)$ corrections are not expected.

There is thus the widely held belief that low-energy effective theories accurately describe Hawking radiation until the black hole's characteristic energy scale $1/a(t)$ becomes too high for them to apply. This suggests that an effective theory valid up to $1\,TeV$ would suffice to describe the decay of a supermassive black hole of 10^{10} solar masses until its Schwarzschild radius shrinks to approximately $a \sim 1\,TeV^{-1} \sim 10^{-19}$ m, which is a tiny fraction $\sim 10^{-22}$ of its initial value.

Loopholes in the nice-slice argument

Consider the time evolution of the Minkowski vacuum in the Minkowski spacetime in an effective theory. There are nice slices that foliate the full Minkowski space, and the effective-theory prediction that the Minkowski vacuum persists under time evolution is reliable. But for any Planck-scale measurement, say, the detection of particles in a region smaller than the Planck length, the prediction of the effective theory is still unreliable.

Therefore, the question is whether the detection of a Hawking particle is a low-energy event. If it is a high-energy event, the nice-slice argument does not justify its prediction by a low-energy effective theory. From the viewpoint of a distant observer, the energy of the Hawking particle is $\sim \mathcal{O}(1/a)$, and the use of effective theories seems to be justified. On the other hand, the detection of a Hawking particle can be viewed as a "delayed measurement"[12] of a tiny region of the vacuum state in the near-horizon region. We can extrapolate the wave packet of a detected Hawking particle backward in time to the near-horizon region, and ask whether the detection of a particle with such a wave packet can be reliably described in the low-energy effective theory in the near-horizon region.

For a Hawking particle detected at a sufficiently late time ($u \geq u_{scr}$), although its Kruskal frequency Ω is not Lorentz invariant, it is associated with a trans-Planckian Lorentz-invariant, as was shown above in Eq. (7.3.4). Thus, while the nice-slice argument is valid for claiming that the Unruh vacuum in the low-energy effective theory remains approximately the same state without becoming highly excited, we still require UV theories for the derivation of Hawking radiation.

It was proposed in Ref. [87] that, even when all intrinsic and extrinsic curvatures are small, quantum gravity effects become important when the hypersurfaces of a foliation are excessively stretched. The trans-Planckian Lorentz-invariant quantity (7.3.4) discussed here can be understood as a physical explanation of this criterion for the relevance of quantum gravity effects.

---

[12] Here we make the analogy with Wheeler's notion of "delayed choice" of the double-slit experiment [86].

7.3.2 *Difficulty in Information Transfer*

The derivation of Hawking radiation in Sect. 7.2.3 suggests that there is no information transfer between Hawking radiation and the collapsing matter in the low-energy effective theory. But even if we consider a complicated UV theory for Hawking radiation, there is a list of reasons why it is still hard for Hawking radiation to carry a substantial portion of the information in the collapsing matter. We recount some of the well-known reasons here.

7.3.2.1 No-Cloning Theorem

As long as the equivalence principle remains valid, the infalling matter should be left undisturbed as it falls across the horizon. Subsequently, the no-cloning theorem [88, 89] does not allow its information to be copied to another quantum system.

The idea of black hole complementarity [90] proposed that the cloning of information is allowed if it cannot be verified by a single observer [91]. However, the compatibility of black hole complementarity with both the equivalence principle and the unitarity of the evaporation process remains an open question [17, 92].

7.3.2.2 Firewall Versus Nonlocality

Apart from the problem mentioned in Sect. 7.2.2.4 that all the information needs to be transmitted within an extremely short period of time [69, 93], there is also a problem with how information can be retrieved from the collapsing matter.

Normally, to probe information at a given length scale L, we need a wave packet of a shorter wavelength, and thus a momentum larger than $1/L$. Hence, to retrieve the information about the wave functions of the quarks in a nucleus, for example, we need a probe at the energy scale of GeV or larger. This background of probes at high energies implies a firewall that burns down all ingoing particles at the horizon. The evaporation is then similar to the burning of a piece of coal, with both the equivalence principle and the adiabatic principle being violated.

Alternatively, information transfer may occur through a nonlocal mechanism [94, 95]. The $ER = EPR$ proposal [96] and the mechanism underlying the entanglement islands [97–102] fall into this category. Naively, these approaches seem to have the advantage of circumventing the AMPS firewall [17]. However, it has been argued in Ref. [103] that gauge/gravity duality might still imply a firewall for large AdS black holes—one that cannot be evaded by merely identifying degrees of freedom inside the black hole with those outside.

In any case, if nonlocal physics is important for the discussion of information in Hawking radiation, why should we believe the prediction of Hawking radiation based on a (local) low-energy effective theory in the first place?

7.4 Breakdown of Low-Energy Effective Theory

In this section, we examine the validity of the low-energy effective theory in describing the black-hole radiation process [31–33]. We shall compute the transition amplitudes of particle creation on top of Hawking radiation due to higher-derivative non-renormalizable interactions during gravitational collapse. We show that these amplitudes increase exponentially over time, becoming significant when the collapsing matter is roughly at the order of a Planck length from the horizon. This exponential growth indicates that the effective-theory description of Hawking radiation breaks down at the scrambling time, necessitating Planckian physics to understand black-hole evaporation [31–33].

7.4.1 Particle Creation by Higher-Derivative Interactions

The original derivation of Hawking radiation can be understood as the transition amplitude of particle creation at the free-field level. Later, it was shown that the corrections due to renormalizable interactions only modify Hawking radiation perturbatively [36, 37, 39]. However, non-renormalizable interactions have not been considered until more recently in Refs. [31–33]. It turns out that higher-derivative non-renormalizable interactions between the collapsing matter and the radiation field induce particle creation that overlaps with the spectrum of Hawking radiation [31–33].

The particle creation due to higher-derivative non-renormalizable interactions is suppressed by certain negative powers of the Planck mass, but its amplitude grows exponentially with time for particles with a given frequency $\sim\mathcal{O}(1/a)$. After the scrambling time, this effect dominates over ordinary Hawking radiation, and the contributions from interactions containing derivatives of higher orders are even larger [31–33]. As a result, the perturbation theory breaks down, and the rate of particle creation can only be reliably derived from the UV theory.

With the particle creation viewed as a scattering process between the radiation field initially in the Minkowski vacuum and the collapsing matter, there is a Lorentz-invariant center-of-mass energy (7.3.4) that grows exponentially with time, as we explained in Sect. 7.3.1.1. This is why the contributions of higher-derivative interactions grow exponentially over time and start to dominate at the scrambling time, when the collapsing matter is approximately $\mathcal{O}(\ell_p)$ away from the horizon [31–33]. The remainder of this section shows this calculation explicitly.

7.4.1.1 Higher-Derivative Non-renormalizable Interactions

In an effective theory, the action must include all interactions compatible with symmetries (see Sect. 12.3 in Ref. [104]). As examples, we consider the following interaction terms [31–33]

$$S_n \equiv \int d^4x \sqrt{-g}\, \frac{\lambda_n}{M_p^{4n-2}}\, R^{\mu_1\nu_1} \cdots R^{\mu_n\nu_n} \left(\nabla_{\mu_1} \cdots \nabla_{\mu_n}\phi\right) \left(\nabla_{\nu_1} \cdots \nabla_{\nu_n}\phi\right), \quad (7.4.1)$$

$$S_{mn} \equiv \int d^4x \sqrt{-g}\, \frac{g_n}{M_p^{2(m+n)}}\, g^{\mu_1\nu_1} \cdots g^{\mu_m\nu_m} g^{\lambda_1\rho_1} \cdots g^{\lambda_n\rho_n} \times$$
$$\times \left(\nabla_{\mu_1} \cdots \nabla_{\mu_m}\psi\right) \left(\nabla_{\lambda_1} \cdots \nabla_{\lambda_n}\psi\right) \left(\nabla_{\nu_1} \cdots \nabla_{\nu_m}\phi\right) \left(\nabla_{\rho_1} \cdots \nabla_{\rho_n}\phi\right), \quad (7.4.2)$$

where λ_n and g_n are dimensionless coupling constants, $R^{\mu\nu}$ the Ricci curvature, ϕ the massless scalar field of Hawking radiation, and ψ represents the field of the collapsing matter.

We note the following important features of these higher-derivative non-renormalizable interactions:

1. In the absence of the collapsing matter, both interaction terms are irrelevant (Recall that the Ricci tensor vanishes for the Schwarzschild solution[13]).
2. It is possible to create ϕ-particles through these interactions.
3. As non-renormalizable interactions, they are suppressed by negative powers of the Planck mass M_p.
4. S_n and S_{mn} lead to stronger interactions for higher energy modes and larger n, m (higher-order derivatives).

When a distant observer detects a particle in the radiation field ϕ at a given frequency $\omega \sim \mathcal{O}(1/a)$, it is equivalent to detecting a wave packet with a blue-shifted central frequency

$$\Omega \sim \omega\, e^{u/2a} \quad (7.4.3)$$

with respect to the Kruskal retarded time U. At a later time (larger u), wave packets with larger Ω would have stronger couplings with the background (Ricci tensor or matter field ψ) through the interaction terms S_n (7.4.1) and S_{mn} (7.4.2).

This expectation can be explicitly verified via a standard S-matrix calculation of the transition amplitude from the initial state of the Unruh vacuum to the final state of multiple particles at large distances [31–33]. In the ensuing discussion, we shall focus specifically on the effects of the interaction term S_n (7.4.1) for simplicity.

Let us consider the final state obtained by adding two additional particles on top of the original Hawking radiation:

$$|f\rangle \equiv b^{\dagger}_{\psi_1} b^{\dagger}_{\psi_2} |0\rangle\,, \quad (7.4.4)$$

where the creation operator corresponding to a wave packet has been defined in Eq. (7.2.52), and the vacuum $|0\rangle$ here refers to the Unruh vacuum (7.2.46). The wave packets ψ_1 and ψ_2 are assumed to have central frequencies $\omega_1 \sim \omega_2 \sim \mathcal{O}(1/a)$ in the same range as ordinary Hawking particles, and these outgoing wave packets are centered around a given retarded time u_0.

[13] At the next order in the ℓ_p^2/a^2 expansion, the back-reaction of the quantum energy-momentum operator introduces a non-zero Ricci tensor outside the shell and leads to qualitatively the same result through the interaction (7.4.1). See Sect. 7.4.2.2.

In the following, as we calculate the amplitude

$$\mathcal{M}_n \equiv i \langle f|S_n|0\rangle \tag{7.4.5}$$

of the transition from the Unruh vacuum to the final state $|f\rangle$ (7.4.4), we shall pay extra attention to the time dependence of this particle-creation mechanism.

7.4.1.2 Order-of-Magnitude Estimate

Let us start with a brief order-of-magnitude estimate of the S-matrix (7.4.5).

In the black-hole background described in Sect. 7.2, the operator S_n (7.4.1) involves R_{VV}, ∇_U, ϕ, and the spacetime integral. The Ricci curvature due to the collapsing matter is

$$R^{UU} \sim R_{VV} \sim 8\pi G_N T_{VV} \sim \mathcal{O}\left(\frac{1}{a^2}\right). \tag{7.4.6}$$

The covariant derivative $-i\nabla_U$ gives the momentum P_U (or the frequency Ω) conjugate to the Kruskal coordinate U. For a Hawking particle with frequency $\omega \sim \mathcal{O}(1/a)$ detected at $u = u_0$ at large distances, each factor of the covariant derivative ∇_U contributes

$$\nabla_U \sim P_U = \Omega \simeq \omega\, e^{u_0/2a} \sim \frac{1}{a}\, e^{u_0/2a}. \tag{7.4.7}$$

The normalization of the field operator ϕ for the Hawking particles is $\phi \sim \mathcal{O}(1/a)$. Finally, assuming that the wave packet of the Hawking particle has a width $\Delta u \sim \mathcal{O}(a)$ along the u direction and that the thickness of the collapsing matter is $\Delta v \sim \mathcal{O}(a)$, the spacetime integral $\int dUdV$ contributes a factor of $a^2\, e^{-u_0/2a}$.

The amplitude of the transition from the Unruh vacuum to the final state (7.4.4) due to the interaction term (7.4.1) is therefore estimated as

$$\mathcal{M}_n \sim \frac{\lambda_n}{M_p^{4n-2}}\left[a^4 e^{-u_0/2a}\right]\frac{1}{a^{2n}}\left(\frac{1}{a}\,e^{u_0/2a}\right)^{2n}\frac{1}{a^2} \sim \lambda_n\left(\frac{1}{a^2 M_p^2}\,e^{u_0/2a}\right)^{2n-1}. \tag{7.4.8}$$

For a coupling $\lambda_n \sim \mathcal{O}(1)$,[14] this amplitude (7.4.8) is large when

$$e^{u_0/2a} > a^2 M_p^2, \tag{7.4.9}$$

[14] In the ℓ_p^2/a^2 expansion, a small constant like 0.0001 is still considered $\mathcal{O}(1)$.

which is equivalent to the condition

$$u_0 > u_{scr} \tag{7.4.10}$$

on the retarded time u_0 when the Hawking particles are detected.[15]

Note that Eq. (7.4.10) coincides with the condition (7.3.5) for an ingoing quantum with $P_V \sim 1/a$ and a Hawking particle with $\omega \sim 1/a$ to have a trans-Planckian center-of-mass energy. In the interaction term S_n (7.4.1), the ingoing quantum is represented by the Ricci tensor R_{VV}, while ∇_U extracts the momentum P_U of the Hawking particle. Thus, the center-of-mass energy squared is reflected in the invariant quantity $R_{VV}\Omega_U^2$ in S_n. The calculation above shows that the contribution of the non-renormalizable interaction S_n becomes large precisely when the center-of-mass energy of this scattering process becomes trans-Planckian.

In short, Eq. (7.4.10) indicates that after the scrambling time u_{scr}, non-renormalizable interactions dominate over renormalizable interactions in terms of their contributions to particle creations in the spectrum of Hawking radiation, and thus the perturbative approximation fails.

7.4.1.3 Derivation of the S-Matrix

Here we present more details about the calculation of the S-matrix (7.4.5) to confirm the order-of-magnitude estimate (7.4.8). The reader may skip this section without missing the physical picture.

The transition amplitude $\mathcal{M}_n$ (7.4.5) from $|0\rangle$ to $|f\rangle$ (7.4.4) is

$$\mathcal{M}_n \equiv i \int d^4x \sqrt{-g}\, \frac{\lambda_n}{M_p^{4n-2}}\, R^{\mu_1\nu_1} \cdots R^{\mu_n\nu_n} \langle f| \left(\nabla_{\mu_1} \cdots \nabla_{\mu_n}\phi\right) \left(\nabla_{\nu_1} \cdots \nabla_{\nu_n}\phi\right) |0\rangle\,. \tag{7.4.11}$$

For a collapsing spherical null shell of radius $R_s(U)$, the Ricci tensor vanishes outside the shell in the Schwarzschild spacetime, and it also vanishes inside the shell in the Minkowski spacetime. The non-vanishing component of the Ricci tensor is given by

$$R_{VV} = \frac{a}{R_s^2(U)}\, \delta_d(V - V_s)\,, \tag{7.4.12}$$

where δ_d is a smooth function that vanishes outside the range $V \in (V_s - d,\, V_s)$ for a collapsing null shell of thickness d in the V direction, and V_s denotes the location of the outer surface of the collapsing shell. We shall assume that $d \lesssim a$ so that we

---

[15] There is a longstanding observation [105–107] that in the presence of a large number N of matter fields in a d-dimensional effective field theory, the quantum gravity cutoff scale $\Lambda \lesssim N^{-1/(d-2)} M_p$ is parametrically lower than the Planck scale. This suggests that the breakdown of the effective field theory description of Hawking radiation at the scrambling time u_{scr} is a conservative estimate, as this breakdown may occur even earlier.

do not have to worry about self-collision inside the shell before it enters the horizon. Recall from the arguments in Sect. 7.2.2.2 that we can also take $V_s = 2a$ without loss of generality.

As other components of the Ricci tensor vanish for a null shell, the first-order tree-level contribution of the interaction term (7.4.1) to the transition from the Unruh vacuum to the final state (7.4.4) can be written as

$$\mathcal{M}_n = i\,(-2)^n \frac{\lambda_n}{M_p^{4n-2}} \int d^4x\, \sqrt{-g}\, (R_{VV})^n \, \langle f|\left(\nabla_U^n \phi\right)^2 |0\rangle. \tag{7.4.13}$$

The matrix element $\langle f|\left(\nabla_U^n \phi\right)^2 |0\rangle$ above is essentially the product of two terms of the form $\langle 0|b_{\psi_i}\left(\nabla_U^n \phi\right)|0\rangle$ for $i = 1, 2$. Using the formulas in Sect. 7.2.3.3, we can derive

$$\langle 0|b_{\psi}\left(\nabla_U^n \phi\right)|0\rangle \simeq \frac{1}{4\pi\,(2a)^n\, R_s(U(u))} F_{\psi}(u - u_0)\, e^{nu/2a}\,, \tag{7.4.14}$$

where $U(u)$ (7.2.11) is the Kruskal coordinate U expressed as a function of u, and

$$F_{\psi}(u) \equiv \int_0^{\infty} \frac{d\omega}{\sqrt{\omega}} \frac{f_{\omega_0}^*(\omega) A_n(\omega)}{1 - e^{-4\pi a\omega}} e^{i\omega u}\,, \tag{7.4.15}$$

$$A_n(\omega) \equiv (n - 1 + 2ia\omega)(n - 2 + 2ia\omega)\cdots(1 + 2ia\omega)(2ia\omega)\,, \tag{7.4.16}$$

with f_{ω_0} being the profile function in the definition (7.2.52) of the wave packet ψ.

As we are interested in Hawking radiation, we focus on a small U-interval (large u_0) when the collapsing shell is just about to enter the horizon, and thus $R_s(U) \simeq a$ is a good approximation. Using Eqs. (7.4.14)–(7.4.16), we can simplify Eq. (7.4.13) as [33]

$$\mathcal{M}_n \simeq \frac{i\lambda_n}{M_p^{4n-2}} \frac{(-1)^n}{2\pi d^{n-1}(2a)^n} G_{\psi_1\psi_2}\, e^{(2n-1)u_0/2a}\,, \tag{7.4.17}$$

where

$$G_{\psi_1\psi_2} \equiv \int_{-\infty}^{\infty} du\, \frac{e^{(2n-1)u/2a}}{a^{2n}} F_{\psi_1}(u) F_{\psi_2}(u) \tag{7.4.18}$$

is a constant depending on the explicit forms of the profile functions characterizing the wave packets ψ_1 and ψ_2 of the particles in the final state $|f\rangle$ (7.4.4).

Notice that the factor $G_{\psi_1\psi_2}$ is independent of u_0, the time at which the particles are detected. As a result, the transition amplitude $\mathcal{M}_n$ has the time dependence of

$$\mathcal{M}_n \propto e^{(2n-1)u_0/2a}\,. \tag{7.4.19}$$

The proportionality constant is determined by Eqs. (7.4.17) and (7.4.18) for given profile functions defining the outgoing wave packets ψ_1 and ψ_2. Assuming that all characteristic length scales in the problem (e.g. the width of the collapsing shell) is $\sim \mathcal{O}(a)$, dimensional analysis demands the order-of-magnitude of the proportionality constant to be $\sim 1/(aM_p)^{4n-2}$, in agreement with Eq. (7.4.8) derived above.

We refer to Refs. [31–33] for more details about the calculation of the transition amplitudes of particle creation. The calculation for the interaction term (7.4.2) can be found in Ref. [33]. We now proceed to comment on the generalization to generic higher-derivative interactions below in Sect. 7.4.2.1.

7.4.2 *Implications of the S-Matrix Calculation*

In the above, we have calculated the transition amplitude for the creation of particles on top of Hawking radiation due to a specific higher-derivative non-renormalizable interaction S_n (7.4.1). In the following, we comment on the generalizations of the result above to other interactions and its implications for the information paradox.

7.4.2.1 Generic Higher-Derivative Interactions

It is now straightforward to extend the result above to generic higher-derivative interactions. For each derivative ∇_U, we get an exponential factor $e^{u_0/2a}$ as in Eq. (7.4.7). There is a suppressing factor of $e^{-u_0/2a}$ from the measure dU in the integration. Therefore, for an interaction with m derivatives, there is an overall exponential factor $e^{(m-1)u_0/2a}$ to enhance the production of Hawking particles at $u = u_0$.

As non-renormalizable interactions are always suppressed by $1/M_p$ to some power k, we need

$$u_0 > \frac{2k}{(m-1)} a \log\left(a/\ell_p\right) \tag{7.4.20}$$

for the contribution to the amplitude of particle creation to be large (For S_n, $k = 4n - 2$ and $m = 2n$ according to Eq. (7.4.1)). Regardless of the numbers k and m, as long as they are finite, the right-hand side is always parametrically the order of magnitude of the scrambling time $\sim u_{scr}$ (7.2.16). As a result, one can conclude that essentially all higher-derivative interactions contribute to the breakdown of the low-energy effective theory beyond the scrambling time.

7.4.2.2 Contribution From Back-Reaction

We mentioned in Sect. 7.2.2.3 that there is always a negative ingoing energy flux $\langle T_{VV}\rangle$ (7.2.25) in the traditional model to balance the outgoing energy flux of Hawking radiation for energy conservation. Through the semi-classical Einstein equation (7.2.22), its back-reaction leads to a non-vanishing Ricci tensor

$$\langle R_{VV} \rangle \sim \mathcal{O}(\ell_p^2/a^4)\,. \tag{7.4.21}$$

Although this is an order of ℓ_p^2/a^2 smaller than the contribution (7.4.6) from the collapsing matter, the higher-derivative interaction term (7.4.1) gets non-trivial contributions from the bulk spacetime outside the collapsing shell. This leads to a larger number k in Eq. (7.4.20), but for any finite k, it is still at the order of magnitude of the scrambling time when these higher-derivative interactions lead to $\mathcal{O}(1)$ corrections to Hawking radiation.

7.4.2.3 UV Theories and Information Paradox

Even if we have complete information about all non-renormalizable interactions, higher-order interaction terms S_n (7.4.1) with larger n have exponentially larger contributions to the amplitude (7.4.8) of particle creation after the scrambling time. Therefore, a perturbative expansion of the UV theory in powers of $1/M_p$ breaks down. This is an $\mathcal{O}(1)$ effect that has to be properly dealt with prior to examining the information paradox. In this sense, one might be tempted to say that the information paradox is already resolved because the standard description of Hawking radiation using effective field theories is unreliable after the scrambling time. On the other hand, as we discussed in Sect. 7.3.2, as long as the information is required to be transmitted into Hawking radiation, there are still puzzles even when UV theories are relevant.

7.4.2.4 Possible Fates of Hawking Radiation

Although the perturbative expansion of the amplitude displays an exponentially large transition rate of particle creation from higher-order terms, it does not necessarily imply that there is a large enhancement of Hawking radiation in a non-perturbative formulation of the UV theory. For instance, if we introduce an exponential suppression factor $e^{-\ell_p^2 k^2}$ to all interaction terms in momentum space, the interactions with the background would be suppressed in the UV limit.

Given a self-consistent UV-complete theory that resolves the information paradox, there are two logical possibilities for its non-perturbative (in contrast with the $1/M_p$ expansion) prediction of Hawking radiation:

1. Some UV mechanism of particle creation dominates over Hawking radiation so that information can be transferred.
2. Some UV mechanism suppresses Hawking radiation so that the evaporated mass is a negligible portion of the initial black hole mass. There is no need for information transfer.

Naively, the creation of Hawking particles through trans-Planckian interactions between the collapsing matter and the radiation opens the window for information transfer from the collapsing matter to the radiation. However, if higher-derivative interactions are capable of producing trans-Planckian particles, there is in principle a danger of violating the adiabatic principle. In Minkowski space, energy-momentum conservation restricts the production of trans-Planckian particles. But in our expanding universe, we should probably have already seen such an effect. The absence of such events, or the validity of the adiabatic principle reflected in observations, is an indication that trans-Planckian particles are not produced this way.

In this sense, and also in the context of the discussions in Sect. 7.3.2, the information paradox persists even after demonstrating that Hawking radiation is modified at the $\mathcal{O}(1)$ level by UV physics. It remains problematic if we insist that (almost) all information about the collapsing matter is released through Hawking radiation.

Owing to this, the second option mentioned above presents a safer alternative. If all higher-derivative interactions align to cancel the original Hawking radiation before the Page time, the black hole is essentially classical and there would be no paradox. The only challenge is determining whether such a UV theory exists. We will examine two toy models that exhibit precisely this feature in Sect. 7.5.

7.4.2.5 Firewall?

Assuming that some UV mechanism dominates Hawking radiation with a large production rate of particles at large distances, does it imply a firewall around the horizon due to the large blue shift? If we trace the particles created by the interaction S_n (7.4.1) backward to the near-horizon region, their wave packets are composed of extremely high-frequency modes in terms of the Kruskal coordinates. Therefore, naively speaking, the particle creation that we discussed in Sects. 7.4.1.2–7.4.1.3 implies a firewall around the horizon [31, 32]. However, the situation could be more subtle than that [33].

As we mentioned in Sect. 7.2.3.2, there is a large uncertainty $\Delta P_U > P_U$ (7.2.41) in the momentum P_U conjugate to the Kruskal coordinate U for a Hawking particle with a well-defined momentum $P_u \sim \mathcal{O}(1/a)$. With a large uncertainty ΔP_U, the requirement of energy conservation is relaxed (see Sect. 4 of Ref. [33]). This is why the Unruh vacuum can be turned into a multi-particle state with a large amplitude.

On the other hand, a state with a very large ΔP_U is not a typical particle state. To illustrate the idea, let us consider the following state

$$|\Psi\rangle \equiv \frac{1}{\sqrt{\Lambda}} \int_0^\Lambda d\Omega \, |\Omega\rangle \,, \tag{7.4.22}$$

where Λ is a large UV cutoff of this superposition of energy eigenstates. This state has a large $\Delta P_U \sim \Lambda$ with a large central value $P_U = \Lambda/2$. Naively, it should be identified with a high-energy particle state. However, the probability of detecting such a particle at any given energy Ω_0 over a width $\Delta\Omega_0$ is

$$\frac{\Delta\Omega_0}{\Lambda}, \tag{7.4.23}$$

which goes to 0 as $\Lambda \to \infty$. For a state with the upper bound $\Lambda = M_p$, the probability of finding a particle with energy between 0 and 1 TeV is as small as 10^{-16}.

Nevertheless, the total energy flux of such a non-particle state becomes very large due to the large blue-shift factor after the scrambling time. This means that the adiabatic principle will be violated, and there is a firewall. To avoid the violation of the adiabatic principle, the contributions of the higher-derivative non-renormalizable interactions have to sum up with a large cancellation.

7.4.2.6 Scrambling Time Versus Page Time

The "scrambling time" was first named to refer to the lower bound on the order of magnitude of the minimal time it takes a quantum system to scramble (thermalize) information, and black holes are said to saturate the bound as "fast scramblers" [34]. For large black holes ($a \gg \ell_p$), the scrambling time is of order $\mathcal{O}(a \log(a/\ell_p))$, which is orders of magnitude shorter than the Page time $\sim \mathcal{O}(a^3/\ell_p^2)$.

While the information paradox becomes significant around Page time, new physics related to black-hole evaporation has typically been proposed to emerge at or near Page time. For example, loop corrections to correlation functions in a black-hole background exhibit secular growth [40, 108], signaling the breakdown of perturbation theory around the Page time. Similarly, the firewall scenario [17] is concerned with the Page time.

In contrast to these proposals, we propose below that Hawking radiation ceases around the scrambling time. The mass lost to Hawking radiation within the scrambling time is negligible, scaling as $\mathcal{O}(\log(a/\ell_p)/a)$, which is a tiny fraction $\sim \mathcal{O}(\ell_p^2 \log(a/\ell_p)/a^2)$ of the black hole's total mass.

7.5 UV Models of Hawking Radiation

In this section, we investigate Hawking radiation in two UV-modified quantum field theories of massless fields that have been studied in the literature. The first model is based on the generalized uncertainty principle (Sect. 7.5.2), and the second model is rooted in the universal UV suppression characteristic of string field theories (Sect. 7.5.3). We demonstrate that in both models, Hawking radiation ceases around the scrambling time.

These models exhibit nonlocality, though restricted to trans-Planckian modes. While nonlocality in the IR is strongly constrained by observations, nonlocality in the UV appears to be a prevalent feature across quantum gravity theories. Both models incorporate a "minimal length" characteristic, as widely anticipated in quantum gravity theories. Furthermore, the nonlocalities in these models share the common feature of a larger spatial extension at higher momenta inspired by string theory.

7.5.1 Nonlocality in String Theory

To motivate the UV-modified models for Hawking radiation, we first discuss the nonlocality inherent in string theory.

In the study of quantum gravity, there is the general expectation that spacetime as we know it breaks down at scales shorter than the Planck length. This implies that excitations with wavelengths shorter than the Planck length or string length[16] should be eliminated in quantum gravity, leading to some form of nonlocality. Indeed, nonlocality is a common feature in quantum gravity theories, while locality is expected only at low energies.

For example, studies of string scattering amplitudes in the high-energy limit showed that strings cannot probe distances shorter than the string length [109–112]. To explore smaller distances, one is tempted to increase the momentum of the probe string, but this also leads to higher energy and a longer string extension.

As a manifestation of nonlocality in quantum gravity and string theory, the generalized uncertainty principle (GUP) [109–113] is proposed as

$$\Delta x\, \Delta p \geq \frac{1}{2}\left(1 + \ell_s^2\, \Delta p^2\right), \tag{7.5.1}$$

where $\ell_s = \sqrt{\alpha'}$ denotes the string length.[17] It is an extension of the Heisenberg uncertainty principle that effectively characterizes the nonlocality inherent in string theory. It also accounts for the presence of a minimal length scale $\Delta x \geq \ell_s$ as motivated by a variety of quantum gravity considerations [114–121].

Another proposal is the space-time uncertainty relation (STUR) [122–125]

$$\Delta x\, \Delta t \geq \ell_s^2\,. \tag{7.5.2}$$

Roughly speaking, as the uncertainty in energy or momentum increases, the uncertainty in the size of the string grows as well.

Both relations (7.5.1) and (7.5.2) have been argued to arise from the stringy nature of interactions in string theory. They are actually interconnected. Combining the GUP (7.5.1) with the usual uncertainty relation $\Delta E\, \Delta t \geq 1$ under the assumption

[16] We will not distinguish the Planck length from the string length in this chapter for simplicity in discussion as they are of the same order in the ℓ_p^2/a^2 expansion.

[17] Strictly speaking, there are two types of uncertainty at play here. The first term $1/2$ on the right-hand side of Eq. (7.5.1) arises from the standard uncertainty relation due to the wave nature of particles. The second term originates from the spatial extension characteristic of strings. When strings are used to probe spacetime geometry, both types of uncertainty contribute to the overall uncertainty in the spacetime geometry.

$$\Delta E \sim \Delta p\,, \tag{7.5.3}$$

we derive

$$\Delta x \gtrsim \ell_s^2\,\Delta p \sim \ell_s^2\,\Delta E \geq \frac{\ell_s^2}{\Delta t}\,, \tag{7.5.4}$$

which is simply the spacetime uncertainty relation (7.5.2).

For an outgoing massless field fluctuation with $E = p$ and satisfying Eq. (7.5.3), both the GUP and the STUR imply

$$\Delta x \gtrsim \ell_s^2\,\Delta E\,. \tag{7.5.5}$$

For an order-of-magnitude estimate, when $\Delta E \sim E$ or $\Delta E \gtrsim E$ (as is the case for Hawking particles in the near-horizon region shown in Eq. (7.2.41)), we find that $\Delta x > \ell_s^2 E$. This uncertainty $\ell_s^2 E$ exceeds the wavelength $\sim 1/E$ only when the energy of the particle is trans-Planckian: $E > \ell_s^{-1}$. Thus, even though the spatial uncertainty Δx can be macroscopic, it remains strictly confined to trans-Planckian degrees of freedom.

Equation (7.5.5) implies that probing a structure of length scale L requires not only selecting a probe with a wavelength shorter than L but also ensuring that the probe's energy does not exceed L/ℓ_s^2. This condition sets a minimum length scale $\Delta x \sim \ell_s$ for defining spacetime geometry, thus limiting the smallest resolvable scale to ℓ_s.

As shown in Sect. 7.2.3.2, in the freely falling frame we have $\Delta E \simeq \Delta P_U \gg \langle P_U \rangle$, and thus Eq. (7.5.5) leads to

$$\Delta x \gg \ell_s^2\,\langle P_U \rangle \sim \ell_s^2\,\langle P_u \rangle\, e^{u/2a} \sim \frac{\ell_s^2}{a}\, e^{u/2a} \tag{7.5.6}$$

for a Hawking particle with $\langle P_u \rangle \sim \mathcal{O}(1/a)$. According to Eq. (7.5.6), the uncertainty Δx grows exponentially for Hawking particles at late times. Specifically, Δx becomes much larger than the Schwarzschild radius a after the scrambling time $u \geq 2a \log(a^2/\ell_s^2)$.

The potential significance of nonlocal effects in string theory for the black-hole information problem has also been suggested in Refs. [72, 94, 95, 126–129].

7.5.2 *Hawking Radiation Under GUP*

In this section, we examine how the GUP influences late-time Hawking radiation [13].

As discussed in Sect. 7.3.1.1, the presence of the collapsing matter and the quantum energy-momentum tensor breaks local Lorentz symmetry in the near-horizon region. Although this breaking is weak with a small curvature $\sim \mathcal{O}(1/a^2)$, it is crucial to consider its effects on Hawking particles after the scrambling time as we have demonstrated in Sect. 7.4.

Rather than examining the effects of trans-Planckian interactions between the collapsing matter and the outgoing radiation field in a fundamental UV theory, it is simpler to focus on UV modifications to the wave equation of the outgoing field in a preferred reference frame.

To illustrate this, we first implement the GUP in the Vaidya coordinates, showing that Hawking radiation ceases at the scrambling time. As we will discuss in Sect. 7.5.2.6, selecting other freely falling frames yields the same conclusion [13]. Thus, the termination of Hawking radiation at the scrambling time is a robust prediction of the GUP, independent of the choice of the local frame.

7.5.2.1 Low-Energy Effective Description

We consider a black hole formed from the collapse of a light-like infalling matter shell. The background geometry was reviewed in Sect. 7.2.2. The Schwarzschild metric outside the shell and the Minkowski space inside the shell can be combined in the ingoing Vaidya metric as

$$ds^2 = -\left(1 - \Theta(v)\,\frac{a}{x+a}\right) dv^2 + 2dv\,dx\,, \tag{7.5.7}$$

where $v \equiv t + r_*$ (7.2.9) is the light-cone time, and $x \equiv r - a$ represents the radial coordinate from the horizon. To streamline our analysis, we focus on the s-wave sector of the radiation field and omit the angular part of the metric. Without loss of generality, we assume that the trajectory of the null shell coincides with the time slice $v = 0$ (see Fig. 7.5). Outside the shell ($v > 0$), the metric describes a Schwarzschild geometry, with the event horizon located at $x = 0$. In the region inside the shell ($v < 0$), the spacetime reduces to flat Minkowski space.

Adopting the line element (7.5.7) in the near-horizon region where $x \equiv r - a \ll a$, the free-field action S_0 of the radiation field in the low-energy effective theory is

$$S_0[\phi] = -\frac{1}{2}\int dv \int dx\,[\partial_x\,\phi(v,x)]\left\{2\partial_v + \left[1 + \Theta(v)\left(\frac{x}{a} - 1\right)\right]\partial_x\right\}\phi(v,x)\,. \tag{7.5.8}$$

Via the Fourier transform

$$\phi(v,x) = \int_{-\infty}^{\infty} \frac{dp}{\sqrt{2\pi}}\,\Phi(v,p)\,e^{ipx}\,, \tag{7.5.9}$$

the Klein-Gordon inner product on a constant-v hypersurface of this background can be expressed as

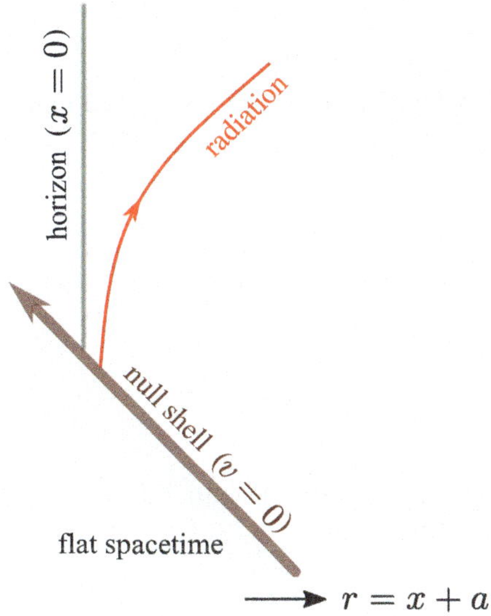

Fig. 7.5 A thin null shell with a flat interior collapses to form a black hole. The calculation of Hawking radiation involves tracing an outgoing wave packet backward in time and decomposing it in terms of the Minkowski modes inside the shell

$$\langle \phi_1 \, , \phi_2 \rangle \equiv -i \int_{-\infty}^{\infty} dx \, \phi_1^*(v, x) \, \overleftrightarrow{\partial_x} \, \phi_2(v, x) = 2 \int_{-\infty}^{\infty} dp \, \Phi_1^*(v, p) \, p \, \Phi_2(v, p) \, . \tag{7.5.10}$$

In terms of the Fourier modes, the action can be brought to the form

$$\begin{aligned} S_0[\Phi] &= \int dv \int_{-\infty}^{\infty} dp \, \Phi^*(v, p) \, p \left[i\partial_v - \frac{p}{2} - \Theta(v) \left(\frac{\hat{x}}{2a} \, p - \frac{p}{2} \right) \right] \Phi(v, p) \\ &= \frac{1}{2} \int dv \left\langle \Phi(v, p) \, , \left[i\partial_v - \frac{p}{2} - \Theta(v) \left(\frac{\hat{x}}{2a} \, p - \frac{p}{2} \right) \right] \Phi(v, p) \right\rangle , \end{aligned} \tag{7.5.11}$$

where $\hat{x} \equiv i\partial_p$.

7.5.2.2 Implementing GUP

Following the proposal of Ref. [130], we incorporate the GUP (7.5.1) into the radial coordinate x of the background (7.5.7) by deforming the commutation relation between x and the radial momentum p as [131]

$$[\hat{x} \, , \hat{p}] = i \left(1 + \ell_s^2 \, \hat{p}^2 \right) , \tag{7.5.12}$$

which implies the desired uncertainty relation

$$\Delta x\, \Delta p \geq \frac{1}{2}\left[1 + \ell_s^2\, (\Delta p)^2 + \ell_s^2\, \langle p\rangle^2\right] \geq \frac{1}{2}\left[1 + \ell_s^2\, (\Delta p)^2\right]. \tag{7.5.13}$$

This modification can be realized in momentum space via the representation

$$\hat{x} \equiv i\left(1 + \ell_s^2\, p^2\right) \partial_p\,, \qquad \hat{p} = p\,. \tag{7.5.14}$$

To incorporate the GUP into this system, we take the action (7.5.11) and modify the definition of $\hat{x}$ and the inner product. We utilize the deformed momentum-space representation of $\hat{x}$ (7.5.14) while modifying the inner product (7.5.10) accordingly as

$$\langle \Phi_1\,, \Phi_2\rangle_{\text{GUP}} \equiv 2\int_{-\infty}^{\infty} \frac{dp}{1 + \ell_s^2\, p^2}\, \Phi_1^*(v, p)\, p\, \Phi_2(v, p)\,. \tag{7.5.15}$$

This ensures the Hermiticity of $\hat{x}$ [131] and thus preserves the reality of the GUP-modified action

$$S_{\text{GUP}}[\Phi] = \frac{1}{2}\int dv \left\langle \Phi(v, p)\,, \left[i\partial_v - \frac{p}{2} - \Theta(v)\left(\frac{\hat{x}}{2a}\, p - \frac{p}{2}\right)\right] \Phi(v, p)\right\rangle_{\text{GUP}} \tag{7.5.16}$$

for the radiation field. As ϕ is by assumption a real scalar, we shall assume the reality condition $\Phi^*(v, p) = \Phi(v, -p)$.

In the presence of a minimal uncertainty $\Delta x \geq \ell_s$, the conventional description of the field in terms of the radial coordinate x is no longer available. We view Eq. (7.5.16) as the starting point for this investigation.

Although this realization of the GUP (7.5.1) violates Lorentz symmetry, this is justified for the following reasons. First, as it was demonstrated in Sect. 7.4, the local Lorentz symmetry in the derivation of Hawking radiation is lost due to the presence of the collapsing matter and the back-reaction of the quantum energy-momentum tensor. There is thus a preferred reference frame.

Second, there are numerous examples in which the Hawking radiation of the traditional model is only mildly modified by Lorentz-violating UV physics implemented in a freely falling frame (see, e.g. [18, 19, 21–26, 28]). Furthermore, even though the realization of the GUP is mathematically inequivalent for each freely falling frame, we will see below that the conclusion of turning off Hawking radiation around the scrambling time is independent of the choice of the frame [13], including the Vaidya coordinates as a limit (see Sect. 7.5.2.6).

We note that there have been attempts [132–139] in the literature to integrate the GUP into quantum field theories in a Lorentz-covariant manner, which could provide the necessary tools to extend our analysis in a more rigorous framework. It will be interesting to extend the calculation here to a Lorentz-covariant version. Let us also mention that, in Sect. 7.5.3, the spacetime uncertainty relation that will be responsible for turning off Hawking radiation is Lorentz-covariant.

Variation of the action (7.5.16) yields the field equation

$$\left\{ i\partial_v - \frac{p}{2} - \Theta(v) \left[\frac{i\left(1+\ell_s^2\, p^2\right)}{2a}\, \partial_p\, p - \frac{p}{2} \right] \right\} \Phi(v,\, p) = 0 \tag{7.5.17}$$

in the outgoing sector, with the continuity of $\Phi(v, p)$ required across the null shell at $v = 0$.

The field Eq. (7.5.17) has to be supplemented with a suitable boundary condition. According to the definition of the inner product (7.5.15), the time derivative of the norm is

$$\partial_v \left\langle \Psi\, ,\, \Psi \right\rangle_{\mathrm{GUP}} = \frac{\Theta(v)}{2a} \left[\left| p\, \Psi(v,\, p) \right|^2 \right]_{p=-\infty}^{\infty} \tag{7.5.18}$$

for a wave Ψ obeying the outgoing field equation (7.5.17). Therefore, the norm of Ψ is conserved under time evolution outside the matter shell only if the boundary condition

$$\lim_{p \to \pm\infty} p\, \Psi(v,\, p) = 0 \tag{7.5.19}$$

is satisfied. Hence, unitarity demands that the Hilbert space of the quantized theory is only composed of wave functions satisfying this boundary condition.

According to the GUP commutation relation (7.5.12), it is natural to define the conjugate momentum of x as

$$K(p) \equiv \ell_s^{-1} \tan^{-1}\left(\ell_s\, p\right), \tag{7.5.20}$$

so that the usual Heisenberg algebra $[\hat{x}, \hat{K}(p)] = i$ is satisfied. This explains the measure of the inner product (7.5.15):

$$dK = \frac{dp}{1+\ell_s^2\, p^2}\,. \tag{7.5.21}$$

The existence of a minimal length $\Delta x \geq \ell_s$ is now reflected in the UV cutoff of the conjugate momentum:

$$|K| \leq \pi/2\ell_s\,. \tag{7.5.22}$$

In terms of this new momentum K, the free field described by Eq. (7.5.16) exhibits a superluminal dispersion relation $\omega(K) = \tan\left(\ell_s K\right)/2\ell_s$.

While the final outcome of Hawking radiation is independent of the choice of momentum variables, we shall adopt the interpretation where the field is dispersionless ($\omega(p) = p/2$), and take p instead of K as the momentum.

7.5.2.3 Hawking Particle's Wave Packet

In the Schwarzschild spacetime outside the shell ($v > 0$), the monochromatic Hawking modes $\Phi_\omega(v, p)$ with frequency ω (defined with respect to the Eddington time v) can be solved from the wave Eq. (7.5.17) as

$$\Phi_\omega(v, p) = \mathcal{N}_\omega \frac{e^{-i\omega v}}{p - i0^+} \exp\left\{-2ia\omega \log\left[\frac{a\,(p - i0^+)}{\sqrt{1 + \ell_s^2\, p^2}}\right]\right\}, \tag{7.5.23}$$

where $\mathcal{N}_\omega$ is a normalization factor that will be determined later. The $i0^+$-prescription for analytic continuation across the pole (and branch point) at $p = 0$ ensures that the expression (7.5.23) is both analytic and bounded in the lower half of the complex p-plane. This guarantees that Φ_ω describes an outgoing wave with no support behind the horizon ($x < 0$), thereby respecting the causal behavior of the wave equation [140].

As will become clear soon, the difference $\exp[-2\pi a\omega\, \Theta(-p)]$ resulting from the discontinuity across the branch cut of the logarithm in Eq. (7.5.23) gives rise to the Boltzmann factor $e^{-4\pi a\omega}$ characterizing the Hawking temperature.

As stated in Sect. 7.2.3.4, to capture the time dependence of Hawking radiation, rather than working with spacetime-filling waves, we construct a localized wave packet Ψ_{ω_0,u_0} as a superposition of the positive-frequency modes (7.5.23):

$$\Psi_{\omega_0,u_0}(v, p) = \int_0^\infty d\omega\; f_{\omega_0}(\omega)\, e^{i\omega u_0}\, \Phi_\omega(v, p)\,. \tag{7.5.24}$$

Its profile $f_{\omega_0}(\omega)$ in the frequency domain is supposed to be compactly supported around $\omega = \omega_0$ with a narrow width $\ll \omega_0$. Under this assumption, the Hawking wave packet (7.5.24) can be written more explicitly as

$$\begin{aligned}\Psi_{\omega_0,u_0}(v, p) &= \int_0^\infty d\omega\; f_{\omega_0}(\omega)\, \mathcal{N}_\omega \frac{e^{-i\omega(v-u_0)}}{p - i0^+} \exp\left\{-2ia\omega \log\left[\frac{a\,(p - i0^+)}{\sqrt{1 + \ell_s^2\, p^2}}\right]\right\} \\ &\simeq \mathcal{N}_{\omega_0} \frac{\exp[-2\pi a\omega_0\, \Theta(-p)]}{p - i0^+} \int_0^\infty d\omega\; f_{\omega_0}(\omega) \exp\left\{-i\omega\left[v - u_0 + 2a \log|aP(p)|\right]\right\},\end{aligned} \tag{7.5.25}$$

where the correction from the GUP is encoded in the *effective momentum*

$$P(p) \equiv \frac{p}{\sqrt{1 + \ell_s^2\, p^2}}\,. \tag{7.5.26}$$

In deriving Eq. (7.5.25), we took advantage of the compactness of $f_{\omega_0}(\omega)$ to approximate the slowly varying term $\mathcal{N}_\omega \exp[-2\pi a\omega\, \Theta(-p)]$ by a constant evaluated at ω_0.

Defining the following quantities:

$$F(p) \equiv \log\left|aP(p)\right|, \tag{7.5.27}$$

$$\tilde{f}_{\omega_0}(v) \equiv \int_0^\infty \frac{d\omega}{\sqrt{2\pi}}\, f_{\omega_0}(\omega)\, e^{-i\omega v}, \tag{7.5.28}$$

we further simplify Eq. (7.5.25) as

$$\Psi_{\omega_0,u_0}(v,p) \simeq \sqrt{2\pi}\,\mathcal{N}_{\omega_0}\,\frac{\exp\left[-2\pi a\omega_0\,\Theta(-p)\right]}{p-i0^+}\,\tilde{f}_{\omega_0}\big(v-u_0+2aF(p)\big). \tag{7.5.29}$$

From this expression, it is clear that the momentum-space wave packet $\Psi_{\omega_0,u_0}(v,p)$ for a Hawking particle is centered along the characteristic trajectory

$$v + 2aF(p) = \text{constant}. \tag{7.5.30}$$

At sufficiently late times (large v), Eqs. (7.5.26), (7.5.27), and (7.5.30) imply that the central momentum $\bar{p}$ of the wave packet is cis-Planckian, i.e., $\bar{p} \ll \ell_s^{-1}$. In this regime, it is reasonable to approximate $F(p) \simeq \log|ap|$. As a result, the wave packet (7.5.29) eventually evolves in accordance with the low-energy theory. It is then natural to normalize the wave packet (7.5.29) by matching it to the ordinary Hawking wave packet in the unmodified theory at late times.

In the unmodified theory, the wave packet is a superposition of the outgoing plane waves $e^{-i\omega u}/\sqrt{4\pi\omega} = e^{-i\omega(v-2r_*)}/\sqrt{4\pi\omega}$. Tracing these modes backward in time to the near-horizon region where $0 < x \ll a$, the form of each individual mode function in momentum space is given by

$$\int_{-\infty}^{\infty} \frac{dx}{\sqrt{2\pi}}\, e^{-ipx}\left[\Theta(x)\,\frac{e^{-i\omega v}}{\sqrt{4\pi\omega}}\left(\frac{x}{a}\right)^{2ia\omega}\right] = \frac{a}{\pi}\sqrt{\frac{\omega}{2}}\, e^{\pi a\omega}\,\Gamma(2ia\omega)\left\{\frac{e^{-i\omega v}}{p-i0^+}\left[a(p-i0^+)\right]^{-2ia\omega}\right\}. \tag{7.5.31}$$

The counterpart of (7.5.29) in the low-energy theory is

$$\Psi^{(0)}_{\omega_0,u_0}(v,p) \simeq a\sqrt{\frac{\omega_0}{\pi}}\, e^{\pi a\omega_0}\,\Gamma(2ia\omega_0)\,\frac{\exp\left[-2\pi a\omega_0\,\Theta(-p)\right]}{p-i0^+}\,\tilde{f}_{\omega_0}\big(v-u_0+2a\log|ap|\big). \tag{7.5.32}$$

By demanding that the GUP-modified wave packet (7.5.29) matches its unmodified counterpart $\Psi^{(0)}_{\omega_0,u_0}(v,p)$ during the late-time evolution within the IR regime:

$$\Psi_{\omega_0,u_0}(v \gg u_0\,,\,p) \simeq \Psi^{(0)}_{\omega_0,u_0}(v \gg u_0\,,\,p), \tag{7.5.33}$$

one can uniquely fix the p-independent overall constant $\mathcal{N}_\omega$ in the mode functions (7.5.23) as

$$\mathcal{N}_\omega = \frac{a}{\pi}\sqrt{\frac{\omega}{2}}\, e^{\pi a\omega}\,\Gamma(2ia\omega). \tag{7.5.34}$$

As the correspondence with an ordinary Hawking particle is established at late times, the phase factor $e^{i\omega u_0}$ that was included in the definition (7.5.24) of the wave packet

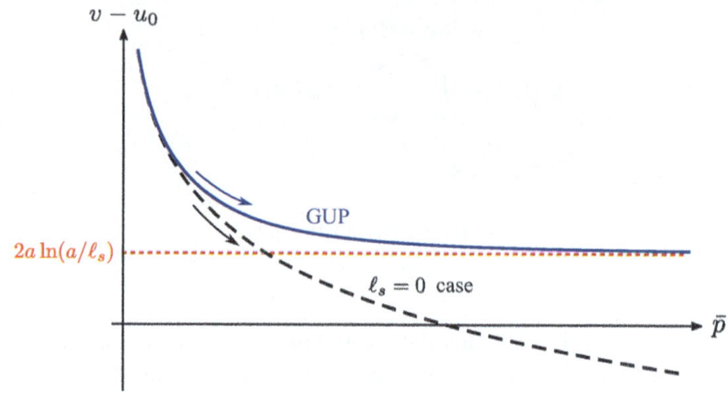

Fig. 7.6 A graph illustrating the positive-p branch of the characteristic trajectory $v - u_0 = -2aF(p)$ (7.5.30) (solid blue curve), alongside its counterpart in the low-energy theory (dashed black curve). Arrows indicate the flow direction toward the past (decreasing v). With the introduction of the GUP, the momentum reaches infinity in a finite time marked by the dotted red line

now takes on a clear interpretation as localizing the wave packet around the Eddington retarded time $u = u_0$.

One can solve for the central momentum $\bar{p}$ of a Hawking wave packet from Eq. (7.5.30) as

$$\bar{p}(v) = \begin{cases} \frac{1}{a} \frac{e^{(u_0 - v)/2a}}{\sqrt{1 - e^{(u_0 - v)/a}\, \ell_s^2/a^2}} & \text{for } \bar{p} > 0 \\ -\frac{1}{a} \frac{e^{(u_0 - v)/2a}}{\sqrt{1 - e^{(u_0 - v)/a}\, \ell_s^2/a^2}} & \text{for } \bar{p} < 0 \end{cases}. \quad (7.5.35)$$

As time v decreases, the magnitude of the central momentum $\bar{p}(v)$ grows sharply due to the rapid decrease of its denominator. An illustration of the characteristic evolution (7.5.30) is presented in Fig. 7.6. During the backward evolution of a wave packet, the blue shift of its central momentum $\bar{p} > 0$ occurs far more rapidly than the exponential growth seen in the low-energy effective theory.

Based on these results, a Hawking particle that reaches the asymptotic region at around the scrambling time $u_0 \sim 2a \log(a/\ell_s)$ would have originated from the collapsing shell at $v = 0$ with a central momentum $|\bar{p}| \gg 1/\ell_s$ greatly exceeding the Planck scale. As it turns out, Eq. (7.5.35) indicates that $|\bar{p}(v)|$ is in fact infinite when $v = u_0 - 2a \log(a/\ell_s)$. Namely, the central momentum would already be blue-shifted to reach the boundary $p = \pm\infty$ within a finite duration of backward propagation. This peculiar feature of the GUP system has profound implications to be discussed in more detail later in this section.

7.5.2.4 Wave Function Inside the Shell

Now that the mode configuration outside the collapsing shell is set up, we proceed to quantize the field inside the shell ($v < 0$). In the flat spacetime inside the shell, the field Eq. (7.5.17) simply reduces to

$$\left(i\partial_v - \frac{p}{2}\right)\Phi(v \le 0\,,\, p) = 0\,, \tag{7.5.36}$$

which suggests the mode expansion

$$\Phi(v \le 0, p) = e^{-ipv/2}\sqrt{\frac{1+\ell_s^2\, p^2}{2|p|}}\left[a_p\,\Theta(p) + a_{-p}^{\dagger}\,\Theta(-p)\right]. \tag{7.5.37}$$

Upon canonical quantization, where the field Φ and its conjugate momentum

$$\Pi(v, p) \equiv \frac{\delta S_{\text{GUP}}}{\delta[\partial_v \Phi(v, p)]} = \frac{2ip}{1+\ell_s^2\, p^2}\,\Phi^*(v, p) \tag{7.5.38}$$

are promoted to operators satisfying the equal-time commutation relation

$$\left[\Phi(v, p), \Pi(v, p')\right] = i\,\delta(p-p')\,, \tag{7.5.39}$$

the creation and annihilation operators obey

$$\left[a_p\,, a_{p'}^{\dagger}\right] = \delta(p-p')\,, \qquad \left[a_p\,, a_{p'}\right] = 0\,, \qquad \left[a_p^{\dagger}\,, a_{p'}^{\dagger}\right] = 0\,. \tag{7.5.40}$$

The quantum state of the field inside the shell is assumed to be the Minkowski vacuum $|0\rangle$ defined by

$$a_p\,|0\rangle = 0 \qquad \forall\, p > 0\,. \tag{7.5.41}$$

7.5.2.5 Derivation of Hawking Radiation Under GUP

In analogy with the approach in Sect. 7.2.3.4, we can associate an annihilation operator b_Ψ to the Hawking particle described by the wave packet Ψ_{ω_0,u_0} (7.5.29). This operator b_Ψ is defined as the inner product between Ψ_{ω_0,u_0} and the field operator Φ:

$$b_\Psi \equiv \left(\Psi_{\omega_0,u_0}\,, \Phi\right)_{\text{GUP}}\,. \tag{7.5.42}$$

The number expectation value $\langle 0 \mid b_\Psi^{\dagger} b_\Psi \mid 0\rangle$ of such Hawking particles in the Minkowski vacuum $|0\rangle$ (7.5.41) is determined by how the operator b_Ψ decomposes into the creation and annihilation operators $(a_p\,, a_p^{\dagger})$ inside the shell. Hence, the calculation of Hawking radiation amounts to tracing the wave packet $\Psi_{\omega_0,u_0}(v, p)$ back in time to the collapsing shell at $v = 0$, where the field $\Phi(v = 0\,, p)$ can be matched with the mode expansion (7.5.37) in Minkowski space.

According to the form (7.5.15) of the inner product under the GUP, Eq. (7.5.42) reads

$$b_\Psi = \int_0^\infty dp \sqrt{\frac{2p}{1+\ell_s^2 p^2}} \left[\Psi^*_{\omega_0,u_0}(0, p)\, a_p - \Psi^*_{\omega_0,u_0}(0, -p)\, a_p^\dagger\right], \tag{7.5.43}$$

which is essentially a decomposition of the wave packet $\Psi_{\omega_0,u_0}(0, p)$ into positive and negative momentum components. It follows that the vacuum expectation value of the number of Hawking particles is given by

$$\begin{aligned}\left\langle 0 \left| b_\Psi^\dagger b_\Psi \right| 0 \right\rangle &= \int_0^\infty dp \, \frac{2p}{1+\ell_s^2 p^2} \left|\Psi_{\omega_0,u_0}(0, -p)\right|^2 \\ &= \left\langle \Psi_{\omega_0,u_0}(0, p), \Theta(-p)\, \Psi_{\omega_0,u_0}(0, p) \right\rangle_{\text{GUP}},\end{aligned} \tag{7.5.44}$$

which is determined by the norm of the negative-momentum components in the wave packet $\Psi_{\omega_0,u_0}(0, p)$ as anticipated.

Plugging the expression (7.5.25) for the wave packet into Eq. (7.5.44), we find

$$\begin{aligned}\left\langle 0 \left| b_\Psi^\dagger b_\Psi \right| 0 \right\rangle &\simeq \left|\mathcal{N}_{\omega_0}\right|^2 e^{-4\pi a\omega_0} \int_0^\infty d\omega \int_0^\infty d\omega' \, f^*_{\omega_0}(\omega)\, f_{\omega_0}(\omega')\, e^{-i(\omega-\omega')u_0} \int_0^\infty dp \, \frac{2\, e^{2ia(\omega-\omega')F(p)}}{p\,(1+\ell_s^2 p^2)} \\ &= 2 \left|\mathcal{N}_{\omega_0}\right|^2 e^{-4\pi a\omega_0} \int_0^\infty d\omega \int_0^\infty d\omega' \, f^*_{\omega_0}(\omega)\, f_{\omega_0}(\omega')\, e^{-i(\omega-\omega')u_0} \int_{-\infty}^{\log(a/\ell_s)} dF \, e^{2ia(\omega-\omega')F},\end{aligned} \tag{7.5.45}$$

from which we observe that the effect of the GUP on Hawking radiation appears in the form of a finite upper bound $\log(a/\ell_s)$ on the F-integration. Subsequently, through a change of variable

$$u \equiv 2aF - u_0\,, \tag{7.5.46}$$

we arrive at [13]

$$\begin{aligned}\left\langle 0 \left| b_\Psi^\dagger b_\Psi \right| 0 \right\rangle &\simeq \frac{a}{\pi} \frac{e^{-4\pi a\omega_0}}{1-e^{-4\pi a\omega_0}} \int_0^\infty d\omega \int_0^\infty d\omega' \, f^*_{\omega_0}(\omega)\, f_{\omega_0}(\omega') \\ &\quad \times \int_{-\infty}^{2a\log(a/\ell_s)-u_0} \frac{du}{2a}\, e^{i(\omega-\omega')u} \\ &= \frac{1}{e^{4\pi a\omega_0}-1} \int_{-\infty}^{2a\log(a/\ell_s)-u_0} du \left| \int_0^\infty \frac{d\omega'}{\sqrt{2\pi}} f_{\omega_0}(\omega')\, e^{-i\omega' u} \right|^2 \\ &= \frac{1}{e^{4\pi a\omega_0}-1} \int_{-\infty}^{2a\log(a/\ell_s)-u_0} du \left| \tilde{f}_{\omega_0}(u) \right|^2,\end{aligned} \tag{7.5.47}$$

where we made use of the definition (7.5.28), as well as the fact that

$$|\mathcal{N}_\omega|^2 = \frac{a}{2\pi} \frac{1}{1-e^{-4\pi a\omega}} \tag{7.5.48}$$

based on Eq. (7.5.34). Note that the Hawking temperature remains the same while the amplitude changes over time.

We thus arrive at a result similar to the scenario involving a non-covariant UV cutoff described in Sect. 7.2.3.5. Specifically, by employing the effective momentum $P(p) \equiv p/\sqrt{1+\ell_s^2\, p^2}$ (7.5.26) and redefining the field as

$$\chi(v, P) \equiv p(P)\, \Phi\big(v, p(P)\big) ,$$

where $p(P)$ is the inverse function of $P(p)$, the action (7.5.16) can be expressed as

$$S_{\text{GUP}}[\chi] = \int dv \int_0^{\ell_s^{-1}} \frac{dP}{P}\, \chi^*(v, P) \left(2i\partial_v - \frac{i}{a}\, \Theta(v) P\, \partial_P \right) \chi(v, P) ,$$

which indeed differs from the low-energy action $S_0[\chi]$ (7.5.11) merely by a UV cutoff ℓ_s^{-1}. Given that the profile function $\tilde{f}_{\omega_0}(u)$ defined in Eq. (7.5.28) is centered around the retarded time $u = 0$ with a width of $\mathcal{O}(a)$, the probability (7.5.47) of detecting Hawking particles centered around the retarded time u_0 diminishes with increasing u_0. Eventually, this probability approaches zero when $u_0 \gtrsim 2a \log(a/\ell_s)$, signifying that Hawking radiation is turned off around the scrambling time.

7.5.2.6 Freely Falling Frames

In the above, we have considered Hawking radiation with the GUP realized in the Vaidya coordinates. The same calculation can be carried out in a generic freely falling frame [13].

Consider a generic freely falling observer traveling radially inward along a time-like geodesic in the Schwarzschild spacetime. We can parametrize the initial speed of the geodesic at the infinite past ($t \to -\infty, r \to \infty$) as

$$\lim_{t \to -\infty} \left(\frac{dr}{dt} \right)^2 = 1 - \gamma , \qquad \text{where } \gamma \in (0, 1] . \tag{7.5.49}$$

By introducing the proper time coordinate

$$\tau(t, r) = \gamma^{-1/2} \left[t + h_\gamma(r) \right] , \tag{7.5.50}$$

where

$$\begin{aligned} h_\gamma(r) &\equiv \int^r dr' \, \frac{\sqrt{1 - \gamma(1 - a/r')}}{1 - a/r'} \\ &= \sqrt{1 - \gamma(1 - a/r)}\, r + \frac{(2 - \gamma)}{2\sqrt{1 - \gamma}}\, a \log \left| \frac{\sqrt{1 - \gamma} + \sqrt{1 - \gamma(1 - a/r)}}{\sqrt{1 - \gamma} - \sqrt{1 - \gamma(1 - a/r)}} \right| \\ &\quad - a \log \left| \frac{1 + \sqrt{1 - \gamma(1 - a/r)}}{1 - \sqrt{1 - \gamma(1 - a/r)}} \right| , \end{aligned} \tag{7.5.51}$$

we can define a class of freely falling frames (τ, r), in which the Schwarzschild line element can be expressed as

$$ds^2 = -d\tau^2 + \gamma\left(dr - v_\gamma(r)\, d\tau\right)^2, \tag{7.5.52}$$

where

$$v_\gamma(r) = -\gamma^{-1/2}\sqrt{1 - \gamma\left(1 - \frac{a}{r}\right)}. \tag{7.5.53}$$

The Eddington-Finkelstein line element that we worked with previously is the limiting case with $\gamma \to 0$.

Following similar calculation steps as in the previous section (with the detailed derivation provided in Ref. [13]), it can be shown that the result (7.5.47) of Hawking radiation with the GUP implemented in the Vaidya coordinates is slightly modified in a generic freely falling frame as

$$\left\langle 0 \left| b_\Psi^\dagger b_\Psi \right| 0 \right\rangle \simeq \frac{1}{e^{4\pi a \gamma^{-1/2}\omega_0} - 1} \int_{-\infty}^{2a\log(a/\ell_s) - u_0} \left| \tilde{f}_{\omega_0}(u) \right|^2 du\,, \tag{7.5.54}$$

where there is a factor $\gamma^{-1/2}$ in front of the central frequency ω_0 defined with respect to the free-fall proper time τ. The conclusion remains the same that Hawking radiation is turned off around the scrambling time $u \sim u_{scr}$ (7.2.16). Notice that the proper time interval $\Delta\tau$ is related to the Schwarzschild time interval Δt via $\Delta\tau = \gamma^{-1/2}\Delta t$ at a fixed radius. Hence, by combining the Lorentz boost factor $\gamma^{-1/2}$ with the frequency ω_0 into the Schwarzschild frequency (defined with respect to t) in the Planck distribution factor in Eq. (7.5.54), we find that the same Hawking temperature is displayed across all freely falling frames.

Since the results of the UV modification by the GUP to Hawking radiation are insensitive to the free parameter γ labeling different freely falling frames, we can infer that this UV effect is robust and insensitive to the details of the gravitational collapse.

7.5.2.7 Comments on Hawking Radiation Under GUP

We showed above that Hawking radiation is terminated around the scrambling time by the GUP implemented in any freely falling frame. In this section, we discuss the mathematics and physics behind this result.

The termination of Hawking radiation can be attributed to the UV cutoff in the Hilbert space, analogous to what was shown in Sect. 7.2.3.5. While the momentum p within the Hilbert space may, in theory, take on arbitrarily large values, the conjugate variable K (7.5.20) is limited to $|K| \leq \pi/(2\ell_s)$. States with $|K| > \pi/(2\ell_s)$ in the near-horizon region, which are necessary for the evolution into Hawking particles post-scrambling time, do not exist within the Hilbert space. This phenomenon is related to the "non-conservation of the norm" of a Hawking wave packet, as highlighted in Refs. [13, 130].

If we start with a state with momentum constrained within $|K| \leq \pi/(2\ell_s)$ on an initial time slice in the past, it will not evolve into a Hawking particle beyond the scrambling time. Conversely, if we consider a Hawking particle that emerges after the scrambling time at a large distance and we allow it to propagate backward in time, it will violate the boundary condition (7.5.19) as demonstrated in Sect. 7.5.2.3. This violation leads to a decrease in the norm during backward evolution until the wave packet falls out of the Hilbert space in the near-horizon region.

The question now is whether it is sensible to say that, at large distances, there are low-energy particle states that cannot be created from the vacuum when these states originate from a small region close to the horizon of a black hole. If this is not acceptable, there is a violation of unitarity. It was suggested in Ref. [130] that the late-time Hawking modes originate from massive string states contained within a "reservoir" of size $\mathcal{O}(\ell_s)$ surrounding the horizon. The origin of the non-conservation of the norm is that these states are excluded from the Hilbert space.[18]

Another possibility is that a proper description of these missing degrees of freedom will become accessible once we achieve a deeper understanding of spacetime. For instance, this includes how to characterize the Schwarzschild background for a trans-Planckian mode with $\Delta x \gg a$.

In any case, recall that the motivation behind the GUP in quantum gravity is to establish a minimal length scale, below which no degrees of freedom exist. A key observation from the GUP (7.5.1) is that the uncertainty Δx in position increases with momentum. Under the influence of the GUP, the width Δp of a Hawking wave packet $\Psi_{\omega_0,u_0}(v, p)$ (7.5.29) in momentum space scales with its central momentum $\bar{p}$ as

$$\Delta p \sim \left(\frac{dF}{dp}\right)^{-1}\Bigg|_{p=\bar{p}} = \bar{p}\left(1 + \ell_s^2\,\bar{p}^2\right). \tag{7.5.55}$$

This implies that at early times in the past, when the Hawking particle state has a highly blue-shifted momentum $\bar{p} \gg a/\ell_s^2$, its wave function not only exhibits a much broader spectrum $\Delta p \gg \bar{p}$ compared to the low-energy theory, but also an enormous spread in position since

$$\Delta x \gtrsim \ell_s^2\,(\Delta p) \gg \ell_s^2\,\bar{p} \gg a\,. \tag{7.5.56}$$

This indicates that the particle state is unlikely to be confined within the near-horizon region and that the uncertainty Δx in its location far exceeds the size of the black hole itself.[19]

---

[18] However, Ref. [130] missed the fact that Hawking radiation stops around the scrambling time because the magnitude of Hawking radiation was not derived.

[19] This feature may seem incompatible with the assumption of the near-horizon approximation $x \ll a$ that we started with in Eq. (7.5.8). That said, this issue was addressed in the appendix of Ref. [13], where a result similar to (7.5.47) was obtained without relying on the assumption of small x.

Notice that in the freely falling frames defined in Eq. (7.5.52), $x = r - a$ measures the proper radial distance from the horizon (up to an $\mathcal{O}(1)$ factor γ) on a constant proper time slice. Therefore, the relation $\Delta x \gg a$ represents a genuinely large distance in length scale.

For this reason, it was proposed in Ref. [13] that highly trans-Planckian momentum states with $p \gg a/\ell_s^2$ should be interpreted as *stringy* states that span macroscopic distances far beyond the characteristic length scale of the black hole. Such states are expected to be ignorant of the geometry of the near-horizon region, or anything that has a characteristic scale $\lesssim \mathcal{O}(a)$. Effectively they propagate in a smooth background of curvature $\ll \mathcal{O}(1/a^2)$. The horizon is invisible to these extremely high-energy modes. With their initial conditions being the Minkowski vacuum in the infinite past, the adiabatic theorem ensures that they remain approximately in the vacuum at a scale $\gg \mathcal{O}(a)$. This means that their quantum states shall agree with the asymptotic vacuum defined for distant observers, rather than the Unruh vacuum in the near-horizon region. Consequently, they do not contribute to Hawking radiation, leading to the termination of Hawking radiation after the scrambling time.

The fact that the vacuum state of a given propagating mode can only be defined in a sufficiently large space is not new: The vacuum state of a fluctuation mode of wavelength L can only be defined in a region much larger than L. Furthermore, this mode is insensitive to spacetime structures at a distance scale much shorter than L. When its wavelength L significantly exceeds the Schwarzschild radius a of a black hole, the black hole can effectively be ignored in the metric. This is why, with the galaxies and black holes smeared out, the Friedmann-Robertson-Walker (FRW) metric is suitable for describing the fluctuations at the scale of the universe.

In addition to the wave nature of a probe, there is also a stringy nature for trans-Planckian modes described above. The abovementioned idea of background smearing that applies to waves can be applied analogously to high-energy stringy probes.

7.5.3 Hawking Radiation in String Field Theory

In this section, we explore the nonlocal structure of string field theory (SFT), aiming to provide a more direct justification from string theory regarding the shutdown of Hawking radiation after the scrambling time.

7.5.3.1 Nonlocality in String Field Theory

A common feature of string field theories [141–144] is the exponential suppression of UV interactions. In d-dimensional Minkowski space $x^\mu = (t, x^i)$, each component field in the theory can be described by an action of the schematic form

$$S_{\text{SFT}}[\phi] = \int d^d x \left[\frac{1}{2} \phi \Box \phi - V(\tilde{\phi}, \partial\tilde{\phi}) \right], \tag{7.5.57}$$

where $\Box \equiv -\partial_0^2 + \partial_i^2$. The field ϕ appears in the interaction vertices $V(\tilde{\phi}, \partial\tilde{\phi})$ only in the nonlocal form

$$\tilde{\phi} \equiv e^{\ell_s^2 \Box/2}\, \phi\,, \tag{7.5.58}$$

while $V(\tilde{\phi}, \partial\tilde{\phi})$ itself contains at most derivative couplings of finite order in $\tilde{\phi}$. Actions of the form (7.5.57) have been employed as toy models to study various aspects of string field theories, such as causality, UV finiteness, and unitarity of the perturbation theory [145–156] .

It is crucial to note that all interaction vertices are dressed with the exponential operator $e^{\ell_s^2 \Box/2}$, so any measurement of field observables via coupling to an Unruh-DeWitt detector [47, 157] would be determined by the Wightman correlator $\left\langle 0 \left| \hat{\tilde{\phi}}(x_1) \cdots \hat{\tilde{\phi}}(x_n) \right| 0 \right\rangle$ for the UV-suppressed field $\tilde{\phi}$ instead of ϕ [14]. It is therefore useful to rewrite the action in terms of $\tilde{\phi}$ as [158]

$$S_{\text{SFT}}[\tilde{\phi}] = \int d^d x \left[\frac{1}{2}\, \tilde{\phi}\, \Box\, e^{-\ell_s^2 \Box}\, \tilde{\phi} - V(\tilde{\phi}, \partial\tilde{\phi}) \right], \tag{7.5.59}$$

which we will refer to as the *stringy model* for convenience.

The interaction terms are now local, and the nonlocality resulting from the operator $e^{-\ell_s^2 \Box}$ can be studied *non-perturbatively* with respect to the string length parameter ℓ_s by focusing on the free-field action:

$$S^{(0)}_{\text{SFT}}[\tilde{\phi}] = \frac{1}{2} \int d^d x\, \tilde{\phi}\, \Box\, e^{-\ell_s^2 \Box}\, \tilde{\phi}\,. \tag{7.5.60}$$

Previous efforts [29, 159] to address the stringy nonlocality have argued that it does not affect Hawking radiation, since transcendental functions of $\Box$ do not change the initial value problem for the original field equation $\Box\phi = 0$ [160, 161]. However, we provide below a recap of Ref. [14] on how a proper non-perturbative treatment of the nonlocal operator $e^{-\ell_s^2 \Box}$ in the stringy model (7.5.60) indicates significant modifications to the amplitude of late-time Hawking radiation.

The behavior of late-time Hawking radiation is encoded in the Wightman function $\langle 0 | \hat{\tilde{\phi}}(x)\, \hat{\tilde{\phi}}(x') | 0 \rangle$ of the field operator $\hat{\tilde{\phi}}(x)$ at arbitrarily short distances. However, the operator $e^{-\ell_s^2 \Box}$ in the action (7.5.60) introduces exponential UV divergences along time-like directions, which is apparent in the two-point correlation function

$$\left\langle \tilde{\phi}(x)\, \tilde{\phi}(x') \right\rangle = \int \frac{d^d p}{(2\pi)^d} \frac{-i e^{-\ell_s^2 p^2}}{p^2 - i\epsilon}\, e^{i p \cdot (x - x')} \tag{7.5.61}$$

from the path-integral formalism. Consequently, some form of analytic continuation, such as Euclideanizing the Minkowski spacetime [146, 148–150, 156], is inevitable to regularize the divergence and extract meaningful results from the stringy model.

In the following, we shall focus on the case of $d = 2$ for simplicity. Recently, a Hamiltonian approach for the stringy model in real time was developed based on consistency with its path-integral correlation functions [14, 162]. The idea is to work with the analytically continued string length parameter:

$$\ell_E^2 = -i\ell_s^2 > 0\,, \tag{7.5.62}$$

and formulate the theory in the light-cone frame

$$x^\pm \equiv t \pm x\,. \tag{7.5.63}$$

This is based on the observation that for modes with fixed retarded frequency p_- with respect to the coordinate x^-, the effect of the nonlocal operator $\exp\left(-i\ell_E^2\,\Box\right) = \exp\left(4i\ell_E^2\,\partial_{x^+}\partial_{x^-}\right)$ simplifies to a finite shift $\exp\left(4\ell_E^2\,p_-\,\partial_{x^+}\right)$ in the light-cone time x^+.[20]

In the continued space of $\ell_E^2 > 0$, the two-point function (7.5.61) can be obtained as

$$\begin{aligned}
&\left\langle\tilde{\phi}(x^+, x^-)\,\tilde{\phi}(y^+, y^-)\right\rangle \\
&= \int_{-\infty}^{\infty} dp_- \int_{-\infty}^{\infty} \frac{dp_+}{2\pi^2}\,\frac{-i\,e^{4i\ell_E^2 p_-\,p_+}}{-4p_-\,p_+ - i\epsilon}\,e^{-ip_-(x^- - y^-)}\,e^{-ip_+(x^+ - y^+)} \\
&= \int_{-\infty}^{\infty} \frac{dp_-}{4\pi p_-}\,e^{-ip_-(x^- - y^-)} \int_{-\infty}^{\infty} \frac{dp_+}{2\pi}\,\frac{i}{p_+ + i\epsilon/p_-}\,e^{-ip_+(x^+ - y^+ - 4\ell_E^2 p_-)} \\
&= \int_0^{\infty} \frac{dp_-}{4\pi p_-}\left[e^{-ip_-(x^- - y^-)}\,\Theta(x^+ - y^+ - 4\ell_E^2 p_-) + e^{ip_-(x^- - y^-)}\,\Theta(y^+ - x^+ - 4\ell_E^2 p_-)\right].
\end{aligned} \tag{7.5.64}$$

The goal is to construct the field operator $\hat{\tilde{\phi}}(x^+, x^-)$ such that the correlation function can be reproduced through the correspondence

$$\begin{aligned}
&\left\langle\tilde{\phi}(x^+, x^-)\,\tilde{\phi}(y^+, y^-)\right\rangle = \left\langle 0\left|\mathcal{T}\left\{\hat{\tilde{\phi}}(x^+, x^-)\,\hat{\tilde{\phi}}(y^+, y^-)\right\}\right|0\right\rangle \\
&\equiv \Theta(x^+ - y^+)\left\langle 0\left|\hat{\tilde{\phi}}(x^+, x^-)\,\hat{\tilde{\phi}}(y^+, y^-)\right|0\right\rangle + \Theta(y^+ - x^+)\left\langle 0\left|\hat{\tilde{\phi}}(y^+, y^-)\,\hat{\tilde{\phi}}(x^+, x^-)\right|0\right\rangle
\end{aligned} \tag{7.5.65}$$

between the path-integral formalism and the vacuum expectation value of time-ordered operators. (We have added hats on operators in the Hamiltonian formalism.) By decomposing the outgoing sector of the field operator $\hat{\tilde{\phi}}(x^+, x^-)$ into Fourier modes $e^{-ip_- x^-}$, we write

$$\hat{\tilde{\phi}}(x^+, x^-) = \int_0^{\infty} \frac{dp_-}{\sqrt{4\pi p_-}}\left[\hat{a}_{p_-}(x^+)\,e^{-ip_- x^-} + \hat{a}_{p_-}^\dagger(x^+)\,e^{ip_- x^-}\right], \tag{7.5.66}$$

[20] Since it is the combination $\ell_s^2\,\partial_{x^+}$ that appears in the exponent, this approach can alternatively be carried out with a real string length parameter ℓ_s while adopting the Euclidean time $x_E^+ = ix^+ \in \mathbb{R}$ [162].

and define the vacuum state $|0\rangle$ by

$$\hat{a}_{p_-}(x^+)\,|0\rangle = 0 \qquad \forall\; p_- > 0\,. \tag{7.5.67}$$

We can then deduce from the correspondence (7.5.65) a self-consistent algebra for the creation and annihilation operators [14]:

$$\left[\hat{a}_{p_-}(x^+)\,,\,\hat{a}^\dagger_{p'_-}(y^+)\right] = \Theta\left(|x^+ - y^+| - 4\ell_E^2 p_-\right)\,\delta(p_- - p'_-)\,, \tag{7.5.68}$$

$$\left[\hat{a}_{p_-}(x^+)\,,\,\hat{a}_{p'_-}(y^+)\right] = \left[\hat{a}^\dagger_{p_-}(x^+)\,,\,\hat{a}^\dagger_{p'_-}(y^+)\right] = 0\,. \tag{7.5.69}$$

Interestingly, the nonlocality of the stringy model is translated to a p_--dependent minimal x^+ separation in the commutation relation of the creation-annihilation operators.

These unequal time commutation relations allow us to construct an operator formalism where the outgoing Wightman function is given by

$$\langle 0|\hat{\tilde{\phi}}(x^+, x^-)\,\hat{\tilde{\phi}}(y^+, y^-)|0\rangle = \int_0^{|x^+ - y^+|/4\ell_E^2} \frac{dp_-}{2\pi}\,\frac{e^{-ip_-(x^- - y^-)}}{2p_-}\,. \tag{7.5.70}$$

This reveals that the light-cone formulation of the stringy model (7.5.60) exhibits nonlocality along light-cone directions, with a UV cutoff

$$p_- \;\leq\; \frac{|x^+ \,-\, y^+|}{4\ell_E^2} \tag{7.5.71}$$

on the retarded frequency, or equivalently, a minimal uncertainty in the advanced light-cone direction:

$$\Delta x^+ \;\geq\; 4\ell_E^2 p_-\,, \tag{7.5.72}$$

where $\Delta x^+ \equiv |x^+ \,-\, y^+|$.

As noticed in Ref. [14], this reflects a *UV/IR connection* in the stringy model, where the correlator between two spacetime points is nonvanishing only when their separation in the light-cone direction x^+ (or x^-) exceeds the nonlocality length scale $4\ell_E^2\, p_-$ (or $4\ell_E^2\, p_+$), which becomes macroscopic for large p_- (or p_+). Approaching the UV limit $p_- \to \infty$ demands the IR limit $|x^+ - y^+| \to \infty$. Moreover, due to the standard uncertainty principle $\Delta x^- \Delta p_- \geq 1$ between conjugate variables, the upper bound (7.5.71) also implies

$$\Delta x^- \;\geq\; \frac{1}{\Delta p_-} \;\gtrsim\; \frac{1}{p_-} \;\geq\; \frac{4\ell_E^2}{\Delta x^+}\,. \tag{7.5.73}$$

Thus, we find that a *Lorentz-invariant* space-time uncertainty relation [14]

$$\Delta x^+ \Delta x^- \gtrsim 4\ell_E^2 \tag{7.5.74}$$

is naturally incorporated into the operator formalism for the stringy model (7.5.60) under a "Wick rotation" of ℓ_s^2.

Due to the nontrivial x^+-dependence of the ladder operators $\hat{a}_{p_-}(x^+)$ and $\hat{a}^\dagger_{p_-}(x^+)$ required by the correspondence (7.5.65) with the path integral result, the field operator $\hat{\tilde{\phi}}(x^+, x^-)$ constructed in Eq. (7.5.66) does not satisfy the equation of motion, i.e. $e^{-i\ell_E^2\Box}\,\Box\hat{\tilde{\phi}} \neq 0$. This is expected in higher-derivative theories of infinite order. The equation of motion should instead be imposed as a constraint defining the physical Hilbert space [162–167]. This feature marks a major difference from the perturbative treatments of nonlocal theories [159, 160, 168].

The x^+-dependence of these operators appears to introduce an infinite number of extra degrees of freedom into the state space of the stringy model—a common feature of theories with infinite time derivatives that typically leads to instabilities or violations of unitarity. However, as shown in Ref. [162], for the particular model under consideration, the physical-state constraint ends up eliminating the negative-norm states and, apart from the decoupled zero-norm states, reducing the physical degrees of freedom to match those of standard local quantum field theory. (See Ref. [162] for more details.)

For our purpose, we only need a free field theory (including a background field interaction). While our calculation of Hawking radiation only relies on the Wightman function, it has been shown that its Hamiltonian formalism can be constructed consistently [162]. The progress is based on a light-cone formulation with a light-cone time coordinate and the analytic continuation of the string tension. This is a remarkable result, as the Hamiltonian formalisms for nonlocal theories are typically pathological [169–171]. After all, when there is nonlocality in the time direction, it is not clear what the meaning of a Hamiltonian formalism is. It remains to be seen whether a complete, consistent Hamiltonian formalism including all component fields in SFT can be achieved by the same approach.

7.5.3.2 Derivation of Hawking Radiation in Stringy Model

We now apply the newly developed operator formalism for the stringy model to the radiation field $\tilde{\phi}$ in a Schwarzschild background. We shall focus on the near-horizon region again, where the Kruskal coordinates U, V are identified with the light-cone coordinates x^-, x^+ above, and the frequency Ω with p_-.

Following the procedure outlined in Sect. 7.2.3.4, a Hawking particle with central frequency $\omega = \omega_0$ localized around the Eddington retarded coordinate $u = u_0$ can be described by the wave packet

$$\psi(u) = \int_0^\infty \frac{d\omega}{\sqrt{\omega}}\, f_{\omega_0}(\omega)\, e^{-i\omega(u-u_0)}\,. \tag{7.5.75}$$

Let $\hat{b}_\omega(v)$ and $\hat{b}^\dagger_\omega(v)$ denote the ladder operators associated with an outgoing Hawking mode $e^{-i\omega u}$ with positive frequency $\omega > 0$ near asymptotic infinity. The annihilation operator $\hat{b}_\psi$ corresponding to the wave packet (7.5.75) is defined as

$$\hat{b}_{\psi}(v) \equiv \int_0^{\infty} d\omega\, f^*_{\omega_0}(\omega)\, e^{-i\omega u_0}\, \hat{b}_{\omega}(v)\,, \tag{7.5.76}$$

where the only deviation from the standard case is the dependence of the operators on the advanced Eddington time v due to the nonlocality in the stringy model.

In the freely falling frame near the horizon, the outgoing sector of the radiation field $\hat{\tilde{\phi}}$ has the Minkowski mode expansion

$$\hat{\tilde{\phi}}(U, V) = \int_0^{\infty} \frac{d\Omega}{\sqrt{4\pi\Omega}} \left[\hat{a}_{\Omega}(V)\, e^{-i\Omega U} + \hat{a}^{\dagger}_{\Omega}(V)\, e^{i\Omega U}\right]. \tag{7.5.77}$$

The quantity of interest is the expectation value of the number operator $\hat{n}_{\psi}(v, v') = \hat{b}^{\dagger}_{\psi}(v)\, \hat{b}_{\psi}(v')$ in the free-fall vacuum state $|0\rangle$ defined by

$$\hat{a}_{\Omega}(V)\, |0\rangle = 0 \qquad \forall\, \Omega > 0\,. \tag{7.5.78}$$

Although the ladder operators acquire time dependence in the stringy model, they are still related by the same Bogoliubov transformation (7.2.47) in the conventional theory:

$$\langle 0|\, \hat{b}^{\dagger}_{\omega}\big(v(V)\big)\, \hat{b}_{\omega'}\big(v'(V')\big)\, |0\rangle = \int_0^{\infty} d\Omega \int_0^{\infty} d\Omega'\, \beta^*_{\omega\Omega}\, \beta_{\omega'\Omega'}\, \langle 0|\, [\hat{a}_{\Omega}(V)\,, \hat{a}^{\dagger}_{\Omega'}(V')]\, |0\rangle\,, \tag{7.5.79}$$

since the Bogoliubov coefficients $\beta_{\omega\Omega}$ are completely determined by the exponential relation $U(u) = -2ae^{-u/2a}$ between the retarded coordinates.

Inserting the free-field commutator (7.5.68)

$$[\hat{a}_{\Omega}(V)\,, \hat{a}^{\dagger}_{\Omega'}(V')] = \delta(\Omega - \Omega')\, \Theta(|V - V'| - 4\ell_E^2\Omega) \tag{7.5.80}$$

into Eq. (7.5.79) leads to the number expectation value [14]

$$\begin{aligned}
&\langle 0|\, \hat{n}_{\psi}\big(v(V), v'(V')\big)\, |0\rangle \\
&= \int_0^{\infty} d\omega \int_0^{\infty} d\omega'\, f^*_{\omega_0}(\omega)\, f_{\omega_0}(\omega')\, e^{i(\omega-\omega')u_0}\, \langle 0|\, \hat{b}^{\dagger}_{\omega}\big(v(V)\big)\, \hat{b}_{\omega'}\big(v'(V')\big)\, |0\rangle \\
&\simeq \frac{a}{\pi} \frac{1}{e^{4\pi a\omega_0} - 1} \int_0^{\infty} d\omega \int_0^{\infty} d\omega'\, f^*_{\omega_0}(\omega)\, f_{\omega_0}(\omega')\, e^{i(\omega-\omega')u_0} \\
&\quad \times \int_0^{|V-V'|/4\ell_E^2} \frac{d\Omega}{\Omega}\, (2a\Omega)^{-2ia(\omega-\omega')}\,.
\end{aligned} \tag{7.5.81}$$

By performing a change of variable $\Omega \mapsto u(\Omega) = 2a \log(2a\Omega)$, the Ω-integral above can be evaluated as

$$\int_0^{|V-V'|/4\ell_E^2} \frac{d\Omega}{\Omega} (2a\Omega)^{-2ia(\omega-\omega')} = \int_{-\infty}^{u_\Lambda(|V-V'|,\ell_E^2)} \frac{du}{2a} e^{-i(\omega-\omega')u} , \tag{7.5.82}$$

where

$$u_\Lambda(|V-V'|, \ell_E^2) \equiv 2a \log\left(\frac{a|V-V'|}{2\ell_E^2}\right). \tag{7.5.83}$$

Ultimately, the physical observables of the original theory are recovered through an analytic continuation $\ell_E^2 \to -i\ell_s^2$ back from the complexified string length. Under this continuation, $u_\Lambda(|V-V'|, \ell_E^2)$ as a function of ℓ_E^2 gets turned into

$$u_\Lambda(|V-V'|, \ell_E^2) \quad \to \quad u_\Lambda(|V-V'|, \ell_s^2) + i\pi a . \tag{7.5.84}$$

The additional term $i\pi a$ introduces an extra factor $e^{\pi a(\omega-\omega')}$ in the integrand of Eq. (7.5.82), which contributes merely a factor of $\mathcal{O}(1)$ to the final outcome (7.5.81) since the compact profile f_{ω_0} restricts both ω and ω' to values near ω_0. Thus, the expectation value becomes

$$\langle 0| \hat{n}_\psi(V, V') |0\rangle \simeq \frac{1}{2\pi} \frac{\omega_0}{e^{4\pi a\omega_0} - 1} \int_{-\infty}^{u_\Lambda(|V-V'|,\ell_s^2)} |\psi(u)|^2 \, du , \tag{7.5.85}$$

which again showcases an amplitude that decays in u_0, this time due to the UV cutoff $\Omega \leq |V - V'| / 4\ell_E^2$. The temperature remains the same as the Hawking temperature (7.2.4).

However, we emphasize that the physical interpretation here differs significantly from a simple truncation of the effective field theory by imposing a UV cutoff on the retarded frequency Ω, as it was done in Sect. 7.2.3.5. Throughout the discussion, we have not excluded the presence of trans-Planckian modes. Instead, as we will argue, these modes simply do not contribute to the Hawking process.

In a physical setting, a stationary observer would detect radiation over a finite duration V_d of light-cone time and measure the expectation value

$$N_\psi(V_\mathrm{d}) \equiv \frac{1}{V_\mathrm{d}^2} \int dV \, s(V) \int dV' \, s(V') \langle 0| \hat{n}_\psi(V, V') |0\rangle , \tag{7.5.86}$$

where the switching function $s(V)$ [172] associated with a particle detector is normalized as

$$\int s(V) \, dV = V_\mathrm{d} . \tag{7.5.87}$$

Since the wave packet $\psi(u)$ (7.5.75) of a Hawking particle is localized around the retarded time $u = u_0$ with a width of $\mathcal{O}(a)$, Eq. (7.5.85) indicates that the mean number $N_\psi(V_\mathrm{d})$ of Hawking particles measured would vanish for sufficiently large u_0 when

$$u_0 \gg u_\Lambda(V_\mathrm{d}\,,\ell_s^2) + \mathcal{O}(a)\,, \tag{7.5.88}$$

where u_Λ is defined in Eq. (7.5.83). In particular, the expression (7.5.83) for $u_\Lambda(V_\mathrm{d}\,,\ell_s^2)$ implies that a Hawking particle emitted on the order of the scrambling time scale with

$$u_0 = 2\,(n+1)\,a\log\left(\frac{a}{\ell_s}\right) \tag{7.5.89}$$

requires a detection range

$$V_\mathrm{d} \gg \frac{\ell_s^2}{a}\,e^{u_0/2a} = \left(\frac{a}{\ell_s}\right)^{n-1} a \tag{7.5.90}$$

that vastly exceeds the size a of the black hole as long as $n > 1$.

The result (7.5.90) is simply a reflection of the UV/IR constraint (7.5.71) for a large $p_- = \Omega$. A Hawking particle with central frequency $\omega_0 \sim 1/a$ at retarded time u_0 (7.5.89) involves modes with trans-Planckian free-fall frequencies

$$\Omega \sim \frac{1}{a}\,e^{u_0/2a} = \left(\frac{a}{\ell_s}\right)^{n-1} \frac{a}{\ell_s^2}\,. \tag{7.5.91}$$

Rather than being confined within a Planckian distance $\sim 1/\Omega$ around the horizon, these trans-Planckian modes should be viewed as extremely long stringy states that span an interval (see Eq. (7.5.71) and identify $|x^+ - y^+|$ with ΔV)

$$\Delta V \;>\; 4\ell_s^2\Omega \sim \left(\frac{a}{\ell_s}\right)^{n-1} a\,, \tag{7.5.92}$$

which far exceeds the size of the black hole for $n > 1$, as illustrated in Fig. 7.7. Equation (7.5.90) is reproduced as the requirement $V_d \gg \Delta V$. That is, the detector has to be switched on for a duration V_d much longer than the uncertainty ΔV.

The black-hole evaporation process can be viewed as a collection of scattering processes. The initial states of the scattering include the ingoing collapsing matter and the Minkowski vacua of outgoing modes. The final states include a black hole and outgoing Hawking particles. In the low-energy approximation, the Schwarzschild geometry accounts for the effect of the exchanges of virtual gravitons.[21] This low-energy effective description fails for Hawking particles after the scrambling time, as we saw in Sect. 7.4. In the high-energy limit, the UV/IR relation (7.5.71) obtained in the stringy model suggests that the underlying geometry is smeared over at a scale $\Delta V \gg a\,(a/\ell_s)^{n-1}$ (7.5.92). From the viewpoint of Hawking particles after the scrambling time, the black hole is smeared by a length scale much larger than a.

---

[21] Recall that the Schwarzschild metric can be reconstructed from a perturbative expansion of scattering amplitudes in Minkowski space in low-energy effective theories [173–178].

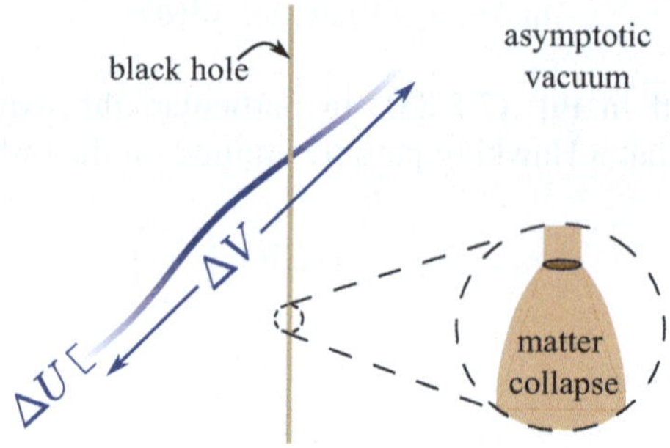

Fig. 7.7 A schematic illustration of a high-frequency outgoing wave packet with a narrow width ΔU. The wave packet behaves like an extended string spanning a range $\Delta V > 4\ell_s^2/\Delta U$. The vertical line represents the worldvolume of matter that eventually collapses into a black hole with a size considerably smaller than ΔV. Since the wave packet cannot be localized near the horizon, it does not contribute to Hawking radiation

For example, consider a solar-mass black hole whose Schwarzschild radius is $3\ km$ and assume that $\ell_s = \ell_p$. For $n = 2$ or equivalently $u_0 \sim 5\ ms$, the required detection period $V_{\rm d}$ has to be much larger than 10^{15} times the size of the observable universe!

Similar to the situation with the GUP, the black-hole background would have had little to no impact on the evolution of these high-frequency modes.[22] These modes remain in the asymptotic vacuum state and do not contribute to particle creation in the first place. As a result, we conclude that Hawking radiation no longer exists after the scrambling time $\mathcal{O}\big(a \log(a/\ell_s)\big)$.

7.6 Comments

We examined two UV models studied in the literature for Hawking radiation but arrived at the surprising conclusion that Hawking radiation is turned off as early as the scrambling time. Why has this simple resolution of the information paradox not been proposed earlier? How does this proposal contradict or align with the existing literature on black-hole evaporation? We comment on related questions in this section.

[22] This statement is supported by demonstrating that the corrections to the correlation function $\left\langle\tilde{\phi}(U, V)\,\tilde{\phi}(U', V')\right\rangle$ due to interactions between high-energy modes and a time-dependent background with certain compact profiles are highly suppressed [162].

7.6.1 Comments on Theoretical Issues

Before commenting on related works in Sect. 7.6.2, we first discuss the theoretical issues involved in our proposal.

7.6.1.1 Robustness of Hawking Radiation

While the magnitude of Hawking radiation is sensitive to UV physics, the Hawking temperature ($T_H = 1/4\pi a$ as given in Eq. (7.2.4)) remains robust against Planck-scale effects, with perturbative corrections of $\mathcal{O}(\ell_p^2/a^2)$ [13].

The mathematical reason is that the Boltzman factor $e^{-4\pi a\omega}$, which appears from the ratio $|\beta_{\omega\Omega}/\alpha_{\omega\Omega}|^2$ of the Bogoliubov coefficients (7.2.48)–(7.2.49), is an infrared feature essentially determined by the analytic continuation of the factor $(2a\Omega - i\epsilon)^{2ia\omega}$ across $\Omega = 0$ from positive to negative values of Ω. The $i\epsilon$-prescription that one should adopt in the Fourier transform of an outgoing Hawking mode $e^{-i\omega u}$ is fixed by the causality and locality of the mode propagation.

Most claims in the literature regarding the robustness of Hawking radiation refer specifically to the robustness of the Hawking temperature. In particular, the insensitivity of the Hawking temperature and its associated thermodynamic properties to the GUP has been noted repeatedly in past studies [179–192]. However, discussions often overlook details about the magnitude of radiation. The time dependence of the magnitude of Hawking radiation has not been thoroughly examined as the temperature has been.

Since the Hawking temperature remains the same, the usual derivation of the Bekenstein-Hawking entropy (7.3.1) from $dS = dM/T_H$ remains valid.

As a dynamic process, a time-dependent magnitude of radiation from the collapsing matter is somewhat expected. This is analogous to the thermal Larmor radiation emitted by a moving point charge, which has an exceedingly small amplitude if the final speed is sufficiently low [193].

7.6.1.2 Local Lorentz Symmetry and Nice Slices

In Sect. 7.3.1, we explained that local Lorentz symmetry and the nice-slice argument have been utilized to claim that UV physics is irrelevant to Hawking radiation, and so Hawking radiation is expected to persist until the black hole becomes microscopic; we also clarified there that these are misconceptions.

In Sect. 7.4, we demonstrated explicitly how the low-energy effective theory fails in S-matrix calculations concerning Hawking radiation when higher-derivative interactions are considered. The nice-slice argument merely ensures that a low-energy effective theory can describe the time evolution of the background geometry defined in the low-energy limit; however, it says nothing about the trans-Planckian scattering involving the background and the Hawking particle. The nice-slice argument cannot ensure that Hawking radiation persists adiabatically.

We have also shown in Sect. 7.5.3 that a Lorentz-invariant uncertainty relation (7.5.74) at the Planck scale inhibits Hawking radiation around the scrambling time. Just like long wavelength modes perceive a smoother background, Eq. (7.5.72) implies that trans-Planckian modes do, too. Instead of saying that local Lorentz symmetry is broken, we would rather say that the spacetime geometry is probe-dependent, and the notion of local Lorentz symmetry is different for different probes.

7.6.2 Comments on Related Works

The information paradox has attracted a lot of attention over the past 50 years. Numerous works have been devoted to this topic. Here we comment on the connections with other works in the literature.

7.6.2.1 AdS/CFT Duality and Entanglement Entropy

The AdS/CFT duality [80, 81] serves as a concrete illustration of the holographic principle of quantum gravity [78, 79], having collected extensive evidence over time. This duality offers profound insights into the physics underlying the Bekenstein-Hawking entropy (7.3.1) and sheds light on the information paradox. On the other hand, while it provides a strong argument for unitarity (no information loss), it does not explicitly reveal the underlying physical mechanism.

The AdS/CFT duality motivated a formula for the generalized fine-grained entropy [194–197]. Recently, with the help of "entanglement islands" [97–102], the generalized entropy formula reproduced the Page curve for an evaporating black hole. However, despite these advancements, the island proposal still faces various criticisms [198–201] and lacks a clear explanation of the physical mechanism for information retrieval [202] or the ultimate fate of the black hole.

Notably, the generalized entropy formula applies to any spacetime geometry, whether it describes a classical black hole or one undergoing continuous evaporation. Regarding the scenario proposed in this chapter, where Hawking radiation terminates near the scrambling time, the entanglement island is irrelevant, as its effects would only become significant if Hawking radiation persists beyond the Page time.

It is important to note that nonlocality is a crucial aspect of the discussion surrounding entanglement islands, as well as many other proposed resolutions to the information paradox. Indeed, the essence of the AdS/CFT duality—the holographic principle [78, 79]—may stem from a type of nonlocality that reveals a UV/IR connection.

In our analysis, we have explored specific forms of nonlocality (GUP and STUR) inspired by string theory studies. However, it remains uncertain whether these formulations accurately reflect the nonlocality inherent in string theory. It will be interesting to see whether they correspond to the same type of nonlocality that contributes to the holographic nature of string theory.

7.6.2.2 Fuzzball

The fuzzball proposal [203, 204] claims that horizons exist only in coarse-grained spacetime geometries (see, for example, Refs. [205, 206] for recent reviews). What we traditionally understand as a black hole is actually an ensemble of horizonless microstate configurations that share the same mass, charge, and angular momentum as a classical black hole.

Since fuzzballs lack horizons, the exponential relation (7.2.11) between U and u is altered near the Schwarzschild radius. Consequently, the conventional derivation of Hawking radiation does not apply. This aspect of the fuzzball paradigm is similar to our proposal in the sense that Hawking radiation occurs only for a brief period. On the other hand, fuzzballs exhibit enhanced radiation [207] due to classical instabilities in spacetime geometry. This situation is analogous to the burning of a piece of coal. The fuzzball's radiation has a fundamentally different origin compared to traditional Hawking radiation.

Both the fuzzball proposal and the approach presented in this chapter are motivated by string theory, though each emphasizes different aspects of it.

In general, whether a collapsing matter shell tunnels into a fuzzball [208, 209] may depend on its initial state. This suggests that one might be able to classify black holes based on their ultimate fates, with possibly multiple scenarios for their final state configurations.

7.6.2.3 Firewall

It was argued that, assuming that information is transmitted from matter to Hawking radiation, there must be a firewall before Page time [17]. As the Unruh vacuum is a pure state, Hawking particles outside the apparent horizon are maximally entangled with their partners inside the black hole, hence the quantum state inside the apparent horizon can no longer be viewed as a vacuum state. At a later time, when the partner emerges from the shrinking apparent horizon, its identity is already determined by its maximally entangled Hawking particle detected in the early radiation. As such, the quantum state at the apparent horizon can no longer be viewed as a pure state. It must be viewed as a particle at the apparent horizon even for freely falling observers, and its frequency is trans-Planckian due to the large blue-shift factor.[23]

This proposal argues that the firewall is a necessity for information to come out in Hawking radiation after the Page time. In our proposal, Hawking radiation stops

[23] The firewall can also be argued from the AdS/CFT duality [103] without using the entanglement argument.

well before the Page time and information stays inside the black hole, so there is no need for a firewall or the violation of the equivalence principle.

7.6.2.4 Massive Remnants

Our proposal bears resemblance to the concept of "*massive remnants*" [210], which suggests that Hawking radiation terminates at a time well beyond the Page time, leaving behind a macroscopic remnant characterized by an entropy density limited by the Planck density. This type of remnant was proposed to retain only a small fraction $\sim (\ell_p M)^{-1/3}$ of its initial mass (but still macroscopic). It does not resolve the conflict between the Bekenstein-Hawking entropy and the entanglement entropy that arises around the halfway point of evaporation (see Sect. 7.3).

A similar idea is the memory burden effect [211], which suggests that the further decay of a black hole is suppressed beyond the Page time. Ultimately, these proposals require a mechanism to stop Hawking radiation well before the black hole becomes microscopic. In out approach, this early termination is accomplished through the nonlocal nature of strings.

7.6.2.5 Analogue Gravity

The results from analogue gravity have contributed to the belief in the adiabaticity of Hawking radiation. The original concept emerged from Unruh's sonic model [212] of a black hole, which is characterized by an inhomogeneous fluid flow and a sonic horizon. In this analogy, the spontaneous emission of sound waves from the acoustic horizon can be likened to Hawking radiation [213]. The breakdown of the continuous description of the fluid, caused by atomic effects at submolecular scales, allows analogue black holes to be used as a platform to investigate Hawking radiation and its connections to short-distance physics.

Nevertheless, in much of the analogue gravity literature [214], the introduction of new UV physics is primarily limited to modifications of the dispersion relation in the UV regime. Although Lorentz symmetry may be broken, such modifications are relatively mild compared to a UV cutoff or nonlocality. Consequently, it is plausible that many studies report only a slightly modified version of Hawking radiation.

An exception is recently reported in Ref. [12], which presents a UV dispersion relation that alters the propagation of the Hawking modes, making them sensitive to the boundary conditions at the black hole singularity. Under specific boundary conditions, this modification also leads to the termination of Hawking radiation.

7.6.2.6 Eternal Black Holes

The information paradox is initially posed in the context of a dynamical black hole, as discussed throughout this work. However, an analogous issue arises for an eternal black hole in AdS space [215]. In AdS_4, if Hawking radiation does not carry information and proceeds as the low-energy effective theory predicts, a bulk correlation function should decay exponentially over time. The paradox lies in the fact that boundary correlators in the CFT on S^3, which are linked to the bulk correlators via the extrapolate dictionary, can only decay to a minimum value $e^{-c\,\mathcal{S}_{\rm BH}}$, where c is a constant of order one and $\mathcal{S}_{\rm BH}$ is the Bekenstein-Hawking entropy (7.3.1).

Our proposal offers a potential resolution to this discrepancy by suggesting that the effects of Hawking radiation may effectively be disregarded. In this scenario, nearly all information remains within the black hole (which only has a horizon in the low-energy effective theory), which possesses a degeneracy of approximately $e^{c\,\mathcal{S}_{\rm BH}}$. Consequently, it is natural for the correlation function to asymptotically approach $e^{-c\,\mathcal{S}_{\rm BH}}$ as $t \to \infty$.

7.6.2.7 Cosmology

A challenge analogous to the trans-Planckian problem in Hawking radiation also arises in cosmological inflation, where low-energy effective theories are utilized to derive primordial cosmological density perturbations despite the involvement of trans-Planckian modes [216, 217].

A UV cutoff that halts Hawking radiation at the scrambling time is translated to a UV cutoff in the static patch of de Sitter space that stops the Gibbons-Hawking radiation [218]. A bound ($t \sim H^{-1}\log(H^{-1}M_p)$) analogous to the scrambling time has been established as the time scale of the validity of effective field theory in de Sitter space [218], agreeing with the predictions from the Trans-Planckian Censorship Conjecture (TCC) [219, 220]. Turning off Hawking radiation at the scrambling time is thus closely related to the TCC.

The TCC suggests that the universe avoids revealing details about trans-Planckian modes to the macroscopic world. This idea aligns with the intuition of a minimal uncertainty in the spacetime geometry. In Ref. [221], the spacetime uncertainty relation is examined in the context of the TCC.

In the early universe, primordial black holes form during the Big Bang. Smaller black holes, typically assumed to evaporate completely, radiate at higher temperatures. If Hawking radiation stops in the early stages and these black holes survive, their implications for cosmology could be significant as they may act as dark matter candidates. Further exploration of this new mass window for primordial black holes inspired by string theory considerations would be intriguing. Similar studies motivated by the memory burden effect that suppresses black-hole decay near the halfway point of evaporation have been considered in Refs. [222–224].

7.7 Conclusion

In the string-motivated models discussed in Sect. 7.5, the most important ingredient that turns off Hawking radiation is the nonlocality exhibiting a UV/IR connection: there is an IR uncertainty in the UV limit. Although the nonlocality is restricted to the UV regime, it has an IR effect. It is reminiscent of the UV/IR connections observed in T-duality, noncommutative field theories, AdS/CFT duality, etc. in string theory. This UV/IR connection is the reason why the quantum modes corresponding to would-be Hawking particles after the scrambling time have spatial uncertainties too large to fit inside the near-horizon region, causing Hawking radiation to cease.

Hawking radiation can be viewed as a scattering process between the radiation field and the collapsing matter through gravitational interactions. In the UV models discussed in Sects. 7.5.2 and 7.5.3, the larger nonlocality associated with higher-frequency modes is expected to result in a smeared geometry. When this nonlocality for trans-Planckian modes surpasses the size of the black hole, the black hole is smeared out as a smooth background (see Fig. 7.8), leading to a cessation of Hawking radiation.

From the perspective of a UV theory that may radically differ from our conventional understanding of spacetime, geometric notions such as the Schwarzschild metric only emerge as effective descriptions in the low-energy regime. They are low-energy effective concepts suitable only for low-energy probes. In string field theories, the nonlocality in spacetime originates from exponentially suppressed interactions in the UV limit. This feature of a weaker gravity in the high-energy limit is shared by other theories, such as asymptotically safe gravity [225–228]. Our idea is also reminiscent of gravity's rainbow [229], that is, the spacetime geometry depends on the energy scale of the propagating modes, but we are free from the dangers of violating Lorentz symmetry in flat space [230]. In our proposal, there is no correction to the Minkowski spacetime.[24]

This chapter proposes an alternative resolution to the black hole information paradox. Instead of focusing on the details of Hawking radiation at late times to look for the desired information transmission from the collapsing matter, we argue that Hawking radiation is essentially suppressed before the Page time, stopping around the scrambling time due to the dominance of trans-Planckian effects. This suppression is arguably a consequence of the fundamental nonlocality present in UV-complete theories of quantum gravity such as string theory.

As black holes can approximately be treated as classical objects in this scenario, all paradoxes associated with Hawking radiation are alleviated. This perspective offers a conceptually straightforward resolution to the information paradox.

This claim is supported in this chapter with the following discussions:

1. Analyzing the breakdown of the low-energy effective theory: We demonstrate that higher-derivative non-renormalizable interactions, always present in any

[24] Hence, we also do not expect significant modifications to the Unruh effect in Minkowski spacetime, which is consistent with the findings of Ref. [231].

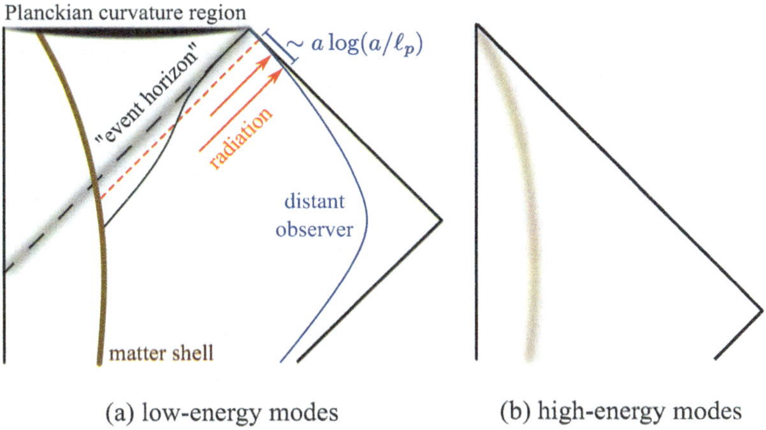

Fig. 7.8 Panel (**a**) shows the Penrose diagram of a black hole with Hawking radiation terminated at an early stage marked by the dashed red line. The background is defined in the low-energy limit. While the spacetime metric may not be well-defined for trans-Planckian modes, panel (**b**) schematically illustrates the Penrose diagram of a highly smeared black-hole geometry, representing the background on which the vacuum states for extremely high-energy modes with $\Delta x \gg a$ are defined

UV completion of general relativity, lead to exponentially growing contributions to particle creation around the scrambling time, invalidating perturbative calculations of Hawking radiation (see Sect. 7.4).

2. Modeling UV physics using GUP and SFT: Two concrete UV models, one incorporating the generalized uncertainty principle (GUP) and the other based on string field theory (SFT), are analyzed to show that trans-Planckian modes effectively do not "see" the black hole geometry and thus do not contribute to late-time Hawking radiation. The suppression of Hawking radiation is attributed to the inherent nonlocality in these models (see Sect. 7.5).
3. Avoiding firewalls and other exotic physics: The challenges associated with transmitting information from the collapsing matter to the radiation, such as the firewall or violation of the no-cloning theorem (see Sect. 7.3), are avoided.

A UV suppression and a corresponding uncertainty relation are important in our proposal. In lower (two or three) spacetime dimensions, a UV-finite theory does not always need a UV suppression factor like $e^{\ell_s^2 \Box}$, and a spacetime uncertainty relation may not be present. Therefore, the problem of Hawking radiation in lower dimensions [232, 233] can be fundamentally different from higher dimensions.

The early cessation of Hawking radiation, driven by trans-Planckian nonlocality, offers a conceptually simple resolution to the information paradox, suggesting that black-hole evaporation is a far less dramatic process than previously assumed and that most information remains within the black hole. We need a nonlocal effect to exclude trans-Planckian modes in a small region near the horizon, but we do not need a nonlocal mechanism for information transfer. In addition to offering the simplest resolution to the information paradox, this work also sheds light on our understanding of the nonlocal UV/IR feature of quantum gravity.

Acknowledgements This work is an invited chapter for the edited book "The Black Hole Information Paradox" (Eds. Ali Akil and Cosimo Bambi, Springer Singapore, expected in 2025). We thank Emil Akhmedov, Robert Brandenberger, Ronny Chau, Chong-Sun Chu, Nick Dorey, Shinji Mukohyama, Toshifumi Noumi, Nobuyoshi Ohta, Domenico Orlando, Susanne Reffert, Yuki Yokokura, and Cheng-Tsung Wang for valuable discussions. This work is supported in part by the Ministry of Science and Technology, R.O.C. (MOST 110-2112-M-002 -016-MY3), and by the National Taiwan University. H.K. thanks Prof. Shin-Nan Yang and his family for their kind support through the Chin-Yu chair professorship. H.K. is partially supported by the Japan Society of Promotion of Science (JSPS), Grants-in-Aid for Scientific Research (KAKENHI) Grants No. 20K03970 and 18H03708, by the Ministry of Science and Technology, R.O.C. (MOST 111-2811-M-002-016), and by the National Taiwan University.

References

1. S.W. Hawking, Breakdown of predictability in gravitational collapse. Phys. Rev. D **14**, 2460 (1976). https://doi.org/10.1103/PhysRevD.14.2460
2. S.W. Hawking, The unpredictability of quantum gravity. Commun. Math. Phys. **87**, 395 (1982). https://doi.org/10.1007/BF01206031
3. S.W. Hawking, Black hole explosions. Nature **248**, 30 (1974). https://doi.org/10.1038/248030a0
4. S.W. Hawking, Particle creation by black holes. Commun. Math. Phys. **43**, 199 (1975). https://doi.org/10.1007/BF02345020
5. S.D. Mathur, The information paradox: a pedagogical introduction. Class. Quantum Gravity **26**, 224001 (2009). https://doi.org/10.1088/0264-9381/26/22/224001. arXiv:0909.1038
6. S.R. Coleman, Black holes as red herrings: topological fluctuations and the loss of quantum coherence. Nucl. Phys. B **307**, 867 (1988). https://doi.org/10.1016/0550-3213(88)90110-1
7. J. Polchinski, A. Strominger, A possible resolution of the black hole information puzzle. Phys. Rev. D **50**, 7403 (1994). https://doi.org/10.1103/PhysRevD.50.7403. arXiv:hep-th/9407008
8. J.A. Wheeler, An analysis of example, in *Relativity, Groups and Topology*, eds. by B. DeWitt, C. DeWitt (Gordon and Breach, 1974), p. 408
9. P. Chen, Y.C. Ong, D.H. Yeom, Black hole remnants and the information loss paradox. Phys. Rep. **603**, 1 (2015). https://doi.org/10.1016/j.physrep.2015.10.007. arXiv:1412.8366
10. S. Barman, G.M. Hossain, C. Singha, Is Hawking effect short-lived in polymer quantization? J. Math. Phys. **60**, 052304 (2019). https://doi.org/10.1063/1.5063401. arXiv:1707.03605
11. P.M. Ho, H. Kawai, UV and IR effects on Hawking radiation. JHEP **03**, 002 (2023). https://doi.org/10.1007/JHEP03(2023)002. arXiv:2207.07122
12. E.T. Akhmedov, T.L. Chau, P.M. Ho, H. Kawai, W.H. Shao, C.T. Wang, UV dispersive effects on Hawking radiation. Phys. Rev. D **109**, 025001 (2024). https://doi.org/10.1103/PhysRevD.109.025001. arXiv:2307.12831
13. T.L. Chau, P.M. Ho, H. Kawai, W.H. Shao, C.T. Wang, Hawking radiation under generalized uncertainty principle. Eur. Phys. J. C **83**, 1118 (2023). https://doi.org/10.1140/epjc/s10052-023-12302-3. arXiv:2309.01638
14. P.M. Ho, Y. Imamura, H. Kawai, W.H. Shao, A stringy effect on Hawking radiation. JHEP **12**, 122 (2023). https://doi.org/10.1007/JHEP12(2023)122. arXiv:2309.12926
15. D.N. Page, Information in black hole radiation. Phys. Rev. Lett. **71**, 3743 (1993). https://doi.org/10.1103/PhysRevLett.71.3743. arXiv:hep-th/9306083

16. D.N. Page, Average entropy of a subsystem. Phys. Rev. Lett. **71**, 1291 (1993). https://doi.org/10.1103/PhysRevLett.71.1291. arXiv:gr-qc/9305007
17. A. Almheiri, D. Marolf, J. Polchinski, J. Sully, Black holes: complementarity or firewalls? JHEP **02**, 062 (2013). https://doi.org/10.1007/JHEP02(2013)062. arXiv:1207.3123
18. W.G. Unruh, Sonic analog of black holes and the effects of high frequencies on black hole evaporation. Phys. Rev. D **51**, 2827 (1995). https://doi.org/10.1103/PhysRevD.51.2827. arXiv:gr-qc/9409008
19. R. Brout, S. Massar, R. Parentani and P. Spindel, Hawking radiation without transPlanckian frequencies. Phys. Rev. D **52**, 4559 (1995). https://doi.org/10.1103/PhysRevD.52.4559. arXiv:hep-th/9506121
20. N. Hambli, C.P. Burgess, Hawking radiation and ultraviolet regulators. Phys. Rev. D **53**, 5717 (1996). https://doi.org/10.1103/PhysRevD.53.5717. arXiv:hep-th/9510159
21. S. Corley, T. Jacobson, Hawking spectrum and high frequency dispersion. Phys. Rev. D **54**, 1568 (1996). https://doi.org/10.1103/PhysRevD.54.1568. arXiv:hep-th/9601073
22. S. Corley, T. Jacobson, Lattice black holes. Phys. Rev. D **57**, 6269 (1998). https://doi.org/10.1103/PhysRevD.57.6269. arXiv:hep-th/9709166
23. S. Corley, Computing the spectrum of black hole radiation in the presence of high frequency dispersion: an Analytical approach. Phys. Rev. D **57**, 6280 (1998). https://doi.org/10.1103/PhysRevD.57.6280. arXiv:hep-th/9710075
24. Y. Himemoto, T. Tanaka, A generalization of the model of Hawking radiation with modified high frequency dispersion relation. Phys. Rev. D **61**, 064004 (2000). https://doi.org/10.1103/PhysRevD.61.064004. arXiv:gr-qc/9904076
25. T. Jacobson, D. Mattingly, Hawking radiation on a falling lattice. Phys. Rev. D **61**, 024017 (2000). https://doi.org/10.1103/PhysRevD.61.024017. arXiv:hep-th/9908099
26. W.G. Unruh, R. Schutzhold, On the universality of the Hawking effect. Phys. Rev. D **71**, 024028 (2005). https://doi.org/10.1103/PhysRevD.71.024028. arxiv:gr-qc/0408009
27. I. Agullo, J. Navarro-Salas, G.J. Olmo, L. Parker, Insensitivity of Hawking radiation to an invariant Planck-scale cutoff. Phys. Rev. D **80**, 047503 (2009). https://doi.org/10.1103/PhysRevD.80.047503. arXiv:0906.5315
28. A. Coutant, R. Parentani, S. Finazzi, Black hole radiation with short distance dispersion, an analytical S-matrix approach. Phys. Rev. D **85**, 024021 (2012). https://doi.org/10.1103/PhysRevD.85.024021. arXiv:1108.1821
29. N. Kajuri, D. Kothawala, Universality of Hawking radiation in non local field theories. Phys. Lett. B **791**, 319 (2019). https://doi.org/10.1016/j.physletb.2019.03.006. arXiv:1806.10345
30. J. Boos, V.P. Frolov, A. Zelnikov, Ghost-free modification of the Polyakov action and Hawking radiation. Phys. Rev. D **100**, 104008 (2019). https://doi.org/10.1103/PhysRevD.100.104008. arXiv:1909.01494
31. P.M. Ho, Y. Yokokura, Firewall from effective field theory. Universe **7**, 241 (2021). https://doi.org/10.3390/universe7070241. arXiv:2004.04956
32. P.M. Ho, From uneventful Horizon to firewall in D-dimensional effective theory. Int. J. Mod. Phys. A **36**, 2150145 (2021). https://doi.org/10.1142/S0217751X21501451. arXiv:2005.03817
33. P.M. Ho, H. Kawai, Y. Yokokura, Planckian physics comes into play at Planckian distance from horizon. JHEP **01**, 019 (2022). https://doi.org/10.1007/JHEP01(2022)019. arXiv:2111.01967
34. Y. Sekino, L. Susskind, Fast scramblers. JHEP **10**, 065 (2008). https://doi.org/10.1088/1126-6708/2008/10/065. arXiv:0808.2096
35. D. Harlow, Jerusalem lectures on black holes and quantum information. Rev. Mod. Phys. **88**, 015002 (2016). https://doi.org/10.1103/RevModPhys.88.015002. arXiv:1409.1231
36. D.A. Leahy, W.G. Unruh, Effects of a $\lambda\phi^4$ interaction on black hole evaporation in two-dimensions. Phys. Rev. D **28**, 694 (1983). https://doi.org/10.1103/PhysRevD.28.694
37. W.G. Unruh, N. Weiss, Acceleration radiation in interacting field theories. Phys. Rev. D **29**, 1656 (1984). https://doi.org/10.1103/PhysRevD.29.1656
38. A.D. Helfer, Quantum character of black holes. arXiv:gr-qc/0503053

39. M. Frasca, Hawking radiation and interacting fields. Eur. Phys. J. Plus **132**, 467 (2017). https://doi.org/10.1140/epjp/i2017-11732-1. arXiv:1412.1955
40. E.T. Akhmedov, H. Godazgar, F.K. Popov, Hawking radiation and secularly growing loop corrections. Phys. Rev. D **93**, 024029 (2016). https://doi.org/10.1103/PhysRevD.93.024029. arXiv:1508.07500
41. S. Iso, Hawking radiation, gravitational anomaly and conformal symmetry: the origin of universality. Int. J. Mod. Phys. A **23**, 2082 (2008). https://doi.org/10.1142/S0217751X08040627. arXiv:0804.0652
42. T. Jacobson, Black hole radiation in the presence of a short distance cutoff. Phys. Rev. D **48**, 728 (1993). https://doi.org/10.1103/PhysRevD.48.728. arXiv:hep-th/9303103
43. A.D. Helfer, Do black holes radiate? Rept. Prog. Phys. **66**, 943 (2003). https://doi.org/10.1088/0034-4885/66/6/202. arXiv:gr-qc/0304042
44. C. Barcelo, L.J. Garay, G. Jannes, Sensitivity of Hawking radiation to superluminal dispersion relations. Phys. Rev. D **79**, 024016 (2009). https://doi.org/10.1103/PhysRevD.79.024016. arXiv:0807.4147
45. R. Brout, S. Massar, R. Parentani, P. Spindel, A primer for black hole quantum physics. Phys. Rept. **260**, 329 (1995). https://doi.org/10.1016/0370-1573(95)00008-5. arXiv:0710.4345
46. V.P. Frolov, I.D. Novikov (eds.), *Black Hole Physics: basic Concepts and New Developments* (Springer, Dordrecht, 1998). https://doi.org/10.1007/978-94-011-5139-9
47. W.G. Unruh, Notes on black hole evaporation. Phys. Rev. D **14**, 870 (1976). https://doi.org/10.1103/PhysRevD.14.870
48. B.A. Juárez-Aubry, J. Louko, Quantum fields during black hole formation: how good an approximation is the Unruh state? JHEP **05**, 140 (2018). https://doi.org/10.1007/JHEP05(2018)140. arXiv:1804.01228
49. P.C.W. Davies, S.A. Fulling, Radiation from a moving mirror in two-dimensional space-time conformal anomaly. Proc. Roy. Soc. Lond. A **348**, 393 (1976)
50. P.C.W. Davies, S.A. Fulling, W.G. Unruh, Energy momentum tensor near an evaporating black hole. Phys. Rev. D **13**, 2720 (1976). https://doi.org/10.1103/PhysRevD.13.2720
51. S.M. Christensen, S.A. Fulling, Trace anomalies and the Hawking effect. Phys. Rev. D **15**, 2088 (1977). https://doi.org/10.1103/PhysRevD.15.2088
52. P. Candelas, Vacuum polarization in schwarzschild space-time. Phys. Rev. D **21**, 2185 (1980). https://doi.org/10.1103/PhysRevD.21.2185
53. R. Parentani, T. Piran, The internal geometry of an evaporating black hole. Phys. Rev. Lett. **73**, 2805 (1994). https://doi.org/10.1103/PhysRevLett.73.2805. arXiv:hep-th/9405007
54. S. Ayal, T. Piran, Spherical collapse of a massless scalar field with semiclassical corrections. Phys. Rev. D **56**, 4768 (1997). https://doi.org/10.1103/PhysRevD.56.4768. arXiv:gr-qc/9704027
55. A. Fabbri, S. Farese, J. Navarro-Salas, G.J. Olmo, H. Sanchis-Alepuz, Semiclassical zero-temperature corrections to Schwarzschild spacetime and holography. Phys. Rev. D **73**, 104023 (2006). https://doi.org/10.1103/PhysRevD.73.104023. arXiv:hep-th/0512167
56. A. Fabbri, S. Farese, J. Navarro-Salas, G.J. Olmo, H. Sanchis-Alepuz, Static quantum corrections to the Schwarzschild spacetime. J. Phys. Conf. Ser. **33**, 457 (2006). https://doi.org/10.1088/1742-6596/33/1/059. arXiv:hep-th/0512179
57. C. Barcelo, S. Liberati, S. Sonego, M. Visser, Fate of gravitational collapse in semiclassical gravity. Phys. Rev. D **77**, 044032 (2008). https://doi.org/10.1103/PhysRevD.77.044032. arXiv:0712.1130
58. P.M. Ho, Y. Matsuo, On the near-horizon geometry of an evaporating black hole. JHEP **07**, 047 (2018). https://doi.org/10.1007/JHEP07(2018)047. arXiv:1804.04821
59. P.M. Ho, Y. Matsuo, S.J. Yang, Vacuum energy at apparent horizon in conventional model of black holes. arXiv:1904.01322
60. H. Kawai, Y. Matsuo, Y. Yokokura, A self-consistent model of the black hole evaporation. Int. J. Mod. Phys. A **28**, 1350050 (2013). https://doi.org/10.1142/S0217751X13500504. arXiv:1302.4733

61. H. Kawai, Y. Yokokura, Phenomenological description of the interior of the Schwarzschild black hole. Int. J. Mod. Phys. A **30**, 1550091 (2015). https://doi.org/10.1142/S0217751X15500918. arXiv:1409.5784
62. H. Kawai, Y. Yokokura, Interior of black holes and information recovery. Phys. Rev. D **93**, 044011 (2016). https://doi.org/10.1103/PhysRevD.93.044011. arXiv:1509.08472
63. P.M. Ho, Comment on self-consistent model of black hole formation and evaporation. JHEP **08**, 096 (2015). https://doi.org/10.1007/JHEP08(2015)096. arXiv:1505.02468
64. P.M. Ho, The absence of horizon in black-hole formation. Nucl. Phys. B **909**, 394 (2016). https://doi.org/10.1016/j.nuclphysb.2016.05.016. arXiv:1510.07157
65. S.L. Braunstein, S. Pirandola, K. Życzkowski, Better late than never: information retrieval from black holes. Phys. Rev. Lett. **110**, 101301 (2013). https://doi.org/10.1103/PhysRevLett.110.101301. arXiv:0907.1190
66. V.P. Frolov, G.A. Vilkovisky, spherically symmetric collapse in quantum gravity, in *Second Seminar on Quantum Gravity*, vol. 10 (1981)
67. T.A. Roman, P.G. Bergmann, Stellar collapse without singularities? Phys. Rev. D **28**, 1265 (1983). https://doi.org/10.1103/PhysRevD.28.1265
68. S.A. Hayward, Formation and evaporation of regular black holes. Phys. Rev. Lett. **96**, 031103 (2006). https://doi.org/10.1103/PhysRevLett.96.031103. arXiv:gr-qc/0506126
69. P.M. Ho, Y. Matsuo, Y. Yokokura, Distance between collapsing matter and apparent horizon in evaporating black holes. Phys. Rev. D **104**, 064005 (2021). https://doi.org/10.1103/PhysRevD.104.064005. arXiv:1912.12863
70. S. Mukohyama, Comments on entanglement entropy. Phys. Rev. D **58**, 104023 (1998). https://doi.org/10.1103/PhysRevD.58.104023. arXiv:gr-qc/9805039
71. G.T. Horowitz, J.M. Maldacena, The black hole final state. JHEP **02**, 008 (2004). https://doi.org/10.1088/1126-6708/2004/02/008. arxiv:hep-th/0310281
72. D.A. Lowe, J. Polchinski, L. Susskind, L. Thorlacius, J. Uglum, Black hole complementarity versus locality. Phys. Rev. D **52**, 6997 (1995). https://doi.org/10.1103/PhysRevD.52.6997. arXiv:hep-th/9506138
73. J. Polchinski, String theory and black hole complementarity, in *STRINGS 95: Future Perspectives in String Theory*, p. 417 (1995). arXiv:hep-th/9507094
74. G. t Hooft, On the quantum structure of a black hole. Nucl. Phys. B **256**, 727 (1985). https://doi.org/10.1016/0550-3213(85)90418-3
75. T. Jacobson, Black hole evaporation and ultrashort distances. Phys. Rev. D **44**, 1731 (1991). https://doi.org/10.1103/PhysRevD.44.1731
76. J.D. Bekenstein, Black holes and entropy. Phys. Rev. D **7**, 2333 (1973). https://doi.org/10.1103/PhysRevD.7.2333
77. J.D. Bekenstein, Generalized second law of thermodynamics in black hole physics. Phys. Rev. D **9**, 3292 (1974). https://doi.org/10.1103/PhysRevD.9.3292
78. G. t Hooft, Dimensional reduction in quantum gravity. Conf. Proc. C **930308**, 284 (1993). arXiv:gr-qc/9310026
79. L. Susskind, The World as a hologram. J. Math. Phys. **36**, 6377 (1995). https://doi.org/10.1063/1.531249. arxiv:hep-th/9409089
80. J.M. Maldacena, The large N limit of superconformal field theories and supergravity. Adv. Theor. Math. Phys. **2**, 231 (1998). https://doi.org/10.4310/ATMP.1998.v2.n2.a1. arXiv:hep-th/9711200
81. E. Witten, Anti-de Sitter space and holography. Adv. Theor. Math. Phys. **2**, 253 (1998). https://doi.org/10.4310/ATMP.1998.v2.n2.a2. arXiv:hep-th/9802150
82. A. Strominger, C. Vafa, Microscopic origin of the Bekenstein-Hawking entropy. Phys. Lett. B **379**, 99 (1996). https://doi.org/10.1016/0370-2693(96)00345-0. arXiv:hep-th/9601029
83. S.S. Gubser, I.R. Klebanov, A.M. Polyakov, Gauge theory correlators from noncritical string theory. Phys. Lett. B **428**, 105 (1998). https://doi.org/10.1016/S0370-2693(98)00377-3. arXiv:hep-th/9802109
84. A. Dabholkar, J. Gomes, S. Murthy, Nonperturbative black hole entropy and Kloosterman sums. JHEP **03**, 074 (2015). https://doi.org/10.1007/JHEP03(2015)074. arXiv:1404.0033

85. A. Almheiri, T. Hartman, J. Maldacena, E. Shaghoulian, A. Tajdini, The entropy of Hawking radiation. Rev. Mod. Phys. **93**, 035002 (2021). https://doi.org/10.1103/RevModPhys.93.035002. arXiv:2006.06872
86. J. Wheeler, The "past" and the "delayed-choice" double-slit experiment, in ed. by A.R. Marlow, *Mathematical Foundations of Quantum Theory* (Academic Press, Cambridge, MA., 1978), p. 9
87. S.D. Mathur, Resolving the black hole information paradox. Int. J. Mod. Phys. A **15**, 4877 (2000). arXiv:gr-qc/0007011
88. W.K. Wootters, W.H. Zurek, A single quantum cannot be cloned. Nature **299**, 802 (1982). https://doi.org/10.1038/299802a0
89. D. Dieks, Communication by EPR devices. Phys. Lett. A **92**, 271 (1982). https://doi.org/10.1016/0375-9601(82)90084-6
90. L. Susskind, L. Thorlacius, J. Uglum, The Stretched horizon and black hole complementarity. Phys. Rev. D **48**, 3743 (1993). https://doi.org/10.1103/PhysRevD.48.3743. arXiv:hep-th/9306069
91. L. Susskind, L. Thorlacius, Gedanken experiments involving black holes. Phys. Rev. D **49**, 966 (1994). https://doi.org/10.1103/PhysRevD.49.966. arXiv:hep-th/9308100
92. R. Bousso, G. Penington, Islands far outside the horizon. arXiv:2312.03078
93. P.M. Ho, Y. Matsuo, Y. Yokokura, Analytic description of semiclassical black-hole geometry. Phys. Rev. D **102**, 024090 (2020). https://doi.org/10.1103/PhysRevD.102.024090. arXiv:1912.12855
94. S.B. Giddings, Nonviolent nonlocality. Phys. Rev. D **88**, 064023 (2013). https://doi.org/10.1103/PhysRevD.88.064023, arXiv:1211.7070
95. S.B. Giddings, Y. Shi, Effective field theory models for nonviolent information transfer from black holes. Phys. Rev. D **89**, 124032 (2014). https://doi.org/10.1103/PhysRevD.89.124032. arXiv:1310.5700
96. J. Maldacena, L. Susskind, Cool horizons for entangled black holes. Fortsch. Phys. **61**, 781 (2013). https://doi.org/10.1002/prop.201300020. arXiv:1306.0533
97. G. Penington, Entanglement wedge reconstruction and the information paradox. JHEP **09**, 002 (2020). https://doi.org/10.1007/JHEP09(2020)002. arXiv:1905.08255
98. A. Almheiri, N. Engelhardt, D. Marolf, H. Maxfield, The entropy of bulk quantum fields and the entanglement wedge of an evaporating black hole. JHEP **12** (2019) 063. https://doi.org/10.1007/JHEP12(2019)063. arXiv:1905.08762
99. A. Almheiri, R. Mahajan, J. Maldacena, Y. Zhao, The page curve of Hawking radiation from semiclassical geometry. JHEP **03**, 149 (2020). https://doi.org/10.1007/JHEP03(2020)149. arXiv:1908.10996
100. A. Almheiri, R. Mahajan, J. Maldacena, Islands outside the horizon. arXiv:1910.11077
101. G. Penington, S.H. Shenker, D. Stanford, Z. Yang, Replica wormholes and the black hole interior. JHEP **03**, 205 (2022). https://doi.org/10.1007/JHEP03(2022)205. arXiv:1911.11977
102. A. Almheiri, T. Hartman, J. Maldacena, E. Shaghoulian, A. Tajdini, Replica Wormholes and the entropy of Hawking radiation. JHEP **05**, 013 (2020). https://doi.org/10.1007/JHEP05(2020)013. arXiv:1911.12333
103. D. Marolf, J. Polchinski, Gauge/gravity duality and the black hole interior. Phys. Rev. Lett. **111**, 171301 (2013). https://doi.org/10.1103/PhysRevLett.111.171301. arXiv:1307.4706
104. S. Weinberg, The Quantum Theory of Fields. Vol. 1: foundations (Cambridge University Press, 2005). https://doi.org/10.1017/CBO9781139644167
105. G. Veneziano, Large N bounds on, and compositeness limit of, gauge and gravitational interactions. JHEP **06**, 051 (2002). https://doi.org/10.1088/1126-6708/2002/06/051. arxiv:hep-th/0110129
106. G. Dvali, Black holes and large N species solution to the hierarchy problem. Fortsch. Phys. **58**, 528 (2010). https://doi.org/10.1002/prop.201000009. arXiv:0706.2050
107. S. Caron-Huot, Y.Z. Li, Gravity and a universal cutoff for field theory. arXiv:2408.06440
108. C.P. Burgess, J. Hainge, G. Kaplanek, M. Rummel, Failure of perturbation theory near horizons: the Rindler example. JHEP **10**, 122 (2018). https://doi.org/10.1007/JHEP10(2018)122. arXiv:1806.11415

109. D. Amati, M. Ciafaloni, G. Veneziano, Can space-time be probed below the string size? Phys. Lett. B **216**, 41 (1989). https://doi.org/10.1016/0370-2693(89)91366-X
110. M. Fabbrichesi, G. Veneziano, Thinning out of relevant degrees of freedom in scattering of strings. Phys. Lett. B **233**, 135 (1989). https://doi.org/10.1016/0370-2693(89)90629-1
111. K. Konishi, G. Paffuti, P. Provero, Minimum physical length and the generalized uncertainty principle in string theory. Phys. Lett. B **234**, 276 (1990). https://doi.org/10.1016/0370-2693(90)91927-4
112. R. Guida, K. Konishi, P. Provero, On the short distance behavior of string theories. Mod. Phys. Lett. A **6**, 1487 (1991). https://doi.org/10.1142/S0217732391001603
113. G. Veneziano, An Enlarged Uncertainty Principle From Gedanken String Collisions? Conf. Proc. C **8903131**, 86 (1989)
114. M. Maggiore, A generalized uncertainty principle in quantum gravity. Phys. Lett. B **304**, 65 (1993). https://doi.org/10.1016/0370-2693(93)91401-8. arXiv:hep-th/9301067
115. L.J. Garay, Quantum gravity and minimum length. Int. J. Mod. Phys. A **10**, 145 (1995). https://doi.org/10.1142/S0217751X95000085. arXiv:gr-qc/9403008
116. F. Scardigli, Generalized uncertainty principle in quantum gravity from micro-black hole Gedanken experiment. Phys. Lett. B **452**, 39 (1999). https://doi.org/10.1016/S0370-2693(99)00167-7. arXiv:hep-th/9904025
117. R.J. Adler, D.I. Santiago, On gravity and the uncertainty principle. Mod. Phys. Lett. A **14**, 1371 (1999). https://doi.org/10.1142/S0217732399001462. arXiv:gr-qc/9904026
118. S. Capozziello, G. Lambiase, G. Scarpetta, Generalized uncertainty principle from quantum geometry. Int. J. Theor. Phys. **39**, 15 (2000). https://doi.org/10.1023/A:1003634814685. arXiv:gr-qc/9910017
119. F. Scardigli, R. Casadio, Generalized uncertainty principle, extra dimensions and holography. Class. Quant. Grav. **20**, 3915 (2003). https://doi.org/10.1088/0264-9381/20/18/305. arXiv:hep-th/0307174
120. A.F. Ali, S. Das, E.C. Vagenas, Discreteness of space from the generalized uncertainty principle. Phys. Lett. B **678**, 497 (2009). https://doi.org/10.1016/j.physletb.2009.06.061. arXiv:0906.5396
121. S. Hossenfelder, Minimal length scale scenarios for quantum gravity. Living Rev. Rel. **16**, 2 (2013). https://doi.org/10.12942/lrr-2013-2. arXiv:1203.6191
122. T. Yoneya, Duality and indeterminacy principle in string theory, in *Symposium in Honor of Professor Nishijima: wandering in the Fields*, p. 419 (1987)
123. T. Yoneya, On the interpretation of minimal length in string theories. Mod. Phys. Lett. A **4**, 1587 (1989). https://doi.org/10.1142/S0217732389001817
124. T. Yoneya, Schild action and space-time uncertainty principle in string theory. Prog. Theor. Phys. **97**, 949 (1997). https://doi.org/10.1143/PTP.97.949. arXiv:hep-th/9703078
125. T. Yoneya, String theory and space-time uncertainty principle. Prog. Theor. Phys. **103**, 1081 (2000). https://doi.org/10.1143/PTP.103.1081. arXiv:hep-th/0004074
126. D.A. Lowe, The Planckian conspiracy: string theory and the black hole information paradox. Nucl. Phys. B **456**, 257 (1995). https://doi.org/10.1016/0550-3213(95)00521-9. arXiv:hep-th/9505074
127. S.B. Giddings, D.J. Gross, A. Maharana, Gravitational effects in ultrahigh-energy string scattering. Phys. Rev. D **77** (2008) 046001. https://doi.org/10.1103/PhysRevD.77.046001. arXiv:0705.1816
128. M. Dodelson, E. Silverstein, String-theoretic breakdown of effective field theory near black hole horizons. Phys. Rev. D **96**, 066010 (2017). arXiv:1504.05536
129. M. Dodelson, E. Silverstein, Longitudinal nonlocality in the string S-matrix. arXiv:1504.05537
130. R. Brout, C. Gabriel, M. Lubo, P. Spindel, Minimal length uncertainty principle and the transPlanckian problem of black hole physics. Phys. Rev. D **59**, 044005 (1999). https://doi.org/10.1103/PhysRevD.59.044005. arXiv:hep-th/9807063
131. A. Kempf, G. Mangano, R.B. Mann, Hilbert space representation of the minimal length uncertainty relation. Phys. Rev. D **52** (1995) 1108. https://doi.org/10.1103/PhysRevD.52.1108. arXiv:hep-th/9412167

132. S. Hossenfelder, Interpretation of quantum field theories with a minimal length scale. Phys. Rev. D **73**, 105013 (2006). https://doi.org/10.1103/PhysRevD.73.105013. arXiv:hep-th/0603032
133. Y. Shibusa, Supersymmetric field theory based on generalized uncertainty principle. Int. J. Mod. Phys. A **22**, 5279 (2007). https://doi.org/10.1142/S0217751X07037913. arXiv:0704.1545
134. M. Kober, Gauge theories under incorporation of a generalized uncertainty principle. Phys. Rev. D **82**, 085017 (2010). https://doi.org/10.1103/PhysRevD.82.085017. arXiv:1008.0154
135. M. Kober, Electroweak theory with a minimal length. Int. J. Mod. Phys. A **26**, 4251 (2011). https://doi.org/10.1142/S0217751X11054413. arXiv:1104.2319
136. V. Husain, D. Kothawala, S.S. Seahra, Generalized uncertainty principles and quantum field theory. Phys. Rev. D **87**, 025014 (2013). https://doi.org/10.1103/PhysRevD.87.025014. arXiv:1208.5761
137. M. Faizal, A.F. Ali and A. Nassar, AdS/CFT correspondence beyond its supergravity approximation. Int. J. Mod. Phys. A **30**, 1550183 (2015). https://doi.org/10.1142/S0217751X15501833. arXiv:1405.4519
138. V. Todorinov, P. Bosso, S. Das, Relativistic generalized uncertainty principle. Annals Phys. **405**, 92 (2019). https://doi.org/10.1016/j.aop.2019.03.014. arXiv:1810.11761
139. P. Bosso, S. Das, V. Todorinov, Quantum field theory with the generalized uncertainty principle I: Scalar electrodynamics. Annals Phys. **422**, 168319 (2020). https://doi.org/10.1016/j.aop.2020.168319. arXiv:2005.03771
140. T. Damour, R. Ruffini, Black hole evaporation in the Klein-Sauter-Heisenberg-Euler formalism. Phys. Rev. D **14**, 332 (1976). https://doi.org/10.1103/PhysRevD.14.332
141. M. Kaku, K. Kikkawa, The field theory of relativistic strings. I. trees. Phys. Rev. D **10**, 1110 (1974). https://doi.org/10.1103/PhysRevD.10.1110
142. E. Witten, Noncommutative geometry and string field theory. Nucl. Phys. B **268**, 253 (1986). https://doi.org/10.1016/0550-3213(86)90155-0
143. B. Zwiebach, Closed string field theory: quantum action and the B-V master equation. Nucl. Phys. B **390**, 33 (1993). https://doi.org/10.1016/0550-3213(93)90388-6. arXiv:hep-th/9206084
144. C. de Lacroix, H. Erbin, S.P. Kashyap, A. Sen, M. Verma, Closed superstring field theory and its applications. Int. J. Mod. Phys. A **32**, 1730021 (2017). https://doi.org/10.1142/S0217751X17300216. arXiv:1703.06410
145. E.T. Tomboulis, Nonlocal and quasilocal field theories. Phys. Rev. D **92**, 125037 (2015). https://doi.org/10.1103/PhysRevD.92.125037. arXiv:1507.00981
146. R. Pius, A. Sen, Cutkosky rules for superstring field theory. JHEP **10**, 024 (2016). https://doi.org/10.1007/JHEP10(2016)024. arXiv:1604.01783]
147. C.D. Carone, Unitarity and microscopic acausality in a nonlocal theory. Phys. Rev. D **95**, 045009 (2017). https://doi.org/10.1103/PhysRevD.95.045009 [arXiv:1605.02030]
148. F. Briscese, L. Modesto, Cutkosky rules and perturbative unitarity in Euclidean nonlocal quantum field theories. Phys. Rev. D **99**, 104043 (2019). https://doi.org/10.1103/PhysRevD.99.104043. arXiv:1803.08827
149. P. Chin, E.T. Tomboulis, Nonlocal vertices and analyticity: Landau equations and general Cutkosky rule. JHEP **06**, 014 (2018). https://doi.org/10.1007/JHEP06(2018)014. arXiv:1803.08899
150. R. Pius, A. Sen, Unitarity of the box diagram. JHEP **11**, 094 (2018). https://doi.org/10.1007/JHEP11(2018)094. arXiv:1805.00984
151. L. Buoninfante, G. Lambiase, A. Mazumdar, Ghost-free infinite derivative quantum field theory. Nucl. Phys. B **944**, 114646 (2019). https://doi.org/10.1016/j.nuclphysb.2019.114646. arXiv:1805.03559
152. C. De Lacroix, H. Erbin, A. Sen, Analyticity and crossing symmetry of superstring loop amplitudes. JHEP **05**, 139 (2019). https://doi.org/10.1007/JHEP05(2019)139. arXiv:1810.07197
153. L. Buoninfante, A. Ghoshal, G. Lambiase, A. Mazumdar, Transmutation of nonlocal scale in infinite derivative field theories. Phys. Rev. D **99**, 044032 (2019). https://doi.org/10.1103/PhysRevD.99.044032. arXiv:1812.01441

154. A.S. Koshelev, A. Tokareva, Unitarity of Minkowski nonlocal theories made explicit. Phys. Rev. D **104**, 025016 (2021). https://doi.org/10.1103/PhysRevD.104.025016. arXiv:2103.01945
155. H. Erbin, A.H. Fırat, B. Zwiebach, Initial value problem in string-inspired nonlocal field theory. JHEP **01**, 167 (2022). https://doi.org/10.1007/JHEP01(2022)167. arXiv:2111.03672
156. L. Buoninfante, Contour prescriptions in string-inspired nonlocal field theories. Phys. Rev. D **106**, 126028 (2022). https://doi.org/10.1103/PhysRevD.106.126028. arXiv:2205.15348
157. B.S. DeWitt, *Quantum Gravity: the New Synthesis* (Cambridge University Press, 1980), p. 680
158. W. Siegel, Stringy gravity at short distances. arXiv:hep-th/0309093
159. N. Kajuri, Unruh effect in nonlocal field theories. Phys. Rev. D **95**, 101701 (2017). https://doi.org/10.1103/PhysRevD.95.101701. arXiv:1704.03793
160. D.G. Barci, L.E. Oxman, M. Rocca, Canonical quantization of nonlocal field equations. Int. J. Mod. Phys. A **11**, 2111 (1996). https://doi.org/10.1142/S0217751X96001061. arXiv:hep-th/9503101
161. N. Barnaby, N. Kamran, Dynamics with infinitely many derivatives: the initial value problem. JHEP **02**, 008 (2008). https://doi.org/10.1088/1126-6708/2008/02/008. arXiv:0709.3968
162. C.H. Chang, P.M. Ho, I.K. Lee, W.H. Shao, Towards Hamiltonian formalism for string field theory and nonlocality. arXiv:2412.02577
163. J. Llosa, J. Vives, Hamiltonian formalism for nonlocal Lagrangians. J. Math. Phys. **35**, 2856 (1994). https://doi.org/10.1063/1.530492
164. R.P. Woodard, A Canonical formalism for Lagrangians with nonlocality of finite extent. Phys. Rev. A **62**, 052105 (2000). https://doi.org/10.1103/PhysRevA.62.052105. arXiv:hep-th/0006207
165. J. Gomis, K. Kamimura, J. Llosa, Hamiltonian formalism for space-time noncommutative theories. Phys. Rev. D **63**, 045003 (2001). https://doi.org/10.1103/PhysRevD.63.045003. arXiv:hep-th/0006235
166. C. Heredia, J. Llosa, Non-local Lagrangian mechanics: Noether's theorem and Hamiltonian formalism. J. Phys. A **54**, 425202 (2021). https://doi.org/10.1088/1751-8121/ac265c. arXiv:2105.10442
167. C. Heredia, J. Llosa, Nonlocal Lagrangian fields: Noether's theorem and Hamiltonian formalism. Phys. Rev. D **105**, 126002 (2022). https://doi.org/10.1103/PhysRevD.105.126002. arXiv:2203.02206
168. T.C. Cheng, P.M. Ho, M.C. Yeh, Perturbative approach to higher derivative and nonlocal theories. Nucl. Phys. B **625**, 151 (2002). https://doi.org/10.1016/S0550-3213(02)00020-2. arXiv:hep-th/0111160
169. M. Ostrogradsky, Mémoires sur les équations différentielles, relatives au problème des isopérimètres. Mem. Acad. St. Petersbourg **6**, 385 (1850)
170. D.A. Eliezer, R.P. Woodard, The problem of nonlocality in string theory. Nucl. Phys. B **325**, 389 (1989). https://doi.org/10.1016/0550-3213(89)90461-6
171. R.P. Woodard, Ostrogradsky's theorem on Hamiltonian instability. Scholarpedia **10**, 32243 (2015). https://doi.org/10.4249/scholarpedia.32243. arXiv:1506.02210
172. S. Takagi, Vacuum noise and stress induced by uniform acceleration: Hawking-Unruh effect in Rindler manifold of arbitrary dimension. Prog. Theor. Phys. Suppl. **88**, 1 (1986). https://doi.org/10.1143/PTP.88.1
173. Z. Bern, C. Cheung, R. Roiban, C.H. Shen, M.P. Solon, M. Zeng, Scattering amplitudes and the conservative Hamiltonian for binary systems at third post-Minkowskian order. Phys. Rev. Lett. **122**, 201603 (2019). https://doi.org/10.1103/PhysRevLett.122.201603. arXiv:1901.04424
174. Z. Bern, C. Cheung, R. Roiban, C.H. Shen, M.P. Solon, M. Zeng, Black hole binary dynamics from the double copy and effective theory. JHEP **10**, 206 (2019). https://doi.org/10.1007/JHEP10(2019)206. arXiv:1908.01493
175. G.U. Jakobsen, Schwarzschild-Tangherlini metric from scattering amplitudes. Phys. Rev. D **102**, 104065 (2020). https://doi.org/10.1103/PhysRevD.102.104065. arXiv:2006.01734
176. S. Mougiakakos, P. Vanhove, Schwarzschild-Tangherlini metric from scattering amplitudes in various dimensions. Phys. Rev. D **103**, 026001 (2021). https://doi.org/10.1103/PhysRevD.103.026001. arXiv:2010.08882

177. M. Driesse, G.U. Jakobsen, G. Mogull, J. Plefka, B. Sauer, J. Usovitsch, Conservative black hole scattering at fifth post-Minkowskian and first self-force order. Phys. Rev. Lett. **132**, 241402 (2024). https://doi.org/10.1103/PhysRevLett.132.241402. arXiv:2403.07781
178. P.H. Damgaard, K. Lee, Schwarzschild black hole from perturbation theory to all orders. Phys. Rev. Lett. **132**, 251603 (2024). https://doi.org/10.1103/PhysRevLett.132.251603. arXiv:2403.13216
179. G. Amelino-Camelia, M. Arzano, Y. Ling, G. Mandanici, Black-hole thermodynamics with modified dispersion relations and generalized uncertainty principles. Class. Quant. Gravity **23**, 2585 (2006). https://doi.org/10.1088/0264-9381/23/7/022. arXiv:gr-qc/0506110
180. B. Majumder, Black hole entropy with minimal length in tunneling formalism. Gen. Rel. Gravity **45**, 2403 (2013). https://doi.org/10.1007/s10714-013-1581-2. arXiv:1212.6591
181. Y.G. Miao, Y.J. Zhao, S.J. Zhang, Maximally localized states and quantum corrections of black hole thermodynamics in the framework of a new generalized uncertainty principle. Adv. High Energy Phys. **2015**, 627264 (2015). arXiv:1410.4115
182. P. Wang, H. Yang, S. Ying, Quantum gravity corrections to the tunneling radiation of scalar particles. Int. J. Theor. Phys. **55**, 2633 (2016). https://doi.org/10.1007/s10773-015-2898-2. arXiv:1410.5065
183. D. Chen, H. Wu, H. Yang, S. Yang, Effects of quantum gravity on black holes. Int. J. Mod. Phys. A **29**, 1430054 (2014). https://doi.org/10.1142/S0217751X14300543. arXiv:1410.5071
184. P. Bargueño, E.C. Vagenas, Semiclassical corrections to black hole entropy and the generalized uncertainty principle. Phys. Lett. B **742**, 15 (2015). https://doi.org/10.1016/j.physletb.2015.01.016. arXiv:1501.03256
185. B. Mu, P. Wang, H. Yang, Minimal length effects on tunnelling from spherically symmetric black holes. Adv. High Energy Phys. **2015**, 898916 (2015). https://doi.org/10.1155/2015/898916. arXiv:1501.06025
186. I. Sakalli, A. Övgün, K. Jusufi, GUP assisted Hawking radiation of rotating acoustic black holes. Astrophys. Space Sci. **361**, 330 (2016). https://doi.org/10.1007/s10509-016-2922-x. arXiv:1602.04304
187. G. Gecim, Y. Sucu, The GUP effect on Hawking radiation of the 2 + 1 dimensional black hole. Phys. Lett. B **773**, 391 (2017). https://doi.org/10.1016/j.physletb.2017.08.053. arXiv:1704.03536
188. G. Lambiase, F. Scardigli, Lorentz violation and generalized uncertainty principle. Phys. Rev. D **97**, 075003 (2018). https://doi.org/10.1103/PhysRevD.97.075003. arXiv:1709.00637
189. T. Kanazawa, G. Lambiase, G. Vilasi, A. Yoshioka, Noncommutative Schwarzschild geometry and generalized uncertainty principle. Eur. Phys. J. C **79**, 95 (2019). https://doi.org/10.1140/epjc/s10052-019-6610-1
190. L. Buoninfante, G.G. Luciano, L. Petruzziello, Generalized uncertainty principle and corpuscular gravity. Eur. Phys. J. C **79**, 663 (2019). https://doi.org/10.1140/epjc/s10052-019-7164-y. arXiv:1903.01382
191. M.A. Anacleto, F.A. Brito, G.C. Luna, E. Passos, The generalized uncertainty principle effect in acoustic black holes. Annals Phys. **440**, 168837 (2022). https://doi.org/10.1016/j.aop.2022.168837. arXiv:2112.13573
192. M.A. Anacleto, F.A. Brito, E. Passos, Modified metrics of acoustic black holes: a review. Biophys. J. **7**, 000245 (2023). https://doi.org/10.23880/psbj-16000245. arXiv:2306.03077
193. E. Ievlev, M.R.R. Good, Thermal larmor radiation. PTEP **2024**, 043A01 (2024). https://doi.org/10.1093/ptep/ptae042. arXiv:2303.03676
194. S. Ryu, T. Takayanagi, Holographic derivation of entanglement entropy from AdS/CFT. Phys. Rev. Lett. **96**, 181602 (2006). https://doi.org/10.1103/PhysRevLett.96.181602. arXiv:hep-th/0603001
195. V.E. Hubeny, M. Rangamani, T. Takayanagi, A Covariant holographic entanglement entropy proposal. JHEP **07**, 062 (2007). https://doi.org/10.1088/1126-6708/2007/07/062. arXiv:0705.0016
196. T. Faulkner, A. Lewkowycz, J. Maldacena, Quantum corrections to holographic entanglement entropy. JHEP **11**, 074 (2013). https://doi.org/10.1007/JHEP11(2013)074. arXiv:1307.2892

197. N. Engelhardt, A.C. Wall, Quantum extremal surfaces: holographic entanglement entropy beyond the classical regime. JHEP **01**, 073 (2015). https://doi.org/10.1007/JHEP01(2015)073. arXiv:1408.3203
198. H. Geng, A. Karch, Massive Islands. JHEP **09**, 121 (2020). https://doi.org/10.1007/JHEP09(2020)121. arXiv:2006.02438
199. H. Geng, A. Karch, C. Perez-Pardavila, S. Raju, L. Randall, M. Riojas et al., Inconsistency of Islands in theories with long-range gravity. JHEP **01**, 182 (2022). https://doi.org/10.1007/JHEP01(2022)182. arXiv:2107.03390
200. H. Geng, Graviton mass and entanglement islands in low spacetime dimensions. arXiv:2312.13336
201. F. Landgren, A. Shekar, Islands and entanglement entropy in d-dimensional curved backgrounds. arXiv:2401.01653
202. B. Guo, M.R.R. Hughes, S.D. Mathur, M. Mehta, Contrasting the fuzzball and wormhole paradigms for black holes. Turk. J. Phys. **45**, 281 (2021). https://doi.org/10.3906/fiz-2111-13. arXiv:2111.05295
203. S.D. Mathur, The Fuzzball proposal for black holes: an elementary review. Fortsch. Phys. **53**, 793 (2005). https://doi.org/10.1002/prop.200410203. arXiv:hep-th/0502050
204. K. Skenderis, M. Taylor, The fuzzball proposal for black holes. Phys. Rept. **467**, 117 (2008). https://doi.org/10.1016/j.physrep.2008.08.001. arXiv:0804.0552
205. I. Bena, E.J. Martinec, S.D. Mathur, N.P. Warner, Snowmass white paper: micro- and macro-structure of black holes. arXiv:2203.04981
206. I. Bena, E.J. Martinec, S.D. Mathur, N.P. Warner, Fuzzballs and microstate geometries: black-hole structure in string theory. arXiv:2204.13113
207. B.D. Chowdhury, S.D. Mathur, Radiation from the non-extremal fuzzball. Class. Quantum Gravity **25**, 135005 (2008). https://doi.org/10.1088/0264-9381/25/13/135005. arXiv:0711.4817
208. S.D. Mathur, Tunneling into fuzzball states. Gen. Rel. Gravity **42**, 113 (2010). https://doi.org/10.1007/s10714-009-0837-3. arXiv:0805.3716
209. I. Bena, D.R. Mayerson, A. Puhm, B. Vercnocke, Tunneling into microstate geometries: quantum effects stop gravitational collapse. JHEP **07**, 031 (2016). https://doi.org/10.1007/JHEP07(2016)031. arXiv:1512.05376
210. S.B. Giddings, Black holes and massive remnants. Phys. Rev. D **46**, 1347 (1992). https://doi.org/10.1103/PhysRevD.46.1347. arXiv:hep-th/9203059
211. G. Dvali, L. Eisemann, M. Michel, S. Zell, Black hole metamorphosis and stabilization by memory burden. Phys. Rev. D **102**, 103523 (2020). https://doi.org/10.1103/PhysRevD.102.103523. arXiv:2006.00011
212. W.G. Unruh, Experimental black hole evaporation. Phys. Rev. Lett. **46**, 1351 (1981). https://doi.org/10.1103/PhysRevLett.46.1351
213. L.J. Garay, J.R. Anglin, J.I. Cirac, P. Zoller, Black holes in Bose-Einstein condensates. Phys. Rev. Lett. **85**, 4643 (2000). https://doi.org/10.1103/PhysRevLett.85.4643. arXiv:gr-qc/0002015
214. C. Barcelo, S. Liberati, M. Visser, Analogue gravity. Living Rev. Rel. **8**, 12 (2005). https://doi.org/10.12942/lrr-2005-12. arXiv:gr-qc/0505065
215. J.M. Maldacena, Eternal black holes in anti-de Sitter. JHEP **04**, 021 (2003). https://doi.org/10.1088/1126-6708/2003/04/021. arxiv:hep-th/0106112
216. J. Martin, R.H. Brandenberger, The TransPlanckian problem of inflationary cosmology. Phys. Rev. D **63**, 123501 (2001). https://doi.org/10.1103/PhysRevD.63.123501. arXiv:hep-th/0005209
217. R.H. Brandenberger, J. Martin, Trans-Planckian issues for inflationary cosmology. Class. Quantum Gravity **30**, 113001 (2013). https://doi.org/10.1088/0264-9381/30/11/113001. arXiv:1211.6753
218. M. Blamart, S. Laliberte, R. Brandenberger, TCC bounds on the static patch of de Sitter space. JHEP **05**, 193 (2023). https://doi.org/10.1007/JHEP05(2023)193. arXiv:2301.02741

219. A. Bedroya, C. Vafa, Trans-Planckian Censorship and the Swampland. JHEP **09**, 123 (2020). https://doi.org/10.1007/JHEP09(2020)123. arXiv:1909.11063
220. A. Bedroya, R. Brandenberger, M. Loverde, C. Vafa, Trans-Planckian Censorship and inflationary cosmology. Phys. Rev. D **101**, 103502 (2020). https://doi.org/10.1103/PhysRevD.101.103502. arXiv:1909.11106
221. R. Brandenberger, P.M. Ho, H. Kawai, W.H. Shao, Stringy spacetime uncertainty principle and a modified trans-Planckian censorship criterion. Phys. Rev. D **109**, 083503 (2024). https://doi.org/10.1103/PhysRevD.109.083503. arXiv:2401.04097
222. A. Alexandre, G. Dvali, E. Koutsangelas, New mass window for primordial black holes as dark matter from the memory burden effect. Phys. Rev. D **110**, 036004 (2024). https://doi.org/10.1103/PhysRevD.110.036004. arXiv:2402.14069
223. V. Thoss, A. Burkert, K. Kohri, Breakdown of hawking evaporation opens new mass window for primordial black holes as dark matter candidate. Mon. Not. Roy. Astron. Soc. **532**, 451 (2024). https://doi.org/10.1093/mnras/stae1098. arXiv:2402.17823
224. G. Dvali, J.S. Valbuena-Bermúdez, M. Zantedeschi, Memory burden effect in black holes and solitons: implications for PBH. Phys. Rev. D **110**, 056029 (2024). https://doi.org/10.1103/PhysRevD.110.056029. arXiv:2405.13117
225. A. Bonanno, M. Reuter, Renormalization group improved black hole space-times. Phys. Rev. D **62**, 043008 (2000). https://doi.org/10.1103/PhysRevD.62.043008. arXiv:hep-th/0002196
226. M. Reuter, F. Saueressig, *Quantum Gravity and the Functional Renormalization Group: the Road towards Asymptotic Safety* (Cambridge University Press, 2019)
227. A. Bonanno, A. Eichhorn, H. Gies, J.M. Pawlowski, R. Percacci, M. Reuter et al., Critical reflections on asymptotically safe gravity. Front. Phys. **8**, 269 (2020). https://doi.org/10.3389/fphy.2020.00269. arXiv:2004.06810
228. F. Saueressig, *The Functional Renormalization Group in Quantum Gravity*. arXiv:2302.14152
229. J. Magueijo, L. Smolin, Gravity's rainbow. Class. Quantum Gravity **21**, 1725 (2004). https://doi.org/10.1088/0264-9381/21/7/001. arXiv:gr-qc/0305055
230. S. Hossenfelder, Bounds on an energy-dependent and observer-independent speed of light from violations of locality. Phys. Rev. Lett. **104**, 140402 (2010). https://doi.org/10.1103/PhysRevLett.104.140402. arXiv:1004.0418
231. H. Hata, H. Oda, S. Yahikozawa, String field theory in Rindler space-time and string thermalization. Prog. Theor. Phys. **96**, 985 (1996). https://doi.org/10.1143/PTP.96.985. arXiv:hep-th/9512206
232. C.G. Callan, Jr., S.B. Giddings, J.A. Harvey, A. Strominger, Evanescent black holes. Phys. Rev. D **45**, R1005 (1992). https://doi.org/10.1103/PhysRevD.45.R1005. arXiv:hep-th/9111056
233. S.B. Giddings, W.M. Nelson, Quantum emission from two-dimensional black holes. Phys. Rev. D **46**, 2486 (1992). https://doi.org/10.1103/PhysRevD.46.2486. arXiv:hep-th/9204072

Chapter 8
The Case For Black Hole Remnants: A Review

Yen Chin Ong

Abstract It has been almost 40 years since the proposal of the idea that Hawking radiation of black holes does not lead to a complete evaporation but rather a "remnant" state. Though traditionally viewed with great criticisms especially from the high energy physics community, in recent years, various approaches have demonstrated that black hole remnants remain a viable possibility. In this review, which is primarily aimed as an introduction to the subject, we will discuss some possible routes to forming remnants and their respective properties and challenges.

8.1 Introduction: The Puzzles of Black Hole Evaporation

Classical black holes are "black" in the sense that nothing, not even light, can escape from inside the horizon. Anything that falls into a black hole is forever lost to the outside world. This naive picture is drastically changed after Hawking showed that black holes can radiate quantum mechanically. For an asymptotically flat Schwarzschild black hole, the Hawking temperature[1] is given by $T = 1/8\pi M$. This is inversely proportional to the black hole mass M. Therefore, as the black hole loses mass and becomes smaller, it also becomes hotter. The temperature is not bounded above and so diverges in the limit $M \to 0$. The obvious question is: what is the end state of Hawking evaporation? The usual model of Hawking evaporation is governed by the simple ODE of the form

$$\frac{dM}{dt} = -CAT^4, \tag{8.1}$$

[1] We will use the units $G = \hbar = c = 4\pi\epsilon_0 = k_B = 1$.

Y. C. Ong (✉)
Center for Gravitation and Cosmology, College of Physical Science and Technology, Yangzhou University, Yangzhou, China

Shanghai Frontier Science Center for Gravitational Wave Detection, School of Aeronautics and Astronautics, Shanghai Jiao Tong University, Shanghai, China
e-mail: ycong@yzu.edu.cn

A. Akil and C. Bambi (eds.), *The Black Hole Information Paradox*, Springer Series in Astrophysics and Cosmology, https://doi.org/10.1007/978-981-96-6170-1_8

where C is some constant related to the particle species, A the area of the black hole and T the temperature. The right hand side is just the Stefan-Boltzmann law for blackbody emission. The cross sectional area is not simply A as it should be determined by the gray-body factor, and even in the geometric optical limit the length scale is determined by the impact parameter of the photon sphere instead of just the Schwarzschild radius. Still, modulo these nitty-gritty details, the qualitative feature can be understood by Eq. (8.1). Since $T \sim 1/M$ and $A \sim M^2$, we see that $dM/dt \sim -M^{-2}$, which we can solve to find that a black hole whose initial mass is M will evaporate completely in a finite time that scales as $t_{\text{life}} \sim M^3$. In general relativity, black holes have "no hair", which means that there is no other physical parameters that characterize a black hole other than the mass, electric (and hypothetically, magnetic) charge, and angular momentum. Including charge and angular momentum changes the history of the evolution somewhat, but since angular momentum and charge are always radiated away more efficiently than mass loss (in the standard picture of Hawking evaporation), the end state is always a Schwarzschild black hole, so it suffices to consider that as far as remnants are concerned.

A model is of course, only as good as the assumptions we put into it. The above conclusion of a complete evaporation is based on the validity of Eq. (8.1), i.e., on the validity of the Stefan-Boltzmann law. Indeed, it is commonly accepted that at sufficiently high energy scale (i.e., high temperature; i.e., small black holes), new physics would likely kick in. This could happen before the Planck scale, and around the Planck scale we would expect quantum gravity effects to become important. Therefore, this already means that the end state of black hole evaporation is not well understood, and on this ground there is no reason to prefer a complete evaporation over the possibility that evaporation somehow effectively stops, becoming a "remnant". By remnant, we mean a state of black holes at late time of its evolution, whose lifetime can be infinite, or finite but longer than M^3, say M^4.

The aim of this review is to provide an overview of black hole remnants that is accessible to colleagues and students who are familiar with black holes but know little about remnants. There is already a more comprehensive review which I co-authored with Yeom and Chen almost 10 years ago [1], and readers are encouraged to refer to that for more details. In this review, I will instead mention some new progress and insights that have been made since then, focusing on (1) how remnants can possibly form and (2) traditional criticisms against remnants and why they do not rule out remnants. Finally we will end with some discussions.

The possibility of a black hole remnant was proposed already at least as far back as 1987, in the work of Aharonov et al. [2], on the ground that Hawking evaporation should satisfy unitarity of quantum evolution.[2] They called remnants "Planckons". Indeed, microscopic black holes may look like an "elementary particle" at least from the view of an exterior observer (the interior of a small black hole may still be

[2] Essentially, the problem is that if a pure state forms a black hole, while the entirety of quantum state remains pure, the interior and exterior configurations are mixed. Since the interior states presumably cannot escape the black hole, they somehow "disappear" if the black hole completely evaporates. This implies that we have a pure-to-mixed evolution, violating unitarity.

macroscopically large; see below). Some authors have interpreted a black hole to be a collection of many microscopic "black hole atoms" [3]. However, the remnant scenario would be deemed "impossible" by later works that instead prefer to maintain unitarity by extracting quantum information from the Hawking radiation itself (see, e.g., the review [4]). In order to better appreciate the issue, we now give a quick summary of the infamous "information paradox" and the "firewall" or AMPS paradox [5, 6] that came after it; see also Chaps. 12 and 13 of this book.

The story started with the realization that black holes have an entropy, characterized by the Bekenstein-Hawking formula $S = A/4$. Black holes need to have entropy otherwise the entropy content of the universe will decrease if we throw things into black holes. The second law of thermodynamics is preserved since the black hole grows afterward. Classically, the area cannot decrease, but when Hawking radiation is included, the *generalized second law* (GSL) still holds: the entropy of the entire system still increases since the emitted Hawking radiation also contains entropy. However, because black holes have no hair, it seems that no information about the stuff that falls into a black hole is retained, or at least not accessible from the observer outside. This is classically acceptable. The problem is when a black hole evaporates—what happens to the information of everything that has fallen inside? The Hawking radiation is thermal and so does not appear to contain any information. If the black hole eventually completely disappears, then we have lost all the information inside, and this violates unitarity of quantum mechanics (in this context it can be thought of as "conservation of information"). A possible way out of this conundrum is to allow the information to leak out via Hawking radiation, but in a very subtle manner—by quantum entanglement. Thus in order to recover the information one has to collect all the Hawking particles and decode the information somehow via a super-quantum computer, impossible in practice but all that is required for unitarity is that this is possible in principle (just like burning a book and recovering the information therein from the ashes is possible in principle, but no one can do it in practice!).

However, the AMPS paradox pointed out that doing so would lead to another curious issue, namely that at late time (after the black hole entropy has been halved, called the *Page time* [7–9]), the black hole horizon ceases to be "uneventful" and would instead be replaced by a high energy region, dubbed a "firewall". Any observer that falls through a black hole in classical general relativity would just pass right through the horizon (a mathematical surface!), but once the firewall sets in, an infalling observer would now be incinerated at the firewall. The reason for the existence of the firewall is as follows. Hawking particles are formed by pair-production from the vacuum "near"[3] the horizon, one of which falls into the black hole while the other is emitted to infinity. The usual argument is that (1) the pair-produced particles are maximally entangled, and that (2) information only starts to be released after Page time. The Hawking particles emitted after Page time needs to purify the earlier Hawking radiation, and therefore they cannot be maximally entangled with their

[3] This is not completely correct. For a Schwarzschild black hole the particles can be emitted in a diffused region called "quantum atmosphere" that extends some $O(1)$ to $O(10)$ distance away from the horizon [10].

own partner that falls into the black hole. This is due to the monogamy of quantum information: a particle cannot be maximally entangled with two particles at the same time. Without the maximal entanglement between the outgoing and ingoing pairs, the field configuration across the horizon becomes discontinuous. This in turn means that the Hamiltonian (energy) of the field diverges across the horizon (since it contains terms like $\partial^2\phi$, that involves the *derivative* of the field); this is the firewall. An observer that falls into a sufficiently old black hole would encounter the firewall and be destroyed (astrophysical black holes are not yet old enough). Such a drastic departure from the "no drama" scenario of general relativity[4] led to a fierce debate in the community as to what has possibly gone wrong in the analysis.

Indeed, there are assumptions that go into the AMPS paradox, and "resolutions" of the paradox by challenging these premises can be found throughout the literature. For example, the requirement for maximal entanglement has been challenged [12, 13]. For our purpose, however, it is the "central dogma" [14]—the claim that a black hole behaves like a quantum system whose number of degrees of freedom is proportional to the horizon area—that requires a closer examination (see, e.g., [15] for some criticisms of the dogma). The dogma is related to known results that there exists an upper bound for how much information one can squeeze into a finite size region with a given energy, namely the Bekenstein bound [16] ($S \leqslant 2\pi RE$) and its various more recent generalizations. Thus, if the central dogma holds, and if Hawking evaporation does not contain any information, then problems already arise when a black hole becomes sufficiently small (after Page time), regardless of whether there is a complete evaporation at the end. This is because the horizon would eventually be too small to bound the large amount of information inside. In particular this would mean that the remnant picture *for solving the information paradox* is incorrect.[5] However, as we will discuss later on, it could be that the central dogma is not correct, and the remnant picture may still well be viable. But for now, let us begin by explaining how one can obtain remnants in various quantum gravitational models.

---

[4] The situation is even worse than initially expected. If firewalls can in fact exist, then there is a nonzero probability that a firewall can occur arbitrarily far away from the horizon, a "naked firewall" [11]. Due to the divergence in energy density, a firewall is essentially a singularity; a naked firewall is therefore as bad as a naked singularity. We will have more discussions later on naked singularity.

[5] Note that remnants themselves are not incompatible with the central dogma in the sense that if the latter is correct, then a microscopic remnant can only have a tiny amount of information, and therefore cannot solve the information paradox. For the same reason, the firewall would arise as in the standard picture; see [1]. On the other hand, this also means that if one's goal is to solve the information paradox, it is *not enough* to just come up with a model in which there is a remnant at the end of the evaporation—one has to assume the failure of the central dogma independently of the existence of the remnant. It is this assumption that is at the heart of the debate.

8.2 How to Obtain Remnants

A black hole becomes a remnant when it stops evaporating. For example, this can happen if it becomes an extremal black hole. In this context, by an extremal black hole we mean the temperature of the black hole is zero (these two notions do not always coincide). Note that there is no need for it to become exactly extremal: just by asymptotically approaching an extremal state the black hole evaporation time scale becomes indefinitely long. Zero temperature extremal black holes are black holes whose inner horizon coincides with the outer horizon. In classical general relativity, one needs to include either a charge or angular momentum in order for a black hole to have two horizons. However, if one envisages removing the black hole singularity, then it is common to end up with "regular black holes" with two horizons,[6] even without charge or angular momentum. Such black holes can then radiate indefinitely as they approach the zero temperature limit (whereas for the charged and/or rotating case in general relativity, the charge and angular momentum are typically shed faster than the mass and so the final state is still the Schwarzschild solution). Bardeen black hole [19], non-commutative geometry black hole [20], asymptotically safe gravity black hole [21], and two-dimensional dilaton black hole [22] are just some examples.

Note, however, inner horizons are themselves highly unstable and would likely collapse into another singularity. This seems to defeat the whole point of regularizing the singularity in the first place. Worse, the end state of this instability is unknown, so we cannot say whether there will still be a remnant afterward. One recent progress is to construct black holes with "inner extremal horizon", with zero surface gravity, and therefore avoids the classical blueshift instability. Unfortunately this does not seem to avoid the Aretakis instability (classical perturbation does not decay) [23–26]. In addition, semi-classically the inner extremal horizon remains unstable [27]. This way of obtaining a remnant is therefore quite challenging conceptually, but clearly more studies are required, as the properties of inner horizon of a black hole are still unclear in general [28], and the possibility that black holes can asymptote towards some extremal configurations and thus become an effective remnant cannot yet be ruled out [29]. Note that extremal black hole is not the only possible end state, as some regular black holes can evolve into a horizonless object [30–32].

Another possibility is to consider quantum gravitational effect that kicks in when the black hole is sufficiently small. Note the subtle difference: in the contexts of regular black holes, while the classical singularity could be removed by some kind of quantum gravity effect, remnants are the result of the existence of extremal state. The inner horizon—which would not be there without the quantum gravitational effect regularizing the singularity—exists already when the black hole is still astronomi-

[6] This is not necessarily the case. For example, in [17], the authors constructed a regular black hole by considering a massive source in a quantum superposition of locations. The black hole only has a single horizon, with the zero-temperature extremal case separating the black hole configuration with a (horizonless) wormhole configuration. For another example, see [18], in which a single-horizon regular black hole whose singularity is replaced by a bounce was constructed via an effective polymerization scheme in the Loop Quantum Gravity approach.

cally large. Thus, these extremal remnants are not *directly* caused by any quantum gravitational correction, only indirectly so.[7] A more direct influence from quantum gravity can happen at or near the Planck scale, without necessarily modifying the black hole solution when M is still large, and this can be studied phenomenologically by employing the "generalized uncertainty principle" (GUP).

Let us restore $\hbar$, c and G for the moment for clarity. The most basic form of GUP is given by[8]

$$\Delta x \Delta p \geqslant \frac{1}{2}\left(\hbar + \alpha L_p^2 \frac{\Delta p^2}{\hbar}\right), \tag{8.2}$$

where L_p is the Planck length, and α is a dimensionless parameter, typically taken to be of order unity. We will first discuss the case in which $\alpha > 0$. From Eq. (8.2) we can obtain the following inequality:

$$\frac{\hbar}{\alpha L_p^2}\left[\Delta x\left(1-\sqrt{1-\frac{\alpha L_p^2}{\Delta x^2}}\right)\right] \leqslant \Delta p \leqslant \frac{\hbar}{\alpha L_p^2}\left[\Delta x\left(1+\sqrt{1-\frac{\alpha L_p^2}{\Delta x^2}}\right)\right]. \tag{8.3}$$

The presence of the square root makes it clear that GUP imposes a minimum uncertainty in the position:

$$\Delta x_{\text{min}} = \sqrt{\alpha} L_p. \tag{8.4}$$

Hawking temperature is modified under GUP. For an asymptotically flat Schwarzschild black hole this modification was obtained heuristically in [33] by considering the characteristic energy of the Hawking particle as $E = pc$ and by identifying Δx with the horizon scale $\Delta x \sim 2GM/c^2$. This gives (upon multiplying with a normalization constant prefactor $1/2\pi$ in order to obtain the correct temperature when $\alpha \to 0$):

$$T_{\text{GUP}} = \frac{Mc^2}{\pi \alpha}\left(1-\sqrt{1-\frac{\alpha \hbar c}{4GM^2}}\right). \tag{8.5}$$

Note that the minus sign before the square root is chosen (instead of the positive one; c.f. Eq. (8.3)) so that the correct Schwarzschild temperature is obtained in the $\alpha \to 0$ limit. The problem with such a heuristic treatment is that we cannot be sure of its validity. After all, while the method works for the Schwarzschild case, it does *not* lead to the correct known temperature[9] for the Kerr or Reissner-Nordström black hole [34] (though intriguingly, how it fails may be related to black hole thermodynamics [34]), so how could we be confident that it works when there is a GUP correction? For other criticisms for heuristic treatments of GUP, see [35–37].

---

[7] Admittedly, what qualifies as "direct" is rather subjective.

[8] The notation Δx^n is a short hand of $(\Delta x)^n$. Similarly for Δp^n.

[9] By "known temperature" we mean the expression obtained using standard methods like Hawking's original Bogoliubov transformation, the Euclidean Wick rotation method, quantum anomaly calculations, or even tunneling method. They all give the same temperature expression.

Equation (8.5), if indeed correct, gives rise to a peculiar behavior that the temperature becomes nonzero but finite at the minimum mass[10]

$$M_{\min} = \frac{\sqrt{\alpha} M_p}{2}, \tag{8.6}$$

where M_p denotes the Planck mass. The fact that there is a nonzero temperature yet the black hole no longer evaporates (otherwise the square root term in Eq. (8.5) becomes imaginary) could be a sign that modifying the Hawking temperature without modifying the Schwarzschild metric may be problematic. Furthermore, if we look at the temperature of a Reissner-Nordström black hole in standard general relativity, there is also a square root involved, namely a factor of $\sqrt{M^2 - Q^2}$ (in Planck units). In this case the temperature tends to zero as $M \to Q$, so the $M = Q$ state is an extremal black hole. Still, even then, there are considerable interests in the possibility of a naked singularity formation. That is, whether some kind of perturbation can bring Q to exceed M. Crucially, note that $\sqrt{M^2 - Q^2} \notin \mathbb{R}$ does not imply there is no state such that $Q > M$, only that such a state would no longer be a black hole. Therefore, in the GUP case discussed above, it is actually not necessary that the $T \neq 0$ remnant "ceases radiating" because the temperature becomes imaginary. Rather, precisely because $T \neq 0$, there is a risk that it may drive the evolution further beyond a black hole stage into another horizonless configuration, if not a naked singularity. Whether this is the case would require a careful investigation.

Of course, the whole idea of a classical, well-defined, metric tensor might already fail near the Planck scale, so it is not clear if any modified metric is trustworthy (worse still, different metrics can be obtained by relying on different heuristic "derivations" [38]). Nevertheless, we note that the entropy of the black hole, which can be obtained by integrating the first law of black hole thermodynamics, can be shown to satisfy (returning to Planck units hereinafter)

$$S_{\text{GUP}} = 4\pi M^2 - \frac{1}{2}\pi\alpha \ln M + \sum_{n=1}^{\infty} c_n (4\pi M^2)^{-n} + \text{const.} \tag{8.7}$$

The logarithmic correction term is consistent with many approaches of quantum gravity (though the sign differs from one approach to another), which gives us some confidence that despite the heuristic argument, Eq. (8.5) may be correct.

Surprisingly, the problem of the final non-evaporating but nonzero temperature black hole can be avoided. This is due to another effect of GUP that can prevent black hole evaporation even when the black hole is still astronomically large (and therefore when the temperature remains largely unmodified), just after a scrambling time [39]. If correct, this renders the aforementioned Planckian size black hole remnant irrelevant except for black holes that were formed to be microscopic in the beginning. For a Schwarzschild black hole, the scrambling time [40, 41] is $r_S \log(r_S/L_p)$, where $r_S = 2M$ denotes the horizon; it is the time scale for information to be scrambled

[10] This can also be obtained from Eq. (8.4) by identifying Δx with the horizon scale $2GM/c^2$.

over the horizon after being dropped into a black hole. Note that this time scale is far shorter than $\sim r_S^3$, the lifetime of a black hole. Thus, if a large black hole stops evaporating after the scrambling time, then it becomes a macroscopic remnant and the peculiarity mentioned above never arises. Intuitively this result can be appreciated as follows. First, we recall the heuristic derivation of Hawking temperature above, in which the Hawking temperature scale is roughly set by Δp, and by the uncertainty principle $\Delta x \sim 1/\Delta p$, we can trace Δx back to the horizon scale, i.e., Hawking radiation comes from the region around the black hole. However, with GUP, it can be shown that beyond the scrambling time, $\Delta p \gg \alpha/L_p^2$, which in turn implies that $\Delta x \gg r_S$, which cannot be interpreted as particle creation by the black hole [39]. See also [42], in which the same conclusion was reached using a string theoretic approach. These results are also reviewed in Chap. 7 of this book. Note that if correct, this also avoids the firewall problem since the black hole never becomes small enough to get past the Page time scale. The remnant is essentially a *classical* black hole. To contrast with this scrambling time scale remnant, in another recent approach, it is argued that a black hole stops evaporating after it has emitted about half of its initial mass, due to "memory burden effect" [43], in which a system with high capacity of information storage (such as a black hole) is stabilized by the load of information it carries. Thus we see that it is not necessarily the case that remnants are microscopic.

It is almost poetic that a quantum gravitational correction like the one in GUP, which we usually think can only be important when the black hole is small, can give rise to a drastic deviation from semi-classical picture even when the black hole is still huge (and even renders the black hole effectively classical with no Hawking radiation). Yet, this is not the first time we have seen such an unexpected behavior. For another example, consider the Chandrasekhar limit of a white dwarf star. White dwarfs are stable because the electron degenerate pressure counteracts gravitational collapse. This pressure is inherently quantum mechanical in nature, it arises from the jittering motion (large Δp) of the particles inside the star, which increases as the star becomes smaller (small Δx). Once GUP is taken into account, it seems that the Chandrasekhar limit is destroyed [44, 45]. Dynamical stability analysis can restore an upper mass limit for white dwarfs for a certain range of the GUP parameter [46], which explains why in astrophysical observation we do not observe arbitrarily large white dwarfs. Still, here too we see that a supposedly small quantum gravity effect can lead to big changes in physics well outside the quantum gravity regime.

In [47] I considered Hawking evaporation under the assumption that the GUP parameter could be negative. In such a scenario there is no longer a minimum mass, and so the black hole can continue to evaporate indefinitely. However, its lifetime turns out to be infinite, so we still have a remnant that asymptotes to zero mass. There is, however, a peculiar property—the zero mass black hole (or rather, no more black hole!) has a finite nonzero temperature in the negative GUP parameter case. This is as weird, if not weirder than, the positive GUP parameter case in which the non-zero temperature remnant supposedly does not radiate. Fortunately, since this bizarre state cannot be reached in a finite time, it is not physically relevant as the end state of an evaporating black hole.[11] See also [48] for more discussions.

So far we have discussed extremal black holes and the GUP approaches that led to remnants. In recent years, another promising possibility has come to light: a *white hole* remnant, which is a scenario supported by loop quantum gravity [49–55], see the recent review [56] for details (and [57] for a nice summary). The basic idea is simple: the static Schwarzschild solution does not distinguish the direction of time, and indeed in the maximally extended spacetime we see that there are black hole *and* white hole, described by the same metric tensor. Classically, we do not usually care about the white hole part because it is unphysical (in the sense that being a time-reversed version of black hole, a white hole cannot be produced by gravitational collapse). However, quantum mechanically one should consider a superposition of both black and white holes, and the possible tunneling from the black hole state to the white hole one. This probability becomes significant as the black hole becomes sufficiently small. In this scenario, a black hole with initial mass M evaporates with lifetime that scales like M^3 as in the standard Hawking picture. However, before it completely evaporates away, it tunnels into the white hole state, which then becomes a remnant that slowly radiates away. The time scale for this diffuse emission is about M^4 [55]. Such a white hole remnant does not live indefinitely. However, like the more traditional black hole remnant it can conceivably still solve the information paradox, as we will discuss in the next section.

There are of course other routes that can possibly lead to black hole remnants, but more studies are required to check these claims. For example, many quantum gravity theories predict that at sufficiently high energy the effective dimension of spacetime is 2 instead of 4 (see [58] for a review; also see [59] and the references therein). In [60] it was proposed that if the early 2-dimensional universe is governed by a dilatonic gravity theory, a black hole can exist. However, as the universe expanded and cooled down, the spacetime transited into 3 dimensions (before it becomes 4). In 3-dimensions however, if there is no negative cosmological constant, then there can be no black hole solution [61]. Thus, the black holes from the 2-dimensional epoch ceased its evaporation and became "stuck" as remnants. This picture is however, somewhat problematic. What is the geometry of a black hole remnant once it crossed the dimensional transitions? Most remnants *are* themselves black holes, so if no black hole solution is allowed, what are those remnants, which in any case, have two dimensions less than the surrounding space? Also, most quantum gravity results indicate that spacetime dimension does not change, but rather the spectral dimension or thermal dimension does (i.e., spacetime is only *effectively* 2-dimensional at high energy scale). In this case there is no problem with black holes with effective dimension being 3. While the scenario above concerns the dimensional transition of the entire universe, we can also consider the dimensional transition of the black hole

---

[11] There is still one subtle issue: a zero mass black hole should be easily produced by quantum fluctuations, so why don't we see them? The reason may have to do with their nonzero temperature, which in a consistent quantum gravity treatment may contribute to a nonzero effective energy-momentum tensor and this then gives it an effective mass after all.

itself. As the black hole evaporates and gets hotter, eventually its spectral dimension reduces to 2. It is shown in [59] that provided that the solution still only has 1 horizon, the lifetime is only slightly longer than the standard M^3, namely $M^3 + cM$ term, where c is a constant. Therefore it seems that the spectral dimension route does not lead to remnants. However, in the case of *thermal dimension* [62, 63], which arises from the modification of the dispersion relation, whether remnants occur or not likely depend on the dispersion relation. Also related to dispersion relation is the idea of "gravity's rainbow" [64], which does lead to black hole remnants [65].

8.3 Why Traditional Criticisms Against Remnants are Not Convincing

There are two major issues concerning remnants, which led to their viability being questioned by the mainstream high energy physics community. The two problems are the *species problem*[12] and the *entropy problem*, which are closely related. We begin with the species problem. Consider again the no-hair theorem of black holes. Once a black hole is formed, there is no way to tell from the outside what was the stuff that collapsed to form the black hole in the first place. The possibility seems to be endless, each of these different internal states gives rise to a distinct "species". Now, consider the probability of pair-producing remnants from quantum fluctuation of the vacuum or in a high energy experiment such as a collider. For each species, whose mass is about the Planck mass, the pair-production probability is tiny. However, since there are infinitely many species, the total production rate seems to be unbounded. This therefore seems to predict that tiny black holes should be copiously produced from the vacuum all the time, as long as the available energy exceeds twice the remnant mass (given that they are Planckian, this is a relatively low energy requirement). The fact that we do not observe them therefore seemingly rules out black hole remnants. This criticism is, however, over-simplistic, as we do not know if the number of internal states is actually infinite. In fact, Giddings had argued that it is finite [68, 69]. Even if there is indeed an infinitely many possible internal states, the usual physics of *particle* pair-production may be quite misleading when applied to remnants. Even though they may appear particle-like with mass the order of the Planck mass, they still have a vast interior structure. The vast amount of space inside a black hole is a result that follows from standard general relativity, not any exotic physics.

For any star of radius r, given that it is spherical, we can deduce that its volume has to be fixed, given by $\frac{4}{3}\pi r^3$. However, a black hole is a highly curved spacetime manifold; its interior is nontrivial. This should be well-known but somehow the literature is still confusticated by statements about the "central singularity" of a

[12] Not to be confused with another "species problem" [66, 67] (though they may be related): if Bekenstein-Hawking entropy is entanglement entropy, it should depend on the species of particles/fields available, so why is it only proportional to the area?

Schwarzschild black hole, as if $r = 0$ is at the geometric center of a sphere.[13] What happens in the Schwarzschild case is that the singularity is spacelike, an event in the future of the in-falling observer. The interior spacetime is best viewed as a special case of Kantowski-Sachs cosmological spacetime, which is anisotropic [71, 72]. In particular, inside the horizon, a line of constant r, θ, ϕ has a proper length that goes to infinity as the singularity is approached, hence "there is no dearth of space inside" of a black hole, in the words of Ted Jacobson [73].

A notion of interior volume, independent of coordinates, was defined by Christodoulou and Rovelli [74], which for a Schwarzschild black hole, is best computed in the Eddington-Finkelstein coordinates, grows linearly in advanced time v as

$$\begin{aligned} \text{Vol.} &\sim \int^{v} \int_{S^2} \max\left(r^2\sqrt{\frac{2M}{r} - 1}\right) \sin\theta \, d\theta d\phi dv \\ &= 3\sqrt{3}\pi M^2 v. \end{aligned} \tag{8.8}$$

Therefore, we see that in classical general relativity, an old Schwarzschild black hole has a larger interior volume compared to a young one, even if they have the same mass. Even more surprisingly, the volume continues to grow even if the black hole is losing mass via Hawking evaporation [75, 76]. A remnant therefore has a vast interior volume, although this is not visible from the outside. Thus, even if we can create mini black holes with powerful high energy collision, these newly formed black holes are *not* the same as a remnant.

As argued in [56, 77]—see also Chap. 6 in this book—black holes may harbor more states than those surface states that account for the Bekenstein-Hawking entropy, and as the black hole shrinks to Planckian size remnant, the information remains inside until finally (after a lifetime that scales as M^4) the remnant decays into a large number of low energy photons. The huge number is required if they were to carry all the information inside. In such a scenario, the usual Hawking radiation need not carry any information. In order to produce a remnant, one has to consider essentially the reverse process: focusing a large number of low energy photons. This explains why it is actually very difficult to form a remnant in high energy physics experiments, or even spontaneously from the vacuum. This is similar to the results in [78], in which it was argued that the most likely history of a fluctuation out of equilibrium is simply the CPT conjugate of the most likely way a system relaxes back to equilibrium. This means that though given enough time, systems near equilibrium can fluctuate into lower-entropy states, it is exceedingly unlikely to fluctuate out an egg; the most likely way to get an egg formed from fluctuation is for the shards of egg shells and runny yolk and egg white to gradually formed and re-assemble into an egg, i.e., the reverse process of breaking an egg. In general, the most likely way to get into a low entropy state is via a sequence of individually unlikely events. For our case, a black hole remnant decays into a large number of photons at the end of

[13] The "$r =$ const." surfaces inside the black hole are sometimes referred to as "T-spheres" of the "T-region", especially in older texts [70] to emphasize the distinction.

its long lifetime. By the second law of thermodynamics, the collection of photons must have a higher entropy state than the remnant itself. Thus, the most likely way for a remnant to be spontaneously created would be for a large number of low energy photons to somehow gather together by chance. This itself is already exceedingly unlikely to happen, so to spontaneously produce a remnant directly from fluctuation is even more unlikely.

In order for remnants to be viable, one has to also address the *entropy problem*. As previously mentioned, the question was how does a "small" remnant possibly store enough entropy. We have already seen that the remnant is not small, quite to the contrary. However, the entropy problem also concerns what is the nature of the Bekenstein-Hawking entropy, namely whether it reflects all the degrees of freedom inside the black hole. If so, then it does not matter how large the interior is, it cannot contain that much information for some reason. This is the position of the "central dogma"—that the number of states inside of black hole goes like $e^{A/4}$. The issue of the interpretation of the Bekenstein-Hawking entropy is of course an old problem (see, e.g., [79]), but the view that the Bekenstein-Hawking entropy measures all the available states (the "strong form" interpretation [80]) became a dominant view due partly to the success of holography (AdS/CFT correspondence). Still, it could be that the Bekenstein-Hawking entropy measures only the "surface states" (the "weak form" interpretation [80]) that are relevant for some physical purposes, while the total number of states inside the black hole far exceeds $e^{A/4}$. The area is relevant for black hole thermodynamics, and by extension holography, but there can be other "non-thermodynamic states" inside. Indeed, as mentioned in [56] the entanglement (von Neumann) entropy is only guaranteed to be equal or smaller than the thermodynamic entropy for ergodic systems, but black holes are not ergodic.

There are other subtleties that one should consider. As mentioned, the large volume inside a black hole is crucial for understanding a remnant. An intuitive picture can also be found in [56], namely that the mass of the black hole is essentially its energy. If the remnant is Planckian, it has very low energy, thus in order to account for huge entropy (large number of states), we need a lot of low energy, i.e., long wavelength particles, inside. This can only be possible if the interior volume is huge, which is indeed the case. Thus, treating remnants as point particles cannot possibly work from the onset, and it is that assumption that led to the idea of remnants being prematurely rejected. Another issue, as already pointed out in [80], is that the species problem rests on an effective field theory argument (which is required to estimate the production cross section), which in turn means that there is a cutoff scale $\lambda_c \gg L_p$ above which the remnant is treated as a point particle. However, due to the highly curved spacetime, it is nontrivial to define such an effective field theory. Indeed, due to gravitational redshift/blueshift, a longer wavelength far away from the black hole, is short close to it, and long again in the interior, and hence a fixed cutoff does not seem to make much sense. This is not to say that we cannot have an effective field theory on curved spacetime at all, but it would be too premature to rule out remnant based on an argument that is itself rather non-rigorous.

Another subtlety concerns the notion of entropy itself being observer-dependent (which is not really a surprise because it is well-known that temperature is also observer-dependent—the Hawking temperature being valid for asymptotic observers). In particular, consider a wave packet of fixed energy in flat spacetime. We can increase the entropy without adding more energy by introducing more particle species. Suppose there are some huge number n of free scalar fields. Then the entropy grows like $\log n$. If we go to the Rindler wedge instead, then the von Neumann entropy of the mixed state, though increases as $\log n$, is bounded above by E/T [81] (note that this is essentially the Bekenstein bound, up to a constant prefactor). This at least suggests that we should be careful about *whose* entropy are we talking about. Perhaps the interior of a remnant can have vastly more entropy than its surface bound, without being in tension with the fact that an asymptotic observer only sees the surface entropy. In addition, even the interpretation of the Bekenstein bound requires some care; in [82] Hayden and Wang argued that unlike classical bits and qubits, zero-bits[14] and their associated information processing capability are generally not constrained by the Bekenstein bound. In other words, we must understand the various subtleties of entropy before we can definitely rule out remnants.

8.4 Discussion: Prospects for Black Hole Remnants

As we have seen, there are various ways one could obtain black hole (or white hole) remnants. It should be emphasized that the GUP approach, though relatively simple, still requires more careful investigations [35–37]. After all, there are many versions of GUP in the literature, which should we trust? Even for a fixed GUP, it is not clear what should the approach be for black hole thermodynamics. For example, should we modify only the Hawking temperature and the entropy, or also modify the metric tensor? If the latter, how do we obtain the metric? No unique, well accepted, answers are available to these questions in the literature [38], and as such the conclusions obtained about remnants are also subject to a certain degree of skepticism. The possibility that black holes can quantum tunnel into a white hole remnant, obtained mostly from loop quantum gravity approach, seems more promising.

In principle remnants can still "resolve" the information paradox, but one has to be careful of some subtle aspects. For one, whether there is a remnant or not is quite independent from whether Hawking radiation encodes information. If information is still carried out via Hawking emission, then the firewall paradox remains because Page time sets in much earlier than the remnant stage [1] if the remnant is Planckian. However, the requirement that information be carried out by Hawking radiation rests upon the assumption of the central dogma, which as we have discussed, need not

[14] A zero-bit is a special case of α-bits, introduced in [83]. They correspond to the ability to perform quantum error correction on arbitrary bounded dimension subspaces, with the real number $0 \leqslant \alpha \leqslant 1$ characterizing the size of the subspace. The standard quantum error correction corresponds to $\alpha = 1$, while $\alpha = 0$ is related to quantum identification, a form of equality testing for quantum states. In the context of black holes, if the central dogma holds, the zero-bits are revealed after the Page time [84]. The information scrambling of Hayden-Preskill protocol [40] also corresponds to the $\alpha = 0$ case [84].

hold. Therefore, Hawking radiation can be information free like initially thought, with the information somehow contained within the large interior of the remnant. If the remnant slowly emits diffuse particles and completely evaporates in a far longer time scale (at least M^4), then eventually all the information is released back to the exterior universe and unitarity is preserved. The large interior allows these particles to be soft particles that carry little energy, which also explains why remnants are not copiously produced in everyday high energy experiments—any mini-black hole we can produce in future laboratory is a completely different object from a remnant. Much of the discussions here are not new, unfortunately the current state remains more or less the same, with many colleagues still unaware of the counter-arguments against the species and entropy problems.

The interior volume of black holes did however, gained some attention in the high energy community due to it possibly being a holographic dual to complexity (the so-called "Complexity-Volume conjecture" or just CV-conjecture [85–91]), yet how this might relate to a remnant is still unclear. In [92], Susskind proposed the idea that black hole complexity cannot grow without bound. Instead, after some time that scales as $\exp(S)$, where S is the Bekenstein-Hawking entropy,[15] the quantum state of the black hole now also contains a significant white hole component. This is the "gray hole" phase,[16] during which the complexity fluctuates around a constant value. After a time of $t \sim \exp[\exp(S)]$ the "gray hole" becomes a white hole, during which the complexity decreases. Note that this is very similar to the black-to-white hole transition we have discussed, but the time scale involved is extremely different. The possibility that even astrophysical black holes are "gray", i.e., in the quantum superposition of black and white hole states has also been discussed in [95], in which the authors also pointed out it is possible to arrange the "black-to-white hole transition" to have zero action—so that it will not be subject to destructive interference in the Feynman path integral.

We should also not rule out the possibility that information escapes into "another universe", interpreted broadly. For example, many of the regular black hole solutions have de Sitter core or other geometries. An interior geometry which asymptotes to $dS_2 \times S_2$ in the future is obtained in [96], which essentially is like another universe whose spacetime dimension becomes effectively 2-dimensional (the size of the S_2 is fixed and thus eventually "compactified" relative to the dS_2 part). Even under the usual Hawking evaporation without remnant, we already mentioned that the interior volume continues to grow. Therefore, if the black hole completely evaporates away, the interior spacetime must pinch off its parent (our universe) and subsequently becomes a new baby universe, taking away all the information with it (assuming that Hawking radiation is information free) [97]. The idea that black holes lead to a new universe and thus avoiding the information loss problem is of course, not new either; see, e.g., [98]. The pinching-off of a baby universe is sometimes referred to as the black hole "emitting" a baby universe. In [99], based on a calculation

[15] This time scale, exponential in the entropy, is also the time scale that is required for one to decode the information contained in Hawking radiation [93], assuming the latter contains information.

[16] A different kind of "gray hole" is considered in [94].

carried out in Jackiw-Teitelboim (JT) gravity, it was argued that a very old black hole can tunnel to a white hole or even a firewall by emitting a large baby universe. However, the probability for tunneling to a white hole is proportional to $t^2 \exp(-2S)$, where t is the age of the black hole, which again differs from the loop quantum gravity tunneling probability of $\exp(-S)$ [56] (granted that the gravity theories are different). In addition, the loop quantum gravity tunneling does not "emit" away the large interior; the interior just slowly shrinks throughout the white hole phase until the remnant completely disappears via diffuse emission.

It is also interesting to compare the remnant picture with that of the usual remnant-free end stage of Hawking evaporation. Note that as a Schwarzschild black hole evaporates, it gets smaller, which in turn means that the curvature (say, the Kretschmann scalar) at and near its horizon is becoming larger. In a truly quantum picture M cannot be continuously changed (although semi-classically we model the evaporation with a differential equation of dM/dt), so at the final stage the Planckian black hole is expected to dissolve away into a collection of particles and we are left with a flat spacetime. Still, this is only an expectation. How do we know that the large spacetime curvature does not "tear" spacetime enough to lead to a naked singularity? Certainly such a possibility has been hinted at in [100, 101]. Including higher curvature terms (quadratic gravity) also suggests such a possibility [102], though it is hard to say what the full quantum gravity result would be. We normally expect that the end state is flat spacetime because it is the smooth limit of $M \to 0$, but it is *not* a smooth limit for the curvature. Of course, naked singularity as the end state will be bad, and by the (weak) cosmic censorship conjecture it is natural to think that this does not happen. Perhaps remnants also play the role of a cosmic censor. Still, this does not completely avoid the problem, as we will discuss below.

In the remnant picture, if the lifetime is finite, the same thing happens towards the end. We do not know for sure whether quantum gravity can indeed "resolve" singularity completely,[17] or replace it with a milder, quantum version (see [106, 107] for my related reviews). In fact, if a remnant has more entropy than the Bekenstein-Hawking entropy, then a version of singularity theorem can be proved [108]. This means that there is still the possibility that the end state is either a singularity or something else that can only be described by quantum gravity. As Penrose wrote in [109]:

> It is hard to avoid the conclusion that the endpoint of the Hawking evaporation of a black hole would be a naked singularity—or at least something that one [sic] a classical scale would closely resemble a naked singularity.

So both the standard picture and the remnant picture may encounter singularity at the end if they are not sufficiently resolved in quantum gravity.

---

[17] Note that GUP may or may not remove the classical singularity. The remnant obtained in [33] still considers the same classical form of Schwarzschild geometry, so the singularity is untouched (here the "minimal length" is reflected in the size of the remnant, not in "smearing out" the singularity. For a review about minimal length, see [103]). Though in other models, one could also employ GUP to obtain regular black holes [104, 105].

Even if the lifetime of the black hole is infinite, one cannot avoid this problem as long as the mass becomes Planckian. This is because naked timelike or null singularities are bad in classical general relativity not because of the singularities themselves, but because their presence renders the theory non-predictive. Without knowing the properties of the singularities, one cannot know what influences it can have on spacetime region even far away from them. Note that this problem does not only arise when the curvature is mathematically divergent, it already arises for large enough curvatures. Usually we expect general relativity to be a good theory when the curvature scale is above the Planck scale (for the subtleties, see [38]). Thus, if black holes become Planckian in size (whether there is a Planck size remnant or not), cosmic censorship will be violated in spirit. Again, it is worth emphasizing that stable remnants (those with infinite lifetime) cannot serve as a cosmic censor; it effectively becomes a "Planckian curvature source", as bad as a naked singularity. This is not necessarily a disaster, it just means that the end state of Hawking evaporation cannot be understood semi-classically. Nevertheless, we should keep in mind that one of the criticisms against firewall is that it is essentially a naked singularity [11]. Should we prefer remnants over firewalls if they have the same problems? Perhaps the former is more palatable because it only arises for Planckian black holes, whereas the latter can occur even for large black holes.

Remnants that stop evaporating after either the Page time or scrambling time is effectively classical and this issue never arises. One might think that this problem also does not arise for extremal black hole remnants if they asymptote to extremal state when still being macroscopic in size. However, this is not the case, as various studies have suggested that near-extremal black holes are highly quantum object and thus the result cannot be trusted [110–116]. In fact, an extremal black hole can be effectively a singular object [110, 112, 113].

To conclude, no argument against remnants holds up to close scrutiny, and thus the option that remnants can avoid information paradox remains viable. Given that Planck size remnants may also serve as a dark matter candidate (see, e.g., [51, 117–119])—a possibility already noted by Aharonov et al. [2]—it would be interesting to further investigate their properties under different models. For example, recently there was a concern that some asymptotically extremal black hole remnants may be too fast (due to the recoil of the emitted radiation) [119] and thus too warm to be dark matter, but subsequent discussions had reassured that it is still viable [119].

In the complete evaporation picture, the lower bound of the primordial black hole mass is set by the Hawking evaporation rate, namely that black holes cannot be too small or otherwise they would have completely evaporated by now. This yields a lower bound of about 10^{17} g. With remnants, this bound has to be modified, depending on the model. For example in [56], the white hole remnant is estimated to be in the mass range of 10^{10} and 10^{15} g, and Schwarzschild radius in the range 10^{-18} and 10^{-13} cm. These bounds are obtained from the requirement that the white hole remnant lifetime is longer than the Hubble time, and its progenitor black hole's lifetime is shorter than the Hubble time, namely $M^3 < t_{\text{Hubble}} < M^4$. If, however, black holes already stop evaporating after the Page time or even the scrambling time, then again we must re-check the bounds. In [119], it is shown that if the remnant forms after about half of its mass is evaporated (the "memory burden model"), then we could have light primordial black holes below 10^{10} g serving as a dark matter candidate. It has recently been pointed out that the evaporation rate for cosmological black holes cannot be treated the same way as isolated black holes in asymptotically flat spacetimes. The consequence is that the decay of primordial black holes occurs faster for larger masses but the decay rate reduces for lower mass [119]. Some mass ranges of remnants can hopefully be tested in the future via gravitational wave signature by, e.g., the Einstein Telescope. One such scenario was studied in [119], in which it was pointed out that if the remnant mass is around 5×10^5 g, then there will be a cosmological gravitational wave signal at frequencies around 100 Hz. The aforementioned memory-burden proposal can potentially also be tested via gravitational waves [119]. What is the actual mass scale for a black hole transitioning into a remnant is therefore an important question that requires more studies.

Acknowledgements We acknowledge support from the National Centre of Competence in Research SwissMAP and the ETH Zurich Quantum Center.

References

1. P. Chen, Y.C. Ong, D.-H. Yeom, Black hole remnants and the information loss paradox. Phys. Rept. **603**, 1 (2015). https://www.sciencedirect.com/science/article/abs/pii/S0370157315004391?via%3Dihub. arXiv:1412.8366
2. Y. Aharonov, A. Casher, S. Nussinov, The unitarity puzzle and planck mass stable particles. Phys. Lett. B **191**, 51 (1987). https://www.sciencedirect.com/science/article/abs/pii/0370269387913207?via%3Dihub
3. G. Volovik, Tsallis-Cirto entropy of black hole and black hole atom. arXiv:2409.15362
4. J. Polchinski, The black hole information problem, in *New Frontiers in Fields and Strings* (World Scientific, 2017), pp. 353–397. arXiv:1609.04036
5. A. Almheiri, D. Marolf, J. Polchinski, J. Sully, Black holes: complementarity or firewalls? JHEP **02**, 062 (2013). https://doi.org/10.1007/JHEP02(2013)062. arXiv:1207.3123
6. A. Almheiri, D. Marolf, J. Polchinski, D. Stanford, J. Sully, An apologia for firewalls. JHEP **09**, 018 (2013). https://link.springer.com/article/10.1007/JHEP09(2013)018. arXiv:1304.6483
7. D.N. Page, Average entropy of a subsystem. Phys. Rev. Lett. **71**, 1291 (1993). https://doi.org/10.1103/PhysRevLett.71.1291. arXiv:gr-qc/9305007
8. D.N. Page, Information in black hole radiation. Phys. Rev. Lett. **71**, 3743 (1993). https://doi.org/10.1103/PhysRevLett.71.3743. arXiv:hep-th/9306083
9. D.N. Page, Time dependence of hawking radiation entropy. https://iopscience.iop.org/article/10.1088/1475-7516/2013/09/028. arXiv:1301.4995
10. S.B. Giddings, Hawking radiation, the Stefan-Boltzmann law, and unitarization. Phys. Lett. B **754**, 39 (2016). https://www.sciencedirect.com/science/article/pii/S0370269316000022?via%3Dihub. arXiv:1511.08221
11. P. Chen, Y.C. Ong, D.N. Page, M. Sasaki, D.-H. Yeom, Naked black hole firewalls. Phys. Rev. Lett. **116**(16), 161304 (2016). https://journals.aps.org/prl/abstract/10.1103/PhysRevLett.116.161304. arXiv:1511.05695
12. D.-C. Dai, State of a particle pair produced by the Schwinger effect is not necessarily a maximally entangled Bell state. Phys. Rev. D **100**(4), 045015 (2019). https://journals.aps.org/prd/abstract/10.1103/PhysRevD.100.045015. arXiv:1908.01005

13. D.-C. Dai, The spin correlation of fermion pairs created by a Kerr black hole gravitational potential. Eur. Phys. J. C **84**(9), 928 (2024). https://link.springer.com/article/10.1140/epjc/s10052-024-13288-2. arXiv:2409.05051
14. A. Almheiri, T. Hartman, J. Maldacena, E. Shaghoulian, A. Tajdini, The entropy of hawking radiation. Rev. Mod. Phys. **93**(3), 035002 (2021). https://journals.aps.org/rmp/abstract/10.1103/RevModPhys.93.035002. arXiv:2006.06872
15. L. Buoninfante, F. Di Filippo, S. Mukohyama, On the assumptions leading to the information loss paradox. JHEP **10**, 081 (2021). https://doi.org/10.1007/JHEP10(2021)081. arXiv:2107.05662
16. Jacob D. Bekenstein, Universal upper bound on the entropy-to-energy ratio for bounded systems. Phys. Rev. D **23**, 287 (1981). https://journals.aps.org/prd/abstract/10.1103/PhysRevD.23.287
17. A. Akil, M. Cadoni, L. Modesto, M. Oi, A.P. Sanna, Semiclassical spacetimes at super-Planckian scales from delocalized sources. Phys. Rev. D **108**(4), 044051 (2023). https://journals.aps.org/prd/abstract/10.1103/PhysRevD.108.044051. arXiv:2211.01657
18. A. Peltola, G. Kunstatter, Effective polymer dynamics of D-dimensional black hole interiors. Phys. Rev. D **80**, 044031 (2009). https://journals.aps.org/prd/abstract/10.1103/PhysRevD.80.044031. arXiv:0902.1746
19. M. Akbar, S. Nema, S.A. Hussein, Thermodynamics of the Bardeen regular black hole. Chinese Phys. Lett. **29**, 070401 (2012). https://iopscience.iop.org/article/10.1088/0256-307X/29/7/070401
20. P. Nicolini, A. Smailagic, E. Spallucci, Noncommutative geometry inspired Schwarzschild black hole. Phys. Lett. B **632**, 547 (2006). https://www.sciencedirect.com/science/article/pii/S0370269305016126. arXiv:gr-qc/0510112
21. B. Koch, F. Saueressig, black holes within asymptotic safety. Int. J. Mod. Phys. A **29**(8), 1430011 (2014). https://www.worldscientific.com/doi/abs/10.1142/S0217751X14300117. arXiv:1401.4452
22. D.A. Easson, Hawking radiation of nonsingular black holes in two dimensions. JHEP **02**, 037 (2003). https://iopscience.iop.org/article/10.1088/1126-6708/2003/02/037. arXiv:hep-th/0210016
23. S. Aretakis, Horizon instability of extremal black holes. Adv. Theor. Math. Phys. **19**, 507 (2015). https://www.intlpress.com/site/pub/pages/journals/items/atmp/content/vols/0019/0003/a001/. arXiv:1206.6598
24. J. Lucietti, H.S. Reall, Gravitational instability of an Extreme Kerr black hole. Phys. Rev. D **86**, 104030 (2012). https://journals.aps.org/prd/abstract/10.1103/PhysRevD.86.104030. arXiv:1208.1437
25. J. Lucietti, K. Murata, H.S. Reall, N. Tanahashi, On the horizon instability of an extreme Reissner-Nordström black hole. JHEP **03**, 035 (2013). https://link.springer.com/article/10.1007/JHEP03(2013)035. arXiv:1212.2557
26. T. Katagiri, M. Kimura, The Aretakis constants and instability in general spherically symmetric extremal black hole spacetimes: higher multipole modes, late-time tails, and geometrical meanings. Phys. Rev. D **105**(6), 064062 (2022). https://journals.aps.org/prd/abstract/10.1103/PhysRevD.105.064062. arXiv:2112.09832
27. T. McMaken, Semiclassical instability of inner-extremal regular black holes. Phys. Rev. D **107**(12), 125023 (2023). https://journals.aps.org/prd/abstract/10.1103/PhysRevD.107.125023. arXiv:2303.03562
28. X.Y. Chew, D.-H. Yeom, Insights and guidelines on the cauchy horizon theorems. J. Korean Phys. Soc. **85**(12), 1050 (2024). arXiv:2308.09225
29. F. Di Filippo, S. Liberati, M. Visser, Fully extremal black holes: a black hole graveyard? Int. J. Mod. Phys. D **33**, 15, 2440005 (2024). https://www.worldscientific.com/doi/10.1142/S0218271824400054. arXiv:2405.08069
30. R. Carballo-Rubio, F. Di Filippo, S. Liberati, M. Visser, A connection between regular black holes and horizonless ultracompact stars. JHEP **08**, 046 (2023). https://link.springer.com/article/10.1007/JHEP08(2023)046. arXiv:2211.05817

31. R. Carballo-Rubio, F. Di Filippo, S. Liberati, M. Visser, Singularity-free gravitational collapse: from regular black holes to horizonless objects. arXiv:2302.00028
32. V. Vertogradov, Dynamical black holes: apparent horizon versus energy conditions. arXiv:2410.10582
33. R.J. Adler, P. Chen, D.I. Santiago, The generalized uncertainty principle and black hole Remnants. Gen. Rel. Grav. **33**, 2101 (2001). https://link.springer.com/article/10.1023/A:1015281430411. arXiv:gr-qc/0106080
34. M.R.R. Good, Y.C. Ong, Hawking temperature and the inverse-radius scale of the horizon. Ann. Phys. **474** (2025) 169919. https://www.sciencedirect.com/science/article/pii/S0003491624003269. arXiv:2407.21114
35. Y.C. Ong, Schwinger pair production and the extended uncertainty principle: can heuristic derivations be trusted? Eur. Phys. J. C **80**(8), 777 (2020). https://link.springer.com/article/10.1140/epjc/s10052-020-8363-2. arXiv:2005.12075
36. Y.C. Ong, A critique on some aspects of GUP effective metric. Eur. Phys. J. C **83**, 209 (2023). https://link.springer.com/article/10.1140/epjc/s10052-023-11360-x. arXiv:2303.10719
37. P. Bosso, G.G. Luciano, L. Petruzziello, F. Wagner, Class. Quantum Gravity **40**(19), 195014 (2023). https://iopscience.iop.org/article/10.1088/1361-6382/acf021. arXiv:2305.16193
38. J.O. Weatherall, Where does general relativity break down?. Phils. Sci. **90**(5), 1342 (2023). https://www.cambridge.org/core/journals/philosophy-of-science/article/where-does-general-relativity-break-down/29A158BBF428113F6AB9F2AFF5B81DCE. arXiv:2204.03869
39. T.-L. Chau, P.-M. Ho, H. Kawai, W.-H. Shao, C.-T. Wang, Hawking radiation under generalized uncertainty principle. Eur. Phys. J. C **83**(12), 1118 (2023). https://link.springer.com/article/10.1140/epjc/s10052-023-12302-3. arXiv:2309.01638
40. P. Hayden, J. Preskill, Black holes as mirrors: quantum information in random subsystems. JHEP **09**, 120 (2007). https://iopscience.iop.org/article/10.1088/1126-6708/2007/09/120. arXiv:0708.4025
41. Y. Sekino, L. Susskind, Fast scramblers. JHEP **10**, 065 (2008). https://iopscience.iop.org/article/10.1088/1126-6708/2008/10/065. arXiv:0808.2096
42. P.-M. Ho, Y. Imamura, H. Kawai, W.-H. Shao, A stringy effect on hawking radiation. Eur. Phys. J. C **83**(12), 1118 (2023). https://link.springer.com/article/10.1007/JHEP12(2023)122JHEP **12** (2023) 122. https://link.springer.com/article/10.1140/epjc/s10052-023-12302-3. arXiv:2309.12926
43. G. Dvali, J.S. Valbuena-Bermúdez, M. Zantedeschi, Memory burden effect in black holes and solitons: implications for PBH. Phys. Rev. D **110**(5), 056029 (2024). https://journals.aps.org/prd/abstract/10.1103/PhysRevD.110.056029. arXiv:2405.13117
44. R. Rashidi, Generalized uncertainty principle removes the Chandrasekhar limit. Annals Phys. 374, 434 (2016). https://linkinghub.elsevier.com/retrieve/pii/S0003491616301890. arXiv:1512.06356
45. Y.C. Ong, Generalized uncertainty principle, black holes, and white dwarfs: a tale of two infinities. JCAP **09**, 015 (2018). https://iopscience.iop.org/article/10.1088/1475-7516/2018/09/015. arXiv:1804.05176
46. A. Mathew, M.K. Nandy, Existence of Chandrasekhar's limit in GUP white dwarfs. R. Soc. Open Sci. **8**, 210301 (2021). https://royalsocietypublishing.org/doi/10.1098/rsos.210301. arXiv:2002.08360
47. Y.C. Ong, Zero mass remnant as an asymptotic state of hawking evaporation. JHEP **10**, 195 (2018). https://link.springer.com/article/10.1007/JHEP10(2018)195. arXiv:1806.03691
48. Y. Yao, M.-S. Hou, Y.C. Ong, A complementary third law for black hole thermodynamics. Eur. Phys. J. C **79**, 513 (2019). https://link.springer.com/article/10.1140/epjc/s10052-019-7003-1. arXiv:1812.03136
49. M. Christodoulou, F. D'Ambrosio, Characteristic time scales for the geometry transition of a black hole to a white hole from spinfoams. Class. Quantum Gravity **41**(19), 195030 (2024). https://iopscience.iop.org/article/10.1088/1361-6382/ad6059. arXiv:1801.03027

50. E. Bianchi, M. Christodoulou, F. D'Ambrosio, H.M. Haggard, C. Rovelli, White holes as remnants: a surprising scenario for the end of a black hole. Class. Quantum Gravity **35** (2018) 22, 225003. https://iopscience.iop.org/article/10.1088/1361-6382/aae550. arXiv:1802.04264
51. C. Rovelli, F. Vidotto, Small black/white hole stability and dark matter. Universe **4**(11), 127 (2018). https://www.mdpi.com/2218-1997/4/11/127. arXiv:1805.03872
52. P. Martin-Dussaud, C. Rovelli, Evaporating black-to-white hole. Class. Quantum Gravity **36**(24), 245002 (2019). https://iopscience.iop.org/article/10.1088/1361-6382/ab5097. arXiv:1905.07251
53. F. D'Ambrosio, M. Christodoulou, P. Martin-Dussaud, C. Rovelli, F. Soltani, The end of a black hole's evaporation—Part I. Phys. Rev. D **103**(10), 106014 (2021). https://journals.aps.org/prd/abstract/10.1103/PhysRevD.103.106014. arXiv:2009.05016
54. F. Soltani, C. Rovelli, P. Martin-Dussaud, The end of a black hole's evaporation—Part II. Phys. Rev. D **104**(6), 066015 (2021). https://journals.aps.org/prd/abstract/10.1103/PhysRevD.104.066015. arXiv:2105.06876
55. S. Kazemian, M. Pascual, C. Rovelli, F. Vidotto, Diffuse emission from black hole remnants. Class. Quantum Gravity **40**(8), 087001 (2023). https://iopscience.iop.org/article/10.1088/1361-6382/acc232. arXiv:2207.06978
56. C. Rovelli, F. Vidotto, Planck stars, White Holes, Remnants and Planck-mass quasi-particles. The quantum gravity phase in black holes' evolution and its manifestations. arXiv:2407.09584
57. P. Martin-Dussaud, How the black-to-white hole scenario resolves the information loss paradox. arXiv:2411.03125
58. S. Carlip, Dimension and dimensional reduction in quantum gravity. Class. Quantum Gravity **34**(19), 193001 (2017). https://iopscience.iop.org/article/10.1088/1361-6382/aa8535. arXiv:1705.05417
59. F. Saueressig, A. Khosravi, Black hole remnants from dynamical dimensional reduction?. Phys. Rev. D **105**(6), 066015 (2022). https://journals.aps.org/prd/abstract/10.1103/PhysRevD.105.066015. arXiv:2111.05018
60. J. Mureika, Primordial black hole evaporation and spontaneous dimensional reduction. Phys. Lett. B **716**, 171 (2012). https://www.sciencedirect.com/science/article/pii/S0370269312008660?via%3Dihub. arXiv:1204.3619
61. D. Ida, No black hole theorem in three-dimensional gravity. Phys. Rev. Lett. **85**, 3758 (2000). https://journals.aps.org/prl/abstract/10.1103/PhysRevLett.85.3758. arXiv:gr-qc/0005129
62. G. Amelino-Camelia, F. Brighenti, G. Gubitosi, G. Santos, Thermal dimension of quantum spacetime. Phys. Lett. B **767**(48) (2017). https://www.sciencedirect.com/science/article/pii/S0370269317300680?via%3Dihub. arXiv:1602.08020
63. I.P. Lobo, G.B. Santos, Thermal dimensional reduction and black hole evaporation. Phys. Lett. B **817**, 136272 (2021). https://linkinghub.elsevier.com/retrieve/pii/S0370269321002124. arXiv:2009.08556
64. J. Magueijo, L. Smolin, Gravity's rainbow. Class. Quantum Gravity **21**, 1725 (2004). arXiv:gr-qc/0305055
65. A.F. Ali, M. Faizal, M.M. Khalil, Remnant for all black objects due to gravity's rainbow. Nucl. Phys. B **894**, 341 (2015). https://www.sciencedirect.com/science/article/pii/S0550321315000966?via%3Dihub. arXiv:1410.5706
66. J.D. Bekenstein, Do we understand black hole entropy?. arXiv:gr-qc/9409015
67. G. Dvali, C. Gomez, D. Lust, Black hole quantum mechanics in the presence of species. Fortsch. Phys. **61**, 768 (2013). https://onlinelibrary.wiley.com/doi/10.1002/prop.201300002. arXiv:1206.2365
68. S. Giddings, Comments on information loss and remnants. Phys. Rev. D **49**, 4078 (1994). https://journals.aps.org/prd/abstract/10.1103/PhysRevD.49.4078. arXiv:hep-th/9310101
69. S. Giddings, Why aren't black holes infinitely produced?. Phys. Rev. D **51**, 6860 (1995). https://journals.aps.org/prd/abstract/10.1103/PhysRevD.51.6860. arXiv:hep-th/9412159
70. V.A. Ruban, Spherically symmetric T-models in the general theory of relativity. Gen. Rel. Gravity 33, 375 (2001). https://link.springer.com/article/10.1023/A:1002779702615 (Translated from original Russian, 1969)

71. R. Doran, F.S.N. Lobo, P. Crawford, Interior of a Schwarzschild black hole revisited. Found. Phys. **38**, 160 (2008). https://link.springer.com/article/10.1007/s10701-007-9197-6. arXiv:gr-qc/0609042
72. R.W. Brehme , Inside the black hole. Am. J. Phys. **45**, 423 (1977). https://pubs.aip.org/aapt/ajp/article-abstract/45/5/423/1050671/Inside-the-black-hole?redirectedFrom=fulltext
73. T. Jacobson, Introduction to quantum fields in curved spacetime and the hawking effect, in *Lectures on Quantum Gravity, Series of the Centro De Estudios Científicos*, eds. by A. Gomberoff, D. Marolf (Springer, Boston, MA). https://doi.org/10.1007/0-387-24992-3_2. arXiv:gr-qc/0308048
74. M. Christodoulou, C. Rovelli, How big is a black hole?. Phys. Rev. D **91**(6), 064046 (2015). https://journals.aps.org/prd/abstract/10.1103/PhysRevD.91.064046. arXiv:1411.2854
75. Y.C. Ong, The persistence of the large volumes in black holes. Gen. Relativ. Gravity **47**, 88 (2015). https://link.springer.com/article/10.1007/s10714-015-1929-x. arXiv:1503.08245
76. M. Christodoulou, T. De Lorenzo, On the volume inside old black holes. Phys. Rev. D **94**(10), 104002 (2016). https://journals.aps.org/prd/abstract/10.1103/PhysRevD.94.104002. arXiv:1604.07222
77. C. Rovelli, Black holes have more states than those giving the Bekenstein-Hawking entropy: a simple argument. arXiv:1710.00218
78. A. Aguirre, S.M. Carroll, M.C. Johnson, Out of equilibrium: understanding cosmological evolution to lower-entropy states. JCAP **02**, 024 (2012). https://iopscience.iop.org/article/10.1088/1475-7516/2012/02/024. arXiv:1108.0417
79. T. Jacobson, D. Marolf, C. Rovelli, Black hole entropy: inside or out?. Int. J. Theor. Phys. **44**, 1807 (2005). https://link.springer.com/article/10.1007/s10773-005-8896-z,. arXiv:hep-th/0501103
80. S. Hossenfelder, L. Smolin, Conservative solutions to the black hole information problem. Phys. Rev. D **81**, 064009 (2010). https://journals.aps.org/prd/abstract/10.1103/PhysRevD.81.064009. arXiv:0901.3156
81. D. Marolf, D. Minic, S. Ross, Notes on spacetime thermodynamics and the observer-dependence of entropy. Phys. Rev. D **69**, 064006 (2004). https://journals.aps.org/prd/abstract/10.1103/PhysRevD.69.064006. arXiv:hep-th/0310022
82. P. Hayden, J. Wang, What exactly does Bekenstein bound?. arXiv:2309.07436
83. P. Hayden, G. Penington, Approximate quantum error correction revisited: introducing the alpha-bit. Commun. Math. Phys. **374**(2), 369 (2020). https://link.springer.com/article/10.1007/s00220-020-03689-1. arXiv:1706.09434
84. P. Hayden, G. Penington, Learning the alpha-bits of black holes. JHEP **12**, 007 (2019). https://link.springer.com/article/10.1007/JHEP12(2019)007. arXiv:1807.06041
85. L. Susskind, Computational complexity and black hole horizons. Fortsch. Phys. **64**, 24 (2016). https://onlinelibrary.wiley.com/doi/10.1002/prop.201500092. arXiv:1402.5674
86. L. Susskind, Addendum to computational complexity and black hole horizons. Fortsch. Phys. **64**, 44 (2016). https://onlinelibrary.wiley.com/doi/10.1002/prop.201500093. arXiv:1403.5695
87. D. Stanford, L. Susskind, Complexity and shock wave geometries. Phys. Rev. D **90**(12), 126007 (2014). https://journals.aps.org/prd/abstract/10.1103/PhysRevD.90.126007. arXiv:1406.2678
88. L. Susskind, Entanglement is not enough. Fortsch. Phys. **64**, 49 (2016). https://onlinelibrary.wiley.com/doi/10.1002/prop.201500095. arXiv:1411.0690
89. M. Alishahiha, Holographic complexity. Phys. Rev. D **92**(12), 126009 (2015). https://journals.aps.org/prd/abstract/10.1103/PhysRevD.92.126009. arXiv:1509.06614
90. R.-Q. Yang, Upper bound about cross-sections inside black holes and complexity growth rate. Phys. Rev. D **102**(10), 106001 (2020). https://journals.aps.org/prd/abstract/10.1103/PhysRevD.102.106001. arXiv:1911.12561
91. B. McInnes, Evidence that the rate of evolution of a black hole interior has a holographic dual. Nucl. Phys. B **1004**, 116567 (2024). https://www.sciencedirect.com/science/article/pii/S0550321324001330. arXiv:2312.05731

92. L. Susskind, The typical-state paradox: diagnosing horizons with complexity. Fortsch. Phys. **64**, 84 (2016). https://onlinelibrary.wiley.com/doi/10.1002/prop.201500091. arXiv:1507.02287
93. D. Harlow, P. Hayden, quantum computation versus firewalls. JHEP **06**, 085 (2013). https://link.springer.com/article/10.1007/JHEP06(2013)085. arXiv:1301.4504
94. H. Singh, M.K. Nandy, Black hole to grey hole metamorphosis in the deep quantum regime. Eur. Phys. J. C **84**(11), 1233 (2024). https://link.springer.com/article/10.1140/epjc/s10052-024-13636-2
95. R. Gaur , M. Visser, Black holes, white holes, and near-horizon physics. JHEP **05**, 172 (2024). https://doi.org/10.1007/JHEP05(2024)172. arXiv:2304.10692
96. M. Bouhmadi-López, C.-Y. Chen, X.Y. Chew, Y.C. Ong, D.-H. Yeom, Regular black hole interior spacetime supported by three-form field. Eur. Phys. J. C **81**(4), 278 (2021). https://link.springer.com/article/10.1140/epjc/s10052-021-09080-1. arXiv:2005.13260
97. R. Parentani, T. Piran, The internal geometry of an evaporating black hole. Phys. Rev. Lett. **73**, 2805 (1994). https://journals.aps.org/prl/abstract/10.1103/PhysRevLett.73.2805. arXiv:hep-th/9405007
98. D.A. Easson, R.H. Brandenberger, Universe generation from black hole interiors. JHEP **06**, 024 (2001). https://iopscience.iop.org/article/10.1088/1126-6708/2001/06/024. arXiv:hep-th/0103019
99. D. Stanford, Z. Yang, Firewalls from wormholes. arXiv:2208.01625
100. J.G. Russo, On black hole singularities in quantum gravity. Phys. Lett. B **10**, 35 (1994). https://www.sciencedirect.com/science/article/abs/pii/0370269394911290
101. W.A. Hiscock, Models of evaporating black holes. I. Phys. Rev. D **23**, 2813 (1981). https://journals.aps.org/prd/abstract/10.1103/PhysRevD.23.2813
102. A. Bonanno, S. Silveravalle, Black holes at a crossroads: late-stage evaporation in quadratic gravity. arXiv:2409.16690
103. S. Hossenfelder, Minimal length scale scenarios for quantum gravity. Living Rev. Relativity **16**, 2 (2013). https://link.springer.com/article/10.12942/lrr-2013-2. arXiv:1203.6191
104. L. Xiang, Y. Ling, Y.G. Shen, Singularities and the finale of black hole evaporation. Int. J. Mod. Phys. D **22**, 1342016 (2013). https://www.worldscientific.com/doi/abs/10.1142/S0218271813420169. arXiv:1305.3851
105. Y. Ling, M.-H. Wu, Regular black holes with sub-planckian curvature. Class. Quantum Gravity **40**(7), 075009 (2023). https://iopscience.iop.org/article/10.1088/1361-6382/acc0c9. arXiv:2109.05974
106. Y.C. Ong, On black hole thermodynamics, singularity, and gravitational entropy. Gen. Relativ. Gravity **54**, 1329 (2022). https://link.springer.com/article/10.1007/s10714-022-03008-0. arXiv:2210.16856
107. Y.C. Ong, Spacetime singularities and cosmic censorship conjecture: a review with some thoughts. Int. J. Mod. Phys. A **35**, 2030007 (2020). https://www.worldscientific.com/doi/abs/10.1142/S0217751X20300070. arXiv:2005.07032
108. R. Bousso, A. Shahbazi-Moghaddam, Singularities from entropy. Phys. Rev. Lett. **128**(23), 231301 (2022). arXiv:2201.11132
109. R. Penrose, The question of cosmic censorship. J. Astrophys. Astronomy **20**, 233 (1999). https://adsabs.harvard.edu/full/1999JApA...20..233P
110. A. Barvinsky, S. Das, G. Kunstatter, Discrete spectra of charged black holes. Found. Phys. **32**, 1851 (2002). https://link.springer.com/article/10.1023/A:1022314632647. arXiv:hep-th/0209039
111. D. Marolf, The dangers of extremes. Gen. Rel. Gravity **42**, 2337 (2010). https://link.springer.com/article/10.1007/s10714-010-1027-z. arXiv:1005.2999
112. G.T. Horowitz, M. Kolanowski, J.E. Santos, Almost all extremal black holes in ads are singular. JHEP **01**, 162 (2023). https://link.springer.com/article/10.1007/JHEP01(2023)162. arXiv:2210.02473
113. G.T. Horowitz, M. Kolanowski, G.N. Remmen, J.E. Santos, Extremal Kerr black holes as amplifiers of new physics. Phys. Rev. Lett. **131**(9), 091402 (2023). https://journals.aps.org/prl/abstract/10.1103/PhysRevLett.131.091402. arXiv:2303.07358

114. G.J. Turiaci, New insights on near-extremal black holes. arXiv:2307.10423
115. Y.C. Ong, A maximum force perspective on black hole thermodynamics, quantum pressure, and near-extremality. Eur. Phys. J. C **83**(11), 1068 (2023). https://link.springer.com/article/10.1140/epjc/s10052-023-12253-9. arXiv:2309.04110
116. Y.C. Ong, Black holes, complex curves, and graph theory: revising a conjecture by Kasner. Nucl. Phys. B **1008**, 116711 (2024). https://www.sciencedirect.com/science/article/pii/S0550321324002773?via%3Dihub. arXiv:2409.08236
117. J.H. MacGibbon, Can planck-mass relics of evaporating black holes close the universe?. Nature **329**, 308. (1987) https://www.nature.com/articles/329308a0
118. P. Chen, R.J. Adler, Black hole remnants and dark matter. Nucl. Phys. Proc. Suppl. **124**, 103 (2003). https://www.sciencedirect.com/science/article/abs/pii/S0920563203020887?via%3Dihub. arXiv:gr-qc/0205106
119. Y. C. Ong, The case for black hole remnants: a review. arXiv preprint arXiv:2412.00322 (2024)

Chapter 9
Page Curve of Average Subsystem Entropy

Oscar C. O. Dahlsten

Abstract The Page curve is a curve of average subsystem entropy as a function of subsystem size. The curve starts from 0, rises, peaks near the maximal possible entropy when the subsystem makes up half of the total system, and then falls back to 0. We here describe subsystem entropy, averaging over quantum states, and why the curve rises and falls in that manner. We also discuss the connection between the curve and the black hole information paradox.

9.1 Introduction

The Page curve concerns the expected value of subsystem entropy. More specifically, for a bipartite pure state $|\Psi\rangle_{AB}$, there is an entropy of subsystem A for example, $S(\rho_A)$ where ρ_A is the reduced state on A. The Page curve shows the average $S(\rho_A)$, averaged over pure bipartite states, as a function of $\log d_A$ where d_A is the dimension of system A. The curve is depicted in Fig. 9.1. The curve likely gained its name because of the 1993 papers Refs. [12, 13]. Similar results concerning the curve were obtained by Lubkin around 1978 in Ref. [9] and consistent results were derived in Ref. [8] by Lloyd and Pagels around 1988. Page's early work appears to have been motivated by wishing to show that entropy does not necessarily change monotonically as subsystems radiate out as suggested by so-called semi-classical Hawking radiation calculations. The curve and related concepts also appear in quantum thermodynamics and quantum information science.

This chapter aims to give an introduction to the curve. The reader is assumed to have graduate knowledge of quantum theory. The emphasis will be on the quantum information science side, reflecting the expertise of the author. The chapter is far from an exhaustive review of the literature. (For a more developed review on the expected entropy of subsystems see e.g. Ref. [3]).

O. C. O. Dahlsten (✉)
City University of Hong Kong, Kowloon Tong, Hong Kong SAR, China
e-mail: oscar.dahlsten@cityu.edu.hk

A. Akil and C. Bambi (eds.), *The Black Hole Information Paradox*, Springer Series in Astrophysics and Cosmology, https://doi.org/10.1007/978-981-96-6170-1_9

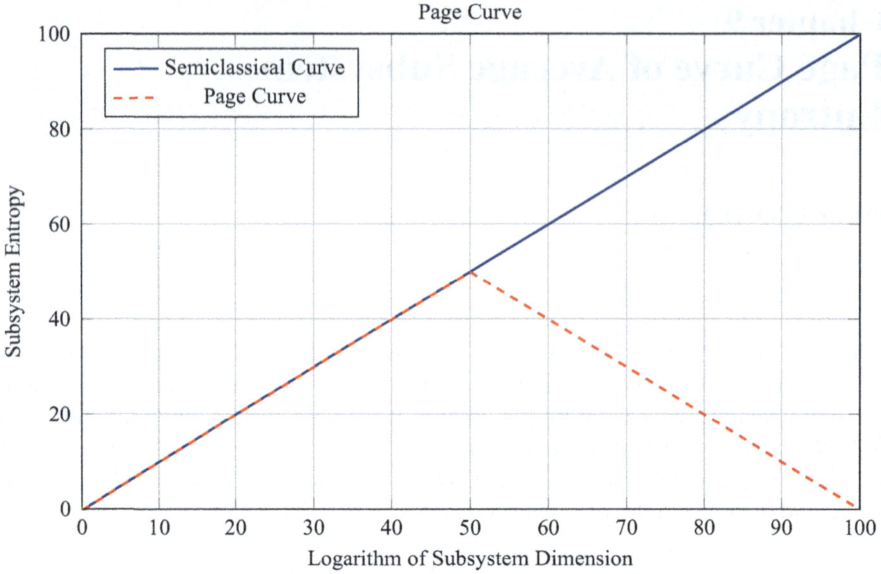

Fig. 9.1 **Page curve**. The entropy of a subsystem, averaged over the uniform distribution over pure quantum states, rises with the subsystem size, peaks when the subsystem is half of the total subsystem (50/100 subsystems in this case) and then falls off in a symmetric manner down to 0. The Page Curve contrasts with the behaviour of the entropy curve of black hole radiation derived semiclassically by Hawking and others. The latter curve increases monotonically as the radiation increases, suggesting a black hole evaporation leaves behind high entropy radiation rather than a pure state radiation field

We proceed as follows. We begin with entropy and subsystem entropy. We then describe averaging over states, before showing an example of how one can derive the curve. Finally we give a discussion on the relation to black holes, thermodynamics and quantum information science.

9.2 Subsystem Entropy

In this section, we first discuss the entropy of quantum states, including Renyi entropy, and then the entropy of subsystems.

9.2.1 Entropy of Quantum States

The entropy of a quantum state ρ is a single real number assigned to the state. An important family of possible entropy choices are the quantum Renyi entropies

$$S_q(\rho) = \frac{1}{1-q} \log \sum_{i=1}^{d} \lambda_i^q, \tag{9.1}$$

where λ_i denotes the i-th eigenvalue of the density matrix ρ, i.e. $\rho = \sum_{i=1}^{d} \lambda_i \mid i \rangle \langle i \mid$. The most commonly employed entropy

$$S_{q\to 1}(\rho) := S_1(\rho) := S(\rho) = \sum_{i=1}^{d} \lambda_i \log \frac{1}{\lambda_i}, \tag{9.2}$$

is often termed the *von Neumann entropy* after a thermodynamical argument leading von Neumann to this definition (see Ref. [14] for the argument). In information theory the logarithm log(.) is normally taken to base 2 and in thermodynamics to base e = 2.72.... Furthermore in thermodynamics the entropy is by definition also multiplied by Boltzmann's constant k_B, becoming $k_B \sum_{i=1}^{d} \lambda_i \ln \frac{1}{\lambda_i}$. For our purposes here, it will also be useful to note that, from Eq. 9.2,

$$S_2 = -\log \sum_{i=1}^{d} \lambda_i^2, \tag{9.3}$$

where $\sum_i \lambda_i^2 \in [\frac{1}{d}, 1]$ is commonly termed the *purity* of the state ρ.

The entropy can be understood in multiple consistent manners:

- $S(\rho) \in [0, \log d]$ quantifies our lack of information about the preparation of the quantum system. $S(\rho) = 0$ if and only if the state has form $\rho = \mid i \rangle \langle i \mid$ for some state $\mid i \rangle$ which then has eigenvalue 1. $S(\rho) = \log d$ for $\rho = \sum_{i=1}^{d} \frac{1}{d} \mid i \rangle \langle i \mid$ which corresponds to no knowledge of the preparation, the maximally mixed state. $S(\rho)$ can be interpreted as quantifying where on the scale between these two extremes the state ρ lies.
- $S(\rho)$ is the classical entropy $H(\vec{\lambda}) = \sum_i p_i \log \frac{1}{p_i}$ of a measurement in the eigenbasis of ρ whereby $\lambda_i = \mathrm{tr}(\rho \mid i \rangle \langle i \mid) =: p_i$ for $\rho = \sum_i \lambda_i \mid i \rangle \langle i \mid$.
- $S(\rho)$ is the classical entropy H of measurement in a given basis *minimised over all bases*:

$$S(\rho) = \min_{\{\mid j \rangle \langle j \mid\}_{j=1}^{d}} H(\vec{p}), \tag{9.4}$$

where $\vec{p}$ is the vector of probabilities with entries $p_j = \mathrm{tr}(\rho \mid j \rangle \langle j \mid)$.

For a tutorial with more on entropies see e.g. Ref. [2].

9.2.2 Entropy of a Subsystem

The entropy of a subsystem A of a joint system AB is given by $S(\rho_A) = S(\mathrm{tr}_B \rho_{AB})$ where tr_B is the partial trace over B, the quantum method of taking a marginal distribution on A. $S(\rho_A) \in [0, \log d_A]$ where d_A is the state dimension of A. $\log d_A$ can be interpreted as the number of qubits, n_A, making up A. n_A qubits of dimension 2 have dimension $d_A = 2^{n_A}$ such that $n_a = \log d_A$.

The subsystem entropy of a pure state can be understood via the singular value decomposition of a matrix representing the state. The singular value decomposition of a complex matrix M is that $M = UDV$ where U and V are unitary and D is diagonal with non-negative real entries. At the component level $M_{ij} = \sum_k U_{ik} D_{kk} V_{kj} := U_{ik}\mu_k V_{kj}$. Thus, a pure state on AB can be rewritten as

$$|\Psi\rangle_{AB} = \sum_{i,j,k} M_{ij} \,|\, i\rangle_A \,|\, j\rangle_B = \sum_{i,j} U_{ik}\mu_k V_{kj} \,|\, i\rangle \,|\, j\rangle = \sum_{i,j,k} \mu_k (U_{ik} \,|\, i\rangle)(V_{kj} \,|\, j\rangle) := \sum_k \mu_k \left|\tilde{k}\right\rangle_A \left|\tilde{k}\right\rangle_B, \tag{9.5}$$

which is termed the *Schmidt decomposition* of $|\,\Psi\rangle_{AB}$ with Schmidt coefficients μ_k. It follows that

$$\rho_A = \sum_k \mu_k^2 \left|\tilde{k}\right\rangle_A \left\langle\tilde{k}\right|_A \tag{9.6}$$

and

$$\rho_B = \sum_k \mu_k^2 \left|\tilde{k}\right\rangle_B \left\langle\tilde{k}\right|_B . \tag{9.7}$$

Thus A and B share the same eigenvalues, μ_k^2 and accordingly $S(\rho_A) = S(\rho_B)$, for this case of a pure bipartite joint state.

$S(\rho_A)$ is, in the case of a pure bipartite state, the standard measure of the entanglement between A and B. A quick way to understand why is as follows. By definition, entangled states are those which are not product states (or probabilistic mixtures thereof). Pure product states have Schmidt form $|\,1\rangle \,|\,1\rangle$. On the other extreme, the maximally entangled state is by definition $\sum_{k=1}^{d} \frac{1}{\sqrt{d_A}} \,|\,k\rangle \,|\,k\rangle$. $S(\rho_A) \in [0, \log d_A]$, can be interpreted as quantifying where on the scale between these two extremes the state on AB is. $S(\rho_A) = 0$ for product states and $S(\rho_A) = \log d_A$ for the maximally entangled state. (For mixed states, entanglement is quantified by other measures that distinguish between quantum and classical correlations—see e.g. Ref. [16]).

9.3 Averaging Over Quantum States

In this section, we first argue that the uniform distribution for fair averaging over states should be unitarily invariant. We then say more about what that unitarily invariant distribution is.

9.3.1 Uniform Distribution Should Be Unitarily Invariant

For fair averaging over quantum states we require a uniform distribution over quantum states. A uniform distribution can be defined via invariance under some permutation or more general group operation. For example for a coin with heads H or tails T, [p(H), p(T)] is uniform if and only if it is invariant under a permutation of p(H) and p(T). Now instead of H and T let the microstates be pure quantum states. There is a continuum of these so we are looking for a uniform probability density over those states.

To visualise distributions over pure quantum states consider the 3-dimensional real space Bloch sphere representation of a qubit, where ρ is represented via the expectation value vector $(\langle X\rangle, \langle Y\rangle, \langle Z\rangle)^T$ (T is the tranpose). Recall that the Bloch vector uniquely determines the normalised density matrix since the Paulis plus the identity, $\{\mathbb{1}, X, Y, Z\}$, form a basis for Hermitian matrices such that

$$\rho = \xi_0 \mathbb{1} + \xi_1 X + \xi_2 Y + \xi_3 Z; \tag{9.8}$$

$$\mathrm{tr}(\rho \mathbb{1}) = 2\xi_0 = 1;\ \langle X\rangle = \mathrm{tr}(\rho X) = 2\xi_1 \text{ etc..} \tag{9.9}$$

Valid qubit states respect

$$\langle X\rangle^2 + \langle Y\rangle^2 + \langle Z\rangle^2 \leq 1, \tag{9.10}$$

with equality for pure states which thus lie on the surface of a sphere. It turns out that each point on the sphere surface (for a single qubit) is a valid state. $\langle X\rangle^2 \in [0, 1]$ can be interpreted as the predictability of the observable X which has eigenvalues ± 1, being 0 for $p(1) = p(-1)$. Thus Eq. 9.10 is sometimes interpreted as an uncertainty relation, bounding the joint predictability of the three Paulis [1].

The analogue of a microstate permutation is now a a rotation of the sphere. That rotation is equivalent to a unitary evolution U being applied on the standard complex vector representation of the state. The analogue of a uniform distribution is then a distribution which is invariant under any unitary: *the unitarily invariant distribution*. This distribution can be visualised as a uniform density over the Bloch sphere surface. If the distribution were not uniform, rotations of the sphere would alter it, as depicted in Fig. 9.2.

The above discussion motivates using unitary invariance to define a uniform distribution (called measure) over quantum states. More concretely, for whatever parameterisation $(\theta_1, \theta_2, ...)$ is used for the state, the joint density over those parameters $\rho(\theta_1, \theta_2, ...) := \rho(|\phi\rangle)$ should respect

$$\rho(|\phi\rangle) = \rho(U|\phi\rangle)\ \forall U \mid U^\dagger U = \mathbb{1}. \tag{9.11}$$

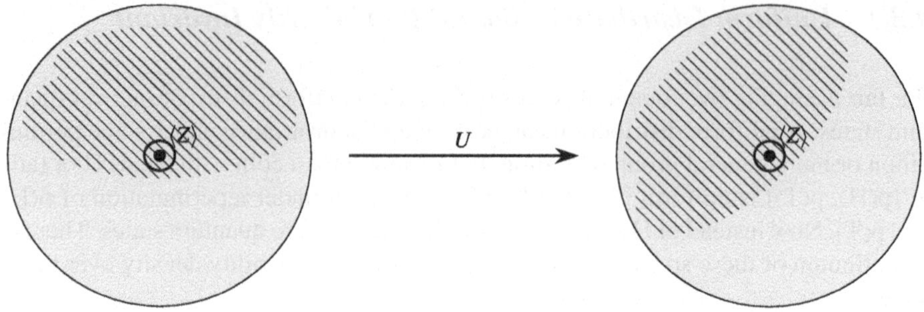

Fig. 9.2 Non-uniform density on Bloch sphere. Toy picture of how non-uniform density of pure states changes under unitary evolution. The circle represents the Bloch sphere real vector representation of a qubit as seen from above the z-axis. The shading indicates density. The unitary acts as a rotation, altering the overall density. In the case depicted, the density changes under rotation and is thus not unitarily invariant. In contrast, a uniform density would be rotationally invariant

9.3.2 Unitarily Invariant Distribution Over Pure States

To gain intuition about how to pick states randomly uniformly, consider firstly picking real vectors in a manner such that the density is invariant under orthogonal matrices. In particular, consider again the case of a single qubit and the Bloch sphere, wherein states are real-vectors and unitaries act as orthogonal matrices. Let for notational simplicity $x := \langle X \rangle$ etc.. We want a method for picking the real vectors $(x, y, z)^T$ with probability density respecting $\rho(x, y, z) = \rho(x', y', z')$ if $(x', y', z')^T = \mathcal{O}(x, y, z)^T$ where $\mathcal{O}$ is an orthogonal matrix. Thus $\rho(x, y, z)$ should at most depend on the radius $r = (x^2 + y^2 + z^2)^{1/2}$ since otherwise we could alter the density via some $\mathcal{O}$ that keeps r constant but alters the angles. A simple way to have $\rho(x, y, z)$ only depending on the radius is to pick x, y, z independently from the normal distribution such that $\rho(x, y, z) = \mathcal{N} \exp(-x^2) \exp(-y^2) \exp(-z^2) = \mathcal{N} \exp(-r^2)$, where $\mathcal{N}$ is the normalisation of the joint probability distribution. This is then an orthogonally invariant density (often called measure). A subtlety is that for pure quantum states we also require r = 1. We then add a second step to alter r, by a deterministic many-to-one map $x \rightarrow r^{-1}x$ etc.. The density of the altered values is by inspection still orthogonally invariant since it is independent of the angles.

Now consider the most common parametrisation of the pure quantum state as a complex vector $|\phi\rangle$ with entries $\phi_j = a_j + ib_j$. It turns out that picking the state from the unitarily invariant distribution means picking all the real components a_j and b_j for all j independently from the normal distribution and finally normalising the vector [4], similar to the above simple case of real numbers.

Consider further the distribution over individual observable expectation values $\langle g_i \rangle$. If the state is picked at random, $\langle g_i \rangle$, or any function thereof such as $\langle g_i \rangle^2$ is picked at random. Recall that $\langle g_i \rangle^2 \in [0, 1]$ can be interpreted as the predictability of g_i for g_i having eigenvalues ± 1. There is an associated 'expectation value of the expectation value', such as $\mathbb{E}(\langle g_i \rangle^2)$. These values, being functions of the density over states should be invariant for observables connected by unitaries e.g. $\mathbb{E}(\langle X\rangle^2) = \mathbb{E}(\langle Z\rangle^2)$. If that equality were not the case, we could alter the density over state parameters $\langle X\rangle$ and $\langle Z\rangle$ by applying the unitary that connects the two observables ($Z = HXH$ where H is the Hadamard unitary). This equality of expected predictability across unitarily related observables is helpful in calculations and in understanding why expected subsystem entropy is high.

9.4 Average Subsystem Entropy

We are now ready to discuss the average subsystem entropy. We give an example of one way to derive the expected purity of a subsystem. We argue that the total predictability is limited and that the expected share thereof which the local observables have small, such that the local purity is small.

We shall focus on the predictability of Pauli matrix observables (generalised to n qubits). This is an approach that the author and collaborators have found convenient for understanding and for undertaking calculations. We used a similar approach to show that states generated by random collections of 2-qubit unitaries of restricted (poly(n)) length generate approximately the same average subsystem entropy [11] as those picked from the unitarily invariant distribution. Furthermore, we used a similar approach to show that there is a classical analogue of states typically being maximally entangled (more details below).

9.4.1 Total Predictability Limited

The predictability of observables is limited [1]. Consider the implication for states written in a Pauli basis for Hermitian matrices $\rho_{AB} = \sum_i \xi_i g_i$ where $g_i \in \{X, Y, Z, \mathbb{1}\}^{\otimes n}$. We can, similarly to the above single qubit case, interpret $\langle g_i\rangle^2 \in [0, 1]$ as the predictability of g_i. The purity determines the total predictability in that

$$\mathrm{tr}(\rho^2) = d\sum_{i=0}^{d^2-1} \xi_i^2 = \frac{1}{d}\sum_{i=0}^{d^2-1}\langle g_i\rangle^2 = \frac{1}{d}\left(1+\sum_{i=1}^{d^2-1}\langle g_i\rangle^2\right), \tag{9.12}$$

where $d = 2^n$ and we used $\mathrm{Tr}(g_j g_k) = d\delta_{jk}\quad \forall j, k$. Furthermore, the first element $g_0 = I_{2^n}$ has a fixed coefficient $\xi_0 = 1/d$ since $\mathrm{tr}\rho = 1$.

Equation 9.12 by inspection implies, for pure states where $\mathrm{tr}(\rho^2) = 1$,

$$\sum_{i=1}^{d^2-1}\langle g_i\rangle^2 = d - 1, \tag{9.13}$$

which is a strong limitation since the algebraic maximum of the left-hand-side is $d^2 - 1$. Thus the total predictability is strongly limited.

9.4.2 Expected Local Predictability a Small Share of Total

There is a system AB comprising a number $n = n_A + n_B$ of individual qubit systems. We can break Eq. 9.13 into one contribution from local observables on A and another from the rest. Local observables on A (B) correspond to measuring A (B) alone and take form $g_A \otimes \mathbb{1}_B$ ($\mathbb{1}_A \otimes g_B$). Thus,

$$\sum_{g_i | g_i = g_i^{(A)} \otimes \mathbb{1}_B} \langle g_i \rangle^2 + \sum_{g_i | g_i \neq g_i^{(A)} \otimes \mathbb{1}_B} \langle g_i \rangle^2 = d - 1, \tag{9.14}$$

The first term associated with local observables on A can be termed the local predictability on A and equals $\sum_i \langle g_i^{(A)} \rangle^2$ since $\langle g_i^{(A)} \rangle = \langle g_i^{(A)} \otimes \mathbb{1}_B \rangle$.

There are many more global observables than local. Global Observables are not of that local form, e.g. $X_A \otimes Z_B$ can be called global. When we use the Pauli bases for systems composed of many qubits $\{X, Y, Z, \mathbb{1}\}^{\otimes n}$, we have four choices for each qubit so 4^n basis elements. There are similarly 4^{n_A} basis elements local on A so the ratio of local on A to global is $4^{n_A - n}$. This ratio decreases exponentially in n for all n_A, since for the largest A of interest, $n_A = n/2$ the ratio becomes $4^{-n/2}$.

Consider the expected share of predictability split in a similar manner. As argued above, for unitarily invariant statistics

$$\mathbb{E}(\langle g_i \rangle^2) = \mathbb{E}(\langle g_j \rangle^2) := \mathbb{E}(\langle g \rangle^2) \quad \text{for } i, j \neq 0. \tag{9.15}$$

Equations 9.13 and 9.15 together give

$$(d^2 - 1)\mathbb{E}(\langle g \rangle^2) = d - 1, \tag{9.16}$$

such that

$$\mathbb{E}(\langle g \rangle^2) = \frac{d - 1}{d^2 - 1}. \tag{9.17}$$

Thus the expected predictability on A,

$$\mathbb{E}\left(\sum_{g_i | g_i = g_i^{(A)} \otimes \mathbb{1}_B} \langle g_i \rangle^2 \right) = (d_A^2 - 1)\mathbb{E}(\langle g \rangle^2) = \frac{(d_A^2 - 1)(d - 1)}{d^2 - 1}, \tag{9.18}$$

which is small compared to the total predictability of $(d^2 - 1)\mathbb{E}(\langle g \rangle^2) = d - 1$.

9.4.3 Expected Local Purity Low, Entropy High

The expected purity on A is closely related to the expected predictability of Eq. 9.18. From Eq. 9.12

$$\mathbb{E}\left(\mathrm{tr}(\rho_A^2)\right) = \frac{1}{d_A}\left(1 + \mathbb{E}\left(\sum_{i\neq 0}\langle g_i\rangle^2\right)\right) = \frac{1}{d}\left(1 + \frac{(d_A^2-1)(d-1)}{d^2-1}\right) \quad (9.19)$$

Equation (9.19) indeed simplifies to Lubkin's expression (derived differently) [9]:

$$\mathbb{E}\left(\mathrm{tr}(\rho_A^2)\right) = \frac{d_A + d_B}{d_A d_B + 1}, \quad (9.20)$$

where total system dimension $d = d_A d_B$. We see that $\mathbb{E}\left(\mathrm{tr}(\rho_A^2)\right) = 1$ (maximal purity) when $d_A = d$ or $d_B = d$ and moreover, by differentiation one sees $\mathbb{E}\left(\mathrm{tr}(\rho_A^2)\right)$ has a minimal value, for fixed d at $d_A = \sqrt{d}$. Thus the entropy, which essentially has an inverse behaviour to the purity has the qualitative behaviour of rising and falling in Fig. 9.1. A quick way to relate purity $\mathrm{tr}(\rho^2)$ to entropy is to note that $S_2 = -\log \mathrm{tr}(\rho^2)$. (A subtlety is that there is a spread around the average but that spread can be shown to be heavily limited [9]).

Similar arguments can be made for the expected subsystem purity/entropy of classical systems [10]. For the case of classical probability distributions on bits we would drop X and Y so ρ is diagonal and there are 2^n global observables and 2^{n_A} on A. The total predictability, which may be limited, is again shared out between observables and similar arguments apply [10]. One may say that, whether in quantum or classical systems, when there is limited predictability, it tends to be found in global observables, i.e. in correlations, rather than local observables.

9.5 Discussion

9.5.1 Expected Subsystem Entropy and Black Holes

Page's seminal work [13] applied expected subsystem entropy ideas to black hole thermodynamics. Page modelled formation and evaporation as a unitary process. The entropy of the radiation then typically increases until what is now termed the Page time (when the dimensions of the inside and outside are equal), then decreases to zero. This model has information preservation, due to the reversibility of unitary evolution, while appearing random/thermal locally. This resolves the apparent paradox where calculations showing entropy increase in reduced states of subsystems might suggest information loss, by noting that information can be preserved in correlations even when subsystems appear maximally mixed.

If the essential picture of black holes radiating away is correct and the radiation entropy follows something like the Page curve, the generalised 2nd law hypothesized by Hawking [7] that $S + \frac{A}{4}$ never decreases where S is the entropy of matter outside the black holes and A is the sum of the surface areas of the event horizons would, at least taken on face value, be violated, as A goes to 0 during evaporation and S goes to 0 at the end of the Page curve. It should finally be noted that Page [13] acknowledges that there is a key question mark around how the subsystems could exit the black hole to facilitate this curve, a question that remains an active area of research.

9.5.2 Expected Subsystem Entropy as an Approach to Thermodynamics

As noted by Feynman "*This fundamental law [that systems are in a Gibbs thermal state] is the summit of statistical mechanics, and the entire subject is either the slide-down from this summit, as the principle is applied to various cases, or the climb-up to where the fundamental law is derived and the concepts of thermal equilibrium and temperature T clarified*" [5].

The mathematical fact that most pure bipartite quantum states are nearly maximally entangled provides an alternative approach to justify the thermal state assumption [6, 8, 9, 15]. For the simplest case where $H = 0$, the Gibbs state reduces to $\rho_{th}(\beta, H) = \frac{\mathbb{1}_{d_A}}{d_A}$. This matches the reduced state $\text{tr}_B \rho_{AB}$ when ρ_{AB} is pure and maximally entangled. The typically maximal entanglement then implies that the system should typically be in a thermal state.

For non-trivial Hamiltonians, the argument is more subtle. One may pick a state from a uniform distribution over a restricted state space wherein the total energy on AB is restricted to a particular value. Then local observables are still not expected to have predictability such that the local reduced state is the same as though we had taken a maximally mixed state in the restricted state space, as in the more common derivation of the Gibbs state (see e.g. Ref. [3] for references). Such arguments give an alternative justification for the Gibbs state.

References

1. Č. Brukner, A. Zeilinger, Information invariance and quantum probabilities. Found. Phys. **39**, 677–689 (2009)
2. O. Dahlsten. Tutorial 1: entropy and majorization in generalized probabilistic theories. https://youtu.be/fXGUiXur1kU?feature=shared. Accessed Feb 2025 (YouTube Video)
3. O.C.O. Dahlsten, C. Lupo, S. Mancini, A. Serafini, Entanglement typicality. J. Phys. A: Math. Theor. **47**(36), 363001 (2014)
4. P. Diaconis, What is a random matrix. Notices AMS **52**(11), 1348–1349 (2005)
5. R.P. Feynman. *Statistical Mechanics: a Set of Lectures* (CRC press, 2018)

6. J. Gemmer, A. Otte, G. Mahler, Quantum approach to a derivation of the second law of thermodynamics. Phys. Rev. Lett. **86**(10), 1927 (2001)
7. S.W. Hawking, Particle creation by black holes. Commun. Math. Phys. **43**(3), 199–220 (1975)
8. S. Lloyd, H. Pagels, Complexity as thermodynamic depth. Ann. Phys. **188**(1), 186–213 (1988)
9. E. Lubkin, Entropy of an n-system from its correlation with a k-reservoir. J. Math. Phys. **19**(5), 1028–1031 (1978)
10. M.P. Müller, O.C.O. Dahlsten, V. Vedral, Unifying typical entanglement and coin tossing: on randomization in probabilistic theories. Commun. Math. Phys. **316**, 441–487 (2012)
11. R. Oliveira, O.C.O. Dahlsten, M.B. Plenio, Generic entanglement can be generated efficiently. Phys. Rev. Lett. **98**(13), 130502 (2007)
12. D.N. Page, Average entropy of a subsystem. Phys. Rev. Lett. **71**(9), 1291 (1993)
13. D.N. Page, Information in black hole radiation. Phys. Rev. Lett. **71**(23), 3743 (1993)
14. A. Peres, *Quantum Theory: concepts and Methods*, vol. 72 (Springer, 1997)
15. S. Popescu, A.J. Short, A. Winter, Entanglement and the foundations of statistical mechanics. Nat. Phys. **2**(11), 754–758 (2006)
16. V. Vedral, M.B. Plenio, M.A. Rippin, P.L. Knight, Quantifying entanglement. Phys. Rev. Lett. **78**(12), 2275 (1997)

Chapter 10
Unitary Evaporation and AdS/CFT

Masamichi Miyaji

Abstract AdS/CFT is the duality between quantum gravity on anti-de Sitter (AdS) spacetime and conformal field theory (CFT) on the AdS boundary. Since the CFT is unitary, the dual quantum gravity on AdS should also be unitary, and thus should not have the information paradox at all. However, what is truly important to understand is rather *how and why* the information paradox can be resolved in quantum gravity, and AdS/CFT plays the key role in such direction as it offers a controlled calculable setup. Such studies would lead us to a deeper understanding of the quantum structure of black holes. The recent remarkable progress in this direction is the discovery of the necessity of including *Euclidean wormholes* into the gravitational path integral, which renders the entropy of the Hawking radiation unitary. This result stems from the two key ingredients in AdS/CFT, the Ryu-Takayanagi formula, and the entanglement wedge reconstruction. This chapter introduces the fundamentals of AdS/CFT, the Ryu-Takayanagi formula, and the entanglement wedge reconstruction. We explain why we need an external reservoir to pose the information paradox in AdS/CFT, and why the inclusion of the Euclidean wormholes to the gravitational path-integral renders the radiation entropy unitary. We also explain why the Euclidean wormholes are tied to the ensemble average interpretation, which introduces new problems, such as the factorization problem and fluctuations in the averaged quantities.

10.1 Introduction

AdS/CFT correspondence [3, 70, 98, 141, 142], which is a special version of *holography* [132, 133, 136], is one of the most powerful tools to study non-perturbative quantum gravity. AdS/CFT states that a quantum gravity theory or a superstring theory on an anti-de Sitter (AdS) space $AdS_{d+1} \times M$ has a dual description in terms of d dimensional conformal field theory (CFT) at the boundary of the AdS space. Assuming this conjecture is true, the real-time evolution of gravity in AdS can be described by the real-time unitary evolution in the boundary CFT. Thus we can con-

M. Miyaji (✉)
Yukawa Institute for Theoretical Physics, Kyoto University, Kyoto, Japan
e-mail: masamichi.miyaji@yukawa.kyoto-u.ac.jp

A. Akil and C. Bambi (eds.), *The Black Hole Information Paradox*, Springer Series in Astrophysics and Cosmology, https://doi.org/10.1007/978-981-96-6170-1_10

clude that the seemingly non-unitary evaporation process of a black hole, the loss of information in black hole evaporation found by Hawking [77], should also be unitary in AdS. We thus can learn that the black hole evaporation is unitary due to some *unknown mechanism*. On the other hand, it does not mean we do not need to worry about the information paradox anymore, because we still do not understand *why* the process is unitary and *how* the information fallen behind the horizon will be contained in the Hawking radiation.

AdS/CFT promises that unitarity can be explained for black holes in AdS. Indeed, the recent progress in the *island formula* and the *replica wormholes* [13–17, 20, 25, 38–41, 61, 67, 80, 114, 116, 120], has enabled us to compute the entropy of Hawking radiation which is consistent with unitary evaporation in various AdS black holes, by including *Euclidean wormholes* [45, 65, 66] in the gravitational path integral. In other words, we have understood how to calculate the information measures gravitationally, and we confirmed that they are consistent with the unitary evaporation. We have also understood how to recover the interior information from the black hole interior, such as via the Petz map. Despite these results, we have not yet come to understand what is really happening at the black hole horizon. We have not understood how the interior information can leak outside the black hole, as the information leakage suggests. Such an investigation is definitely the next important step toward the full understanding of the black hole information paradox as well as testing the theory of quantum gravity.

The aim of this chapter is to overview this progress and clarify the statements above. For this purpose, we first review how AdS/CFT was derived from superstring theory, which is widely believed to be a concrete realization of unitary quantum gravity, in Sect. 10.2. In Sect. 10.3, we will briefly discuss black holes in AdS, highlighting technical difficulty of formulating the information paradox in this set-up. In Sect. 10.4, we couple AdS gravity with an external reservoir and the double-holography model, in order to accomplish this task. Section 10.5 explains the Ryu-Takayanagi formula and the entanglement wedge reconstruction, which are the basis of the island formula and the replica wormhole explained in Sect. 10.6. In Sect. 10.7, the caveat in the gravitational path-integral with Euclidean wormholes, such as ensemble averaging and factorization are discussed. We conclude with future directions in Sect. 10.8.

We emphasize that the list of references included in this chapter is far from complete, due to the vast number of studies. We highly recommend that readers refer to the references therein, for a complete list of literature.

10.2 AdS/CFT and Holography

AdS/CFT correspondence [98] is a concrete realization of *holography* [132, 136], the conjecture that a quantum gravity theory in a $d+1$-dimensional spacetime with a d-dimensional boundary, is equivalent to the boundary quantum mechanics without gravity. In other words, the information of the $d+1$-dimensional gravity theory is contained in the lower dimensional quantum field theory like a hologram. The first implication of holography is that the entropy of quantum gravity in a region R, cannot exceed the Hilbert space dimension of quantum mechanics at the boundary ∂R. If the boundary quantum mechanics is an ordinary quantum field theory (QFT), then its entropy should be proportional to the volume Area$[\partial R]$ of the $d-1$-dimensional boundary ∂R. Thus the entropy of the quantum gravity in d-dimensional region R is at most proportional to [30, 31]

$$S_R \propto \frac{\text{Area}[\partial R]}{l_P^{d-1}}, \tag{10.1}$$

here we set the UV cut-off of the boundary QFT at the Planck length scale l_P. Note that this behavior is unique to gravity, as the entropy of usual matter is proportional to the volume. This implies that degrees of freedom in gravity are highly non-locally stored in a subregion. Holography is conjectured to hold in various spacetime backgrounds, such as de Sitter space and flat space [129, 130]. Among them, AdS/CFT is the most established and well-tested. In the following, we will review how AdS/CFT arises from string theory.

Superstring theory is a theory of strings and branes [117]. Massless fluctuations of these solitonic objects give massless quantum fields at low energy, including graviton and gauge fields. The closed string coupling g_s and open string coupling g_s^2 controls string interactions. Excited strings have an energy of order l_s^{-1}, where l_s is the string length scale, and modes below this energy scale are massless supergravity fields.

The remarkable feature of string theory is that strings can couple only with particular backgrounds, imposing constraints on the spacetime dimension of the background spacetime, giving equation of motion for metric and fields. Indeed, the worldsheet Weyl gauge invariance fixes the dimension of the background spacetime. This dimension is called *critical dimension* and is 10 in type I, II-A, II-B, and heterotic superstring theories for the background Minkowski spacetime. It is also remarkable that the worldsheet Weyl gauge invariance imposes the equation of motion for the massless fields, giving rise to the Einstein field equation for the metric fluctuation. The low-energy effective action for this equation of motion is the corresponding supergravity action.

Since we would like to consider AdS/CFT in $\text{AdS}_5 \times S^5$, we will consider type II-B superstring theory. The type II-B superstring theory contains closed strings as well as D-branes. D-branes are heavy non-perturbative solitonic objects from a closed string viewpoint. However, the remarkable finding was that the fluctuation of D-branes can also be understood as the motion of open strings whose endpoints are attached to D-branes. The worldsheet Weyl gauge invariance of open string constrains the background fields on the D-brane, giving the low-energy effective action for these background fields by the Dirac-Born-Infeld (DBI) action [1, 93].

The important point is that the low energy fluctuation near the stack of N D3-branes decouples from closed superstrings propagating in the flat 10-dimensional Minkowski space when $l_s \to 0$. The action for the low energy fluctuations of the stack of N D3-branes can be obtained by taking the leading order in $\alpha' := l_s^2$ expansion of the DBI action. The effective action is then given by

$$S = -\frac{1}{2\pi g_s}\int d^4x\, \mathrm{Tr}\Big[\frac{1}{4}F_{\mu\nu}F^{\mu\nu} + \sum_{i=1,\ldots,6}\frac{1}{2}\partial^\mu X^i \partial_\mu X^i + \sum_{ij}[X^i, X^j]^2 + \cdots\Big], \tag{10.2}$$

where the fields take value in $U(N)$. This action is the $\mathcal{N}=4$ $U(N)$ super-Yang-Mills theory with 't Hooft coupling $\lambda = g_{YM}^2 N = 2\pi g_s N$ [135].

In the previous paragraph, we described the stack of parallel D3-branes from the open string theory viewpoint. We can also describe this system from the closed string theory viewpoint, whose low energy effective field theory is the type II-B supergravity in 10d, which takes the following form in the Einstein-frame

$$S_{\mathrm{IIB}} = \frac{1}{(2\pi)^7 l_P^8}\int d^{10}x\sqrt{-g}\Big[R - \frac{1}{2}\nabla_M\phi\nabla^M\phi - \frac{2g_s^2}{5!}F_{(5)MNPQR}F_{(5)}^{MNPQR} + \cdots\Big], \tag{10.3}$$

where $F_{(5)MNPQR}$ is the Ramond-Ramond 5-form field strength and dilaton ϕ. $l_P = g_s^{1/4} l_s$ is the 10-dimensional Planck length. From the supergravity perspective, the stack of parallel D3-branes in Minkowski spacetime is a heavy object that is charged under the Ramond-Ramond field strength and preserves 16 supersymmetry. The stack of D3-branes can be described by the extremal black 3-brane solution of the supergravity,

$$\begin{aligned} ds^2 &= H(r)^{-1/2}\eta_{\mu\nu}dx^\mu dx^\nu + H(r)^{1/2}\sum_{i=1}^{6}\delta_{ij}dy^i dy^j,\ H(r) = 1 + \left(\frac{L}{r}\right)^4, \\ \phi(r) &= 0,\ r^2 = \sum_{i=1}^{6}(y^i)^2,\ L^4 = 4\pi g_s N\alpha'^2, \\ A_{(4)} &= H(r)^{-1}dx^0\wedge dx^1\wedge dx^2\wedge dx^3,\ F_{(5)} = N(\mathrm{vol}(S^5) + *\mathrm{vol}(S^5)), \end{aligned} \tag{10.4}$$

where $r\to 0$ corresponds to the near horizon region and $r=\infty$ to the asymptotically flat region, see Fig. 10.1. The curvature scale L of this metric must be sufficiently

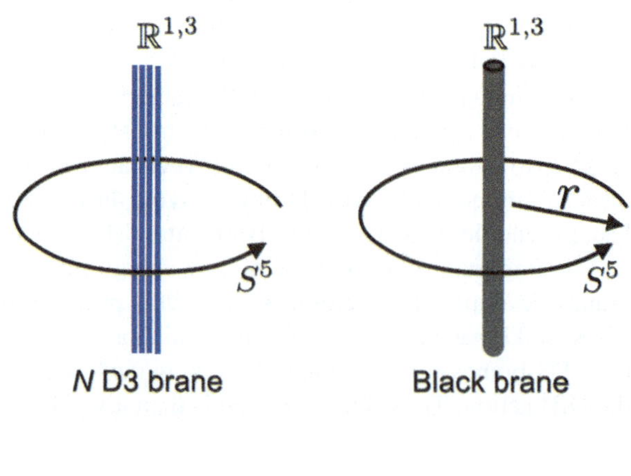

Fig. 10.1 The stack of N D3-branes (left) and its description in the type II-B supergravity as the BPS saturated black brane

large compared to the string length scale l_s, in order to suppress massive string effects so that the black hole can be described purely by supergravity. This solution also preserves half of the supersymmetry of the flat space, namely 16 supersymmetry. This requirement that $L \gg l_s$ implies

$$g_s N \gg 1. \tag{10.5}$$

From a gauge theory perspective, this is equivalent to having strong 't Hooft coupling. Since the string coupling g_s needs to be small, we also need

$$N \gg 1, \tag{10.6}$$

which is the large N limit in the gauge theory perspective.

The near horizon region $r/L \to 0$ of this black hole is described by

$$\begin{aligned} ds^2 &= \frac{r^2}{L^2}\eta_{\mu\nu}dx^\mu dx^\nu + \frac{L^2}{r^2}\delta_{ij}dx^i dx^j \\ &= \frac{L^2}{z^2}(\eta_{\mu\nu}dx^\mu dx^\nu + dz^2) + L^2 ds^2_{S^5}, \ z = \frac{L^2}{r}, \end{aligned} \tag{10.7}$$

which is the 5-dimensional *anti de Sitter (AdS) space* cross 5-sphere $AdS_5 \times S^5$. While the region far away from the horizon is 10-dimensional Minkowski space. The black hole horizon $r = 0$ now is at $z = \infty$, reflecting the fact that the near horizon region of an extremal black hole is infinitely long. The AdS space has a conformal boundary at $z = 0$ ($r = \infty$), which was originally a region far away from the black 3-brane horizon, which connects the flat Minkowski region at infinity and the near horizon AdS region. Again, the closed string excitation in the AdS region is decoupled with the closed string propagating in the flat Minkowski region.

Thus, by taking $g_s N \to \infty$ and $g_s \to 0$, we have two different descriptions of the same system after removing the supergravity on flat Minkowski space from both sides; one is the super Yang-Mills theory at large N with strong 't Hooft coupling $\lambda = 2\pi g_s N \gg 1$, and the other is type II-B superstring theory on $AdS_5 \times S^5$ with AdS radius $L = (4\pi g_s N)^{1/4} l_s$ large compared to the string scale and weak string coupling $g_s \ll 1$. AdS/CFT correspondence states that these two theories are equivalent.

So far we have only seen the equivalence at large $N \gg 1$ and strong coupling $g_s N \gg 1$ in type II-B superstring on $AdS_5 \times S^5$. However, there is a variety of evidence that shows that this duality holds even for finite N and finite coupling $g_s N$. It was soon recognized that this duality holds in other theories and other AdS backgrounds [4]. More recently, we have come to understand the exact equivalence of two sides in two-dimensional dilaton gravity, namely Jackiew-Teitelboim gravity on AdS_2 and a matrix integral [124].

Let us explain the fundamental formula of AdS/CFT, the Gubser-Klebanov-Polyakov-Witten (GKPW) formula [70, 141] which states that

$$Z_{\mathrm{CFT}_d}\left[e^{\int d^dx\,\phi_0(x)O(x)}\right] = Z_{AdS_{d+1}}\left[\lim_{z\to 0} z^{\Delta-d}\phi(z,x) = \phi_0(x)\right], \tag{10.8}$$

where O is a CFT operator with conformal dimension Δ. The left-hand side is the CFT partition function with source $e^{\int d^dx\,\phi_0(x)O(x)}$, and the right-hand side is the gravity partition function with *Dirichlet* boundary conditions for the bulk field $\phi(z,x)$. Using this relation, it is possible to compute matrix elements between any CFT states and any correlation function. When the bulk side is weakly coupled, we can approximate the right-hand side by saddle point with the above boundary condition

$$Z_{\mathrm{CFT}_d}\left[e^{\int d^dx\,\phi_0(x)O(x)}\right] \approx \underset{\phi_{\mathrm{saddle}}}{\mathrm{Min\,Ext}}\left[e^{-S_{\mathrm{gravity}}[\phi_{\mathrm{saddle}}]}\right]. \tag{10.9}$$

This relation enables us to probe strongly coupled CFT from classical general relativity.

10.3 Black Hole and the Evaporation in AdS

In AdS/CFT, black holes in AdS are dual to highly excited states in CFT. To study the statistical property of these excited states, it is very convenient to consider the canonical ensemble and the microcanonical ensemble. In the canonical ensemble, we are interested in the thermal CFT partition function with inverse temperature β. Suppose the space-like section of the 4d CFT is S^3, then the thermal CFT partition function is the CFT path integral on $S^3 \times S^1_\beta$, with periodic boundary condition for bosons and anti-periodic boundary for fermions. The solution of the bulk equation of motion satisfying the periodicity and the spherical symmetry is either the *Schwarzschild-AdS*

$$ds^2 = f(r)d\tau_E^2 + \frac{dr^2}{f(r)} + r^2 ds_{S^3}^2,\ f(r) = 1 + \frac{r^2}{L^2} - \frac{\mu}{r^2},\ \mu := \frac{8G_N M}{3\pi}, \tag{10.10}$$

or the *thermal AdS*

$$ds^2 = (1 + \frac{r^2}{L^2})d\tau_E^2 + \frac{dr^2}{1+\frac{r^2}{L^2}} + r^2 ds_{S^3}^2. \tag{10.11}$$

Here $\tau_E \sim \tau_E + \beta$. The Schwarzschild-AdS black hole has the horizon at $r = r_+$, which is the solution of $f(r_+) = 0$. This Euclidean black hole metric can have a codimension-2 conical defect in the τ_E direction for general μ at $r = r_+$. The conical deficit angle is given by

$$\Delta\theta = 2\pi - \beta r_+\left(\frac{1}{L^2} + \frac{\mu}{r_+^4}\right). \tag{10.12}$$

Such a conical defect does not satisfy the equation of motion and is not a saddle point, thus we require the absence of such a defect. This condition $\Delta\theta = 0$ can be written as, in terms of r_+

$$\beta = \frac{2\pi r_+ L^2}{2r_+^2 + L^2}. \tag{10.13}$$

Or expressing r_+ in terms of β only

$$r_+ = \frac{\pi L^2}{2\beta}\left[1 \pm \sqrt{1 - \frac{2\beta^2}{\pi^2 L^2}}\right]. \tag{10.14}$$

The plus sign has a larger horizon and corresponds to *large black hole*, whereas the minus sign case corresponds to *small black hole*. In the canonical ensemble, the large black hole dominates over the small black hole. The small black holes are unstable as they have negative specific heat. Indeed, the small black hole becomes lighter as the temperature increases, whereas the large black hole becomes heavier. At the high-temperature limit, the small black hole does not feel the AdS cosmological constant and behaves in the same way as the Schwarzschild black hole in the flat space.

After imposing the condition (10.14), we can analytically continue the geometry to Lorentzian signature $\tau_E \to it_L$, and the codimension-2 surface defined by $r = r_+$ becomes indeed the event horizon of the black hole. Using (10.9), we can evaluate the free energy, expectation value for energy, and entropy. The result of the entropy is

$$S = \frac{\text{Area}[r = r_+]}{4G_N}, \tag{10.15}$$

which is the Bekenstein-Hawking formula. The large Schwarzschild-AdS black hole has lower free energy than the thermal AdS and thus is dominant at high temperature $\beta \leq \frac{2\pi L}{3} = \beta_c$. At the critical temperature $\beta = \beta_c$, a phase transition called the *Hawking-Page transition* occurs [78], and at a lower temperature the thermal AdS becomes the dominant configuration. The thermal AdS is the AdS space filled with a gas of gravitons and has no event horizon. The entropy is zero at the tree level, of order $O(G_N^{-1})$. At one loop, the gas of gravitons contributes to the entropy of order $O(G_N^0)$ both in the thermal AdS and in the large Schwarzschild-AdS black hole.

What are the roles played by the Hawking radiation in the large Schwarzschild-AdS black hole? First, we must note that the massless Hawking radiation emitted at the event horizon will eventually hit the AdS boundary and will be reflected back to the black hole in a finite time, see Fig. 10.2. Thus, in the canonical ensemble, the system will eventually reach an equilibrium between the black hole evaporation and the infalling matter. Since the entropy of matter in this equilibrium is of order $O(G_N^0)$, the radiation entropy cannot exceed the black hole entropy.

How about the microcanonical ensemble? Let's consider AdS black holes in $AdS_5 \times S^5$ [3]. In this case, at high energy, we have a 5-dimensional large Schwarzschild-AdS black hole, which is uniform over S^5. As we lower the temperature, instead of having the Hawking-Page transition, the 5-dimensional large

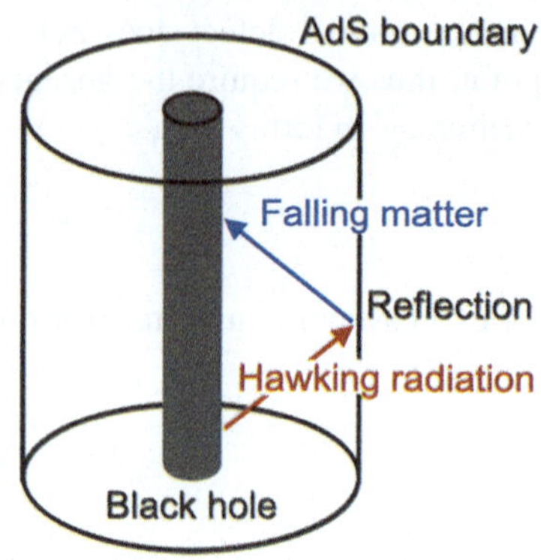

Fig. 10.2 The Hawking radiation from the AdS black hole. The asymptotic boundary of the AdS spacetime with Dirichlet boundary condition reflects the Hawking radiation and the radiation falls behind the horizon

Schwarzschild-AdS black hole transitions into a small ten-dimensional black hole which is localized in S^5, due to the Gregory-Laflamme instability [68]. This 10-dimensional black hole will be in equilibrium with the radiation after a small amount of evaporation [82]. Again, the entropy of matter is small compared to the black hole entropy. As we further lower the temperature, the 10-dimensional black hole transitions into the gas of strings and the horizon disappears.

Thus, testing the black hole information paradox in AdS directly in the micro-canonical or canonical ensemble is a non-trivial task. The entropy of matter is too small to cause a tension between the black hole entropy and the entropy of the Hawking radiation. An alternative approach to test unitarity is to restrict the problem to the information recovery exclusively. Namely, we can ask whether the information fallen behind the horizon can be recovered from the Hawking radiation, by studying the auto-correlation function $\langle O(t)O(0)\rangle_\beta$. Such a two-point function decays exponentially as $\langle O(t)O(0)\rangle_\beta \sim e^{-at/\beta}$ from the classical black hole geometry. However, if this behavior persists forever, the observer at the late time would not be able to recover the initial information. It is also inconsistent with the finiteness of the black hole entropy S, which implies the long time average of the two-point function to be $O(e^{-cS})$.[1] This question was asked by Maldacena [99] and is called the *Maldacena information paradox*, which was later addressed by incorporating effects from *baby universe emission* by Saad [123].

10.4 Reservoir in AdS/CFT

As we have seen in the previous section, it is difficult to formulate the information paradox of entropy of the Hawking radiation in AdS/CFT. To circumvent this problem, we introduce an external reservoir system that can collect sufficient amount of Hawking radiation, whose entropy is comparable to that of the black hole. For

[1] The autocorrelation function can be thought as the overlap between two random states. On average, such overlap is of order $O(e^{-S/2})$.

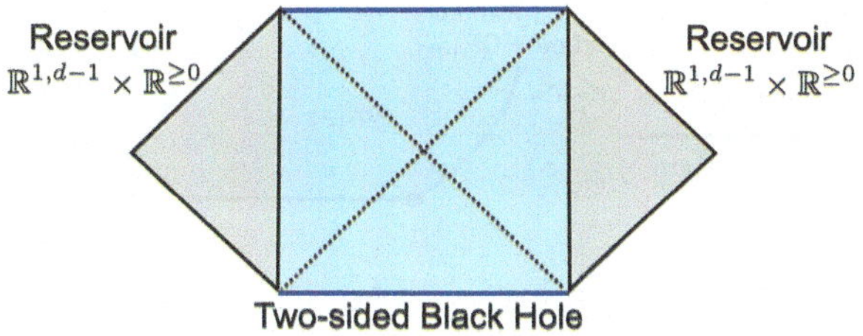

Fig. 10.3 AdS two-sided black hole coupled with two reservoirs at the two AdS boundaries via transparent boundary condition

this purpose, we couple the dual CFT to an external system, allowing it to collect the Hawking radiation. There are various ways to realize such a setup. The common setup is to glue the boundary of AdS to that of a half of a $d+1$-dimensional Minkowski space $\mathbb{R}^{1,d-1} \times \mathbb{R}^{\geq 0}$, see Fig. 10.3. The boundary condition imposes the *transparent* boundary condition for the matter and the usual *Dirichlet* boundary condition for the metric. In this set-up, the Minkowski space has the same matter content as the gravity on the AdS, while it does not contain the graviton.

Since the QFT on the half Minkowski space has an infinite number of degrees of freedom, we can now probe the finiteness of the black hole entropy. Suppose that the Hamiltonian for the CFT is H_{CFT_d} and for the Minkowski is H_{R}. Then the total Hamiltonian is

$$H_{\text{total}} = H_{\text{CFT}_d} + H_{\text{Int}} + H_{\text{R}}, \tag{10.16}$$

with an interaction H_{int} which couples d-dimensional CFT and $d+1$-dimensional QFT on $\mathrm{M}^{\text{half}}_{d+1}$. This construction of the reservoir, however, is known to introduce mass to the graviton [60]. This mass is proportional to the coupling strength of H_{Int}. We note that unlike in flat space, the van Dam-Veltman-Zakharov discontinuity [139, 145] which states that massless limit of massive gravity does not reproduce correlators of massless gravity, is absent in AdS space [87, 89, 119] and in de Sitter space [81], as long as the mass approaches to zero fast enough compared to the cosmological constant. This implies that we can recover the standard correlators of massless AdS gravity in the weak interaction limit.

A convenient model of a black hole coupled with a reservoir is the *double-holography model* [5, 6, 16, 19, 20, 28, 38, 40, 41, 44, 63, 80, 120, 131, 134], which is a special version of the AdS/BCFT correspondence [58, 87, 88, 137] see Fig. 10.4. In the double-holography model, we assume that the boundary conformal field theory (BCFT) on half Minkowski space $\mathbb{R}^{1,d-1} \times \mathbb{R}^{\geq 0}$ whose Hamiltonian is H_{total} is holographically dual to AdS_{d+1} gravity with d-dimensional Karch-Randall brane [87, 88] where the AdS spacetime terminates. On this d-dimensional Karch-Randall brane, nearly massless graviton lives, with the correct number of degrees of freedom for d-dimensional diffeomorphism, and with extra degrees of freedom being gauged away by brane bending effect [87, 89]. The metric on the brane is $d+1$-dimensional AdS, obeying the action given by the Einstein-Hilbert action at leading order. In the double-holography model, we further assume that the bulk CFT_{d+1} on the Karch-

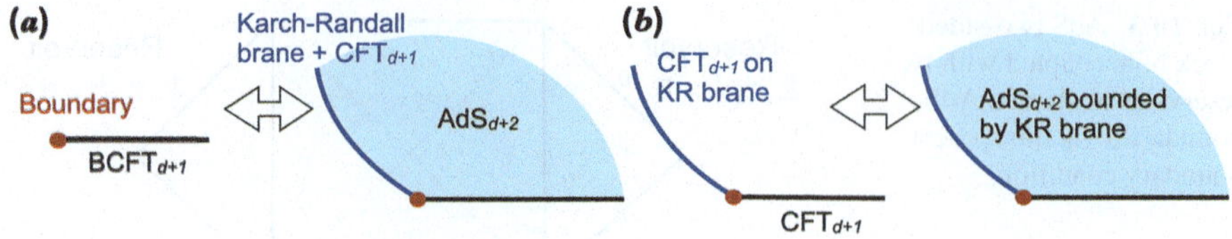

Fig. 10.4 Two different types of holography in the double-holography model. Note that on the dual gravity side, **a** contains the KR brane while **b** does not

Randall brane is dual to AdS_{d+2} whose boundary is the union of the Karch-Randall brane and the half-Minkowski space [16, 87].

Thus, we have three descriptions of the same theory: (1) the boundary description: BCFT_{d+1} on $\mathbb{R}^{1,d-1} \times \mathbb{R}^{\geq 0}$, (2) the bulk description: $d + 2$-dimensional AdS gravity with the Karch-Randall brane at the boundary, (3) the intermediate effective description: CFT_{d+1} on $\text{AdS}_{d+1} \cup \mathbb{R}^{1,d-1} \times \mathbb{R}^{\geq 0}$ plus $d + 1$-dimensional AdS gravity. However, we must keep in mind that the intermediate description is only an effective description valid for small fluctuations of the brane, and we need to take a large tension limit to obtain a nearly massless graviton.

10.5 Ryu-Takayanagi Formula

One of the most profound findings in the AdS/CFT dictionary is the Ryu-Takayanagi formula [57, 95, 111, 121, 122], followed by the discovery of the entanglement wedge reconstruction [11, 12, 48, 71, 85, 113]. These findings turned out to be the foundation for the derivation of the Page curve [51] and the information recovery from the black hole [79]. In this section, we will explain these fundamental results.

Consider a holographic CFT_d that is dual to classical Einstein gravity on AdS_{d+1}. Let us assume Euclidean signature. The Ryu-Takayanagi formula states that the entanglement entropy of subregion A in the boundary is holographically given by

$$S_A = \underset{\gamma_A}{\text{Min}}\left[\frac{\text{Area}[\gamma_A]}{4G_N^{(d+1)}}\right], \tag{10.17}$$

where γ_A is co-dimension 2 surface anchored from ∂A and is homologous to A, see Fig. 10.5. When γ_A minimizes the area, such surface is called *Ryu-Takayanagi surface*, and the formula (10.17) is the *Ryu-Takayanagi formula*. We can include quantum corrections from the bulk matter, which is a bulk low-energy effective field theory. In this case, the formula is given by [55, 57]

$$S_A = \underset{\gamma_A}{\text{Min}}\left[\frac{\text{Area}[\gamma_A]}{4G_N^{(d+1)}} + S^{\text{matter}}_{\text{EW}[\gamma_A]}\right], \tag{10.18}$$

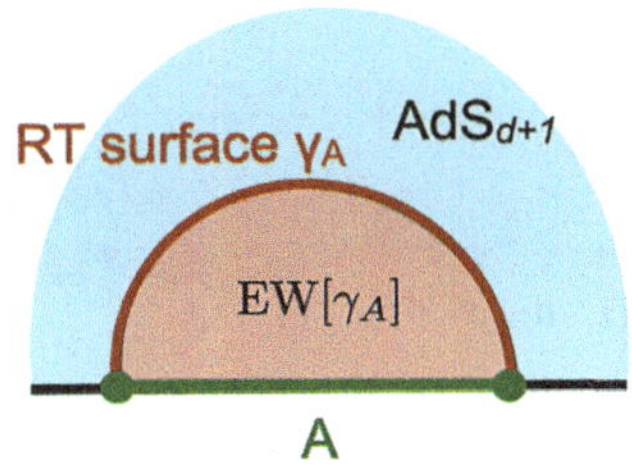

Fig. 10.5 The Ryu-Takayanagi surface γ_A in AdS_{d+1} for boundary subregion A. The region bounded by γ_A and the boundary subregion A is called the entanglement wedge of A, denoted as $\text{EW}[\gamma_A]$

where the bulk subregion $\text{EW}[\gamma_A]$ is co-dimension 1 and is bounded by $A \cup \gamma_A$. When the right-hand side is minimized, the bulk subregion $\text{EW}[\gamma_A]$ is called *entanglement wedge* of A.

In QFT, it is possible to compute entanglement entropy in several cases, using *replica trick*. In the replica trick, one computes

$$S[\rho] = -\text{Tr}\Big[\rho \log \rho\Big] = \lim_{n\to 1} \frac{\log \text{Tr}[\rho^n]}{1-n} = \lim_{n\to 1} S_n[\rho], \tag{10.19}$$

where $S_n[\rho]$ is the *Renyi* entropy. One computes $\text{Tr}[\rho^n]$ for positive integer n and analytically continue to real n. In QFT, the matrix element of reduced density matrix $\rho_A = \text{Tr}_{\bar{A}}[|\Omega\rangle\langle\Omega|]$ of the vacuum $|\Omega\rangle$ is given by

$$\langle \phi_- | \rho_A | \phi_+ \rangle = \int \mathscr{D}\phi \Big|_{\phi(x,\pm 0)=\phi_\pm(x),\ x\in A} e^{-S} =$$

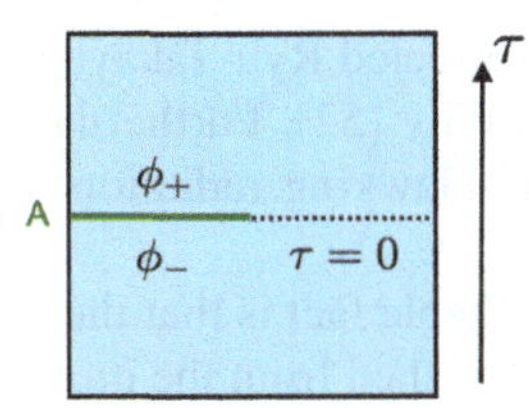

. (10.20)

Thus the replica partition function is the usual partition function on the n-sheeted replica manifold M^n

$$\mathrm{Tr}[\rho_A^n] = \int \Pi_{i=1}^n \mathscr{D}\phi^{(i)}\Big|_{\phi^{(i)}(x,-0)=\phi^{(i+1)}(x,+0),\ x\in A} e^{-S} =$$

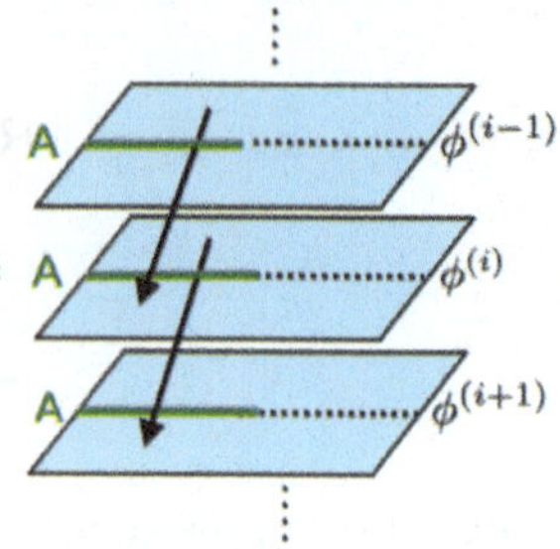

, (10.21)

which has conical excess angle $2\pi(n-1)$ at ∂A. The CFT replica partition function can also be computed in AdS/CFT setup, which leads to the derivation of the Ryu-Takayanagi formula [95]. This can be accomplished by considering the AdS spacetime with the replica manifold at the boundary. If the dual geometry is replica symmetric, we can take the orbifold to obtain the original manifold M^1 and remove the conical excess angle. Now the bulk has a co-dimension-2 defect γ anchored from ∂A, which has conical deficit angle $2\pi(1-\frac{1}{n})$. Since we have n copies of M^1, the action of this defect is proportional to $n \times 2\pi(1-\frac{1}{n})\frac{\mathrm{Area}[\gamma]}{G_N}$. When $n > 1$ is close to 1, we can ignore the backreaction to the background metric from the defect, and the replica partition function is now evaluated as

$$\mathrm{Tr}[\rho_A^n] \underset{n\to 1}{\propto} \underset{\gamma}{\mathrm{Min}}\, e^{-(n-1)\frac{\mathrm{Area}[\gamma_A]}{4G_N}}, \tag{10.22}$$

which immediately leads to the Ryu-Takayanagi formula. Using identical gravitational path integral for bulk matter as well as higher derivative gravity, one can obtain a quantum-corrected Ryu-Takayanagi formula as well as a generalization to higher derivative gravity [52]. Furthermore, the replica wormhole, which gives the Page curve for the Hawking radiation, is also based on the identical line of derivation [15, 116].

The remarkable fact is that the effective field theory (EFT) states inside $\mathrm{EW}[\gamma_A]$ can be distinguished from the quantum state on A if these EFT states have the same background metric. This can be seen by considering the difference between the bulk EFT state's relative entropy and the relative entropy of microscopic states on the boundary. It was found that these two are approximately equal in holography at large N [85],

$$S(\rho_{\mathrm{EW}}^{\mathrm{EFT}:a}|\rho_{\mathrm{EW}}^{\mathrm{EFT}:b}) \approx S(\rho_A^a|\rho_A^b), \tag{10.23}$$

called JLMS formula. This equality implies that there is approximately no loss of information when we extract the information of the bulk state $\rho_{\mathrm{EW}}^{\mathrm{EFT}:a}$ from the boundary reduced density matrix ρ_A.

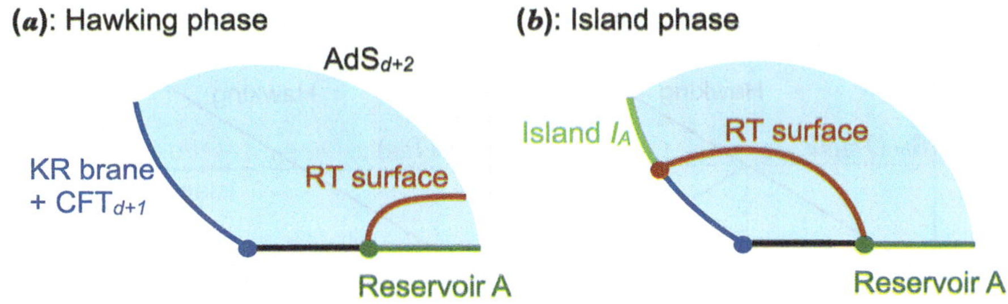

Fig. 10.6 Two phases of the Ryu-Takayanagi surface in the double-holography model. **a** In the Hawking phase, the Ryu-Takayanagi surface does not end on the Karch-Randall (KR) brane. The entanglement entropy of the reservoir A can increase arbitrarily, violating the upper bound from the black hole entropy thus the information paradox. **b** The Ryu-Takayanagi surface now ends on the KR brane. The region on the KR brane which is homologous to the reservoir A is the island I_A of A

10.6 Island Formula and Euclidean Wormhole

In this section, we consider the entropy of the reservoir in the double-holography model [16]. We will see that the application of the Ryu-Takayanagi formula to the double-holography model immediately leads to the unitary Page curve.

We consider the entropy of subregion A in $\mathbb{R}^{1,d-1} \times \mathbb{R}^{\geq 0}$ in the double-holography model, see Fig. 10.6. We first use the Ryu-Takayanagi formula for A using the duality between BCFT_{d+1} and AdS_{d+2}. The crucial point is that the Ryu-Takayanagi surface γ_A can end on the brane. The endpoint ∂I_A is the boundary of subregion I_A in the brane, which is called *island* of A. This endpoint ∂I_A contributes to the holographic entanglement entropy. Assuming the leading brane action is the Einstein-Hilbert action on AdS_{d+1} with Newton constant $G_N^{(d+1)}$, the total holographic entanglement entropy is

$$S_A = \underset{\gamma_A,\, I_A}{\mathrm{Min}}\left[\frac{\mathrm{Area}[\gamma_A]}{4G_N^{(d+2)}} + \frac{\mathrm{Area}[\partial I_A]}{4G_N^{(d+1)}}\right]. \tag{10.24}$$

Note that here both I_A and γ_A are variables in the minimization, and the island can be empty, in which case $\mathrm{Area}[\partial I_A] = 0$. Next, we rewrite the above expression in terms of EFT on AdS_{d+1}, so that we obtain an expression that is valid for more generic gravity coupled with a reservoir, without the dual AdS_{d+2}. For this purpose we consider the CFT_{d+1} living on $\mathrm{AdS}_{d+1} \cup (\mathbb{R}^{1,d-1} \times \mathbb{R}^{\geq 0})$, which is dual to AdS_{d+2} gravity. If we generalize the Ryu-Takayanagi formula for CFT on the general curved background as in [54, 109], we arrive at

$$S^{\mathrm{matter}}_{I_A \cup A} = \underset{\gamma_A}{\mathrm{Min}}\left[\frac{\mathrm{Area}[\gamma_A]}{4G_N^{(d+2)}}\right], \tag{10.25}$$

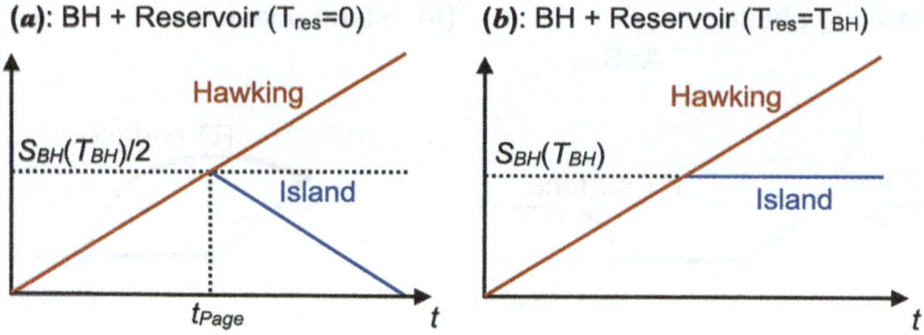

Fig. 10.7 Schematic figure of the entropy of the reservoir, comparing Hawking's calculation and the island formula. T_{res} is the temperature of the reservoir, and we draw **a** $T_{\text{res}} = 0$ in which case the black hole evaporates into nothing and thus the final entropy should be 0, and **b** $T_{\text{res}} = T_{\text{BH}}$ case so that the black hole and the reservoir will be in equilibrium at this temperature, and the entropy computed by Hawking grows linearly in time, which saturates at the black hole entropy which is the upper bound dictated by the unitarity

with I_A being fixed. Since we can perform minimization with fixed I_A inside the parenthesis in (10.24), we can rewrite (10.24) as

$$S_A = \underset{I_A}{\text{Min}}\left[S^{\text{matter}}_{I_A \cup A} + \frac{\text{Area}[\partial I_A]}{4G_N^{(d+1)}}\right] = \underset{I_A}{\text{Min}}\ S^{\text{gen}}_{A\cup I_A}, \tag{10.26}$$

where $S^{\text{gen}}_{A\cup I_A}$ is the *generalized entropy* [55] of region $A \cup I_A$ (excluding area contribution from non-gravitating region A). The formula (10.26) is now written in terms of $d+1$-dimensional quantity only, and the variable in the minimization is I_A. This formula is known as the *island formula,* which reproduces the unitary Page curve of the Hawking radiation as we explain shortly.

The important new ingredient in the island formula is the introduction of the island I_A, which typically contains the interior region of the black hole. Without I_A, the entropy of the reservoir is always given by S_A^{matter}, which can grow arbitrarily large, and the growth of the entropy exceeds the black hole entropy, giving rise to the black hole information paradox. However, now we must minimize over configurations with the presence of I_A as well. Consider an entangled pair in EFT between I_A and A. This pair contributes to S_A^{matter} but not to $S^{\text{matter}}_{I_A\cup A}$. Thus the emergence of the island I_A reduces the matter entropy, at the price of $\text{Area}[\partial I_A]/(4G_N^{(d+1)})$. In particular, once I_A contains the portion of the black hole interior which is entangled with A, the increase of S_A cannot be greater than $\text{Area}[\partial I_A]/(4G_N^{(d+1)}) \approx S_{\text{BH}}$. This implies we can reproduce the Page curve and thus do not have the information paradox in the entropy of the Hawking radiation, see Fig. 10.7.

Furthermore, we can generalize the JLMS formula (10.27) to the island as

$$S(\rho^{\text{EFT}:a}_{I_A}|\rho^{\text{EFT}:b}_{I_A}) \approx S(\rho^a_A|\rho^b_A), \tag{10.27}$$

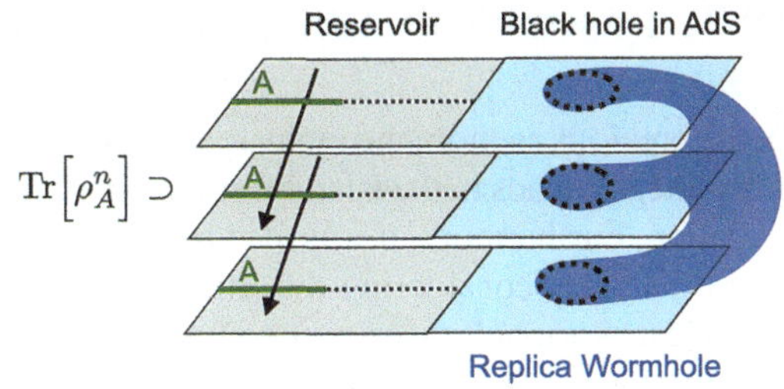

Fig. 10.8 The replica wormhole configuration which appears in the gravitational path-integral for Tr[ρ_A^n]. In general, we can also have other replica wormhole configurations that only connect a subset of full replica sheets

which implies that the information on the island can be recovered from the Hawking radiation in the reservoir [114, 116]. In other words, the island is in the entanglement wedge of A. This indicates that the information thrown into the black hole does escape the interior as Hawking radiation after the Page time, through an unknown mechanism that has not been understood to date.

The island formula is known to hold not only in the double-holography model but in more generic models of gravity, derived through the introduction of the *Euclidean wormholes* or the *replica wormholes*. In the gravitational path-integral, the *Hartle-Hawking prescription* [73] dictates us to sum over all possible spacetime configurations with the given boundary condition. In particular, when there are multiple boundaries, we need to sum over all possible spacetime configurations including connected spacetimes, namely spacetimes connected via Euclidean wormholes, see Fig. 10.8. These Euclidean wormholes play the central role in the replica trick calculation of the entropy of the Hawking radiation, as the island formula can be derived by including these Euclidean wormhole contributions [15, 116]. In the path-integral of Tr[ρ_A^n], we have n AdS boundaries where the n AdS spacetimes can end, and as the Hartle-Hawking prescription dictates that we need to include Euclidean wormholes which connect these AdS boundaries through the bulk spacetime. Using similar argument as in the derivation of the Ryu-Takayanagi formula [95], by taking orbifold of the replica symmetric wormhole configuration, we can arrive at the island formula (10.26) [15, 116].

The island formula is believed to be correct in more general setups. It has been applied not only for AdS black holes, but also for black holes in various backgrounds including asymptotically flat space [18, 75, 76], de Sitter space [21, 22, 42, 62, 74, 116], and Friedmann-Lemaître-Robertson-Walker spacetimes [35, 74]. The island formula can be extended to other information measures which can be computed via a replica trick. Such as reflected entropy [7, 36, 96], Renyi entropy [90, 116], the capacity of entanglement [90, 91], the negativity [26, 53], and the coherent information [23, 110].

10.7 Ensemble in AdS/CFT

In the previous section, the inclusion of the Euclidean wormholes to the gravitational path-integral leads to the island formula, deriving the unitary entropy of the Hawking radiation. Furthermore, it was also found that Maldacena's information paradox can be resolved by considering the Euclidean wormholes [123]. Why the inclusion of Euclidean wormholes is so successful in reproducing results consistent with unitarity, and why the previous calculations without wormholes were wrong even qualitatively?

Consider low-energy effective field theory orthonormal states $|\alpha_i\rangle$, $(i = 1, \ldots, k)$ and their embeddings $|\Psi_{\alpha_i}\rangle$ into the non-perturbative Hilbert space of the microscopic theory. The important point is that the orthonormality of EFT states is not exact, although the error is suppressed by the inverse of the Hilbert space dimension e^S of the microscopic theory so that it is almost invisible. The overlap of the microscopic states is schematically given by

$$\langle \Psi_{\alpha_i} | \Psi_{\alpha_j} \rangle = \delta_{ij} + e^{-S/2} R_{ij}, \tag{10.28}$$

where R_{ij} is a $O(1)$ matrix element with $R_{ii} = 0$. The conventional calculation has been capturing the first term only, however, the second subleading term is essential in order to probe the finite dimensionality of the gravity Hilbert space. Indeed, the entropy of Hawking radiation crucially relies on the finite dimensionality of the gravity Hilbert space. The key point is that the gravitational path-integral including the Euclidean wormholes can capture the *statistics* of this second term, as the *ensemble average over microscopic gravity theories* [33, 34, 100, 118, 124, 127, 140].

To understand this statement, let us consider a microscopic boundary Hamiltonian H which has a gravity dual on AdS. We would like to learn about the gravity dual of each boundary Hamiltonian. This is a highly non-trivial task because the bulk theory may contain various complicated, theory-specific excitations and branes. Instead of studying such microscopic theory, it is convenient to study the statistical behavior of the *ensemble* of Hamiltonians. When we tune the probability distribution $e^{-V(H)}$ of the Hamiltonian appropriately, it turns out that the products of averaged partition functions are dual to the gravitational path-integral with Euclidean wormholes as

$$\mathbb{E}[Z(H,\beta_1)Z(H,\beta_2)\cdots] = \int \mathcal{D}H\, e^{-V(H)} \mathrm{Tr}[e^{-\beta_1 H}]\mathrm{Tr}[e^{-\beta_2 H}]\cdots$$

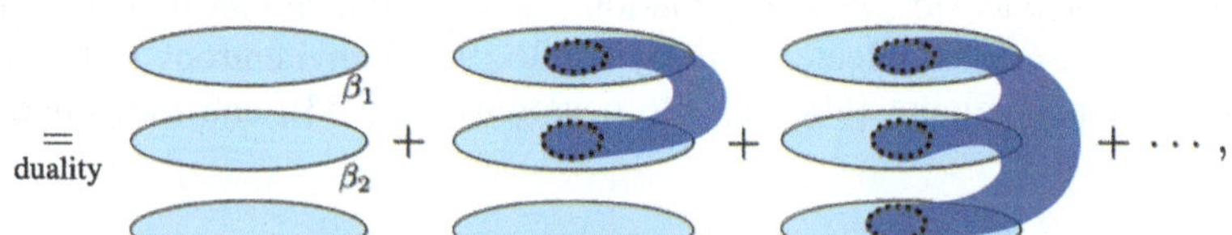

(10.29)

here the last line is the sum over topologies, and $\mathbb{E}[\cdots]$ denotes the ensemble averaging over Hamiltonians. This highly non-trivial relation between ensemble of theories and smooth Euclidean wormhole geometry was found to be true in some important examples in lower dimensional gravity, such as the two-dimensional dilaton gravity called Jackiew-Teitelboim gravity [8, 56, 69, 72, 84, 86, 92, 97, 102, 103, 124, 138, 143, 144], and three-dimensional topological gravity theories [2, 27, 46, 104] or AdS_3 pure gravity [49, 50]. However, in dimensions higher than three, it is still an open problem how this ensemble of theories is even possible. For example, there is no room for any random coupling in $\mathcal{N} = 4$ super Yang-Mills theory in four dimensions, as it is the unique maximally supersymmetric Yang-Mills theory in four dimensions for a given gauge group. See also [112, 125, 126].

The reason why the ensemble averaging and the Euclidean wormhole are related is as follows. Suppose we consider two AdS boundaries. The Hartle-Hawking prescription dictates us to include the Euclidean wormhole which connects two AdS boundaries β_1 and β_2. The implication of the dual CFT is that the product of two partition functions never *factorizes* [100, 107]

$$Z(\beta_1, \beta_2) \neq Z(\beta_1)Z(\beta_2). \tag{10.30}$$

This is impossible if there were no correlation between two boundaries β_1 and β_2. As there is no causal correlation between the two boundaries,[2] we can conclude that the correlation originates from the ensemble averaging, in which case we indeed have

$$\mathbb{E}[Z(H, \beta_1)Z(H, \beta_2)\ldots] \neq \mathbb{E}[Z(H, \beta_1)]\mathbb{E}[Z(H, \beta_2)]\ldots. \tag{10.31}$$

In other words, the gravitational path-integral with a connected Euclidean wormhole is dual to an ensemble averaging over boundary theories.

Let us describe how the Euclidean wormhole captures the off-diagonal components in (10.28). For simplicity let us take $\mathbb{E}[R_{ij}] = 0$. Then the product of overlaps is given by

$$\mathbb{E}[\langle\Psi_{\alpha_i}|\Psi_{\alpha_j}\rangle\langle\Psi_{\alpha_{j'}}|\Psi_{\alpha_{i'}}\rangle] = \delta_{ij}\delta_{j'i'} + e^{-S}\mathbb{E}[R_{ij}R_{j'i'}]$$

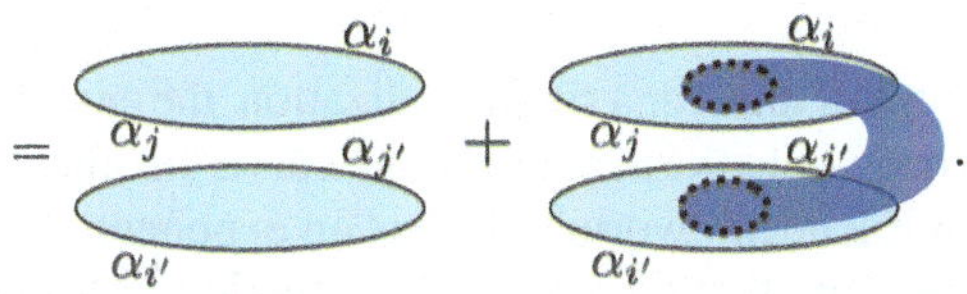

(10.32)

Since the first term in the second line is equal to $\delta_{ij}\delta_{j'i'}$, we can see that the Euclidean wormhole corresponds to the two-point function of the off-diagonal component R_{ij}. The important point is that the Euclidean wormhole contribution in (10.32) is nonzero in general, in particular when $i = i'$ and $j = j'$ [116]. If we had omitted such Euclidean wormhole contribution, we would have arrived at the wrong conclusion that the states $|\Psi_{\alpha_i}\rangle$ and $|\Psi_{\alpha_j}\rangle$ with $i \neq j$ are orthogonal non-perturbatively. Such wrong assumption would lead to the inconsistency that the EFT Hilbert space

[2] When we introduce explicit interaction between two CFTs as in [59, 101], they are now in causal contact and the dual side can even have a traversable wormhole.

dimension in the black hole is larger than $e^{S_{\rm BH}}$, as we can consider orthogonal states $|\Psi_{\alpha_i}\rangle$ $(i = 1, \ldots, k)$ with $k \gg e^{S_{\rm BH}}$. Such inconsistency can be observed, for example, by considering entanglement between the Hawking radiation and the interior EFT state. Such entangled state can schematically be written as

$$|\Phi\rangle = \frac{1}{\sqrt{k}} \sum_{i=1}^{k} |\Psi_{\alpha_i}\rangle \otimes |i\rangle_{\rm rad}. \tag{10.33}$$

If the interior EFT states are orthogonal to each other, assuming $\langle i|j\rangle_{\rm rad} = \delta_{ij}$, the entanglement entropy of the radiation is $\log k$, which contradicts with the dimensionality of the black hole Hilbert space. As demonstrated in [116], once we consider the gravitational path-integral with Euclidean wormholes that can account for the non-orthogonality, we can indeed reproduce the expected entropy of the radiation $S_{\rm rad} \approx S_{\rm BH}$, which is indeed consistent with the unitarity of the black hole evaporation as well as the dimensionality of the black hole Hilbert space.

So far, we have seen how the unitary Page curve and the reconstruction of the interior information can be explained by considering the gravitational path integral with Euclidean wormholes, which computes ensemble-averaged quantities. It is important to ask whether this unitarity holds for *individual theory in the ensemble*. This is not guaranteed automatically, as the gravitational path-integral over Euclidean wormholes only computes ensemble-averaged quantities, and in principle, the typical value can have a large deviation from the average. The most well-known examples are the ramp and the plateau of the spectral form factor, which has order-one signal to noise ratio [47]. Fortunately, in the gravitational path-integral, it is possible to show that such a deviation in the Page curve and the interior reconstruction are typically exponentially suppressed by the black hole entropy [32, 110]. Namely, we have the following entropy fluctuation

$$\mathbb{E}[S_R^2] - \mathbb{E}[S_R]^2 = O(e^{-2S_{\rm BH}}). \tag{10.34}$$

This result implies that, for a single realization, the entropy of the radiation is *typically* exponentially close to the averaged Page curve. In other words, the gravitational path integral with Euclidean wormholes has an exponentially suppressed error.

Finally, we note that we can study a single realization of the ensemble by concentrating on a super-selection sector called α-sector [24, 29, 45, 105, 106]. In each *alpha* sector, the multi-boundary partition function and correlation functions factorize. This implies the absence of the correlation between different boundaries, thus each α-sector corresponds to a single realization of the ensemble.

10.8 Future Directions

AdS/CFT provides a powerful tool for studying how the apparent loss of unitarity can be resolved in quantum gravity. So far, we have understood that the inclusion of the Euclidean wormholes is a convenient method to capture the information recovery by evaluating the quantum information theoretic measures. However, this is only the beginning of understanding the black hole interior quantitatively, as we have many important unsolved questions about quantum black holes, quantum cosmology, and quantum gravity. We still do not fully understood how interior information is recovered by an external observer or what happens at the event horizon and the interior. In particular, the existence of the firewalls at the horizon [9, 10], has recently been studied from the novel Euclidean wormhole point of view by constructing the bulk non-perturbative Hilbert space [83, 108, 128], which may have potential connections to the algebra of observables of black holes [37, 94, 115]. It is also intriguing to understand how the emergence of the event horizon is related to the boundary quantum chaos [43, 64]. It is also important to understand the relation between the ensemble-averaged boundary theory and gravitational path integral over Euclidean wormholes in higher dimensions, which is still far from clear at present. In short, AdS/CFT continues to be one of the most important tools to investigate the non-perturbative properties of quantum gravity.

Acknowledgements We are grateful to Raphael Bousso, Venkatesa Chandrasekaran, Satoshi Iso, Chitraang Murdia, Yasunori Nomura, Pratik Rath, Tadakatsu Sakai, Masaki Shigemori, Tadashi Takayanagi and Zhenbin Yang for the discussions. MM is supported by JSPS KAKENHI Grant-in-Aid for Early-Career Scientists (24K17044).

References

1. A. Abouelsaood, et al., Open strings in background gauge fields. Nucl. Phys. B **280**, 599–624 (1987). https://doi.org/10.1016/0550-3213(87)90164-7
2. N. Afkhami-Jeddi, et al., Free partition functions and an averaged holographic duality. JHEP **01**, 130 (2021). https://doi.org/10.1007/JHEP01(2021)130. arXiv: 2006.04839 [hep-th]
3. O. Aharony, et al., Large N field theories, string theory and gravity. Phys. Rept. **323**, 183–386 (2000). https://doi.org/10.1016/S0370-1573(99)00083-6. arXiv: hep-th/9905111
4. O. Aharony, et al., N = 6 superconformal Chern-Simons-matter theories, M2-branes and their gravity duals. JHEP **10**, 091 (2008). https://doi.org/10.1088/1126-6708/2008/10/091. arXiv: 0806.1218 [hep-th]
5. I. Akal, et al., Codimension two holography for wedges. Phys. Rev. D **102**(12), 126007 (2020). https://doi.org/10.1103/PhysRevD.102.126007. arXiv: 2007.06800 [hep-th]
6. I. Akal, et al., Holographic moving mirrors. Class. Quant. Grav. **38**(22), 224001 (2021). https://doi.org/10.1088/1361-6382/ac2c1b. arXiv: 2106.11179 [hep-th]
7. C. Akers, et al., The Page curve for reflected entropy. JHEP **06**, 089 (2022). https://doi.org/10.1007/JHEP06(2022)089. arXiv: 2201.11730 [hep-th]
8. A. Almheiri, J. Polchinski, Models of AdS_2 backreaction and holography. JHEP **11**, 014 (2015). https://doi.org/10.1007/JHEP11(2015)014. arXiv: 1402.6334 [hep-th]

9. A. Almheiri, et al., An apologia for firewalls. JHEP **09**, 018 (2013). https://doi.org/10.1007/JHEP09(2013)018. arXiv: 1304.6483 [hep-th]
10. A. Almheiri, et al., Black holes: complementarity or firewalls? JHEP **02**, 062 (2013). https://doi.org/10.1007/JHEP02(2013)062. arXiv: 1207.3123 [hep-th]
11. A. Almheiri, X. Dong, D. Harlow, Bulk locality and quantum error correction in AdS/CFT. JHEP **04**, 163 (2015). https://doi.org/10.1007/JHEP04(2015)163. arXiv:1411.7041 [hep-th]
12. A. Almheiri, X. Dong, B. Swingle, Linearity of holographic entanglement entropy. JHEP **02**, 074 (2017). https://doi.org/10.1007/JHEP02(2017)074. arXiv:1606.04537 [hep-th]
13. A. Almheiri, R. Mahajan, J. Maldacena, Islands outside the horizon (2019). arXiv: 1910.11077 [hep-th]
14. A. Almheiri, et al., The entropy of bulk quantum fields and the entanglement wedge of an evaporating black hole. JHEP **12**, 063 (2019). https://doi.org/10.1007/JHEP12(2019)063. arXiv: 1905.08762 [hep-th]
15. A. Almheiri, et al., Replica wormholes and the entropy of Hawking radiation. JHEP **05**, 013 (2020). https://doi.org/10.1007/JHEP05(2020)013. arXiv: 1911.12333 [hep-th]
16. A. Almheiri, et al., The Page curve of Hawking radiation from semiclassical geometry. JHEP **03**, 149 (2020). https://doi.org/10.1007/JHEP03(2020)149. arXiv: 1908.10996 [hep-th]
17. A. Almheiri, R. Mahajan, J.E. Santos, Entanglement islands in higher dimensions. SciPost Phys. **9**(1), 001 (2020). https://doi.org/10.21468/SciPostPhys.9.1.001. arXiv: 1911.09666 [hep-th]
18. T. Anegawa, N. Iizuka, Notes on islands in asymptotically flat 2d dilaton black holes. JHEP **07**, 036 (2020). https://doi.org/10.1007/JHEP07(2020)036. arXiv: 2004.01601 [hep-th]
19. D. Bak, et al., Unitarity of entanglement and islands in two-sided Janus black holes. JHEP **01**, 155 (2021). https://doi.org/10.1007/JHEP01(2021)155. arXiv: 2006.11717 [hep-th]
20. V. Balasubramanian, et al., Geometric secret sharing in a model of Hawking radiation. JHEP **01**, 177 (2021). https://doi.org/10.1007/JHEP01(2021)177. arXiv: 2003.05448 [hep-th]
21. V. Balasubramanian, A. Kar, T. Ugajin, Islands in de Sitter space. JHEP **02**, 072 (2021). https://doi.org/10.1007/JHEP02(2021)072. arXiv: 2008.05275 [hep-th]
22. V. Balasubramanian, A. Kar, T. Ugajin, Entanglement between two disjoint universes. JHEP **02**, 136 (2021). https://doi.org/10.1007/JHEP02(2021)136. arXiv: 2008.05274 [hep-th]
23. V. Balasubramanian, et al., Quantum chaos and the complexity of spread of states. Phys. Rev. D 106(4), 046007 (2022). https://doi.org/10.1103/PhysRevD.106.046007. arXiv: 2202.06957 [hep-th]
24. V. Balasubramanian, et al., Factorization of the Hilbert space of eternal black holes in general relativity (2024). arXiv: 2410.00091 [hep-th]
25. V. Balasubramanian, et al., Microscopic origin of the entropy of black holes in general relativity. Phys. Rev. X **14**(1), 011024 (2024). https://doi.org/10.1103/PhysRevX.14.011024. arXiv: 2212.02447 [hep-th]
26. J.K. Basak, et al., Islands for entanglement negativity. SciPost Phys. **12**(1), 003 (2022). https://doi.org/10.21468/SciPostPhys.12.1.003. arXiv: 2012.03983 [hep-th]
27. N. Benjamin, et al., Narain to Narnia. Commun. Math. Phys. **390**(1), 425–470 (2022). https://doi.org/10.1007/s00220-021-04211-x. arXiv: 2103.15826 [hep-th]
28. A. Bhattacharya, et al., Islands and complexity of eternal black hole and radiation subsystems for a doubly holographic model. JHEP **05**, 135 (2021). https://doi.org/10.1007/JHEP05(2021)135. arXiv: 2103.15852 [hep-th]
29. J. Boruch, et al., How the Hilbert space of two-sided black holes factorises (2024). arXiv: 2406.04396 [hep-th]
30. R. Bousso, A covariant entropy conjecture. JHEP **07**, 004 (1999). https://doi.org/10.1088/1126-6708/1999/07/004. arXiv: hep-th/9905177
31. R. Bousso, The holographic principle. Rev. Mod. Phys. **74**, 825–874 (2002). https://doi.org/10.1103/RevModPhys.74.825. arXiv: hep-th/0203101
32. R. Bousso, M. Miyaji, Fluctuations in the entropy of Hawking radiation. Phys. Rev. D **109**(2), 026006 (2024). https://doi.org/10.1103/PhysRevD.109.026006. arXiv: 2307.13920 [hep-th]
33. R. Bousso, M. Tomašević, Unitarity from a smooth horizon? Phys. Rev. D **102**(10), 106019 (2020). https://doi.org/10.1103/PhysRevD.102.106019. arXiv: 1911.06305 [hep-th]

34. R. Bousso, E. Wildenhain, Gravity/ensemble duality. Phys. Rev. D **102**(6), 066005 (2020). https://doi.org/10.1103/PhysRevD.102.066005. arXiv: 2006.16289 [hep-th]
35. R. Bousso, E. Wildenhain, Islands in closed and open universes. Phys. Rev. D **105**(8), 086012 (2022). https://doi.org/10.1103/PhysRevD.105.086012. arXiv: 2202.05278 [hep-th]
36. V. Chandrasekaran, M. Miyaji, P. Rath, Including contributions from entanglement islands to the reflected entropy. Phys. Rev. D **102**(8), 086009 (2020). https://doi.org/10.1103/PhysRevD.102.086009. arXiv: 2006.10754 [hep-th]
37. V. Chandrasekaran, G. Penington, E. Witten, Large N algebras and generalized entropy. JHEP **04**, 009 (2023). https://doi.org/10.1007/JHEP04(2023)009. arXiv: 2209.10454 [hep-th]
38. H.Z. Chen, et al., Information flow in black hole evaporation. JHEP **03**, 152 (2020). https://doi.org/10.1007/JHEP03(2020)152. arXiv: 1911.03402 [hep-th]
39. H.Z. Chen, et al., Quantum extremal islands made easy, Part I: Entanglement on the brane. JHEP **10**, 166 (2020). https://doi.org/10.1007/JHEP10(2020)166. arXiv: 2006.04851 [hep-th]
40. H.Z. Chen, et al., Quantum extremal islands made easy, Part II: Black holes on the brane. JHEP **12**, 025 (2020). https://doi.org/10.1007/JHEP12(2020)025. arXiv: 2010.00018 [hep-th]
41. H.Z. Chen, et al., Evaporating black holes coupled to a thermal bath. JHEP **01**, 065 (2021). https://doi.org/10.1007/JHEP01(2021)065. arXiv: 2007.11658 [hep-th]
42. Y. Chen, V. Gorbenko, J. Maldacena, Bra-ket wormholes in gravitationally prepared states. JHEP **02**, 009 (2021). https://doi.org/10.1007/JHEP02(2021)009. arXiv: 2007.16091 [hep-th]
43. Y. Chen, H.W. Lin, S.H. Shenker, BPS chaos (2024). arXiv: 2407.19387 [hep-th]
44. J. Chu, F. Deng, Y. Zhou, Page curve from defect extremal surface and island in higher dimensions. JHEP **10**, 149 (2021). https://doi.org/10.1007/JHEP10(2021)149. arXiv: 2105.09106 [hep-th]
45. S.R. Coleman, Black holes as red herrings: topological fluctuations and the loss of quantum coherence. Nucl. Phys. B **307**, 867–882 (1988). https://doi.org/10.1016/0550-3213(88)90110-1
46. S. Collier, A. Maloney, Wormholes and spectral statistics in the Narain ensemble. JHEP **03**, 004 (2022). https://doi.org/10.1007/JHEP03(2022)004. arXiv: 2106.12760 [hep-th]
47. J.S. Cotler, et al., Black holes and random matrices. JHEP **05**, 118 (2017). https://doi.org/10.1007/JHEP05(2017)118. arXiv: 1611.04650 [hep-th] [Erratum: JHEP **09**, 002 (2018)]
48. J. Cotler, et al., Entanglement wedge reconstruction via universal recovery channels. Phys. Rev. X **9**(3), 031011 (2019). https://doi.org/10.1103/PhysRevX.9.031011. arXiv: 1704.05839 [hep-th]
49. J. Cotler, K. Jensen, AdS_3 gravity and random CFT. JHEP **04**, 033 (2021). https://doi.org/10.1007/JHEP04(2021)033. arXiv: 2006.08648 [hep-th]
50. G. Di Ubaldo, E. Perlmutter, AdS_3/RMT_2 duality. JHEP **12**, 179 (2023). https://doi.org/10.1007/JHEP12(2023)179. arXiv: 2307.03707
51. D.N. Page, Average entropy of a subsystem. Phys. Rev. Lett. **71**, 1291–1294 (1993). https://doi.org/10.1103/PhysRevLett.71.1291. arXiv: gr-qc/9305007
52. X. Dong, Holographic entanglement entropy for general higher derivative gravity. JHEP **01**, 044 (2014). https://doi.org/10.1007/JHEP01(2014)044. arXiv: 1310.5713 [hep-th]
53. X. Dong, S. McBride, W.W. Weng, Replica wormholes and holographic entanglement negativity. JHEP **06**, 094 (2022). https://doi.org/10.1007/JHEP06(2022)094. arXiv: 2110.11947 [hep-th]
54. W. Donnelly, V. Shyam, Entanglement entropy and $T\overline{T}$ deformation. Phys. Rev. Lett. **121**(13), 131602 (2018). https://doi.org/10.1103/PhysRevLett.121.131602. arXiv: 1806.07444 [hep-th]
55. N. Engelhardt, A.C. Wall, Quantum extremal surfaces: holographic entanglement entropy beyond the classical regime. JHEP **01**, 073 (2015). https://doi.org/10.1007/JHEP01(2015)073. arXiv: 1408.3203 [hep-th]
56. J. Engelsöy, T.G. Mertens, H. Verlinde, An investigation of AdS_2 backreaction and holography. JHEP **07**, 139 (2016). https://doi.org/10.1007/JHEP07(2016)139. arXiv: 1606.03438 [hep-th]
57. T. Faulkner, A. Lewkowycz, J. Maldacena, Quantum corrections to holographic entanglement entropy. JHEP **11**, 074 (2013). https://doi.org/10.1007/JHEP11(2013)074. arXiv: 1307.2892 [hep-th]

58. M. Fujita, T. Takayanagi, E. Tonni, Aspects of AdS/BCFT. JHEP **11**, 043 (2011). https://doi.org/10.1007/JHEP11(2011)043. arXiv: 1108515 [hep-th]
59. P. Gao, D.L. Jafferis, A.C. Wall, Traversable wormholes via a double trace deformation. JHEP **12**, 151 (2017). https://doi.org/10.1007/JHEP12(2017)151. arXiv: 1608.05687 [hep-th]
60. H. Geng, A. Karch, Massive islands. JHEP **09**, 121 (2020). https://doi.org/10.1007/JHEP09(2020)121. arXiv: 2006.02438 [hep-th]
61. H. Geng, et al., Information transfer with a gravitating bath. SciPost Phys. **10**(5), 103 (2021). https://doi.org/10.21468/SciPostPhys.10.5.103. arXiv: 2012.04671 [hep-th]
62. H. Geng, Y. Nomura, H.-Y. Sun, Information paradox and its resolution in de Sitter holography. Phys. Rev. D **103**(12), 126004 (2021). https://doi.org/10.1103/PhysRevD.103.126004. arXiv: 2103.07477 [hep-th]
63. H. Geng, et al., Entanglement phase structure of a holographic BCFT in a black hole background. JHEP **05**, 153 (2022). https://doi.org/10.1007/JHEP05(2022)153. arXiv: 2112.09132 [hep-th]
64. E. Gesteau, H. Liu, Toward stringy horizons (2024). arXiv: 2408.12642 [hep-th]
65. S.B. Giddings, A. Strominger, Axion induced topology change in quantum gravity and string theory. Nucl. Phys. B **306**, 890–907 (1988). https://doi.org/10.1016/0550-3213(88)90446-4
66. S.B. Giddings, A. Strominger, Loss of incoherence and determination of coupling constants in quantum gravity. Nucl. Phys. B **307**, 854–866 (1988). https://doi.org/10.1016/0550-3213(88)90109-5
67. K. Goto, T. Hartman, A. Tajdini, Replica wormholes for an evaporating 2D black hole. JHEP **04**, 289 (2021). https://doi.org/10.1007/JHEP04(2021)289. arXiv: 2011.09043 [hep-th]
68. R. Gregory, R. Laflamme, Black strings and p-branes are unstable. Phys. Rev. Lett. **70**, 2837–2840 (1993). https://doi.org/10.1103/PhysRevLett.70.2837. arXiv: hep-th/9301052
69. D. Grumiller, W. Kummer, D.V. Vassilevich, Dilaton gravity in two-dimensions. Phys. Rept. **369**, 327–430 (2002). https://doi.org/10.1016/S0370-1573(02)00267-3. arXiv: hep-th/0204253
70. S.S. Gubser, I.R. Klebanov, A.M. Polyakov, Gauge theory correlators from noncritical string theory. Phys. Lett. B **428**, 105–114 (1998). https://doi.org/10.1016/S0370-2693(98)00377-3. arXiv: hep-th/9802109
71. D. Harlow, The Ryu–Takayanagi formula from quantum error correction. Commun. Math. Phys. **354**(3), 865–912 (2017). https://doi.org/10.1007/s00220-017-2904-z. arXiv: 1607.03901 [hep-th]
72. D. Harlow, D. Jafferis, The factorization problem in Jackiw-Teitelboim gravity. JHEP **02**, 177 (2020). https://doi.org/10.1007/JHEP02(2020)177. arXiv: 1804.01081 [hep-th]
73. J.B. Hartle, S.W. Hawking, Wave function of the universe. Phys. Rev. D **28**, 2960–2975 (1983). Ed. by Li-Zhi Fang, R. Ruffini. https://doi.org/10.1103/PhysRevD.28.2960
74. T. Hartman, Y. Jiang, E. Shaghoulian, Islands in cosmology. JHEP **11**, 111 (2020). https://doi.org/10.1007/JHEP11(2020)111. arXiv: 2008.01022 [hep-th]
75. T. Hartman, E. Shaghoulian, A. Strominger, Islands in asymptotically flat 2D gravity. JHEP **07**, 022 (2020). https://doi.org/10.1007/JHEP07(2020)022. arXiv: 2004.13857 [hep-th]
76. K. Hashimoto, N. Iizuka, Y. Matsuo, Islands in Schwarzschild black holes. JHEP **06**, 085 (2020). https://doi.org/10.1007/JHEP06(2020)085. arXiv: 2004.05863 [hep-th]
77. S.W. Hawking, Particle creation by black holes. Commun. Math. Phys. **43**, 199–220 (1975). Ed. by G.W. Gibbons, S.W. Hawking. https://doi.org/10.1007/BF02345020 [Erratum: Commun. Math. Phys. **46**, 206 (1976)]
78. S.W. Hawking, D.N. Page, Thermodynamics of black holes in anti-De Sitter space. Commun. Math. Phys. **87**, 577 (1983). https://doi.org/10.1007/BF01208266
79. P. Hayden, J. Preskill, Black holes as mirrors: quantum information in random subsystems. JHEP **09**, 120 (2007). https://doi.org/10.1088/1126-6708/2007/09/120. arXiv: 0708.4025 [hep-th]
80. J. Hernandez, R.C. Myers, S.-M. Ruan, Quantum extremal islands made easy. Part III. Complexity on the brane. JHEP **02**, 173 (2021). https://doi.org/10.1007/JHEP02(2021)173. arXiv: 2010.16398 [hep-th]

81. A. Higuchi, Forbidden mass range for spin-2 field theory in De Sitter space-time. Nucl. Phys. B **282**, 397–436 (1987). https://doi.org/10.1016/0550-3213(87)90691-2
82. G.T. Horowitz, Comments on black holes in string theory. Class. Quant. Grav. **17**, 1107–1116 (2000). Ed. by O. Lechtenfeld et al. https://doi.org/10.1088/0264-9381/17/5/320. arXiv: hep-th/9910082
83. L.V. Iliesiu, et al., On the non-perturbative bulk Hilbert space of JT gravity (2024). arXiv: 2403.08696 [hep-th]
84. R. Jackiw, Lower dimensional gravity. Nucl. Phys. B **252**, 343–356 (1985). Ed. by R. Baier, H. Satz. https://doi.org/10.1016/0550-3213(85)90448-1
85. D.L. Jafferis, et al., Relative entropy equals bulk relative entropy. JHEP **06**, 004 (2016). https://doi.org/10.1007/JHEP06(2016)004. arXiv: 1512.06431 [hep-th]
86. C.V. Johnson, Nonperturbative Jackiw-Teitelboim gravity. Phys. Rev. D **101**(10), 106023 (2020). https://doi.org/10.1103/PhysRevD.101.106023. arXiv: 1912.03637 [hep-th]
87. A. Karch, L. Randall, Locally localized gravity. JHEP **05**, 008 (2001). Ed. by M.J. Duff, J.T. Liu, J. Lu. https://doi.org/10.1088/1126-6708/2001/05/008. arXiv: hep-th/0011156
88. A. Karch, L. Randall, Open and closed string interpretation of SUSY CFT's on branes with boundaries. JHEP **06**, 063 (2001). https://doi.org/10.1088/1126-6708/2001/06/063. arXiv: hep-th/0105132
89. A. Karch, E. Katz, L. Randall, Absence of a VVDZ discontinuity in AdS (AdS). JHEP **12**, 016 (2001). https://doi.org/10.1088/1126-6708/2001/12/016. arXiv: hep-th/0106261
90. K. Kawabata, et al., Probing Hawking radiation through capacity of entanglement. JHEP **05**, 062 (2021). https://doi.org/10.1007/JHEP05(2021)062. arXiv: 2102.02425 [hep-th]
91. K. Kawabata, et al., Replica wormholes and capacity of entanglement. JHEP **10**, 227 (2021). https://doi.org/10.1007/JHEP10(2021)227. arXiv: 2105.08396 [hep-th]
92. A. Kitaev, S. Josephine Suh, The soft mode in the Sachdev-Ye-Kitaev model and its gravity dual. JHEP **05**, 183 (2018). https://doi.org/10.1007/JHEP05(2018)183. arXiv: 1711.08467
93. R.G. Leigh, Dirac-Born-Infeld action from Dirichlet sigma model. Mod. Phys. Lett. A **4**, 2767 (1989). https://doi.org/10.1142/S0217732389003099
94. S. Leutheusser, H. Liu, Causal connectability between quantum systems and the black hole interior in holographic duality. Phys. Rev. D **108**(8), 086019 (2023). https://doi.org/10.1103/PhysRevD.108.086019. arXiv: 2110.05497 [hep-th]
95. A. Lewkowycz, J. Maldacena, Generalized gravitational entropy. JHEP **08**, 090 (2013). https://doi.org/10.1007/JHEP08(2013)090. arXiv: 1304.4926 [hep-th]
96. T. Li, J. Chu, Y. Zhou, Reflected entropy for an evaporating black hole. JHEP **11**, 155 (2020). https://doi.org/10.1007/JHEP11(2020)155. arXiv: 2006.10846 [hep-th]
97. D. Louis-Martinez, J. Gegenberg, G. Kunstatter, Exact Dirac quantization of all 2-D dilaton gravity theories. Phys. Lett. B **321**, 193–198 (1994). https://doi.org/10.1016/0370-2693(94)90463-4. arXiv: gr-qc/9309018
98. J.M. Maldacena, The large N limit of superconformal field theories and supergravity. Adv. Theor. Math. Phys. **2**, 231–252 (1998). https://doi.org/10.4310/ATM1998.v2.n2.a1. arXiv: hep-th/9711200
99. J.M. Maldacena, Eternal black holes in anti-de Sitter. JHEP **04**, 021 (2003). https://doi.org/10.1088/1126-6708/2003/04/021. arXiv: hep-th/0106112
100. J.M. Maldacena, L. Maoz, Wormholes in AdS. JHEP **02**, 053 (2004). https://doi.org/10.1088/1126-6708/2004/02/053. arXiv: hep-th/0401024
101. J. Maldacena, X.-L. Qi, Eternal traversable wormhole (2018). arXiv: 1804.00491 [hep-th]
102. J. Maldacena, D. Stanford, Remarks on the Sachdev-Ye-Kitaev model. Phys. Rev. D **94**(10), 106002 (2016). https://doi.org/10.1103/PhysRevD.94.106002. arXiv: 1604.07818 [hep-th]
103. J. Maldacena, D. Stanford, Z. Yang, Conformal symmetry and its breaking in two dimensional nearly anti-de-Sitter space. PTEP **2016**(12), 12C104 (2016). https://doi.org/10.1093/ptep/ptw124. arXiv: 1606.01857 [hep-th]
104. A. Maloney, E. Witten, Averaging over Narain moduli space. JHEP **10**, 187 (2020). https://doi.org/10.1007/JHEP10(2020)187. arXiv: 2006.04855 [hep-th]

105. D. Marolf, H. Maxfield, Transcending the ensemble: baby universes, spacetime wormholes, and the order and disorder of black hole information. JHEP **08**, 044 (2020). https://doi.org/10.1007/JHEP08(2020)044. arXiv: 2002.08950 [hep-th]
106. D. Marolf, H. Maxfield, Observations of Hawking radiation: the Page curve and baby universes. JHEP **04**, 272 (2021). https://doi.org/10.1007/JHEP04(2021)272. arXiv: 2010.06602 [hep-th]
107. D. Marolf, J.E. Santos, AdS Euclidean wormholes. Class. Quant. Grav. **38**(22), 224002 (2021). https://doi.org/10.1088/1361-6382/ac2cb7. arXiv: 2101.08875 [hep-th]
108. M. Miyaji, Non-perturbative discrete spectrum of interior length and timeshift in two-sided black hole, JHEP **04**, 190 (2025). https://doi.org/10.1007/JHEP04(2025)190 [hep-th]
109. M. Miyaji, T. Takayanagi, Surface/state correspondence as a generalized holography. PTEP **2015**(7), 073B03 (2015). https://doi.org/10.1093/ptep/ptv089. arXiv: 1503.03542 [hep-th]
110. M. Miyaji, K. Yano, Fluctuation in the fidelity of information recovery from Hawking radiation. Phys. Rev. D **109**(8), 086011 (2024). https://doi.org/10.1103/PhysRevD.109.086011. arXiv: 2311.02881 [hep-th]
111. T. Nishioka, S. Ryu, T. Takayanagi, Holographic entanglement entropy: an overview. J. Phys. A **42**, 504008 (2009). https://doi.org/10.1088/1751-8113/42/50/504008. arXiv: 0905.0932 [hep-th]
112. Y. Nomura, Interior of a unitarily evaporating black hole. Phys. Rev. D **102**(2), 026001 (2020). https://doi.org/10.1103/PhysRevD.102.026001. arXiv: 1911.13120 [hep-th]
113. F. Pastawski, et al., Holographic quantum error-correcting codes: toy models for the bulk/boundary correspondence. JHEP **06**, 149 (2015). https://doi.org/10.1007/JHEP06(2015)149. arXiv: 1503.06237 [hep-th]
114. G. Penington, Entanglement wedge reconstruction and the information paradox. JHEP **09**, 002 (2020). https://doi.org/10.1007/JHEP09(2020)002. arXiv: 1905.08255 [hep-th]
115. G. Penington, E. Witten, Algebras and states in JT gravity (2023). arXiv: 2301.07257 [hep-th]
116. G. Penington, et al., Replica wormholes and the black hole interior. JHEP **03**, 205 (2022). https://doi.org/10.1007/JHEP03(2022)205. arXiv: 1911.11977 [hep-th]
117. J. Polchinski, Dirichlet Branes and Ramond-Ramond charges. Phys. Rev. Lett. **75**, 4724–4727 (1995). https://doi.org/10.1103/PhysRevLett.75.4724. arXiv: hep-th/9510017
118. J. Pollack, et al., Eigenstate thermalization and disorder averaging in gravity. Phys. Rev. Lett. **125**(2), 021601 (2020). https://doi.org/10.1103/PhysRevLett.125.021601. arXiv: 2002.02971 [hep-th]
119. M. Porrati, No van Dam-Veltman-Zakharov discontinuity in AdS space. Phys. Lett. B **498**, 92–96 (2001). https://doi.org/10.1016/S0370-2693(00)01380-0. arXiv: hep-th/0011152
120. M. Rozali, et al., Information radiation in BCFT models of black holes. JHEP **05**, 004 (2020). https://doi.org/10.1007/JHEP05(2020)004. arXiv: 1910.12836 [hep-th]
121. S. Ryu, T. Takayanagi, Aspects of holographic entanglement entropy. JHEP **08**, 045 (2006). https://doi.org/10.1088/1126-6708/2006/08/045. arXiv: hep-th/0605073
122. S. Ryu, T. Takayanagi, Holographic derivation of entanglement entropy from AdS/CFT. Phys. Rev. Lett. **96**, 181602 (2006). https://doi.org/10.1103/PhysRevLett.96.181602. arXiv: hep-th/0603001
123. P. Saad, Late time correlation functions, baby universes, and ETH in JT gravity (2019). arXiv: 1910.10311 [hep-th]
124. P. Saad, S.H. Shenker, D. Stanford, JT gravity as a matrix integral (2019). arXiv: 1903.11115 [hep-th]
125. P. Saad, et al., Wormholes without averaging. JHEP **09**, 133 (2024). https://doi.org/10.1007/JHEP09(2024)133. arXiv: 2103.16754 [hep-th]
126. J.-M. Schlenker, E. Witten, No ensemble averaging below the black hole threshold. JHEP **07**, 143 (2022). https://doi.org/10.1007/JHEP07(2022)143. arXiv: 2202.01372 [hep-th]
127. D. Stanford, More quantum noise from wormholes (2020). arXiv: 2008.08570 [hep-th]
128. D. Stanford, Z. Yang, Firewalls from wormholes (2022). arXiv: 2208.01625 [hep-th]
129. A. Strominger, The dS/CFT correspondence. JHEP **10**, 034 (2001). https://doi.org/10.1088/1126-6708/2001/10/034. arXiv: hep-th/0106113

130. A. Strominger, Lectures on the infrared structure of gravity and gauge theory (2017). arXiv: 1703.05448 [hep-th]. ISBN: 978-0-691-17973-5
131. J. Sully, M. Van Raamsdonk, D. Wakeham, BCFT entanglement entropy at large central charge and the black hole interior. JHEP **03**, 167 (2021). https://doi.org/10.1007/JHEP03(2021)167. arXiv: 2004.13088 [hep-th]
132. L. Susskind, The world as a hologram. J. Math. Phys. **36**, 6377–6396 (1995). https://doi.org/10.1063/1.531249. arXiv: hep-th/9409089
133. L. Susskind, E. Witten, The holographic bound in anti-de Sitter space (1998). arXiv: hep-th/9805114
134. K. Suzuki, T. Takayanagi, BCFT and Islands in two dimensions. JHEP **06**, 095 (2022). https://doi.org/10.1007/JHEP06(2022)095. arXiv: 2202.08462 [hep-th]
135. G. 't Hooft, A planar diagram theory for strong interactions. Nucl. Phys. B **72**, 461 (1974). Ed. by J.C. Taylor. https://doi.org/10.1016/0550-3213(74)90154-0
136. G. 't Hooft, Dimensional reduction in quantum gravity. Conf. Proc. C **930308**, 284–296 (1993). arXiv: gr-qc/9310026
137. T. Takayanagi, Holographic dual of BCFT. Phys. Rev. Lett. **107**, 101602 (2011). https://doi.org/10.1103/PhysRevLett.107.101602. arXiv: 1105.5165 [hep-th]
138. C. Teitelboim, Gravitation and Hamiltonian structure in two space-time dimensions. Phys. Lett. B **126**, 41–45 (1983). https://doi.org/10.1016/0370-2693(83)90012-6
139. H. van Dam, M.J.G. Veltman, Massive and massless Yang-Mills and gravitational fields. Nucl. Phys. B **22**, 397–411 (1970). https://doi.org/10.1016/0550-3213(70)90416-5
140. M. Van Raamsdonk, Comments on wormholes, ensembles, and cosmology. JHEP **12**, 156 (2021). https://doi.org/10.1007/JHEP12(2021)156. arXiv: 2008.02259 [hep-th]
141. E. Witten, Anti-de Sitter space and holography. Adv. Theor. Math. Phys. **2**, 253–291 (1998). https://doi.org/10.4310/ATM1998.v2.n2.a2. arXiv: hep-th/9802150
142. E. Witten, Anti-de Sitter space, thermal phase transition, and confinement in gauge theories. Adv. Theor. Math. Phys. **2**, 505–532 (1998). Ed. by L. Bergstrom, U. Lindstrom. https://doi.org/10.4310/ATM1998.v2.n3.a3. arXiv: hep-th/9803131
143. E. Witten, Matrix models and deformations of JT gravity. Proc. Roy. Soc. Lond. A **476**(2244), 20200582 (2020). https://doi.org/10.1098/rspa.2020.0582. arXiv: 2006.13414 [hep-th]
144. Z. Yang, The quantum gravity dynamics of near extremal black holes. JHEP **05**, 205 (2019). https://doi.org/10.1007/JHEP05(2019)205. arXiv: 1809.08647 [hep-th]
145. V.I. Zakharov, Linearized gravitation theory and the graviton mass. JETP Lett. **12**, 312 (1970)

Chapter 11
The Fuzzball Paradigm

Samir D. Mathur and Madhur Mehta

Abstract We describe the puzzles that arise in the quantum theory of black holes, and explain how they are resolved in string theory. We review how the Bekenstein entropy is obtained through the count of brane bound states. We describe the fuzzball construction of black hole microstates. These states have no horizon and radiate from their surface like a normal body, so there is no information puzzle. We explain how the semiclassical approximation is violated in gravitational collapse even though curvatures are low at the classical horizon. This violation happens because the collapse leads to a stretching of space that is *fast*: light does not have time to travel across the collapsing region to establish the 'vecro' correlations needed in the quantum gravitational vacuum. These vecro correlations arise from the existence of virtual fuzzball fluctuations in the gravitational vacuum, and are significant because of the large degeneracy of fuzzball states implied by the Bekenstein entropy. We conjecture that similar effects of fast expansion may be responsible for effects like dark energy and the Early Dark Energy postulated to explain the Hubble tension.

11.1 The Puzzles

The quantum theory of black holes presents us with three interrelated puzzles:

(A) *The entropy puzzle*: Gedanken experiments suggest that black holes have an entropy [1]

$$S_{bek} = \frac{c^3}{\hbar}\frac{A}{4G}. \tag{11.1}$$

But it has been argued that 'black holes have no hair', so what is this entropy counting? At the other extreme, the bags-of-gold construction by Wheeler suggests that we can put an infinite amount of entropy inside the horizon.

S. D. Mathur (✉) · M. Mehta
Department of Physics, The Ohio State University, Columbus, OH, USA
e-mail: mathur.16@osu.edu

M. Mehta
e-mail: mehta.493@osu.edu

A. Akil and C. Bambi (eds.), *The Black Hole Information Paradox*, Springer Series in Astrophysics and Cosmology, https://doi.org/10.1007/978-981-96-6170-1_11

Thus, the puzzle is: *Does the entropy* (11.1) *correspond to a count of black hole microstates through* $S_{bek} = \log \mathcal{N}_{micro}$, *in the way it would for a normal system described by statistical mechanics?*

(B) *The information paradox*: In classical general relativity, matter collapses to the center of the black hole, leaving a vacuum around the horizon. Hawking [2, 3] found that in quantum theory, this vacuum region is unstable to the creation of entangled pairs. One member of the pair (which we will call b) escapes to infinity as 'Hawking radiation'. The other member (which we will call c) has negative energy and falls into the hole, lowering its mass. The two members of the pair are in an entangled state, which we can schematically write as

$$|\psi\rangle_{pair} = \frac{1}{\sqrt{2}}\left(|0\rangle_b|0\rangle_c + |1\rangle_b|1\rangle_c\right). \tag{11.2}$$

Thus, there is a monotonically growing entanglement between the emitted radiation and the remaining hole, leading to a sharp puzzle at the endpoint of evaporation. If the hole evaporates away, then we are left with radiation that is entangled, but there is nothing that this radiation is entangled *with*. Then this radiation cannot be described by *any* pure state in quantum theory, but only by a density matrix; thus we have a loss of quantum unitarity in the process of black hole formation and evaporation. This difficulty is known as the black hole information paradox.

To evade this difficulty, one may postulate that quantum gravity effects prevent the further evaporation of the hole once the hole evaporates down to Planck size. The initial matter making the hole, and all the negative energy quanta $\{c_i\}$ are then trapped in a tiny 'remnant'. Since we could have started with a black hole of arbitrarily large size and evaporated down to Planck scale, the Planck mass remnant should be able to hold an arbitrarily large number of internal states. One model that has been proposed for remnants is in terms of a 'baby universe' which is attached to the rest of spacetime through a Planck sized neck.

But remnants do not appear to be possible in string theory. For one, we believe we know all the states at the Planck scale; we just have a few strings and branes. For another, the conjecture of AdS/CFT duality rules out remnants. Suppose we have an object with energy E_{grav} sitting at the center of an AdS_5 space with curvature radius R_{AdS}. The dual theory is an $SU(N)$ gauge theory, with a finite value of N: we have $N \sim (R_{AdS}/l_s)^4 \sim (R_{AdS}/l_p)^4$ (we have assumed that the string coupling is a fixed number not scaling with N; thus in these estimates we are using $l_s \sim l_p$). The gauge theory lives on a sphere S^3 whose radius can be taken to be any length R_{CFT}. Then the remnant in the gravity theory with $E_{grav} \sim m_p$ is described in this gauge theory by a state of energy

$$E_{CFT} \sim \frac{R_{AdS}E_{grav}}{R_{CFT}} \sim \frac{N^{\frac{1}{4}}}{R_{CFT}}. \tag{11.3}$$

The gauge theory has finite N and lives on a finite volume sphere with radius R_{CFT}. In the finite energy range $E \lesssim E_{CFT}$, this gauge theory has only a finite number of allowed states. Thus, AdS/CFT duality does not allow the gravity theory to have an infinite number of states for a remnant with $m \lesssim m_p$.

Given this, how do we resolve the problem of monotonically rising entanglement in string theory? Some people had initially hoped that the puzzle might be resolved through 'small corrections'. Suppose there is a small correction of order ε to the state of each emitted pair; such small corrections to Hawking's computation can always arise from hitherto unknown quantum gravity effects. The number of emitted quanta is large: $N_{quanta} \sim (M/m_p)^2$, where M is the black hole mass. Thus, if $\varepsilon N_{quanta} \gtrsim 1$, one might hope that subtle correlations among this large number of emitted quanta might remove the monotonically growing entanglement that Hawking observed in his leading order calculations.

But the 'Small corrections theorem' [4] shows that such a resolution is not possible. Suppose we assume

(i) There is small correction to the state of each pair $\{b_i, c_i\}$

$$|\psi\rangle_{pair} = \frac{1}{\sqrt{2}}\left(|0\rangle_{b_i}|0\rangle_{c_i} + |1\rangle_{b_i}|1\rangle_{c_i}\right) + |\delta\psi_i\rangle, \qquad \Big||\delta\psi_i\rangle\Big| < \varepsilon, \qquad \varepsilon \ll 1. \tag{11.4}$$

(ii) There is no significant change to the state of the quanta $\{b_i\}$ after they recede sufficiently far from the hole.

Then the entanglement $S_{ent}(N)$ of the radiation with the hole at step N must keep growing monotonically as

$$S_{ent}(N+1) > S_{ent}(N) + \log 2 - 2\varepsilon. \tag{11.5}$$

In other words, small corrections to the low energy dynamics at the horizon cannot resolve the difficulty; we need order *unity* corrections.

Thus, our puzzle is: *How should we resolve the puzzle of monotonically rising entanglement?*

(C) *Breakdown of the semiclassical approximation:* Consider a shell of mass M that is collapsing to form a black hole. Curvatures are low at the horizon scale

$$\mathcal{R} \sim \frac{1}{(GM)^2} \ll \frac{1}{l_p^2}, \tag{11.6}$$

so one expects that usual low energy semiclassical dynamics will hold, and the shell will pass smoothly through the horizon radius. If the shell goes through the horizon, how does its information ever come out? Points on the shell that are inside the horizon cannot send light signals to the horizon or exterior of the hole. Thus even though new physics can occur at the singularity, how can this new physics modify the pair creation at the horizon and resolve Hawking's paradox?

Thus, the puzzle is: *What can lead to a breakdown of semiclassical dynamics at the horizon where curvatures are low?*

In this article, we will describe how string theory resolves these puzzles. The resolution of puzzles (B) and (C) is obtained through the *fuzzball paradigm*, which describes the structure and dynamics of black hole states in the theory. Before proceeding, we summarize the resolutions obtained in this paradigm.

In string theory, the Bekenstein entropy does correspond to a count of microstates in the theory. Unlike the classical hole, which has all its mass at $r = 0$, we find that these microstates are horizon-size quantum objects called fuzzballs. The fuzzballs have no horizon and radiate from their surface like a normal body; thus there is no analogue of the monotonically rising entanglement (11.5).

But why does gravitational collapse lead to fuzzballs rather than the semiclassical hole? It is natural to expect that semiclassical dynamics will break down when curvatures exceed the Planck scale: $\mathscr{R} \gtrsim l_p^{-2}$. But as we noted above, this will not help us to resolve the information paradox; we need order unity corrections to dynamics at the *horizon*. We will find that there is indeed a second mode of breakdown of the semiclassical approximation, due to an effect that can be traced back to the large value of the Bekenstein entropy (11.1).

Fuzzballs are bound states of the elementary excitations in string theory. The existence of a bound state in a field theory leaves an imprint on the correlations in the *vacuum*. This effect is usually small in a normal quantum field theory, which has a limited set of bound states. But the correlations created by fuzzballs are particularly significant, because (i) fuzzballs have an extended structure (with size of order horizon radius), and (ii) fuzzballs have a very high degeneracy (given by the Bekenstein entropy) which increases with the size of the fuzzball. These correlations created by virtual excitations of fuzzballs are termed 'vecro correlations', for reasons we will see later.

The existence of these vecro correlations in the gravitational vacuum leads to a breakdown of the semiclassical approximation in the process of gravitational collapse. This process of collapse is 'fast', in the sense that light does not have time to travel across the collapsing region to establish the vecro correlations appropriate to the vacuum. Thus, we do not reach the semiclassical vacuum solution inside the horizon region; instead we transition to a linear superposition of fuzzball states.

11.2 Black Holes in String Theory

The first difficulty we encounter in quantizing gravity is that loop amplitudes diverge, and cannot be renormalized. In string theory, on the other hand, we find that loop amplitudes are finite. This taming of divergences can be traced back to the fact that string theory has an effectively infinite number of particle species: as we go to higher energies, we find new 'particles' arising as new vibration modes of the string.

The second issue we must deal with in any theory of quantum gravity is the diverging value for the vacuum energy of quantum fields. In string theory, we assume supersymmetry of the underlying theory, so that at we have flat Minkowski space as a solution of the theory.

The most remarkable feature of string theory is that it is *unique*; i.e., it has no free parameters. The theory must live in 9 + 1 dimensions, and the fundamental objects—strings and branes—are uniquely determined along with their tensions and interactions. Interestingly, this unique and elaborate structure can be obtained, in retrospect, in the following way. The classical supergravity theory has certain discrete symmetries called T and S dualities. If we require that these symmetries continue to hold at the quantum level, we obtain string theory.

11.2.1 Black Hole Entropy

To make black holes in string theory, we must consider a bound state of the elementary objects in the theory. It is reasonable to think that the microstates of black holes involve strong quantum gravity effects, and that it would be hard to construct and count these microstates. But it turns out that in string theory, at least for extremal holes (holes with mass M equaling the charge Q), there is a way to count these states without understanding their explicit structure. The coupling in the theory is given by the value of a scalar field (the dilaton). Thus, we can study the bound states at weak coupling or at strong coupling. Extremal holes are supersymmetric states of the theory, and an index which counts the states must remain the same as we change the coupling. It is relatively straightforward to count the states at weak coupling. This index argument then yields the count of states at strong coupling, where the bound state is expected to yield a black hole.

Let us now turn to the bound states, which will give us our black hole. We compactify some spatial directions; it is convenient to study the theory in 4 + 1 noncompact directions to start with, and then extend the constructions to the case of 3 + 1 noncompact directions.

Suppose we wrap a string around a compact circle. From the viewpoint of the dimensionally reduced theory, this string would appear as a point mass in the noncompact directions. The string also carries a charge, with strings winding one way around the circle having an opposite charge to strings winding the other way. Thus, in the dimensionally reduced theory, we obtain an object with mass m_1 and charge q_1. Here $m_1 = 2\pi RT$, where T is the tension of the string and R is the radius of the compact circle. The subscript 1 corresponds to the string being a 1-dimensional object, or '1-brane'; more specifically, this string is called an NS1 brane. The string winding around the circle has $M = Q$ (mass equaling charge in appropriate units); this implies that the wound string gives a supersymmetric state of the theory.

To make a black hole with a large mass, we must take a bound state of a large number n_1 of strings. Such a bound state of NS1 branes has a simple description: we just have a string that winds n_1 times around the circle before closing on itself. Thus, we obtain an object with mass $n_1 m_1$ and charge $n_1 q_1$.

It turns out however that such a mass source does not create a good black hole. The tension of the string causes the size of the compact circle to shrink to zero at the location where the string is wound. In the dimensionally reduced theory, we find that the horizon area is zero: the horizon coincides with the singularity. Thus, the Bekenstein entropy is $S_{bek} = \frac{A}{4G} = 0$. This is at least consistent with the microscopics: the multi-wound string has just one state, so the microscopic entropy is $S_{micro} = \log 1 = 0$.[1]

To get a black hole we must add excitations which tend to *increase* the size of this compact circle. Consider a graviton moving around this circle, at a fixed location in the noncompact directions. This is a mode carrying momentum, so we call it a P excitation. In the dimensionally reduced theory, such a Kaluza-Klein mode carries a mass $m_p = \frac{1}{R}$ and a charge $q_p = \pm\frac{1}{R}$, where R is the radius of the compact circle and the two possible signs of q_p correspond to the two directions in which the graviton can orbit the circle. The bound state of n_p such gravitons is a single graviton carrying n_p units of momentum, and in the dimensionally reduced theory corresponds to an extremal object with $M = Q = \frac{n_p}{R}$. We see that M decreases as R is increased, so the momentum modes tend to expand the compact circle.

Thus, we try to make a bound state carrying both winding and momentum charges. The corresponding hole is called the 2-charge extremal hole, because it carries two types of charges—one from winding strings and one from momentum modes. Suppose we assume that at strong coupling, the bound state of these charges forms a spherically symmetric solution of the supergravity equations, carrying the above mass and charges. We seek to measure the horizon area of this solution and thus get S_{bek} for the hole.

It turns out that if we use the leading order supergravity action, we still get a vanishing horizon area. But the effective action arising from string theory has higher derivative terms of the form $\alpha'\mathscr{R}^2$. Taking these into account, we find that there is indeed a small horizon (of radius $\alpha'^{\frac{1}{2}}$), and we can compute the Bekenstein-Wald entropy [5] of this horizon. Consider type IIB string theory with the compactification $\mathscr{M}_{9,1} \to \mathscr{M}_{3,1} \times K3 \times T^2$ where $K3$ is a Calabi-Yau 4-manifold. One finds that [6]

$$S_{bek-wald} = 4\pi\sqrt{n_1 n_p}. \tag{11.7}$$

We now wish to see if this gravity result can be reproduced from the microscopic description of the bound state. At weak coupling, it is easy to understand our bound state. We have already seen that the bound state of n_1 strings is a single string with winding n_1; this string has a total length $L_T = 2\pi R n_1$. The momentum-carrying modes bound to this string take the form of traveling waves on the string. The waves on a string of length L_T carry momentum in units of $2\pi/L_T$. Thus, we write the total momentum on the string as

$$P = \frac{n_p}{R} = \frac{2\pi n_1 n_p}{L_T}. \tag{11.8}$$

[1] Actually the spin of the string makes it a multiplet of 256 states, but we will regard $S_{micro} = \log 256 \approx 0$ since this entropy does not scale with the number n_1 of strings in the bound state.

A quantum in the k th harmonic carries a momentum $p = \frac{2\pi k}{L_T}$. If we have N_k units of the excitation in harmonic k, then we will get

$$\sum_{k=1}^{\infty} k N_k = n_1 n_p. \tag{11.9}$$

Note that each different direction of vibration counts as a different 'flavor' of excitation. The different ways of partitioning the total momentum P among these excitations gives the microscopic number of states $\mathcal{N}_{micro}$, and we obtain the microscopic entropy as $S_{micro} = \log \mathcal{N}_{micro}$. We find, for the compactification mentioned above [6–8],

$$S_{micro} = 4\pi \sqrt{n_1 n_p} = S_{bek-wald}. \tag{11.10}$$

The 2-charge extremal hole is also called the 'small black hole' because the horizon radius is $\sim \alpha'^{\frac{1}{2}}$, a microscopic length in the theory. To get a larger horizon, we can add a third kind of charge. We take the compactification $\mathcal{M}_{9,1} \to \mathcal{M}_{4,1} \times \mathcal{M}_4 \times S^1$, where $\mathcal{M}_4$ is a 4-manifold, which can be K3 or T^4. We wrap strings and momentum modes around the S^1 the same way we did for the 2-charge hole, but now also add n_5 NS5-branes wrapping $\mathcal{M}_4 \times S^1$. The corresponding classical hole is a traditional Reissner-Nordstrom hole, whose Bekenstein entropy can be found by using the leading order action $\frac{1}{16\pi G} \int dx \sqrt{-g} \mathcal{R}$; the higher order Wald corrections are subleading. We find

$$S_{bek} = \frac{A}{4G} = 2\pi \sqrt{n_1 n_5 n_p}. \tag{11.11}$$

The microscopic entropy can be again computed at weak coupling, and one finds [9]

$$S_{micro} = 2\pi \sqrt{n_1 n_5 n_p} = S_{bek}. \tag{11.12}$$

Extremal holes do not radiate. To understand radiation from the hole and the resolution of the information paradox, we should look at non-extremal holes. We can make a non-extremal microstate by taking momentum modes P carrying momentum n_p in one direction and also momentum modes $\bar{P}$ carrying momentum $\bar{n}_p$ in the opposite direction. The microstate will no longer be supersymmetric, so we cannot rigorously argue that the count of states at weak coupling must agree with the count at strong coupling. We can avoid violating supersymmetry 'too much' by taking $n_p, \bar{n}_p \ll n_1, n_5$. Then we find a microscopic entropy at weak coupling

$$S_{micro} = 2\pi \sqrt{n_1 n_5} \left(\sqrt{n_p} + \sqrt{\bar{n}_p} \right). \tag{11.13}$$

We compare this microscopic entropy to $S_{bek} = \frac{A}{4G}$, where A is computed using the horizon area of a black hole carrying the mass and charges of the microstates; note that the momentum charge carried by the hole is $n_p - \bar{n}_p$. We again find agreement: $S_{micro} = S_{bek}$ [10].

Remarkably, a natural extension of the weak coupling expression (11.13) continues to give an exact agreement with the Bekenstein entropy even for holes which are arbitrarily far from extremality [11]. There is no clear reason known why this agreement should hold. Nevertheless, in view of all these agreements, it is reasonable to say that in string theory, the Bekenstein-Wald entropy indeed represents a count of microstates of the hole, in exactly the way we understand entropy in statistical physics as a count of microstates.

11.2.2 Radiation from Microstates

Extremal holes cannot radiate, but near-extremal holes radiate, and we can use them to approach the Hawking puzzle. At strong coupling where we expect black holes, we can perform a computation similar to the one that Hawking did, and obtain a radiation rate $\Gamma_H(\omega)$ which depends on the energy ω of the radiated quantum. On the microscopic side, we can look at the weak coupling description which gave the entropy (11.13). In this description we have massless momentum carrying modes P and $\bar{P}$ running in opposite directions along the S^1 on which the branes are wrapped. These modes can collide, cancelling their momentum charge, and leading to the emission of a graviton from the string bound state. The interaction vertex $\hat{V}$ for this process thus gives an operator of the form

$$V(\omega)\hat{a}_P(p)\hat{a}_{\bar{P}}(p)\hat{a}_h^\dagger(\omega), \tag{11.14}$$

where $\hat{a}_P(p)$ annihilates a P excitation with momentum p and energy p, $\hat{a}_{\bar{P}}(p)$ annihilates a $\bar{P}$ mode carrying momentum $-p$ and energy p, and $\hat{a}_h^\dagger(\omega)$ creates a graviton with no momentum along the S^1 and energy $\omega = p + p = 2p$. The radiation rate Γ is then computed by multiplying $V(\omega)$ by the occupation numbers N_p and $\bar{N}_p$ of the relevant P and $\bar{P}$ modes. Schematically:

$$\Gamma_{micro}(\omega) = V(\omega)N_p\left(\frac{\omega}{2}\right)\bar{N}_p\left(\frac{\omega}{2}\right). \tag{11.15}$$

The entropy (11.13) reproduced the Bekenstein entropy when we summed over all the allowed configurations of the P, $\bar{P}$ modes. By the usual arguments of statistical mechanics, we see that the entropy (11.13) will be reproduced by using thermal distributions for N_p, $\bar{N}_p$, with temperatures corresponding to the total energies in the P, $\bar{P}$ sectors. Substituting these thermal distributions for N_p, $\bar{N}_p$ in the radiation rate (11.15), we find [12, 13]

$$\Gamma_{micro}(\omega) = \Gamma_H(\omega), \tag{11.16}$$

where $\Gamma_H(\omega)$ is the radiation rate obtained from the semiclassical Hawking computation. This agreement (11.16) suggests that string theory is on the correct track to understand black hole radiation. But note that the microscopic emission (11.15) arises from a unitary process, with excitations on the brane colliding and emerging as gravitons. By contrast, Γ_H is obtained from the usual process of pair creation from the vacuum, which leads to the information paradox. This suggests there is something wrong with the strong coupling picture of the black hole, and we will now see that such is indeed the case.

11.3 Constructing the Microstates—Fuzzballs

In the above discussion we counted black hole microstates by an indirect argument—using supersymmetry to relate the state count at weak coupling to the count at strong coupling. At weak coupling we could count states, but the states did not make black holes. At strong coupling we just used the standard metric of the hole and computed the value of Bekenstein's area entropy. But this black hole metric has a horizon, and horizons lead to the information paradox. To understand how the paradox is to be resolved, we must construct the microstates at *strong* coupling, and examine their structure. This is what we will proceed to do now. We will start with the simplest black hole: the 2-charge extremal hole with entropy (11.10). The expressions for entropy were similar for all holes, so we expect that any lessons that we extract from the simple 2-charge extremal case would hold, at least qualitatively, for all holes.

Recall that the 2-charge extremal hole was made by winding a string around a compact circle, and adding momentum along this string. All the states of the hole are given by the different traveling waves on this string. Thus, we need to construct the gravity solution created by strings carrying traveling waves.

With a view to study the 3-charge hole later, we compactify spacetime as $\mathcal{M}_{4,1} \times S^1 \times T^4$. The S^1 is parametrized by $0 \le y < 2\pi R$. The T^4 has volume $(2\pi)^4 V$ and is parametrized by $z_a, a = 1, \ldots 4$.

First, consider a static string source, with the string wound along the S^1. We assume that this string source is made up of n_1 fundamental strings, all at the same position. The classical solution for such a string source is known,

$$ds^2_{string} = H(-dudv) + \sum_{i=1}^{4} dx_i dx_i + \sum_{a=1}^{4} dz_a dz_a,$$

$$B_{uv} = -\frac{1}{2}(H-1), \quad e^{2\phi} = H, \quad H^{-1} = 1 + \frac{Q_1}{r^2}, \quad Q_1 = \frac{g^2\alpha'^3}{V} n_1. \quad (11.17)$$

Let us now add a traveling wave to this string. There are two important aspects to note. First, the fundamental string of string theory has no longitudinal vibration modes; it can have only transverse vibrations. Thus, when we add momentum charge P in the form of a traveling wave, the string necessarily bends away from its central position, spreading over some transverse spatial region. In other words, the string does not remain a point source in these transverse directions. Second, the string is multiwound, with winding n_1. Thus, the transverse displacement of the string has to return to its initial value, not after $\Delta y = 2\pi R$ but after $\Delta y = 2\pi R\, n_1$. Under such a wave, the strands separate from each other in general, and we find n_1 string sources, each carrying a wave traveling in the y direction at the speed of light. The metric for such a set of string sources is known:

$$
\begin{aligned}
ds^2_{string} &= H[-du\,dv + K dv^2 + 2\,A_i\,dx_i\,dv] + \sum_{i=1}^{4} dx_i\,dx_i + \sum_{a=1}^{4} dz_a\,dz_a,\\
B_{uv} &= -\frac{1}{2}[H-1], \quad B_{vi} = HA_i, \quad e^{2\phi} = H,\\
H^{-1}(\mathbf{x}, y, t) &= 1 + \sum_s \frac{Q_1^{(s)}}{|\mathbf{x} - \mathbf{F}^{(s)}(t-y)|^2}, \quad K(\mathbf{x}, y, t) = \sum_s \frac{Q_1^{(s)} |\dot{\mathbf{F}}^{(s)}(t-y)|^2}{|\mathbf{x} - \mathbf{F}^{(s)}(t-y)|^2},\\
A_i(\mathbf{x}, y, t) &= -\sum_s \frac{Q_1^{(s)} \dot{F}_i^{(s)}(t-y)}{|\mathbf{x} - \mathbf{F}^{(s)}(t-y)|^2}.
\end{aligned}
\tag{11.18}
$$

In studying black holes, we are interested in the limit $n_1, n_p \gg 1$. In this limit we may replace the sum in (11.18) by an integral

$$
\sum_{s=1}^{n_1} \to \int_{s=0}^{n_1} ds = \int_{y=0}^{2\pi R n_1} \frac{ds}{dy} dy
\tag{11.19}
$$

Since the length of the compact S^1 is $2\pi R$ we have $\frac{ds}{dy} = \frac{1}{2\pi R}$. Also, since the vibration profile is a function of $v = t - y$ we can replace the integral over y by an integral over v. Thus, we have

$$
\sum_{s=1}^{n_1} \to \frac{1}{2\pi R} \int_{v=0}^{L_T} dv,
\tag{11.20}
$$

where $L_T = 2\pi R n_1$ is the total range of the y coordinate on the multiwound string. Finally, note that $Q_1^{(s)} = \frac{Q_1}{n_1}$. With all this, the solution (11.18) becomes

$$
\begin{aligned}
ds^2_{string} &= H[-du\,dv + K\,dv^2 + 2\,A_i\,dx_i\,dv] + \sum_{i=1}^{4} dx_i\,dx_i + \sum_{a=1}^{4} dz_a\,dz_a,\\
B_{uv} &= -\frac{1}{2}[H-1], \quad B_{vi} = HA_i, \quad e^{2\phi} = H,\\
H^{-1} &= 1 + \frac{Q_1}{L_T} \int_0^{L_T} \frac{dv}{|\mathbf{x} - \mathbf{F}(v)|^2},\\
K &= \frac{Q_1}{L_T} \int_0^{L_T} \frac{dv (\dot{F}(v))^2}{|\mathbf{x} - \mathbf{F}(v)|^2},
\end{aligned}
$$

$$A_i = -\frac{Q_1}{L_T}\int_0^{L_T}\frac{dv\dot{F}_i(v)}{|\mathbf{x}-\mathbf{F}(v)|^2}. \tag{11.21}$$

In Fig. 11.1 we give a graphic description of the steps above. If we do not take into account the transverse displacements of the string, we get the 'naive' geometry describing a black hole with horizon (after taking account $\alpha'\mathscr{R}^2$ corrections). Taking into account the actual transverse displacements gives the fuzzball solutions, which have a different geometry for each allowed distribution of momentum on the string.

The solutions (11.21) give the microstates of the 2-charge extremal hole made with string winding charge (usually called NS1) and momentum charge (usually called P) [14]. These are all the microstates where the string is allowed to oscillate in the 4 noncompact directions $\mathbf{x}$. The solutions with vibrations in the compact directions z_a are similar, and were given in [15], and the set of solutions for the heterotic string were given in [16]. Special cases of these metrics were noted earlier. Solutions with $\mathbf{F}(v)$ lying along a circle were found in [17–20]. Solutions with excitations lying in a plane (supertubes) were found in [21].

Let us now analyze the nature of these solutions, and the lessons that we can draw from them about the structure of black holes.

11.4 Lessons from the Fuzzball Construction

Analysis of fuzzball construction above suggests the following nature for black hole microstates in string theory:

(A) **Absence of a horizon**: A little analysis shows that none of the microstates (11.21) have a horizon. Along the curve $\mathbf{x} = \mathbf{F}(v)$, the redshift goes to infinity, but there is no closed trapped surface, and light rays can escape to infinity from any point in these geometries.

(B) **Lack of spherical symmetry**: Note that the solutions (11.21) are not spherically symmetric. If we *impose* spherical symmetry on the low energy gravity solution, then we get the 'naive' metric pictured in Fig. 11.1. This naive metric is not realized in string theory, because it is not possible to add a momentum carrying wave to the string without breaking spherical symmetry. This in turn follows from the fact that the fundamental string has no longitudinal vibration mode; it only allows transverse vibrations, and the polarization of this vibration at any point along the string must break the spherical symmetry in the angular S^3 directions or the symmetry in the T^4.

(C) **The relation between entropy and area**: Given that there is no horizon, one might wonder what happens to the Bekenstein entropy relation (11.1). From Fig. 11.1 we see that the microstates have the structure of flat space at infinity, then a 'neck' which leads to a 'throat', and this throat ends in a 'cap'. The geometries are almost identical everywhere except in the cap; thus this cap contains the detailed information of the choice of microstate. Let us draw a surface that surrounds the region where the typical microstates differ significantly from each other; this surface

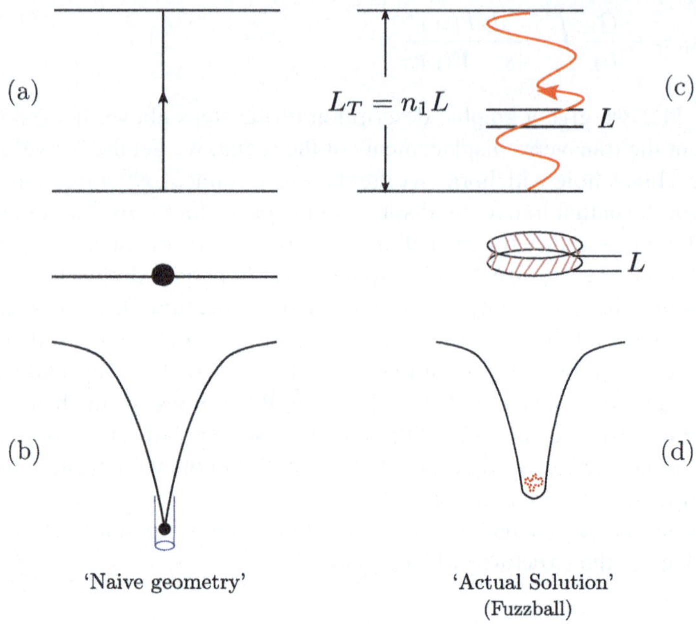

Fig. 11.1 **a** If a string could carry momentum as a longitudinal wave, then it would generate a spherically symmetric geometry. **b** This 'naive' geometry would have a singular horizon at leading order, but α' corrections turn this singularity into a regular horizon. **c** The actual string of string theory has only transverse vibrations, which break spherical symmetry and cause the string to spread out over these transverse directions. **d** The resulting geometry is a 'fuzzball' with no horizon

can be said to enclose the 'information' of the microstate. We find that the area of this surface satisfies [22]

$$\frac{A}{G} \sim \sqrt{n_1 n_p} \sim S_{micro}. \tag{11.22}$$

We can also consider the subset of the microstates which have a large angular momentum J, with $J \gtrsim (n_1 n_p)^{\frac{1}{4}}$. The entropy for such microstates can be computed in a manner similar to (11.9), and we find

$$S_{micro} = 2\pi\sqrt{2}\sqrt{n_1 n_p - J}. \tag{11.23}$$

In this case, the geometries differ from each other over a tube shaped region, so we get the analogue of a black ring. The surface area of this tube is found to satisfy [22]

$$\frac{A}{G} \sim \sqrt{n_1 n_p - J} \sim S_{micro}. \tag{11.24}$$

Thus, it appears that the Bekenstein entropy relation tells us the number of orthogonal gravity solutions that we can fit in a given region. We can investigate this further by restricting attention to only that subset of microstates which 'fit' in a small ball of surface area $A' = \mu A$ where $\mu < 1$ and A is the bounding area of the generic microstate, appearing in (11.22). In other words, we look at states that differ significantly from each other only inside the ball of area A'; these are a special subset of all microstates. The count of these states turns out to give [23]

$$S_{micro} \sim \mu\sqrt{n_1 n_p} \sim \frac{A'}{G}, \tag{11.25}$$

so we again find a Bekenstein-type relation.

Note that the geometries depend on several continuous parameters R, V and g. In the ratio A/G in (11.22) all these variables cancel out, yielding the entropy which depends only on the charges. It is not clear why exactly we have this cancellation, so it is not easy to answer the question of why the surface area is related in the above way to entropy. But these computations do indicate that the gravity solutions we have constructed are packed to maximal density; i.e., we cannot fit more than $Exp[S_{bek}]$ orthogonal gravity states in a region that is bounded by a surface smaller than that given by the relation $A/4G = S_{micro}$.

The computations of S_{micro} in Sect. 11.2.1 were performed at weak coupling, while the fuzzball solutions (11.21) were constructed at strong coupling. In [24], the space of fuzzball solutions (i.e., the solutions obtained at strong coupling) was quantized and shown to yield the entropy S_{micro}.

(D) **3-charge holes**: We have obtained all the microstates of the 2-charge extremal hole, and found that they are all fuzzballs. We now wish to consider more complicated holes like the 3-charge hole. This hole can be made by adding a 5-brane charge to the winding and momentum charges that we took for the 2-charge hole.

Before looking at the 3-charge hole, it is useful to take the 2-charge fuzzball solutions (11.21) and apply some discrete symmetry maps—called S and T dualities—to these solutions. After the dualities, the charges will change as follows

$$\text{NS1 charge} \rightarrow \text{NS5 brane charge}, \quad \text{P charge} \rightarrow \text{NS1 charge}. \tag{11.26}$$

Thus, we can map the fuzzballs (11.21) with NS1-P charges of the fuzzballs to NS5–NS1 charges.[2] Under this map we find that the singularity in (11.21) along the locus $\mathbf{x} = \mathbf{F}(v)$ gets removed, and we obtain metrics that are smooth everywhere. The points along the curve $\mathbf{x} = \mathbf{F}(v)$ become centers of Kaluza-Klein monopoles, which have a nontrivial topology but no singularity anywhere. More generally, microstates can have string and brane sources, as well as nontrivial topological features. They will in general be described by quantum wavefunctionals, and only a subset of fuzzballs can be well-approximated by a classical metric. The important point however is that

[2] Most of the black holes constructions discussed below were done using D5 and D1 branes, but here for simplicity we apply an S-duality map to those constructions and work with the NS5-NS1 bound state.

no microstate has the traditional horizon with the semiclassical vacuum state in its vicinity.

In [25–28], simple examples of extremal solutions were constructed with NS1, NS5 and P charges; i.e. these solutions carried the charges of the Strominger-Vafa black hole. In each case, the solution was a fuzzball; i.e, there was no horizon. Amplitudes computed for particular states in the dual CFT agree with amplitudes computed in the corresponding gravity microstates [29].

In the solutions of [28], four of the directions were compactified to a torus T^4, which was trivially tensored with the other directions. Dimensionally reducing on this T^4 gives a supersymmetric solution in 5 + 1 dimensions, where one of the directions is a S^1 parametrized by $0 < y < 2\pi R$, one direction is time t, and four space directions are noncompact. There was a general result [30] that supersymmetric solutions in 5 + 1 dimensions could be written as a 4-dimensional hyperkahler base times a 1 + 1 dimensional fiber along the directions t, y. In [31], the solutions of [28] were decomposed in this base$\times$fiber fashion, and a surprise was found. Hyperkahler manifolds have signature $++++$, but in this decomposition the hyperkahler base had signature $++++$ in one region (connected to spatial infinity) and signature $----$ in an interior region. The base $\times$ fiber split degenerated at the boundary of these two regions, but the overall 6-dimensional geometry remained smooth everywhere, with signature 5 + 1. Thus, we again have an example of a fuzzball geometry where the compact directions are not trivially fibered with the noncompact directions, though the overall manifold remains regular.

(E) **Finding large sets of fuzzball solutions**: In [32], a general method was found to find extremal solutions of supergravity with this kind of base $\times$ fiber split. This work led to an intense program which was very successful in constructing a large number of families of microstates for the 3-charge extremal hole [33–35]. These constructions describe microstates of black holes, black rings, and combinations of black holes and black rings. Some solutions have brane sources in the form of supertubes in the 'cap' region. The arbitrary 1-dimensional curve $\mathbf{F}(v)$ appearing in the 2-charge extremal solutions (11.21) has been extended to an arbitrary 2-dimensional surface in 3-charge extremal solutions [36, 37]. An important aspect of these constructions was that they succeeded in constructing microstates in the 'black hole regime': while earlier 3-charge fuzzball constructions were typically solutions that were 'overrotating', these solutions have the macroscopic parameters required to describe a nonrotating 3-charge hole. The large number of microstates found in this program went a long way towards establishing the non-uniqueness of the standard black hole geometry in theories with extra dimensions and, in particular, in string theory.

(F) **Non-extremal fuzzballs and Hawking radiation**: Extremal holes do not radiate. Thus while extremal fuzzballs tell us something about the structure of black holes, we should construct non-extremal microstates to understand radiation from the hole and the resolution of the information paradox.

In [38] a family of non-extremal D1D5P microstates were constructed. These solutions were again found to be fuzzballs; i.e., they had no horizon. They did have an ergoregion, and the emission from this ergoregion was computed in [39]. In [40–42] it was found that this emission spectrum was exactly the spectrum of Hawking radiation that was expected from these very non-generic microstates. To understand the nature of this agreement, we proceed in the following steps:

(i) In the CFT, we have an emission vertex $\hat{V}$ (Eq. (11.14)) which causes a transition from excitations on the weakly coupled NS1–NS5 bound state to gravitons that escape this bound state. If we have a generic thermal distribution of excitations on the bound state, then we multiply this vertex by Bose/Fermi distributions as in (11.15) to obtain the emission rate from the generic brane bound state at weak coupling. As noted in Sect. 11.2.2, this emission spectrum exactly matches the spectrum of Hawking radiation from the corresponding black hole.

(ii) With the fuzzball constructions, we have explicit gravity microstates (without horizon) for specific cases, together with knowledge of the corresponding weak coupling CFT state. Thus, we can compute the CFT emission from a specific microstate by multiplying the vertex V with the occupation numbers for excitations on that *particular* microstate. What we find is that this emission rate from the particular microstate at weak coupling exactly matches the emission from the corresponding gravity solution (i.e. fuzzball) obtained at strong coupling.

(iii) In short, the same weakly coupled CFT computation which reproduces the Hawking rate from a black hole, also reproduces the emission from specific fuzzballs. But while Hawking's emission led to a loss of information, the emission from the fuzzball is completely unitary, and we can see the details of the specific fuzzball imprinted on the emission spectrum. Of course, since these fuzzballs are very nongeneric states, the emission profile in these cases is not at all like thermal radiation. But these computations do suggest that when we are able to compute radiation from generic fuzzballs, we will find that the radiation is unitary but still with the emission spectrum found by Hawking.

Other classes of nonextremal microstates have been constructed in [43–45].

(G) **The general nature of fuzzball solutions**: We end this section with some general comments on the nature of fuzzballs. The Buchdahl theorem says that if a ball of perfect fluid has a radius $R < 9/4GM$, then it must collapse into a black hole. On the other hand, we expect fuzzballs to have a surface that is just order Planck length outside the horizon radius [46]. What novel effects in string theory would allow such solutions?

A black hole in 3 + 1 dimensions will have 6 compact directions in string theory. We have already seen above that in fuzzballs, the noncompact directions are not trivially tensored with compact directions. As a toy example, consider 3 + 1 noncompact spacetime with one additional compact circle. Then the vacuum Einstein equations have a solution given by Euclidean Schwarzschild times a trivial a time direction

$$ds^2 = -dt^2 + \left(1 - \frac{r_0}{r}\right) dy^2 + \frac{dr^2}{1 - \frac{r_0}{r}} + r^2(d\theta^2 + \sin^2\theta d\phi^2) \tag{11.27}$$

where $r \geq r_0$ and the radius of the compact circle y is $4\pi r_0$. If we dimensionally reduce on the y circle, we get 3 + 1 spacetime coupled to a scalar field. This scalar field has the usual sign of the stress tensor, and is nonzero everywhere, diverging as $r \to r_0^+$. From a 3 + 1 dimensional perspective, we can ask: why is this distribution of energy density not collapsing inwards under its own gravity?

If we take a different radial function for the same compactification, we can obtain something even more startling: Witten's bubble of nothing [47]. This is also a vacuum solution of the Einstein's equations in 4 + 1 dimensions, and from a 3 + 1 dimensional perspective the matter is a scalar field with the usual stress tensor. Yet instead of collapsing inwards under gravity, the solution is exploding outwards.

As noted in [48], in each of the above cases a key point is that the manifold is not a trivial product of 3 + 1 dimensional space and a circle: the circle degenerates on a S^2, creating a cigar shaped 'cap' at that location. In this situation the stress tensor of the scalar field is highly non-isotropic, with positive pressure in the radial direction and negative pressures in the angular directions; thus the pressure is very different from the isotropic one assumed in Buchdahl's theorem. These toy examples show how the usual intuition of gravity can be violated when we have compact dimensions.[3]

As noted in (F) above, fuzzball solutions in appropriate duality frames have compact directions that are not trivially tensored with noncompact directions. In other duality frames this topological nontriviality shows up as a string or brane source: for example in the NS1-P frame the 2-charge extremal fuzzballs had a string source, while in the NS1-NS5 duality frame this source was replaced by a smooth KK-monopole tube. Thus, it is the special features of string theory which allow fuzzballs to exist.

In [49] it was shown how these features bypass the usual no hair arguments for black holes. In [50] solitonic stars were constructed to model neutral black holes, with compact directions pinching off in the cigar shaped manner mentioned above.

Far outside from the fuzzball, we expect to get the usual metric of the black hole. What happens as we approach the horizon radius? To get an idea of this, let us examine the nature of the extremal fuzzballs which have been constructed. At different locations in the fuzzball, we have a string theory source, which is always a $\frac{1}{2}$-BPS object. The overall configuration is $\frac{1}{4}$-BPS or $\frac{1}{8}$-BPS, depending on the hole whose microstates we are constructing [51]. Some part of these source charges cancel out in the overall charge of the fuzzball; we call these dipole charges. The other parts add up to yield the net charge of the fuzzball. As we approach one of these source charge locations, the geometry becomes very asymmetric: the metric along the brane source shrinks to zero by the tension of the brane, and the metric perpendicular to the brane source expands to infinity.

A similar situation is noted with metrics that have been proposed for some neutral Extremely Compact Objects (ECO) [52]. In the traditional black hole metric, as we approach the horizon in a radial direction, we find Rindler space. In a suitable frame

[3] These toy models cannot be directly realized in string theory because the cigar geometry is not compatible with the presence of fermions in the theory. A fermion that is periodic around the compact circle at infinity will have a singularity at the tip of the cigar.

around any point near the horizon, the time direction and the radial direction make up the right Rindler wedge, while the directions along the horizon separate out with a metric of the form $ds_{\perp}^2 \approx dX_1^2 + \ldots dX_{d-1}^2$. But in the ECO geometry in [52], the metric in some of these horizon directions diverges as we approach the horizon radius, while in other directions it shrinks to zero. This behavior is also observed for the solitonic stars mentioned above. We can regard these local solutions as similar to the vacuum Kasner metrics for cosmology. In the Kasner solutions, the metric has the form

$$ds^2 = -dt^2 + \sum_i t^{2\alpha_i} dX_i^2 \tag{11.28}$$

To get the possible geometries near the fuzzball surface, we replace the t coordinate in the Kasner solutions by the distance from the horizon s. The metric then takes the form

$$ds^2 \approx -s^{2\alpha_0} dt^2 + \sum_i s^{2\alpha_i} dX_i^2 \tag{11.29}$$

We again find the conditions on the exponents that are found for the Kasner case:

$$\alpha_0 + \sum_i \alpha_i = 1, \quad \alpha_0^2 + \sum_i \alpha_i^2 = 1 \tag{11.30}$$

(The metric in [52] for example satisfies these relations.)

For the case of cosmology, the BKL estimate [53] shows that as $t \to 0$, we get a random orientation of the spatial axes determining the directions X_i. That is, these directions and their corresponding exponents change very rapidly from position to position. If a similar situation holds for the metric (11.29), then we will get a large family of allowed solutions as we approach the horizon, and this can be a manifestation of the large entropy of the hole. When compact directions X_i in (11.30) have positive exponents, they shrink towards zero size near the horizon. Branes wrapping these directions can become very light, and the fuzzball surface will become a region with large quantum fluctuations.

11.5 Recovering Black Hole Thermodynamics

Let us summarize what we have seen above. The presence of a horizon leads to the information paradox. The small corrections theorem makes this into a precise conflict: if semiclassical physics around a horizon emerges, even as a leading order approximation, then black hole evaporation must end in either a loss of unitarity or the formation of remnants with unbounded degeneracy. String theory escapes this problem by not allowing the formation of a horizon: if we make a bound state of the fundamental objects in the theory—strings and branes—then this bound state is a horizon sized quantum object—a fuzzball—with no horizon. The fuzzball radiates from its surface like a normal body. Thus, the black hole is just like a 'string star', and there is no information paradox.

But now we are faced with a different question. The traditional semiclassical black hole has an elegant thermodynamics. For example, in 3 + 1 dimensions, the Schwarzschild hole has a temperature

$$T_H = \frac{1}{8\pi GM}, \tag{11.31}$$

and an entropy

$$S_{bek} = \frac{A}{4G}. \tag{11.32}$$

Further, this semiclassical hole radiates like a black body: i.e., the radiation rate $\Gamma_{BH}(l, m, \omega)$ for a partial wave $Y_{l,m}$ of a massless scalar at frequency ω is related to the absorption probability $\mathscr{P}(l, m, \omega)$ by the relation of detailed balance:

$$\Gamma_{BH}(l, m, \omega)d\omega = \frac{\mathscr{P}(l, m, \omega)}{e^{\frac{\omega}{T_H}} - 1}\frac{d\omega}{2\pi}. \tag{11.33}$$

The absorption probability $\mathscr{P}(l, m, \omega)$ depends only on the geometry of the hole outside the horizon.

But if the black hole is just a string star with no horizon, then why should it have the properties (11.31)–(11.33)? Hawking's derivation of the temperature (11.31) relied on his picture of pair creation at the horizon. If we have no horizon, then we resolve the information paradox. But in the process, do we lose the elegance of black hole thermodynamics?

As we will now note, if an object is sufficiently compact (i.e., it has a radius that is sufficiently close to the horizon radius $R = 2GM$), then it *must* have the temperature, entropy, and radiation rate given by black hole thermodynamics [54, 55]. We use the term Extremely Compact Object (ECO) to describe an object with a surface which is a very small distance s outside the horizon radius $r = 2GM$. ECOs have been of interest in general theories of quantum gravity—not just string theory—and the argument below will be a general one which applies to any ECO.

Consider a spherically symmetric ECO with mass M as measured from infinity. To get the essence of the argument, consider an extreme case: the surface of the ECO is at a proper distance $s_{ECO} \sim l_p$ outside the horizon radius $R = 2GM$, and the temperature is $T_{ECO} = 0$. We assume nothing about the dynamics inside the ECO surface, but outside this surface we are allowed to use the semiclassical dynamics of quantum fields on curved space. It is possible to argue that this semiclassical region will have a stress tensor that equals, to leading order, the stress tensor of the Boulware vacuum. This vacuum has a vacuum energy density $\rho_v \sim -\frac{1}{s^4}$, at a distance s from the horizon radius. Integrating ρ_v from the surface of the ECO to infinity, we find that this vacuum energy contributes a mass $M_v \sim -\alpha M$, with α a positive constant of order unity.

Since the mass at infinity is M, the mass inside the ECO surface is $(1+\alpha)M$, which corresponds to a horizon radius $2(1+\alpha)GM$. Thus, the semiclassical region just outside the ECO is deep inside this horizon radius. The geometry in this semi-classical cannot be time independent, due to the 'inward pointing' behavior of light cones inside a horizon. Thus, we conclude that an ECO with surface at $s_{ECO} \sim l_p$ and $T = 0$ cannot exist.

A similar argument works for other values of T_{ECO}, except for $T_{ECO} = T_H$, where T_H is the Hawking temperature. When $T = T_H$, the local blueshifted temperature near the ECO surface is the Unruh temperature

$$T_s = T_U(s) = \frac{1}{2\pi s}. \tag{11.34}$$

The near surface region is filled with a thermal gas at temperature $T(s)$, and if $T(s) = T_U(s)$ then the negative vacuum energy ρ_v is cancelled by this thermal energy. In this case, we bypass the above argument and the ECO can exist. Repeating this analysis with more care, one finds that in 3 + 1 dimensions, if

$$s_{ECO} \ll \left(\frac{M}{m_p}\right)^{\frac{1}{2}} l_p, \tag{11.35}$$

then the ECO cannot exist if $T_{ECO} \neq T_H$. In $d+1$ dimensions, this condition becomes $s_{ECO} \ll \left(\frac{M}{m_p}\right)^{\frac{2}{(d+1)(d-2)}} l_p$.

To make the above argument more precise, we write the metric outside the ECO surface in the standard form

$$ds^2 = -e^{2B(r)}dt^2 + e^{-2A(r)}dr^2 + r^2 d\Omega^2, \tag{11.36}$$

and solve the Tolman-Oppenheimer-Volkoff equation in the region just outside the ECO surface. We allow an arbitrary thermal stress-tensor from thermal excitations which have $p_T(r) = \frac{1}{d}\rho_T(r)$ and from vacuum energy, for which we take $p_v(r) = \frac{1}{d}\rho_v(r)$.[4] The approximation $r \approx 2GM$ allows an explicit solution of the TOV equation, and one finds that if (11.35) holds, then there is no solution where $e^{A(r)}$, $e^{B(r)}$ remain positive throughout the semiclassical region outside the ECO.

This argument tells us that we need $T_{ECO} \approx T_H$, with the approximation becoming better and better as s decreases below the scale on the RHS of (11.35). The general thermodynamic relation then tells us that

$$dS_{ECO} \approx T^{-1}dE = (8\pi GM)dM, \tag{11.37}$$

which integrates to

[4] The tracelessness of the vacuum stress tensor to leading order follows from the anomaly relation giving the trace. The pressure p_v is isotropic in the Boulware vacuum, and we make the assumption that it continues to be isotropic to leading order in the near surface geometry.

$$S_{ECO} \approx 4\pi G M^2 = \frac{A}{4G}. \tag{11.38}$$

This steps (11.37), (11.38) were used to convert Bekenstein's qualitative conjecture $S \sim A/G$ [1] to the precise relation (11.32) after Hawking's discovery of the temperature (11.31); here we are just noting that *any* object with the same $T[M]$ as the black hole would yield the same entropy $S[M]$ as the black hole.

To obtain the radiation rate from an ECO, we first write the semiclassical computation of Hawking radiation in Schwarzschild coordinates. The effective potential $V_{eff}(r)$ for quantum field modes in the black hole geometry vanishes at $r \to 2GM$ and at $r \to \infty$, and has a 'bump' in between. In the near horizon region, we have Rindler modes excited at the local Unruh temperature (11.34). These excitations have a small probability to tunnel through the 'bump' in V_{eff} and escape to infinity; the quanta that escape give Hawking radiation at the temperature (11.31).

An ECO at temperature $T_{ECO} = T_H$ has radiation at the temperature (11.34) in the near surface region. Since the surface of the ECO is close to the horizon radius, the modes of quantum fields encounter virtually the same potential barrier $V_{eff}(r)$ as the modes in the black hole geometry. Thus, the spectrum of quanta escaping the barrier is the same, to leading order, as the spectrum (11.33) of radiation from the black hole.

In short, ECO's satisfying (11.35) must have the same thermodynamic properties (11.31)–(11.33) as the semiclassical black hole.

How compact do we expect fuzzballs to be? In string theory, we have seen that the count of microstates reproduces the Bekenstein entropy $A/4G$. Consider any theory of quantum gravity where the entropy is order A/G. Suppose we wish to reproduce this entropy by having some structure at the horizon. We will now give a heuristic argument which suggests that the object reproducing this entropy must be very compact [46].

Consider the quantum gravitational structure responsible for the entropy. We imagine this structure to be made of Planck sized elementary objects arranged near the horizon. Each such object will cover an area $\sim l_p^2$, so the number of such objects will be

$$N \sim \frac{A}{l_p^2} \sim \left(\frac{M}{m_p}\right)^2. \tag{11.39}$$

If we give each elementary object one bit of entropy (say, a spin which can be up or down) then we recover an entropy $S \sim A/G$, as desired. But the mass of each elementary object will be $m \sim m_p$, giving a total mass

$$M_T \sim N m_p \sim \left(\frac{M}{m_p}\right) M. \tag{11.40}$$

This is far larger than M. How then can we think of the entropy in terms of structure at the horizon?

The key point is that if the microscopic structures mentioned above lie a short distance s outside the horizon radius $R = 2GM$, then they sit at a location with high redshift. Thus, the mass seen from infinity is

$$M_\infty \sim (-g_{tt})^{\frac{1}{2}} M_T \sim \left(\frac{s}{GM}\right) M_T \sim \frac{s}{l_p} M. \tag{11.41}$$

We see that if $s \sim l_p$, then we get $M_\infty \sim M$, as required. Interestingly, this argument gives the same scale $s \sim l_p$ as the required location for the ECO surface in any spacetime dimension $d + 1$.

Thus, on quite general grounds we expect fuzzballs to be very compact, with $s_{ECO} \sim l_p$. This length scale is much smaller than the limit (11.35), and so we expect the fuzzballs will reproduce the black hole thermodynamic relations (11.31)–(11.33) to an excellent approximation.

11.6 Why Is the Semiclassical Approximation Violated?

We have seen that in all cases where we have been able to construct black hole microstates in string theory, these microstates have turned out to be fuzzballs—objects with no horizon. This resolves the information paradox, but at this point one might wonder: why did the classical picture of the hole change so radically in the full quantum gravity theory? The curvature at the horizon of a black hole of radius R is $\mathcal{R} \sim \frac{1}{R^2}$. For a large black hole with $M \gg m_p$, this curvature is very low: $\mathcal{R} \ll l_p^{-2}$. It is generally agreed that semiclassical physics should break down when $\mathcal{R} \gtrsim l_p^{-2}$. But here we are finding a large change all over the interior of the horizon, which is a region of low curvature at all points away from the central singularity. Thus, we are observing a second mode of failure of the semiclassical approximation. What is the trigger for this second mode of failure?

To answer this question, we will proceed in three steps. First, we will use rough estimates to understand why black hole dynamics can differ from dynamics in other situations with low curvature. Next, we consider the quantum gravitational description of a static star, as we reduce its radius towards the horizon radius. We will see that in the limit, $R_{star} \to 2GM$ the virtual fluctuations of fuzzball type configurations become larger and larger. Finally, we will study the dynamics of gravitational collapse to understand why evolution to the semiclassical black hole geometry is not an allowed path in the quantum gravity theory, and the evolution leads to fuzzballs instead.

11.6.1 The Role of Bekenstein's Entropy

A key aspect in understanding the mysteries of the black hole is the large value of the Bekenstein entropy (11.32). This entropy is far larger than the entropy of a gas with the same energy and volume as the hole. Now that we have some understanding of what this entropy represents—it counts horizon sized quantum objects (fuzzballs) that give the microstates of the hole—we can ask what role these microstates play in the dynamics of gravitational collapse.

Consider a shell with energy M that is collapsing to make a black hole. In semi-classical evolution, this shell passes smoothly through its horizon radius $R = 2GM$, and continues inwards to create a singularity at $r = 0$. But looking at the structure of the fuzzball geometries we know, we find that there is a path for the classical shell to tunnel into a fuzzball microstate. The tunneling amplitude will be very small, since we are talking about tunneling between two very different macroscopic states. Let the spacetime dimension be $d + 1$. The probability of tunneling is $P_{tunnel} = |\mathscr{A}|^2$ where $\mathscr{A}$ is the tunneling amplitude. In typical tunneling processes, we can estimate $\mathscr{A}$ as

$$\mathscr{A} \sim e^{-S_{cl}}, \tag{11.42}$$

where S_{cl} is the classical action for the Euclidean gravitational path between the two configurations. To get some estimate of S_{cl}, let us set all length and time scales to be of order of the horizon radius $R \sim (GM)^{\frac{1}{d-2}}$. This gives

$$S_{cl} \sim \frac{1}{G} \int d^{d+1}x \sqrt{-g}\mathscr{R} \sim \frac{1}{G} R^{d+1} \frac{1}{R^2} \sim \left(\frac{R}{l_p}\right)^{d-1}. \tag{11.43}$$

We see that the tunneling probability is indeed small for $R \gg l_p$. But the number of possible fuzzball states we can tunnel to is

$$\mathscr{N} \sim e^{S_{bek}}, \tag{11.44}$$

where

$$S_{bek} \sim \frac{A}{G} \sim \left(\frac{R}{l_p}\right)^{d-1}. \tag{11.45}$$

We see that it is possible to have

$$P_{tunnel}\, \mathscr{N} \sim 1. \tag{11.46}$$

so that the collapsing shell tunnels into fuzzballs in a short time, invalidating the semiclassical approximation [56]. In [57] it was argued that the numerical coefficients in the exponents in P_{tunnel} and $\mathscr{N}$ are such that they actually cancel. In [58] tunneling was considered between the members of a particular family of fuzzball microstates, and the enhancement due to the degeneracy factor analogous to $\mathscr{N}$ was observed.

The argument leading to (11.46) is of course very heuristic; in particular, the simple estimate (11.43) may not give the tunneling rate to very complicated microstates. But (11.46) does indicate that something special can happen for black holes that does not happen for a normal star: 'entropy enhanced tunneling' can invalidate the semiclassical approximation even when the curvature scale is low. The larger we make the black hole, the smaller the curvature around the horizon, but at the same time the larger the Bekenstein entropy. Thus, the violation of the semiclassical approximation happens for all holes, however large they are.

Put another way, in the path integral

$$Z \sim \int D[g] e^{\frac{i}{\hbar} S_{cl}}, \tag{11.47}$$

we usually assume that the measure factor is small for macroscopic systems, and thus we extremise the classical action S_{cl} to get the semiclassical approximation. But the measure factor $D[g]$ involves the degeneracy of states that can contribute to the path integral, so for black holes we may write

$$D[g] \sim e^{S_{bek}}. \tag{11.48}$$

The above estimates then show that the large value of S_{bek} can make the measure term compete with the classical action, thus invalidating the description of the black hole as a semiclassical object.

11.6.2 The VECRO Hypothesis

Let us now see how the above estimates can be used to develop a more explicit picture of gravitational wavefunctional. First consider a star of mass M_{star} and radius R_{star}. In Fig. 11.2a, along the horizontal axis, we depict schematically all the configurations of the gravity theory that are involved in describing the star. Thus, this is a space of large dimension, which for the sake of illustration, we are drawing as a 1-dimensional line. The vertical axis gives the energy of the configuration. Thus, this graph is similar to the potential energy graph $V(x)$ for a quantum mechanics problem.

The central point on the horizontal axis is the classical configuration of the star. Around this point we have the small Gaussian deformations describing the quadratic fluctuations of quantum fields around the classical geometry of the star. Thus, this part of the potential graph is similar to the potential $V(x) = \frac{1}{2}kx^2$ for a harmonic oscillator, except that there are many modes of the quantum fields with such a quadratic potential. These quadratic fluctuations are the ones that lead to Hawking's particle creation under deformation of the geometry. But further away from the configuration depicting the classical star, we must have the fuzzball configurations that describe the $Exp[S_{bek}]$ states that account for the black hole entropy. We focus on fuzzball states that have radius R_{star}, since the wavefunctional of the star can overlap with

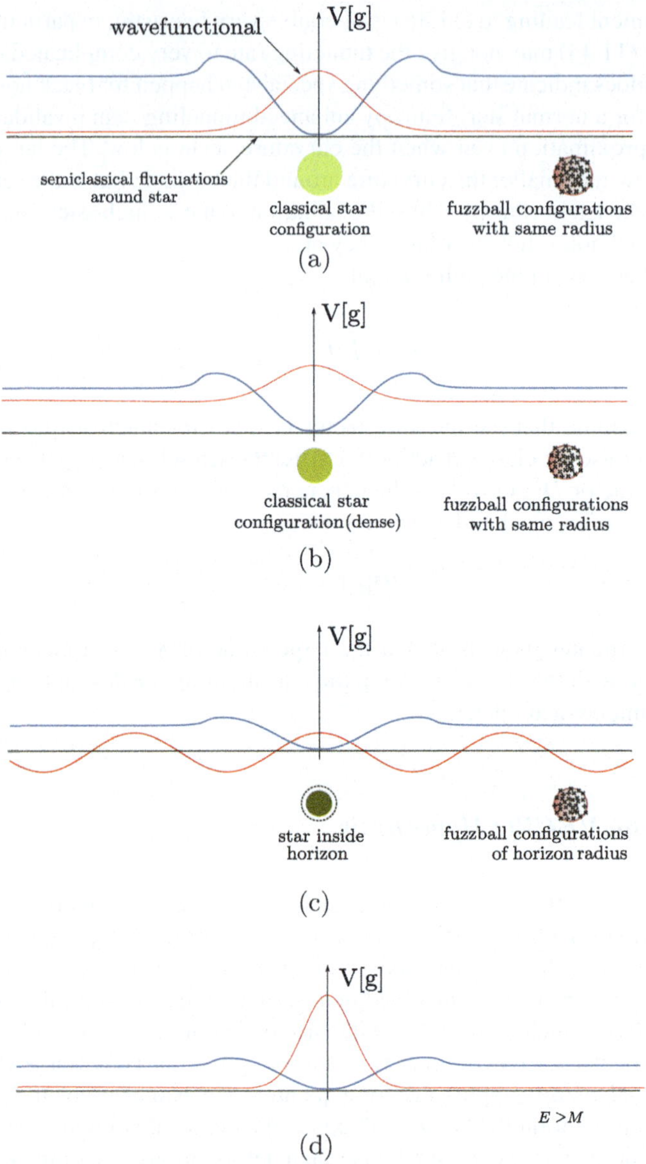

Fig. 11.2 A schematic depiction of the potential function over the space of configurations of the gravitational theory. The thick blue line is the potential and the thin red line is the wavefunctional. **a** When a star is much bigger than its horizon radius, the fuzzballs with the same radius as the star are much heavier than the star, and the wavefunctional has a small support over these fuzzball configurations. **b** For a denser star, this support over fuzzball configurations is larger. **c** For a star inside its horizon radius, the wavefunctional spills over the entire space of fuzzball configurations of mass M_{star}. **d** For a star inside its horizon radius, we attempt to construct a wavefunction that is not allowed to spill over into the fuzzball region. The energy of the wavefunctional exceeds the mass of the star, so such a wavefunctional does not describe a possible state of the collapsing star

these states. Since the star has a mass $M_{star} < 2GR_{star}$, these fuzzball states have a mass M_{fuzz} much larger than the mass M_{star} of the star. This fact is depicted in Fig. 11.2a by the fact that the potential graph in the region of the fuzzball states is high. In between the semiclassical configurations around the star and the fuzzball states, we have drawn a 'bump'; this indicates that there is a barrier to tunneling between the star and the fuzzballs.

Now let us ask: what is the meaning of the evolution described by classical gravity? The classical metric $g[x]$ depicts the peak of a quantum wavefunctional; the wavefunctional itself has a spread around this classical peak. For our star, the wavefunctional is depicted by the dashed line in Fig. 11.2a. Since the fuzzball configurations are also present in our space of gravity configurations, we must allow the wavefunctional to spread over these fuzzball configurations as well. Of course the potential is high at these fuzzball configurations, so the wavefunctional there is very small, and the semiclassical part describing Gaussian fluctuations around the classical configuration gives an adequate approximation to the wavefunctional.

Now consider the situation in Fig. 11.2b. The star still has mass M_{star}, but is denser. The mass of the fuzzballs with radius R_{star} is still more than M_{star} but not too much more. The potential in the fuzzball region is now lower, and the wavefunctional has a larger amplitude over these fuzzball configurations.

In Fig. 11.2c we attempt to shrink the star inside its horizon radius. Now, the wavefunction does not stay confined to the potential well around the classical configurations; it spreads over the vast space of fuzzball configurations. For a toy model of this phenomenon, consider the quantum mechanics problem of a square well of depth V_0 and radius a, in a large number of dimensions d_n. For a particle of mass m, there is a bound state in this well only if

$$V_0 \gtrsim \frac{\hbar^2}{8ma^2} d_n^2 \tag{11.49}$$

If $V_0 \lesssim \frac{\hbar^2}{8ma^2} d_n^2$, then the wavefunction 'pops out' of the well and spreads over the entire spatial region. This is what has happened to our star: while classical gravity would suggest that a star can shrink inside its horizon, we must consider if this classical configuration can really be the peak of a wavefunctional which is localized around this classical configuration. The large dimension of the space of fuzzballs (given through the Bekenstein entropy) makes it impossible to trap the wavefunctional into the Gaussian part of the potential well when the radius of the star reaches $R_{star} \approx 2GM_{star}$, since all the $Exp[S_{bek}]$ states of mass M_{star} are accessible to the wavefunctional. Thus the semiclassical approximation breaks down when the star tries to shrink inside its horizon radius.

Finally, in Fig. 11.2d we depict a state where we force the wavefunctional to be localized in the Gaussian part of the potential even though the star has collapsed through its horizon. While we can certainly consider such a wavefunctional, it has an energy $\langle E\rangle$ which is higher than M_{star}; thus it does not describe a star of mass M_{star} which has collapsed through its horizon. The situation can again be understood with the help of our toy model of the potential well. For, $V_0 \lesssim \frac{\hbar^2}{8ma^2} d_n^2$ we can certainly consider a wavefunction that is localized inside the well. But this wavefunction will not be a bound state with $E < 0$; it will have an expectation value of energy that is higher than the continuum, i.e., we will find $\langle E \rangle > 0$.

To summarize, classical evolution describes the peak of a wavefunctional that in principle spreads over all quantum states. When we try to shrink a star inside its horizon, the wavefunctional does not stay narrowly localized around this classical configuration, so the semiclassical approximation breaks down.

The notion that the fuzzball component of the wavefunctional is important is called the VECRO hypothesis [59]. The term VECRO stands for Virtual Extended Compression-Resistant Objects. The fuzzball tail of the semiclassical wavefunctional consists of 'virtual' fluctuations of 'extended' objects, since the fuzzball configurations are not pointlike particles but rather large complex configurations. As we will see below, these configurations are resistant to compression or extension, a property that is important in their contribution to gravitational dynamics. We will now see the role of the vecro part of the gravitational wavefunctional in the process of black hole formation.

11.6.3 The Dynamics of Gravitational Collapse

Finally, we come to the crucial question: why does gravitational collapse not lead to the semiclassical geometry of the hole? Consider a shell of mass M composed of radially ingoing massless quanta, in $3+1$ dimensional asymptotically flat Minkowski space. By causality, no particle on this shell can receive a signal from any other particle on the shell. Thus, it would seem that each particle will move independently through empty space, crossing the horizon uneventfully and progressing towards the center. New physics can certainly arise at the center since $\mathscr{R} \gtrsim l_p^{-2}$ at the singularity, but by causality this new physics cannot affect the horizon which is spacelike separated from the singularity. It would therefore seem that we will have a semiclassical horizon and the associated information loss problem. What is the way out?

As we will now see, the particles of the shell indeed cannot exchange signals with each other. But once the shell reaches the vicinity of the horizon, the particles of the shell do not leave behind them a vacuum spacetime. In the region of this 'wake' (i.e., the region $r > R_{shell}$) we *can* receive information from a curved segment of the shell. If the shell is at a radius $R_{shell} \gg 2M$, then this 'wake' region is close to the vacuum, as expected from usual semiclassical dynamics. But for $R_{shell} \approx 2GM$, it is important to note that the metric outside the shell is a Schwarzschild metric with mass M. We had noted in Sect. 11.6.2 that the fluctuations of virtual fuzzballs—the vecro component of the wavefunctional—will be larger and larger as the shell approaches the horizon. When the shell crosses into the horizon region, the wake left behind the shell generates fuzzballs rather than the traditional vacuum geometry.

To arrive at this picture of evolution, we need to provide a crucial ingredient: a reason why the 'wake' region behind the shell cannot just be the semiclassical vacuum. We will now give a bit model of the gravitational vacuum which incorporates the vecro hypothesis; with this model we will find that the collapsing shell cannot give the semiclassical black hole geometry but must yield fuzzballs instead. We proceed in the following steps:

(A) We know that the vacuum of quantum theory has fluctuations of electrons and positrons; it is these fluctuations that turn to on-shell particles in the process of Hawking radiation. But the electron and positron also form a bound state—the positronium. Is there any effect of this on-shell bound state on the structure of the vacuum? The answer is yes: the fluctuations where the electron and positron are at a separation of order the positronium radius have a slightly greater amplitude than fluctuations where they do not have this separation. Thus, bound states in the theory lead to correlations among the vacuum fluctuations. The particles of the standard model have only a few bound states at any given energy, and as we consider bound states of higher and higher energies, their fluctuations are quite suppressed anyway. Thus, it may seem that the correlations induced by bound states are not an important aspect of the vacuum wavefunctional.

But could there be some large class of bound states that we have conventionally ignored? Yes: the $Exp[S_{bek}(M)]$ microstates of black holes for each mass M! It is true that the virtual fluctuations corresponding to states of larger mass M will be suppressed, but we have already noted in Sect. 11.6.1 that this suppression can be offset by the larger degeneracy of bound states at larger M. The vecro hypothesis says that there are important correlations in the vacuum at all length scales due to the presence of on-shell black hole microstates in the theory. We will now make a model of these correlations, and see what they do in the process of gravitational collapse.

(B) It is generally agreed that spacetime has violent quantum fluctuations at the Planck scale. Let us make a toy model of these fluctuations; this will enable us to talk about correlations among these fluctuations in a more concrete way. Consider a spacetime with a compact circle, depicted in Fig. 11.3a. Small fluctuations in the radius of this circle describe a massless scalar in the dimensionally reduced theory. But we can also have larger fluctuations which pinch off the circle and create, for example, a KK monopole or antimonopole (Fig. 11.3b). In string theory, the KK monopole forms a 256 dimensional multiplet after we take into account fermion zero modes localized at the monopole. Here we will just assume for simplicity that at the microscopic scale we have excitations (the analogue of KK monopoles) that carry a spin $\frac{1}{2}$ degree of freedom. We can thus imagine a lattice of spin $\frac{1}{2}$ bits, with the proviso that as space expands in volume we can create more lattice sites and when space shrinks in volume we move to a configuration with less lattice sites.

(C) The vacuum is the state with the lowest energy. Because of the Hamiltonian interaction between neighboring spins, the lowest energy state of the theory will be such that the spin at any site is entangled with the spins at nearby sites. This entanglement will capture the correlations that we seek to model. The correlations result from the presence of virtual fuzzballs with all possible sizes. Thus, we should have correlations at all length scales, though the correlations should fall off with distance. Let us make a toy model describing these correlations. First, suppose that we had correlations only among neighboring sites. (These will be correlations arising

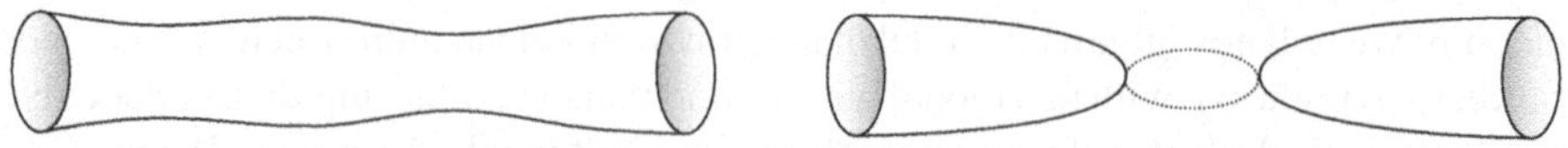

Fig. 11.3 **a** Small fluctuations of a compact circle give a massless scalar in the low energy dimensionally reduced theory. **b** At high energy scales we have larger fluctuations that create topological deformations like KK monopoles and anti-monopoles; these are the bits carrying spin that we use in Fig. 11.4

from the virtual fluctuations of the smallest black holes.) We group the spins in pairs. Consider a pair of neighboring spins called $s_1 s_1'$. Let their singlet state be S_1 and triplet T_1. If we only wished to consider the entanglement arising from the smallest virtual black holes, then we would let the state of this pair be S_1.

Now let us see how we can also allow correlations across longer distances, which would arise from virtual fluctuations of large black hole microstates. We want an entanglement that is large between neighboring spins, and smaller (but not zero) between spins that are further away. We can model this as follows. For the two neighboring spins $s_1 s_1'$ considered above, we take the combined state to be mostly the singlet S_1, but with a small admixture of the triplet T_1

$$|\psi\rangle_{12} = (1 - |\delta|^2) S_1 + \delta\, T_1, \quad \delta \ll 1. \tag{11.50}$$

Consider another similar pair of spins $s_2 s_2'$ that are at a distance D away, and which form a similar spin state: mostly the singlet S_2 and a bit of the triplet T_2. We then entangle the triplet states T_1, T_2 between these two distant pairs of spins to make a singlet. This is a toy model describing a situation where we have a large correlation with nearby spins, as well as a small but nonzero correlations between spins that are further away. By making such a hierarchical set of entanglements between a given spin and other spins at various distances, we can include the effect of vecros (virtual black hole microstates) of all sizes.

(D) Let us now input the energy scales of black hole physics into our model. We have a hierarchical set of correlations. There is an entanglement between the spins in the smallest block of radius 2 lattice sites. Next we group these blocks into larger blocks of radius 4 lattice sites; these blocks in turn are grouped into blocks with radius 8 lattice sites and so on. At each stage of the blocking, we have an entanglement between the blocks that are joining to make the next larger block, of the kind described in (C) above. Consider the blocks of radius R_v, which will correlate with each other to make the next larger block of radius $2R_v$ (the symbol R_v stands vecro radius). These correlations of the regions of radius R_v contribute a (negative) energy, $E(R_v)$ which we set to be of order of the black hole mass for radius R_v. In 3 + 1 dimensions, this will be

$$E(R_v) \sim -\frac{R_v}{G}. \tag{11.51}$$

In terms of energy density, this is

$$\rho(R_v) \sim -\frac{1}{R_v^3}E(R_v) \sim -\frac{1}{GR_v^2}. \tag{11.52}$$

To summarize, the virtual black hole microstates (vecros) of radius $\sim R_v$ lead to correlations in the vacuum that lower the energy density by the amount (11.52). Taking into account the lowering of energy density by all vecros with $0 < R_v < \infty$ brings us down to the net vacuum energy density which is assumed to be zero since we are starting with flat Minkowski space. From all this, we see that if we have a state with the correct correlations at scales $0 < R_v < R_v^{max}$ but not at scales $R_v > R_v^{max}$, then we will get a *positive* energy density of order

$$\rho \sim \frac{1}{G(R_v^{max})^2}. \tag{11.53}$$

(E) Now we come to a crucial point. Suppose the space depicted in Fig. 11.4 expands. The two pairs of spins s_1s_1' and s_2s_2' get separated to a distance $D' > D$, and new spins appear at lattice points in between. To get the optimal correlations that lead to the zero energy vacuum, we will need the triplet T_1 of the first pair to entangle with the triplet T_3 of the new pair s_3s_3' that is a separation D away in the new metric. But now we see an important issue. For the triplet T_1 of the first pair to entangle with the triplet T_3 of the new pair, we must first remove the entanglement between the two original triplets T_1, T_2. We can remove this entanglement by suitable interactions between the two sets of spins s_1s_1' and s_2s_2', but for these interactions to

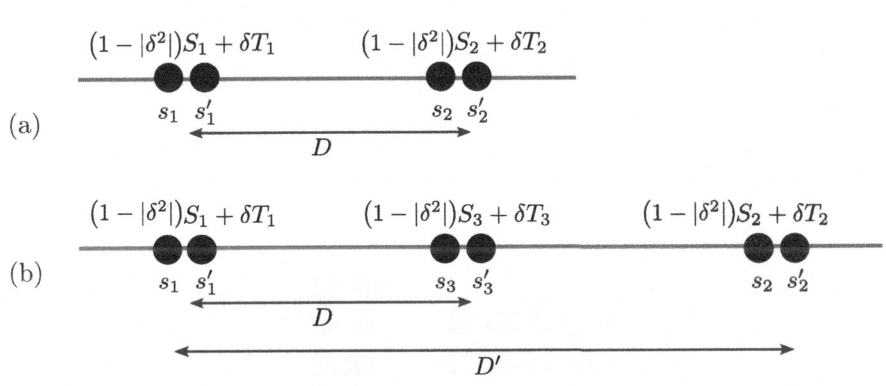

Fig. 11.4 **a** Neighboring spins s_1s_1' form a state that is mostly a singlet S_1 but with a small admixture of the triplet T_1, and similarly for s_2s_2'. Entangling the triplets into a singlet gives a situation where each spin is largely entangled with its nearest neighbor, but has a small entanglement with spins further away. **b** When space expands, we create a new spin pair s_3s_3'. For the stretched space to be a vacuum, we must have the same entanglement structure at separation D as we had in the initial manifold in (**a**)

take place we must have enough time for light to travel between $s_1 s_1'$ and $s_2 s_2'$. We now have two cases:

(i) If the expansion is slow, light will be able to traverse between $s_1 s_1'$ and $s_2 s_2'$ several times in the course of the expansion. Thus, we can reach the entanglement appropriate to the vacuum in the new expanded lattice; i.e., T_1 will disentangle from T_2 and entangle with T_3. Under this adiabatic expansion, the initial zero energy vacuum stretches to the new zero energy vacuum.

(ii) If the expansion is rapid, then light cannot travel between $s_1 s_1'$ and $s_2 s_2'$ during the expansion. Then we cannot remove the entanglement between, T_1, T_2 and thus we cannot entangle T_1, T_3. We do not reach the vacuum state of the expanded manifold. Instead, we get a state with extra energy density of order (11.53), where R_v^{max} is given by the distance light has been able to travel during the expansion.

Note that the two cases of adiabatic and nonadiabatic expansion also arise in the usual study of quantum fields on curved space. What is new here is that there are extra correlations in the vacuum that stretch across all length scales; these correlations arise from virtual black hole microstates (vecros) which are *extended* objects spanning all length scales. The contribution of these correlations is significant because of the large degeneracy of bound states implied by the Bekenstein entropy.

(F) With all this, we can now explain how the semiclassical approximation can break down in black hole formation but not in other places where we have seen general relativity to be valid:

(i) In a gravitational wave, there is no increase in the volume of space; we just deform its shape. Consider a wave traveling in the z direction. When the x direction is compressed, the y direction expands and vice versa. In Fig. 11.5 we see how the microscopic bits can rearrange themselves locally to create this distortion. This distortion does not change overall volume and thus not require the creation or annihilation of bits. Since this rearrangement of bits is at the local level, it can be completed in order of Planck time, which is much smaller than the period of the wave. We do not have any issue of the kind described in

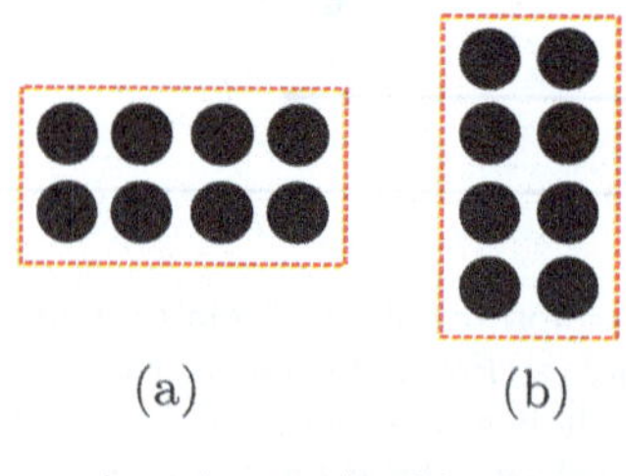

Fig. 11.5 In a gravitational wave, the spin carrying bits of Fig. 11.4 just rearrange themselves locally to create a shape distortion from (**a**) to (**b**); we do not need to create or annihilate bits. Thus, the dynamics of vecros is not involved in this situation, even though the wave moves fast (at the speed of light)

(E)(ii) above. Thus, we do not get any novel effect from vecro dynamics, and general relativity is a valid approximation for waves with $\lambda \gg l_p$.

(ii) Consider the formation of a star from a ball of gas. The particles move much slower than the speed of light, and so light has time to traverse the star multiple times during the star-formation process. The entanglements required to minimize the vacuum energy can be created, and general relativity is again a valid approximation.

(iii) Consider a shell that is collapsing at the speed of light to make a black hole. The black hole puzzle appears strongest when seen in a 'good slicing', so we draw such slices in Fig. 11.6. We see that the slices stretch progressively as the time at infinity evolves. But note that inside the horizon, the analogues of the spins $s_1 s_1'$ in Fig. 11.4 cannot communicate with the analogues of spins $s_2 s_2'$ on the diametrically opposite side, due to the inward pointing structure of light cones. Spins inside the horizon also cannot send signals to the spins just outside. Thus, we will not be able to establish the correlations appropriate to the vacuum spacetime over distances of size $\sim GM$. Thus, in (11.53) we set $R_v^{max} \sim GM$. Using this ρ and a volume $\sim (GM)^3$ for the black hole region, we find an extra energy from the vecro effect on the vacuum

$$E_{extra} \sim \rho (GM)^3 \sim \frac{1}{G(GM)^2}(GM)^3 \sim M. \qquad (11.54)$$

This energy E_{extra} was missing in a semiclassical analysis of collapse. We see that the classical black hole spacetime with stretching slices cannot be a valid on-shell solution of the full quantum gravity theory. The configurations of Fig. 11.6b cost more energy than we have available. This extra energy is carried by virtual excitations of bubbles—KK monopoles and antimonopoles in our toy model—which are separated spatially but have their spins entangled with each other. In a quantum mechanical analogue, these configurations would correspond to points where a particle is 'under a potential barrier'. In Fig. 11.6c we show configurations where the bubbles have linked up into a configuration that describes a fuzzball of mass M; transitioning to one of these configurations is like emerging from under the potential barrier to a new set of allowed states. This is the process of tunneling to fuzzballs.

(G) We have seen above that semiclassical physics breaks down when 'space stretches too fast'. Where else do we find such fast stretching? Consider cosmological expansion. The spatial volume doubles in a time $\sim H^{-1}$, where H is the Hubble constant. If we consider a region with size $L \ll H^{-1}$, then light can traverse this region several times during the doubling time. As we noted above, in this situation the vacuum correlations at scales L will settle down to their vacuum values, and we will get no novel effects from the vecro part of the quantum gravity wavefunctional. But consider regions of size $L \sim H^{-1}$. Here light has barely enough time to cross the region in doubling time, and according to our vecro hypothesis, we expect extra energy density (11.53) with $R_v^{max} \sim H^{-1}$. In [60] it was noted that this extra energy

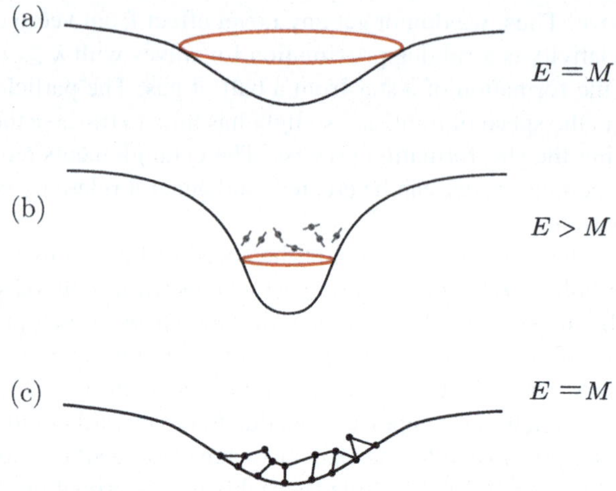

Fig. 11.6 **a** The good slicing of the semiclassical hole manifests a stretching of spatial slices as a shell (indicated by the thick red circle) comes in from infinity. **b** Light signals cannot travel between different points of this stretching region to establish the vecro correlations appropriate for the quantum gravitational vacuum wavefunctional. As a result we have local excitations (indicated by dark dots) that are entangled with each other, in the manner of Fig. 11.4b, and the configuration is 'under the potential barrier' with energy $E > M$ (**c**). The spins in the dark dots can link up in a fuzzball configuration which has energy $E = M$; this is the result of tunneling from the collapsing shell state to a linear combination of fuzzballs

might help resolve issues like the Hubble tension, the cosmological constant, and the energy needed to drive inflation; all of these are effects involving a mysterious source of energy density of the order given by (11.53).

11.6.4 Summary

In this section, we have seen how semiclassical evolution breaks down in the process of black hole formation, even though the curvatures are low in the region around the horizon. This is a second mode of failure of the semiclassical approximation, different from the usual mode where $\mathcal{R} \gtrsim l_p^{-2}$. This second mode involves fast stretching of space, which leads to a non-adiabatic evolution of correlations that arise from virtual black hole microstates. We always knew that the Bekenstein entropy was large, but without an understanding of the microstates counted by this entropy, we could not find the role of this entropy in black hole dynamics. With the fuzzball construction, we now understand that black hole microstates are extended structures of horizon size. The corresponding correlations in the vacuum wavefunctional—vecros—therefore also extend over all length scales, and create extra energy when the expansion is too quick to allow these correlations to settle down to their optimal form [61]. This extra energy creates a potential barrier to evolving along the traditional semiclassical path, and leads to a tunneling to fuzzballs instead. This is the resolution of the information paradox.

Note that we have maintained causality and locality at all steps in our analysis: no signals propagate outside the light cone, and the Hamiltonian of the fundamental theory does not have interactions between spatially distant points. The vecro correlations on all length scales in the Minkowski vacuum arise from this local Hamiltonian, and can exist because this space has an infinitely long past history which allowed these correlations to form. In an expanding cosmology, light has not had time to travel further than the particle horizon, and so there are no vecro correlations outside this distance; this was an essential ingredient in the cosmological analysis of [60].

Correlations across spatial distances also arise in the local, causal theory of a scalar field in Minkowski space, and are reflected in the Wightman functions between spatially separated points. One might then ask if vecro correlations can be seen in quantum fluctuations of the observed low energy quantum fields of the standard model. But in string theory, the standard model fields arise as the analogues of the gravitational waves depicted in Fig. 11.5; the only (irrelevant) difference is that the polarization of these waves is in the compact directions. As we saw in Sect. 11.6.3, part (F)(i), in a gravitational wave we do not change the number of bits involved in the gravitational wavefunctional; we simply rearrange these bits locally. Thus while vecro correlations are in principle just like the correlations across spacelike distances observed in usual quantum fields, they involve a different set of excitations, and affect dynamics only when we have the 'fast stretching' described above.[5]

11.6.5 The Possibility of Fuzzball Complementarity

We still have one question left: what is the fate of an infalling observer? In the classical black hole, the observer would pass uneventfully through the horizon, and feel strong tidal forces only as he approaches the singularity at $r = 0$. In the fuzzball paradigm, the observer would transition to fuzzball degrees of freedom as he reaches the horizon. Is there still some way that he could feel that he is falling through the horizon into a smooth black hole interior?

To understand the answer, it is useful to step back into the history of different ideas and counter-ideas that have been proposed for this 'infall' question.

Starting around 1988, 't Hooft argued in a set of papers that matter falling in towards the horizon of a black hole will bounce off the horizon and return to infinity [62]. It was soon noted that his scattering process used 'post-selection' of states, and was differed from the usual Schrödinger evolution along the 'good slices' of the black hole geometry.

[5] In a rough analogy, we may say that the usual low energy quanta of the standard model are like phonons in a superfluid, while the bits involved in the vecro correlations are like the rotons.

'tHooft's argument suggested that ingoing and outgoing modes were not different parts of a quantum field, but were 'complementary' in the way that position and momentum are complementary in quantum mechanics. Susskind et al. [63] attempted to develop the idea of complementarity with the following postulates. (i) Infalling matter gets absorbed by a surface (called the stretched horizon) just outside $r = 2GM$, and the information in this matter gets radiated back to infinity from this surface. (ii) There is a 'complementary' description in which the infalling matter encounters nothing at the horizon and falls smoothly through. (iii) The two copies thus created—one inside and one outside—appear to violate the 'no cloning' theorem of quantum mechanics. But it is argued that 'no cloning' can be violated here, since an observer cannot measure one copy and then the other because of the constraints presented by the horizon.

It was not clear what to make of this notion of black hole complementarity; it seemed to require postulating new physical principles for which we had no evidence. In particular, it was not clear what mechanism would scramble and return information at the stretched horizon, since at the time there was no evidence of any degrees of freedom that could stand at this location without falling through the horizon. But this proposal of complementarity suggested in an abstract way the notion of holography, where the degrees of freedom of the hole could be regarded as living at the boundary of the hole; i.e., at the horizon.

The years 1995 and 1996 saw rapid progress in understanding black hole entropy and the structure of black holes in terms of bound states of D-branes. In June 1997 it was estimated that the size of these bound states was always of the order of the horizon radius, suggesting that black holes in string theory are normal objects with no horizon [64]. Further studies of brane bound states continued to support this conclusion, leading to the emergence of the fuzzball paradigm discussed in this article [35, 65, 66]. The fuzzball resolution of the information paradox is quite different from Susskind's conjecture of complementarity. Susskind had required that the horizon appear to be a local vacuum in some choice of variables. But with a fuzzball, the low energy modes involved in Hawking radiation do not see a vacuum horizon in any choice of variables.

In November 1997, a concrete realization of holography was obtained in string theory by Maldacena [67–69]. Imagine a graviton falling onto a stack of D3 branes. The graviton gets absorbed by the branes, its energy getting converted to a complicated collection of open strings stretching between the branes. But with a change of variables, these open strings can be written in terms of closed strings. In this dual description, the infalling graviton passes smoothly into an AdS space, encountering no branes in the process. In the open string description—the 'CFT'—the ball of open string excitations spreads with time over the surface of the D-branes. In the dual 'gravity' description, we simply see the graviton fall deeper into empty AdS space. This AdS/CFT duality is a realization of holography, since in the CFT description the degrees of freedom live only on the surface of the branes, while in the gravity description we have an extra radial direction orthogonal to these branes in which the graviton moves.

The conjecture of AdS/CFT duality says that all the gravitational physics inside AdS space can be recast, in different variables, as a field theory without gravity. We can make and evaporate a black hole in AdS space. Since the field theory is unitary, this indicates that black hole evaporation should be unitary as well. But the AdS/CFT conjecture does not tell us the resolution of the information paradox, since it leaves all possibilities open: (i) The black hole microstates are fuzzballs, so there is no pair creation at a horizon (ii) The hole has its semiclassical form, but nonlocal effects (e.g., wormholes) somehow transport information from the singularity to the radiation region (iii) AdS/CFT duality itself breaks down or gets modified when a black hole forms; after all, quantum theory is valid everywhere other than in the black hole Hawking process, so a similar situation could hold for AdS/CFT duality.

Given this, how should we relate the AdS/CFT duality conjecture to black holes? When this duality was initially proposed, the fuzzball idea was still quite new, and many people considered it too radical a change to the structure of the hole. They assumed that in the gravity description, black holes would continue to be described by the semiclassical hole, at least to a first approximation. But then, how do we evade the Hawking puzzle of monotonically growing entanglement?

One suggestion was that the Hawking computation was itself flawed. Hawking did a leading order computation of pair production using quantum fields on curved space. But there can always be small (hitherto unknown) quantum gravity corrections (say $O(\varepsilon)$, $\varepsilon \ll 1$) to this leading order computation. The number of quanta N emitted by the hole is large, with $N \sim (M/m_p)^2$. Even though the correction to the state of each emitted pair would be small, these small quantum gravity effects could induce subtle correlations between the large number of quanta N. If

$$\varepsilon N \gg 1, \tag{11.55}$$

then the large number of these 'small correction' terms might overwhelm the leading order computation and yield an entanglement $S_{ent} = 0$ for the radiation at the endpoint of evaporation. Then there would be no information paradox.

In 2004, Hawing conceded that his argument for information loss might be invalid due to such small corrections [70]. He noted that the Euclidean path integral would have subleading saddle points. A subleading saddle would produce small corrections to Lorentzian evolution, and perhaps the subtle correlations induced thereby would remove the troublesome entanglement.

But most relativists were perplexed by Hawking's retraction. If the correction to each pair is small, then they expected that the correction to the overall entanglement would be small; thus, such small corrections should have no bearing on the original argument of information loss.

In 2009, it was shown that such small corrections *cannot* remove the troublesome entanglement [4]. If the correction to each pair is small as in (11.4), and if there are no nonlocal Hamiltonian interactions linking the black hole to the radiation at infinity, then the entanglement must continue to rise monotonically like (11.5). This 'small corrections theorem' established fuzzballs as the natural resolution of the information paradox: the theorem says that order unity corrections are required at the horizon to resolve the Hawking puzzle, and all the microstates constructed so far in string theory are found to be fuzzballs which have no horizon.

We have noted above that the conjecture of AdS/CFT duality cannot by itself tell us how to resolve the information paradox. But this duality conjecture did suggest a possibility for the *infall* problem. Hawking radiation consists of quanta with energies $E \sim T$, where T is the temperature of the hole. These radiated quanta cannot see a vacuum around the horizon in any set of variables, since quanta emerging from the vacuum have the state (11.4) and not a state that carries information about the hole. But the infall problem addresses observers that fall *into* the hole. The temperature of the hole is very low: the wavelength $\lambda \sim 1/T$ is order the horizon radius. Thus, we can focus on infalling objects which have $E \gg T$. Could it be that in some choice of variables, such infalling objects feel that they are falling through the classical black hole geometry, to a first approximation? This experience of free fall would have to be approximate, since it must fail completely when $E \sim T$; thus the best that we can hope for is that the approximation becomes better and better in the limit $E/T \gg 1$.

More explicitly, the analogy with AdS/CFT would be as follows:

(i) The infalling object reaches the surface of the hole, and its energy gets converted to fuzzball degrees of freedom, yielding a complicated stringy mess. The analogue in AdS/CFT would be that a graviton hits a collection of D3 branes and gets converted to a messy collection of open strings.
(ii) In some other variables, the object falling onto the horizon would appear to continue through the horizon into a black hole interior. The analogue in AdS/CFT would be the gravity description, where the incoming graviton continues to fall smoothly into AdS space.

The black hole case would still differ from AdS/CFT in the following way: the dual smooth infall (ii) emerges only as an approximation that works when $E \gg T$, and fails when, we consider the modes with $E \sim T$ that are involved in Hawking radiation.

The above possibility for infall into a black hole is called 'fuzzball complementarity'. The word 'complementarity' is meant to acknowledge similarities with Susskind's notion of complementarity mentioned above. The word 'fuzzball' emphasizes the difference from Susskind's complementarity. Susskind had imagined that a vacuum region around the horizon would emerge in some variables, and all modes including those with $E \sim T$ could be studied in these variables. In fuzzball complementarity, the modes with $E \sim T$ do not see a vacuum region in *any* choice of variables, and complementarity emerges only as an approximation for $E \gg T$.

In 2012, Almheiri, Marolf, Polchinski and Sully (AMPS) gave a 'firewall' argument to show that Susskind's proposal of complementarity could not work [71]. In the variables described by the stretched horizon, the hole is supposed to radiate like a normal body. For a normal body, in the second half of the evaporation process, the degrees of freedom in the body are almost maximally entangled with the emitted radiation, since the body has fewer bits than the radiation. Consider a hole that is in this second half of its evaporation history. Suppose that in some variables we obtain a horizon with a vacuum region around it, as required by Susskind's complementarity. Then the bits in this vacuum region must be entangled with each other in the form (11.2); i.e., the bits describing excitations b and c are entangled with each other. But the bits in the hole are all supposed to be entangled with the radiation, so they *cannot* entangle with each other (this follows from the 'monogamy of entanglement'). Thus, Susskind's complementarity is not possible.

It had never been clear how Susskind's postulates could be realized, so perhaps there was no great surprise that complementarity could be ruled out. But at first it appeared that the AMPS argument would also rule out the possibility of fuzzball complementarity. The AMPS argument focused on the Hawking modes around the horizon, which have energies $E \sim T$. But if the region around the horizon cannot be a vacuum, then won't every infalling observer interact and scatter off whatever degrees of freedom are present at the fuzzball surface? In that case, it seemed that one could not have fuzzball complementarity either.

In 2014, it was found that there is a flaw in the AMPS argument: while the AMPS argument does rule out Susskind's complementarity, it cannot rule out fuzzball complementarity [72]. To see this flaw, let us recall the argument in more detail.

At first sight, the goal of the AMPS argument seems puzzling. Hawking had shown that if the region around the horizon was a vacuum, then the entanglement of the radiation with the hole would keep growing monotonically. AMPS required that the entanglement should *not* keep rising, and concluded that the region around the horizon *cannot* be a vacuum. Is the AMPS statement not the same as Hawking's statement? Even the line of argumentation was similar: AMPS considered the entangled pairs used by Hawking and described them by the same bit model that was used in the small corrections theorem to make Hawking's argument rigorous.

In fact, AMPS *did* try to do something a little different from Hawking, by making an extra assumption. It is this extra assumption that provides the loophole allowing fuzzball complementarity to exist, so let us recall the assumptions of AMPS in a little more detail.

AMPS required that (i) The entanglement of the radiation with the hole should come down to zero by the end of evaporation, just like it does for a normal body. (ii) Outside a surface at $r = 2GM + \varepsilon$, physics should be 'normal'; i.e., we should just get the usual low energy effective field theory. (iii) The surface at $r = 2GM + \varepsilon$ responds to infalling objects in a causal manner; i.e., it cannot alter its state in response to an infalling particle *before* the infalling particle reaches this surface.

We can now see the difference between what AMPS and Hawking could argue from their respective assumptions. Hawking's argument can indeed be recast as saying that if the entanglement has to come down, then the dynamics around the horizon cannot be that of a vacuum region. But the radiated quanta have a long wavelength $\lambda \sim 1/T \sim 2GM$, so the required departure from semiclassical evolution can be a sharp departure at some location outside the horizon, or a gentle one, say over the region $2GM \lesssim r \lesssim 4GM$. AMPS *postulated* that physics is normal at $r > 2GM + \varepsilon$, so any departure from vacuum dynamics must take place sharply in the region $2GM < r < 2GM + \varepsilon$. In other words, the vacuum around the horizon must be replaced by a sharp 'firewall'.

But instead of focusing on Hawking modes which have $E \sim T$, consider the infall of a shell with $E \gg T$. Since physics is 'normal' at $r > 2GM + \varepsilon$, the usual dynamics of gravity in this region will create a *new* horizon at

$$R_h^{new} = 2G(M + E), \tag{11.56}$$

which is a location outside the stretched horizon at $r = 2GM + \varepsilon$. Light cones 'point inwards' in the region $R_{shell} < r < R_h^{new}$, where R_{shell} is the location of the shell as it falls in. Thus, radiation from the stretched horizon cannot escape to infinity, in contradiction to requirement (i). This difficulty would be avoided if the stretched horizon surface jumped out to a location outside R_h^{new} *before* the infalling shell reached $r = R_h^{new}$, but this is disallowed by postulate (iii). Thus, we see that there is an internal conflict in the assumptions (i)–(iii) used by AMPS: an infalling shell creates a new horizon that causally traps the radiation trying to emerge from the stretched horizon, so information cannot actually emerge like it does for a normal body.

We can now see how fuzzball dynamics can evade the above situation. Consider a fuzzball of mass M, with a surface that is at $2GM + \varepsilon$. Let a shell of energy E fall onto this fuzzball. As the shell approaches the radius R_h^{new} given by (11.56), the fuzzball structure begins to nucleate just *outside* $r = R_h^{new}$. This nucleation is possible because the theory contains extended objects (fuzzballs) of all radii, and the gravity wavefunctional has the tail indicated in Fig. 11.2b over fuzzball configurations of radius $\sim R_h^{new}$. As the shell passes through the radius $r = R_h^{new}$, the virtual fluctuations (vecros) represented by this tail transition to on-shell fuzzballs, with the wavefunctional evolving as in Fig. 11.2c.

Let us return to the issue of how fuzzball complementarity would work. The degrees of freedom that are analogous to the open strings in AdS/CFT are the *new* fuzzballs that are created by the energy of the infalling object—in the above discussion, this infalling object is the shell of energy E. The analogue of the evolution in the CFT would be the continued spreading of the wavefunctional in Fig. 11.2c away from the location of the central potential dip. The graviton falling into AdS in AdS/CFT is analogous to the progressive infall of the shell into the interior geometry of the classical hole.

With this understanding of fuzzball complementarity, we can see how the condition $E \gg T$ bypasses the AMPS argument. The fuzzball of mass M has entropy $S_{bek}[M]$ and therefor a number of possible states

$$N_{states}[M] \sim e^{S_{bek}[M]}. \tag{11.57}$$

After the shell of energy E comes in, the total number of accessible states is

$$N_{states}[M + E] \sim e^{S_{bek}[M+E]}. \tag{11.58}$$

Assuming $E \ll M$ for simplicity, we have

$$S_{bek}[M+E] \approx S_{bek}[M] + \frac{dS_{bek}}{dE}[M]E = S_{bek}[M] + \frac{E}{T}. \tag{11.59}$$

Thus,

$$\frac{N_{states}[M+E]}{N_{states}[M]} \sim e^{\frac{E}{T}}. \tag{11.60}$$

Thus, for $E \gg T$ most of the accessible states are *new* states, not present in the hole of mass M. The AMPS argument against complementarity was based on the assumption that the states of the hole were, after the halfway evaporation point, maximally entangled with the radiation. *But these new states created by the infall are not entangled with anything.* The spread of the wavefunctional over these newly accessible states can give the 'fuzzball complementarity' description of the shell, where the shell appears to fall through the interior of a classical black hole geometry.

In [73] a bit model was given that exhibits fuzzball complementarity.

We thus see that fuzzball complementarity is *possible*; the AMPS argument cannot establish that infallers with energy $E \gg T$ feel a firewall at the horizon. On the other hand, there is no reason that the dynamics of fuzzballs should be such that we *do* get fuzzball complementarity. While the information paradox is a puzzle that needs to be resolved (otherwise we violate quantum theory), there is no a priori requirement on what an infalling observer should feel. In other words, it is completely consistent that the infalling observer smashes onto the surface of a fuzzball, with no feeling on continuing to fall smoothly through in any choice of variables. However, the AMPS argument cannot establish that the infaller feels a firewall, since for an infaller with $E \gg T$ most of the accessible degrees of freedom are not entangled with anything, even if the hole is old (i.e, past the halfway evaporation point).

11.6.6 Wormholes

In 2013, Maldacena and Susskind [74] proposed a picture of the hole where the horizon had the traditional semiclassical form, but tiny wormholes linked the inside and outside members of the entangled pairs (b, c) of (11.4). This proposal led to much confusion. Were the wormholes just an alternative depiction of the fact that the two members of the Hawking pair were entangled? If so, how does wormhole picture say anything new about the entanglement problem? Or were the wormholes indicative of a nonlocal Hamiltonian interaction between the region inside the hole and the radiated quanta near infinity? These two regions are causally separated from each other, and there is no evidence of such large scale violation of causality in string theory.

While the wormhole proposal remained unclear, it led to a belief about black holes which turned out to be incorrect. This belief was as follows. Suppose we accept that the black hole radiates from its surface like a normal body. Fuzzballs give such a behavior for the hole, so we can assume that the black hole is a fuzzball in its exact quantum gravity description. But, according to the above-mentioned belief, there is a ‘code subspace’ out of all the complicated quantum gravitational degrees of freedom which describes the low energy dynamics of the fuzzball, and this low energy dynamics is to a first approximation the usual semiclassical dynamics of the traditional hole.

However, the Effective small corrections theorem [75] showed that such a code subspace cannot exist. This theorem is a small extension of the small corrections theorem, and says the following. Consider the exact quantum gravity Hilbert space of the black hole region, which we can take, for concreteness, as the region $r < 4GM$. Suppose there is a linear subspace of this exact Hilbert space (the code subspace) in which the evolution approximates the usual semiclassical evolution. (We allow the possibility that this code subspace gives the semiclassical evolution for only a few crossing times; after that we may need a different subspace to function as the code subspace.) Finally, we assume that there are no nonlocal Hamiltonian interactions between the hole and the region near infinity. Then the theorem says that with these assumptions, the entanglement S_{ent} in the *exact* theory will have to keep rising monotonically. In other words, if the Page curve is to come down in the exact theory, then we cannot have a code subspace which gives semiclassical dynamics to leading order.

The intuitive idea behind the Effective small corrections theorem is simple. If a semiclassical horizon emerges as a leading order approximation to the exact dynamics, then Hawking pairs will be created in this semiclassical description, with the form (11.4). Then the entanglement S_{ent} of the hole with the radiation will keep rising monotonically in this semiclassical approximation. The small corrections theorem can then be used to prove that the entanglement will keep rising monotonically in the *exact* theory, since evolution in the code subspace of the exact theory departs only in a ‘small’ way from semiclassical evolution.

There were attempts to evade this conclusion by making the code subspace using the bits at infinity along with the bits in the black hole region (the so-called $A = R_B$ models), but in this case one does not obtain semiclassical dynamics around the horizon because the dynamics of the bits at infinity is not the one required by semiclassical dynamics in the horizon region.

11.7 Summary

Hawking’s black hole information paradox posed a remarkable challenge to the unification of general relativity and quantum theory. General relativity predicts that black holes will form, and the evaporation of these holes appears to violate the unitarity of quantum theory. The resolution of the puzzle through string theory is equally remarkable: quantum gravity effects modify the entire region inside the horizon, so that we get a fuzzball—a ball of stringy matter with no horizon. This fuzzball radiates from its surface like a normal body; i.e., not by the creation of pairs from the vacuum. Thus, there is no information paradox.

Some people found this resolution of the puzzle surprising, because one usually assumes that quantum gravity effects will only stretch over distances $\sim l_p$. After all, Planck length is the only length scale that we can make from the fundamental constants $c, \hbar, G$. But a large black hole is made of a large number of elementary quanta N, so we need to ask if the size of a bound state is $\sim l_p$ or $\sim N^\alpha l_p$ for some $\alpha > 0$. The fuzzball constructions indicate that the latter is true, with α such that the radius of the generic bound state is order the horizon radius.

The fuzzball paradigm emerged soon after the discovery of D-branes and the construction of black holes as bound states of branes. In [64] an estimate of the radius of brane bound states gave a result of order of the horizon radius. Over the years several families of brane bound states have been explicitly constructed, and in each case we find a 'fuzzball'; i.e., an object with no horizon. Note that a fuzzball is in general a very quantum object, though we often construct examples of fuzzballs in terms of a classical geometry. An analogy is black body radiation in a box: the general state is very far from classical, with occupation number $\langle n_{\mathbf{k}}\rangle \sim O(1)$ for the typical mode $\mathbf{k}$. But to get an estimate of the energy or pressure of this radiation, we can look at specific states where we place all the energy in a few harmonics with $\langle n_{\mathbf{k}}\rangle \gg 1$; such states can be understood in terms of the classical electromagnetic fields $\mathbf{E}, \mathbf{B}$. The classical fuzzball geometries are the analogues of these latter states, and help us to understand how the no-hair theorems can be broken in string theory to yield horizon-sized structures that describe the microstates of the black hole. But the relevant aspect of all fuzzball solutions—classical or quantum—is that there is no horizon around which the usual low energy semiclassical evolution holds.

Since all the microstates constructed so far have turned out to be fuzzballs, it does not appear plausible that the remaining microstates do have a horizon and that the information puzzle is resolved in string theory by some alternative to the fuzzball paradigm. Nevertheless, a lot of confusion has persisted over the years due to attempts where one tries to maintain the traditional semiclassical horizon and yet get the information to emerge in the Hawking radiation. Some examples of such attempts are the following:

(i) The small corrections idea was based on assuming a semiclassical horizon at leading order, and arguing that 'subleading saddle points' in the Euclidean path integral would introduce delicate correlations in the radiation to remove the problematic entanglement [70]. The 'Small corrections theorem' showed that this idea cannot work [4].

(ii) The idea of complementarity postulated that with one choice of variables the information reflects off a stretched horizon outside the hole, while with another choice we get the usual semiclassical vacuum around the horizon. The AMPS firewall argument [71] showed that such complementarity is not possible.

(iii) Once it became clear that microstates were fuzzballs, there were attempts to argue that the low energy dynamics of fuzzballs would be captured by evolution in a 'code subspace' of the full Hilbert space, and that dynamics in this code subspace would be approximately the traditional semiclassical dynamics around a horizon. The 'Effective small corrections theorem' (an extension of the small corrections theorem) showed that this is not possible [75].

In short, the situation in string theory is the following. There is only *one* exact description of a black hole microstate; this state is in general a very messy quantum fuzzball, though for special states it may be well approximated by a classical geometry without horizon. The fuzzball radiates from its surface like any other quantum body, through quanta of energy, $E \sim T$ where T is the temperature of the hole. The dynamics of these radiated quanta near the fuzzball surface depends on the details of the fuzzball microstate; that is, how these quanta are able to carry the information of the hole. There is *no* description of the microstate in which we obtain the semiclassical horizon dynamics for these $E \sim T$ quanta. The detailed arguments for this are covered in the results (i)–(iii) above, but in essence the reason is simple: if one could replace the fuzzball by an effective semiclassical horizon, then these $E \sim T$ modes will be produced as members of entangled Hawking pairs, and will not carry any information since the vacuum does not have any information about the microstate.

The picture of the vacuum described by the vecro hypothesis shows how the semiclassical approximation can break down in gravitational collapse and create fuzzballs, even though the curvature is low at the horizon. The key is that in this situation, space is stretching too *fast*: light cannot travel across the stretching region to create the correlations required to obtain a vacuum on the stretched manifold. We noted that in other low curvature situations like gravitational waves or star formation, we do not have this fast stretching, and so the semiclassical approximation does not break down. At the scale of the cosmological horizon we again encounter fast stretching, and here we do find the need for unexplained new sources of energy, like dark energy and the Early Dark Energy that might explain the Hubble tension.

One can extract the essential ideas of string theory in an abstract form [76]; it is plausible that these ideas can be incorporated into other alternatives for quantizing gravity. In particular, the vecro picture of the gravitational vacuum can be a starting point for how to think of spacetime in any theory of gravity; this picture is essential to resolving the information paradox while preserving causality and locality.

Acknowledgments We are grateful to all the people who have worked over the years on the black hole information puzzle, and in particular to all those who have contributed to the fuzzball paradigm. This work was supported in part by DOE grant DE-FG02-91ER-40690.

References

1. J.D. Bekenstein, Black holes and entropy. Phys. Rev. D **7**, 2333–2346 (1973). https://doi.org/10.1103/PhysRevD.7.2333
2. S.W. Hawking, Particle creation by black holes. Commun. Math. Phys. **43**, 199–220 (1975). https://doi.org/10.1007/BF02345020 [Erratum: Commun. Math. Phys. **46**, 206 (1976)]

3. S.W. Hawking, Breakdown of predictability in gravitational collapse. Phys. Rev. D **14**, 2460–2473 (1976). https://doi.org/10.1103/PhysRevD.14.2460. https://link.aps.org/doi/10.1103/PhysRevD.14.2460
4. S.D. Mathur, The information paradox: a pedagogical introduction. Class. Quant. Grav. **26**, 224001 (2009). https://doi.org/10.1088/0264-9381/26/22/224001. arXiv:0909.1038 [hep-th]
5. R.M. Wald, Black hole entropy is the Noether charge. Phys. Rev. D **48**, R3427–R3431 (1993). https://doi.org/10.1103/PhysRevD.48.R3427. https://link.aps.org/doi/10.1103/PhysRevD.48.R3427
6. A. Dabholkar, Exact counting of black hole microstates. Phys. Rev. Lett. **94**, 241301 (2005). https://doi.org/10.1103/PhysRevLett.94.241301. arXiv:hep-th/0409148
7. A. Sen, Black hole solutions in heterotic string theory on a torus. Nucl. Phys. B **440**, 421–440 (1995). https://doi.org/10.1016/0550-3213(95)00063-X. arXiv:hep-th/9411187
8. A. Sen, Extremal black holes and elementary string states. Mod. Phys. Lett. A **10**, 2081–2094 (1995). https://doi.org/10.1142/S0217732395002234. arXiv:hep-th/9504147
9. A. Strominger, C. Vafa, Microscopic origin of the Bekenstein-Hawking entropy. Phys. Lett. B **379**, 99–104 (1996). https://doi.org/10.1016/0370-2693(96)00345-0. arXiv:hep-th/9601029
10. C.G. Callan, J.M. Maldacena, D-brane approach to black hole quantum mechanics. Nucl. Phys. B **472**, 591–610 (1996). https://doi.org/10.1016/0550-3213(96)00225-8. arXiv:hep-th/9602043
11. G.T. Horowitz, J.M. Maldacena, A. Strominger, Nonextremal black hole microstates and U duality. Phys. Lett. B **383**, 151–159 (1996). https://doi.org/10.1016/0370-2693(96)00738-1. arXiv:hep-th/9603109
12. S.R. Das, S.D. Mathur, Comparing decay rates for black holes and D-branes. Nucl. Phys. B **478**, 561–576 (1996). https://doi.org/10.1016/0550-3213(96)00453-1. arXiv:hep-th/9606185
13. J. Maldacena, A. Strominger, Black hole greybody factors and d-brane spectroscopy. Phys. Rev. D **55**, 861–870 (1997). https://doi.org/10.1103/PhysRevD.55.861. https://link.aps.org/doi/10.1103/PhysRevD.55.861
14. O. Lunin, S.D. Mathur, AdS/CFT duality and the black hole information paradox. Nucl. Phys. B **623**, 342–394 (2002). https://doi.org/10.1016/S0550-3213(01)00620-4. arXiv:hep-th/0109154
15. O. Lunin, J.M. Maldacena, L. Maoz, Gravity solutions for the D1–D5 system with angular momentum. arXiv:hep-th/0212210
16. I. Kanitscheider, K. Skenderis, M. Taylor, Fuzzballs with internal excitations. JHEP **06**, 056 (2007). https://doi.org/10.1088/1126-6708/2007/06/056. arXiv:0704.0690 [hep-th]
17. V. Balasubramanian, J. de Boer, E. Keski-Vakkuri, S.F. Ross, Supersymmetric conical defects: towards a string theoretic description of black hole formation. Phys. Rev. D **64**, 064011 (2001). https://doi.org/10.1103/PhysRevD.64.064011. arXiv:hep-th/0011217
18. J.M. Maldacena, L. Maoz, Desingularization by rotation. JHEP **12**, 055 (2002). https://doi.org/10.1088/1126-6708/2002/12/055. arXiv:hep-th/0012025
19. O. Lunin, S.D. Mathur, Metric of the multiply wound rotating string. Nucl. Phys. B **610**, 49–76 (2001). https://doi.org/10.1016/S0550-3213(01)00321-2. arXiv:hep-th/0105136
20. E.J. Martinec, W. McElgin, String theory on AdS orbifolds. JHEP **04**, 029 (2002). https://doi.org/10.1088/1126-6708/2002/04/029. arXiv:hep-th/0106171
21. R. Emparan, D. Mateos, P.K. Townsend, Supergravity supertubes. JHEP **07**, 011 (2001). https://doi.org/10.1088/1126-6708/2001/07/011. arXiv:hep-th/0106012
22. O. Lunin, S.D. Mathur, Statistical interpretation of Bekenstein entropy for systems with a stretched horizon. Phys. Rev. Lett. **88**, 211303 (2002). https://doi.org/10.1103/PhysRevLett.88.211303. arXiv:hep-th/0202072
23. S.D. Mathur, Black hole size and phase space volumes. arXiv:0706.3884 [hep-th]
24. V.S. Rychkov, D1–D5 black hole microstate counting from supergravity. JHEP **01**, 063 (2006). https://doi.org/10.1088/1126-6708/2006/01/063. arXiv:hep-th/0512053
25. S.D. Mathur, A. Saxena, Y.K. Srivastava, Constructing 'hair' for the three charge hole. Nucl. Phys. B **680**, 415–449 (2004). https://doi.org/10.1016/j.nuclphysb.2003.12.022. arXiv:hep-th/0311092

26. O. Lunin, Adding momentum to D-1–D-5 system. JHEP **04**, 054 (2004). https://doi.org/10.1088/1126-6708/2004/04/054. arXiv:hep-th/0404006
27. S. Giusto, S.D. Mathur, A. Saxena, Dual geometries for a set of 3-charge microstates. Nucl. Phys. B **701**, 357–379 (2004). https://doi.org/10.1016/j.nuclphysb.2004.09.001. arXiv:hep-th/0405017
28. S. Giusto, S.D. Mathur, A. Saxena, 3-charge geometries and their CFT duals. Nucl. Phys. B **710**, 425–463 (2005). https://doi.org/10.1016/j.nuclphysb.2005.01.009. arXiv:hep-th/0406103
29. A. Galliani, S. Giusto, E. Moscato, R. Russo, Correlators at large c without information loss. JHEP **09**, 065 (2016). https://doi.org/10.1007/JHEP09(2016)065. arXiv:1606.01119 [hep-th]
30. J.B. Gutowski, D. Martelli, H.S. Reall, All supersymmetric solutions of minimal supergravity in six-dimensions. Class. Quant. Grav. **20**, 5049–5078 (2003). https://doi.org/10.1088/0264-9381/20/23/008. arXiv:hep-th/0306235
31. S. Giusto, S.D. Mathur, Geometry of D1–D5-P bound states. Nucl. Phys. B **729**, 203–220 (2005). https://doi.org/10.1016/j.nuclphysb.2005.09.037. arXiv:hep-th/0409067
32. I. Bena, N.P. Warner, One ring to rule them all ... and in the darkness bind them? Adv. Theor. Math. Phys. **9**(5), 667–701 (2005). https://doi.org/10.4310/ATMP.2005.v9.n5.a1. arXiv:hep-th/0408106
33. P. Berglund, E.G. Gimon, T.S. Levi, Supergravity microstates for BPS black holes and black rings. JHEP **06**, 007 (2006). https://doi.org/10.1088/1126-6708/2006/06/007. arXiv:hep-th/0505167
34. I. Bena, N.P. Warner, Bubbling supertubes and foaming black holes. Phys. Rev. D **74**, 066001 (2006). https://doi.org/10.1103/PhysRevD.74.066001. https://link.aps.org/doi/10.1103/PhysRevD.74.066001
35. I. Bena, N.P. Warner, Black holes, black rings and their microstates. Lect. Notes Phys. **755**, 1–92 (2008). https://doi.org/10.1007/978-3-540-79523-0_1. arXiv:hep-th/0701216
36. I. Bena, S. Giusto, R. Russo, M. Shigemori, N.P. Warner, Habemus superstratum! A constructive proof of the existence of superstrata. JHEP **05**, 110 (2015). https://doi.org/10.1007/JHEP05(2015)110. arXiv:1503.01463 [hep-th]
37. I. Bena, S. Giusto, E.J. Martinec, R. Russo, M. Shigemori, D. Turton, N.P. Warner, Smooth horizonless geometries deep inside the black-hole regime. Phys. Rev. Lett. **117**(20), 201601 (2016). https://doi.org/10.1103/PhysRevLett.117.201601. arXiv:1607.03908 [hep-th]
38. V. Jejjala, O. Madden, S.F. Ross, G. Titchener, Non-supersymmetric smooth geometries and D1–D5-P bound states. Phys. Rev. D **71**, 124030 (2005). https://doi.org/10.1103/PhysRevD.71.124030. arXiv:hep-th/0504181
39. V. Cardoso, O.J.C. Dias, J.L. Hovdebo, R.C. Myers, Instability of non-supersymmetric smooth geometries. Phys. Rev. D **73**, 064031 (2006). https://doi.org/10.1103/PhysRevD.73.064031. arXiv:hep-th/0512277
40. B.D. Chowdhury, S.D. Mathur, Radiation from the non-extremal fuzzball. Class. Quant. Grav. **25**, 135005 (2008). https://doi.org/10.1088/0264-9381/25/13/135005. arXiv:0711.4817 [hep-th]
41. B.D. Chowdhury, S.D. Mathur, Pair creation in non-extremal fuzzball geometries. Class. Quant. Grav. **25**, 225021 (2008). https://doi.org/10.1088/0264-9381/25/22/225021. arXiv:0806.2309 [hep-th]
42. B. Chakrabarty, D. Turton, A. Virmani, Holographic description of non-supersymmetric orbifolded D1–D5-P solutions. JHEP **11**, 063 (2015). https://doi.org/10.1007/JHEP11(2015)063. arXiv:1508.01231 [hep-th]
43. S.D. Mathur, D. Turton, Oscillating supertubes and neutral rotating black hole microstates. JHEP **04**, 072 (2014). https://doi.org/10.1007/JHEP04(2014)072. arXiv:1310.1354 [hep-th]
44. G. Bossard, S. Katmadas, D. Turton, Two kissing bolts. JHEP **02**, 008 (2018). https://doi.org/10.1007/JHEP02(2018)008. arXiv:1711.04784 [hep-th]
45. B. Ganchev, S. Giusto, A. Houppe, R. Russo, N.P. Warner, Microstrata. JHEP **10**, 163 (2023). https://doi.org/10.1007/JHEP10(2023)163. arXiv:2307.13021 [hep-th]
46. B. Guo, S. Hampton, S.D. Mathur, Can we observe fuzzballs or firewalls? JHEP **07**, 162 (2018). https://doi.org/10.1007/JHEP07(2018)162. arXiv:1711.01617 [hep-th]

47. E. Witten, Instability of the Kaluza-Klein Vacuum. Nucl. Phys. B **195**, 481–492 (1982). https://doi.org/10.1016/0550-3213(82)90007-4
48. S.D. Mathur, What prevents gravitational collapse in string theory? Int. J. Mod. Phys. D **25**(12), 1644018 (2016). https://doi.org/10.1142/S0218271816440181. arXiv:1609.05222 [hep-th]
49. G.W. Gibbons, N.P. Warner, Global structure of five-dimensional fuzzballs. Class. Quant. Grav. **31**, 025016 (2014). https://doi.org/10.1088/0264-9381/31/2/025016. arXiv:1305.0957 [hep-th]
50. I. Bah, P. Heidmann, Topological stars, black holes and generalized charged Weyl solutions. JHEP **09**, 147 (2021). https://doi.org/10.1007/JHEP09(2021)147. arXiv:2012.13407 [hep-th]
51. V. Balasubramanian, E.G. Gimon, T.S. Levi, Four dimensional black hole microstates: from D-branes to spacetime foam. JHEP **01**, 056 (2008). https://doi.org/10.1088/1126-6708/2008/01/056. arXiv:hep-th/0606118
52. S. Toktarbay, H. Quevedo, A stationary q-metric. Grav. Cosmol. **20**, 252 (2014). https://doi.org/10.1134/S0202289314040136. arXiv:1510.04155 [gr-qc]
53. V.A. Belinskii, E.M. Lifshitz, I.M. Khalatnikov, On a general cosmological solution of the Einstein equations with a time singularity. Zh. Eksp. Teor. Fiz. Ser. **62**, 1606–1613 (1972)
54. S.D. Mathur, M. Mehta, The universality of black hole thermodynamics. Int. J. Mod. Phys. D **32**(14), 2341003 (2023). https://doi.org/10.1142/S0218271823410031. arXiv:2305.12003 [hep-th]
55. S.D. Mathur, M. Mehta, The universal thermodynamic properties of extremely compact objects. Class. Quant. Grav. **41**(23), 235011 (2024). https://doi.org/10.1088/1361-6382/ad869e. arXiv:2402.13166 [hep-th]
56. S.D. Mathur, Tunneling into fuzzball states. Gen. Rel. Grav. **42**, 113–118 (2010). https://doi.org/10.1007/s10714-009-0837-3. arXiv:0805.3716 [hep-th]
57. P. Kraus, S.D. Mathur, Nature abhors a horizon. Int. J. Mod. Phys. D **24**(12), 1543003 (2015). https://doi.org/10.1142/S0218271815430038. arXiv:1505.05078 [hep-th]
58. I. Bena, D.R. Mayerson, A. Puhm, B. Vercnocke, Tunneling into microstate geometries: quantum effects stop gravitational collapse. JHEP **07**, 031 (2016). https://doi.org/10.1007/JHEP07(2016)031. arXiv:1512.05376 [hep-th]
59. S.D. Mathur, The VECRO hypothesis. arXiv:2001.11057 [hep-th]
60. S.D. Mathur, The elastic vacuum. Int. J. Mod. Phys. D **30**(14), 2141001 (2021). https://doi.org/10.1142/S0218271821410017. arXiv:2105.06963 [hep-th]
61. S.D. Mathur, The secret structure of the gravitational vacuum. arXiv:2405.08945 [hep-th]
62. C.R. Stephens, G. 't Hooft, B.F. Whiting, Black hole evaporation without information loss. Class. Quant. Grav. **11**, 621–648 (1994). https://doi.org/10.1088/0264-9381/11/3/014. arXiv:gr-qc/9310006
63. L. Susskind, L. Thorlacius, J. Uglum, The stretched horizon and black hole complementarity. Phys. Rev. D **48**, 3743–3761 (1993). https://doi.org/10.1103/PhysRevD.48.3743. arXiv:hep-th/9306069
64. S.D. Mathur, Emission rates, the correspondence principle and the information paradox. Nucl. Phys. B **529**, 295–320 (1998). https://doi.org/10.1016/S0550-3213(98)00336-8. arXiv:hep-th/9706151
65. S.D. Mathur, The fuzzball proposal for black holes: an elementary review. Fortsch. Phys. **53**, 793–827 (2005). https://doi.org/10.1002/prop.200410203. arXiv:hep-th/0502050
66. B.D. Chowdhury, A. Virmani, Modave lectures on fuzzballs and emission from the D1–D5 system, in *5th Modave Summer School in Mathematical Physics*, vol. 1 (2010). arXiv:1001.1444 [hep-th]
67. J.M. Maldacena, The large N limit of superconformal field theories and supergravity. Adv. Theor. Math. Phys. **2**, 231–252 (1998). https://doi.org/10.4310/ATMP.1998.v2.n2.a1. arXiv:hep-th/9711200
68. S.S. Gubser, I.R. Klebanov, A.M. Polyakov, Gauge theory correlators from noncritical string theory. Phys. Lett. B **428**, 105–114 (1998). https://doi.org/10.1016/S0370-2693(98)00377-3. arXiv:hep-th/9802109

69. E. Witten, Anti-de Sitter space and holography. Adv. Theor. Math. Phys. **2**, 253–291 (1998). https://doi.org/10.4310/ATMP.1998.v2.n2.a2. arXiv:hep-th/9802150
70. S.W. Hawking, Information loss in black holes. Phys. Rev. D **72**, 084013 (2005). https://doi.org/10.1103/PhysRevD.72.084013. arXiv:hep-th/0507171
71. A. Almheiri, D. Marolf, J. Polchinski, J. Sully, Black holes: complementarity or firewalls? JHEP **02**, 062 (2013). https://doi.org/10.1007/JHEP02(2013)062. arXiv:1207.3123 [hep-th]
72. S.D. Mathur, D. Turton, The flaw in the firewall argument. Nucl. Phys. B **884**, 566–611 (2014). https://doi.org/10.1016/j.nuclphysb.2014.05.012. arXiv:1306.5488 [hep-th]
73. S.D. Mathur, A model with no firewall. arXiv:1506.04342 [hep-th]
74. J. Maldacena, L. Susskind, Cool horizons for entangled black holes. Fortsch. Phys. **61**, 781–811 (2013). https://doi.org/10.1002/prop.201300020. arXiv:1306.0533 [hep-th]
75. B. Guo, M.R.R. Hughes, S.D. Mathur, M. Mehta, Contrasting the fuzzball and wormhole paradigms for black holes. Turk. J. Phys. **45**(6), 281–365 (2021). https://doi.org/10.3906/fiz-2111-13. arXiv:2111.05295 [hep-th]
76. S.D. Mathur, How the black hole puzzles are resolved in string theory. Gen. Rel. Grav. **57**(1), 3 (2025). https://doi.org/10.1007/s10714-024-03336-3

Chapter 12
Fast Scrambling and the Emergence of Black Hole Firewalls

Zhi-Wei Wang, Saurya Das, and Samuel L. Braunstein

Abstract Under reasonable assumptions, black holes have been argued to form firewalls, burning up anything crossing their horizons. This argument finds that a firewall would appear very late in a black hole's lifetime, when Hawking radiation has caused the horizon to shrink to one-half its original area. For stellar-mass black holes, the time-scale for this process to produce a firewall surpasses the universe's present age and so no such black hole would currently possess a firewall. However, black holes have recently been conjectured to scramble their interior degrees of freedom with a scrambling time scale comparable to the time it takes light to travel a Schwarzschild radius' distance. Here we prove that local observers will already experience a firewall from the scrambling time onwards after the black hole's formation. Here 'local' means that the observer couples to fewer than one-half the black hole's total interior 'qubits.' Indeed, for observers to fail to be local in this manner, it would mean that they couple to more 'qubits' within such black holes than exist in all the stars of the observable universe. Therefore we find that if black holes are indeed fast scramblers, then every astrophysical black hole in the universe will already have a fully developed firewall for any local physical process.

Z.-W. Wang (✉)
College of Physics, Jilin University, Changchun, People's Republic of China

S. Das
Theoretical Physics Group, Department of Physics and Astronomy, University of Lethbridge, Lethbridge, Alberta, Canada
e-mail: saurya.das@uleth.ca

S. L. Braunstein (✉)
Computer Science, University of York, York, UK
e-mail: sam.braunstein@york.ac.uk

A. Akil and C. Bambi (eds.), *The Black Hole Information Paradox*, Springer Series in Astrophysics and Cosmology, https://doi.org/10.1007/978-981-96-6170-1_12

12.1 The Case for and Against the Black Hole Firewall

According to classical general relativity, black holes are regions from which nothing can escape. However, Hawking's seminal work [11] showed that black holes emit thermal radiation, known as Hawking radiation, due to quantum effects near the event horizon. If a black hole evaporates completely and this radiation is purely thermal, then it can carry no information about the matter that originally collapsed to form the black hole. Hawking therefore argued that black hole evaporation would inexorably lead to a loss of unitarity and a breakdown of quantum mechanics [11]. This is the key reasoning behind the black hole information paradox.

To better appreciate the implications of this paradox, Page considered a simple quantum information theoretic model of 'black hole' evaporation which was manifestly unitary [20]. In its simplest form, this model consisted of starting with a pure-state black hole and randomly selecting a subspace of the black hole interior that will be 'ejected' as the Hawking radiation. Under the assumption of selection via random unitaries, Page's model yields a universal curve for the entropy of the radiation as a function of the size of the radiation subspace [20]. It should be noted, that Page's model implies that the black hole interior Hilbert space is decreasing as the black hole itself evaporates. This behavior is consistent with the so-called 'central dogma' of black hole physics [3], which states that the interior Hilbert space dimension, N, may be quantified by the Bekenstein-Hawking entropy, S, of the black hole's horizon by $N = \exp(S)$, where Boltzmann's constant is taken to be unity.

Page's universal curve for black hole evaporation, known as the Page curve, displays a von Neumann entropy for the radiation that increases by almost exactly one bit for every extra qubit of radiation 'emitted,' until the black hole has shrunk to one-half its original surface area (at the so-called Page time), after which the entropy symmetrically drops by one bit for each additional qubit of radiation [20]. Complete evaporation leads to the radiation having net zero entropy corresponding to a globally pure state.

It will be convenient to call a black hole that has been evaporating for longer than its Page time an 'old' black hole. Similarly, black holes are considered 'young' prior to their reaching their respective Page time. According to this nomenclature, all known astrophysical black holes are believed to be young.

For young black holes, every qubit of Hawking radiation is maximally entangled with some subspace remaining within the black hole. Further, according to the central dogma, the Page time is the point in a black hole's evaporation, where exactly one-half of the number of qubits accounting for the initial black hole Hilbert space have evaporated away as Hawking radiation. At this point, the radiation is maximally entangled with the entirety of the black hole interior Hilbert space. As the black hole transitions to being an old black hole, the remaining black hole interior is maximally entangled with some subspace of the distant Hawking radiation.

According to this reasoning, the reduced state of an old black hole is a maximally-mixed state. Naively, an infinite temperature state will exist within the black hole interior Hilbert space at the Page time.

We can come to a similar conclusion about old black holes without relying on the reduced state of the interior. In particular, according to quantum theory, bipartite entanglement is monogamous [8, 15]. This means that for old black holes, the bipartite entanglement between the black hole interior and some subspace of the radiation precludes the black hole interior being entangled with any other system. In particular, this implies that there can be no trans-event horizon entanglement between the black hole interior and its neighboring exterior from the Page time onwards [4]. Consequently, the quantum fields across the event horizon must be very far from the vacuum state after the Page time [4]; the absence of this entanglement leads to the formation of a high-energy region at the horizon from the Page time, often referred to as a 'firewall' [2, 4].

The presence of a firewall on the black hole horizon would violate Einstein's equivalence principle, a cornerstone of general relativity. The equivalence principle asserts that the experience of an observer freely falling into a black hole should be locally indistinguishable from their freely floating in empty space. Therefore, the results above lead to a manifest failure of the equivalence principle for old black holes and raises another paradox between quantum mechanics and general relativity in describing black holes, known the firewall paradox [2, 4]. Resolving this paradox remains a significant challenge in the quest to reconcile quantum mechanics with general relativity and to develop a consistent theory of quantum gravity.

The original proof of the firewall paradox [2] required the infalling observer to extract enough information from the early Hawking radiation before reaching the horizon so that they could predict what correlations would be expected in the late radiation. However, calculations based on quantum computation show that the time to extract this information is generally longer than the lifetime of the black hole, which forms a potential loophole to the original firewall claim [10].

There have been other proposals for overcoming the firewall paradox, questioning for example, the existence of the tensor product structure typically assumed at the horizon (a structure that rigorously exists for Rindler horizons [18]). For instance, the so-called "ER=EPR" conjecture proposed by Maldacena and Susskind suggests that the entanglement between the outgoing Hawking radiation and the black hole interior could be related to non-traversable wormholes, providing a non-local resolution that preserves the equivalence principle [17]. Additionally, in Refs. [13, 14] it has been noted that momentum kicks produced by the Hawking radiation itself will lead to the black hole evolving into a macroscopic superposition with distinct locations. To counter this, it has been noted that when viewed in the position basis, a firewall would occur in each branch of the wavefunction and hence the firewall phenomenon would be unaffected by such macroscopic superposition [2].

A more sustained critique of the existence of a tensor product factorization between the degrees-of-freedom near a black hole including ones describing the black hole as viewed from the outside and the degrees-of-freedom very far away has been made by Raju and colleagues [16, 21, 23]. They argue that such factorization, known to be facilitated by massive gravitons, fails in the limit of massless gravitons. Notwithstanding this, the majority of the Hawking radiation from a large black hole is in the form of photons, for which measurable entanglement is unproblematic.

Thus, the huge entropy of entanglement built up in the distant Hawking radiation must have entangled partners somewhere in the vicinity of the black hole itself. This again implies a breakdown of 'no drama' in the vicinity of the horizon and hence a firewall [7]. In any case, the current consensus appears to be in favor of the tensor product structure holding at the horizon either exactly or at least to an excellent approximation for large black holes [3].

12.2 Fast Scrambling

Quantum scrambling denotes the dispersion of local quantum information into its neighborhoods and finally the entire system. Within a scrambling time, a quantum state has a random unitary operator applied to it; and each subsequent scrambling time would lead to the application of a new randomly selected unitary operator. Thus, a pure quantum state would be mapped to a random pure state, which would be further mapped to a new random pure state with each additional scrambling time. Indeed, the application of this effect is widely studied in the literature on black hole dynamics. By assuming that the radiation from a black hole is always a subsystem of a random pure state, Page proved that the entropy of the radiation will first increase and then decrease [20]. He assumed that a black hole is a fast scrambler, without providing a concrete realization of it. Following this, Hayden and Preskill, for the first time, explicitly proposed that black holes obey fast random unitary transformations, and showed that old black holes behave as information mirrors [12]. In their work, Hayden and Preskill argued that the scrambling time of a black hole should scale as $O(\sqrt{S}\ \log S)$, where S is the entropy of the black hole [12]. Note that the Page time is $O(M^3)$, while the scrambling time is $O(M\ \log M)$, implying that for a large (astrophysical) black hole ($M \gg 1$ in appropriate units), the latter is far shorter than the former.

By analyzing the spread of perturbations on the stretched horizon [22] of the D0-brane black hole and the ADS black hole, Sekino and Susskind further showed that the scrambling time of a black hole should be $\frac{1}{T_H} \log S$ times a constant, where T_H is the Hawking temperature of the black hole [24, 26]. This scrambling time approximately equals that proposed by Hayden and Preskill for black holes far from extremality, which are the ones relevant to astrophysical black holes in nature. To have a sense of this time scale, consider a solar-mass black hole, its scrambling time is only about 10^{-1} s, while its Page time is about 10^{72} s (which is more than 10^{54} times the current age of the universe) [27]. Indeed, black holes are believed by some to be the fastest scramblers in nature for any system of comparable size [24].

Although the original proofs of the black hole firewall require a black hole older than the Page time, Almheiri et al. speculated that the firewall phenomenon might already exist after the very short scrambling time [2]. However, Susskind disagreed with this conjecture and claimed that they have mistaken the scrambled state to be the generic state [27]. He argued that while a generic state describes a maximally mixed state which corresponds to infinite temperature, a scrambled pure state will still be a pure state globally. Anchored in the principle that quantum scrambling is consistent with unitary transformations, Susskind's perspective—that the global temperature of a quantum system is unaffected by scrambling—intuitively appears straightforward. Yet, our investigations reveal a significant dependence of his argument on the global state of the system [29]. In Sect. 12.3, we utilize a 1 + 1-dimensional lattice quantum field model to conclusively demonstrate that quantum scrambling (of the lattice sites) can indeed lead to an observed local temperature.

Thus, Sect. 12.3 provides an explicit counterexample to Susskind's argument that a random pure state will not be observed to have a temperature. The key insight is that an observer coupled to a limited number of degrees-of-freedom of a globally scrambling state will couple only to the reduced state of the global system. Thus they can see a well-defined finite temperature even for a globally scrambled pure state. In order to explore the importance of the size of the 'neighborhood' to which the observer is coupled in a generic scenario and for which the scrambling is not limited to lattice-site permutations, we turn next to the methods of quantum information. In Sect. 12.4, we begin by reviewing some basic concepts about black holes and quantum fidelity and then in Sect. 12.5, we show that any sufficiently small neighborhood of a scrambled black hole will be infinitesimally close to a maximally mixed quantum state. This implies that locally an infalling observer will experience a high temperature as they pass the horizon. We allow for arbitrary amounts of emitted radiation and a black hole which may be initially pure or mixed. Finally, in Sect. 12.6 we summarize our conclusions.

12.3 Local Temperature of a Scrambled Lattice Quantum Field

In this section, we study a toy model of quantum scrambling to investigate how local observables can exhibit thermal behavior even when the global state remains pure. We consider a class of random unitary operators acting on the state of a lattice quantum field, which serves as an analogue for information scrambling in black holes.

Consider a quantized real massless scalar field on a spatial lattice in $(n+1)$-dimensional Minkowski spacetime, the Lagrangian can be written as

$$
\begin{aligned}
L &= \int \mathrm{d}^n x \, \mathrm{d}t \, \mathscr{L} = \int \mathrm{d}^n x \, \mathrm{d}t \left(\frac{1}{2} \eta^{\mu\nu} \partial_\mu \phi \partial_\nu \phi \right) \\
&= \frac{1}{2} \int \left(\partial_t \phi \partial_t \phi - \sum_{A=1}^{n} \partial_{x^{(A)}} \phi \partial_{x^{(A)}} \phi \right) \mathrm{d}^n x \, \mathrm{d}t,
\end{aligned}
\tag{12.1}
$$

where $\eta^{\mu\nu}$ is the Minkowski metric and we use A to label the spatial dimension. If we define the momentum as $\pi = \frac{\partial \mathscr{L}}{\partial(\partial_t \phi)} = \partial_t \phi = \dot{\phi}$, the Hamiltonian of the massless scalar field at a given time equals

$$H = \int [\dot{\phi}\pi - \mathscr{L}]\,\mathrm{d}^n x = \frac{1}{2}\int \left(\pi^2 + \sum_{A=1}^{n} (\partial_{x^{(A)}}\phi)^2\right) \mathrm{d}^n x, \tag{12.2}$$

where a general spatial coordinate is given by $x = (x^{(A)}) = (x^{(1)}, \ldots, x^{(n)})$.

Let us now consider a hypercubical lattice to discretize the n-dimensional spatial hypersurface. This yields a multi-index $i \ldots j$ to label any lattice point. The Ath coordinate of the ith lattice point (which lies along the Ath axis) is simply given by $x_i^{(A)}$. Thus, the location of a labeled lattice point $i \ldots j$ is given by $x_{i\ldots j} = (x_i^{(1)}, \ldots, x_j^{(n)})$.

For simplicity, we assume that the lattice spacing along each spatial direction is equal, so that $\delta \equiv x_{i+1}^{(1)} - x_i^{(1)} = \cdots = x_{j+1}^{(n)} - x_j^{(n)}$ for arbitrary $i, \ldots, j$. So an elementary spatial hypercube $\Delta_{i\ldots j}$ has one extreme corner at $x_{i\ldots j} = (x_i^{(1)}, \ldots, x_j^{(n)})$ and the opposite corner at $x_{i+1\ldots j+1} = (x_{i+1}^{(1)}, \ldots, x_{j+1}^{(n)})$, with all other corners as expected, so that the volume of the hypercube is given by

$$\int_{\Delta_{i\ldots j}} d^n x = \int_{x_i^{(1)}}^{x_{i+1}^{(1)}} \cdots \int_{x_j^{(n)}}^{x_{j+1}^{(n)}} d^n x = \delta^n \tag{12.3}$$

We now define the discretized scalar field at the lattice point $x_{i\ldots j}$ as

$$\phi_{i\ldots j} \equiv \delta^{-a} \int_{\Delta_{i\ldots j}} \phi(x_{i\ldots j})\, d^n x \simeq \delta^{n-a}\phi(x_{i\ldots j}), \quad \text{for } x_{i\ldots j} \in \Delta_{i\ldots j}\,. \tag{12.4}$$

Similarly, the momentum may be defined as

$$\pi_{i\ldots j} \equiv \delta^{-b} \int_{\Delta_{i\ldots j}} \pi(x_{i\ldots j})\, \mathrm{d}^n x \simeq \delta^{n-b}\pi(x_{i\ldots j}), \quad \text{for } x_{i\ldots j} \in \Delta_{i\ldots j}\,. \tag{12.5}$$

Here we have introduced the parameters a and b to ensure both a simple canonical commutation relation and a simple Hamiltonian for the discrete variables as we shall now see.

With these constructions, the canonical commutation relation may be calculated as

$$\left[\int_{\Delta_{i\ldots j}} \phi(x_{i\ldots j})\mathrm{d}^n x, \int_{\Delta_{i'\ldots j'}} \pi(x_{i'\ldots j'})\mathrm{d}^n x\right] = \delta^{a+b}[\phi_{i\ldots j}, \pi_{i'\ldots j'}]. \tag{12.6}$$

On the other hand, the canonical commutation relation also equals

$$\begin{aligned}\left[\int_{\Delta_{i\ldots j}} \phi(x_{i\ldots j})\,\mathrm{d}^n x, \int_{\Delta_{i'\ldots j'}} \pi(x_{i'\ldots j'})\,\mathrm{d}^n x\right] &= \int_{\Delta_{i\ldots j}}\int_{\Delta_{i'\ldots j'}} \mathrm{d}^n x\,\mathrm{d}^n x\,[\phi(x_{i\ldots j}), \pi(x_{i'\ldots j'})] \\ &= \int_{\Delta_{i\ldots j}}\int_{\Delta_{i'\ldots j'}} \mathrm{d}^n x\,\mathrm{d}^n x\, i\,\delta(x_{i\ldots j} - x_{i'\ldots j'}) \\ &= i\,\delta_{ii'}\cdots\delta_{jj'}\delta^n,\end{aligned} \tag{12.7}$$

where $\hbar = 1$. Comparing Eqs. (12.6) and (12.7) yields

$$[\phi_{i\cdots j}, \pi_{i'\cdots j'}] = i\,\delta_{ii'}\cdots\delta_{jj'} \tag{12.8}$$

providing $a + b = n$. Inserting $b = n - a$ into Eqs. (12.4) and (12.5) yields

$$\phi_{i\cdots j} = \delta^{n-a}\phi(x_{i\cdots j}) \quad \text{and} \quad \pi_{i\cdots j} = \delta^{a}\pi(x_{i\cdots j}). \tag{12.9}$$

With the above results, the Hamiltonian on the lattice may be written as

$$\begin{aligned} H &= \frac{1}{2}\int\left(\pi^2 + \sum_{A=1}^{n}(\partial_{x^{(A)}}\phi)^2\right)\mathrm{d}^n x \\ &= \frac{1}{2}\delta^n \sum_{i\ldots j}\left(\delta^{-2a}\pi_{i\ldots j}^2 + \frac{\delta^{2(a-n)}}{\delta^2}(\phi_{i+1\ldots j} - \phi_{i\ldots j})^2 + \cdots\right). \end{aligned} \tag{12.10}$$

We choose a to ensure that each term in Eq. (12.10) should have the same power of δ, so we have $-2a = 2a - 2n - 2 \Rightarrow a = \frac{n+1}{2}$, $b = \frac{n-1}{2}$. Therefore, the Hamiltonian becomes

$$H = \frac{1}{2\delta}\sum_{i\cdots j}\left(\pi_{i\cdots j}^2 + (\phi_{i+1\cdots j} - \phi_{i\cdots j})^2 + \cdots\right). \tag{12.11}$$

In this work, we consider $n = 1$ as an example, and the study of higher dimensions should be similar. We then have for a lattice of N points

$$H = \frac{1}{2\delta}\sum_{i=1}^{N}\left(\pi_i^2 + (\phi_{i+1} - \phi_i)^2\right) = \frac{1}{2\delta}\left(\pi^{\mathrm{T}}\mathrm{I}_N\pi + \boldsymbol{\phi}^{\mathrm{T}}\cdot V\cdot\boldsymbol{\phi}\right), \tag{12.12}$$

where the field ϕ_i at each lattice site i allows us to form the N-dimensional vector $\boldsymbol{\phi} = (\phi_1, \ldots, \phi_N)$ with conjugate momentum $\boldsymbol{\pi} = (\pi_1, \ldots, \pi_N)$ satisfying $[\phi_i, \pi_{i'}] = i\delta_{ii'}$, with natural units so that $\hbar = 1$. The interaction matrix V here is given by

$$V = \begin{pmatrix} 1 & -1 & 0 & 0 & 0 & 0 \\ -1 & 2 & -1 & 0 & 0 & 0 \\ 0 & -1 & 2 & \ddots & 0 & 0 \\ 0 & 0 & \ddots & \ddots & -1 & 0 \\ 0 & 0 & 0 & -1 & 2 & -1 \\ 0 & 0 & 0 & 0 & -1 & 1 \end{pmatrix}. \tag{12.13}$$

The ground state of the above scalar field in the lattice representation is written as a Gaussian as follows

$$\Psi = \mathcal{N} e^{-\frac{1}{2}\boldsymbol{\phi}^{\mathsf{T}} \cdot Q \cdot \boldsymbol{\phi}}\,, \tag{12.14}$$

where Q is an $N \times N$ matrix to be determined below. Recall that the canonical commutation relation $[x_i, \pi_j] = i\delta_{ij}$ implies $\pi_j = -i\frac{\partial}{\partial x_j}$. Therefore, Eq. (12.12) can be written as

$$H = \frac{1}{2\delta}\left(-\delta_{ij}\frac{\partial}{\partial \phi_i}\frac{\partial}{\partial \phi_j} + \boldsymbol{\phi}^{\mathsf{T}} \cdot V \cdot \boldsymbol{\phi}\right). \tag{12.15}$$

Applying this Hamiltonian operator to the ground state yields

$$\begin{aligned} H\Psi &= \frac{\mathcal{N}}{2\delta}\left(-\delta_{ij}\frac{\partial}{\partial \phi_i}\frac{\partial}{\partial \phi_j} + \boldsymbol{\phi}^{\mathsf{T}} \cdot V \cdot \boldsymbol{\phi}\right) e^{-\frac{1}{2}\boldsymbol{\phi}^{\mathsf{T}} \cdot Q \cdot \boldsymbol{\phi}} \\ &= \frac{\mathcal{N}}{2\delta}\left(Q_{ii} - \delta_{ij} Q_{ik}\phi_k Q_{jl}\phi_l + \boldsymbol{\phi}^{\mathsf{T}} \cdot V \cdot \boldsymbol{\phi}\right) e^{-\frac{1}{2}\boldsymbol{\phi}^{\mathsf{T}} \cdot Q \cdot \boldsymbol{\phi}} \\ &= \frac{1}{2\delta}\left(\operatorname{tr} Q + \boldsymbol{\phi}^{\mathsf{T}} \cdot (V - Q^2) \cdot \boldsymbol{\phi}\right)\Psi. \end{aligned} \tag{12.16}$$

Now since Eq. (12.14) is an eigenstate of H, it follows from Eq. (12.16) that $Q = \sqrt{V}$. Thus, the ground state simply becomes

$$\Psi = \mathcal{N} e^{-\frac{1}{2}\boldsymbol{\phi}^{\mathsf{T}} \cdot \sqrt{V} \cdot \boldsymbol{\phi}}, \tag{12.17}$$

with the corresponding ground state energy

$$E_0 = \frac{1}{2\delta}\operatorname{tr}\sqrt{V}. \tag{12.18}$$

To normalize the ground state, we require $\int \Psi^2 d^N\phi = 1$, i.e.,

$$1 = \mathcal{N}^2 \int e^{-\frac{1}{2}\boldsymbol{\phi}^{\mathsf{T}} \cdot 2\sqrt{V} \cdot \boldsymbol{\phi}} d^N\phi = \mathcal{N}^2 \sqrt{\frac{(2\pi)^N}{\det(2\sqrt{V})}} = \mathcal{N}^2 \sqrt{\frac{\pi^N}{\det\sqrt{V}}}, \tag{12.19}$$

where we have used the result from the multivariate Gaussian integral that

$$\int e^{-\frac{1}{2}\mathbf{x}^{\mathsf{T}} A \mathbf{x} + \mathbf{J}^{\mathsf{T}}\mathbf{x}} d^n x = e^{\frac{1}{2}\mathbf{J}^{\mathsf{T}} A^{-1} \mathbf{J}} \sqrt{\frac{(2\pi)^n}{\det A}}. \tag{12.20}$$

This implies

$$\mathcal{N} = \frac{(\det V)^{\frac{1}{8}}}{\pi^{\frac{N}{4}}}. \tag{12.21}$$

Next, if we randomly permute (scramble) the scalar field ϕ_i at the lattice points, such that $\boldsymbol{\phi} \rightarrow P\boldsymbol{\phi}$, where P is the permutation operator, and repeat the above analysis, we obtain similar results. Namely, the scrambled Hamiltonian becomes

$$H = \frac{1}{2\delta}\left(\pi^{\mathsf{T}}\mathrm{I}\pi + \boldsymbol{\phi}^{\mathsf{T}} \cdot K \cdot \boldsymbol{\phi}\right), \tag{12.22}$$

where $K = PVP^{\mathsf{T}}$ is the new potential matrix in the randomly permuted $\boldsymbol{\phi}$. The scrambled ground state is now

$$\Psi = \frac{(\det K)^{\frac{1}{8}}}{\pi^{\frac{N}{4}}} e^{-\frac{1}{2}\boldsymbol{\phi}^{\mathsf{T}} \cdot \sqrt{K} \cdot \boldsymbol{\phi}}, \tag{12.23}$$

with the ground state energy

$$\begin{aligned} E_0' &= \mathrm{tr}\sqrt{K} = \frac{1}{2\delta}\mathrm{tr}\sqrt{P\sqrt{V}P^{\mathsf{T}}P\sqrt{V}P^{\mathsf{T}}} \\ &= \frac{1}{2\delta}\mathrm{tr}(P\sqrt{V}P^{\mathsf{T}}) = \frac{1}{2\delta}\mathrm{tr}(\sqrt{V}P^{\mathsf{T}}P) \\ &= \frac{1}{2\delta}\mathrm{tr}\sqrt{V}. \end{aligned} \tag{12.24}$$

which is identical to Eq. (12.18). In fact, scrambling the Hamiltonian and ground state simultaneously is akin to using a different coordinate system to describe the same physics, such that the energy does not change. Therefore to measure the scrambling effect, we need to calculate the mean value of the original Hamiltonian Eq. (12.12) in the scrambled ground state Eq. (12.23):

$$\begin{aligned} E = \langle H \rangle &= \int (\Psi^* H \Psi) d^N\phi \\ &= \frac{\mathcal{N}^2}{2\delta}\int \left(\pi^{\mathsf{T}}\mathrm{I}\pi + \boldsymbol{\phi}^{\mathsf{T}}V\boldsymbol{\phi}\right) e^{-\frac{1}{2}\boldsymbol{\phi}^{\mathsf{T}}2\sqrt{K}\boldsymbol{\phi}} d^N\phi \\ &= \frac{\mathcal{N}^2}{2\delta}\int \left(\mathrm{tr}\sqrt{K} + \boldsymbol{\phi}^{\mathsf{T}}(V-K)\boldsymbol{\phi}\right) e^{-\frac{1}{2}\boldsymbol{\phi}^{\mathsf{T}}2\sqrt{K}\boldsymbol{\phi}} d^N\phi \\ &= \frac{1}{2\delta}\left(\mathrm{tr}\sqrt{K} + \mathrm{tr}[(V-K)(2\sqrt{K})^{-1}]\right) \\ &= \frac{1}{2\delta}\left(\mathrm{tr}\sqrt{V} + \frac{1}{2}\mathrm{tr}[(V-K)(\sqrt{K})^{-1}]\right) \end{aligned} \tag{12.25}$$

where $\int \mathbf{x}^{\mathrm{T}} B \mathbf{x} e^{-\frac{1}{2}\mathbf{x}^{\mathrm{T}} A \mathbf{x}} d^n x = \sqrt{\frac{(2\pi)^n}{\det A}} \mathrm{tr}(BA^{-1})$ has been used in moving from the third to the fourth line.

Therefore, the energy difference before and after scrambling equals

$$\Delta E = E - E_0 = \frac{1}{4\delta} \mathrm{tr}[(V - K)(\sqrt{K})^{-1}] , \tag{12.26}$$

where we have used Eqs. (12.18) and (12.25). This gives the total energy difference for the whole lattice field before and after the fast quantum scrambling.

Now, consider a local observer weakly coupled to our field at a single lattice site i. We can now ask what will be the expected energy seen by such a local observer when coupled to the permuted ground state above the actual ground state. The Hamiltonian of the single-site lattice field may be written

$$H_i = \frac{1}{2\delta}(\pi_i^2 + V_{ii}\phi_i^2) = \frac{1}{2\delta}\Big(-\Big(\frac{\partial}{\partial \phi_i}\Big)^2 + 2\phi_i^2\Big), \tag{12.27}$$

where we have used $\pi_i = -i\frac{\partial}{\partial \phi_i}$ and we have assumed that $i \neq 1$ and N for simplicity. Therefore, the energy of this lattice field felt by the observer equals

$$\begin{aligned} E_i^0 = \langle H_i \rangle &= \int (\Psi^* H_i \Psi) d^N \phi \\ &= \frac{\mathcal{N}^2}{2\delta} \int e^{-\frac{1}{2}\boldsymbol{\phi}^{\mathrm{T}}\sqrt{V}\boldsymbol{\phi}} \Big(-\Big(\frac{\partial}{\partial \phi_i}\Big)^2 + 2\phi_i^2\Big) e^{-\frac{1}{2}\boldsymbol{\phi}^{\mathrm{T}}\sqrt{V}\boldsymbol{\phi}} d^N \phi \\ &= \frac{\mathcal{N}^2}{2\delta} \int \Big((\sqrt{V})_{ii} - (\sqrt{V})_{ik}\phi_k\phi_l(\sqrt{V})_{li} + 2\phi_i^2\Big) e^{-\frac{1}{2}\boldsymbol{\phi}^{\mathrm{T}} 2\sqrt{V}\boldsymbol{\phi}} d^N \phi \\ &= \frac{1}{2\delta}\Big((\sqrt{V})_{ii} - (\sqrt{V})_{ik}((2\sqrt{V})^{-1})_{kl}(\sqrt{V})_{li} + 2((2\sqrt{V})^{-1})_{ii}\Big) \\ &= \frac{1}{2\delta}\Big((\sqrt{V})_{ii} - \frac{(\sqrt{V})_{ii}}{2} + ((\sqrt{V})^{-1})_{ii}\Big) \\ &= \frac{1}{2\delta}\Big(\frac{(\sqrt{V})_{ii}}{2} + ((\sqrt{V})^{-1})_{ii}\Big), \end{aligned} \tag{12.28}$$

where, in going from the third to the fourth line, we have used the normalization condition Eq. (12.19) and a result from the multivariate Gaussian integral that

$$\int x_i x_j B_{ij} e^{-\frac{1}{2}\mathbf{x}^{\mathrm{T}} A \mathbf{x} + \mathbf{J}^{\mathrm{T}} \mathbf{x}} d^n x = B_{ij}\left[(A^{-1})_{ji} + (A^{-1})_{jk} J_k (A^{-1})_{il} J_l\right] e^{\frac{1}{2}\mathbf{J}^{\mathrm{T}} A^{-1} \mathbf{J}} \sqrt{\frac{(2\pi)^n}{\det A}}, \tag{12.29}$$

with $\mathbf{J} = 0$ in our case.

Similarly, the ground state of this Hamiltonian $\Psi = \mathcal{N} \exp\left(-\frac{1}{2}\boldsymbol{\phi}^T \cdot \sqrt{V} \cdot \boldsymbol{\phi}\right)$ allows us to easily construct the 'scrambled' ground state by the replacement

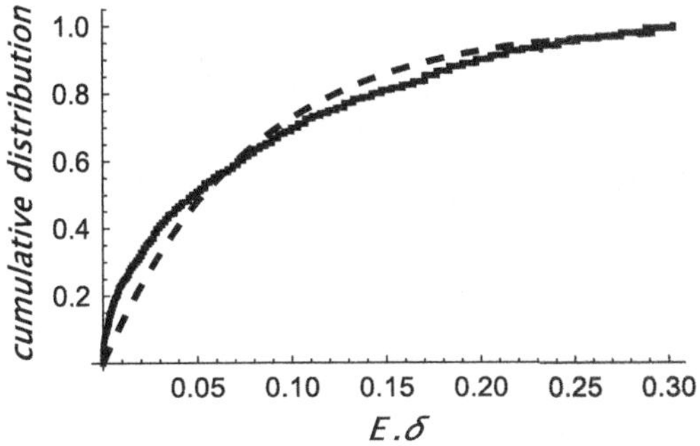

Fig. 12.1 Cumulative distribution for 1000 sampled energies seen by an observer weakly coupled to the scrambled ground state at a single lattice point (solid) on a lattice with $N = 200$ sites. The fit shown is for $k_B T.\delta = 0.0768$ against the cumulative Boltzmann distribution $1 - e^{-E/k_B T}$ (dashed)

$\boldsymbol{\phi} \mapsto P\boldsymbol{\phi}$, for a random permutation operator P. After the scrambling, the energy of the lattice field at the ith lattice equals

$$\begin{aligned} E_i = \langle H_i \rangle &= \int (\Psi^* H_i \Psi) d^N \phi \\ &= \frac{\mathcal{N}^2}{2\delta} \int e^{-\frac{1}{2}\boldsymbol{\phi}^{\mathrm{T}}\sqrt{K}\boldsymbol{\phi}} \Big(-\Big(\frac{\partial}{\partial \phi_i}\Big)^2 + 2\phi_i^2 \Big) e^{-\frac{1}{2}\boldsymbol{\phi}^{\mathrm{T}}\sqrt{K}\boldsymbol{\phi}} d^N \phi \\ &= \frac{\mathcal{N}^2}{2\delta} \int \Big((\sqrt{K})_{ii} - (\sqrt{K})_{ik}\phi_k\phi_l(\sqrt{K})_{li} + 2\phi_i^2 \Big) e^{-\frac{1}{2}\boldsymbol{\phi}^{\mathrm{T}} 2\sqrt{K}\boldsymbol{\phi}} d^N \phi \\ &= \frac{1}{2\delta}\Big((\sqrt{K})_{ii} - (\sqrt{K})_{ik}((2\sqrt{K})^{-1})_{kl}(\sqrt{K})_{li} + 2((2\sqrt{K})^{-1})_{ii} \Big) \\ &= \frac{1}{2\delta}\Big((\sqrt{K})_{ii} - \frac{(\sqrt{K})_{ii}}{2} + ((\sqrt{K})^{-1})_{ii} \Big) \\ &= \frac{1}{2\delta}\Big(\frac{(\sqrt{K})_{ii}}{2} + ((\sqrt{K})^{-1})_{ii} \Big). \end{aligned} \tag{12.30}$$

We can numerically simulate the change $\Delta E_i \equiv E_i - E_i^0$ with respect to the number of contiguous sites being scrambled. Although each permutation yields a single expected energy, we may consider the distribution of these energies across random permutations. Each successive scrambling time induces a new random permutation and consequently a new energy seen by our local observer. By fitting this distribution to the Boltzmann distribution, we may extract a local temperature T as a function of lattice size N (see the cumulative distribution fit for $N = 200$ in Fig. 12.1). Numerical results of randomly scrambled lattices with $N \leq 200$ are shown in Fig. 12.2.

Note that unitary scrambling seamlessly translates pure states into other pure states. Further, a pure quantum state inherently signifies an absence of temperature. Therefore, Susskind postulated that quantum scrambling should be ineffectual in creating any substantial rise in temperature [27]. However, our toy model has shown

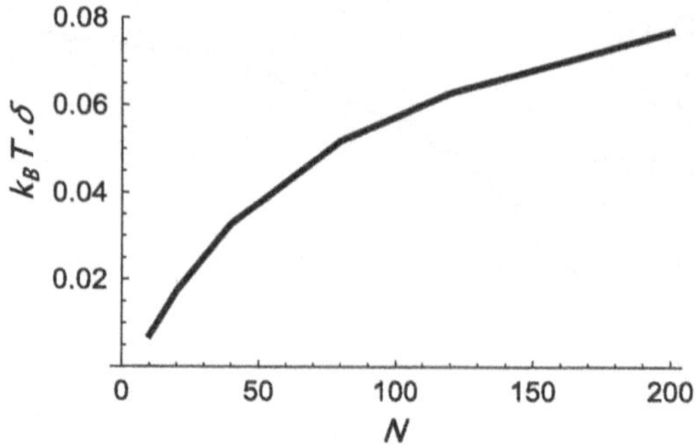

Fig. 12.2 Mean temperature $k_B T$ seen by an observer weakly coupled to the scrambled ground state of a massless scalar field at a single lattice point within a lattice of N sites. This local temperature appears to asymptote to a value approaching the cutoff scale $O(1/\delta)$

that an observer who is only coupled to a limited number of degrees-of-freedom (a limited 'neighborhood') of the global quantum system will see only the reduced state of this system. Thus, a local observer will indeed experience a local temperature.

We find that as the number of lattice sites becomes large, the temperature seen by an observer locally coupled to our scrambled scalar field approaches the cutoff scale $O(1/\delta)$; or within an order-of-magnitude of that scale. Although our simplified model of scrambling involves only the permutation of lattice sites, rather than full-blown random unitary operators applied to the Hilbert space of interest, the emergence of a very high temperature due to scrambling would be expected to continue to hold for other randomization operations, as well as in higher spacetime dimensions.

To move beyond the specifics of the scenario studied in this section in order to derive a generic result we turn in the next sections to the tools of quantum information.

12.4 Local State Encountered by an Infalling Particle

First, consider a black hole that is initially in a quantum state described by a density matrix ρ_0. We assume its Hilbert space dimension to be $N = e^S$, where $S = A/4$ is the Bekenstein-Hawking entropy of the black hole and the Boltzmann constant is set to one. Now, if the evolution of the black hole is unitary, its state after a unitary transformation can be written as

$$\rho_0 \rightarrow U\rho_0 U^\dagger \equiv \rho_U \tag{12.31}$$

where U is a unitary operator, i.e., $U^{\dagger} = U^{-1}$.

Next, consider a small particle (compared to the black hole) falling into this black hole. It passes the horizon and interacts with a small neighborhood surrounding it within the black hole. The quantum state of this neighborhood, which is a tiny subsystem of the black hole, may be obtained by tracing out the other degrees-of-freedom of the black hole. If we assume that the dimension of this subsystem to be n, its state may be expressed as $\mathrm{tr}_{\bar{n}}(\rho_U)$, where $\bar{n}$ represents the degrees-of-freedom of the state complementary (i.e., orthogonal) to degrees-of-freedom contained within this tiny subsystem.

Since we assume that a black hole is a fast scrambler, it follows that the particle arrives at the black hole horizon after the scrambling time, which is believed to be quite small for any reasonably sized black hole [27]. At this stage, the state of the black hole may be calculated as the average over all the unitary transformations [12]. Under these conditions, we will prove that the local quantum state 'observed' by the particle (i.e., with which it directly interacts) is almost a maximally mixed state, thereby having an almost infinite temperature. To show this, we will calculate the fidelity F between $\mathrm{tr}_{\bar{n}}(\rho_U)$ and I_n/n after the scrambling time. In this paper, we define the fidelity of two states characterized by the density matrices ρ and σ as $F(\rho,\sigma) = \mathrm{tr}\sqrt{\sqrt{\rho}\sigma\sqrt{\rho}}$ [19]. It can be shown that $F(\rho,\sigma) = F(\sigma,\rho)$. Since $1 - F(\rho,\sigma) \le \frac{1}{2}\|\rho-\sigma\|_1 \le \sqrt{1-F(\rho,\sigma)^2}$, the fidelity of two states ρ and σ satisfies the inequalities [19]

$$1 - \frac{1}{2}\|\rho-\sigma\|_1 \le F(\rho,\sigma) \le \sqrt{1 - \frac{1}{4}\|\rho-\sigma\|_1^2}\,, \tag{12.32}$$

where the Schatten p-norm of A is defined as $\|A\|_p = \left(\mathrm{tr}(A^{\dagger}A)^{\frac{p}{2}}\right)^{\frac{1}{p}}$. Hölder's inequality implies [6]

$$\|\rho-\sigma\|_1 \le \|\mathrm{I}\|_2\|\rho-\sigma\|_2, \tag{12.33}$$

where I has the same dimensionality as ρ and σ.

Applying Eq. (12.33) to Eq. (12.32) yields

$$1 - \frac{1}{2}\|\mathrm{I}\|_2\|\rho-\sigma\|_2 \le F(\rho,\sigma) \le \sqrt{1 - \frac{1}{4}\|\rho-\sigma\|_1^2}\,. \tag{12.34}$$

Since the lower bound of the fidelity in Eq. (12.34) plays the key role in the following analysis, we will disregard the upper bound from this point onward and only use

$$F(\rho,\sigma) \ge 1 - \frac{1}{2}\|\mathrm{I}\|_2\|\rho-\sigma\|_2\,. \tag{12.35}$$

To study the fidelity between $\mathrm{tr}_{\bar{n}}(\rho_U)$ and I_n/n after scrambling, we insert them into Eq. (12.35). Under the assumption of fast scrambling, the mean fidelity is averaged over all unitary operators of the state of the black hole. We obtain the following relation

$$
\begin{aligned}
F\left(\mathrm{tr}_{\bar{n}}(\rho_U), \frac{\mathrm{I}_n}{n}\right) &\geq 1 - \frac{1}{2}\int_U \|\mathrm{I}_n\|_2 \times \left\|\mathrm{tr}_{\bar{n}}(\rho_U) - \frac{\mathrm{I}_n}{n}\right\|_2 dU \\
&= 1 - \frac{1}{2}\sqrt{n}\int_U \left\|\mathrm{tr}_{\bar{n}}(\rho_U) - \frac{\mathrm{I}_n}{n}\right\|_2 dU.
\end{aligned} \tag{12.36}
$$

12.5 Local Neighborhood State Within Different Black Holes

12.5.1 Pure-State Black Hole with Negligible Radiation

One may first consider a newly formed black hole in a pure quantum state. Since the scrambling time is very short, any small amount of Hawking radiation emitted by the black hole as it scrambles can be safely ignored. For such a black hole, the two norm $\|\mathrm{tr}_{\bar{n}}(\rho_U) - \frac{\mathrm{I}_n}{n}\|_2^2$ that appeared in the fidelity relation Eq. (12.36) may be calculated as

$$
\begin{aligned}
\int_U \left\|\mathrm{tr}_{\bar{n}}(\rho_U) - \frac{\mathrm{I}_n}{n}\right\|_2^2 dU &= \int_U \mathrm{tr}_n\left[\mathrm{tr}_{\bar{n}}(\rho_U)\,\mathrm{tr}_{\bar{n}}(\rho_U) - \frac{2}{n}\,\mathrm{tr}_{\bar{n}}(\rho_U) + \frac{\mathrm{I}_n}{n^2}\right] dU \\
&= \int_U \mathrm{tr}_n\left[\mathrm{tr}_{\bar{n}}(\rho_U)\,\mathrm{tr}_{\bar{n}}(\rho_U)\right] dU - \frac{1}{n} \\
&= \int_U \mathrm{tr}_{n,n}\left[\mathrm{tr}_{\bar{n},\bar{n}}(\rho_U \otimes \rho_U)\,\mathrm{SWAP}_{n,n}\right] dU - \frac{1}{n} \\
&= \mathrm{tr}\left[(\rho \otimes \rho)\int_U \left[(U^\dagger \otimes U^\dagger)\,\mathrm{SWAP}_{n,n}(U \otimes U)\right] dU\right] - \frac{1}{n},
\end{aligned} \tag{12.37}
$$

where $\|A\|_2^2 = \mathrm{tr}(A^\dagger A)$ is applied to the first line, and $\mathrm{tr}(AB) = \mathrm{tr}(A \otimes B\ \mathrm{SWAP}_{A,B})$ is applied in moving from the second to the third line. Here $\mathrm{SWAP}_{A,B}$ is the SWAP operator between the quantum subsystems A and B. Applying the Schur-Weyl duality in Appendix 1 to Eq. (12.37) then yields

$$
\begin{aligned}
&\int_U \left\|\mathrm{tr}_{\bar{n}}(\rho_U) - \frac{\mathrm{I}_n}{n}\right\|_2^2 dU \\
&= \frac{1}{2}\mathrm{tr}\left[(\rho \otimes \rho)\left(\frac{\frac{N}{n} + n}{N+1}(\mathrm{I}_{N,N} + \mathrm{SWAP}_{N,N}) + \frac{\frac{N}{n} - n}{N-1}(\mathrm{I}_{N,N} - \mathrm{SWAP}_{N,N})\right)\right] - \frac{1}{n} \\
&= \mathrm{tr}\left[(\rho \otimes \rho)\left(\frac{\frac{N^2}{n} - n}{N^2 - 1}\mathrm{I}_{N,N} + \frac{Nn - \frac{N}{n}}{N^2 - 1}\mathrm{SWAP}_{N,N}\right)\right] - \frac{1}{n}
\end{aligned}
$$

$$
\begin{aligned}
&= \frac{\frac{N^2}{n}-n}{N^2-1}(\mathrm{tr}\,\rho)^2 + \frac{Nn-\frac{N}{n}}{N^2-1}\mathrm{tr}(\rho^2) - \frac{1}{n} = \frac{\frac{N^2}{n}-n}{N^2-1} + \frac{Nn-\frac{N}{n}}{N^2-1} - \frac{1}{n} \\
&= \frac{Nn^2-N-n^2+1}{(N^2-1)n} = \frac{n^2-1}{(N+1)n} = \frac{1}{N+1}\left(n-\frac{1}{n}\right),
\end{aligned}
\tag{12.38}
$$

where $(\mathrm{tr}\,\rho) = \mathrm{tr}(\rho^2) = 1$ is used in the fourth line because the entire black hole is a pure quantum state.

Since $N \geq n \geq 1$, and from Jensen's inequality $\langle X^2 \rangle \geq \langle X \rangle^2$, Eq. (12.38) may be further simplified as

$$
\int_U \left\| \mathrm{tr}_{\bar{n}}(\rho_U) - \frac{\mathrm{I}_n}{n} \right\|_2 dU < \sqrt{\frac{n}{N}}. \tag{12.39}
$$

Then the fidelity relation in Eq. (12.36) may be written

$$
F\left(\mathrm{tr}_{\bar{n}}(\rho_U), \frac{\mathrm{I}_n}{n}\right) \geq 1 - \frac{n}{2\sqrt{N}}. \tag{12.40}
$$

Recall that for a stellar-mass black hole $N \sim \exp(10^{80})$. This bound determines how close the local substate encountered by an infalling particle is to a maximally mixed state. Since a maximally mixed quantum state corresponds to an infinitely high temperature (see Theorem 12.1), this means that the infalling object will experience a very high temperature as well.

To have some idea about the exact value of this temperature, we now try to write out the expression of this quantum state. The Fidelity $F(\rho, \sigma)$ of a pair of states ρ and σ is the maximum overlap over all purifications $|\psi_\rho\rangle$ and $|\psi_\sigma\rangle$, respectively, of these states, i.e.,

$$
F \equiv F(\rho, \sigma) = \max_{\psi_\rho, \psi_\sigma} \langle \psi_\rho | \psi_\sigma \rangle. \tag{12.41}
$$

Consequently, there exist purifications $|\psi_\rho\rangle$ and $|\psi_\sigma\rangle$ which satisfy [5]

$$
|\psi_\rho\rangle = F|\psi_\sigma\rangle + \sqrt{1-F^2}\,|\psi_\sigma^\perp\rangle, \tag{12.42}
$$

where $|\psi_\sigma^\perp\rangle$ is some quantum state orthogonal to $|\psi_\sigma\rangle$. Taking the partial trace of this pure-state representation yields

$$
\rho = F^2\sigma + O\big(\sqrt{1-F^2}\big), \qquad \text{for } 1 - F^2 \ll 1. \tag{12.43}
$$

Applying this result to Eq. (12.40), we find that typically

$$
\mathrm{tr}_{\bar{n}}(\rho_U) = (1-\epsilon)\frac{\mathrm{I}_n}{n} + O(\sqrt{\epsilon}), \quad \text{for } \epsilon \equiv 1 - F^2 \leq \frac{n}{\sqrt{N}} \ll 1. \tag{12.44}
$$

Theorem 12.1 *A maximally mixed quantum state corresponds to an infinity high temperature for the quantum system.*

Proof In statistical mechanics, a Boltzmann distribution describes the probability distribution of a system for different microstates i with respect to the state's energy E_i and temperature of the system T. The distribution may be written

$$p_i = \frac{1}{Z} e^{-E_i/(k_B T)}, \tag{12.45}$$

where function Z is used to normalized the distribution and hence equals the sum of $e^{-E_i/(k_B T)}$ for all different E_i.

The generalization of Eq. (12.45) to quantum mechanics may be written

$$\hat{\rho} = \frac{1}{Z} \exp\left(-\frac{\hat{H}}{k_B T}\right), \tag{12.46}$$

where $\hat{\rho}$ is the density matrix of the quantum system, and $\hat{H}$ is the system's Hamiltonian operator. Here, the function Z is used to normalize the density matrix of the quantum state and equals the trace of $\exp(-\hat{H}/(k_B T))$.

Now we prove that Eq. (12.46) is the quantum version of the Boltzmann distribution Eq. (12.45). Suppose the basis states of the quantum system are labeled by i and $\{|i\rangle\}$ is a complete basis, then Eq. (12.46) may be written

$$\hat{\rho} = \frac{1}{Z} \exp\left(-\frac{\hat{H}}{k_B T}\right) = \frac{1}{Z} \sum_{i,j} |i\rangle\langle i| \exp\left(-\frac{\hat{H}}{k_B T}\right) |j\rangle\langle j| = \sum_{i,j} p_{i,j} |i\rangle\langle j|, \tag{12.47}$$

where $p_{i,j} = \frac{1}{Z}\langle i| \exp\left(-\frac{\hat{H}}{k_B T}\right)|j\rangle$ is the probability of finding the quantum system in the quantum state $|i\rangle\langle j|$ which has the similar physical interpretation as p_i in the Boltzmann distribution Eq. (12.45).

If we choose the basis $\{|i\rangle\}$ as a complete basis of the energy eigenstates, the $p_{i,j}$ in Eq. (12.47) may be further simplified

$$p_{i,j} = \frac{1}{Z}\langle i| \exp\left(-\frac{\hat{H}}{k_B T}\right) |j\rangle = \frac{1}{Z}\langle i||j\rangle e^{-\frac{E_j}{k_B T}} = \frac{1}{Z} e^{-\frac{E_j}{k_B T}} \delta_{i,j}, \tag{12.48}$$

where E_j is the eigenvalues $|j\rangle$. Inserting Eq. (12.48) into Eq. (12.47) yields

$$\hat{\rho} = \sum_{i,j} \frac{1}{Z} e^{-\frac{E_j}{k_B T}} \delta_{i,j} |i\rangle\langle j| = \sum_i \frac{1}{Z} e^{-\frac{E_i}{k_B T}} |i\rangle\langle i|, \tag{12.49}$$

where $Z = \sum_{i=1}^{N} e^{\frac{-E_i}{k_B T}}$ and N is the total number of the basis eigenvectors. The density matrix is diagonal in this basis and each entry gives the probability for the quantum system with a specific energy.

If we assume the temperature of the system to be high ($\frac{E_i}{k_B T}$ is small), then Eq. (12.49) may be approximated by

$$\begin{aligned}\hat{\rho} &= \sum_i \frac{1}{\sum_{j=1}^{N} e^{\frac{-E_j}{kT}}} e^{-\frac{E_i}{k_B T}} |i\rangle\langle i| \approx \sum_i \frac{1 - \frac{E_i}{k_B T} + O(\frac{1}{k_B T})}{\sum_{n=1}^{N}(1 - \frac{E_j}{k_B T} + O(\frac{1}{k_B T}))} |i\rangle\langle i| \\ &= \sum_i \frac{1 - \frac{E_i}{k_B T} + O(\frac{1}{k_B T})}{N - \sum_{j=1}^{N} \frac{E_j}{k_B T} + O(\frac{1}{k_B T})} |i\rangle\langle i| = \sum_i \frac{1}{N} \frac{1 - \frac{E_i}{k_B T} + O(\frac{1}{k_B T})}{1 - \frac{\langle E\rangle}{k_B T} + O(\frac{1}{k_B T})} |i\rangle\langle i| \\ &= \sum_i \frac{1}{N}\Big(1 + \frac{\langle E\rangle - E_i}{k_B T} + o\big(\frac{1}{k_B T}\big)\Big)|i\rangle\langle i|, \end{aligned} \quad (12.50)$$

where $\langle E\rangle = (\sum_{j=1}^{N} E_j)/N$. When the temperature is infinity high, the quantum state of the system Eq. (12.50) reduces to $\frac{1}{N}\mathrm{I}_N$. Thus a maximally mixed quantum state corresponds to an infinity high temperature for the quantum system. □

For example, in the case of a stellar-mass black hole, the local state described in Eq. (12.44) with which an infalling particle interacts will be in a quantum state nearly indistinguishable from a completely mixed state—a state with infinite temperature—whenever $n \ll \sqrt{N}$.

12.5.2 Pure-State Black Hole with Non-negligible Radiation

We now consider the above black hole that has radiated a non-negligible amount of itself away. If we assume the dimension of the radiation is R, then the dimension of the remaining black hole will be $N_B = N/R$. The quantum state of the remaining black hole may be represented by $\mathrm{tr}_R(\rho_U)$. The key step in our calculation of the 2-norm is to take an average over all possible unitary operators, since black holes are fast scramblers. Now, with part of the initial state radiated outside the horizon, we also take an average over the unitary operators of the remaining black hole U_B. Note that U_B is different from the U, and does not influence the radiation outside the horizon. With this physical picture, the fidelity relation Eq. (12.36) may be written

$$F \geq 1 - \frac{1}{2}\sqrt{n} \int_{U_B} \left\| \mathrm{tr}_{\bar{n}}\left(U_B \mathrm{tr}_R(\rho_U) U_B^{\dagger}\right) - \frac{\mathrm{I}_n}{n} \right\|_2 dU_B. \quad (12.51)$$

Using a similar analysis to that used in Sect. 12.5.1, the 2-norm in Eq. (12.51) may be calculated as

$$
\begin{aligned}
&\int_{U_B}\left\|\mathrm{tr}_{\bar{n}}\left(U_B\mathrm{tr}_R(\rho_U)U_B{}^{\dagger}\right)-\frac{\mathrm{I}_n}{n}\right\|_2^2 dU_B\\
&=\int_{U_B}\mathrm{tr}_n\left[\mathrm{tr}_{\bar{n}}\left(U_B\mathrm{tr}_R(\rho_U)U_B{}^{\dagger}\right)\mathrm{tr}_{\bar{n}}\left(U_B\mathrm{tr}_R(\rho_U)U_B{}^{\dagger}\right)-\frac{2}{n}\mathrm{tr}_{\bar{n}}\left(U_B\mathrm{tr}_R(\rho_U)U_B{}^{\dagger}\right)+\frac{\mathrm{I}_n}{n^2}\right]dU_B\\
&=\int_{U_B}\mathrm{tr}_n\left[\mathrm{tr}_{\bar{n}}\left(U_B\mathrm{tr}_R(\rho_U)U_B{}^{\dagger}\right)\mathrm{tr}_{\bar{n}}\left(U_B\mathrm{tr}_R(\rho_U)U_B{}^{\dagger}\right)\right]dU_B\\
&\quad-\frac{2}{n}\int_{U_B}\mathrm{tr}_{\bar{R}}\left(U_B\mathrm{tr}_R(\rho_U)U_B{}^{\dagger}\right)dU_B+\frac{1}{n}\\
&=\int_{U_B}\mathrm{tr}_{n,n}\left[\mathrm{tr}_{\bar{n},\bar{n}}\left(U_B\mathrm{tr}_R(\rho_U)U_B{}^{\dagger}\right)\otimes\left(U_B\mathrm{tr}_R(\rho_U)U_B{}^{\dagger}\right)\mathrm{SWAP}_{n,n}\right]dU_B\\
&\quad-\frac{2}{n}\int_{U_B}\mathrm{tr}_{\bar{R}}(\mathrm{tr}_R(\rho_U))dU_B+\frac{1}{n}\\
&=\int_{U_B}\mathrm{tr}\left[(U_B\otimes U_B)\left(\mathrm{tr}_R(\rho_U)\otimes\mathrm{tr}_R(\rho_U)\right)(U_B{}^{\dagger}\otimes U_B{}^{\dagger})\left(\mathrm{SWAP}_{n,n}\otimes\mathrm{I}_{\bar{n},\bar{n}}\right)\right]dU_B-\frac{1}{n}\\
&=\mathrm{tr}\left[\left(\mathrm{tr}_R(\rho_U)\otimes\mathrm{tr}_R(\rho_U)\right)\int_{U_B}\left[(U_B{}^{\dagger}\otimes U_B{}^{\dagger})\left(\mathrm{SWAP}_{n,n}\otimes\mathrm{I}_{\bar{n},\bar{n}}\right)(U_B\otimes U_B)\right]dU_B\right]-\frac{1}{n},
\end{aligned}
\tag{12.52}
$$

where $\|A\|_2^2=\mathrm{tr}(A^{\dagger}A)$ is applied to the first step, and $\mathrm{tr}(AB)=\mathrm{tr}(A\otimes B\ \mathrm{SWAP}_{A,B})$ is applied to the third line. Applying the Schur-Weyl duality (see Appendix 1) to Eq. (12.52) yields

$$
\begin{aligned}
&\int_{U_B}\left\|\mathrm{tr}_{\bar{n}}\left(U_B\mathrm{tr}_R(\rho_U)U_B{}^{\dagger}\right)-\frac{\mathrm{I}_n}{n}\right\|_2^2 dU_B\\
&=\frac{1}{2}\mathrm{tr}\left[\left(\mathrm{tr}_R(\rho_U)\otimes\mathrm{tr}_R(\rho_U)\right)\left(\frac{\frac{N_B}{n}+n}{N_B+1}(\mathrm{I}_{N_B,N_B}+\mathrm{SWAP}_{N_B,N_B})\right.\right.\\
&\quad\left.\left.+\frac{\frac{N_B}{n}-n}{N_B-1}(\mathrm{I}_{N_B,N_B}-\mathrm{SWAP}_{N_B,N_B})\right)\right]-\frac{1}{n}\\
&=\mathrm{tr}\left[\left(\mathrm{tr}_R(\rho_U)\otimes\mathrm{tr}_R(\rho_U)\right)\left(\frac{\frac{N_B^2}{n}-n}{N_B^2-1}\mathrm{I}_{N_B,N_B}+\frac{N_Bn-\frac{N_B}{n}}{N_B^2-1}\mathrm{SWAP}_{N_B,N_B}\right)\right]-\frac{1}{n}\\
&=\frac{\frac{N_B^2}{n}-n}{N_B^2-1}\left(\mathrm{tr}\left(\mathrm{tr}_R(\rho_U)\right)\right)^2+\frac{N_Bn-\frac{N_B}{n}}{N_B^2-1}\mathrm{tr}\left(\left(\mathrm{tr}_R(\rho_U)\right)^2\right)-\frac{1}{n}\\
&=\frac{\frac{N_B^2}{n}-n}{N_B^2-1}(\mathrm{tr}(\rho_U))^2+\frac{N_Bn-\frac{N_B}{n}}{N_B^2-1}\mathrm{tr}\left(\left(\mathrm{tr}_R(\rho_U)\right)^2\right)-\frac{1}{n}.
\end{aligned}
\tag{12.53}
$$

Since the interior of and radiation from the black hole are both random density matrices, the term $\mathrm{tr}\left((\mathrm{tr}_R(\rho_U))^2\right)$ in Eq. (12.53) may be calculated as

$$
\begin{aligned}
\left\langle \mathrm{tr}((\mathrm{tr}_R(\rho_U))^2)\right\rangle &= \int_U \mathrm{tr}((\mathrm{tr}_R(\rho_U))^2)dU = \int_U \mathrm{tr}\Big[\mathrm{tr}_R(\rho_U)\otimes \mathrm{tr}_R(\rho_U)\mathrm{SWAP}_{\bar{R},\bar{R}}\Big]dU \\
&= \int_U \mathrm{tr}\Big[\rho_U\otimes\rho_U \mathrm{SWAP}_{\bar{R},\bar{R}}\Big]dU \\
&= \int_U \mathrm{tr}\Big[(U\otimes U)(\rho\otimes\rho)(U^\dagger\otimes U^\dagger)\mathrm{SWAP}_{\bar{R},\bar{R}}\Big]dU \\
&= \mathrm{tr}\Big[(\rho\otimes\rho)\int_U (U^\dagger\otimes U^\dagger)\mathrm{SWAP}_{\bar{R},\bar{R}}(U\otimes U)dU\Big] \\
&= \frac{1}{2}\mathrm{tr}\Big[(\rho\otimes\rho)\Big(\frac{N/\bar{R}+\bar{R}}{N+1}(\mathrm{I}_{N,N}+\mathrm{SWAP}_{N,N}) \\
&\quad +\frac{N/\bar{R}-\bar{R}}{N-1}(\mathrm{I}_{N,N}-\mathrm{SWAP}_{N,N})\Big)\Big] \\
&= \mathrm{tr}\Big[(\rho\otimes\rho)\Big(\frac{NR-N_B}{N^2-1}\mathrm{I}_{N,N}+\frac{NN_B-R}{N^2-1}\mathrm{SWAP}_{N,N}\Big)\Big] \\
&= \frac{NR-N_B}{N^2-1}(\mathrm{tr}(\rho))^2+\frac{NN_B-R}{N^2-1}\mathrm{tr}(\rho^2) = \frac{NR-N_B}{N^2-1}+\frac{NN_B-R}{N^2-1} \\
&= \frac{N_B+R}{N+1},
\end{aligned}
\tag{12.54}
$$

where the Schur-Weyl duality is used in moving from the fourth to the fifth line, $\bar{R}=N_B$ and $N_B=N/R$ are used in moving from the sixth to the seventh line, and $(\mathrm{tr}(\rho))^2=\mathrm{tr}(\rho^2)=1$ is used in the last step.

Applying Eq. (12.54) and $(\mathrm{tr}(\rho_U))^2=1$ to Eq. (12.53) yields

$$
\begin{aligned}
&\int_{U_B}\left\|\mathrm{tr}_{\bar{n}}\left(U_B\mathrm{tr}_R(\rho_U)U_B{}^\dagger\right)-\frac{\mathrm{I}_n}{n}\right\|_2^2 dU_B \\
&= \frac{\frac{N_B^2}{n}-n}{N_B^2-1}+\frac{N_Bn-\frac{N_B}{n}}{N_B^2-1}\frac{N_B+R}{N+1}-\frac{1}{n} \\
&= \frac{(N_B^2-n^2)(N+1)+(N_Bn^2-N_B)(N_B+R)-(N_B^2-1)(N+1)}{(N_B^2-1)(N+1)n} \\
&= \frac{N_B^2n^2+N_BRn^2-Nn^2-N_B^2-N_BR+N-n^2+1}{(N_B^2-1)(N+1)n} \\
&= \frac{N_B^2n^2-N_B^2-n^2+1}{(N_B^2-1)(N+1)n}=\frac{n^2-1}{(N+1)n},
\end{aligned}
\tag{12.55}
$$

where $N=N_BR$ is used in moving from the third to the fourth line.

Since $N\geq n\geq 1$, and by Jensen's inequality Eq. (12.55) may be further simplified as

$$
\int_{U_B}\left\|\mathrm{tr}_{\bar{n}}\left(U_B\mathrm{tr}_R(\rho_U)U_B{}^\dagger\right)-\frac{\mathrm{I}_n}{n}\right\|_2 dU_B<\sqrt{\frac{n}{N}}. \tag{12.56}
$$

Inserting Eq. (12.56) into Eq. (12.51) yields

$$F \geq 1 - \frac{n}{2\sqrt{N}}. \tag{12.57}$$

Note that Eq. (12.57) is identical to Eq. (12.40). So again the quantum state of any local neighborhood encountered by an infalling object typically has the form

$$\mathrm{tr}_{\bar{n}}(U_B \mathrm{tr}_R(\rho_U) U_B{}^{\dagger}) = (1-\epsilon)\frac{\mathrm{I}_n}{n} + O(\sqrt{\epsilon}), \tag{12.58}$$

where $0 \leq \epsilon \leq \frac{n}{\sqrt{N}} \ll 1$. Therefore, infalling objects will observe a very high temperature (see Theorem 12.1). We emphasize that this result does not require the black hole to radiate half of itself away, which means that this phenomenon may occur much earlier than the Page time.

12.5.3 Black Holes with an Initially Impure Quantum State

In the above analysis we have assumed that the initial state of the black hole is a pure state. However, it may seem natural that the quantum state of an astrophysical black hole does not begin as a pure state. Therefore, now we consider a black hole that begins from a generic quantum state ρ_0, with Hilbert space dimension N. The local subsystem observed by the infalling observer with dimension n may be written $\mathrm{tr}_{\bar{n}}(\rho_U)$. The 2-norm in the fidelity relation Eq. (12.36) may be calculated

$$\begin{aligned}
&\int_U \left\| \mathrm{tr}_{\bar{n}}(\rho_U) - \frac{\mathrm{I}_n}{n} \right\|_2^2 dU \\
&= \int_U \left\| \mathrm{tr}_{\bar{n}}(U\rho_0 U^{\dagger}) - \frac{\mathrm{I}_n}{n} \right\|_2^2 dU \\
&= \int_U \mathrm{tr}_n \left[\mathrm{tr}_{\bar{n}}(U\rho_0 U^{\dagger})\, \mathrm{tr}_{\bar{n}}(U\rho_0 U^{\dagger}) - \frac{2}{n}\mathrm{tr}_{\bar{n}}(U\rho_0 U^{\dagger}) + \frac{\mathrm{I}_n}{n^2} \right] dU \\
&= \int_U \mathrm{tr}_n \left[\mathrm{tr}_{\bar{n}}(U\rho_0 U^{\dagger})\, \mathrm{tr}_{\bar{n}}(U\rho_0 U^{\dagger}) \right] dU - \frac{1}{n} \\
&= \int_U \mathrm{tr}_{n,n} \left[\mathrm{tr}_{\bar{n},\bar{n}}(U\rho_0 U^{\dagger}) \otimes (U\rho_0 U^{\dagger})\, \mathrm{SWAP}_{n,n} \right] dU - \frac{1}{n} \\
&= \int_U \mathrm{tr} \left[(U \otimes U)\big(\rho_0 \otimes \rho_0\big)(U^{\dagger} \otimes U^{\dagger})\big(\mathrm{SWAP}_{n,n} \otimes \mathrm{I}_{\bar{n},\bar{n}}\big) \right] dU - \frac{1}{n} \\
&= \mathrm{tr}\left[(\rho_0 \otimes \rho_0) \int_U \Big((U^{\dagger} \otimes U^{\dagger})\big(\mathrm{SWAP}_{n,n} \otimes \mathrm{I}_{\bar{n},\bar{n}}\big)(U \otimes U) \Big) dU \right] - \frac{1}{n},
\end{aligned} \tag{12.59}$$

where $\mathrm{tr}_{\bar{n}}(U\rho_0 U^{\dagger}) = 1$ is applied to the third line. Applying the Schur-Weyl duality to Eq. (12.59) yields

$$\begin{aligned}
&\int_U \left\| \mathrm{tr}_{\bar{n}}(\rho_U) - \frac{\mathrm{I}_n}{n} \right\|_2^2 dU \\
&= \frac{1}{2}\mathrm{tr}\left[(\rho_0 \otimes \rho_0)\left(\frac{\frac{N}{n}+n}{N+1}(\mathrm{I}_{N,N} + \mathrm{SWAP}_{N,N}) + \frac{\frac{N}{n}-n}{N-1}(\mathrm{I}_{N,N} - \mathrm{SWAP}_{N,N})\right)\right] - \frac{1}{n} \\
&= \mathrm{tr}\left[(\rho_0 \otimes \rho_0)\left(\frac{\frac{N^2}{n}-n}{N^2-1}\mathrm{I}_{N,N} + \frac{Nn-\frac{N}{n}}{N^2-1}\mathrm{SWAP}_{N,N}\right)\right] - \frac{1}{n} \\
&= \frac{\frac{N^2}{n}-n}{N^2-1}(\mathrm{tr}\,\rho_0)^2 + \frac{Nn-\frac{N}{n}}{N^2-1}\mathrm{tr}(\rho_0^2) - \frac{1}{n} \\
&= \frac{Nn^2-N}{(N^2-1)n}\mathrm{tr}(\rho_0^2) - \frac{n^2-1}{(N^2-1)n} \\
&= \frac{(N\,\mathrm{tr}(\rho_0^2)-1)(n^2-1)}{(N^2-1)n},
\end{aligned} \tag{12.60}$$

where $\mathrm{tr}\,\rho_0 = 1$ is applied to the fourth line.

Note that as $n \leq N$, then $N^2n^2 - N^2 \leq N^2n^2 - n^2$ and hence $N^2(n^2-1) \leq n^2(N^2-1)$. Further taking $1 < N$, and using Eq. (12.60) and Jensen's inequality we find

$$\int_U \left\| \mathrm{tr}_{\bar{n}}(\rho_U) - \frac{\mathrm{I}_n}{n} \right\|_2 dU \leq \frac{\sqrt{n(N\,\mathrm{tr}(\rho_0^2)-1)}}{N}. \tag{12.61}$$

Inserting Eq. (12.61) into the fidelity relation Eq. (12.36) yields

$$F \geq 1 - \frac{n}{2N}\sqrt{N\,\mathrm{tr}(\rho_0^2)-1}. \tag{12.62}$$

However, identical bounds are found when radiation is allowed for (see Appendix 2). Therefore Eq. (12.62) represents the generic result, with N the dimensionality of the original black hole, or equivalently the product of the dimensionality of the current state of the black hole and that of the radiation.

Since $1/N \leq \mathrm{tr}(\rho_0^2) < 1$ for any impure quantum state, the lower bound to the fidelity for a black hole originating from a non-pure state is larger than that for the pure-state scenarios studied above. Consequently, the reduced state of a neighborhood within an initially impure black hole will be even closer to being a maximally mixed state than for an initially pure state black hole.

Again, the local quantum state with which an infalling object interacts may be written as typically given by

$$\mathrm{tr}_{\bar{n}}(U\rho_0U^\dagger) = (1-\epsilon)\frac{\mathrm{I}_n}{n} + O(\sqrt{\epsilon}), \tag{12.63}$$

where $0 \leq \epsilon \leq \frac{n}{N}\sqrt{N\,\mathrm{tr}(\rho_0^2) - 1} \leq n/\sqrt{N} \ll 1$. Therefore an infalling observer will typically experience even higher temperatures (see Theorem 12.1) as impurity of the quantum state of the initial black hole is increased.

12.6 Discussion

Black holes are conjectured to be fast scramblers; possibly even the fastest scramblers in the universe [12]. The scrambling process itself is conceived of as a random unitary operation on the black hole interior Hilbert space [12, 20]. For a black hole of mass M, across its entire lifetime, during any time interval of duration $O(M \log M)$, a random unitary will operate on this subspace. This time interval is called the scrambling time.

For simplicity, let us suppose the black hole interior is initially pure and let us ignore any evaporation process or new material being added to the black hole. The question arises as to whether the behavior of the quantum state of the black hole should be treated as an ensemble average over the random unitary operators [2], or as a single (though randomly selected) pure state [27]. In the former case, the reduced state of a sufficiently small subsystem would appear to be the generic maximally mixed state corresponding to an infinite temperature [2]. In the latter case, it has been argued that a random pure state is still pure and therefore has zero associated temperature [27].

We have addressed this controversy head on in Sect. 12.3. There we study an explicit model of a $1 + 1$-dimensional quantum field on a lattice undergoing random lattice site permutations (a highly restricted class of random unitary operations on the Hilbert space of the quantum field). We compute the energy of an observer weakly coupled to a single lattice site for a quantum field initially in the ground (vacuum) state. Each random unitary operator yields a distinct energy above the ground state for this observer. After each additional scrambling time, a new random unitary operator will cause our local observer to experience a new local energy. As the total number of scrambled lattice sites increases, the distribution of energies experienced by our local observer is found to be well approximated by a Boltzmann distribution. At each point in time the global quantum state is pure, nevertheless, local behavior is correctly described by the ensemble statistics of the random unitary operators (in this case permutations of lattice sites). The specific temperature calculated in this analysis as seen by our local observer is found to approach the natural cutoff scale of that model.

If the cutoff scale were Planckian, our toy model would generate a scrambling temperature approaching the Planck scale. This is very close to the temperatures found at roughly one Planck time into the Big Bang, which like a black hole is also reputed to have a singularity at its origin. Of course, these similar temperature scales may be just a coincidence caused by our choice of random operations and the underlying Hamiltonian, or it may suggest a previously unforeseen connection to the creation of our universe and black hole physics. In the latter case, one might attempt to model the earliest stages of the Big Bang using scrambling dynamics.

In any case, to overcome the model-dependent limitations associated with relying on any specific dynamics, we consider a more general, information theoretic approach in Sect. 12.5. This allows us to take into account both the inclusion of black hole evaporation but also an initial black hole state which may be anywhere from completely pure to maximally mixed. We find that if an observer is coupled to a sufficiently small 'neighborhood' of the entire interior Hilbert space they will experience a maximally mixed state to a close approximation and therefore a near infinite temperature. We find that a neighborhood is sufficiently small to achieve this high-temperature behavior provided only that its Hilbert space dimensionality, n, satisfies

$$n \leq \varepsilon\sqrt{N}, \tag{12.64}$$

where N is the dimensionality of the *newly formed* pure state black hole. (A looser bound is found when the initial back hole is impure which only strengthens our discussion below, see Sect. 12.5.3 for details.) The prefactor satisfies $\varepsilon \ll 1$; for simplicity we take $\varepsilon = 2^{-10} \simeq 10^{-3}$, though using values of 2^{-100} or 2^{-1000} makes only trivial changes to our discussion below.

Let us take a step back and consider the scenario where an observer has jumped into a newly formed black hole. In such a scenario, we may neglect any Hawking radiation and consider solely the Hilbert space of the black hole interior. The Hilbert space dimensionality of any physically accessible degrees-of-freedom of the interior is assumed to be given by the 'central dogma' of black hole physics as $\exp(S)$ for a black hole with Bekenstein-Hawking entropy S. Applying random unitary operators to this Hilbert space then implies our result of Eq. (12.64), without needing to explicitly assume how the interior Hilbert space is connected to the external universe, say within a tensor product structure. Thus, if fast scrambling means anything at all on this internal Hilbert space, Eq. (12.64) follows.

Note further that the simplicity of Eq. (12.64) belies the counter-intuitive nature of our results, which are technically summarized for initially pure-state black holes in Eq. (12.40) and more generally for initially mixed-state black holes in Eq. (12.62) which gives the lower bounds to the fidelity of the fast-scrambled quantum state of a black hole as compared to the maximally-mixed (infinite-temperature) state. The counter-intuitive result is that the larger the value of N—or in other words, the larger the black hole—the tighter this bound becomes. For astrophysical black holes the fidelity is negligibly close to unity as a consequence. This in turn implies that the infalling observer will experience an enormously large temperature—in effect, a firewall. This result is contrary to expected thinking about black holes, which is that the larger they are, the more classical they should appear.

In particular, rather than quantifying the size of the neighborhood in terms of its dimensionality, it is more physically intuitive to quantify it in terms of qubits. Note that we are not saying that any part of the black hole is actually made up of two-level systems, only that the *number* of two-level systems that could be supported by its dimensionality is a more familiar quantity—analogous to entropy, the number of qubits is additive. So the total number of qubits within the black hole Hilbert space at any time is the sum of the number of qubits within the selected neighborhood and the number still within the black hole, but not within this neighborhood.

Within the language of qubits, Eq. (12.64) states that a neighborhood of the black hole interior is sufficiently small to correspond to an almost infinite-temperature state provided only that

$$\#_{\text{neighborhood}} \le \frac{1}{2} \#_{\text{newly-formed-black-hole}} - 10, \tag{12.65}$$

where we denote the number of qubits within subsystem A by the hashtag $\#_A$. Replacing ε by much tighter values mentioned above only replaces the -10 by -100 or -1000, which as noted are utterly trivial in comparison to values of $\#_{\text{newly-formed-black-hole}} \simeq 10^{80}$ for a stellar mass black hole.

Consider an initially pure-state black hole at least as old as its scrambling time (not very different in order of magnitude than the light-travel time across a Schwarzschild radius' distance in flat spacetime). Equation (12.65) tells us that any body which simultaneously couples to less than one-half of the total number of qubits in the original black hole will experience an extremely high temperature as soon as it passes the horizon. Thus scrambling alone places enormous constraints on the survival of the equivalence principle's claimed prediction of 'no drama' for infalling bodies. As the black hole evaporates this constraint becomes even more difficult to attain.

Consider an initially pure-state black hole that has partially evaporated leaving a fraction, f, of the qubits in the current state of the black hole, which therefore has

$$\#_{\text{BH}} = f \, \#_{\text{newly-formed-black-hole}}, \tag{12.66}$$

qubits remaining (or equivalently, it's area has shrunk to this fraction of its original size). Then Eq. (12.65) becomes

$$\#_{\text{neighborhood}} \le \frac{1}{2f} \#_{\text{BH}} - 10. \tag{12.67}$$

Thus, as $f \to \frac{1}{2}^{+}$, (i.e., as the evaporation approaches the Page time), any infalling body must simultaneously couple to virtually the entire black hole interior to have any hope of experiencing no drama has it passes the horizon. From the Page time onwards, even this is not sufficient and we recover the usual firewall result, though without the need for any decoding or complexity assumption.

In order to preserve the claim of 'no drama' for scrambling black holes even prior to the onset of a full blown firewall, any infalling body must couple to virtually the entire interior Hilbert space of the black hole. Further, it must do so in a uniform manner without random phases appearing in the coupling, whatever the direction of the infalling body. We can envisage only one scenario where this is possible: that the quantum state of the black hole interior is actually described by a Bose-Einstein condensate, so that all interior degrees-of-freedom correspond to excitations of a single Bose-Einstein condensate mode. However, we note, that even with this radical assumption, the onset of a firewall at the Page time will occur regardless of the size of the local neighborhood of the infalling observer.

If the black hole's quantum state is initially partially mixed, the constraints on satisfying the equivalence principle become even more extreme. It is convenient to define the log-purity of the *initial* black hole as

$$\ell \equiv -\log_2(\mathrm{tr}(\rho_0^2) - 1/N) \geq 0, \tag{12.68}$$

where ρ_0 is the initial black hole's density matrix and N its dimensionality. Black hole scrambling therefore implies that a fully developed firewall will be present once the black hole's area has shrunk to the fraction

$$f = \frac{1}{2 - \ell/\#_{\mathrm{BH}}} \geq \frac{1}{2}, \tag{12.69}$$

of its original size. In other words, for a black hole with an initially partially mixed quantum state, the firewall becomes fully developed prior to the Page time regardless of the size of the local neighborhood of the infalling observer.

As noted, the only caveats we can identify to our analysis are either that black holes are described by a Hamiltonian that is totally degenerate with regard to all their interior degrees-of-freedom, wherein the notion of temperature becomes a meaningless concept, or that the quantum state of a black hole interior is described by a Bose-Einstein condensate. Absent these caveats we may interpret our results as proving a new black hole paradox demonstrating the incompatibility between the assumptions of fast scrambling, the equivalence principle's "no drama" at the horizon, and the locality of an infalling observer. Unlike the usual firewall result which holds only for old black holes, our paradox applies to every astrophysical black hole in the universe, which has the implication that either the astrophysical objects observed by LIGO and the EHT are not really black holes but something very similar (and strange), or that there is something wrong with one of the assumptions we make about black hole physics.

Appendix 1

Let us first review a result of the Schur-Weyl duality. Suppose we have a Hermitian X, then Schur-Weyl duality implies that [1]

$$\int_{U(A)} (U^\dagger \otimes U^\dagger)\, X (U \otimes U) dU = \alpha_+ \Pi_+^A + \alpha_- \Pi_-^A\,, \tag{12.70}$$

where

$$\alpha_{\pm} = \frac{\mathrm{tr}\left(X\Pi_{\pm}^{A}\right)}{\mathrm{rank}\left(\Pi_{\pm}^{A}\right)} \quad \text{and} \quad \Pi_{\pm}^{A} = \frac{1}{2}\left(\mathrm{I}_{A,A} \pm \mathrm{SWAP}_{A,A}\right). \tag{12.71}$$

Here $\mathrm{SWAP}_{A,B}$ represents the SWAP operator between the system A and B.

Now suppose that $A = A_1 \otimes A_2$, which is what we will use in this work, then we have

$$\begin{aligned}\mathrm{rank}\Pi_{\pm}^{A} &= \mathrm{rank}\frac{1}{2}\left(\mathrm{I}_{A,A} \pm \mathrm{SWAP}_{A,A}\right)\\ &= \frac{1}{2}(d_A^2 \pm d_A)\\ &= \frac{1}{2}((d_{A_1}d_{A_2})^2 \pm d_{A_1}d_{A_2}),\end{aligned} \tag{12.72}$$

where d_X equals the dimension of the state X. If we suppose X equals SWAP_{A_2,A_2}, then $\mathrm{tr}\left(X\Pi_{\pm}^{A}\right)$ in Eq. (12.71) may be calculated as

$$\begin{aligned}\mathrm{tr}\left(\Pi_{\pm}^{A}\,\mathrm{SWAP}_{A_2,A_2}\right) &= \frac{1}{2}\mathrm{tr}\left[\left(\mathrm{I}_{A,A} \pm \mathrm{SWAP}_{A_1,A_1} \otimes \mathrm{SWAP}_{A_2,A_2}\right)\mathrm{SWAP}_{A_2,A_2}\right]\\ &= \frac{1}{2}\left[\mathrm{tr}\left(\mathrm{I}_{A_1,A_1} \otimes \mathrm{SWAP}_{A_2,A_2}\right) \pm \mathrm{tr}\left(\mathrm{SWAP}_{A_1,A_1} \otimes \mathrm{I}_{A_1,A_2}\right)\right]\\ &= \frac{1}{2}\left(d_{A_1}^2 d_{A_2} \pm d_{A_1}d_{A_2}^2\right).\end{aligned} \tag{12.73}$$

Inserting Eqs. (12.71), (12.72) and (12.73) into Eq. (12.70) yields

$$\begin{aligned}\int_{U(A)}\left(U^{\dagger} \otimes U^{\dagger}\right)X(U \otimes U)dU &= \frac{d_{A_1}^2 d_{A_2} + d_{A_1}d_{A_2}^2}{(d_{A_1}d_{A_2})^2 + d_{A_1}d_{A_2}}\Pi_{+}^{A} + \frac{d_{A_1}^2 d_{A_2} - d_{A_1}d_{A_2}^2}{(d_{A_1}d_{A_2})^2 - d_{A_1}d_{A_2}}\Pi_{-}^{A}\\ &= \frac{d_{A_1} + d_{A_2}}{d_{A_1}d_{A_2} + 1}\Pi_{+}^{A} + \frac{d_{A_1} - d_{A_2}}{d_{A_1}d_{A_2} - 1}\Pi_{-}^{A}\\ &= \frac{d_{A_1} + d_{A_2}}{d_A + 1}\Pi_{+}^{A} + \frac{d_{A_1} - d_{A_2}}{d_A - 1}\Pi_{-}^{A}.\end{aligned} \tag{12.74}$$

Appendix 2

Similarly to Sect. 12.5.2, we can also consider black holes originating from a generic black hole that has non-negligible radiation. If we still assume the initial dimension of the entire black hole state to be N and the dimension of the radiation is R, then the dimension of the remaining black hole N_B will equal N/R. The quantum state of the remaining black hole may be represented by $\mathrm{tr}_R(\rho_U)$. The key step in our calculation of the 2-norm is to take an average over all possible unitary operators, since black holes are fast scramblers. Now, with part of the initial state radiated outside the horizon, we first take an average over the unitary operators of the remaining black hole U_B. Here U_B is different from the U, and does not influence the radiation outside the horizon. With this physical picture, the fidelity will have a range of

$$F \geq 1 - \frac{1}{2}\sqrt{n} \int_{U_B} \left\| \mathrm{tr}_{\bar{n}}\left(U_B \mathrm{tr}_R(\rho_U) U_B^\dagger\right) - \frac{\mathrm{I}_n}{n} \right\|_2 dU_B. \tag{12.75}$$

The 2-norm in Eq. (12.75) may be calculated

$$\begin{aligned}
&\int_{U_B} \left\| \mathrm{tr}_{\bar{n}}\left(U_B \mathrm{tr}_R(\rho_U) U_B^\dagger\right) - \frac{\mathrm{I}_n}{n} \right\|_2^2 dU_B \\
&= \int_{U_B} \mathrm{tr}_n\Big[\mathrm{tr}_{\bar{n}}\left(U_B \mathrm{tr}_R(\rho_U) U_B^\dagger\right) \mathrm{tr}_{\bar{n}}\left(U_B \mathrm{tr}_R(\rho_U) U_B^\dagger\right) - \frac{2}{n}\mathrm{tr}_{\bar{n}}\left(U_B \mathrm{tr}_R(\rho_U) U_B^\dagger\right) + \frac{\mathrm{I}_n}{n^2}\Big] dU_B \\
&= \int_{U_B} \mathrm{tr}_n\left[\mathrm{tr}_{\bar{n}}\left(U_B \mathrm{tr}_R(\rho_U) U_B^\dagger\right) \mathrm{tr}_{\bar{n}}\left(U_B \mathrm{tr}_R(\rho_U) U_B^\dagger\right)\right] - \frac{2}{n}\mathrm{tr}_{\bar{R}}\left(U_B \mathrm{tr}_R(\rho_U) U_B^\dagger\right) dU_B + \frac{1}{n} \\
&= \int_{U_B} \mathrm{tr}_{n,n}\left[\mathrm{tr}_{\bar{n},\bar{n}}\left(U_B \mathrm{tr}_R(\rho_U) U_B^\dagger\right) \otimes \left(U_B \mathrm{tr}_R(\rho_U) U_B^\dagger\right) \mathrm{SWAP}_{n,n}\right] dU_B - \frac{2}{n} + \frac{1}{n} \\
&= \int_{U_B} \mathrm{tr}\left[(U_B \otimes U_B)\left(\mathrm{tr}_R(\rho_U) \otimes \mathrm{tr}_R(\rho_U)\right)(U_B^\dagger \otimes U_B^\dagger)\left(\mathrm{SWAP}_{n,n} \otimes \mathrm{I}_{\bar{n},\bar{n}}\right)\right] dU_B - \frac{1}{n} \\
&= \mathrm{tr}\Big[\left(\mathrm{tr}_R(\rho_U) \otimes \mathrm{tr}_R(\rho_U)\right) \int_{U_B} \left[(U_B^\dagger \otimes U_B^\dagger)(\mathrm{SWAP}_{n,n} \otimes \mathrm{I}_{\bar{n},\bar{n}})(U_B \otimes U_B)\right] dU_B\Big] - \frac{1}{n},
\end{aligned} \tag{12.76}$$

where $\|A\|_2 = \mathrm{tr}(A^\dagger A)$ is applied to the first step, and $\mathrm{tr}(AB) = \mathrm{tr}(A \otimes B\, \mathrm{SWAP}_{A,B})$ is applied in moving from the third to the fourth line. Applying the Schur-Weyl duality in Appendix 1 to Eq. (12.76) yields

$$\begin{aligned}
&\int_{U_B} \left\| \mathrm{tr}_{\bar{n}}\left(U_B \mathrm{tr}_R(\rho_U) U_B^\dagger\right) - \frac{\mathrm{I}_n}{n} \right\|_2^2 dU_B \\
&= \frac{1}{2}\mathrm{tr}\Big[\left(\mathrm{tr}_R(\rho_U) \otimes \mathrm{tr}_R(\rho_U)\right) \\
&\quad \times \Big(\frac{\frac{N_B}{n} + n}{N_B + 1}(\mathrm{I}_{N_B,N_B} + \mathrm{SWAP}_{N_B,N_B}) + \frac{\frac{N_B}{n} - n}{N_B - 1}(\mathrm{I}_{N_B,N_B} - \mathrm{SWAP}_{N_B,N_B})\Big)\Big] - \frac{1}{n} \\
&= \mathrm{tr}\Big[\left(\mathrm{tr}_R(\rho_U) \otimes \mathrm{tr}_R(\rho_U)\right)\Big(\frac{\frac{N_B^2}{n} - n}{N_B^2 - 1}\mathrm{I}_{N_B,N_B} + \frac{N_B n - \frac{N_B}{n}}{N_B^2 - 1}\mathrm{SWAP}_{N_B,N_B}\Big)\Big] - \frac{1}{n} \\
&= \frac{\frac{N_B^2}{n} - n}{N_B^2 - 1}\left(\mathrm{tr}\left(\mathrm{tr}_R(\rho_U)\right)\right)^2 + \frac{N_B n - \frac{N_B}{n}}{N_B^2 - 1}\mathrm{tr}\left(\left(\mathrm{tr}_R(\rho_U)\right)^2\right) - \frac{1}{n} \\
&= \frac{\frac{N_B^2}{n} - n}{N_B^2 - 1}(\mathrm{tr}(\rho_U))^2 + \frac{N_B n - \frac{N_B}{n}}{N_B^2 - 1}\mathrm{tr}\left(\left(\mathrm{tr}_R(\rho_U)\right)^2\right) - \frac{1}{n} \\
&= \frac{\frac{N_B^2}{n} - n}{N_B^2 - 1} + \frac{N_B n - \frac{N_B}{n}}{N_B^2 - 1}\mathrm{tr}\left(\left(\mathrm{tr}_R(\rho_U)\right)^2\right) - \frac{1}{n},
\end{aligned} \tag{12.77}$$

where we have used $(\mathrm{tr}(\rho_U))^2 = 1$ in the last step.

Since black holes are believed to be fast quantum scrambler, a black hole should already finish its quantum scrambling before it radiates too much state away. Thus. the term $\mathrm{tr}\big((\mathrm{tr}_R(\rho_U))^2\big)$ in Eq. (12.77) may be calculated

$$
\begin{aligned}
&\Big\langle \mathrm{tr}\big((\mathrm{tr}_R(\rho_U))^2\big)\Big\rangle \\
&= \int_U \mathrm{tr}((\mathrm{tr}_R(\rho_U))^2)dU = \int_U \mathrm{tr}\Big[\mathrm{tr}_R(\rho_U)\otimes \mathrm{tr}_R(\rho_U)\mathrm{SWAP}_{\bar{R},\bar{R}}\Big]dU \\
&= \int_U \mathrm{tr}\Big[\rho_U \otimes \rho_U \mathrm{SWAP}_{\bar{R},\bar{R}}\Big]dU = \int_U \mathrm{tr}\Big[(U\otimes U)(\rho\otimes\rho)(U^\dagger \otimes U^\dagger)\mathrm{SWAP}_{\bar{R},\bar{R}}\Big]dU \\
&= \mathrm{tr}\Big[(\rho\otimes\rho)\int_U (U^\dagger\otimes U^\dagger)\mathrm{SWAP}_{\bar{R},\bar{R}}(U\otimes U)dU\Big] \\
&= \frac{1}{2}\mathrm{tr}\Big[(\rho\otimes\rho)\Big(\frac{N/\bar{R}+\bar{R}}{N+1}(\mathrm{I}_{N,N}+\mathrm{SWAP}_{N,N}) + \frac{N/\bar{R}-\bar{R}}{N-1}(\mathrm{I}_{N,N}-\mathrm{SWAP}_{N,N})\Big)\Big] \\
&= \mathrm{tr}\Big[(\rho\otimes\rho)\Big(\frac{NR-N_B}{N^2-1}\mathrm{I}_{N,N} + \frac{NN_B-R}{N^2-1}\mathrm{SWAP}_{N,N}\Big)\Big] \\
&= \frac{NR-N_B}{N^2-1}(\mathrm{tr}(\rho))^2 + \frac{NN_B-R}{N^2-1}\mathrm{tr}(\rho^2) \\
&= \frac{NR-N_B+(NN_B-R)\mathrm{tr}(\rho^2)}{N^2-1},
\end{aligned}
\tag{12.78}
$$

where the Schur-Weyl duality is used in moving from the fourth to the fifth line, $\bar{R} = N_B$ and $N_B = N/R$ are used in moving from the fifth to the sixth line, and $(\mathrm{tr}(\rho))^2 = 1$ is applied to the seventh line.

Applying Eq. (12.78) to Eq. (12.77) yields

$$
\begin{aligned}
&\int_{U_B} \Big\| \mathrm{tr}_{\bar{n}}\Big(U_B \mathrm{tr}_R(\rho_U) U_B{}^\dagger\Big) - \frac{\mathrm{I}_n}{n}\Big\|_2^2 dU_B \\
&= \frac{\frac{N_B^2}{n} - n}{N_B^2-1} + \frac{N_B n - \frac{N_B}{n}}{N_B^2-1}\frac{NR-N_B+(NN_B-R)\mathrm{tr}(\rho^2)}{N^2-1} - \frac{1}{n} \\
&= \frac{1-n^2}{(N_B^2-1)n} + \frac{N_B n^2 - N_B}{(N_B^2-1)n}\frac{NR-N_B+(NN_B-R)\mathrm{tr}(\rho^2)}{N^2-1} \\
&= \frac{(n^2-1) - N_B^2(n^2-1) + N(N_B^2-1)(n^2-1)\mathrm{tr}(\rho^2)}{(N_B^2-1)(N^2-1)n} \\
&= \frac{-(n^2-1)+N(n^2-1)\mathrm{tr}(\rho^2)}{(N^2-1)n} \\
&= \frac{(N\mathrm{tr}(\rho^2)-1)(n^2-1)}{(N^2-1)n},
\end{aligned}
\tag{12.79}
$$

where $N = N_B R$ is used in moving from the third to the four line line. As $n \le N$, then $N^2n^2 - N^2 \le N^2n^2 - n^2$ and hence $N^2(n^2-1) \le n^2(N^2-1)$. Thus, Eq. (12.79) satisfies

$$\int_{U_B} \left\| \mathrm{tr}_{\bar{n}}\left(U_B \mathrm{tr}_R(\rho_U) U_B{}^{\dagger} \right) - \frac{\mathrm{I}_n}{n} \right\|_2^2 dU_B = \frac{(N\mathrm{tr}(\rho^2) - 1)(n^2 - 1)}{(N^2 - 1)n} \leq \frac{(N\mathrm{tr}(\rho^2) - 1)n}{N^2}. \tag{12.80}$$

Applying Eq. (12.80) to Eq. (12.75) yields

$$F \geq 1 - \frac{n}{2N}\sqrt{N\mathrm{tr}(\rho^2) - 1}. \tag{12.81}$$

References

1. A. Abeyesinghe, I. Devetak, P. Hayden, A. Winter, Proc. R. Soc. A **465**, 2537–2563 (2009)
2. A. Almheiri, D. Marolf, J. Polchinski, J. Sully, JHEP **02**, 062 (2013). https://doi.org/10.1007/JHEP02(2013)062. arXiv:1207.3123 [hep-th]
3. A. Almheiri, T. Hartman, J. Maldacena, E. Shaghoulian, A. Tajdini, Rev. Mod. Phys. **93**(3), 035002 (2021). https://doi.org/10.1103/RevModPhys.93.035002. arXiv:2006.06872 [hep-th]
4. S.L. Braunstein. arXiv:0907.1190 [quant-ph]
5. S.L. Braunstein, A.K. Pati, Phys. Rev. Lett. **98**, 080502 (2007). https://doi.org/10.1103/PhysRevLett.98.080502. arXiv:gr-qc/0603046 [gr-qc]
6. S.L. Braunstein, S. Pirandola, K. Życzkowski, Phys. Rev. Lett. **110**(10), 101301 (2013). https://doi.org/10.1103/PhysRevLett.110.101301. arXiv:0907.1190 [quant-ph]
7. S.L. Braunstein, S. Pirandola, Philos. Trans. A **376**, 20170324 (2018)
8. V. Coffman, J. Kundu, W.K. Wootters, Phys. Rev. A **61**, 052306 (2000). https://doi.org/10.1103/PhysRevA.61.052306. arXiv:quant-ph/9907047 [quant-ph]
9. P.C.W. Davies, S.A. Fulling, W.G. Unruh, Phys. Rev. D **13**, 2720–2723 (1976). https://doi.org/10.1103/PhysRevD.13.2720
10. D. Harlow, P. Hayden, JHEP **06**, 085 (2013). https://doi.org/10.1007/JHEP06(2013)085. arXiv:1301.4504 [hep-th]
11. S.W. Hawking, Commun. Math. Phys. **43**, 199–220 (1975) [Erratum: Commun. Math. Phys. **46**, 206 (1976)]. https://doi.org/10.1007/BF02345020
12. P. Hayden, J. Preskill, JHEP **09**, 120 (2007). https://doi.org/10.1088/1126-6708/2007/09/120. arXiv:0708.4025 [hep-th]
13. S.D.H. Hsu. arXiv:1302.0451 [hep-th]
14. S.D.H. Hsu. arXiv:1308.5686 [hep-th]
15. M. Koashi, A. Winter, Phys. Rev. A **69**(2), 022309 (2004). https://doi.org/10.1103/PhysRevA.69.022309
16. A. Laddha, S.G. Prabhu, S. Raju, P. Shrivastava, SciPost Phys. **10**(2), 041 (2021). https://doi.org/10.21468/SciPostPhys.10.2.041. arXiv:2002.02448 [hep-th]
17. J. Maldacena, L. Susskind, Fortsch. Phys. **61**, 781–811 (2013). https://doi.org/10.1002/prop.201300020. arXiv:1306.0533 [hep-th]
18. B. Michel. arXiv:1612.03158 [hep-th]
19. M.A. Nielsen, I.L. Chuang (Cambridge University Press, 2010). https://doi.org/10.1017/cbo9780511976667
20. D.N. Page, Phys. Rev. Lett. **71**, 3743–3746 (1993). https://doi.org/10.1103/PhysRevLett.71.3743. arXiv:hep-th/9306083 [hep-th]
21. K. Papadodimas, S. Raju, JHEP **10**, 212 (2013). https://doi.org/10.1007/JHEP10(2013)212. arXiv:1211.6767 [hep-th]
22. R.H. Price, K.S. Thorne, Phys. Rev. D **33**, 915–941 (1986). https://doi.org/10.1103/PhysRevD.33.915

23. S. Raju, Class. Quant. Grav. **39**(6), 064002 (2022). https://doi.org/10.1088/1361-6382/ac482b. arXiv:2110.05470 [hep-th]
24. Y. Sekino, L. Susskind, JHEP **10**, 065 (2008). https://doi.org/10.1088/1126-6708/2008/10/065. arXiv:0808.2096 [hep-th]
25. L. Susskind, L. Thorlacius, J. Uglum, Phys. Rev. D **48**, 3743–3761 (1993). https://doi.org/10.1103/PhysRevD.48.3743. arXiv:hep-th/9306069 [hep-th]
26. L. Susskind. arXiv:1101.6048 [hep-th]
27. L. Susskind. arXiv:1208.3445 [hep-th]
28. K.S. Thorne, R.H. Price, D.A. Macdonald
29. Z.W. Wang, S. Das, S.L. Braunstein. arXiv:2206.02053 [gr-qc]
30. J. Watrous (Cambridge University Press, 2018). https://doi.org/10.1017/9781316848142

Chapter 13
The Firewall Paradox Is Wigner's Friend Paradox

Ladina Hausmann and Renato Renner

Abstract Quantum gravity thought experiments, such as the firewall paradox, probe the compatibility of quantum theory with general relativity. The usual conclusions of these experiments rely on the universality of quantum theory—the assumption that we can always apply the theory to any physical system, including black holes. Similarly, non-gravitational thought experiments from quantum foundations, like Wigner's friend paradox, also rely on the universality assumption. They examine scenarios involving multiple physicists, each applying quantum theory to different physical systems, which may themselves consist of other physicists. A key lesson learnt from these thought experiments is that combining the different physicists' perspectives is far from straightforward. Even a seemingly intuitive rule—that if physicist Alice knows that physicist Bob knows a measurement yielded result x then Alice also knows this measurement has result x—leads to contradictions. In this chapter, we revisit quantum gravity thought experiments like the firewall paradox and point out that they also crucially depend on such combination rules, some of which have been implicit so far. They can, therefore, be understood as gravitational versions of Wigner's friend thought experiments. This correspondence enables the transfer of insights between quantum foundations and gravity. In particular, the lesson from quantum foundations—that the universality of quantum theory conflicts with seemingly self-evident rules physicists use to combine their perspectives—applies equally to the firewall paradox. Questioning the validity of these rules thus offers a new avenue for resolving it.

L. Hausmann (✉) · R. Renner
Institute for Theoretical Physics, ETH Zürich, Zürich, Switzerland
e-mail: hladina@phys.ethz.ch

R. Renner
e-mail: renner@ethz.ch

A. Akil and C. Bambi (eds.), *The Black Hole Information Paradox*, Springer Series in Astrophysics and Cosmology, https://doi.org/10.1007/978-981-96-6170-1_13

13.1 Introduction

Imagine two quantum physicists, Alice and Bob. Alice is enclosed in an isolated laboratory, where she performs a projective measurement on a qubit. From her perspective, after observing the outcome of her measurement, the qubit is in a pure state. Meanwhile, Bob, who stays outside, describes the entire laboratory—including everything inside it—as a big quantum system. From his perspective, this system undergoes a unitary time evolution, which generally entangles the qubit with other systems within the laboratory. Thus, the two physicists arrive at incompatible conclusions about the final state of the qubit: For Alice, it is in a pure state, whereas, for Bob, it is entangled with other systems. This is the basic version of Wigner's friend paradox [48].[1]

Thought experiments in quantum gravity, like the Xeroxing paradox [25, 42] and the firewall paradox [2, 30], similarly involve different physicists adopting different perspectives. For example, Alice, who is freely falling into a black hole, has access to the interior, whereas Bob, who remains outside, has access to the Hawking radiation emitted after Alice reaches the horizon. Furthermore, as in Wigner's friend paradox, the conclusions that Alice and Bob draw about the state of certain systems, such as the radiation modes in "the zone", are incompatible.

In this chapter, we aim to turn these superficial similarities between quantum foundations and quantum gravity considerations into a deeper correspondence. For this, we explore recent experiments that extend Wigner's friend paradox, developed within the field of quantum foundations [6, 11, 16, 21] (for reviews see [12, 34, 39]). They rely on the principle that "physicists are physical".[2] This principle was already used in the basic version: it allowed Bob to apply quantum theory to Alice.

Wigner's friend experiments, like quantum gravity thought experiments, test the universality of quantum theory—the assumption that we can always use quantum theory to describe any physical system. Combined with the aforementioned principle, universality implies that quantum physicists can use quantum theory to describe other quantum physicists. As we will explain in Sect. 13.2, this test shows that the universality of quantum theory is incompatible with seemingly innocent assumptions about how different physicists combine their conclusions.

An example of such a rule is the following [21]. Suppose that Bob, using quantum theory, concluded that another physicist, Charly, using quantum theory, concluded that the result of a measurement of a qubit is 0. The assumption is now that Bob can conclude that the measurement result is 0.

---

[1] The physicist in Alice's role is commonly referred to as "the Friend" and the one in Bob's role as "Wigner". In quantum foundations, they are sometimes called "agents", a term associated with additional assumptions like "free will", which play no role in our arguments. In gravity, Alice and Bob are also called "observers", but this notion degrades them to a too passive role. Here, we want to emphasize the aspect that they can interact with physical systems and use a physical theory to make predictions. Therefore, we refer to them as "physicists".

[2] The similarity to Landauer's slogan "Information is physical" [28] is deliberate.

Combining information acquired by different physicists with different perspectives is also relevant to quantum gravity thought experiments, as we will discuss in Sect. 13.3. In situations where these pieces are not operationally accessible to a single physicist, as in the Xeroxing paradox, this issue is explicitly recognized as problematic and referred to as *black hole complementarity* [41–43]. However, in other situations, particularly in the firewall paradox, the rules for combining information appeared so innocent that they remained implicit and unexamined. Making these rules explicit reveals that the firewall paradox belongs to same class of thought experiments as Wigner's friend paradox—the class of thought experiments where the universality of quantum theory, together with a rule for combining information held by different physicists, leads to a contradiction.[3]

13.2 Wigner's Friend Thought Experiments

For a long time, the discussion around Wigner's friend paradox was mostly philosophical, and the implications for concrete physics questions were unclear. Recent extensions of the experiment, proposed in [6, 11, 21], changed this as they test specific properties of a physical theory, similarly to Bell experiments testing local causality [4, 19]. One of these properties is the universal applicability of the theory—the tenet that any physicist can apply the theory to describe any systems around them. Wigner's friend thought experiments push the universality of quantum theory to its limits, by applying the theory to its users.

Physicists are part of the world and, therefore, physical systems, which may again be described by other physicists. We frame this as a principle, which any fundamental physical theory, like quantum theory, must satisfy.

Principle (PP) — Physicists are physical!
A physicist can use the theory to describe another physicist who uses the theory.

The principle treats physicists as objects of study. At first sight, this may appear problematic, potentially involving vague notions like "free will", "consciousness", or "the mind".[4] However, we avoid this vagueness by adopting an operational approach and treating physicists as information-processing devices, for example,

[3] Rules for combining information are also relevant in other theories, such as classical thermodynamics (see [27] for more examples). An example is Maxwell's demon paradox [31]. It features a physicist, P, who holds the usual coarse-grained thermodynamic description of a gas, and a demon, D, who can observe individual gas molecules. A careless combination of these perspectives leads to the conclusion that P would observe a violation of the second law of thermodynamics [5].

[4] Indeed, Wigner's original conclusion was that a "being with a consciousness must have a different role in quantum mechanics than the inanimate measuring device" [48].

Turing machines connected to experimental devices.[5] The physicists' task is then captured by an algorithm for processing data and generating predictions using specific rules.[6]

Quantum States Cannot Be Agreed On …

We are now ready to examine the basic version of Wigner's friend paradox in more detail. The paradox involves two physicists, Alice and Bob, who follow the instructions specified in Protocol 13.1.

Protocol 13.1 (Wigner's friend paradox [48] *with minor modifications)*

Setup

- Bob is outside a laboratory. He has full quantum control over the laboratory as well as any system initially entangled with it. He can send a qubit Q into the laboratory and isolate the laboratory from its environment.
- Alice is part of the laboratory and can control the qubit Q she receives from the outside. We denote the part of the laboratory without the qubit by L_A.

Protocol

1. Bob prepares L_A in a pure state $|\varphi\rangle_{L_A}$.[7]
2. Bob prepares Q in state

$$|\bar{0}\rangle_Q = \sqrt{\tfrac{1}{2}}\left(|0\rangle_Q + |1\rangle_Q \right) \tag{13.1}$$

   and sends it to Alice into the laboratory.
3. Bob isolates the laboratory.
4. Alice measures Q projectively in the computational basis, $\{|0\rangle, |1\rangle\}$, and records the outcome.
5. Alice infers the quantum state of Q.

---

[5] Principle (PP) may be stated in a way reminiscent of the Church-Turing thesis [13, 45]: A computer programmed with the rules of the theory can be modelled within the theory by another computer.

[6] Principle (PP) also plays a central role in resolving Maxwell's demon paradox described in Footnote 3. On the one hand, to fulfil his task, the demon D needs to process information gained by observing the gas molecules. On the other hand, D, and in particular his memory containing the record of his observations, must be modelled as a physical system. In particular, in P's coarse-grained description, D is itself subject to the laws of thermodynamics. Taking this into account, no violation of the second law occurs [5].

6. Bob infers the quantum state of the composite system $L_A Q$.

Principle (PP) plays a key role in this experiment: Alice is, at the same time, the subject using quantum theory and the object described by it.[8] This differs from standard experiments, where physicists are exterior to the experimental setup described by the theory.

Protocol 13.1 requires both Alice and Bob to reason about the experiment, using the rules of quantum theory. However, their descriptions encompass different systems. Therefore, we take precautionary measures: We always declare explicitly from whose perspective a system is described. For example, we write ρ_Q^{Alice} for the state Alice assigns to the qubit Q. Additionally, we state the assumption that Alice and Bob are justified in their use of quantum theory explicitly.

Assumption (Q) *Any physicist can apply quantum theory*[9] *to any physical system.*[10]

Before proceeding with the analysis of Protocol 13.1, we highlight two key aspects of a theory's universality. Each aspect corresponds to one of the words "any" in Assumption (Q). First, any physicist can apply the theory, here, quantum theory. In particular, Alice, who is in an isolated laboratory, can apply quantum theory. Second, any system can be described by the theory. In particular, Bob can describe Alice as a quantum system.

The first aspect of universality is often left implicit, although there are firm views on it. Proponents of many-worlds [46] or Bohmian mechanics [44], for instance, might proclaim that quantum theory can ultimately be applied only from a perspective outside the universe.[11] A similar view is common in AdS/CFT, where the

---

[7] If one assumes that the total state is pure, Bob can achieve this by an appropriate measurement of the systems initially entangled with the laboratory. Such a measurement does not act on the laboratory itself. In particular, by non-signalling, it does not harm Alice. The same is also true for the remaining steps. For example, Bob can isolate the laboratory without acting on Alice.

[8] That Alice is a physicist, able to use quantum theory, is the crucial distinction between Wigner's friend and Schrödinger's cat experiment. (Unlike the cat, she is also not subjected to a lethal dose of hydrocyanic acid [40].)

[9] By quantum theory, we mean the postulates stated in [33].

[10] One might wonder whether this leads to problems if the system is not closed, such as a qubit A coupled to an environment B. Suppose that, in a particular run of an experiment, a physicist P_1 has acquired information about B that allows him to predict that AB evolves into $\rho_{AB}^{P_1} = |0\rangle\langle 0|_A \otimes |\varphi\rangle\langle\varphi|_B$. Another physicist P_2, who holds a description of qubit A only, may conclude that A evolves to $\rho_A^{P_2} = \frac{1_A}{2}$. This mixed state reflects the fact that P_2 has no knowledge about the environment B, and the coupling between A and B made this uncertainty propagate into A. However, there is no contradiction between P_1 and P_2's conclusions, as there exists a joint state $\psi_{AB} = |0\rangle\langle 0|_A \otimes |\varphi\rangle\langle\varphi|_B$ they could both agree on, according to the definition given below.

[11] These approaches require that quantum theory be applied to a closed system. But when a physicist P interacts with a system S, any closed system that includes S will necessarily also include P. Consequently, P must refer to a hypothetical outside perspective onto this closed system. In many-worlds this perspective is captured by "the wave function of the universe". But this leads to an information-theoretic problem: P is fundamentally lacking the information to infer this wave function, because his access is limited to a single branch of it.

boundary theory of the AdS universe is regarded as the ultimate authority. In contrast, Assumption (Q) asserts that quantum theory must be usable by the physicists inside the universe. This is a natural and hard-to-dispute requirement—after all, the theory was discovered and validated through countless experiments by such physicists.

We are now ready to analyse the descriptions of the qubit Q that Alice and Bob arrive at by the end of Protocol 13.1. After having performed her projective measurement, and conditioned on the recorded outcome, Alice's state will be either $|0\rangle_Q$ or $|1\rangle_Q$, i.e., ρ_Q^{Alice} is pure. Meanwhile, for Bob, who does not know Alice's measurement outcome, Q will be in a mixed state, i.e., $\rho_Q^{\mathrm{Bob}} = \frac{\mathbb{1}_Q}{2}$. Furthermore, Bob can apply quantum theory to the entire laboratory $L_A Q$, which, for him, is in a pure state. Because it is isolated and thus evolves unitarily, it will remain pure. But since the marginal state ρ_Q^{Bob} is mixed, he concludes that Q is entangled with the rest of the laboratory L_A.

One may now ask whether Alice's and Bob's conclusions can be combined, in the sense that there exists a quantum state, ψ, on which both of them can agree. We say that a physicist P *can agree on* ψ if, for any finite-dimensional system S described by P, there exists information such that, conditioned on this information, the state ρ_S^{P} that P assigns to S equals ψ_S. It is not hard to convince oneself that this corresponds to the condition $\mathrm{supp}(\psi_S) \subseteq \mathrm{supp}(\rho_S^{\mathrm{P}})$, where $\mathrm{supp}(\sigma)$ denotes the support of the operator σ. Note that, if ρ_S^{P} is pure then this condition implies $\psi_S = \rho_S^{\mathrm{P}}$.

Let us apply this definition to Alice's and Bob's descriptions. Our analysis above showed that for Alice the qubit Q is in a pure state ρ_Q^{Alice}, whereas for Bob the joint system QL_A is in an entangled state $\rho_{QL_A}^{\mathrm{Bob}}$, which is also pure. As we just noted, the state ψ would need to satisfy both $\psi_Q = \rho_Q^{\mathrm{Alice}}$ and $\psi_{QL_A} = \rho_{QL_A}^{\mathrm{Bob}}$. This is impossible because, according to ψ, the qubit Q would at the same time be pure and entangled with L_A. Therefore, there is no state ψ on which both Alice and Bob can agree. We formulate this conclusion as a no-go theorem:

Theorem 13.1 (State agreement paradox) *The following assumptions are incompatible:*

- ***Executability**: Physicists can execute Protocol 13.1.*
- ***Universal applicability of quantum theory**: Assumption (Q) holds.*
- ***State agreement**: For any two physicists* P *and* P′ *who use quantum theory there exists a state* ψ *on which both* P *and* P′ *can agree.*

If Alice and Bob want to agree on the quantum state of Q, then at least one of them has to admit a mistake. We now discuss the two possible options.[12]

If Alice is correct and Bob is wrong, this indicates that physics imposes a constraint on at least one of the first two assumptions, preventing Bob from describing and controlling large systems such as Alice's laboratory. Justifying such a constraint is the program pursued by objective collapse theorists [22]. They propose to modify quantum theory and supplement it with a collapse mechanism. The argument leading

[12] Combinatorially, there are three, but when Alice and Bob are both wrong, this is akin to admitting that quantum theory is not applicable from the perspective of any physicist.

to Theorem 13.1 can be used to impose bounds on the strength of this mechanism: it must be strong enough to prevent the conclusions Bob draws about Alice, who may be a quantum computer programmed with the rules of quantum theory.[13] Conversely, the collapse mechanism must be weak enough to not be ruled out by current experiments. This leaves only a small window of possible collapse models, which, in view of recent progress in quantum technologies, is likely to close completely.

If Alice is wrong and Bob is correct, then this indicates a mechanism that prevents physicists in isolated laboratories, like Alice, from applying quantum theory. This option entails another modification of Assumption (Q). For example, one could demand that only physicists who are not in superposition can use quantum theory. However, this is problematic, as we cannot determine whether we are in such a situation: Consider a powerful alien physicist who would describe our entire galaxy and any systems entangled with it. From the alien's perspective, we, together with Alice and Bob, could be in a superposition of performing the experiment or not performing it. Hence, the modified rule would not allow us to use quantum theory. Nonetheless, this is the approach taken by many-worlds and Bohmian mechanics. They indeed consider an ultimate outside perspective.[14]

The first two assumptions can be experimentally tested and are not unlikely to be confirmed.[15] This would mean we have to abolish the third assumption of Theorem 13.1. Then, both Alice and Bob can be correct even though they do not agree in their state assignment. Consequently, their states cannot represent an objective property of the qubit Q. But if states are not objective, is there anything that we can firmly say about the world?

We may hope that, even if states are not objective, it is still possible to agree on things that can be directly observed—measurement outcomes. We will explore this possibility with more elaborate Wigner's friend experiments. But before doing so, we highlight two information-theoretic aspects of quantum theory.

The first is the *state update rule*. The state assignment of a physicist P changes as she obtains new information about a system, such as the outcome of a measurement. Let ρ^{P}_{QS} be the state that P assigns to a joint system QS. If P measures the subsystem Q and obtains outcome x, then the *updated state* that P assigns to S is

$$\sigma_S^{\mathrm{P}|Q=x} \propto \mathrm{tr}_Q(|x\rangle\langle x|_Q\, \rho^{\mathrm{P}}_{QS}) \tag{13.2}$$

where $|x\rangle\langle x|_Q$ is the projector corresponding to this measurement outcome. Note that we extended the superscript to keep track of the outcome, which is now stored in P's memory.

[13] Although the algorithm that implements the rules is classical and therefore can run on a classical computer, Bob needs to have quantum control over the computer.

[14] While reasoning about such a perspective might be of interest for theological reasons, it stands in stark contrast to the fact that quantum theory was discovered and experimentally confirmed by us who inhabit the universe.

[15] Such experimental tests still require some assumptions, e.g., that if another physicist falsifies Assumption (Q), we should also consider this a falsification of Assumption (Q).

The second concept is the *Heisenberg cut* of a physicist [26]. It defines the boundary separating the systems that a physicist models within the theory from those she does not. Alice's Heisenberg cut, for instance, surrounds the qubit Q, whereas Bob's cut surrounds the entire laboratory.

In a universal theory like quantum theory, the location of the Heisenberg cut is not determined by the theory. Indeed, Assumption (Q) allows each physicist to choose her cut arbitrarily. However, to avoid problems of self-reference, we will not consider scenarios where a physicist is enclosed within her own Heisenberg cut, as this would require her to describe herself.

... Outcomes Are Not Objective ...

Deutsch proposed an extension of Wigner's experiment [16], Protocol 13.2, in which a qubit is measured in two complementary bases. This, by itself, is not in conflict with quantum theory, as the measurement outcome is only acessible to a single physicist. However, we can use this experiment to prove a stronger no-go result, Theorem 13.2, which questions the objectivity of measurement outcomes.

To this end, we first need to define what we mean by *objective measurement outcomes*. The idea is to use the following operational criterion: if measurement outcomes are objective, then there should be a rule which any physicist can use to consistently update her description of the experiment. Specifically, whenever a physicist's description is initially compatible with quantum theory, updating it with the objective outcomes must preserve this compatibility.

Protocol 13.2 (Deutsch's extension of Wigner's friend paradox [16]*)*

Setup

- Bob is outside a laboratory, over which he has full quantum control. He can send a qubit Q into the laboratory and isolate the laboratory from its environment.
- Alice is part of the laboratory and can control the qubit Q she receives from the outside. We denote the part of the laboratory without the qubit by L_A.

Protocol

1. Bob prepares Q and sends it to Alice into the laboratory.
2. Bob isolates the laboratory.

3. Alice measures Q in the computational basis and records the outcome in register R_A.
4. Bob reverses the unitary evolution of $L_A Q$ back to step 2.[16]
5. Bob measures Q in the diagonal basis, defined by

$$|\bar{0}\rangle_Q = \sqrt{\tfrac{1}{2}}\left(|0\rangle_Q + |1\rangle_Q\right) \text{ and } |\bar{1}\rangle_Q = \sqrt{\tfrac{1}{2}}\left(|0\rangle_Q - |1\rangle_Q\right), \quad (13.3)$$

and records the outcome in R_B.

Let us assume by contradiction that there exists a rule for updating a description such that it incorporates both the outcome observed by Alice, r_A, and the outcome observed by Bob, $\bar{r}_B$. This rule may be represented as a function that takes as input any possible initial descriptions in the form of a quantum state ρ_{QS} of the qubit Q and a reference system called S, as well as the values of the outcomes r_A and $\bar{r}_B$,

$$\mathscr{R} : (\rho_{QS}, r_A, \bar{r}_B) \mapsto (P^{\mathscr{R}}_{R_A, R_B}(r_A, \bar{r}_B), \sigma^{\mathscr{R}}_{S|R_A=r_A, R_B=\bar{r}_B}). \quad (13.4)$$

The outcome consists of two pieces. The first, $P^{\mathscr{R}}_{R_A, R_B}(r_A, \bar{r}_B)$, is the probability that the outcome pair $(r_A, \bar{r}_B)$ occurs, which is needed to allow updating if one has access to only one of the outcomes r_A or $\bar{r}_B$. The second, $\sigma^{\mathscr{R}}_{S|R_A=r_A,\, R_B=\bar{r}_B}$, is the updated quantum description. The only requirement we impose on the update rule $\mathscr{R}$ is that, it is compatible with the quantum-theoretic update rule (13.2). For example, because Alice has access to r_A, we require that

$$\forall r_A : \quad \sum_{\bar{r}_B} P^{\mathscr{R}}_{R_A, R_B}(r_A, \bar{r}_B)\, \sigma^{\mathscr{R}}_{S|R_A=r_A, R_B=\bar{r}_B} = \mathrm{tr}_Q(|r_A\rangle\langle r_A|_Q\, \rho_{QS}). \quad (13.5)$$

Here, $|r_A\rangle\langle r_A|_Q$ is the projector onto a computational basis state, corresponding to Alice's measurement.

To show that such an update rule $\mathscr{R}$ cannot exist, we rewrite it as the function

$$\mathscr{F} : \rho_{QS} \mapsto \sigma_{R_A R_B S} = \sum_{r_A, \bar{r}_B} P^{\mathscr{R}}_{R_A R_B}(r_A, \bar{r}_B)\, |r_A\rangle\langle r_A|_{R_A} \otimes |\bar{r}_B\rangle\langle \bar{r}_B|_{R_B} \otimes \sigma^{\mathscr{R}}_{S|R_A=r_A,\, R_B=\bar{r}_B}. \quad (13.6)$$

Note that, while $\mathscr{F}$ is not necessarily a linear map, its output is by construction a valid quantum state. We can, therefore, rewrite both the compatibility condition for updating on r_A, Eq. (13.5), and the compatibility condition for updating on r_B as

[16] This is a highly invasive operation on Alice. One may be worried that the argument presented here depend on Alice surviving this operation [1]. However, the argument does not rely on Alice being a physicist after this step.

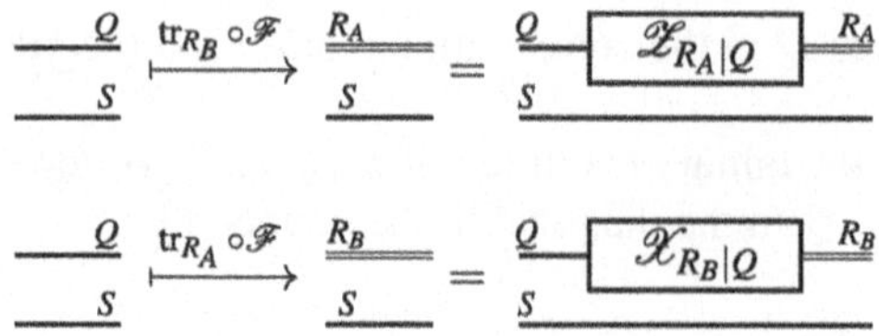

(13.7)

where $\mathscr{Z}_{R_A|Q}$ and $\mathscr{X}_{R_B|Q}$ are completely positive trace-preserving maps corresponding to the measurements in the computational and diagonal basis, respectively.

But the two conditions in Eq. (13.7) cannot hold simultaneously. To see this, we may choose the reference system S to be a qubit maximally entangled with Q, i.e., $\rho_{QS} = |\psi\rangle\langle\psi|_{QS}$ with $|\psi\rangle_{QS} \propto |0\rangle_Q |0\rangle_S + |1\rangle_Q |1\rangle_S$. In this case, $\mathscr{F}$ produces the output $\sigma_{R_A R_B S} := \mathscr{F}(\rho_{QS})$. We now calculate the conditional entropies of the outcomes Z and X of a measurement of S in the computational basis and in the diagonal basis, conditioned on R_A and R_B, respectively. Because these entropies only depend on the marginals $\sigma_{R_B S}$ and $\sigma_{R_A S}$, we can use Eq. (13.7) to see that

$$H(Z|R_A) = H(X|R_B) = 0. \tag{13.8}$$

But this contradicts strong subadditivity and the entropic Heisenberg uncertainty principle [29], which assert that

$$H(Z|R_A) + H(X|R_B) \geq H(Z|R_A R_B) + H(X|R_A R_B) \geq 1. \tag{13.9}$$

Therefore, our assumption that an update function $\mathscr{R}$ exists must be wrong.[17]

We summarize this conclusion as another no-go theorem.[18]

Theorem 13.2 (Objective outcome paradox) *The following assumptions are incompatible:*

- ***Executability**: Physicists can execute Protocol 13.2.*
- ***Universal applicability of quantum theory**: Assumption (Q) holds.*
- ***Objective measurement outcomes**: A physicist's description of a physical system can be updated with all measurement outcomes ever observed.*

The purpose of our no-go theorems is to show that certain assumptions about physics, which a priori sound reasonable, cannot hold. The first two assumptions of Theorem 13.2 capture the universality of quantum theory, whereas the third can

[17] This violation—expressed by the discrepancy between Eqs. (13.8) and (13.9)—is robust, meaning that even an approximate version of $\mathscr{R}$ cannot exist.

[18] Similar no-go theorems have been proposed in [6, 11]. These works consider a different experiment which, in contrast to ours, relies on the idea that measurement settings can be chosen freely. Their notion of objectivity is then defined based on free choice: they demand that, conditioned on any choice of the measurement setting, there is a joint probability distribution of all measurement outcomes that satisfies certain assumptions. Specifically, these are no-superdeterminism and locality, analogously to the assumptions that enter Bell's theorem.

be regarded as a rule to combine information held by different physicists. We have already described earlier, after Theorem 13.1, that the validity of the first two assumptions is, in principle, experimentally testable. Provided that this test will confirm them, Theorem 13.2 implies that the third assumption is wrong.

Giving up the third assumption—the objectivity of measurement outcomes—might seem like a radical move. However, the assumption refers to outcomes that are not in general accessible to the same physicist, so from a purely operational point of view, there is no problem. Protocol 13.2 illustrates this inaccessibility. If Bob asked Alice to tell him r_A, he would break the laboratory's isolation, making its evolution non-unitary, and thus change his other conclusions. Conversely, the reversal of the time evolution of the laboratory will erase Alice's memory of r_A. Hence, even if Bob told her the outcome $\bar{r}_B$ after his final measurement, she no longer knows r_A.[19] Consequently, the assumption of objective measurement outcomes has no operational relevance—any argument defending it would be purely philosophical.

...and Even Communication Cannot Help

Expecting that all measurement outcomes—even those which are not operationally accessible—can be consistently incorporated in any physicist's description might have still been too much to ask. Here, we derive another no-go result, the *quantum collaboration paradox*, which we will state as Theorem 13.3. It replaces the assumption that measurement outcomes are objective by a weaker and more operational consistency assumption, phrased as Assumption (C) below. To emphasize the operational nature of the argument, we present it by referring to a game, the *complementarity game*.

Complementarity Game *Consider N collaborating physicists and a referee. The game is played as follows.*

1. *The physicists indicate a qubit Q to the referee.*
2. *The physicists issue a pair of predictions* $(P, \bar{P})$ *to the referee.*
3. *Depending on a fair coin, the referee chooses one of two possible tests:*
   a. *Measure Q in the computational basis and check if P equals the result.*
   b. *Measure Q in the diagonal basis and check if* $\bar{P}$ *equals the result.*

The N collaborating physicists win if the test of the referee succeeds.

In Protocol 13.3, we describe a strategy, which is based on the universality of quantum theory and a consistency assumption, that allows the players to win the complementarity game with probability 1. This will lead to a contradiction, as according to quantum theory the maximal probability of winning this game is $\frac{1}{2} + \frac{1}{\sqrt{8}} < 1$ [36, Eq. (17) with $m = 2$].

[19] This can be seen as a form of complementarity, which we will discuss further in Sect. 13.3.

Protocol 13.3 (Wigner's friend strategy)

Setup

- Darwin is outside two laboratories, L_A and L_C. He has full quantum control over L_C and can send a qubit Q_A into L_A and a qubit Q_C into L_C.
- Bob is outside the laboratories L_A and L_C and has full quantum control over L_A.
- Alice is part of laboratory L_A and can control the qubit Q_A.
- Charly is part of laboratory L_C and can control the qubit Q_C.

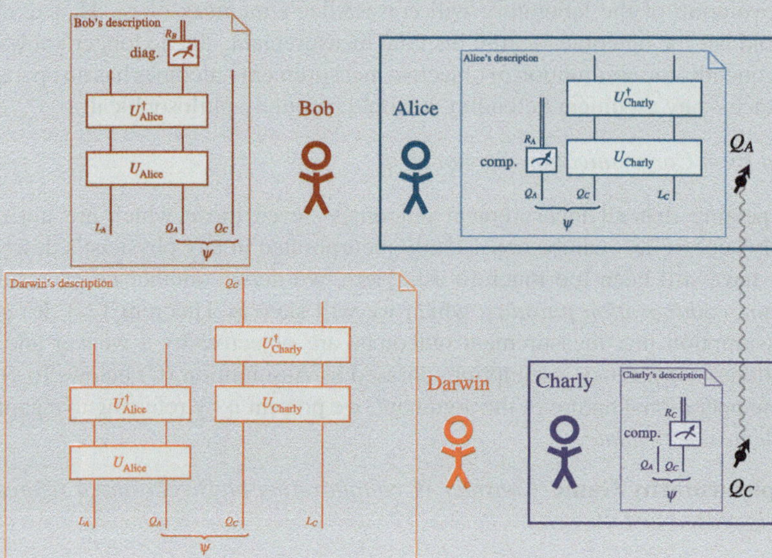

Protocol

1. Darwin prepares Q_A and Q_C in state[20]

$$|\psi\rangle^{\text{all}}_{Q_A Q_C} = \sqrt{\tfrac{1}{3}}\left(|0\rangle_{Q_A}|0\rangle_{Q_C} + |0\rangle_{Q_A}|1\rangle_{Q_C} + |1\rangle_{Q_A}|0\rangle_{Q_C}\right) \quad (13.10)$$

and sends them to Alice and Charly, respectively.
2. Alice measures Q_A in the computational basis and records the outcome in register R_A.
3. Charly measures Q_C in the computational basis and records the outcome in R_C.
4. Bob reverses the evolution of $L_A Q_A$ back to the beginning of step 2.
5. Bob measures Q_A in the diagonal basis, records the outcome in R_B, and communicates it to Darwin.
6. Bob continues if the outcome is $\bar{1}$, else he restarts the experiment from step 1.
7. Darwin reverses the evolution of $L_C Q_C$ back to the beginning of step 3.
8. Bob indicates the qubit Q_C to the referee.
9. Bob issues the pair of predictions $(P = 1, \bar{P} = \bar{0})$ to the referee.

We analyse Protocol 13.3 from Bob's perspective. Bob will also reason about the viewpoints of other physicists, like Charly. He does this by simulating their reasoning using the information he has about them.

Bob's reasoning: Bob chooses his Heisenberg cut to surround the joint system $L_A Q_A Q_C$ but not himself. Note that this is a valid choice, as it surrounds the laboratory L_A and the qubit Q_A, which Bob must act on to reverse its time evolution. Furthermore, for Bob, the physicists Charly and Darwin are classical.

After step 4, the state he assigns to the subsystem $Q_A R_C$ is

$$\rho^{\text{Bob}}_{Q_A R_C} = \frac{1}{3}\left(2\,|\bar{0}\rangle\langle\bar{0}|_{Q_A} \otimes |0\rangle\langle 0|_{R_C} + |0\rangle\langle 0|_{Q_A} \otimes |1\rangle\langle 1|_{R_C}\right) \quad (13.11)$$

Once Bob has indicated a qubit to the referee, i.e., he has reached step 8, his measurement result is $\bar{1}$. The update rule (13.2) yields

$$\rho^{\text{Bob}|Q_A=\bar{1}}_{Q_C} = 6\,\mathrm{tr}_{Q_A}\left(|\bar{1}\rangle\langle\bar{1}|\,\rho^{\text{Bob}}_{Q_A R_C}\right) = |1\rangle\langle 1|_{Q_C}\,. \quad (13.12)$$

Bob is thus certain that Charly's measurement outcome was 1. Note that, from Eq. (13.12), it also follows that the probability of Bob getting outcome $\bar{1}$ is strictly larger than 0. This confirms that the protocol will indeed advance to step 8 after finite time.

[20] This state is common knowledge to all physicists because we assume that all of them are provided with a description of the protocol.

Next Bob simulates Charly's reasoning using the information that Charly observed outcome 1.

Bob's simulation of Charly's reasoning: Charly chooses his Heisenberg cut to surround $Q_A Q_C$ and such that all physicists are classical. Having observed the measurement outcome 1, she can apply the update rule (13.2) to the state (13.10), making her assign the state $|0\rangle_{Q_A}$ to Q_A. She is thus certain that Alice's outcome is 0. To proceed, Charly simulates Alice's reasoning using this information.

Charly's simulation of Alice's reasoning: Alice applies quantum theory to $Q_A L_C Q_C$[21]. Having observed the measurement outcome 0, Alice can apply the update rule (13.2) to (13.10), which leads her to assign the state $|\bar{0}\rangle_{Q_C}$ to Q_C. She is thus certain that a measurement of the referee in the diagonal basis will yield outcome $\bar{0}$.

Based on this analysis, Bob has information about the information his fellow physicists possess. We summarize it here.[22]

1. Bob is certain that Charly's measurement has outcome 1
2. Bob is certain that Charly is certain that Alice's measurement has outcome 0.
3. Bob is certain that Charly is certain that Alice is certain that the referee's measurement, if the diagonal basis was chosen, has outcome $\bar{0}$.[23]

These statements can be experimentally tested. For example, Bob could, in principle, abort the game before the Darwin undoes Charly's measurement, break the isolation of Charly's laboratory, and ask her what she knows about Alice's measurement. If quantum theory is correct, Bob would find statement 2 of the summary confirmed by this test.

It seems Bob can use statement 3 to be certain that with prediction $\bar{P} = \bar{0}$ he will win the game if the referee measures in the diagonal basis. However, to arrive at this conclusion, he needs an additional assumption:

---

[21] The system L_C needs to be surrounded by Alice's Heisenberg cut, as she makes a statement about Q_C after the referee has reversed Charly's measurement. One could, rightfully, be worried that Charly cannot do this simulation as she needs to simulate herself. One way to avoid this problem is for Alice to run the simulation of what she would do for all her possible measurement outcomes and communicate the result to Charly.

[22] The derivation of the statements below requires Bob to combine the statements derived above using standard logic. There are researchers who think logic needs to be modified when applied in the context of quantum experiments [3, 38]. However, the language in which papers are written, even those who deny logic, use some form of logic. Hence, Principle (PP) implies that physicists who argue that quantum theory breaks logic cannot themselves use logic in their arguments.

[23] Those worried about the counterfactual nature of this statement may consider a similar game where the referee's tests are deterministic. Here, the goal is to predict a pair of values (c, d) derived from the outcome m of the referee's measurement by taking the logical AND with the bit that determines whether the measurement is in the computational ($b = 0$) or the diagonal ($b = 1$) basis. Specifically, $(c, d) = (\neg b \text{ AND } \neg m, b \text{ AND } m)$. Alice then predicts d, and the statement reads: Bob is certain that Charly is certain that Alice is certain that d is 0.

Assumption (C) *Let* P_1 *be a physicist who describes another physicist* P_2 *as a classical information-processing system.*[24]
If P_1 *makes the statement "I am certain that* P_2 *is certain, based on reasoning using the same theory as me, that measurement M has outcome x", then* P_1 *can also make the statement "I am certain that measurement M has outcome x".*

Using Assumption (C), Bob can now make the desired conclusion: He is certain that, with the prediction $\bar{P} = \bar{0}$ he will win the game if the referee measures in the diagonal basis.

To gain information about the other outcome of the referee, Bob simulates Darwin using the information that the outcome $\bar{1}$ was communicated to him.[25]

Bob's simulation of Darwin's reasoning: Darwin chooses his Heisenberg cut to surround $L_A Q_A L_C Q_C$ and such that Bob is classical. As Darwin has been communicated the outcome of Bob's measurement, he is certain that, Bob is certain that his measurement outcome is $\bar{1}$. By Assumption (C), Darwin is certain that Bob's measurement outcome is $\bar{1}$. The update rule then yields

$$|\psi'\rangle_{Q_C}^{\text{Darwin}|Q_A=\bar{1}} = \sqrt{6}\left({}_{Q_A}\langle\bar{1}|\,|\psi\rangle_{Q_A Q_C}^{\text{Bob}}\right) = |1\rangle_{Q_C}\,. \tag{13.13}$$

Therefore, Darwin is certain that the referee's measurement of Q_C in the computational basis will yield outcome 1.

Bob can now again apply Assumption (C) and, combined with his previous reasoning, conclude that he is certain that, with the pair of predictions $(P, \bar{P}) = (1, \bar{0})$, the game is won.

Does this mean that the Complementarity Game is *really* won? It has been shown that in quantum theory the Complementarity Game has a maximal winning probability of $\frac{1}{2} + \frac{1}{\sqrt{8}} < 1$ [36, Eq. (17) with $m = 2$]. The test of the referee will thus fail with non-zero probability. If it fails, we immediately have a contradiction with Bob's conclusion that the game is won with certainty. In the other case, if the game is won, this instance of the experiment does not yield a contradiction. However, we can provoke a contradiction with certainty by repeating the game until the physicists lose the game. According to Assumption (Q), this will happen after finitely many repetitions.

---

[24] We say that P_1 describes a system as *classical* if there exists an orthonormal basis with respect to which all states that P_1 assigns to the system are diagonal. For a typical system that strongly interacts with its environment, this condition is met if the Heisenberg cut is placed between system and environment—the system has decohered.

[25] One may wonder why Bob does not directly predict the referee's measurement of Q_C. The reason is that an accurate prediction requires a description of the unitary evolution of Charly, yet Bob must consider Charly as classical to enable the previous use of Assumption (C).

We conclude from this analysis that the assumptions on which Bob's reasoning were based are contradictory. We summarize this finding with the following no-go theorem.[26]

Theorem 13.3 (Quantum collaboration paradox) *The following assumptions are incompatible:*

- ***Executability****: Physicists can execute Protocol 13.3.*
- ***Universality of quantum theory****: Assumption (Q) holds.*
- ***Consistency of knowledge****: Assumption (C) holds.*

Compared to the previous no-go theorems, the quantum collaboration paradox imposes a significantly stronger constraint: either we give up one of the first two assumptions, the consequences of which we already discussed after Theorem 13.1, or we give up Assumption (C). Contrary to the assumptions featured in Theorems 13.1 and 13.2—the state agreement assumption and the objective outcome assumption, respectively—Assumption (C) is operational. So, can we abolish it?

Abolishing it completely—without replacement—would have disastrous consequences. The reason is that it is employed extensively. For example, experimentalists gain data and communicate them to their theorist colleagues. The theorists communicate their conclusions based on this data back to the experimentalists. Without Assumption (C), this collaboration would not be possible.

So, does the quantum collaboration paradox mean that we need to distrust all results from experiments we did not perform ourselves? Luckily, this is not the case. At least for experiments performed today, there is a way out: All physicists simply have to agree on a common Heisenberg cut and regard themselves as one big "meta-physicist". However, this solution is unsatisfactory because there is no fundamental principle that would determine the location of this cut. Fixing a cut would be analogous to postulating an ether in spacetime! Indeed, as we shall see, when we move to experiments involving physicists crossing the event horizon of a black hole, a common Heisenberg cut does not generally exist.

The quantum collaboration paradox poses problems even if there is only a single physicist performing experiments, as even then, we employ a variant of Assumption (C). Suppose that, yesterday, you prepared n spin particles oriented in different directions, and you measure them today. To derive a statement about the time evolution of the spins, you need to combine your knowledge about their initial orientations with the knowledge of the measurement outcomes. But the former is no longer directly accessible—at best, you find it somewhere in your lab notebook.[27] Thus, the knowledge about the initial spin orientations has been communicated to you over time via

[26] This no-go theorem was first proposed in [21]. In [21], the executability was not phrased as a separate assumption, but instead captured by making the statement conditional on the executablity of the experiment. Conversely, the requirement that a physicist should not arrive at contradictory conclusions was stated explicitly as an assumption, called (S).

[27] We assume n is too large for you to remember all the spin orientations. But even if you could, your brain would just take the role of the notebook.

a physical system. Crucially, that we can use these notebook entries from yesterday in our reasoning today is an assumption—very much in the spirit of Assumption (C).

For these reasons, we need to face the challenging task of modifying Assumptions (Q) and (C) in such a way that they are weak enough to provoke the contradiction in Protocol 13.3, but still strong enough to be usable for all practical purposes, e.g., when we communicate with other physicists.

Assumptions (Q) and (C) have been discussed intensively in the foundations literature (see [34] and references therein). Some argue that the assumptions are unjustified, others contend some of them as unnecessary, while still others propose specific modifications [14, 17, 20, 32, 35, 37, 47], intended to circumvent the quantum collaboration paradox. We present here a list of such arguments and then comment on them in the following paragraphs.

First, it is argued that Assumption (C) is obviously too strong because, to accept someone else's result, one would also need to trust them to obtain the result properly and honestly report the result. Second, it is claimed that Assumption (C) is unnecessary because, in all experiments performed today, the measurement result is communicated and not obtained by inference. Third, it is argued that Assumption (C) can only be expected to hold for outcomes of measurements which have already been performed, and that restricting the assumption accordingly would avoid the contradiction. Fourth, it has been noted that, in the argument leading to the quantum collaboration paradox, one uses conclusions of physicists who will later be subject to a measurement. Hence, it is suggested that excepting these physicists from the applicability of Assumptions (Q) and (C) removes the problem.[28] Fifth, it is claimed that constraining the use of Assumptions (Q) and (C) to situations where no physicists are in superposition avoids any contradictions but still allows their use for all practical purposes.

To address the first argument: Trust in a physicist's abilities is not the point in Assumption (C). For Charly to be able to apply Assumption (C), she needs to be certain that Alice is certain about the measurement outcome and that Alice has come to this conclusion using the same theory that Charly uses. If Charly was not convinced that Alice would tell the truth or apply the theory correctly, then this would correspond to Alice applying a different theory as Charly. Hence, in this case, Charly could not apply Assumption (C). But this does not affect its validity.

To address the second argument: Intuitively, it seems that communication and inference are two different concepts. However, what we classify as communication is also a form of inference. When Alice sends Bob a letter telling him the result of her measurement, then by reading the letter Bob measures it and infers that the result he read must be the result Alice got.

To address the third argument: Such a restriction would lead to severe practical limitations. If an engineer is certain that tomorrow the Gotthard Base Tunnel will not

[28] This concern has been summarized by Scott Aaronson as: "It's hard to think when someone Hadamards your brain" [1]. However, in order to avoid the contradiction of the quantum collaboration paradox, one would have to argue that "It's even hard to think when your brain will be Hadamarded in the future." (See [15] for a discussion of this issue.)

collapse, then without Assumption (C) you could not use this prediction to decide whether you risk the travel. It also defies the point of predictions, as the purpose of a prediction is to make a usable statement about a measurement before it happens.

To address the fourth argument: It is correct that a modified version of no-go Theorem 13.3, in which Assumptions (Q) and (C) apply only to physicists who will never be measured, does not hold. However, such a modification makes the assumptions too weak to still be usable in standard situations. For example, it would limit communication, as communication necessarily counts as a measurement, thus disallowing the use of Assumptions (Q) and (C). Additionally, a physicist can usually not decide whether she is even allowed to use Assumptions (Q) and (C), as the constraint of the assumptions depends on the future. This argument is discussed in more detail in [15].

To address the fifth argument: It is not clear what is meant by there being no physicist in superposition. Even in today's experiments, a physicist P_1, who describes another physicist P_2 together with P_2's environment from the outside, would typically conclude that these systems are entangled. But this means that P_1's description of P_2 involves a superposition of states. Hence, a corresponding constraint on Assumptions (Q) and (C) would almost always apply, preventing their use even for all practical purposes.

13.3 Black Holes

The aim of this section is to argue that the findings of Sect. 13.2 on Wigner's friend experiments, which did not involve any spacetime considerations, are relevant for quantum gravity. Like Wigner's friend experiments, thought experiments in gravity involve different physicists with different perspectives. To combine them, rules such as Assumption (C) are needed. Therefore, these rules also play a role in quantum gravity puzzles.

To illustrate this, we consider specific thought experiments that closely match the Wigner's friend thought experiments described in Sect. 13.2, see Table 13.1. We chose these examples for concreteness, but readers who disagree with the gravity assumptions should not be deterred. We expect our central conclusion—that rules like Assumption (C) are essential for analysing puzzles in quantum gravity—holds generically.

Wigner's Friend Implies Complementarity ...

To start, we consider an extension of the thought experiment proposed by Hayden and Preskill [25], which we describe in Protocol 13.4. As we shall see below, this protocol can be regarded as the black hole analogue of Protocol 13.2 and leads to a no-go theorem analogous to Theorem 13.2. Furthermore, we will argue that the original protocol of Hayden and Preskill leads to the analogue of Theorem 13.1.

Table 13.1 Correspondence between non-gravitational Wigner's friend and black hole thought experiments, considered in quantum foundations and quantum gravity, respectively. All these experiments depend on an assumption for how different physicists combine their perspectives. The underlying assumption is indicated in the left column

Assumption	Wigner's friend	Black hole
State agreement	Theorem 13.1	Hayden-Preskill experiment [25]
Objective outcomes	Theorem 13.2	Theorem 13.4
Assumption (C)	Theorem 13.3	Theorem 13.5

Protocol 13.4 (Deutsch-type extension of the Hayden-Preskill experiment [25]*)*

Setup

- Alice is freely falling into a black hole older than the Page time and carries a qubit Q.
- Bob is an outside observer of the black hole and has collected all Hawking radiation R since the black hole has formed.

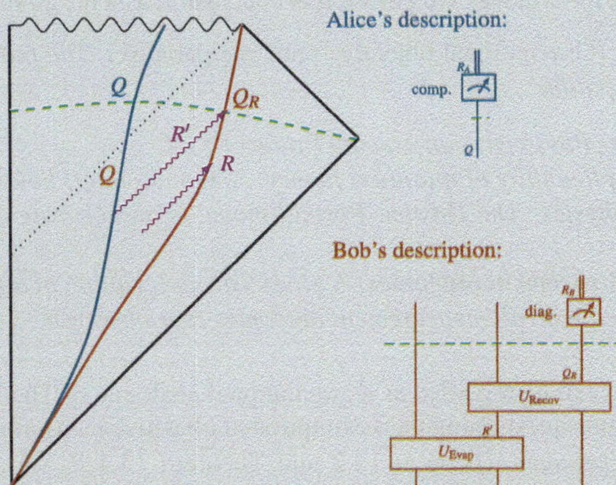

Protocol

- Once Alice is behind the event horizon she measures the qubit Q in the computational basis and records the outcome in register R_A.
- Bob continues collecting Hawking radiation R' until he can reconstruct Q.
- Bob reconstructs the qubit Q from the radiation RR', labelling it Q_R.
- Bob measures Q_R in the diagonal basis and records the outcome in R_B.

We analyse Protocol 13.4 in the Hayden-Preskill model [25]. In this model, for an outside physicist, the evolution of a black hole over a scrambling time corresponds to a Haar-typical unitary. It follows from quantum information theoretic arguments (decoupling theorems [18]) that, for an old black hole, Bob can reconstruct qubit Q from the radiation RR', as required by Protocol 13.4.

With Protocol 13.4, Alice and Bob have achieved the same task as with Protocol 13.2: A given qubit has been measured in two incompatible bases. Alice measured a qubit in the computational basis, Bob has reversed that measurement and measured the qubit in the diagonal basis. Another similarity to Protocol 13.2 is that no physicist can know both outcomes: Alice is in a black hole and, after Bob reconstructed Q_R, it is too late for him to jump into the black hole to meet Alice before she hits the singularity. The difference to Protocol 13.2 is that Bob does not need to isolate a laboratory but uses a black hole to ensure that he can reverse Alice's measurement. Despite this difference, it is crucial that Alice obeys Principle (PP)—in Bob's perspective, she is an ordinary physical system. In particular, her measurement is part of the evolution of the black hole, so nothing special, like a "state collapse", occurs. Therefore, her measurement does not inhibit Bob from reconstructing Q out of the Hawking radiation.

With the same argument as in the analysis of Protocol 13.2, i.e., employing strong subadditivity and the entropic uncertainty relation, one can show that the objectivity of measurement outcomes is in conflict with Assumption (Q) and the gravitational assumptions we have made. We phrase this conclusion as a no-go result.

Theorem 13.4 (Gravitational objective outcome paradox) *The following assumptions are incompatible:*

- ***Executability**: Physicists can execute Protocol 13.4.*
- ***Universal applicability of quantum theory**: Assumption (Q) holds.*[29]
- ***Black hole physics**: The Hayden-Preskill model is an accurate description of a black hole.*[30]
- ***Objective measurement outcomes**: A physicist's description of a physical system can be updated with all measurement outcomes ever observed.*

This theorem can be regarded as a gravitational analogue of Theorem 13.2: both are no-go theorems questioning the assumption of objective measurement outcomes. The difference between Theorem 13.2 and Theorem 13.4 is, of course, that the executability of Protocol 13.4 hinges on the controllability of concrete objects like black holes instead of an abstract notion like an isolated laboratory. This is an instance of the general idea that, whenever a physicist is tasked to perform quantum operations on a macroscopic system, he could execute this task by throwing it into a black hole and act on its Hawking radiation.

---

[29] Despite the similarity of this argument and the one leading to Theorem 13.2, there is a subtle difference in the use of Assumption (Q). In Theorem 13.2, Assumption (Q) could have been weakened such that a physicist can only apply quantum theory to systems she can access. Here, such a weaker variant may not apply, as it is unclear whether there is a system S which both Alice and Bob can access once Bob has reconstructed Q_R.

[30] This assumption must only be guaranteed until a short time after the Page time. In particular, nothing needs to be assumed about the time when the black hole reaches Planck scale.

We remark, that the analysis of the original Hayden-Preskill protocol leads to a statement analogous to Theorem 13.1. It differs from Protocol 13.4 in that neither Alice nor Bob measures their qubit. Bob merely reconstructs Q_R from the Hawking radiation and Alice carries the qubit Q. At the end of the protocol, Q and its copy Q_R exist on the same Cauchy slice, but they are not both accessible to a single physicist. We may now assume by contradicition that there exists a joint quantum description both Alice and Bob could agree on, and consider a reference system S initially maximally entangled with Alice's qubit. For Alice, this entanglement will persist, but Bob's recovered qubit would also need to be maximally entangled with S. This violates monogamy of entanglement. Consequently, there is no joint state of $Q Q_R S$ Alice and Bob could agree on.

One of the favoured resolutions of the contradictions arising in the Hayden-Preskill scenario is *black hole complementarity* [41–43], i.e., the idea that there does not need to exist a consistent joint description among physicists who cannot communicate. In other words, the conclusion was that there is no state agreement and that measurement outcomes are not objective. These conclusions match our no-go Theorems 13.1 and 13.2 which arose from analysing the foundations of bare quantum theory without any gravity considerations. This is rather remarkable: thought experiments in unrelated fields—quantum gravity and quantum foundations—led to the same insight.[31]

...and That It Is Not Enough

To further explore the correspondence between Wigner's friend and gravity thought experiments, we proceed with the firewall paradox [2, 30] and show that its conclusions can be phrased as a no-go theorem analogous to the quantum collaboration paradox. To stress the argument's operational character and similarity to Protocol 13.3, we phrase the firewall paradox as a strategy to win the Complementarity Game with certainty. The concrete strategy is specified by Protocol 13.5.

Protocol 13.5 (The firewall strategy [2, 30]*)*

Setup

- Alice is freely falling into a black hole older than the Page time.
- Bob is an outside observer to the black hole who has collected all Hawking radiation R emitted since the black hole has formed, and stored it in his quantum computer. For him, the joint state of the black hole and R is pure.
- The referee is told to fall into the black hole and a qubit Q_{Ref} in the vacuum just outside the horizon B is indicated to him.

[31] In Raphael Bousso's words: "gravity [acted] as an oracle" for quantum theory [9].

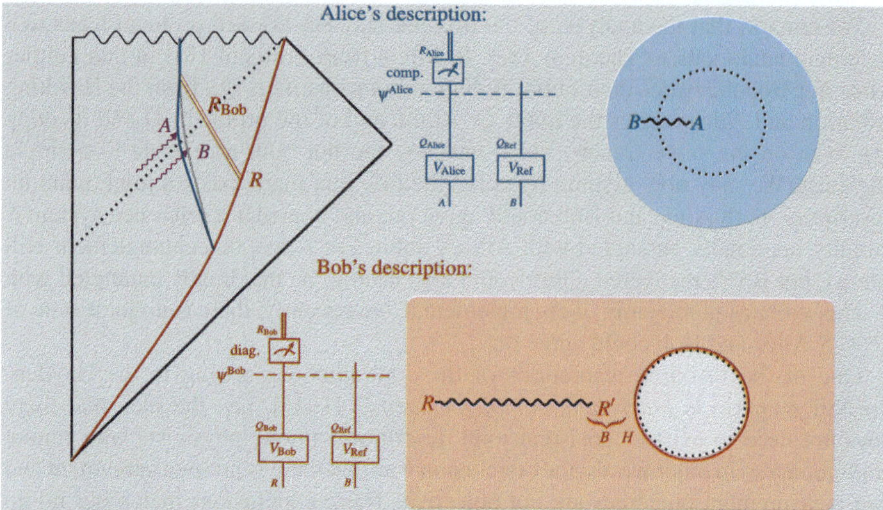

Protocol[32]

- Alice distils a qubit Q_{Alice} from the modes A inside the horizon that is maximally entangled with Q_{Ref}.
- Alice measures Q_{Alice} in the computational basis and records the outcome m_{Alice} in register R_{Alice}.
- Bob distils a qubit Q_{Bob} from the collected Hawking radiation R that is maximally entangled with Q_{Ref}.
- Bob measures Q_{Bob} in the diagonal basis and records the outcome $\bar{m}_{\text{Bob}}$ in R_{Bob}.
- Bob sends R_{Bob} to Alice into the black hole.
- Alice reads the register R_{Bob}.
- Alice issues the pair of predictions $(P = m_{\text{Alice}}, \bar{P} = \bar{m}_{\text{Bob}})$ to the referee.

We analyse Protocol 13.5 from Alice's viewpoint based on the following assumptions:[33]

(G_1) For a freely falling physicist, all fields at the horizon are in the Minkowski vacuum state.

(G_2) For an outside physicist, the evolution of the black hole and all fields (hence also the Hawking radiation) from the creation to the asymptotic future is a Haar-typical unitary on a finite-dimensional Hilbert space.

---

[32] It may appear questionable if this protocol is executable. For example, Alice and Bob blatantly violate monogamy of entanglement. However, to establish such a violation, Alice's and Bob's view need to be combined, which is precisely what we want to explore.

[33] The statements we will derive from these assumptions are robust, i.e., it suffices that the assumptions hold approximately.

(G_3) For an outside physicist, the fields outside the stretched horizon are described by effective field theory.

Alice's reasoning: Alice is a freely falling observer. She has access to field modes B just outside and field modes A just inside the horizon. By assumption (G_1), for her these modes are in a Minkowski vacuum, which is maximally entangled across the horizon. In particular, it is also maximally entangled with Q_{Ref}. Therefore, Alice can distil a qubit Q_{Alice} from the vacuum just inside the horizon, which is maximally entangled with Q_{Ref}:

$$|\psi\rangle^{\text{Alice}}_{Q_{\text{Alice}}Q_{\text{Ref}}} = \sqrt{\tfrac{1}{2}}\left(|0\rangle_{Q_{\text{Alice}}}|0\rangle_{Q_{\text{Ref}}} + |1\rangle_{Q_{\text{Alice}}}|1\rangle_{Q_{\text{Ref}}}\right) \tag{13.14}$$

Upon obtaining result m_{Alice} from her measurement of Q_{Alice}, she can apply the state update rule (13.2) to this state, here expressed as a circuit,

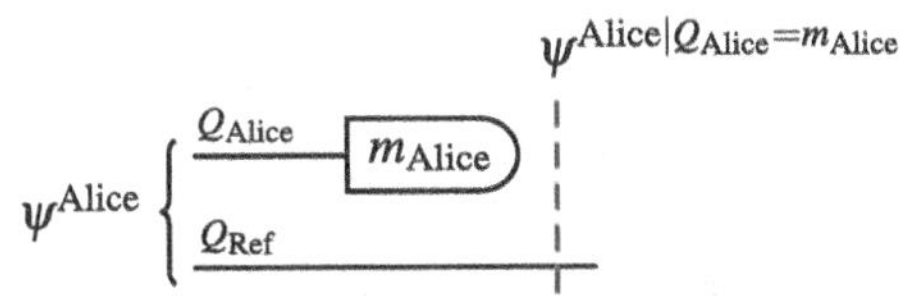

(13.15)

Using Assumption (Q), she can now predict the referee's outcome if he measures in the computational basis: she is certain that this outcome is m_{Alice}.

Upon reading the register R_{Bob} she got from Bob, Alice concludes that Bob is certain his outcome is $\bar{m}_{\text{Bob}}$. Using this information, she simulates Bob.

Alice's simulation of Bob's reasoning: Bob, who is an outside physicist, describes the early radiation R, emitted before Alice has jumped in, and the rest of the black hole, R', which still needs to evaporate. By assumption (G_3), for him, R' consists of two parts: the degrees of freedom at the stretched horizon H and the modes of Hawking radiation in "the zone" B.

By assumption (G_2), the state of the radiation in the asymptotic future is typical. We now regard the radiation as a bipartite system consisting of the early and late radiation RR'. As the black hole is older than the Page time, the Hilbert space dimension of R is larger than that of R'. Because this state is pure by the setup of the experiment, it follows from decoupling theorems that R' is maximally entangled with R. Because Q_{Ref} is by definition a subsystem of B and hence also of R', it is maximally entangled with R. Therefore, Bob can distil a qubit Q_{Bob} from R that is maximally entangled with Q_{Ref},

$$|\psi\rangle^{\text{Bob}}_{Q_{\text{Bob}}Q_{\text{Ref}}} = \sqrt{\tfrac{1}{2}}\left(|0\rangle_{Q_{\text{Bob}}}|0\rangle_{Q_{\text{Ref}}} + |1\rangle_{Q_{\text{Bob}}}|1\rangle_{Q_{\text{Ref}}}\right). \tag{13.16}$$

He measures Q_{Bob} in the diagonal basis and stores the outcome $\bar{m}_{\text{Bob}}$ in R_{Bob}. Applying the state update rule (13.2) to this state, here expressed as a circuit,

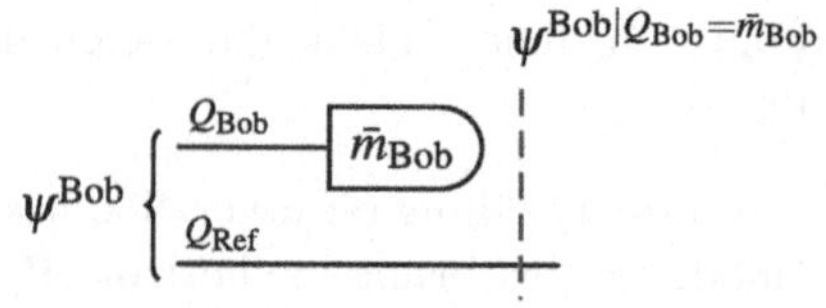

(13.17)

Using Assumption (Q), Bob can now predict the referee's outcome if the referee measures in the computational basis: Bob is certain that this outcome is $\bar{m}_{\text{Bob}}$.

With this simulation of Bob, Alice concludes that she is certain that Bob is certain that, if the referee measures in the diagonal basis, the outcome is $\bar{m}_{\text{Bob}}$. Therefore, using Assumption (C), she is certain that $\bar{P} = \bar{m}_{\text{Bob}}$ will win the game.

We have now reached a conclusion similar to the one for Protocol 13.3: the strategy defined by Protocol 13.5 wins the Complementarity Game with certainty, which contradicts quantum theory.[34] In summary, our formulation of the firewall paradox as a game leads to the following no-go theorem.

Theorem 13.5 (Firewall paradox) *The following assumptions are incompatible:*

- ***Executability****: Physicists can execute Protocol 13.5.*
- ***Universal applicability of quantum theory****: Assumption (Q) holds.*
- ***Black hole physics****: Assumptions (G_1), (G_2), and (G_3) hold.*
- ***Consistency of knowledge****: Assumption (C) holds.*

One might wonder why Assumption (C) must be explicitly stated, as it is only known to fail in a Wigner's friend experiment. However, black holes may be used as a model of an isolated laboratory, so we actually are in a Wigner's friend situation. This suggests that Assumption (C) cannot simply be dismissed.

One may still ask whether the contradiction that lead to Theorem 13.5 can be reached without invoking Assumption (C). This, however, is impossible in the above version of the thought experiment: Bob, as the outside observer, cannot describe the interior modes—they are inside the black hole. Alice, as a freely falling observer, will end up in the black hole, so she cannot describe the late radiation R'. But because her prediction $\bar{P}$ is derived from Bob's analysis of $R'R$, she needs to simulate Bob and thus use Assumption (C).

The purpose of the firewall paradox [2] was to show that, in contrast to the Xeroxing paradox, a contradiction arises even if one assumes black hole complementarity. Specifically, the contradiction is constructed in such a way that it is testable by a single physicist. For example, in the version considered in [10], Alice takes over the tasks Bob was supposed to do. This is possible because Q_{Bob} can be distilled and measured before Alice jumps into the black hole. But because assumptions (G_2) and (G_3) require an outside perspective, their use needs to be justified. This is done by

---

[34] To reach an operational contradiction, one may follow the same argument as described in the analysis of Protocol 13.3. However, because the test is performed in the interior, it is more challenging to decide from the outside whether it has succeeded. But Bob may still obtain this information by reconstructing the referee from the late Hawking radiation.

arguing that Alice, at the time when she uses these assumptions, still has the choice to stay outside. Therefore, her conclusions must agree with what she would conclude if she changed her mind to stay outside and perform a test of the assumptions (see [7, 8, 24] for the history of this argument).

This justification implicitly uses an extra Assumption (E), which ensures a statement of the following form: suppose a physicist has the possibility to perform experiment Exp_1, but actually performs experiment Exp_2; then the conclusions that she would have drawn, had she performed Exp_1, also hold for Exp_2. This assumption has a similar flavour to Assumption (C). But, in contrast to Assumption (C), which considers different physicists in the same experiment, Assumption (E) considers the same physicist in different experiments. To summarise, if one replaces multiple physicists by a single physicist, one faces the challenge of finding a variant of assumption (E) that allows the above conclusion, yet is not obviously wrong.[35]

13.4 Discussion

Non-gravitational Wigner's friend thought experiments teach us an important lesson: the universality of quantum theory, Assumption (Q), conflicts with assumptions that are needed for combining information held by different physicists—the state agreement assumption, the objective outcome assumption, or Assumption (C). This is made precise by no-go Theorems 13.1–13.3.

A similar lesson was, independently, learnt from quantum gravity thought experiments. In these experiments, contradictions arise when conclusions obtained by physicists with different perspectives are combined. This was recognized and lead to the paradigm of black hole complementarity. Comparing to Wigner's friend experiments, black hole complementarity corresponds to giving up the state agreement assumption.

In this chapter, we have deepened this correspondence. The key finding is that gravity thought experiments can be viewed as instances of Wigner's friend thought experiments, with a black hole serving as a perfectly isolated laboratory. We made this correspondence precise by relating no-go theorems originating in Wigner's friend considerations to no-go theorems concerning black holes, as summarized in Table 13.1.

Our finding sheds new light on quantum gravity thought experiments. We uncover that the firewall paradox relies on Assumption (C), which by the quantum collaboration paradox, Theorem 13.3, is in conflict with another assumption of quantum gravity, the universality of quantum theory. This opens a new avenue for resolving puzzles like the firewall paradox: resolve the quantum collaboration paradox!

[35] Consider the following realistic example: Suppose you are in bed and want to know if your cat is in the kitchen. You know that if you go to the kitchen, you will find the cat there (expecting you to feed her). But you decide to stay in bed. Assumption (E) now implies that the cat is in the kitchen. But, as every cat owner knows, this conclusion is false.

If successful, this would show that the firewall paradox is not genuinely rooted in gravity.

Resolving the quantum collaboration paradox means finding replacements for Assumptions (Q) and (C) that are usable for all practical purposes but do not lead to a contradiction. For example, inspired by Principle (PP), Assumption (Q) could be modified such that only physicists who hold a large enough physical reference frame are eligible to apply the rules of quantum theory.[36] In Protocol 13.3, for instance, Charly, who simulates Alice, must hold a reference for Alice. However, because Alice describes Charly, Alice needs a reference for Charly. This leads to a recursive situation: Charly's reference must serve as a reference for Alice's reference for Charly.

The close correspondence between quantum foundations and quantum gravity thought experiments offers an opportunity for the two communities to learn from each other. Thought experiments in quantum gravity may inspire strategies to resolve the quantum collaboration paradox. For example, modelling isolated systems as black holes might reveal features of isolated laboratories not obvious in bare quantum theory. Conversely, concepts such as Principle (PP), which are crucial to understand Wigner's friend thought experiments in quantum foundations, could be utilized in the study of thought experiments in quantum gravity.[37]

Acknowledgements We acknowledge support from the National Centre of Competence in Research SwissMAP and the ETH Zurich Quantum Center.

References

1. S. Aaronson, It's hard to think when someone Hadamards your brain. https://scottaaronson.blog/?p=3975. Accessed 11 Dec 2024
2. A. Almheiri, D. Marolf, J. Polchinski, J. Sully, Black holes: complementarity or firewalls? J. High Energy Phys. **2013**(2) (2013). https://doi.org/10.1007/JHEP02(2013)062
3. F. Atzori, E. Rebufello, M. Violaris, L.T. Knoll, A. Alhajri, A. Avella, M. Gramegna, C. Marletto, V. Vedral, F. Piacentini, I.P. Degiovanni, M. Genovese, Universal quantum theory contains twisted logic (2024). arXiv:2409.20480
4. J.S. Bell, On the Einstein Podolsky Rosen paradox. Phys. Phys. Fizika **1**(3), 195 (1964). https://doi.org/10.1103/PhysicsPhysiqueFizika.1.195
5. C.H. Bennett, The thermodynamics of computation—a review. Int. J. Theor. Phys. **21**(12), 905–940 (1982)
6. K.-W. Bong, A. Utreras-Alarcón, F. Ghafari, Y.-C. Liang, N. Tischler, E.G. Cavalcanti, G.J. Pryde, H.M. Wiseman, A strong no-go theorem on the Wigner's friend paradox. Nat. Phys. **16**(12), 1199–1205 (2020). https://doi.org/10.1038/s41567-020-0990-x
7. R. Bousso, Observer complementarity upholds the equivalence principle (2012). arXiv:1207.5192v1

---

[36] This idea has been discussed widely in the quantum foundations community but only recently been considered in quantum gravity [23].

[37] Indeed, for the firewall paradox, ideas in the spirit of Principle (PP) have already been used to propose a solution [49, 50].

8. R. Bousso, Complementarity is not enough. Phys. Rev. D **87**, 124023 (2013). https://doi.org/10.1103/PhysRevD.87.124023
9. R. Bousso, Gravity as an oracle, in *Talk at the Quantum Information and Gravity Workshop in Heidelberg, Germany* (2024)
10. R. Bousso, Firewalls from general covariance (2025). arXiv:2502.08724
11. Č. Brukner, *On the Quantum Measurement Problem* (Springer International Publishing, Cham, 2017), pp. 95–117. https://doi.org/10.1007/978-3-319-38987-5_5
12. J. Bub, Understanding the Frauchiger-Renner argument. Found. Phys. **51**(2), 36 (2021). https://doi.org/10.1007/s10701-021-00420-5
13. A. Church, An unsolvable problem of elementary number theory. Am. J. Math. **58**(2), 345–363 (1936). http://www.jstor.org/stable/2371045
14. J.B. DeBrota, C.A. Fuchs, R. Schack, Respecting one's fellow: Qbism's analysis of Wigner's friend. Found. Phys. **50**(12) (2020). https://doi.org/10.1007/s10701-020-00369-x
15. L. del Rio, R. Renner, Reply to: Quantum mechanical rules for observed observers and the consistency of quantum theory. Nat. Commun. **15**(1), 3024 (2024). https://doi.org/10.1038/s41467-024-47172-0
16. D. Deutsch, Quantum theory as a universal physical theory. Int. J. Theor. Phys. **24**(1), 1–41 (1985). https://doi.org/10.1007/BF00670071
17. A. Di Biagio, C. Rovelli, Stable facts, relative facts. Found. Phys. **51**(1), 30 (2021). https://doi.org/10.1007/s10701-021-00429-w
18. F. Dupuis, M. Berta, J. Wullschleger, R. Renner, One-shot decoupling. Commun. Math. Phys. **328**(1), 251–284 (2014). https://doi.org/10.1007/s00220-014-1990-4
19. A. Einstein, B. Podolsky, N. Rosen, Can quantum-mechanical description of physical reality be considered complete? Phys. Rev. **47**, 777–780 (1935). https://doi.org/10.1103/PhysRev.47.777
20. M. Federico, P. Grangier, A contextually objective approach to the extended Wigner's friend thought experiment (2023). arXiv:2301.03016
21. D. Frauchiger, R. Renner, Quantum theory cannot consistently describe the use of itself. Nat. Commun. **9**(1) (2018). https://doi.org/10.1038/s41467-018-05739-8
22. G.C. Ghirardi, A. Rimini, T. Weber, Unified dynamics for microscopic and macroscopic systems. Phys. Rev. D, **34**, 470–491 (1986). https://doi.org/10.1103/PhysRevD.34.470
23. H. Gomes, S. Langenscheidt, D. Oriti, Boundaries, frames and gravitational limits to objectivity (2024). arXiv:2412.00993
24. D. Harlow, Complementarity, not firewalls (2012). arXiv:1207.6243
25. P. Hayden, J. Preskill, Black holes as mirrors: quantum information in random subsystems. J. High Energy Phys. **2007**(09), 120–120 (2007). https://doi.org/10.1088/1126-6708/2007/09/120
26. W. Heisenberg, Ist eine deterministische Ergänzung der Quantenmechanik möglich?, in *Wolfgang Pauli: Wissenschaftlicher Briefwechsel mit Bohr, Einstein, Heisenberg u.a. Band II: 1930–1939*, ed. by K. von Meyenn (Springer Berlin Heidelberg, 1985), pp. 409–418
27. C.L. Jones, M.P. Mueller, On the significance of Wigner's friend in contexts beyond quantum foundations (2025). arXiv:2402.08727
28. R. Landauer, Information is physical. Phys. Today **44**(5), 23–29 (1991). https://doi.org/10.1063/1.881299. https://pubs.aip.org/physicstoday/article-pdf/44/5/23/8304036/23_1_online.pdf
29. H. Maassen, J.B.M. Uffink, Generalized entropic uncertainty relations. Phys. Rev. Lett. **60**, 1103–1106 (1988). https://doi.org/10.1103/PhysRevLett.60.1103
30. S.D Mathur, The information paradox: a pedagogical introduction. Class. Quantum Gravity **26**(22), 224001 (2009). https://doi.org/10.1088/0264-9381/26/22/224001
31. J.C. Maxwell, *Theory of Heat*, Chapter 12. Longmans (Green, and Co, London, 1871)
32. V. Narasimhachar, Agents governed by quantum mechanics can use it intersubjectively and consistently (2020). arXiv:2010.01167
33. M.A. Nielsen, I.L. Chuang, *Quantum Computation and Quantum Information* (Cambridge University Press, Cambridge, 2012). https://doi.org/10.1017/CBO9780511976667

34. N. Nurgalieva, R. Renner, Testing quantum theory with thought experiments. Contemp. Phys. **61**(3), 193–216 (2021). https://doi.org/10.1080/00107514.2021.1880075
35. A.P. Polychronakos, Quantum mechanical rules for observed observers and the consistency of quantum theory. Nat. Commun. **15**(1), 3023 (2024). https://doi.org/10.1038/s41467-024-47170-2
36. J.M. Renes, Better bounds on optimal measurement and entanglement recovery, with applications to uncertainty and monogamy relations. Phys. Rev. A **96**, 042328 (2017). https://doi.org/10.1103/PhysRevA.96.042328
37. J.M. Renes, Consistency in the description of quantum measurement: quantum theory can consistently describe the use of itself (2021). arXiv:2107.02193
38. S. Samuel, The Frauchiger-Renner Gedanken experiment: flaws in its analysis—how logic works in quantum mechanics (2023). arXiv:2208.00060
39. D. Schmid, Y. Yīng, M. Leifer, A review and analysis of six extended Wigner's friend arguments (2024). arXiv:2308.16220
40. E. Schrödinger, Die gegenwärtige Situation in der Quantenmechanik. Naturwissenschaften **23**(48), 807–812 (1935)
41. C.R. Stephens, G. 't Hooft, B.F. Whiting, Black hole evaporation without information loss. Class. Quantum Gravity **11**(3), 621–647 (1994). https://doi.org/10.1088/0264-9381/11/3/014
42. L. Susskind, L. Thorlacius, Gedanken experiments involving black holes. Phys. Rev. D **49**(2), 966–974 (1994). https://doi.org/10.1103/physrevd.49.966
43. L. Susskind, L. Thorlacius, J. Uglum, The stretched horizon and black hole complementarity. Phys. Rev. D **48**(8), 3743–3761 (1993). https://doi.org/10.1103/physrevd.48.3743
44. S. Teufel, D. Dürr, *Bohmian Mechanics* (Springer, 2009). https://doi.org/10.1007/b99978_8
45. A.M. Turing, On computable numbers, with an application to the Entscheidungsproblem. Proc. Lond. Math. Soc. **s2–42**(1), 230–265 (1937). https://doi.org/10.1112/plms/s2-42.1.230
46. L. Vaidman, Many-worlds interpretation of quantum mechanics, in *The Stanford Encyclopedia of Philosophy*, ed. by E.N. Zalta, Fall 2021 edn. (Metaphysics Research Lab, Stanford University, 2021)
47. V. Vilasini, M.P. Woods, A general quantum circuit framework for extended Wigner's friend scenarios: logically and causally consistent reasoning without absolute measurement events (2024). arXiv:2209.09281
48. E.P. Wigner, *Remarks on the Mind-Body Question* (William Heinemann, 1962), pp. 284–302
49. B. Yoshida, Firewalls vs. scrambling. J. High Energy Phys. **2019**(10), 132 (2019). https://doi.org/10.1007/jhep10(2019)132
50. B. Yoshida, Observer-dependent black hole interior from operator collision. Phys. Rev. D **103**, 046004 (2021). https://doi.org/10.1103/PhysRevD.103.046004

Chapter 14
Planckian Discreteness and the Resolution of the BH Information Puzzle

Alejandro Perez and Sami Viollet

In memory of Sami, up in his mountains.

Abstract A scenario addressing the black hole information paradox is proposed, based on assumptions about the ultraviolet (UV) completion of quantum gravity. This scenario is exemplified in loop quantum gravity and is physically motivated independently of any specific approach by black hole thermodynamics. The central idea is that traditional descriptions of black hole formation and evaporation, which rely on smooth field theories, are inadequate near singularities where quantum gravity effects dominate. In this strong quantum gravity regime, Planck-scale discreteness provides a vast reservoir of degrees of freedom for information purification, aligning with energy conservation and avoiding the inconsistencies of remnant scenarios. This perspective challenges foundational assumptions in effective field theory, suggesting that information degradation occurs through decoherence with a pre-geometric quantum substrate. The scenario is demonstrated through a series of practical models in quantum mechanics, discrete quantum field theory, and simple quantum gravity toy models. The work emphasizes that resolving the paradox requires recognizing quantum discreteness and posits that this insight could have significant observational consequences, extending beyond the theoretical discussion of the Hawking puzzle.

A. Perez (✉) · S. Viollet
Centre de Physique Théorique, Aix Marseille Université, Université de Toulon, CNRS, Marseille, France
e-mail: perez@cpt.univ-mrs.fr

S. Viollet
e-mail: sami.viollet@cpt.univ-mrs.fr

A. Akil and C. Bambi (eds.), *The Black Hole Information Paradox*, Springer Series in Astrophysics and Cosmology, https://doi.org/10.1007/978-981-96-6170-1_14

14.1 Introduction

In this chapter, we propose a scenario to resolve the information puzzle in black hole formation and evaporation. This is based on the idea that there are more degrees of freedom at the fundamental scale than can be captured by any effective field theory. While our ideas originate from the framework of loop quantum gravity, our proposed solution does not rely on its specific details.

Just as standard thermodynamics has provided important insights into the microscopic nature of matter and influenced the development of statistical mechanics and quantum mechanics, our scenario can be motivated by the principles of black hole thermodynamics. We view the discussion in this chapter as exploratory, guiding us toward greater understanding of black hole formation and evaporation while also potentially offering clues for developing a consistent theory of quantum gravity.

The chapter is organized as follows. In the next section we give a very brief (and very incomplete) account of the history of the subject from the perspective that we propose. In Sect. 14.3 we review the motivation provided by some features of loop quantum gravity. In Sect. 14.4 we discuss several formal issues that are important in order to setup the problem. In Sect. 14.6 we provide the physical motivation that derives from black hole thermodynamics. In Sect. 14.7 we present the scenario. In Sect. 14.8 we illustrate how the key features of the scenario can be realized in simple models allowing for analytic computations. The confrontation of our scenario with the generally covariant timeless nature of quantum gravity is discussed in Sect. 14.9. We conclude with a general discussion in Sect. 14.10 and provide a biased list of desired features for a convincing resolution in Sect. 14.11.

14.2 A Brief Historic Account

We start by recalling historical results that set the context where the present proposal was born:

- 1975: Hawking showed in [1] that a black hole spontaneously emits a thermal radiation for an observer located on $\mathscr{I}^+$.
- 1975: Wald and Parker showed in [2, 3] that a scalar field initially in a vacuum state on $\mathscr{I}^-$ remains pure on the Cauchy surface formed by the horizon and $\mathscr{I}^+$, and that Hawking's result is retrieved by tracing out the modes on the horizon.
- 1976: Hawking argued in [4] that the emission of thermal radiation by a black hole causes its mass to decrease until it completely evaporates, transforming an initially pure state into a mixed state, suggesting a violation of unitarity and a loss of information.
- 1979: Page showed in [5] by a qualitative argument that the back reaction of the Hawking radiation could induce quantum fluctuation of the geometry at a time much shorter than the evaporation time, suggesting that Hawking's semi classical treatment leading to loss of information is not allowed.

- 1980: Bekenstein argued in [6] that, for a generic system, the thermodynamic entropy is bounded by $2\pi RE$, where R represents the system's characteristic size and E its energy. This bound was later generalized by Bekenstein and others to apply to a broader class of systems, forming the foundation of what would become the so-called *central dogma* [7].
- 1983: Banks, Susskind and Peskin argued in [8] that an evolution transforming a pure state into a mixed state comes into conflict with the conservation of energy, which has motivated approaches proposing an early time purification of Hawking radiation.
- 1993: Page suggested in [9] that the information could gradually escape from the black hole during the entire evaporation process, in a way that the entanglement entropy of the radiation would follow the so called Page curve.
- 1993: t'Hooft argued in [10] that all the information contained in our $3 + 1$ dimensional world is encoded on a $2 + 1$ dimensional surface, which is at the origin of the holographic principle.
- 1995: Wald and Unruh criticized Banks, Susskind and Peskin's result in [11] by arguing that the formation and the evaporation of a black hole is not a Markovian process (which was a hypothesis made by Banks, Susskind and Peskin), and showed that for a non Markovian process the result obtained by Banks Peskin and Susskind does not hold.
- 1996: Strominger and Vafa derived in [12] the Bekenstein formula for black hole entropy by a counting of microscopic states corresponding to five dimensional black hole in the framework of string theory. This result is one of the motivations for the central dogma.
- 1997: Maldacena conjectured in [13] the so called AdS/CFT correspondence which provide a framework to describe AdS spacetime in an holographic way, and therefore illustrating the holographic principle.
- 2012: Unruh proposed in [14] a counter example to Banks, Susskind and Peskin's result by constructing a model where a system with a finite number of degrees of freedom evolves from a pure state to a mixed state while conserving its energy.
- 2014: The present scenario is proposed [15].
- 2015: Hotta, Schutzhold and Unruh studied in [16] the radiation produced by an accelerating mirror in a $1 + 1$ D spacetime, which is analogous to the Hawking radiation. They showed that the partner modes purifying this radiation is in a state undistinguished from the vacuum state by local measurements, meaning that the correlations with the thermal emissions need not be associated with any energy transport.
- 2019: Wald argued in [17] that the mechanism mentioned in the previous item is not possible for $3 + 1$ D black holes because entanglement between Hawking radiation with vacuum fluctuations requires in this case the emission of as many late time particles (distinguishable from the vacuum) as Hawking particles.

14.3 Motivation from Loop Quantum Gravity

Loop quantum gravity [18–20] is a background-independent approach to the quantization of gravity, formulated within the Hamiltonian framework. As such, it aims to describe the quantum theory of dynamical processes occurring in the universe considered as a whole. Although this approach can be adapted to the treatment of local regions with boundaries [21] and asymptotic idealizations with customary boundary conditions, we will argue here that the problem of black hole formation and evaporation requires an understanding of the quantum dynamics in the bulk. More importantly, it involves regions where the classical theory predicts the appearance of singularities, which represent the apparent sink of information inside a black hole. In this sense, a Hamiltonian (spacetime-encompassing) dynamical treatment provides a natural framework for addressing the problem of unitarity and information in gravitational collapse.

The emphasis on the relevance of singularities inside black holes is one of the key features that distinguish the present approach from others discussed in this book. Another central point of departure from other proposals is the assumption that the fundamental nature of quantum gravity is discrete. This discreteness may resemble the approach proposed by the formalism of loop quantum gravity, or it may take an (as yet) unknown form that could be realized differently in the future. This assumption is also strongly suggested by the physics of black hole thermodynamics.

Discreteness implies the existence of degrees of freedom at the Planck scale that are not captured by the smooth field representation used in the low-energy description provided by standard quantum field theory. We will argue that, although the continuous idealization may be a satisfactory effective description away from singularities, near black hole singularities, effective field theory fails because microscopic degrees of freedom cannot be ignored.

Loop quantum gravity provides a formalism that is well-suited for the analysis of this question and, at the same time, serves as an example of a quantum theory in which physics is discrete at the Planck scale. For these reasons, loop quantum gravity is the natural framework for discussing the scenario we propose. However, one should keep in mind that, although strongly motivated by loop quantum gravity, the ideas we present may find a more precise realization in a future UV completion of a quantum gravity theory, which could emerge from further developments in loop quantum gravity or arise independently.

The current state of development of loop quantum gravity does not yet allow for a precise treatment of the question of black hole formation and evaporation. We believe this remains an open challenge for all approaches to quantum gravity at present. In this regard, the discussion of the fate of information serves as a testing ground for the theory. In our view, the fact that this question cannot be separated from the issue of what replaces the singularities of the classical theory in the quantum theory provides a strong motivation for considering the information puzzle as an exercise that should yield non-trivial insights into the nature of the consistent theory we all aspire to construct.

Loop quantum gravity may not be complete, but it offers the bold proposal (suggested by the application of a background-independent Hamiltonian quantization of general relativity) that, at the Planck scale, matter and geometry are described by fundamental building blocks in a background-independent fashion. The smooth spacetime geometry with smooth matter fields is expected to emerge from the combinatorial dynamics of microscopic elementary tiles. Under suitable circumstances (but not under all circumstances, see Sect. 14.5.3), an effective smooth description of long-wavelength physics is possible, where the fundamental granularity can be ignored. However, such an effective description is bound to fail in situations involving singularities, such as in cosmology or gravitational collapse. The case of black holes represents the idealized situation where this failure is most apparent. From the perspective of distant observers, the dynamics seems to reduce to the description of low-energy excitations. These range from gravitational and matter radiation in the early stages of black hole formation to quantum Hawking radiation during evaporation. However, the dynamics inside the black hole drive the system into the strong UV quantum regime. In this regime, if discreteness is truly part of the physics of quantum gravity, the existence of fundamental building blocks cannot be ignored. Any effective smooth description would fail to capture this essential aspect. Since the problem of unitarity is fundamentally about the quantum entanglement between the external degrees of freedom and those falling inside, it is clear that any serious discussion of the puzzle must take the strong quantum gravity region near the singularity into account (if quantum gravity resembles what loop quantum gravity proposes discreteness will have to play a central role).

We emphasize that other approaches to quantum gravity, in which the smooth field-theoretic description is expected to emerge from some form of coarse-graining of discrete building blocks at the fundamental Planck scale, might also serve as the theoretical foundation for the scenario we describe. Loop quantum gravity provides motivational value, and its framework helps produce (as we will see in Sect. 14.8) simple toy models of the mechanism we propose. It is also possible that other approaches, such as the causal set approach [22], causal dynamical triangulations [23], or the group field theory approach [24], could similarly provide a platform for realizing the ideas we present. More generally, we assume that this key feature will be present in the consistent formulation of quantum gravity—a theory that we still lack.

14.4 Considerations Framing the Discussion

In the following few sections we discuss a few facts that are important to keep in mind when discussing the problem at hand.

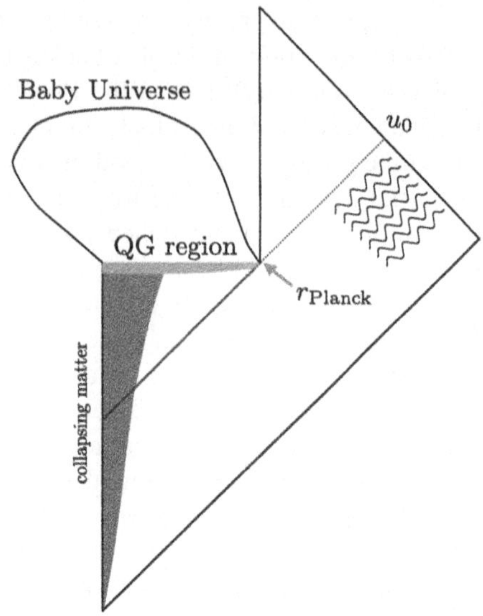

Fig. 14.1 In order to simplify the discussion we ignore the baby universe scenario and focus on the global structure represented in Fig. 14.2. *Reused from Alejandro Perez and Daniel Sudarsky, AVS Quantum Sci. 4 (4), 10.1116/5.0091962, (2022) with permission of AIP Publishing*

14.4.1 The Global Spacetime Features

Despite the limitations of the Penrose diagram representation of the evaporation process (discussed in Sects. 14.4.3 and 14.5.1), we will use it to illustrate qualitative features. In particular, we will assume that quantum gravity does not lead to the picture represented in Fig. 14.1. Specifically, we assume that the interior singularity (or strong quantum gravity region) does not evolve into a separate, future-causally disconnected region from the black hole exterior. This assumption is made to simplify the discussion, but it is not strictly necessary. The fate of the information proposed in our scenario can be adapted to the formation of a baby universe if that is what the quantum dynamics actually leads to. In such a situation, a unitary account would indeed exist at the fundamental level, but information would effectively be lost to outside observers. What we assume here is that the future of the quantum gravity region includes the exterior spacetime. This is a statement made at the quantum level. Specifically, we mean that the future state is described by a quantum superposition of spacetime geometries and matter, involving both the outside degrees of freedom and the future of what would traditionally be called the singularity (which, in the usual sense, becomes naked). The situation is illustrated in Fig. 14.2.

14.4.2 Sensible BH Definitions Are Teleological

In the context of asymptotically flat spacetimes, the black hole region is defined as $\text{BH} = \mathscr{M} - J^-(\mathscr{I}^+)$, where $J^-(\mathscr{I}^+)$ denotes the chronological past of future null

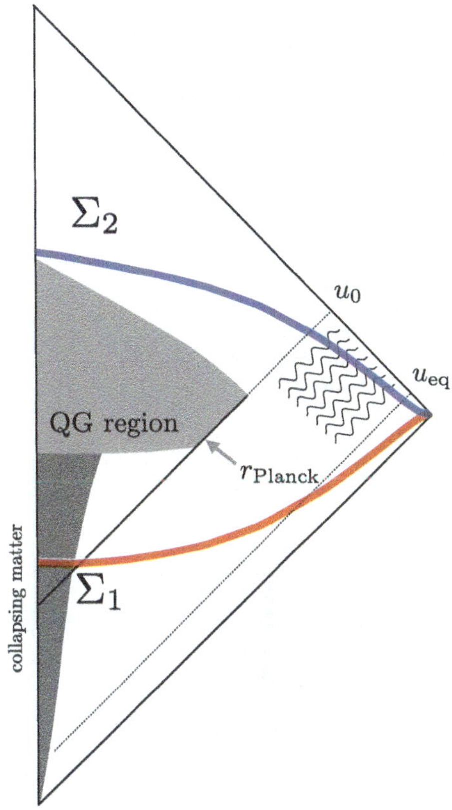

Fig. 14.2 The global structure we assume allows the quantum gravity region to exit the black hole region at the end of evaporation. The light gray region in the figure represents the would-be-singularity as it might be viewed from a classical perspective. This causal structure corresponds to the general one associate with standard remnant scenarios, while also allows for the type of alternative we will present here. For the sake of clarity and based on the proposals we will discuss in detail, this framework will serve as the main focus of this article. Past and future generalized Cauchy surfaces, labeled Σ_1 and Σ_2, are also shown. *Reused from Alejandro Perez and Daniel Sudarsky, AVS Quantum Sci. 4 (4), 10.1116/5.0091962, (2022) with permission of AIP Publishing*

infinity $\mathscr{I}^+$ (in the notation of [25]). One of the first difficulties encountered is that this definition no longer works when quantum effects are taken into account, as real black holes evaporate due to Hawking radiation.

Under certain circumstances, the most appropriate spacetime representation of the process appears to be globally hyperbolic (as depicted in Fig. 14.2), in which case the previous definition would imply that there is no black hole at all. However, there are regions in the spacetime representation where curvatures approach the Planck scale. A definition of the black hole region, which coincides with the standard classical definition in the far past but remains valid globally, is given by

$$\mathrm{BH} = \mathscr{M} - J^-_{\mathrm{LC}}(\mathscr{I}^+),$$

where $J^-_{\mathrm{LC}}(\mathscr{I}^+)$ denotes the low-curvature past of $\mathscr{I}^+$, defined by the collection of events in the bulk that can be connected to $\mathscr{I}^+$ via causal curves (timelike or null) for which the local curvature is below the Planck scale. One possibility is to require that the Kretschmann scalar

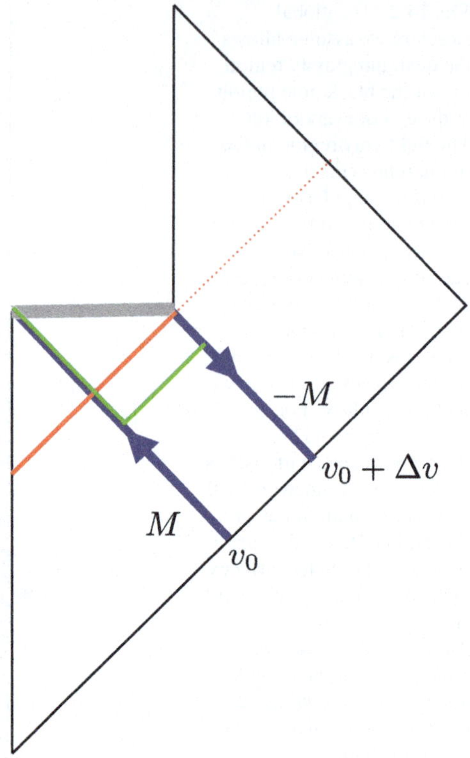

Fig. 14.3 A Vaidya black hole formed and annihilated via null shock waves of mass M and $-M$ sent at advanced time v_0 and $v_0 + \Delta v$ respectively. The dynamical horizon is represented in green. Our definition of event horizon in red. For large Δv dynamical and event horizon have areas approaching exponentially at $v = v_0$. However, the two differ by about $\Delta r \approx 2M$ at $v = v_0 + \Delta v$. *Reused from Alejandro Perez and Daniel Sudarsky, AVS Quantum Sci. 4 (4), 10.1116/5.0091962, (2022) with permission of AIP Publishing*

$$K \equiv R_{abcd} R^{abcd} \leq C\ell_p^{-4}$$

for some constant C of order unity.

The dependence of the definition on the constant C (or, eventually, on the choice of a different curvature scalar) is not a significant limitation for the discussion of information. Different choices are expected to lead to black hole regions that coincide, up to small Planck-scale corrections.

Quasi-local definitions based on the notion of apparent horizons, such as the *dynamical horizon* [26] or the analogous notion of holographic screens (introduced later in [27]), all tend to agree in the far quasi-stationary past, where the evaporation process is slow. However, they can differ significantly in generic situations, as illustrated in the extreme case shown in Fig. 14.3. Dynamical horizons and holographic screens can be timelike or spacelike, depending on the dynamical situation. In contrast, the horizon in our definition is always null, and thus generally defines the notion of a causal barrier that is relevant for the discussion of the information puzzle.

14.4.3 The Tension Between Discreteness and Asymptotic Descriptions

The bulbous bow was a groundbreaking engineering advancement that significantly reduced drag on cargo ships. This design features a bulb at the front of the ship that generates a wave specifically shaped to counteract the wave produced by the ship's hull. From a distance, the two waves interfere destructively, resulting in minimal energy loss to surface waves. This means that, due to conservation of energy, drag is reduced.

Drag is related to friction, and there is a more detailed explanation involving the interactions between the ship's surface and water molecules. However, this detailed view is complex and not particularly useful for engineers. The description from a distance is sufficient for designing an effective bulbous bow because it focuses only on energy conservation. However, this asymptotic description is not enough if one wishes to demonstrate the unitarity of evolution, which requires the inclusion of molecular degrees of freedom.

This familiar physical situation is analogous, in some respects, to the problem we address in this chapter. While a simple description of various features of black hole formation and evaporation is available in terms of asymptotic concepts at infinity, such a description cannot be entirely accurate because certain regions in the bulk are subject to strong quantum gravity effects, especially if the quantum gravity theory predicts discreteness at the Planck scale. This is particularly relevant for regions where classical gravity predicts singularities. Additionally, as we will discuss, the asymptotic description may fail for regions in the future of the singularity where, in classical terms, it becomes naked: a strong quantum gravity fluctuating region that is causally connected to distant observers.

Moreover, asymptotic infinity is defined through conformal compactification, which brings infinity to a finite distance in the conformal metric. This operation is extremely useful in classical theory but conflicts with a fundamental understanding of the microscopic discrete structures underlying the UV completion proposed by quantum gravity approaches, such as loop quantum gravity. More precisely, in classical or semiclassical situations, dynamical fields admit a representation at infinity through mathematical techniques that are well-suited only for smooth fields. Consequently, any emphasis on describing the problem of black hole formation and evaporation solely from the perspective of asymptotic observers will inherently overlook the types of degrees of freedom that we argue are essential.[1] This is not to say that discussions based on asymptotic concepts are not useful; however, one should be mindful of their limitations.

The problem of information is not only about how to purify Hawking radiation. A more fundamental aspect of the question is what happens, for instance, to the Encyclopaedia Britannica if it falls into a black hole. This question is rarely addressed in discussions about unitarity because Hawking radiation represents a more clearly

[1] Quasi-local treatments involving finite regions, currently under active investigation [28], might not face this issue, as they should remain sensitive to discreteness.

defined and idealized situation (after all, it is only about the scattering of the vacuum in a black hole spacetime, or so one might think). However, no complete resolution can be achieved without addressing this more basic question.

Now, it seems quite obvious that boundaries and asymptotic regions cannot help with this issue if they do not include some account of the story involving the strong quantum gravity regime of the singularity where the Encyclopaedia falls. The question also concerns the fate of the universe as we evolve backward in time toward the primordial singularity (as suggested by the classical picture of standard cosmology). Thus, asymptotic boundaries cannot assist with such a question, and if they were to help in some hypothetical treatment involving boundaries, they should at least provide a way to understand the situation physically in their absence. Boundaries exist only in our notebooks and mathematical idealizations. One should keep in mind that such idealizations come with intrinsic limitations.

14.5 Remarks on the Semiclassical Picture

14.5.1 Large Quantum Fluctuations in the Future

The usual accounts of black hole evaporation rely on the validity of Hawking radiation calculations (which are performed on a fixed, non-dynamic background) and the semiclassical approximation. However, at least the second assumption has been shown to be inadequate by a simple argument presented by Don Page in [5]. In that paper, Page applies momentum conservation to the black hole and quantum radiation to estimate the uncertainty in the position of the black hole as it recoils due to the emission of individual Hawking radiation particles.

The uncertainty in position approaches the order of the Schwarzschild radius M when $\delta M/M > (m_p/M)^{2/3}$ (for a solar mass black hole, this occurs when $\delta M_\odot \approx 10^{-25} M_\odot$ [5]). By the time the black hole completes its evaporation, this uncertainty grows to the order of M^2. It is important to note that there is no quantitative way to perform this calculation within the standard framework of quantum field theory, as the results clearly exceed the limits of perturbation theory.

An important aspect of this situation is that fluctuations are controlled and confined to the UV neighborhood of the black hole horizon when traced back to the past, where a well-defined semiclassical (coarse-grained) notion of the horizon exists. In other words, uncertainties diminish in the past, allowing discussions of black hole entropy and thermodynamics to be framed within a semiclassical geometry for macroscopic black holes. A diagram representing this spacetime is shown in Fig. 14.4.

If we were to represent the situation with respect to a common asymptotic structure, the end of evaporation would appear as a superposition of final bursts from the black hole, with significant uncertainties in the Bondi parameters. Specifically, $\Delta(u_0 - u_{\rm eq}) \approx M^2$ and $\delta \equiv \Delta(u_0 - u_{\rm eq})/(u_0 - u_{\rm eq}) \approx 1/M$ (as indicated in Fig. 14.2). Even if late fluctuations prevent a faithful single spacetime representation, the

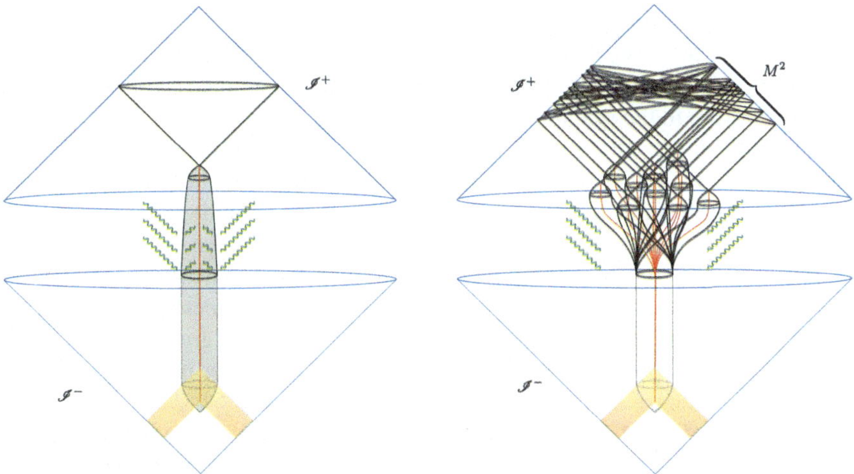

Fig. 14.4 On the left panel we represent the semiclassical expectation of the black hole evaporation process for a spherical black hole (this is a Penrose diagram with one extra dimension). On the right panel we represent the expected situation according to Page's analysis. The back reaction of the particles emitted in the Hawking radiation on the black hole leads to a superposition of black hole geometries where the uncertainty in the black hole position is as large as M^2 in Planck units by the end of evaporation. *Reprinted from Encyclopedia of Mathematical Physics (2nd edition) Volume 1, 2025, Pages 531–541, Black Hole Entropy and Planckian Discreteness, Copyright (2024), with permission from Elsevier*

Penrose diagram still provides a useful framework for discussing macroscopic black holes.

We will also see that, in addition to the large fluctuations in the center-of-mass parameters of the evaporating black hole, significant fluctuations are expected when considering the UV microscopic structure of the final state. Indeed, it is in this microscopic structure that the information needed to purify the apparently mixed Hawking radiation can be found in the current scenario.

14.5.2 *Semiclassical Description: Very Low Mass Is Available for Purification*

Figure 14.2 represents the qualitative features of the spacetime describing black hole formation and evaporation. As discussed in Sect. 14.5.1, no faithful spacetime representation of the future part of the spacetime is expected to exist due to large deviations from semiclassicality. The picture in the future should rather be replaced by a superposition of very different spacetime geometries, roughly represented in Fig. 14.4. Nevertheless, the representation used for the present discussion remains useful, particularly for macroscopic black holes. From the results of Page mentioned in the previous section, the quantum fluctuation in the time of evaporation is given by $\delta \equiv \Delta(u_0 - u_{\rm eq})/(u_0 - u_{\rm eq}) \approx 1/M$ (in the notation of Fig. 14.2), and thus the fluctuations (which remain important and should not be ignored) would have a less predominant role in the graphical representation.

These fluctuations are assumed to be small in the far past by the choice of the initial state: the starting point is semiclassical. We assume that the matter initially forms a diluted cloud in a geometry that is well represented by Minkowski spacetime. The matter in this initial state falls in and eventually produces a black hole via gravitational collapse. One could replace the matter in this excited state by an initially excited state of very weak gravitational waves, focused in a way that produces a black hole in the future, or by both matter and geometry in a state leading to gravitational collapse (black holes can be produced out of pure geometry with matter in an initial vacuum state).

The initially diluted matter cloud (or focused gravitational radiation, or both) becomes denser over time and eventually collapses under gravity. Trapped surfaces form, and if the initial cloud satisfies classical energy conditions, singularities predicted by singularity theorems appear [29]. These energy conditions are expected to be significantly violated by quantum effects near the singularity, where a full quantum gravity treatment will be necessary. They also occur near the black hole horizon, contributing to the gradual decrease in horizon area associated with Hawking radiation. At this point, a suitably defined version of the cosmic censorship conjecture [30] implies the formation of a black hole, which we assume starts with an initially macroscopic mass $M \ll m_p$. The ADM mass being constant, we refer to the Bondi mass $M(u)$, which changes with a suitably defined retarded time u at $\mathscr{I}^+$.

The formation process can be quite intricate and dynamic. During this time, both gravitational and matter radiation are emitted. Eventually, the system reaches a quasi-stationary state where all radiation related to the gravitational collapse has escaped to $\mathscr{I}^+$. This occurs at a late retarded (equilibrium) time $u = u_{\rm eq}$ in the Bondi (rest) frame of the black hole [31, 32], which is well-defined during this phase. Since the spacetime closely resembles a stationary, asymptotically flat spacetime, there is a unique Bondi rest frame up to time translations, fixed by an arbitrary choice of Bondi cut $u = 0$. Once this is set, the Bondi system can be extended to the whole of $\mathscr{I}^+$. The initial mass of the black hole, as invoked in the previous paragraph, is thus $M = M(u_{\rm eq})$.

Therefore, in the present qualitative account, we assume that for sufficiently late times, $u \geq u_{\rm eq}$ and up to the retarded time u_0 (which marks the end of evaporation), the spacetime geometry can be accurately approximated by a member of the Kerr-Newman family of (quasi) stationary black hole solutions. In this scenario, the mass $M(u)$, angular momentum $J(u)$, and electromagnetic charge $Q(u)$ evolve according to the relevant Bondi balance equations at $\mathscr{I}^+$, with fluxes governed by the Hawking radiation spectrum [1]. Such a quasi-stationary description breaks down due to large fluctuations (see Sect. 14.4.3) and strong quantum gravity effects toward the end of evaporation. However, the conservation issues we now invoke should admit a suitable translation to the full quantum account, at least until near the end of evaporation.

Thus, as far as energy balance is concerned, the quasi-stationary description should hold until the mass $M(u)$ approaches the Planck mass m_p near the retarded time u_0 over a very long time $u_0 - u_{\text{eq}} \approx M^3$. At this point, Hawking's calculations can no longer be relied upon, as the curvature near the black hole horizon reaches the Planck scale. A complete quantum gravity description becomes necessary at this stage.

However, the key point is that, regardless of the specifics of this final phase of evaporation, the total energy available—measured by distant Bondi observers—is limited to roughly the Planck mass. Therefore, for $u \geq u_0$, the spacetime geometry and the state of the matter fields and geometry should be well approximated by a quantum superposition of states, all of which are extremely close to the flat vacuum state with excitations that, together, cannot contain much more energy than the Planck mass.

Throughout this time, Hawking particles have been emitted to $\mathscr{I}^+$ in a spectrum close to thermal. Large quantum fluctuations are expected in the later stages of this process, even while the black hole remains macroscopic. This suggests that deviations from Hawking's calculations are likely. Some approaches to the information problem argue that the radiation spectrum is only approximately thermal, proposing that the true state is pure and arises from the unitary evolution of the initial state, meaning no information is lost.

We disagree with this conclusion, especially if it is reached without considering the strong quantum gravity regime near the singularity. The key issue is that even before fluctuations cast doubt on Hawking's calculations, information is already lost to the singularity. This occurs because, while the black hole is macroscopic (and before fluctuations become significant), the emitted Hawking particles are maximally correlated with partners that fall into the strong quantum gravity region we call the singularity.

Furthermore, even at later times, this aspect of the Hawking radiation will still hold true. A hypothetical observation (in the sense of a measurement in quantum mechanics) would cause the quantum system to collapse into a semiclassical spacetime containing a macroscopic black hole, thereby eliminating the quantum fluctuations. This means that throughout the entire evaporation process, as long as $M(u) \gg m_p$, Hawking particles are emitted in maximal correlation with partners falling into the singularity. As a result, the final state of the Hawking radiation must be well described by a highly mixed state with significant entropy $S > A/4$, where A is the area of the black hole at u_{eq} (which is roughly M^2 in order of magnitude).

The challenge of the information problem is to identify the degrees of freedom needed to purify this mixed state for $u > u_0$ with a minimal energy budget of the order of $m_p \ll M$.

14.5.3 On the Impossibility of an Effective Field Theory Description

The effective field theory (EFT) paradigm is crucial for understanding models in quantum field theory and their physical implications. Due to the presence of ultraviolet (UV) divergences in interacting quantum field theories, the EFT perspective is essential for extracting meaningful physics from the Standard Model of particle physics, as well as from extensions that may involve certain situations involving gravity.

The key idea behind effective field theories is the separation of scales. This means we can ignore high-energy (UV) degrees of freedom in order to focus on the low-energy phenomena of interest, while still preserving unitarity. This concept is best illustrated through the Wilsonian approach to renormalization, which reorganizes the path integral representation of quantum dynamics. More precisely, the path integral is recast in terms of low-energy (infrared, or IR) degrees of freedom, with a new effective action that encodes the effects of the UV physics that has been integrated out. This process does not eliminate any degrees of freedom; it simply changes how the integration field variables are interpreted. However, this view imposes a sharp constraint on the use of the effective quantum theory, which can only be used to calculate transition amplitudes between the associated low-energy states.

Thus, while effective field theory is important for interpreting particle physics (and certain special situations beyond), it cannot address the unitarity problem in black hole formation and evaporation. The reason is that the separation of scales necessary for effective field theory breaks down in the context of gravitational collapse. During Hawking evaporation, low-energy degrees of freedom (the escaping Hawking partners) are strongly correlated with high-energy degrees of freedom that are formed as they fall into the singularity (see Sect. 14.7.1 for discussion). To analyze unitarity, we must consider both extremes of the scale spectrum simultaneously. This is beyond the domain of applicability of the effective field theory approach.

14.5.4 Phenomenological Implications

As argued here, discreteness at the Planck scale provides a natural resolution of the information paradox, while simultaneously opening the door to vast possibilities in phenomenology that need to be explored in greater depth. For example, if low-energy degrees of freedom can entangle with the microscopic ones, diffusion could generically occur under suitable conditions. It has been suggested that such diffusion may have observable consequences in cosmology, ranging from providing a microscopic explanation for the origin of dark energy [33–35] to potentially addressing current tensions in observational data [36–38]. The existence of granularity at the Planck scale can also provide a natural scenario for the origin of structure in the early universe, as well as suggest viable candidates for dark matter [39, 40] and their independent observation [41]. This underscores the importance of discussing the problem of unitarity in black hole formation and evaporation, which might initially seem unrelated to more immediate physical applications.

14.6 Motivation from Black Hole Thermodynamics

The mathematical models that define our successful physical theories are all reversible. This means they can predict the future based on initial conditions, while the past can be uniquely reconstructed from the future. The memory of the initial conditions is preserved in the dynamics, maintaining their information content. This holds true for classical mechanics, field theory, quantum mechanics, and quantum field theory, provided we do not consider the collapse of the wave function. In other words, it remains valid as long as we do not make an external measurement. In quantum mechanics, this reversibility arises from the fact that evolution is governed by a unitary operator, which can always be reversed using its adjoint transformation.[2]

14.6.1 The Physical Arrow of Time

In systems with many degrees of freedom, an effective form of irreversibility can arise when information degrades and becomes inaccessible because the observer cannot distinguish between different configurations due to their limited probing abilities. This often occurs when microscopic degrees of freedom are involved. An iconic example is the burning of a newspaper. Common sense suggests that the information it contained is simply lost. However, as physicists, we would argue that the information is merely degraded or hidden (to the extent that it becomes practically unrecoverable) within the vast number of microscopic variables that describe the paper's molecular structure. One could contend that the words in the newspaper remain encoded in the correlations between the molecular degrees of freedom in the combustion gases diffusing into the atmosphere. Information remains, but it is transferred to increasingly larger regions of phase space in an unbounded universe, with degrees of freedom that are inaccessible to macroscopic observers (due to coarse graining). While the physicist cannot prove this explicitly, the quantum version aligns with the strongly held principle of unitarity.

This effective irreversibility is familiar from the second law of thermodynamics, which states that in systems with a large number of degrees of freedom, entropy can only increase. At first glance, this seems to conflict with Liouville's theorem,

---

[2] This is particularly explicit in the case of generally covariant theories like general relativity or quantum gravity, where, due to diffeomorphism invariance, all dynamics are captured by a complete set of Dirac observables, which are constants of motion. We will discuss this point further in Sect. 14.9.

which implies that the volume occupied by a phase space distribution representing the system is preserved during evolution. However, the apparent conflict is resolved when we introduce the concept of coarse graining. More precisely, suitable initial conditions (as defined by the observer) come with some uncertainty due to limited measurement capabilities. In the newspaper example, these initial conditions correspond to the macroscopic arrangement of ink particles forming words on the paper before it burns. These initial conditions, and their associated uncertainties, can be idealized as a distribution in phase space that occupies a regular initial volume. As the paper burns, the apparent phase space volume (though not the actual volume, which remains constant) seems to grow from the coarse-grained perspective of the observer, who can no longer distinguish between points in phase space that the system truly occupies and those nearby points where the system is absent. As emphasized by Penrose, this phenomenon, which we interpret as the second law of thermodynamics, occurs in our part of the universe due to the very special initial conditions of our distant past (entropy can increase because it is low in the past).

We argue that the general outline of this story remains the same when considering black hole formation and evaporation. Once the physics of quantum gravity is properly taken into account (especially near regions where general relativity predicts singularities), the fate of information in black hole formation and evaporation may be explained in a manner similar to the familiar newspaper example.

In the case of black holes, the singularity represents the point where initially low-energy degrees of freedom (described by quantum field theory modes from a macroscopic perspective) are forced to interact with the fundamental components of gravity and matter at the Planck scale. This is the scale at which effective quantum field theory is no longer a reliable description. It cannot be, if Planckian physics is discrete, or more generally, if there is no well-defined notion of the metric at the Planck scale. The singularity in classical gravity is analogous to the fire in the previous example, as it enables the system to explore a vast portion of phase space that was previously inaccessible in a low-curvature, low-energy configuration.[3] We will argue that information is not destroyed or lost; rather, it is degraded into correlations with Planck-scale microscopic degrees of freedom, which are hidden from macroscopic observers who describe the dynamics using the limited tools of effective quantum field theory.

A remarkable indication of the relevance of this perspective is that, according to general relativity combined with quantum field theory (in the regime where both are expected to be accurate), large, isolated black holes are well represented as thermodynamic systems in equilibrium. The no-hair theorem states that classical black holes (formed from standard model matter) are completely characterized by only three macroscopic parameters: mass M, angular momentum J, and electric charge Q. The various ways these black holes can form from the gravitational collapse of suitable initial conditions shape the radiation flux of matter and gravity at $\mathscr{I}^+$ before the black hole becomes stationary. Meanwhile, the "inside" degrees of freedom, located beyond the event horizon and falling toward the interior singularity, seem to be lost

[3] We will introduce an example of this via a toy model in Sect. 14.8.3.

to outside observers. These degrees of freedom evolve into a region where general relativity becomes unpredictable, and their future behavior can only be understood through a complete quantum gravity description.

In quantum mechanics, stationary black holes—formed from gravitational collapse as described by classical physics and governed by the no-hair theorem—are almost in equilibrium but never truly so. When quantum effects are considered, they emit particles at the Hawking temperature, slowly losing energy, especially when they are large. If disturbed, black holes return to a new state of near-equilibrium, and this process follows the first law of black hole thermodynamics:

$$\delta M = T\delta S_{\rm bh} + \Omega\delta J + \Phi\delta Q, \tag{14.1}$$

with entropy given by $S_{\rm bh} = \frac{A}{4\ell_p^2}$, where A is the horizon area. Under such perturbations, it is expected that the total entropy of the outside universe can only increase. Specifically, according to the so-called generalized second law (GSL[4])

$$\delta S = \delta S_{\rm matter} + \frac{\delta A}{4} \geq 0, \tag{14.2}$$

where $\delta S_{\rm matter}$ represents the Von Neumann entropy of whatever is outside the black hole (including, for instance, the emitted radiation). If black holes are thermal systems with macroscopic entropy that is comparable (according to the GSL) to the coarse-grained entropy of material systems, then an obvious question arises: what are the microscopic degrees of freedom responsible for this entropy? Loop quantum gravity (see [44, 45] and references therein) offers a tentative answer to this question.

If the microscopic geometry and matter Planck scale elements are responsible for black hole entropy, then, much like the molecules in a newspaper, they hold the key to understanding the information problem in black hole evaporation. This is the central idea behind the perspective presented here, inspired by loop quantum gravity.

14.6.2 Black Hole Entropy and a Key Assumption About Quantum Gravity

Assuming the radiation process to be quasi-stationary—i.e. that the state of the macroscopic black hole during evaporation can be well approximated by that of a stationary black hole in general relativity—the entropy of the whole of the Hawking radiation produced matches the initial black hole entropy $A/4$. In the spherically symmetric case, this follows immediately from (14.1). More precisely, one has $dS/dM = T^{-1} = 8\pi M$ (where we used $J = 0$ and the corresponding value of the

[4] The GSL has been proven in the context of quantum field theory on curved spacetimes via perturbations of a black hole background in [42]. The proof can be generalized in terms of an equivalent statement involving free energy and observables defined at infinity [43].

Hawking temperature for a spherical black hole). Integrating, it follows that

$$\Delta S = \int_M^0 8\pi \overline{M} d\overline{M} = \frac{A}{4}. \tag{14.3}$$

The previous calculation can be generalized to the rotating case if, in addition, we use conservation of angular momentum at $\mathscr{I}^+$ plus the precise angular momentum dependence of the grey-body factors. The left hand side of Eq. (14.3) can be also expressed as the Von Neumann entropy of the reduces density matrix of the state ρ_{out} of the quantum fields at $\mathscr{I}^+$ when ignoring the degrees of freedom after the retarded time u_0 of complete evaporation, namely

$$\Delta S = -\text{Tr}[\rho_{\text{out}} \log(\rho_{\text{out}})] = \frac{A}{4} + \text{negligible correction}. \tag{14.4}$$

This is the entanglement entropy in the quantum fields when tracing out the degrees of freedom registered on the portion Σ_2 of the Cauchy surface depicted in Fig. 14.5 that is inside of the black hole. The negligible correction comes from a contribution of the boundary separating the inside from the outside that in Σ_2 has a size the order of the cutoff (see (14.5) below). The inner part of Σ_2 refers to the last moment before reaching the quantum area where spacetime representations break down due to Planckian effects. In this case, we're extending the semiclassical approximation to its limits, similar to what was done in the earlier calculation of radiation entropy (14.3).

The equivalence between thermodynamic entropy and entanglement can be understood from the fact that Hawking radiation is produced in perfectly correlated pairs of outgoing particles: one travels to infinity, while the other falls into the black hole singularity. This is a specific instance of a broader phenomenon of particle creation in curved spacetimes [2] and by external potentials. These correlations—predicted by standard quantum field theory—ensure the unitarity of evolution in Fock space for Cauchy surfaces before the quantum gravity region (such as Σ_1 in Fig. 14.5). One also expects unitarity to hold when extending this evolution to the "latest possible time" Σ_2. We neglect possible quantum field theory contributions near the dotted null line. While quantum gravity effects may play a role here, they are unlikely to drastically change the overall picture as far as the radiation degrees of freedom are concerned. Unitarity is maintained by the entanglement between the outside and inside regions of these Cauchy surfaces, meaning that the radiation entropy represents both standard thermodynamic entropy and entanglement entropy between the black hole's exterior and interior.

This perspective suggests that the outside entropy, which satisfies the GSL (14.2), is simply entanglement entropy, not just at the moment Σ_2, but for all Cauchy surfaces that have the quantum gravity region in their future (such as Σ_1 in Fig. 14.5). We will see below that this view allows us to define black hole entropy as a measure of the (local, or would-be divergent) entanglement between the degrees of freedom just outside and just inside the black hole's event horizon [46].

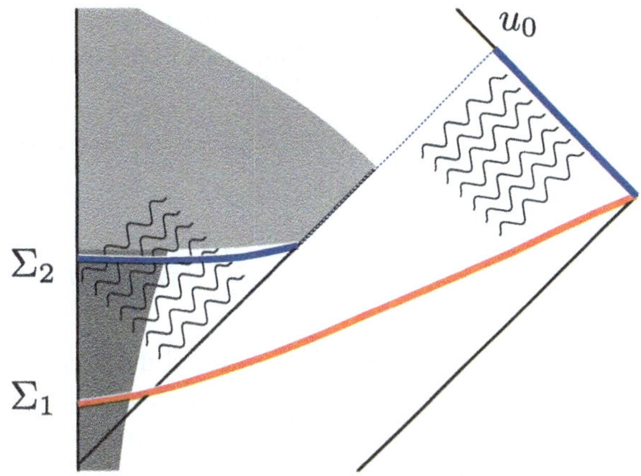

Fig. 14.5 Two 'instants' represented by suitable Cauchy surfaces relevant for the discussion of entropy and information. *Reused from Alejandro Perez and Daniel Sudarsky, AVS Quantum Sci. 4 (4), 10.1116/5.0091962, (2022) with permission of AIP Publishing*

Specifically, for a pure state $\rho = |0\rangle \langle 0|$ of a quantum field on a Cauchy surface like Σ_1 in Fig. 14.3, the entanglement entropy between the inside and outside of Σ_1 with respect to the black hole horizon diverges because of ultra-local entanglement across the horizon in local quantum field theory. To manage these "high-energy" or UV contributions, we introduce a cut-off length scale ϵ. One finds [47]:

$$\begin{aligned} S_{\text{ent}} &= -\text{Tr}[\rho_{\text{out}} \log(\rho_{\text{out}})] \\ &= f_1 \frac{A}{\epsilon^2} + f_2 \log(A\epsilon^{-2}) + S_{\text{rad}}, \end{aligned} \tag{14.5}$$

where f_1 and f_2 are constants, and S_{rad} is a finite term representing the entropy of the radiation outside the black hole. This term is expected to vanish in a suitable vacuum state if Σ_1 is early enough that it doesn't contain any Hawking particles. The term proportional to the black hole area A, which diverges as ϵ^{-2}, is the one we want to associate with black hole entropy. However, this term is not only divergent but also ambiguous in quantum field theory, as it depends on the regularization method used and the number of degrees of freedom involved (the so-called species problem.[5])

In addition to leading to the correct formal dependence on A, there's reason to believe that the UV divergent term, once made finite by including missing quantum gravitational degrees of freedom, captures the true notion of black hole entropy that fits into black hole thermodynamics. We can express this idea as:

$$f_1 \frac{A}{\epsilon^2} \quad \underset{\text{quantum gravity}}{\longrightarrow} \quad \frac{A}{4\ell_p^2}. \tag{14.6}$$

[5] The quantum field theory fact that the entanglement entropy (14.5) scales with the number of species considered.

Support for this conjecture comes from studying perturbations of the Hartle-Hawking state of an eternal Schwarzschild black hole [48, 49]. By perturbing the definition in (14.5), we find:

$$\delta S_{\text{ent}} = \frac{\delta A}{4\ell_p^2} + \delta S_{\text{rad}}, \tag{14.7}$$

where

$$\delta S_{\text{rad}} = \frac{1}{T}\int_{\mathcal{I}^+} \langle \delta T_{\mu\nu}\rangle\, \xi^\mu dS^\nu = \frac{\delta E_{\text{rad}}}{T}. \tag{14.8}$$

Here, ξ^a is the stationary Killing field of the background geometry, and expectation values are taken in the Hartle-Hawking state. The second term on the right-hand side of Eq. (14.8) is the thermodynamic entropy carried by the radiation. This relation highlights the direct connection between entanglement entropy and thermodynamic entropy in the context of black holes.

This argument provides the necessary Bekenstein-Hawking horizon area contribution, as it emerges from the semiclassical Einstein equations via the Raychaudhuri equation, which describes the perturbation of the black hole horizon's generators sourced by $\langle \delta T_{\mu\nu}\rangle$. Including gravity helps resolve both the divergence and ambiguity issues, including the species problem. Notably, any dependence on the number of particle species cancels out in this analysis, as their number influences both the entropy calculation and the change in area in the same way.

More generally, we can support the validity of the conjecture (14.6) by examining the UV structure of entanglement in any Hadamard state (states whose two-point function resembles the UV structure of the Minkowski vacuum and is used to renormalize the energy-momentum tensor in semiclassical gravity) in quantum field theory, along with a suitable assumption about the granularity at the Planck scale (for example, loop quantum gravity provides such a framework).

Specifically, if we assume that the background geometry closely resembles that of a stationary black hole (such as a Kerr-Newman black hole), there exist local observers just outside and near the horizon who perceive the horizon as being at rest, essentially a "rest frame of the horizon." These observers have four-velocities given by $u^a = \chi^a/\sqrt{|\chi \cdot \chi|}$, where χ^a are the Killing generators of the horizon, defined as:

$$\chi^a = \xi^a + \Omega\psi^a,$$

with ξ^a being the time translational Killing field and ψ^a the axial symmetry Killing field (where Ω is the angular velocity of the horizon in the first law (14.1)). According to local quantum field theory, these stationary observers will detect a thermal environment with a local temperature $T_{\text{local}} = 2\pi/\ell + \mathcal{O}(\ell)$, where ℓ is the space-like distance to the horizon in a background assumed to be stationary in the past. As $\ell \to 0$, the temperature increases, and the UV structure of quantum states suggests that, in this limit, all available degrees of freedom must behave as if they are excited at infinite temperature, so the state appears locally like the vacuum to freely falling observers.

Thus, for the specific observers defined by the time translational symmetry near the horizon (the natural rest frame), the state must resemble an infinite temperature state to avoid the presence of a firewall at the horizon. As a result, in the limit $\ell \to 0$, all degrees of freedom must be equally probable (according to the density matrix defined for these preferred observers) and maximally correlated with their partner modes on the other side.

Under such conditions of equipartition of probability, the UV contribution to the entanglement entropy (the divergent part) must satisfy:

$$f_1 \frac{A}{\epsilon^2} \underset{\text{Hadamard}}{\longrightarrow} \log(N), \tag{14.9}$$

where N represents the number of degrees of freedom characterizing the infinitely thin atmosphere close to the horizon (the UV modes), which we can consider as horizon states. Here, N corresponds to the dimension of the Hilbert space of these horizon states (assumed to be finite due to discreteness, that is the case in loop quantum gravity []).

The number N is divergent in an effective quantum field theory because it does not account for the fine structure of quantum geometry. We expect this seemingly divergent number to actually be limited by the area of the black hole (the entanglement surface) in a suitable theory of quantum gravity. This is true in models of quantum geometry that come from loop quantum gravity [50], where the dimension of the allowed part of the Hilbert space, constrained by the horizon's area being a macroscopic number A, is given by $N \approx \exp \eta A$. Here, η is a dimensionless number. Thus, we find:

$$f_1 \frac{A}{\epsilon^2} \underset{\substack{\text{in loop} \\ \text{quantum gravity}}}{\longrightarrow} \eta A. \tag{14.10}$$

In loop quantum gravity, the number of microscopic states of the spacetime geometry with a black horizon of fixed macroscopic area can be counted. This number is finite, which is directly related to the theory's prediction of a fundamental granular structure, where spacetime emerges from fundamental geometric building blocks. The constraint that the horizon area must be A limits the number of possible combinations and produces a finite entropy, as given by (14.10).

It is important to note that, despite the valuable insights from these results, the meaning and computation of black hole entropy in loop quantum gravity remain partly unresolved. The main limitation is the insufficient control over the physical states in the quantum gravity Hilbert space corresponding to a specific macroscopic black hole, including all relevant degrees of freedom.

This situation should generally arise if the smooth spacetime geometry and fields of general relativity emerge from a more fundamental Planckian physics that is discrete at its core. Loop quantum gravity [19, 51] and causal set theory [22] are examples of mathematical models where this idea is partially realized. In these approaches, smooth fields and geometry are expected to emerge as effective descriptions of the collective behavior of microscopic building blocks under certain conditions.

14.6.3 The Page Curve, the Holographic Principle, and the Central Dogma

Every dogma has its day.

The current intuition that thermodynamic entropy represents a measure of the number of internal states of a system (valid in familiar non-relativistic systems) does not seem to apply to black holes or strongly gravitating systems. Nevertheless, the historical evolution of the concept of black hole entropy—starting with the seminal paper by Bekenstein [52] and followed by his proposal of entropy bounds and their generalizations—has elevated these properties (whose known precise versions are properties of standard quantum field theory [53, 54]) to the idea of a *holographic principle*. In the case of black holes, this principle motivates the postulate of the so-called ***central dogma***: *the number of internal degrees of freedom of a black hole is proportional to its horizon area.*

An important consequence of the central dogma is that it implies deviations from the thermality of Hawking radiation when the black hole is still macroscopic—by the so-called Page time [9] when the entanglement entropy between the outside radiation and the interior of the black hole should start decreasing, as illustrated by the celebrated Page curve. This requires significant departures from standard quantum field theory on curved spacetimes, even in situations where these deviations are unexpected (low curvatures, low energies).

Quantum field theory, applied within its expected regime of validity, suggests that internal degrees of freedom in a black hole must correlate with the Hawking radiation outside. Since this radiation can be very large, depending on the black hole's past history, the internal degrees of freedom would need to be unbounded. This aligns with the semiclassical description, where very old black holes can conceal a large volume of space behind their horizons [55]. This is also consistent with the fact that semiclassical causality implies that for Cauchy surfaces Σ_1 and Σ_2, as in Fig. 14.5, located to the past of the singularity, the domain of dependence of the interior region of Σ_2 contains the domain of dependence of the interior of any prior Cauchy surface Σ_1 (even when the horizon of Σ_2 has a smaller area) [56], and with the general perspective taken in this chapter.

In our view, black hole entropy measures the number of degrees of freedom that interact with the exterior, as suggested by the entanglement entropy interpretation of the Bekenstein-Hawking entropy we advocate in Sect. 14.6.2. This is particularly clear from the validity of the generalized second law, formulated and proven in [42]. In particular, the form of the proof admits a reverse interpretation, where the external degrees of freedom are exchanged with those inside the black hole [43]. The number of degrees of freedom inside the black hole is as unbounded by the presence of the black hole horizon as the number of degrees of freedom outside.

For all these reasons, we do not adhere to the central dogma. Our philosophy is based instead on the belief that one can trust quantum field theory and general relativity as long as the situation is within their expected regime of validity. All radical deviations[6] from what is well understood occur (in our scenario) near the Planckian regime close to singularities and their future domain of dependence. The discussion in this chapter shows that a consistent account of unitarity is possible without the holographic principle. We take this as sufficient reason for skepticism regarding the central dogma.

14.7 Discreteness and Unitarity: A Scenario

While the goal mentioned earlier remains an open question, a common feature among these approaches is that a classical solution, characterized by a smooth geometry that satisfies Einstein's equations and related field equations, does not uniquely identify a microscopic state in the fundamental quantum theory. This suggests that the presence of a complex microscopic structure introduces new degrees of freedom that cannot be captured by a smooth effective field theory.

For example, the quantum state of flat spacetime with matter in the Minkowski vacuum emerges from a superposition of fundamental building blocks at the Planck scale, where the arrangement of these pieces is not strictly determined by the macroscopic flatness condition. Importantly, this means that what we might think of as the "ground state" would actually be highly degenerate, contrary to typical quantum field theory expectations. This fundamental degeneracy is the origin of black hole entropy in loop quantum gravity, where a single classical black hole corresponds to many microscopic states. Consequently, this degeneracy should also be crucial for understanding unitarity in black hole evaporation.

Currently, the status of loop quantum gravity (and, to our knowledge, other approaches to quantum gravity) does not allow for a detailed dynamic analysis of black hole formation and evaporation. However, considering this issue from the loop quantum gravity perspective reveals a new way to think about the fate of information.

To clarify this idea, we will present a series of formal equations meant to illustrate concepts rather than provide precise predictions. These equations should be viewed with caution, as they might be inaccurate in some cases—for example, when we describe the final state of Hawking radiation as a thermal state at a constant temperature, ignoring that the temperature changes as the black hole shrinks. In other

[6] One radical departure from standard quantum field theory is that, in order to accommodate a sufficient number of purifying degrees of freedom with low mass, the vacuum must be highly degenerate, with a nontrivial entropy density capacity. This implies an infinite density of states near the vacuum, invalidating certain effective field theory arguments [57] in favor of holography. This is suggested by the structure of nonperturbative quantum gravity and realized in simple models, such as the one we present in Sect. 14.8.3.

cases, they may only roughly express properties of quantum gravity that could only be expressed precisely if we had the complete formalism of quantum gravity.

For simplicity, we assume that the Hilbert space of quantum gravity can be described as the tensor product of the quantum geometry Hilbert space and the matter Hilbert space:

$$\mathscr{H} = \mathscr{H}_{\text{geo}} \otimes \mathscr{H}_{\text{matter}}.$$

This is certainly an oversimplification, as one would expect matter and geometry to emerge from a unified fundamental structure where the associated Hilbert space might not necessarily have such a tensor product structure.[7]

In loop quantum gravity, there are infinitely many quantum geometry states for each classical solution, differing in Planck-scale degrees of freedom that smooth field observables do not detect. These degrees of freedom can be seen as defects in the fabric of quantum gravity, which are not detectable using low-energy probes (see Sect. 14.8.3 for a concrete realization in a toy model).

For flat Minkowski spacetime, we assume that the different quantum geometry states, which can be viewed as flat Minkowski states, belong to a countable set. We can label them as:

$$|\text{flat}, n\rangle \in \mathscr{H}_{\text{geo}},$$

where n is a natural number. Each quantum geometry state corresponds to a vacuum matter state:

$$|0, n\rangle \in \mathscr{H}_{\text{matter}}.$$

The combined states

$$|\Psi_{0,n}\rangle = |\text{flat}, n\rangle \otimes |0, n\rangle$$

are physical states that satisfy all the dynamical equations of quantum gravity. These states are vacuum states with no wavelike matter excitations and are labeled "flat" because they satisfy:

$$\widehat{M}_{\text{ADM}}\, |\text{flat}, n\rangle \otimes |0, n\rangle = 0, \tag{14.11}$$

where $\widehat{M}_{\text{ADM}}$ refers to the ADM mass operator. Thus, the states $|\Psi_{0,n}\rangle$ represent the degenerate "vacua" of quantum gravity.

[7] Such a tensor product structure is probably not present in the fundamental theory if matter and geometry are emergent. Another instance where the tensor product is an oversimplification is in the assumption of the local structure of field theories, where the Hilbert space of $\mathscr{H}_{A+B}$ is assumed to be $\mathscr{H}_A \otimes \mathscr{H}_B$. This relationship is known to fail in standard gauge theories due to the presence of constraints, and the failure is expected to be more significant in gravitational theories [58]. In this respect, it is important to point out the remarkable properties of localization in quantum gravity that are made apparent in the recent work [59]. On general grounds, this is related to the known fact that in quantum gravity, physical degrees of freedom are highly non-local—something that ought to play a role in a refined discussion of the present perspective but would not change the overall picture dramatically. In turn, the microscopic degeneracy of the quantum states representing the low-energy physics (advocated here) also modifies their conclusions, as it violates some of the assumptions made [57].

The motivation for having a multiplicity of states that satisfy the low curvature requirement in Eq. (14.11) comes from loop quantum gravity in several ways. First, by studying quantum gravitational degrees of freedom at the Planck scale (quantum geometry), we find that the discrete nature of these quantum states introduces new degrees of freedom that do not have a smooth classical counterpart. Additional evidence comes from analyzing simplified models [60], where we can rigorously demonstrate the analog of (14.11). In these cases, the Planck-scale degrees of freedom appear as defects in the fabric of geometry, which are not necessarily propagating.[8]

Another piece of evidence for these hidden defect-like degrees of freedom comes from examining the area spectrum in loop quantum gravity. This degeneracy in coarse-grained geometry is crucial for calculating black hole entropy. A single classical isolated horizon, defined by its macroscopic area, can have many microscopic eigenstates of area A, with eigenvalues that grow exponentially with A, specifically as $\exp(\eta A)$ (see [44, 45] and references therein). However, this large number of microscopic states cannot be interpreted classically. The isolated horizon's phase space, defined by its boundary conditions, implies the absence of local excitations on the horizon [62]. Therefore, these new degrees of freedom act like quantum hair for black holes, where one macroscopic black hole corresponds to a vast number of microstates that differ only by tiny details noticeable only at the fundamental scale, rather than by wavy modes like gravitons propagating in the continuous description.

Additional motivation for the previous example comes from analyzing certain mini-superspace models. In these cases, the equivalent of Eq. (14.11) appears in symmetry-reduced models for quantum cosmology and black holes (as we will explicitly review in Sect. 14.8). Here, the infinite degrees of freedom of quantum gravity are simplified to a finite number through symmetry assumptions.

Despite their simplicity, these toy models, based on a similar algebra of phase space variables as full loop quantum gravity, exhibit two important interrelated features. First, the structure of the Hilbert space resolves the Big Bang singularity (or the singularity inside black holes) using a well-defined notion of quantum dynamics, as shown in Fig. 14.5. Second, loop quantization results in a much larger Hilbert space that includes new hidden degrees of freedom, which lack a low-energy interpretation.

In cosmology, this means that the solutions to the Hamiltonian constraint are infinitely degenerate. Since there isn't a clear definition of energy in cosmology, one cannot write the equivalent of (14.11). However, in a related framework called unimodular gravity [63–66], energy is well-defined as a conserved quantity acting like the cosmological constant.

In the singularity-avoiding quantization [67, 68], the eigenstates of the cosmological constant are also infinitely degenerate, reflecting the spirit of (14.11). For spherically symmetric black hole models, the ADM mass can be loop-quantized, demonstrating that it is indeed infinitely degenerate, with degeneracy represented by

[8] One possible example is the degeneracy of volume eigenstates in loop quantum gravity, where each volume eigenvalue (representing a basic volume of quantum space) has a double degeneracy, resulting in a bit of degeneracy for each building block that makes up continuous geometry [61].

quantum numbers that have no classical counterpart, but instead, as envisaged above, are represented by microscopic defects.

We assume that the initial state is a tensor product of the geometry and matter degrees of freedom. While this assumption isn't strictly necessary (see [48] for more details), it simplifies the notation without affecting the main ideas. However, it's important to note that this assumption idealizes the fact that the initial state must be special and "low entropy". For simplicity, we can assume that the initial quantum geometry state is represented by $|\text{flat}, 0\rangle$ in the decomposition of (14.11). We can always define a basis of macroscopically equivalent vacua to meet this requirement. Thus, we can express the initial state as:

$$|\Psi_{\text{initial}}\rangle = \underbrace{|\text{flat}, 0\rangle}_{\text{quantum geometry}} \otimes \underbrace{\hat{C}(f)\,|0_{\text{in}}, 0\rangle}_{\text{matter fields}} \tag{14.12}$$

This represents an initial geometry well approximated by a low-curvature (or zero-curvature) background $|\text{flat}, 0\rangle$, with matter encoded in the state $\hat{C}(f)\,|0_{\text{in}}, 0\rangle$. This state represents a diluted semiclassical matter distribution with a profile f, which will eventually collapse to form the black hole. The operator $\hat{C}(f)$ acts on the vacuum compatible with the quantum geometry state.

This is a very specific state, and reality may not perfectly fit such a simple product state. However, it captures the idea that gravitational collapse is a highly asymmetric process, where the past is particularly special (and low entropy) [69, 70]. This special nature of the past, reflected in the initial condition, captures a key physical aspect of the system, albeit in a highly idealized way.

The history of black hole formation and evaporation is well illustrated by Penrose's depiction of irreversibility. Before a black hole forms, our system typically explores larger portions of the available phase space. This involves the non-gravitational degrees of freedom in the semiclassical cloud f, its self-interactions, and interactions with gravity (producing gravitational waves and matter radiation). The initial cloud evolves into a star that eventually collapses to form a black hole, which behaves as a thermal system in quasi-thermal equilibrium, slowly evaporating into Hawking radiation. A key point of the present perspective is that the proposed existence of underlying microscopic Planckian granularity can only accentuate such time asymmetry.

During black hole formation, a singularity develops within the collapsing body, and thus a vast new portion of phase space opens up for the relevant degrees of freedom to explore. To describe gravitational collapse and the formation of what we call a singularity, we must also explain what happens beyond the event horizon in the Planckian curvature region. This region extends beyond classical descriptions, representing a quantum gravity domain where dynamics should remain well-defined. The singularity marks a quantum gravity zone where tidal forces and densities reach Planckian levels, connecting the system to the Planck scale and its microscopic degrees of freedom—similar to how a lighter can set a newspaper on fire in our earlier example.

Gravitational collapse triggers interactions with the Planckian regime inside the black hole's horizon, and the consideration of dynamics in such an extreme regime becomes essential for any unitary description of black hole formation and evaporation.

14.7.1 Strong Quantum Fluctuations Near the Singularity

To understand the fate of information during black hole evaporation, we need to consider the existence of Planckian degrees of freedom. This is evident for the degrees of freedom in the initially diluted semiclassical cloud that later becomes a collapsing star, eventually entering the quantum gravity region where general relativity predicts a singularity.

This also applies to the dynamics of the vacuum state $|0_{in}, 0\rangle$, which interacts with the low-curvature gravitational field, leading to Hawking pairs and the emission of Hawking radiation to $\mathscr{I}^+$. Black hole evaporation means that the strong field region near the singularity (produced by the collapse of the energy-momentum in the initial cloud forming the black hole) must gradually soften by the backreaction of the Hawking radiation as the evaporation takes place: the singularity itself evaporates in a sense. Understanding this process requires a full quantum gravitational treatment, as the interaction between the matter forming the black hole and the Hawking particles occurs only within the quantum gravity region.

This distinction is highlighted by the fact that gravitational collapse happens quickly, with matter reaching the singularity on timescales comparable to the initial black hole mass M, while Hawking radiation is a slower process, characterized by timescales of about M^3.

We should not overlook the non-local aspects of quantum field theory. For example, vacuum polarization contributions to $\langle T_{ab}\rangle$ can create "negative energy densities" as measured by observers falling with the collapsing matter, particularly near the singularity [15, 71, 72]. However, these effects are significant close to the singularity, where quantum field theory in curved spacetime is unreliable. Therefore, a complete quantum gravity description of this high-curvature regime is essential.

Concentrating on Hawking radiation, two key aspects are important for our discussion, both well-established in semiclassical physics. First, Hawking radiation is produced through particle creation caused by the interaction of matter fields with the curved spacetime background. Second, the backreaction of this radiation drives the evolution of the black hole's geometry and its eventual evaporation.

From the general theory of particle production in a gravitational field, we know that particles are produced in maximally entangled pairs (see, for example, [2]). In this specific case, this means that for each outgoing Hawking particle emitted to $\mathscr{I}^+$, there is a corresponding particle falling into the singularity that is maximally entangled with its Hawking partner. Even if the Hawking particle takes an infinite amount of time to reach $\mathscr{I}^+$, its partner inside the black hole hits the singularity after a finite amount of time.

Therefore, addressing questions about unitarity—particularly regarding the fate of the entanglement between these two particles—require a quantum gravity description of the dynamics within the quantum gravity region. This reinforces the idea that unitarity is fundamentally a quantum gravity issue.

It's worth noting that many analyses in the literature discussing information tend to avoid considering dynamics near the singularity. For the reasons mentioned, we believe such discussions cannot provide a serious resolution to the apparent paradoxes involved (recall Sect. 14.6.3).

We can argue that the Hawking particles falling into the singularity strongly interact with the microscopic Planckian structure of spacetime, especially as they become infinitely blue-shifted during their journey. To explore this, let's focus on one of these particles and ignore the backreaction of the Hawking radiation. As mentioned earlier, the time it takes for the partner particle to hit the singularity is very short, so we can treat the particle as propagating freely along a geodesic in a fixed black hole geometry.

For simplicity, we will consider a massless particle (like a photon or graviton) in a Schwarzschild background. Let k^a denote the four-momentum of the particle. The symmetries of the background lead to the conservation of angular momentum $L \equiv k^a \psi_a$ and the Killing energy $E \equiv k^a \xi_a$, where ψ^a and ξ^a are the rotational and stationary Killing fields of Schwarzschild.

While one might argue that these quantities are only approximately conserved because the black hole is evaporating, the time scales for evaporation of macroscopic black holes are much longer than the time scales involved in the fall of the particles. Therefore, this approximation is reasonable for sufficiently massive black holes.

Using these constants of motion, we can calculate the frequency measured by a radially falling observer moving toward the singularity (for example, those normal to the $r =$ constant hypersurfaces). The result is:

$$\omega^2(r) = \frac{\ell^2}{r^2} + \frac{r}{2M} E^2 + \mathcal{O}\left(\frac{r^2}{M^2}\right). \tag{14.13}$$

The divergence as we approach $r = 0$ is generic for any regular observer measuring ω. The important point is that the Hawking spectrum includes particles with non-zero angular momentum L [73]. This means that the frequency of these particles is infinitely blue-shifted for local freely falling observers, which is expected due to angular momentum conservation and the shrinking length of the orbits of ψ. Only perfectly spherically symmetric modes with $L = 0$ would become infrared at the singularity. However, this conclusion is not very significant if the black hole rotates, or if we consider that exact spherical symmetry inside a black hole is extremely rare.

As a result, the partners falling into the black hole develop Planckian wavelengths near the quantum gravity region, making interactions between the Hawking particles and the microscopic Planckian structure unavoidable.

A clear sign that the backreaction of the background to the presence of Hawking radiation becomes important near the singularity, regardless of how semiclassical the black hole appears from the outside, comes from the behavior of the renormalized expectation value of the energy-momentum tensor in the quantum state associated with the Hawking radiation. This can be calculated in an effective 2D model by focusing on $\ell = 0$ modes and neglecting backscattering [74]. The result shows that energy densities and pressure contributions from the quantum state diverge as we approach the singularity, specifically as $r \to 0$. These divergences are not due to the approximations used; they are also known to occur in the 4d case, even though we cannot provide analytic expressions for it. This divergence in the source of the semiclassical gravity description indicates that a complete quantum gravitational treatment is essential.

Additionally, similar to Page's argument about the external development of large fluctuations (as shown in Fig. 14.4), a full quantum gravitational treatment of individual contributions from modes in Hawking radiation—each with different quantum numbers—leads to the conclusion that strong quantum fluctuations will inevitably arise near the singularity. This means that the singularity becomes a region where effective field theory cannot be applied and where a semiclassical geometric interpretation is not possible (the grey region in Fig. 14.5).

14.7.2 Dynamics of Information in the Strong Quantum Gravity Regime

With this preliminary discussion in place, we can now describe the scenario proposed in [15]. As mentioned at the beginning of Sect. 14.5.2, the best way to understand black hole formation and evaporation is from the perspective of observers at infinity. This is why we will use the Penrose diagram in Fig. 14.8 as our framework.

Our first assumption is that there is a well-defined evolution across the quantum region inside the black hole, allowing us to represent the situation as shown in Fig. 14.8 [75–77]. In this context, we can use a "scattering theory" approach, where an in-state evolves into an out-state, even though, as we will discuss, the results cannot be expressed in the language of effective quantum field theory.

A maximally entangled Hawking pair (a, b) is created through the interaction of the gravitational field and the vacuum (see Fig. 14.6). Particle b escapes to $\mathscr{I}^+$ as Hawking radiation, carrying positive Killing energy. Meanwhile, the entangled partner a falls into the singularity with negative Killing energy, which reduces the Bondi mass of the black hole, though the details of this interaction can only be understood through a full quantum gravity calculation.

Energy conservation at $\mathscr{I}^+$ means that particle a effectively "annihilates part of the mass" contributed by the matter source that formed the black hole (the initial cloud f in (14.12)). The maximal entanglement between the pair (a, b) cannot be maintained due to a property of entanglement known as monogamy. Specifically, the quantum gravity interaction between a and the Planckian degrees of freedom in the quantum gravity region leads to this entanglement being transferred to the microscopic degrees of freedom.

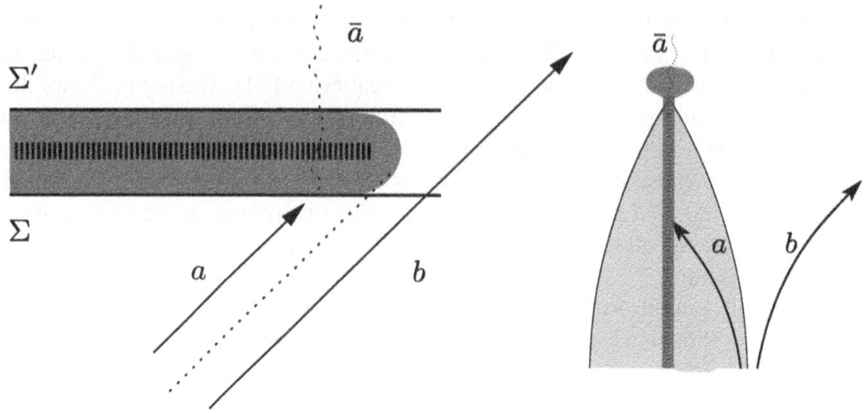

Fig. 14.6 A maximally entangled Hawking pair (a, b) is created through the interaction of the gravitational field and the vacuum. In the quantum gravity region, the entanglement between a and b is transferred to a Planckian defect a and b. The defect a emerges from the quantum region into a macroscopically flat vacuum environment. This transfer occurs due to the monogamy of entanglement, while the lack of energy in the final degree of freedom results from the infinite degeneracy of the flat vacuum in quantum gravity. The left image illustrates the process in a conformal representation. The right image shows the process with a more accurate depiction of the black hole horizon area. *Reused from Alejandro Perez and Daniel Sudarsky, AVS Quantum Sci. 4 (4), 10.1116/5.0091962, (2022) with permission of AIP Publishing*

In a theory where the condition (14.11) and energy conservation at $\mathscr{I}^+$ hold, only about one Planck mass remains at the end of the evaporation process. This suggests that the entanglement must be transferred to very low-energy excitations. Given the vast number of "defect-like" degrees of freedom at the Planck scale, it becomes clear that these correlations will likely shift to the defects that define the Planckian microscopic degeneracy of the vacuum, as suggested by (14.11) and the accompanying discussion.

Now, recall the structure of the Minkowski vacuum state $|0\rangle$ of a quantum field when expressed in terms of the modes that correspond to Rindler accelerated observers. These observers have their own notion of positive frequency. The vacuum state can be written as:

$$|0\rangle = \prod_k \left(\sum_n \exp\left(-n\frac{\pi\omega_k}{a}\right) |n, k\rangle_R \otimes |n, k\rangle_L \right), \tag{14.14}$$

where $|n, k\rangle_L$ and $|n, k\rangle_R$ represent the particle modes as seen by an accelerated observer with uniform acceleration a, on the left and right sides of the Rindler wedge, with n particles in the mode with wave number k [78].

From this expansion, we can see that even though we have a pure state (if we define the density matrix as $|0\rangle\langle 0|$), tracing over one of the two wedges produces a reduced density matrix that represents a thermal state with temperature $T = \frac{a}{2\pi}$.

Building on the analogy from the previous expression, the idea we propose about the purification of information in black hole (BH) evaporation can be roughly represented (this is not a precise equation) as:

$$|\Psi_{\text{final}}\rangle = \mathbf{U}\,|\text{flat}, 0\rangle \otimes \hat{C}(f)\,|0_{\text{in}}, 0\rangle = \prod_k \left(\sum_n \exp\left(-\frac{\beta}{2} n\omega_k\right) |\text{flat}, n\rangle \otimes a_k^{\dagger}\,|0_{\text{out}}, n\rangle \right), \tag{14.15}$$

where the unitary evolution operator $\mathbf{U}$ acts on the initial state (for example, defined on Σ_1 in Fig. 14.2). Here, $|k, n\rangle \equiv a_k^{\dagger}\,|0_{\text{out}}, n\rangle$ represents a Hawking particle in the out-vacuum $|0_{\text{out}}, n\rangle$, which is compatible with the flat quantum geometry state $|\text{flat}, n\rangle$.

The final state on the right is a pure state where the initial correlations between Hawking partners (essentially vacuum quantum field correlations in $|0_{\text{in}}, 0\rangle$) have been transferred to correlations between the Hawking quanta $|k, n\rangle$ and the many quantum flat geometries $|\text{flat}, n\rangle$ (this final state can be thought of as reflecting the state on Σ_2 in Fig. 14.2).

At the end of the evaporation, there are no localized remnants hiding the vast degeneracy; instead, there is a large superposition of states that are distinct in the fundamental quantum gravity Hilbert space but appear the same to low-energy observers (as illustrated in Fig. 14.8). These degrees of freedom cannot be captured by any effective description using smooth fields (such as effective quantum field theory) because they are fundamentally discrete in nature.

It's important to highlight a key feature of the degrees of freedom entangled with Hawking radiation. These degrees of freedom carry nearly zero or exactly zero energy, as stated in Eq. (14.11). However, they should not be viewed as low-energy particle-like excitations linked to very long wavelengths. Instead, they are "exotic" degrees of freedom that do not fit into the usual framework of effective field theories, meaning they don't follow the Einstein-Planck relation $E = \hbar\omega$ with a corresponding wavelength $\lambda \sim 1/E$.

To illustrate this, consider a familiar example: a non-relativistic charged particle in a two-dimensional infinite conductor within a uniform magnetic field. The energy levels are given by Landau levels, $E_n = \hbar\omega_{\text{B}}(n + 1/2)$, where ω_{B} is the Bohr magneton frequency, and these levels are infinitely degenerate. In this case, we have variables that are cyclic and do not appear in the Hamiltonian, allowing us to create wave packets that are localized in one variable without changing the system's energy. This scenario, studied in full detail in Sect. 14.8.2, exemplifies a potential loss of information similar to what we propose here.

Returning to Hawking radiation, these degrees of freedom are described as Planckian defects but don't actually carry Planckian energy. The relationship $E \sim \hbar\omega$ applies only under certain conditions, which are often met but not always. In quantum gravity, particularly in the context of loop quantum gravity, the defects we discuss arise from building a background geometry with elementary Planckian units. This construction reveals degrees of freedom that have no weight. While the usual uncertainty principle might hold for some emerging degrees of freedom, evidence suggests that this is not true for all the degrees of freedom involved in quantum gravity.

At this stage of developing different approaches to quantum gravity, it's not possible to test the proposed scenario in a fully realistic gravitational collapse situation. However, we can show that the essential elements needed for this scenario are already present in existing models.

We've noted the significant number of microscopic states compatible with a single macroscopic black hole geometry, as seen in black hole entropy calculations based on the isolated horizon boundary condition. Similar degeneracies of micro-states also appear in simplified models of black holes and cosmology. In these cases, it can be shown that a decoherence process occurs during evolution across the would-be singularity, where low-energy macroscopic degrees of freedom interact with microscopic "defect-like" degrees of freedom.

To start this process, we need to choose an initial state carefully so that macroscopic features are not entangled with the microscopic defects. Specifically, we select a "special" initial state that makes the entanglement entropy, defined as the Von Neumann entropy of the reduced density matrix after tracing over the defects, equal to zero. It turns out that, generically, regardless of the initial state and the matter involved, entanglement with the microscopic degrees of freedom develops as we evolve through the quantum region. This causes the entanglement entropy to grow: if we assume that macroscopic observers are not sensitive to the microscopic defects, pure states evolve into mixed states for them. However, if we consider all degrees of freedom, the evolution remains unitary across the quantum gravity region.

This idea can also be applied to spherically symmetric and stationary black hole models studied in loop quantum gravity. These models have microscopic degrees of freedom similar to those in cosmological models, and even though they don't account for black hole evaporation, they do allow for considering Hawking radiation from a test field, making the analogy with real physics clearer. Technically, the situation resembles the cosmological case because the interior of such black hole models is similar to a homogeneous (but not isotropic) cosmological model, known as a Kantowski-Sachs model.

Even when the homogeneity is disrupted by matter falling into the black hole, the model remains a useful approximation. This is especially true for the spherically symmetric Hawking partner a, which can be effectively modeled as a stationary matter contribution due to the extreme redshift behavior of the $L = 0$ modes. Thus, the quantum dynamics of both the geometry and the falling partner a can be represented using Kantowski-Sachs models. The concepts of entanglement monogamy and nontrivial quantum interactions between a and the microscopic degrees of freedom in the quantum gravity region lead to the type of decoherence central to our scenario.

14.7.3 *Information Density Near the Past and the Future of the Singularity*

The renormalization of the energy-momentum tensor in the in-vacuum state (leading to Hawking radiation at $\mathscr{I}^+$) on a fixed background black hole spacetime has some ambiguities and is difficult to compute explicitly in 4d [79, 80]. However, 2d models—which neglect the non-spherical field modes and backscattering—provide qualitatively reliable information about certain semiclassical questions. In particular, the renormalization of the energy-momentum tensor for a conformally invariant scalar field can be achieved explicitly [74].

Focusing on the black hole horizon with normal k^a shows that $\langle T_{ab}\rangle k^a k^b < 0$, i.e., there is a flux of negative energy across the horizon that violates the energy condition of the classical area theorem and is thus responsible for the black hole's evaporation.

This feature makes the in-going Vaidya metric

$$ds^2 = -\left(1 - \frac{2M(v)}{r}\right) dv^2 + 2dvdr + r^2 d\Omega^2 \tag{14.16}$$

with its $T_{ab} = \dot{M}/(4\pi r^2) dv_a dv_b$, a natural model to describe a poor's man version of the semiclassical back reaction of the Hawking radiation on the black hole geometry. More precisely, if one assumes that, after being created by a shockwave with initial mass M_0, the mass evolves according to the Hawking radiation flux [81]

$$\frac{dM}{dv} = -\frac{|\beta|}{M^2}, \tag{14.17}$$

with β order one (if all the other quantities are written in Planck units). We are interested in the spacelike surface $r = 1$ (Planck radius) to the past of the singularity extending from the event of the formation of the singularity when $M = M_0$ (at the event where the shockwave collapses to zero radius) and where we set the advanced time $v = 0$ up to advanced time v. Let us call this hyper-surface Σ (see Fig. 14.6), and write its volume element

$$dV_\Sigma = 4\pi\sqrt{2M(v) - 1}dv = -\frac{4\pi}{|\beta|}\sqrt{2M - 1}\, M^2 dM, \tag{14.18}$$

where we used the flux law (14.17) to parametrize the evolution directly in terms of the mass M at advanced time v. In the case of a real black hole the generalized second law [42, 43] requires that

$$\delta\left(\frac{A(v)}{4} + S_\Sigma\right) \geq 0, \tag{14.19}$$

where S_Σ is the entropy of the Hawking radiation falling into the singularity (maximally correlated to the radiation at $\mathscr{I}^+$). In the reversible (least entropic) scenario $dS_\Sigma = -dA/4 = -8\pi M dM$. Using the previous two equations we can calculate the minimal entropy density on Σ

$$\frac{dS_\Sigma}{dV_\Sigma} = \frac{2|\beta|}{M\sqrt{2M-1}}. \tag{14.20}$$

We conclude that the entropy density near the singularity (when approached from its past) grows monotonically as the black hole evaporates reaching a maximum Planckian bit-density (of order one in Planck units) at $M = 1$. This implies that the total information is extremely diluted for a macroscopic black hole in the portions of Σ away from the final evaporation event.[9]

We now try to estimate the spacetime volume to the future of the singularity needed to hold enough bits to purify the Hawking radiation. Without a solid fundamental understanding, we can use here an heuristic estimate based on the the idea that one can code a bit of information in each Planck volume (this would be the case in loop quantum gravity where volume quantum numbers have a two-fold degeneracy [61]). This allows us to gather some geometric information about the transition from the past of the quantum gravity region, represented in a time slice Σ (see Fig. 14.7), to its future where the black hole has evaporated. In this future area, the quantum gravity region becomes "visible" to outside observers and represents a kind of "naked" singularity.

The question arises: how large must this "naked" singularity region be, as measured on a time slice Σ' (see Fig. 14.7). To address this, we first note that the volume of the shaded region in the past slice Σ depends on the specifics of the evaporation process and quantum gravity effects in the non-semiclassical regime. However, simple dimensional analysis—along with the idea that this volume should diverge as $\hbar \to 0$ (the classical limit)—suggests that for a large initial mass M, the volume behaves like:

$$V(\Sigma) \propto M^{3+\alpha},$$

where the proportionality constant and $\alpha > 0$ depend on the dynamics inside (for example, $\alpha = 1/2$ when modeling evaporation with an advanced Vaidya metric as (14.18) implies).

Next, we can estimate the volume of the shaded region in the flat hypersurface Σ' by ensuring that it has at least as many volume "bits" as needed to purify the

[9] This is not surprising as the Hawking quanta are emitted very slowly for macroscopic black holes over the M^3 lifetime of the black hole. Some authors argue that the quantum gravity evolution across the singularity can therefore be well approximated by symmetry reduced models [82] or black hole to white hole transitions [83] where a few degrees of freedom are involved. We disagree with this perspective on at least two grounds: on the one hand the sparse nature of the Hawking flux does not imply that its quantum back reaction would be negligible near the singularity (recall discussion in Sect. 14.7.1), on the other hand, there are no symmetric microscopic states in the quantum gravity framework justifying the validity of the mini superspace models at the Planck scale.

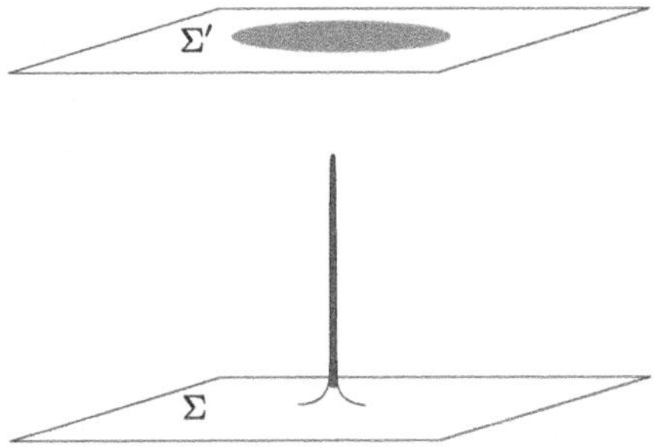

Fig. 14.7 A qualitative representation of the Riemannian geometry of Σ and Σ' of Fig. 14.6. The shaded regions are those 'touching' the quantum region. *Reused from Alejandro Perez and Daniel Sudarsky, AVS Quantum Sci. 4 (4), 10.1116/5.0091962, (2022) with permission of AIP Publishing*

radiation emitted during the Hawking evaporation. For a "simple" black hole—one that forms quickly and remains isolated throughout evaporation—the equality in Eq. (14.3) holds. Thus, the number of bits is given by the initial black hole entropy $A(M)$, where $A(M)$ is the area of the black hole at retarded time u_0 (see Fig. 14.7). This leads to:

$$V(\Sigma') \propto M^2.$$

From this, we find a characteristic size $L = V(\Sigma')^{1/3} \approx 10^{-11}\left(\frac{M}{M_\odot}\right)^{2/3}$ meters for the naked "would-be-singularity." For a black hole with $M = 10^{15}\,g$ (like primordial black holes that have just finished evaporating), this gives a size of about $10^{-2}\ell_{LHC}$, where ℓ_{LHC} is the smallest scale detectable by the Large Hadron Collider today.

The lifetime $\tau = u_2 - u_1$ of the naked region, as seen from $\mathscr{I}^+$, would also be of a similar order in geometric units. Thus, while the "naked-singularity" region is macroscopic compared to the fundamental Planck scale, the size of the naked event can be quite small on cosmological or particle physics scales. The number of degrees of freedom needed to purify the Hawking radiation occupies a tiny volume in relation to that available inside a black hole, i.e., $V(\Sigma') \ll V(\Sigma)$. Consequently, the final quantum gravity event where purification occurs could be a very brief and tiny process. This estimate assumes the black hole has a simple formation history, but the extent of this region can vary widely depending on that history, reflecting the idea that information is not destroyed.

Finally, note that in this discussion we have so far neglected an important source of quantum fluctuations in the final evaporation event: the effect of the recoil of Hawking particles, reviewed in Sect. 14.5.1. Including this effect, the final state is expected to be a superposition of quantum gravity regions Σ' (as described here) with a space-time location distributed in an astronomically large region of size M^2 (as illustrated in Fig. 14.4).

14.7.4 On the Irreversibility of the Evaporating Process

When looking at the final pure state $|\Psi_{\text{final}}\rangle$ defined in Eq. (14.15), we can express the pure final density matrix:

$$\rho_{\text{final}} = |\Psi_{\text{final}}\rangle \langle \Psi_{\text{final}}| = \mathbf{U} |\Psi_{\text{initial}}\rangle \langle \Psi_{\text{initial}}| \mathbf{U}^\dagger,$$

which shows the unitary evolution of the initial density matrix made from $|\Psi_{\text{initial}}\rangle$ in Eq. (14.12). If we trace out the defect degrees of freedom in the geometry, we find that $\rho_{\text{final}}^{\text{reduced}}(\text{matter})$ is the mixed state representing the Hawking radiation when ignoring the microscopic de degrees of freedom that purify it (which low-energy observers naturally do). The Von Neumann entropy $S(\rho_{\text{final}}^{\text{reduced}}(\text{matter}))$ represents the thermal entropy of the Hawking radiation, satisfying the generalized second law in Eq. (14.2).

Alternatively, starting from the final pure state ρ_{final}, we can trace out the matter degrees of freedom to get $\rho_{\text{final}}^{\text{reduced}}(\text{flat})$. This makes sense because late observers at $\mathscr{I}^+$ who are in retarded time $u > u_0$ may not even know about the prior existence and evaporation of the black hole. Their relevant physics can be described by either the pure density matrix ρ_{final} or the mixed one $\rho_{\text{final}}^{\text{reduced}}(\text{flat})$. The entropy of this mixed state is given by

$$S(\rho_{\text{final}}^{\text{reduced}}(\text{flat})) = S(\rho_{\text{final}}^{\text{reduced}}(\text{matter})), \tag{14.21}$$

due to the purity of the original state. From their viewpoint, assuming they can access the fundamental degrees of freedom, the state representing flat spacetime is mixed and contains a certain amount of entropy because of the black hole's evaporation history. This leads us to conclude that as black holes form and evaporate in our universe, the state of spacetime ages, with its age measured by the value of the entropy in Eq. (14.21).

14.7.5 On the Relation with Traditional Remnant Scenarios

An alternative to the scenario we present here is the possibility that the purifying degrees of freedom are associated with internal degrees of freedom of a particle-like remnant of the black hole (see [84] and references therein). It has been proposed that the information in the remnant could leak out via the radiation of very low energy particles, with total energy $E_{\text{late}} \approx m_p$ available for purification. From this it is possible to roughly estimate the time needed for purification to be of the order of $\tau_p \approx (M/m_p)^4$. This suggests that the point-like remnant lasts for a very long time and has a huge internal degeneracy. It is believe that this is problematic in effective quantum field theory due to the high probability for pair production associated with such objects [85]. However, it's unclear why this effective field theory description would lead to these inconsistencies.

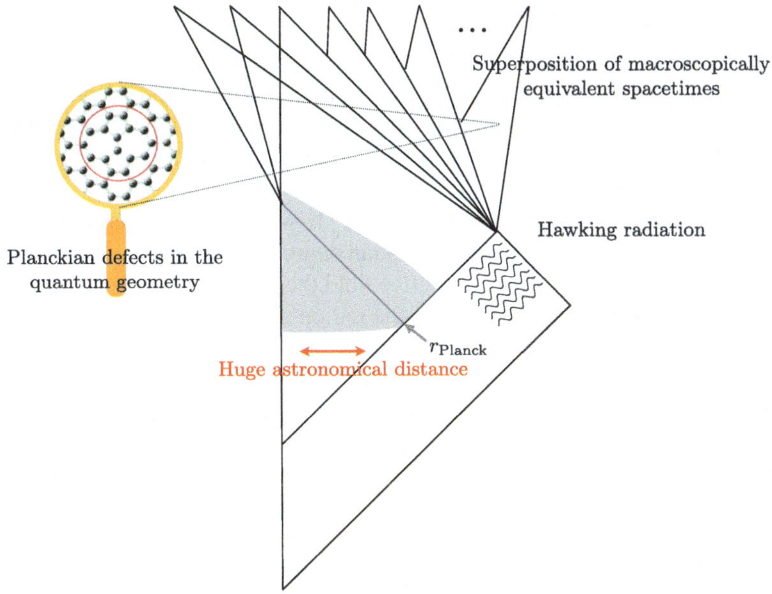

Fig. 14.8 Penrose diagram illustrating the scenario. The shaded area represents the "would-be-singularity", where there are significant fluctuations in geometry and fields. Here, the low-energy degrees of freedom from Hawking pairs interact with fundamental Planckian degrees of freedom. *Reused from Alejandro Perez and Daniel Sudarsky, AVS Quantum Sci. 4 (4), 10.1116/5.0091962, (2022) with permission of AIP Publishing*

These issues with high production probabilities arise only if we treat the remnant as a point particle in effective field theory, which seems questionable given the complex internal geometry of a black hole. If remnants exist, they would likely have a significant internal structure with a non-classical geometry (all this is related to our discussion in Sect. 14.5.3).

Note that the scenario described here is not that of a traditional remnant. The purifying degrees of freedom are not confined to a single point-like object; instead, they are spread out into the future of the quantum gravity region, forming the microscopic structure of the final state. These degrees of freedom cannot be captured by effective field theory; they carry information but have (close to) zero energy, recall Eq. (14.11), yet they are microscopic granular defects. Their existence is expected in loop quantum gravity and in any quantum gravity framework where low-energy spacetime geometry arises from the collective behavior of Planckian structures (Fig. 14.8)

14.8 Educated Models

The general scheme we propose is based on the hypothesis (supported by certain approaches to quantum gravity) that the nature of the fundamental degrees of freedom is discrete. Even though such discreteness is a basic feature of loop quantum gravity, the latter approach plays only a motivational role in our scenario (our arguments remain valid even if loop quantum gravity is shown to be incorrect). The central assumption is that, on the one hand, Planckian degrees of freedom vastly outnumber those that can be accounted for in effective field theory, and on the other hand, they cannot be neglected, as the dynamics of the region near the singularity are essential in the present discussion.

Under low-energy circumstances, the collective behavior of the fundamental building blocks can be described in terms of the mathematical continuous idealization of smooth quantum fields and spacetime geometry. Gravitational collapse and black hole formation lead to the breaking of the continuum approximation near the singularity. On entropic grounds, decoherence with the underlying fundamental granularity degrades information that can no longer be recovered in the continuum approximation. Energy conservation requires that such decoherence occurs with almost no dissipation, which we argue is possible if the microscopic structure of spacetime and matter allows for a large ground-state microscopic degeneracy.

One may ask whether such features can actually be realized in more familiar systems where explicit calculations are available, or where there is a link to actual, familiar physical systems. Here, we briefly review a series of models that illustrate that this is indeed the case. We begin with the simple Unruh model of interaction between two particles and a hidden (baby-universe) sector. In Sect. 14.8.2, we show how a charged particle in a constant magnetic field interacting with another external particle provides a concrete (and more realistic) example of decoherence without dissipation; and in Sect. 14.8.3, we construct a discrete model with a $1+1$ field-theoretic continuum limit that exhibits the desired behavior. We end this section with examples where the hidden sector emerges from the quantization of geometry *á la* loop quantum gravity, first in cosmology (Sect. 14.8.4), and then for spherically symmetric black holes (Sect. 14.8.5).

14.8.1 Decoherence Without Dissipation: The Unruh Model

The article [14] discusses a common belief in the study of the quantum behavior of black holes, suggested by Banks et al. [8], which is the following: black holes, by destroying the coherence of quantum states outside them, should lead to energy non-conservation. Over the years, this idea has become a profound belief for some people. However, Unruh and Wald [11] have pointed out some flaws in the arguments leading to this affirmation. Unruh proposes in this paper a simple quantum mechanical model illustrating that decoherence can occur without dissipation. The model consists of two particles that scatter by interacting through a potential term that depends on a "hidden system". The latter is more precisely a spin $1/2$ bath, where spins represent generic two-level quantum systems. During the process, the energy carried by the two particles remains conserved, while the reduced density matrix (obtained by ignoring the spin bath) evolves from a pure state to a mixed one due to the entanglement that the interaction created with the hidden sector. This model is inspired by the spin decoherence concept from condensed matter physics, as explored by Stamp and Prokof'ev [86]. In the case of black holes, this hidden system could be, according to the author, regarded "as some sort of baby universe", but the idea naturally adapts to what we propose in this chapter.

Concretely, the Hamiltonian is

$$\hat{H} = \frac{\hat{p}_1^2 + p_2^2}{2m} + \delta(\hat{x}_1 - \hat{x}_2) \sum_i \hat{s}_i^z, \tag{14.22}$$

where $\hat{p}_1$ and $\hat{x}_1$ (respectively $\hat{p}_2$ and $\hat{x}_2$) is the momentum and position operator of the first (respectively second) particle, and $\hat{s}_i^z$ is the z component of the ith spin operator. The two particles only interact when they come into contact. Moreover, we can see that the first term, corresponding to the particles energy before and after interaction, commutes with the total Hamiltonian. Their energy is therefore conserved.

An initial product state of the form

$$|\psi\rangle = \int dx_1 dx_2\, \lambda_1(x_1)\lambda_2(x_2)\, |x_1\rangle\, |x_2\rangle \prod_i |s_i^x\rangle\,, \tag{14.23}$$

where λ_1 and λ_2 are Gaussian functions, is such that the reduced density matrix obtained from tracing out the spins is pure. A simple calculation shows that after the scattering, the reduced density matrix representing the particle ends up in a mixed state. This indicates that it undergoes decoherence while still conserving its energy. Thus, decoherence does not necessarily require energy loss. This feature is applicable to familiar matter systems, like the one discussed in this paper, and may also extend to gravitational black hole systems with no inconsistency.

A key feature of Unruh's example is the large degeneracy of the asymptotic energy spectrum (note that the spin state does not contribute to energy in (14.22) when the particles are away from each other). This is a common feature of all the examples we will study in the rest of this section. The hidden sector, however, is very abstract in this simplistic model and has no realistic interpretation. The nature of the interaction is somewhat mysterious, with the spins manifesting their presence via a contact interaction between the particles. We show next that a very concrete and realistic quantum mechanical model can be realized via interactions with a charged particle in a constant magnetic field. In Sect. 14.8.3, we show that a lattice model with a $1+1$ dimensional quantum field theory as the continuum limit can be coupled consistently with a local hidden sector in a way that preserves the basic features of Unruh's toy model in the context of infinitely many degrees of freedom.

14.8.2 Decoherence Without Dissipation: Quantum Memory in the Landau Model

In the previous section, we reviewed a simple example introduced by Unruh, of a system with a hidden sector where information can be transferred without any energy cost. The strength of the model lies in its simplicity, while its weakness is the artificial nature of the hidden sector. After briefly mentioning it in Sect. 14.7.2, we now return to the study of a charged particle in a constant magnetic field, an elementary physical system that plays a central role in understanding a remarkably wide spectrum of physical phenomena, ranging from Landau diamagnetism [87], the quantum Hall effect [88], topological insulators [89], effective relativistic electronic properties in graphene [90], and more. We refer to this as the Landau model and demonstrate that it also represents an ideal physical system that can produce decoherence without dissipation.

We consider a charged particle confined in the $z = 0$ plane in $\mathbb{R}^3$ moving under the influence of a uniform magnetic field $\mathbf{B} = B\hat{z}$. In a gauge where $\mathbf{A} = -By\hat{x}$ the Hamiltonian becomes

$$H = \frac{1}{2m}\left[\left(p_x + \frac{eB}{c}y\right)^2 + p_y^2\right]. \tag{14.24}$$

As p_x is conserved (a constant of motion) we can write the previous Hamiltonian in the Landau form

$$H = \frac{p_y^2}{2m} + \frac{1}{2}m\omega^2(y - y_0)^2, \tag{14.25}$$

where we introduced $\omega = \frac{|e|B}{mc}$ and $y_0 = -\frac{cp_x}{eB}$. The dynamics of this system is therefore identical to the one of an harmonic oscillator. The Schroedinger equation is therefore

$$\left[\frac{p_y^2}{2m} + \frac{1}{2}m\omega(y - y_0)^2 - E_n\right]|\psi\rangle = 0, \tag{14.26}$$

which is solved by

$$\langle x, y|n, p_x\rangle = \left(\frac{m\omega}{\pi(2^n n!)^2}\right)^{1/4} e^{ixp_x} e^{-\frac{m\omega(y-y_0)^2}{2}} H_n(\sqrt{m\omega}(y - y_0)), \tag{14.27}$$

where H_n denotes the Hermite polynomials of order n, and

$$E_n = \omega\left(n + \frac{1}{2}\right), \tag{14.28}$$

called the Landau levels. Each energy level is infinitely degenerate. One can understand this from the fact that the charged particle classically moves around a circle whose center has coordinates (x_0, y_0) given by constants of motion (the second of which is given explicitly above, the first is $x_0 = p_y eB + x$ which can be explicitly shown to commute with H). The degeneracy corresponds then to placing the orbit at different locations which does not change the energy. One can easily show that the degeneracy is proportional to the area of the plane if the plane is finite (see [91] for details).

Not only are the energy levels degenerate, but they are also separated by an energy gap. It is therefore easy to imagine an interaction with another degree of freedom in a scattering process that would produce nontrivial entanglement between the interacting dynamical quantities without changing the energy of the scattering degree of freedom, provided the interaction is soft enough to avoid a transition between the Landau levels of the charged particle.

An explicit solvable situation consists of a free particle moving in the z direction (idealizing a plane wave-packet in $\mathbb{R}^3$ transversal to the plane where the Landau system is) with a contact interaction

$$H_{\text{int}} = \delta(Z)F(p_x), \tag{14.29}$$

where $F(p_x)$ is some given function of p_x and Z is the coordinate of the free outside particle. The total Hamiltonian is

$$H_{\text{tot}} = H + \frac{P_Z^2}{2M} + \delta(Z)F(p_x). \tag{14.30}$$

The dynamics of the plane-wave packet along the z-axis can be reduced to a one-dimensional problem with a delta function potential. The interaction term does not commute with the operator defining one of the coordinates of the center of the Landau orbits, x_0 (previously a constant of motion). The passage of the plane-wave particle will change the value of x_0, thereby producing entanglement with the charged particle. However, this will not change the energy of the Landau particle, thus conserving both the initial and final kinetic energies of the external particle.

Explicitly, solutions of the Schroedinger equation can be written as

$$|n, P_Z, p_x; t\rangle = e^{-iE(n,P_Z)t} \, |n, p_x\rangle \, |P_Z, p_x\rangle \,, \tag{14.31}$$

where $E(n, P_Z) = \hbar\omega(n + 1/2) + P_Z^2/(2M)$, $|n, p_x\rangle$ is given by (14.31) and the $|P_Z, p_x\rangle$ satisfying the eigenvalue equation (for the incident particle degrees of freedom)

$$\left[-\frac{1}{2M}\frac{\partial^2}{\partial Z^2} + \delta(Z)F(p_x)\right]\langle Z|P_Z, p_x\rangle = \left(\frac{P_Z^2}{2M}\right)\langle Z|P_Z, p_x\rangle \,, \tag{14.32}$$

which reduces the dynamics to that of the outside particle scattering in a one-dimensional Dirac delta potential. Elementary methods lead to the solution

$$\langle Z|P_Z, p_x\rangle = \begin{cases} e^{iP_Z Z} + \frac{1}{\frac{iP_Z}{MF(p_x)}-1} e^{-iP_Z Z} & \text{for } Z < 0 \\ \frac{1}{1-\frac{MF(p_x)}{iP_Z}} e^{iP_Z Z} & \text{for } Z > 0, \end{cases} \tag{14.33}$$

where there is a discontinuity in the Z-derivative of the wave function due to the presence of the delta function potential in the interaction.

One can superimpose stationary states (14.31) to form wave packets which are suitable for interpreting some of the important features of the physics involved. For instance, consider a superposition of states with $n = 0$ (the ground states in the charged particle degree of freedom) as

$$|\psi(t)\rangle = \int a_{x_0,y_0}(p_x) A_{Z_0,P_Z^0}(P_Z)\,|0, P_Z, p_x; t\rangle\, dp_x dP_Z, \tag{14.34}$$

with amplitudes $a_{x_0,y_0}(p_x)$ and $A_{Z_0,P_Z^0}(P_Z)$ such that the wave function of the full system $\langle x, y, Z|\psi(0)\rangle$ is initially peaked at some (x_0, y_0) in the 2-dimensional plate (as far as the degrees of freedom of the charged particle are concerned) while the outside particle wave function is peaked at $Z_0 \gg \Delta Z$ far to the right incident toward the plate with $P_Z^0 < 0$.

Under such conditions, the incident particle will move in time toward the plate in a manner practically indistinguishable from a regular free particle (as long as $\langle Z(t)\rangle$ remains much larger than the width of the packet, ΔZ the reduced dynamic is unitary). An observer, who is unaware of the presence of the plate (very far to the left in his lab), could describe this particle as a free particle and conclude (after performing suitable experiments with that particle) that unitarity is respected during this phase of the dynamics far to the right of the plate.

Notice that the non-locality of quantum mechanics implies that, no matter how far we are to the right, the exact solutions of the Schroedinger equation 'know' about the presence of the plate, as solutions depend explicitly on the quantum numbers of the charged particle (see discussion in Sect. 14.9). There are complex experiments that the observer could perform to detect such dependence, but in practice, there is no reason to consider performing them, as such observations would not seem necessary or natural from the experimenter perspective.

We chose a very specific type of interaction to make the problem explicitly solvable. However, the qualitative features of the previous model are generic, as can be seen from the perturbative treatment of a general interaction Hamiltonian of this form.

$$\alpha H_{\text{int}} = \alpha H_{\text{int}}(x, y; \mathbf{X}), \tag{14.35}$$

where x, y are the coordinates of the charged particle in the plate, $\mathbf{X}$ the coordinates of an exterior particle, and $\alpha \ll 1$ a dimensionless coupling. One makes the ansatz

$$|\Psi(t)\rangle = \int \sum_n c_n(p_x, \mathbf{P}, t) e^{-i\left(\omega(n+\frac{1}{2})+\frac{\mathbf{P}^2}{2M}\right)}\,|n, p_x\rangle\,|\mathbf{P}\rangle\, dp_x d^3\mathbf{P}, \tag{14.36}$$

where $|n, p_x\rangle$ is given by (14.31), and $\langle \mathbf{X}|\mathbf{P}\rangle = e^{i\mathbf{P}.\mathbf{X}}$. The Schroedinger equation can be written in integral form (suitable for perturbation theory)

$$c_n(p_x, \mathbf{P}, t) = ic_n(p_x, \mathbf{P}, t_0) - i\int \sum_{n'} \langle n, p_x, \mathbf{P}|\,\alpha H_{\text{int}}\,|n', p'_x, \mathbf{P}'\rangle \tag{14.37}$$

$$\times \int_{t_0}^{t} c_{n'}(p'_x, \mathbf{P}', t')e^{-it'\left(\omega_H(n'-n)+\frac{\mathbf{P}'^2-\mathbf{P}^2}{2M}\right)}\,|n', p'_x, \mathbf{P}'\rangle\, dp'_x d^3\mathbf{P}'dt'.$$

Taking the following initial conditions at $t_0 \to -\infty$

$$c_n(p_x, \mathbf{P}, -\infty) = \delta_{n,0}\delta(p_x - p_x^{\text{in}})\delta(\mathbf{P} - \mathbf{P}_{\text{in}}). \tag{14.38}$$

The transition amplitude (to first order in α) from $t = -\infty$ to $t = \infty$ is given by

$$\mathcal{M}_{|0, p_x^{\text{in}}, \mathbf{P}_{\text{in}}\rangle \to |n, p_x^{\text{out}}, \mathbf{P}_{\text{out}}\rangle}$$

$$= -i\,\langle n, p_x^{\text{out}}, \mathbf{P}_{\text{out}}|\, H_{\text{int}}\,|0, p_x^{\text{in}}, \mathbf{P}_{\text{in}}\rangle \int_{-\infty}^{+\infty} e^{-it\left(-\omega n+\frac{\mathbf{P}_{\text{in}}^2-\mathbf{P}_{\text{out}}^2}{2M}\right)}dt. \tag{14.39}$$

The transition amplitude vanishes if $\omega_H n - \frac{\mathbf{P}_0{}^2-\mathbf{P}^2}{2M} \neq 0$ as implied by energy conservation. This implies, in particular, that if $\frac{\mathbf{P}_0{}^2}{2M} < \omega$ (lower than the energy gap), the charged particle in the plate must remain in the ground state, and thus the interaction takes place elastically, without dissipation. Meanwhile, a nontrivial entanglement with the constants of motion at zero order, (x_0, y_0), will generally occur.

14.8.3 *Decoherence Without Dissipation: A Field Theoretic Model*

The previous two sections present systems with finitely many degrees of freedom. Here, we introduce a simple one-dimensional lattice model with a continuum field-theoretic limit in which the phenomenon of decoherence without energy cost persists. Aside from the inherent Lorentz invariance violation induced by the lattice, this model describes a quantum field theory involving a hidden sector that cannot be accounted for as an effective field theory.

Consider a chain of N identical oscillator of mass M connected by springs of spring constant K. The distance at rest between the oscillator is noted a, so the total length of the chain is $L = Na$. The displacement of the jth oscillator with respect to its equilibrium position at time t is noted $\varphi_j(t)$. The Hamiltonian is

$$H^{(0)} = \sum_{j=1}^{N}\left[\frac{p_j^2}{2M} + \frac{K}{2}(\varphi_j - \varphi_{j+1})^2\right], \tag{14.40}$$

and the equations of motion can be written as

$$-\frac{M}{Ka^2}\frac{d^2\varphi_j(t)}{dt^2}+\frac{(\varphi_{j+1}(t)+\varphi_j(t)-2\varphi_j(t))}{a^2}=0, \tag{14.41}$$

We impose periodic boundary condition $\varphi_1=\varphi_{N+1}$ for simplicity. The system is solved by decomposing it into normal modes

$$\varphi_j(t)=Ae^{i(qja-\omega_{q'}t)}, \quad \text{with} \quad q=\frac{2\pi n}{Na}, \quad n\in\left[-\frac{N}{2},\frac{N}{2}\right]\subset\mathbb{Z}, \tag{14.42}$$

and dispersion relation

$$\omega_q^2=2\frac{K}{M}(1-\cos(qa))=4\frac{K}{M}\sin^2\left(\frac{qa}{2}\right). \tag{14.43}$$

The range $q\in\left[-\frac{\pi}{a},\frac{\pi}{a}\right]$ is known as the first Brillouin zone (BZ). The limit $a\to 0$ while holding fixed $c^2\equiv Ka^2/M$ leads to the $1+1$ scalar field equation

$$\left(-\frac{1}{c^2}\frac{\partial^2}{\partial t^2}+\frac{\partial^2}{\partial x^2}\right)\varphi(t,x)=0 \tag{14.44}$$

for phonons with propagation velocity c. Deviations from linear dispersion relations caused by the lattice have been suggested in the literature as potential Lorentz invariance violations related to quantum gravity. In the present model, the source is clear: the background lattice provides a preferred notion of 'rest'.

The quantization is made by replacing the dynamical variables φ_j and p_j by operators satisfying the usual commutation relations

$$[\varphi_i,\varphi_j]=0 \qquad [p_i,p_j]=0 \qquad [\varphi_i,p_j]=i\delta_{i,j}. \tag{14.45}$$

The Fourier transformed[10] operators satisfy

$$[\varphi_q,\varphi_{q'}]=0 \quad [p_q,p_{q'}]=0 \quad [\varphi_q,p_{q'}]=\frac{La}{4\pi^2}i\delta_{q,-q'}. \tag{14.48}$$

[10] The discrete Fourier is defined as

$$f_q=\sum_{j=1}^{N}\frac{a}{2\pi}e^{-iqja}f_j \Leftrightarrow f_j=\sum_{q\in \mathrm{BZ}}\frac{2\pi}{L}e^{iqja}f_q, \tag{14.46}$$

with completeness relations

$$\sum_{j=1}^{N}e^{i(q-q')ja}=N\delta_{q,q'} \quad \text{and} \quad \sum_{q\in \mathrm{BZ}}e^{iq(j-j')a}=N\delta_{j,j'}. \tag{14.47}$$

The reality conditions imply $\varphi_q^\dagger = \varphi_{-q}$ and $p_q^\dagger = p_{-q}$. Introducing creation and annihilation operators

$$a_q := \frac{2\pi}{\sqrt{2M\omega_q La}}(M\omega_q\varphi_q + ip_q), \tag{14.49}$$

which satisfy the usual commutation relations $\left[a_{q'}, a_q^\dagger\right] = \delta_{q,q'}$. The phonon free Hamiltonian (14.40) becomes

$$H^{(0)} = \sum_q \omega_q\, a_q^\dagger a_q, \tag{14.50}$$

in its normal ordered version (which we chose for simplicity). We modify the free phonon Hamiltonian by adding additional (microscopic) degrees of freedom interacting with the oscillators according to

$$H = \sum_q \omega_q a_q^\dagger a_q + \frac{\alpha}{N}\sum_{j,q,q'} S_{z_j} e^{-i(q+q')ja}\sqrt{\omega_q\omega_{q'}}a_{-q'}^\dagger a_q, \tag{14.51}$$

where $\mathbf{S}_j$ are spins located at each lattice site and we assume $\alpha \ll 1$. The interaction has been suitably engineered to feet our illustrating purposes.

The system is simple enough to allow for a perturbative resolution of its dynamics (see [92] for details). One of its key features is that its ground state (as well as all energy levels) has a degeneracy of 2^N at zeroth order in α. Interestingly, a significant degree of degeneracy remains even when the interaction is taken into account. This is the hallmark of the model, which sharply contrasts with what is typically assumed in the context of quantum field theories (where the uniqueness of the vacuum is one of the central axioms of the algebraic approach to quantum field theory).

In particular, the ground state of this model remains degenerate to all orders in perturbation theory. The system exhibits residual entropy in the thermodynamic sense, violating some versions of the third law. This contradicts assumptions commonly made in standard quantum field theory, which are critical to certain approaches to the information problem [57]. This property is expected in quantum gravity, as argued in Sect. 14.6.2, and plays a role in the aging of the vacuum discussed in Sect. 14.7.4.

One can study the Von Newman entropy of the phonon subsystem—when tracing out the microscopic spins—in the idealized case of an incident phonon with momentum q and frequency $\omega \equiv \omega_q$. A lengthy perturbation theory calculation up to second order in α yields [92]

$$S(t,\omega) = \frac{2\alpha^2}{N}\left(\sum_{q'}\omega_{q'}\omega\left(\frac{\sin\left(\frac{t}{2}(\omega_{q'}-\omega)\right)}{\omega_{q'}-\omega}\right)^2 - (\omega t)^2\right) + \mathcal{O}(\alpha^3) \tag{14.52}$$

The Von Neumann entropy can grow in time and the time with a possible maximum given by $S \leq S_{\max} = N \log(2)$ given by the entropy of the maximally mixed state. This bound tends to infinity when $N \to \infty$ which would seem to suggest that the presence of the defects might be out of control in the continuum limit.

However, a more closer look shows the desired behaviour in the continuum. We note that the continuum limit can be taken in two inequivalent ways: One can first consider the limit $N \to \infty$ while keeping the maximal frequency ω_p constant. In this limit the sum (14.52) becomes and integral which (by appropriate change of variables can be written as)

$$S(t, \omega) = \frac{4\alpha^2 \omega}{\pi \omega_p} F(t, \omega), \tag{14.53}$$

with

$$F(t, \omega) = \int_0^{\pi/2} d\phi \, \omega_p^2 \sin(\phi) \left(\frac{\sin\left(\frac{t}{2}(\omega_p \sin(\phi) - \omega)\right)}{\omega_p \sin(\phi) - \omega} \right)^2. \tag{14.54}$$

The second way is the standard continuum limit where $N \to \infty$ and $a \to 0$ while $Na = L$ is kept constant. In this limit the maximal frequency ('Planck frequency') $\omega_p \to \infty$. The entropy can be approximated for large times $\omega t \approx 1/\alpha \gg 1$ as

$$S(t, \omega) \approx 2\alpha^2 \left(\frac{\omega}{\omega_p} \right) \omega t, \tag{14.55}$$

which tends to *zero* in the standard continuum limit. Thus we see that in the second version of the continuum limit (which we can call *naive*) the effect of the microscopic degrees of freedom disappear. A naive analysis that would not take into account the existence of an underlying discrete structure would lead to the standard unitarity of the effective description that ignores the spin defects. While the full theory that includes the spin defects in the Hilbert space is unitary, unitarity will seem to fail in the effective quantum field theory treatment when the energy scales approach the fundamental scale (e.g. as near a singularity inside a black hole).

14.8.4 Apparent Information Loss Due to Planckian Discreteness in Cosmology

In the previous three sections, we presented simple examples in which information is degraded through decoherence with degrees of freedom that do not contribute to the energy budget, thereby causing no dissipation. The first example by Unruh is a mathematical construction; the second example, by contrast, is closer to a real physical situation; and the third incorporates local degrees of freedom in the continuum limit. However, none of these examples relate in any obvious way to the gravitational realm, where we would like this phenomenon to apply. In this section, we show that it is straightforward to find a quantum gravity model exhibiting the desired behavior within the simplified context of quantum symmetry-reduced models of cosmology.

To illustrate our point, we present here a loop quantum cosmology model defined within the framework of unimodular gravity (following [67, 68]). The main reason for this choice is that, in the cosmological Friedmann-Lemaître-Robertson-Walker (FLRW) context, it effectively addresses the problem of time. Specifically, the theory provides a preferred time evolution and is described by a Schroedinger-like equation in which the universe's states evolve using a unitary operator. The line element is written as

$$ds^2 = -N^2 dt^2 + a^2(t) d\mathbf{x}^2. \tag{14.56}$$

Einstein-Hilbert action supplemented with the unimodular constraint becomes

$$S = -\kappa^{-1} \int \left(\sqrt{|g|} R + \lambda(\sqrt{|g|} - 1)\right) dx^4 \tag{14.57}$$

where $\kappa = 16\pi G$, and λ is a Lagrange multiplier imposing the unimodular constraint $\sqrt{|g|} - 1 = 0$, and we have put an overall minus sign in front of the action for later convenience. Specializing to the FLRW metric (14.56) one gets

$$\begin{aligned} S &= \kappa^{-1} V_0 \int \left(6\frac{a\dot{a}^2}{N} - \lambda(N|a|^3 - 1)\right) dt \\ &= \kappa^{-1} V_0 \int 6a^4 \dot{a}^2 ds \\ &= \frac{1}{2} \int m\dot{x}^2 ds, \end{aligned} \tag{14.58}$$

where total derivative terms have been eliminated, the 3-volume V_0 of a fiducial cell has been introduced, the constraint $N = |a|^{-3}$ has been solved in the second line, and in the third line we introduced the variable $x = a^3 V_0 / l_p^2$ (representing the physical volume of the fiducial cell with units of length l_p), which casts the action of pure gravity in the form of that of a non-relativistic particle with mass.

$$m \equiv \frac{\ell_p^3}{12\pi V_0} m_p, \tag{14.59}$$

The minus sign in front of (14.58) was chosen so that the kinetic term of the particle analog has the usual sign. Also notice that if we use comoving time $d\tau = -ds/|a|^3$ we have that the momentum relates directly to the Hubble rate, namely $p = -3m \frac{1}{|a|} \frac{da}{d\tau} = -3m\ \mathrm{sign}(a) H$. We introduce a scalar field as a matter model with the action

$$S_M = \frac{1}{2}\int \sqrt{|g|}\left(\nabla_a\phi\nabla^a\phi + U(\phi)\right). \tag{14.60}$$

Restricting to the zero mode $\phi(s)$ the action of gravity coupled to the scalar field is

$$S(x,\phi) = \int \left(\frac{1}{2}m\left(\frac{dx}{ds}\right)^2 - \frac{x^2}{2V_0}\ell_p^4\left(\frac{d\phi}{ds}\right)^2 - V_0 U(\phi)\right) ds, \tag{14.61}$$

which, in Hamiltonian form, becomes

$$S(x,\phi) = \int p\frac{dx}{ds} + p_\phi\frac{d\phi}{ds} - \left(\frac{p^2}{2m} - \frac{V_0 p_\phi^2}{2\ell_p^4 x^2} - V_0 U(\phi)\right) ds, \tag{14.62}$$

where

$$p_\phi = -\frac{\ell_p^4 x^2}{V_0}\frac{d\phi}{ds}. \tag{14.63}$$

For simplicity we will assume $U(\phi) = 0$ so that the Hamiltonian of our cosmological model becomes

$$H \equiv \frac{p^2}{2m} - \frac{V_0 p_\phi^2}{2\ell_p^4 x^2}. \tag{14.64}$$

For further use it is convenient to denote $H_g = p^2/(2m)$ the classical geometry contribution to the Hamiltonian, and $H_\phi = V_0 p_\phi^2/(2\ell_p^4 x^2)$ the matter part.
Quantization: The main input—leading to the discreteness of the variable x in analogy to the full loop quantum gravity theory—is to chose a representation of the canonical variables where the action of translations in configuration space are not weakly continuous. This is the consequence of working with a (non separable) Hilbert space $\mathscr{H}$ with inner product

$$\langle\psi_1|\psi_2\rangle = \lim_{\Delta\to+\infty}\frac{\int_{-\Delta}^{+\Delta}\overline{\psi_1(p)}\psi_2(p)dp}{2\Delta}. \tag{14.65}$$

As a consequence plane waves $\langle p|x\rangle = \exp(ipx)$ for $x \in \mathbb{R}$ are normalizable forming an orthonormal basis. One can check that

$$\langle x|y\rangle = \lim_{\Delta\to\infty}\frac{\sin(\Delta(x-y))}{\Delta(x-y)} = \delta_{x,y}, \tag{14.66}$$

with $\delta_{x,y}$ the Kronecker delta with $\delta_{x,x} = 1$ and zero when $x \neq y$. This implies that there is no translation infinitesimal generator p (as $|x\rangle$ and $|x+\epsilon\rangle$ are orthogonal states for all $\epsilon \neq 0$) but only the operators corresponding to finite x translations exist [76, 93]. It is still convenient to write the translation operators as

$$\exp(i2kp)\Psi(x) = \Psi(x-4k). \tag{14.67}$$

Consider the one parameter family of lattices Γ_λ^ϵ given by the form $x = 4n\lambda\ell_p + \epsilon$ with $\epsilon \in [0, 4\lambda\ell_p)$ and $n \in \mathbb{N}$. If we focus on the class of finite translations mapping points of Γ_λ^ϵ into themselves (i.e. those of the form (14.67) with $k = q\lambda\ell_p$ and $q \in \mathbb{Z}$) then we can introduce 'plane wave' states $|p_0; \Gamma_\lambda^\epsilon\rangle \in \mathscr{H}$ whose wave functions are given by $\Psi_{p_0}(x) \equiv \langle x|p_0; \Gamma_\lambda^\epsilon\rangle = \exp\left(-i\frac{p_0 x}{2}\right)\delta_{\Gamma_\lambda^\epsilon}(x)$ where $\delta_{\Gamma_\lambda^\epsilon}(x) = 1$ for $x \in \Gamma_\lambda^\epsilon$ and vanishes otherwise. Using (14.67) it is immediate to check that these are eigenstates of the finite translations preserving the lattice Γ_λ^ϵ, namely

$$\exp(i2kb)\,|p_0; \Gamma_\lambda^\epsilon\rangle = \exp\left(i2kp_0\right)|p_0; \Gamma_\lambda^\epsilon\rangle\,. \tag{14.68}$$

Notice, that the eigenvalues are independent of the parameter ϵ. i.e. they are infinitely degenerate (a key point in what follows).

The gravity part of the Hamiltonian (14.64) cannot be written as such in the quantum theory. In order for it to be a well defined operator in $\mathscr{H}$ a 'regularization' in terms of finite translation operators is needed. Choosing $k \equiv \pm 2\lambda\ell_p$ in (14.67), where λ is the area gap in Planck units [93], one can write

$$\widehat{H}_{\mathrm{g}} \equiv \frac{1}{2m}\frac{\sin^2\left(\lambda\ell_p\, p\right)}{\lambda^2\ell_p^2}, \tag{14.69}$$

which coincides with the geometry term in (14.64) to leading (zero) order in ℓ_p^2. States (14.68) with $k = k_\lambda \equiv \lambda\ell_p$ diagonalize (14.69), i.e., they are eigenstates of the cosmological constant

$$\widehat{H}_{\mathrm{g}}\,|p_0; \Gamma_\lambda^\epsilon\rangle = E_\lambda(p_0)\,|p_0; \Gamma_\lambda^\epsilon\rangle\,, \tag{14.70}$$

with eigenvalues $E_\lambda(p_0) = \sin^2\left(\lambda\ell_p\, p_0\right)/(2m\lambda^2\ell_p^2)$ (Fig. 14.9).

For the dynamics produced by (14.64) the scalar field momentum p_ϕ is a constant of motion and we have that the conserved Hamiltonian satisfies $H \to H_{\mathrm{g}} = \Lambda V_0/\kappa$ when $x \to \infty$ with Λ the value of the cosmological constant that dominates when the universe is large. Explicitly the cosmological constant depends on the asymptotic value of p_0 and it is given by

$$\Lambda_\lambda(p_0) = \frac{\kappa}{2V_0 m}\frac{\sin^2\left(\lambda\ell_p\, p_0\right)}{\lambda^2\ell_p^2}, \tag{14.71}$$

which is positive definite and bounded from above by the maximum value $\Lambda_{\max} = 96\pi^2/(\ell_p^2\lambda^2)$. Negative Λ solutions are possible when matter is added [94].

Notice that the energy eigenvalues do not depend on $\epsilon \in [0, 4\lambda\ell_p)$. Thus, the energy levels are infinitely degenerate. This is not a peculiarity of our model but a general property of the representation of the canonical commutation relations used in loop quantum cosmology. The infinite degeneracy of the energy eigenvalues should

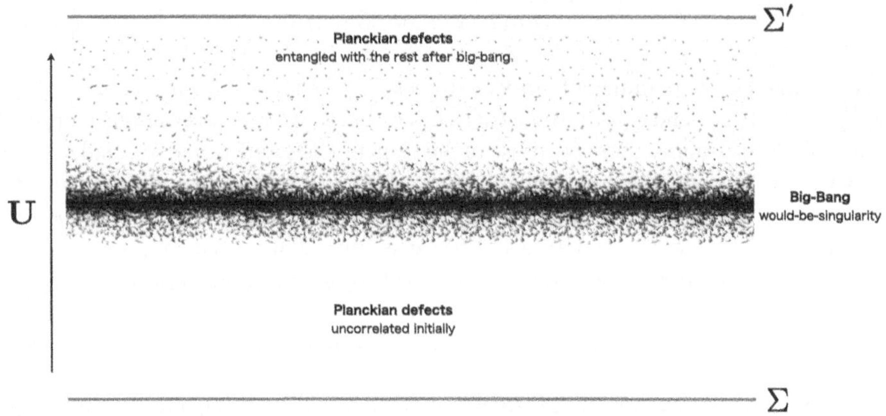

Fig. 14.9 The universe evolves across the strong quantum gravity region (the singularity from the classical perspective) unitarily. The process is fundamentally quantum and can only be fully understood through superpositions of different spacetime geometries differing in their microscopic Planck scale details. Information is lost in coarse grained descriptions. *Reused from Alejandro Perez and Daniel Sudarsky, AVS Quantum Sci. 4 (4), 10.1116/5.0091962, (2022) with permission of AIP Publishing*

be understood as indicating the existence of additional quantum degrees of freedom in loop quantum cosmology (for more discussion of this point see [68]).

This is a consequence of the specific representation we have used for quantization. If we had followed the standard Schroedinger representation with $p = -i\hbar\partial/\partial x$, we would have found only the two-fold degeneracy $E(p) = E(-p)$, corresponding to the two distinct physical situations associated with a contracting and/or expanding universe. The large abundance of additional degrees of freedom in the present model is due to the underlying (microscopic) discreteness of the volume of the universe (14.66).

Low-energy observers (defined in this toy context as those who would use the Wheeler-DeWitt quantization in quantum cosmology) are insensitive to the vast degeneracy of energy eigenstates (cosmological constant here). All these infinitely many states for a single eigenvalue of Λ in the loop quantum cosmology representation must be considered equivalent, except for the two-fold degeneracy mentioned above. In other words, coarse observers can only distinguish the value of the cosmological constant (energy) and whether the universe is expanding or contracting. Such coarse-graining generally leads to the type of decoherence associated with an apparent loss of information, as we show below.

But first, let us briefly discuss the effect of coupling the previous pure-gravity model to matter. The matter contribution $H_\phi = V_0 p_\phi^2/(2\ell_p^4 x^2)$ to the Hamiltonian is regular for all lattices Γ_λ^ϵ with $\epsilon \neq 0$ (those that avoid the origin) [68] (for a discussion of a more sophisticated quantization avoiding the singularity at $x = 0$, see [93]). As the momentum p_ϕ is a constant of motion, quantum dynamics for the geometry sectors corresponds to the scattering of a particle in a 1-dimensional lattice with a $1/x^2$ potential. Such a situation can actually be dealt with numerically using basic methods [68]. We will simplify the discussion by replacing this potential with a $\delta_{x,0}$ potential, as the qualitative results are the same.

Therefore, in the toy model the Hamiltonian becomes

$$H \equiv \frac{1}{2m}\frac{\sin^2\left(\lambda\ell_p\, p\right)}{\lambda^2\ell_p^2} - p_\phi^2\delta_{x,0}, \tag{14.72}$$

We have added by hand a 'matter-term' that switches on only when the universe evolves through the *would-be-singularity* at the zero volume state. The key feature of the $\hat{H}_{\rm int}$ is that—as the more realistic matter $1/x^2$ Hamiltonian—it breaks translational invariance and in that it will lead to different dynamical evolution for different ϵ-sectors. For large universes $x \to \pm\infty$, the state of the universe is described by the eigenstates of the pure gravity Hamiltonian (14.70). Consider, the superposition of two such eigenstates but supported on different lattices Γ_λ^ϵ with $\epsilon = 0$ for the first one and $\epsilon = 2\lambda\ell_p$ for the second one. The eigenstates of (14.68) with eigenvalues $\exp(i2kp)$ will be denoted

$$|p,1\rangle \equiv |p;\Gamma_\lambda^0\rangle\,, \quad \text{and} \quad |p,2\rangle \equiv |p;\Gamma_\lambda^{2\lambda\ell_p}\rangle\,. \tag{14.73}$$

We refer to the underlying lattices as Γ^1 and Γ^2. A key point is that states on Γ^2 (superpositions of $|p,2\rangle$) can move freely because this lattice does not include the point $x = 0$. In contrast, states on Γ^1 (superpositions of $|p,1\rangle$) are influenced by the interaction at the big bang. This qualitatively reflects the same property seen with the scalar field $1/x^2$ potential, which arises from the unavoidable breaking of translation invariance in the matter Hamiltonian.

Therefore, the non-trivial scattering problem concerns only states on the lattice $\Gamma^1 = \{x = 2n\lambda\ell_p \mid n \in \mathbb{Z}\}$. The solution of the scattering problem for a universe initially contracting from the right is

$$|\psi_k\rangle = |x\rangle \begin{cases} e^{-i\frac{k}{2}x} + A(k)\, e^{i\frac{k}{2}x} & (x \geq 0) \\ B(k)\, e^{-i\frac{k}{2}x} & (x \leq 0), \end{cases} \tag{14.74}$$

where $x \in \Gamma^1$, and

$$A(k) = \frac{-i\Theta(k)}{1+i\Theta(k)}, \qquad B(k) = \frac{1}{1+i\Theta(k)}, \qquad \Theta(k) \equiv \frac{32\pi m p_\phi^2}{\sin(2k\lambda\ell_p)}.$$

Wave packets can then be constructed in terms of appropriate superpositions of these 'plane-wave' states. If one takes the initial state in-state of the form (valid for early times)

$$|\psi_{\rm in}, t\rangle = \frac{\pi}{\sqrt{2}\lambda\ell_p}\int dp\big(|p,1\rangle\,\psi(p) + |p,2\rangle\,\psi(p)\big)e^{-iE_\lambda(p)t}, \tag{14.75}$$

where $\psi(p)$ is a wave function picked at some $b = p_0$ value and $x = x_0$ and $E_\lambda(p)$ is given by (14.70). Notice that we are superimposing two wave packets supported on lattices Γ^1 and Γ^2 respectively.

Assuming that $\psi(p)$ is highly peaked at some p_0 so that we can substitute the integration variables p and p' with p_0 and obtain a finite-dimensional representation of the reduced density matrix after the scattering (this step is rather formal; it involves an approximation, but it helps visualize the result). Then, the reduced density matrix, representing the state as seen by coarse observers (only sensitive to p), is defined by tracing over the discrete degree of freedom labeling the components of the state in either the Γ^1 or the Γ^2 lattices, namely

$$\langle p|\,\rho^{\mathrm{R}}_{\mathrm{out}}\,|p'\rangle \equiv \sum_{i=1}^{2} \langle p, i|\Psi_{\mathrm{in}}, \infty\rangle \,\langle \Psi_{\mathrm{in}}, \infty|p', i\rangle = \frac{1}{2}\begin{pmatrix} 1+|B(p_0)|^2 & \overline{A}(-p_0)B(p_0) \\ A(-p_0)\overline{B(p_0)} & |A(-p_0)|^2 \end{pmatrix}, \tag{14.76}$$

which is mixed while initially pure, namely

$$\langle p|\,\rho^{\mathrm{R}}_{\mathrm{in}}\,|p'\rangle \equiv \sum_{i=1}^{2} \langle p, i|\Psi_{\mathrm{in}}, 0\rangle \,\langle \Psi_{\mathrm{in}}, 0|p', i\rangle = \begin{pmatrix} 1 & 0 \\ 0 & 0 \end{pmatrix}. \tag{14.77}$$

Thus, the entanglement entropy jumps according to

$$\delta S = \log(2) - \frac{3\lambda^2}{128\pi^2\gamma^2\mu^2}\Lambda \ell_p^2 + \mathcal{O}(\Lambda^2 \ell_p^4), \tag{14.78}$$

where the result is valid for small Λ. The entropy jumps at the big bang no matter how small the scalar curvature of the initial state is. Low-energy observers (insensitive to the UV details) would conclude that pure states evolve into mixed states.

The sudden jump here is just a result of the overly simplified matter contribution to the Hamiltonian (14.72). In a more accurate model using the free scalar field Hamiltonian (14.64), the entanglement entropy increases gradually as we get closer to the big bang. Near the singularity, the matter Hamiltonian changes more rapidly with x, leading to a faster growth in entropy because quantum states on different ϵ-sectors experience stronger variations in the interaction potentials (see Fig. 14.10). This effect is influenced by the curvature of the semiclassical state. Specifically, when the universe is large and Λ is low, decoherence becomes significant closer to the big bang compared to higher Λ states. This is because the wave function oscillates faster for higher p, highlighting the differences in the matter Hamiltonian across two different ϵ-sectors. Both features are confirmed by numerical solutions of the Hamiltonian (14.64) [94].

In conclusion, the quantization methods of loop quantum gravity applied to cosmology create a Hilbert space that is much larger than that of the Wheeler-DeWitt representation. As a result, physical states labeled by macroscopic quantum numbers have an extra degeneracy related to microscopic or Planckian degrees of freedom.

Fig. 14.10 In the upper panel we represent the $1/x^2$ potential representing the coupling with a massless scalar field in lattice Γ^1 (in blue) and Γ^2 (in yellow). As the interaction is not translational invariant states supported on different lattices evolve differently. In the lower panel our simplistic (yet exactly solvable) model capturing the same qualitative physics

This is thought to hold true in the full theory as well, where states corresponding to flat spacetime are expected to be degenerate.

These microscopic degrees of freedom are responsible for black hole entropy in this approach [45, 50]. This reservoir of quantum numbers remains hidden in any effective quantum field theory approach to black hole evaporation and may help resolve the information puzzle [15, 45]. Our example shows that entanglement with the microscopic structure leads to the apparent loss of information for low-energy observers, even though the fundamental theory is unitary.

These results, when applied to black hole formation and evaporation, suggest a potential resolution to the information paradox that avoids problematic features like firewalls [95, 96] and the risks of information cloning associated with holographic scenarios [97]. In our model, decoherence occurs without diffusion [14], which helps sidestep the usual challenges [8] related to energy conservation in the purification process. This is achieved within a concrete quantum gravity framework, avoiding the issues faced by the quantum field theory approach [16, 17]. In the next chapter, we will demonstrate how these concepts can be applied in a simplified model of a quantum black hole.

14.8.5 Apparent Information Loss Due to Planckian Discreteness in a Black Hole Model

In this last example, we present a toy model of a black hole radiating a Hawking pair of particles, where one can track the backreaction on the black hole geometry of the inner member of the pair as it falls into the singularity, as well as the transfer of information to the microscopic degrees of freedom nearby. This is a model that is the closest to the real physical situation one would like to describe in a (yet-to-come) full theory of quantum gravity.

Here we briefly report the construction of the model introduced in [98] (refer to that paper for more details). The first step towards canonical quantization is to define the Hamiltonian description of the model. It is well known that for a spherically symmetric and static spacetime, the line element can be written without any loss of generality as

$$ds^2 = -f(r)dt^2 + h(r)dr^2 + r^2 d\Omega^2 \,. \tag{14.79}$$

It follows that the Einstein-Hilbert action (with the appropriate boundary term that renders it differentiable: the usual Gibbons-Hawking-York boundary term) becomes

$$S_{\text{geo}} = \frac{1}{16\pi\ell_p^2}\left[\int_R d^4x\sqrt{-g}R + 2\int_{\partial R} K\right] = \frac{T_0}{2\ell_p^2}\int dr\left(\sqrt{fh} + \sqrt{\frac{f}{h}} + \frac{\dot{f}r}{\sqrt{fh}}\right), \tag{14.80}$$

where $\ell_p^2 = G$ (in $\hbar = 1$ units) and T_0 is a infra-red cut off in the Killing time t necessary to make the action of the symmetry reduced model finite (this is the analog of V_0 in the cosmology model (14.58) of the previous section, for more details see [98]). The matter model is massless scalar field with action

$$S_m = -\frac{1}{2}\int_R d^4x\sqrt{-g}\partial_a\phi\partial^a\phi = -2\pi T_0\int dr r^2\dot{\phi}^2\sqrt{\frac{f}{h}}. \tag{14.81}$$

The conjugate momenta to f, h and ϕ are given by

$$\mathrm{p}_f = \frac{T_0}{2\ell_p^2}\frac{r}{\sqrt{fh}} \quad , \quad \mathrm{p}_h = 0 \quad \text{and} \quad \mathrm{p}_\phi = -4\pi r^2 T_0\sqrt{\frac{f}{h}}\dot{\phi}. \tag{14.82}$$

The first two equations turn our to be second class constraints [98] which, when solved, and replaced in the Hamiltonian $H = \dot{f}\mathrm{p}_f + \dot{h}\mathrm{p}_h + \dot{\phi}\mathrm{p}_\phi - \mathscr{L}_\phi - \mathscr{L}_{\text{geo}}$, becomes

$$\mathrm{H}_2 = -\frac{1}{r}\left(f\mathrm{p}_f + \frac{1}{16\pi\ell_p^2}\frac{\mathrm{p}_\phi^2}{f\mathrm{p}_f} + \frac{T_0^2}{4\ell_p^4}\frac{r^2}{\mathrm{p}_f}\right). \tag{14.83}$$

In the *deep interior region* $r \ll M$, where M is the mass scale defined by the corresponding black hole solution perturbed by the presence of matter, the last term becomes negligible [98]. Thus we arrive at the simplified solvable model with Hamiltonian

$$\mathrm{H}_{\mathrm{di}} = -\frac{1}{r}\left(f\mathrm{p}_f + \frac{1}{16\pi\ell_p^2}\frac{\mathrm{p}_\phi^2}{f\mathrm{p}_f}\right). \tag{14.84}$$

This toy theory reflects the dynamics of the leading order in an expansion near the singularity. In order to recover the structure of a gauge theory, with a clear analogy with the full theory of loop quantum gravity, it is convenient to 'reparametrize' the system by promoting the area radius r to a degree of freedom with conjugate momentum p_r and add a scalar constraint $\mathrm{C} = \mathrm{p}_r - \mathrm{H}_2 = 0$. Introducing the *deep interior* (canonically conjugate) variables

$$m = -f\mathrm{p}_f \quad \text{and} \quad \mathrm{p}_m = -\log(-f)\,, \tag{14.85}$$

and adopting the area a of the surfaces of constant r as time

$$a = 4\pi r^2 \quad \text{and} \quad \mathrm{p}_a = \frac{\mathrm{p}_r}{8\pi r}\,. \tag{14.86}$$

The Hamiltonian constraint describing the deep interior regime is

$$\mathrm{C}_a = \mathrm{p}_a + \frac{1}{2a}\left(m + \frac{\mathrm{p}_\phi^2}{16\pi\ell_p^2 m}\right) \approx 0. \tag{14.87}$$

The previous constraint defines the classical dynamical equation of the model. In the quantum theory—and in the loop representation that mimics the one used in the full loop quantum gravity framework—the area variable will evolve in discrete steps (if we use the representation of canonical variables inspired by loop quantum gravity and already used in the cosmology setting of the previous section). In the Schroedinger representation there is no area discreteness and no microscopic degrees of freedom, and the previous constraint is the Schroedinger (Wheeler-DeWitt) equation

$$\left(-i\frac{\partial}{\partial a} + \frac{1}{2a}\left(m + \frac{\mathrm{p}_\phi^2}{16\pi\ell_p^2 m}\right)\right)|\Psi\rangle = 0 \tag{14.88}$$

imposed on physical states.

Instead, we will work with the type of representation we already used in the previous section, which in the present case (as in the full theory of loop quantum gravity) exhibits area quantization. This is analogous to what is done in the full theory of loop quantum gravity that uses the Ashtekar-Barbero connection variables [99, 100] as the starting point. There, the connection is not represented as a fundamental operator but only its holonomy (an exponentiated version of the connection in essence) in the definition of the Hilbert space [101]. More precisely, one replaces the kinematical inner product in the Schroedinger representation by (exactly as in the quantum cosmology example; recall Eq. (14.65)).

$$\langle\psi_1,\psi_2\rangle = \lim_{\Delta\to+\infty} \frac{\int_{-\Delta}^{+\Delta}\left(\int_{-\infty}^{+\infty}\overline{\psi_1(m,\mathrm{p}_\phi,\mathrm{p}_a)}\psi_2(m,\mathrm{p}_\phi,\mathrm{p}_a)dmd\mathrm{p}_\phi\right)d\mathrm{p}_a}{2\Delta}. \quad (14.89)$$

With this inner product, periodic functions of p_a with arbitrary period are normalizable and the conjugate a-representation acquires the property of discreteness in a way that closely mimics the structure of the fundamental theory of loop quantum gravity [93]. In particular, eigenstates of $\widehat{a}$ exist

$$\widehat{a}\,|a\rangle = a\,|a\rangle \quad (14.90)$$

with $a \in \mathbb{R}$. These states form an orthonormal basis with inner product

$$\langle a, a'\rangle = \delta_{a,a'}, \quad (14.91)$$

just as in (14.66). Discreteness of the spectrum of $\widehat{a}$ comes at the price of changing the kinematical Hilbert space structure in a way that prevents the infinitesimal translation operator $\widehat{\mathrm{p}}_a$ from existing (this is because the inner product (14.89) is not weakly continuous under the action of translations: according to (14.91), an infinitesimally translated state is orthogonal to the original state). Instead, only finite translations (quasi-periodic functions of p_a) can be represented as unitary operators in the polymer Hilbert space. Their action on the a-basis is given by

$$\widehat{e^{i\lambda \mathrm{p}_a}}\psi(m,\mathrm{p}_\phi,a) = \psi(m,\mathrm{p}_\phi,a+\lambda\ell_p). \quad (14.92)$$

As the classical Hamiltonian depends on p_a, quantization requires regularization. As in the cosmology example (and the full theory), regularization is an ambiguous procedure. However, as we only want to use the model for qualitative purposes of illustration, we are not really concerned here with the issue of ambiguities.[11]

We proceed as follows. In order to maintain geometric compatibility with the Schroedinger (14.88) equation, one exponentiates the second term in the classical Hamiltonian (14.87), obtaining the well-known unitary evolution operator producing finite area evolution. The quantum constraint preserving the geometric consistency with the Schroedinger equation is

[11] In the full theory, a new perspective on the regularization issue has been introduced motivated by the notion of generalized gauge covariant Lie derivatives [102] and their geometric interpretation allowing for the introduction of a natural regularization (and subsequent) anomaly-free quantization of the Hamiltonian constraint [103]. Even when the procedure does not eliminate all ambiguities of quantization (choices are available in the part of the quantum constraint responsible for propagation [104]), the new technique reduces drastically some of them in the part of the Hamiltonian that is more stringently constrained by the quantum algebra of surface deformations.

$$\underbrace{\exp(i\lambda \mathrm{p}_a)}_{\substack{\text{finite areatime} \\ \text{translation}}} |\psi\rangle - \overbrace{\exp\left(\frac{i}{2}\log\left(\frac{a+\lambda\ell_p^2}{a}\right)\left(m+\frac{\mathrm{p}_\phi^2}{16\pi\ell_p^2 m}\right)\right)}^{\substack{\text{finite areatime} \\ \text{unitary evolution operator}}} |\psi\rangle = 0, \quad (14.93)$$

whose action is well defined in the polymer representation and whose solutions are easily found (by acting on the left with $\langle m, \mathrm{p}_\phi, a|$) to be wave functions satisfying the discrete dynamics given by

$$\psi(m, \mathrm{p}_\phi, a + \lambda\ell_p^2) = e^{\frac{i}{2}\log\left(\frac{a+\lambda\ell_p^2}{a}\right)\left(m+\frac{\mathrm{p}_\phi^2}{16\pi\ell_p^2 m}\right)}\psi(m, \mathrm{p}_\phi, a). \quad (14.94)$$

The physical Hilbert space is defined via the usual inner product at fixed (discrete) time a via

$$\langle\psi_1(a)|\psi_2(a)\rangle_{\text{phys}} = \int_{-\infty}^{+\infty}\int_{-\infty}^{+\infty} \overline{\psi_1(m, \mathrm{p}_\phi, a)}\psi_2(m, \mathrm{p}_\phi, a)dmd\mathrm{p}_\phi, \quad (14.95)$$

which is independent of the lattice sites as required (a property identified with the unitarity of the dynamics generated by the quantum constraint). More precisely the physical inner product is a constant of the quantum motion associated to the full history represented by the lattice $\Gamma^{\epsilon,\lambda}$ as implied by unitarity. Explicitly one has

$$\langle\psi_1(a)|\psi_2(a)\rangle_{\text{phys}} = \langle\psi_1(a+\lambda)|\psi_2(a+\lambda)\rangle_{\text{phys}} \quad \forall \quad a \in \Gamma^{\epsilon,\lambda}. \quad (14.96)$$

The one step evolution (14.94) can be composed to produce the arbitrary initial to final area evolution

$$\psi(m, \mathrm{p}_\phi, \epsilon + n\lambda) = \left(\frac{\epsilon + n\lambda}{\epsilon + q\lambda}\right)^{\frac{im}{2}\left(1+\frac{\mathrm{p}_\phi^2}{16\pi\ell_p^2 m^2}\right)} \psi(m, \mathrm{p}_\phi, \epsilon + q\lambda), \quad (14.97)$$

for arbitrary integers $n, q > 1$.

Modelling the emission of a Hawking pair: In order to simplify as much as we can the description of the matter part, we will consider that the *inside* matter can be in two different orthogonal states $|+\rangle$ or $|-\rangle \in \mathscr{H}_{in}$ (with subscript in standing for the *inside* Hilbert space of the scalar degrees of freedom), which are eigenvectors of $\hat{\mathrm{p}}_\phi$ with eigenvalues p_+ and p_-, respectively. On the other hand, we will consider that the outgoing part of the matter can be in two different orthogonal states $|+\rangle$ or $|-\rangle \in \mathscr{H}_{out}$ (with subscript *out* standing for the *outside* Hilbert space of the scalar degrees of freedom). As the outside outgoing partner evolves in regions of low curvature, where the type of effects we are focusing on should be very small, we model such evolution by a trivial Hamiltonian. This simple description of the matter is sufficient to write down a state for the total matter where the ingoing particle and the outgoing one are initially maximally entangled, and to study how the dynamics modifies these correlations.

We can now write down the total Hilbert space describing the geometrical and the matter degrees of freedom. It can be written as

$$\mathscr{H}_{total} = \underbrace{\mathscr{H}_m \otimes \mathscr{H}_\epsilon}_{\text{geometry}} \otimes \overbrace{\mathscr{H}_{in} \otimes \mathscr{H}_{out}}^{\text{matter}}, \tag{14.98}$$

where we have organized the Hilbert space of geometry as the tensor product of normalizable functions $\mathscr{H}_m$ of the configuration variable m, defined in (14.85), on a given lattice times the separable truncation $\mathscr{H}_\epsilon = \mathbb{C}^{n_\epsilon}$ of the epsilon sector where we assume that a finite number n_ϵ of lattices $\Gamma^{\epsilon,\lambda}$—introduced in the text below Eq. (14.66)—are allowed. Concretely, and for simplicity, we will take $n_\epsilon = 2$ with two alternative values $\epsilon_\pm$. With such simple choices we have that $\mathscr{H}_\epsilon = \mathscr{H}_{in} = \mathscr{H}_{out} = \mathbb{C}^2$, i.e., with the exception of $\mathscr{H}_m$ that is infinite dimensional, the other quantum numbers have been truncated to standard q-bits. Thus, a basis of this Hilbert space is defined by the set of vectors $|m, \varepsilon, i, o\rangle$ with $m \in \mathbb{R}^+$, ε (encoding the microscopic ϵ sectors), i (encoding the *inside* particle state) and o (encoding the *outside* state) spin-like quantum numbers taking values $\pm$. The inner product between basis states is

$$\langle m_1, \varepsilon_1, i_1, o_1 | m_2, \varepsilon_2, i_2, o_2\rangle = \delta(m_1 - m_2)\delta_{\varepsilon_1,\varepsilon_2}\delta_{i_1,i_2}\delta_{o_1,o_2}\,. \tag{14.99}$$

After having defined the truncation of the Hilbert space we introduce the initial state representing a toy version of the in-vacuum in the context of Hawking radiation, namely

$$|\psi_0\rangle = \sum_{\substack{i=\pm\\ \varepsilon=\pm}} \int dm\ \psi_0(m, a_0 + \epsilon_\varepsilon)|m, \varepsilon, i, i\rangle, \tag{14.100}$$

and the normalization condition $\langle\psi_0|\psi_0\rangle = 1$ implying

$$\int dm\ |\psi_0(m, a_0 + \epsilon_\varepsilon)|^2 = \frac{1}{4}\,. \tag{14.101}$$

The key feature of the initial state is that it represents a state where the excitations in $\mathscr{H}_{in}$ and $\mathscr{H}_{out}$ are maximally correlated as in the standard in-vacuum. We can then apply the evolution law (14.93) to this state, a shift by $n\lambda$ is given by

$$
\begin{aligned}
|\psi\rangle &= \sum_{\substack{i=\pm\\ \varepsilon=\pm}} \int dm\, e^{\frac{i}{2}\log\left(1+\frac{n\lambda}{a_0+\epsilon_\varepsilon}\right)\left(m+\frac{p_I^2}{16\pi\ell_p^2 m}\right)} \psi_0(m, a_0+\epsilon_\varepsilon)|m, \epsilon_\varepsilon, i, i\rangle \\
&= \sum_{\substack{i=\pm\\ \varepsilon=\pm}} \int dm\, C(m, \epsilon_\varepsilon, i, n)\, |m, \epsilon_\varepsilon, i, i\rangle\,,
\end{aligned}
\tag{14.102}
$$

where, in order to simplify the notation, we introduced

$$
C(m, \epsilon_\varepsilon, i, n) = e^{\frac{i}{2}\log\left(1+\frac{n\lambda}{a_0+\epsilon_\varepsilon}\right)\left(m+\frac{p_I^2}{16\pi\ell_p^2 m}\right)} \psi_0(m, a_0+\epsilon_\varepsilon)\,. \tag{14.103}
$$

Equation (14.102) describes the quantum gravitational dynamics of the initial state (of maximally correlated Hawking partners). This dynamics encodes the backreaction of the Hawking particles with the quantum geometry exactly (in the sense of our simplified model). It only remains to study the dynamical evolution of the entanglement between the relevant degrees of freedom as the *inside* partner evolves toward the strong quantum gravity regime (the singularity) inside the black hole.

Dynamics of correlation measures: Stating with the previous initial state and evolving in $a = n\lambda\ell_p^2$, for $n\lambda\ell_p^2 \in [0, a_0 - 2\lambda\ell_p^2]$ toward the internal singularity one observes the following behaviour of the relevant correlation and entanglement measures (for a definition and properties see Appendix 1 and 2).

1. If one traces out $\mathscr{H}_m \otimes \mathscr{H}_\epsilon$ (the geometry sector) to find the reduced density matrix representing the quantum field theory sector describing the Hawking pairs, one finds that the initially pure density matrix (representing the typically assumed closeness of the effective quantum field theory formulation) quickly becomes mixed. The associated off-diagonal terms decay exponentially. This reveals the decoherence produced by the geometry environment, with a non-trivial contribution from the discreteness (microscopic) degrees of freedom, as the following features demonstrate.
2. The Von Neumann entropy of the reduced density matrix obtained when tracing out over the microscopic degrees of freedom $\mathscr{H}_\epsilon$ grows as the inside particle approaches the singularity [105]. The entanglement with the Planckian degrees of freedom grows.
3. The mutual information between the (initially maximally correlated) inside and the outside Hawking excitations in the matter sector

$$
I_{\rm in,out}(|\psi\rangle\,\langle\psi|) = S(\rho^{\rm in}) + S(\rho^{\rm out}) - S(\rho^{\rm in,out}) \tag{14.104}
$$

decreases as the inside partner approaches singularity. This is to be expected, as the back reaction of the inside partner on the quantum geometry transfers information to the geometry degrees of freedom in $\mathscr{H}_{in}$. From such a general perspective, the result is anticipated; it is a straightforward implication of the monogamy of entanglement in quantum mechanics. The non-trivial feature here, which will become explicit in the next item, is that essential correlations are established with the microscopic degrees of freedom (modeled here by the ϵ sectors).

4. The mutual information between the ingoing and the ϵ degrees of freedom

$$I_{\text{in},\epsilon}(|\psi\rangle\langle\psi|) = S(\rho^{\text{in}}) + S(\rho^{\epsilon}) - S(\rho^{\text{in},\epsilon}) \tag{14.105}$$

grows with time.

5. The mutual information between the subsystem corresponding to the outside partner and ϵ

$$I_{\text{out},\epsilon}(|\psi\rangle\langle\psi|) = S(\rho^{\text{out}}) + S(\rho^{\epsilon}) - S(\rho^{\text{out},\epsilon}). \tag{14.106}$$

also grows. The outside partner escaping to infinity develops correlations with the quantum degrees of freedom of the geometry, even if it never comes into direct interaction with them. This is a consequence of the monogamy of entanglement. Indeed it turns out that (due to unitarity and the initial maximal entanglement in the Hawking pairs)

$$I_{\text{in},\epsilon}(|\psi\rangle\langle\psi|) = I_{\text{out},\epsilon}(|\psi\rangle\langle\psi|). \tag{14.107}$$

Correlations between outside partner and the microscopic ϵ sector appear at the same rate as those between the inside partner and the ϵ sector.

The results in this section consistently demonstrate how the information encoded in the initial correlations between the Hawking partners in $\mathscr{H}_{in} \otimes \mathscr{H}_{out}$ is transferred to the other degrees of freedom involved. This is expected due to the gravitational interaction between the matter sector and the spacetime geometry. What is particularly interesting and novel in our case is the generic transfer of entanglement to the 'quantum hair' represented in our simple model by $\mathscr{H}_{\epsilon}$. This indicates that, in a process where the mass of the black hole would completely disappear via Hawking evaporation (a process that our simple model cannot account for), unitarity could be maintained via the purification of Hawking radiation facilitated by degrees of freedom like the ϵ sector in this toy model. Despite the simplistic nature of the model, we observe that it complies with the expected behavior of the backreaction: the lower the mass (the smaller m_0), or equivalently the smaller the black hole, the stronger the development of entanglement between the matter and the defects in the quantum geometry (stronger interaction with the Planckian granularity). The effect also grows with the particle momentum p_ϕ and dispersion σ, as expected.

Quantum hair in the mass operator (the vacuum case): As mentioned at the end of Sect. 14.8.4, the huge degeneracy introduced by the presence of the ϵ sectors plays an important role in the spectrum of physical observables. The mass of the black hole (in the vacuum case) is given by (see [98] for details)

$$M = \alpha(a) m^2 e^{\mathrm{p}_m}, \tag{14.108}$$

in terms of the variables (14.85), where $\alpha(a) \equiv 2\sqrt{4\pi}\ell_p^4/(T_0^2\sqrt{a})$. For the ordering $m \exp \mathrm{p}_m m$ the eigenstates are

$$|M, \epsilon\rangle = \sum_{a \in \Gamma^{\epsilon,\lambda}} \int \phi_M(\mathrm{p}_m, a)|\mathrm{p}_m\rangle|a\rangle d\mathrm{p}_m, \tag{14.109}$$

where the sum runs over the discrete lattice $\Gamma^{\epsilon,\lambda}$ when introducing the dynamical constraint (14.93), with

$$\phi_M(\mathrm{p}_m, a) = \sqrt{\frac{2\sqrt{M}}{\alpha(a)}}e^{-\frac{\mathrm{p}_m}{2}} J_1\left(2\sqrt{\frac{M}{\alpha(a)}}e^{-\frac{\mathrm{p}_m}{2}}\right), \tag{14.110}$$

where J_1 is a Bessel function. One can explicitly verify that the quantum dynamics (14.93) preserves the eigenstates. The spectrum of the mass operator is continuous. Moreover, the wave functions of the form (14.110) can be supported on lattices with different values of ϵ. Thus eigenstates of the mass should then be denoted $|M, \epsilon\rangle$ with orthogonality relation

$$\langle M, \epsilon|M', \epsilon'\rangle_{\text{phys}} = \delta(M, M')\delta_{\epsilon,\epsilon'}, \tag{14.111}$$

where $\delta_{\epsilon,\epsilon'}$ is the Kronecker delta symbol. The existence of such a large degeneracy is a generic feature of the polymer (or loopy) representation. The mass operator is hugely degenerate—as the vacuum in our field theory model of Sect. 14.8.3. The model violates the usual assumption of the uniqueness of the vacuum in background dependent treatments of quantum field theory and might reflect a generic feature present in the full quantum gravity context. These conclusions are independent of factor ordering ambiguities.

14.9 The Timeless Perspective

Any acceptable approach to quantum gravity that is consistent with general covariance is timeless at the fundamental level. This is formally clear from the canonical (Dirac) perspective, which implies that the total Hamiltonian of the universe is zero, and from the fact that no preferred notion of time can be defined *a priori*. This feature is, in a way, already present in the classical theory of generally relativistic systems (where a spacetime description is still available at the price of diffeomorphism invariance), but it becomes more severe in the quantum theory, where fundamentally spacetime itself loses its meaning. This is often referred to as *the problem of time*.

Thus, any dynamical account of any system in terms of spacetime, Penrose diagrams, asymptotic boundary conditions, etc., must be regarded as an effective description where some sort of approximation is at play. We believe that a dynamical description in a framework where we can talk about spacetime is necessary when invoking a solution to the black hole information problem (see further discussion in Sect. 14.11). Nevertheless, whatever such (effective) spacetime description may be, it should connect with some form of a more fundamental timeless version in agreement with quantum general covariance.

Even though in the case of general relativity the problem of time is clearly more severe, a cartoon version of it is the difficulty that arises when trying to describe dynamics in a closed quantum mechanical system without making reference to any outside clock. As energy defines a superselected sector in such a context, the physics of such a (toy model) closed universe is captured by the eigenstates of the Hamiltonian with a definite energy E and the (timeless) time-independent Schroedinger equation

$$(H - E)\,|\psi\rangle = 0, \tag{14.112}$$

which is the analog of the (time-less) Wheeler-DeWitt constraint equation

$$H(N, \mathbf{N})\,|\psi\rangle = 0, \tag{14.113}$$

for all lapse $N(x)$ and shift $\mathbf{N}(x)$ fields and $H(N, \mathbf{N})$ the Hamiltonian constraint of quantum gravity.

Even when it would be hard to describe the scenario we propose in great detail in such a timeless fashion, it is easy to illustrate it in the context of one of the simple toy models we introduced in Sect. 14.8. A case which is particularly adapted for the discussion is the Landau model presented in Sect. 14.8.2. The point can be illustrated with the following normalized state[12] (representing a maximal loss of information example)

$$|\psi(t)\rangle \equiv \frac{1}{\sqrt{2}} e^{-iEt} \left(|0, p_x^1\rangle\,|P_Z\rangle + |0, p_x^2\rangle\,|-P_Z\rangle\right)$$

where states on the right are written using the notation of Eq. (14.31), and the four terms in the linear combination have the same energy $E = P_Z^2/(2M) + \hbar\omega/2$ and p_x^1 and p_x^2 are two different values of the constant of motion p_x so that $\langle 0, p_x^\alpha | 0, p_x^\beta\rangle = \delta^{\alpha\beta}$. For this reason, the time dependent phase drops out in the expression of the density matrix of this model universe so that $\rho = |\psi\rangle\,\langle\psi|$ is time independent.

If we take as the analog of the quantum field theory low energy degrees of freedom the possible values $\pm P_Z$ for the momentum of the scattered particle, and the p_x as the analog of the hidden degrees of freedom, we observe that there is a timeless measure of the decoherence process—responsible for the illusion of information loss in the quantum field theory sector—defined for example in terms of the (timeless) reduced density matrix

[12] There is a technical issue here of no real importance in the fact that plane wave states are not normalizable. One would need to work with superpositions of plane waves, but this would take us out of the single eigenstate scenario that leads to a timeless situation. A solution would be to take the degree of freedom Z and compactify it into a circle. One can show that such regularization makes the calculations in this example correct. Wave packets in the p_x degree of freedom (not appearing in the Hamiltonian) can be formed without interference with the closeness of our toy universe.

$$\rho_{\mathrm{QFT}} \equiv \sum_{\alpha} \langle 0, p_x^{\alpha}|\rho|0, p_x^{\alpha}\rangle = \frac{1}{2}|P_Z\rangle\langle P_Z| + \frac{1}{2}|-P_Z\rangle\langle -P_Z| \quad (14.114)$$

which mixed and has a timeless maximal entropy $S_{\mathrm{QFT}} = \log(2)$. The quantum field theory observers interact without dissipation with the p_x sector and conclude that their quantum field theory (timeless energy measure, which is the analog of the ADM mass) is $E_{\mathrm{ADM}} = P_Z^2/(2M)$ and remain insensitive to the true degeneracy of the universe Hamiltonian (responsible of the apparent information loss) encoded in the p_x (hidden) degree of freedom.

The timeless version of black hole formation and evaporation, according to our scenario, would have to be captured in a timeless physical state $|\psi\rangle$ satisfying the Wheeler-DeWitt constraint (14.113) and involving Planckian degrees of freedom, which cannot be accounted for in terms of low-energy smooth quantum field theory quantum fields. The reduced density matrix, constructed by tracing out the microscopic degrees of freedom (or perhaps more appropriately via a suitable coarse-graining mechanism), will be timeless yet mixed, with a measure of its mixed nature given, for instance, by a suitable notion of entropy. In the idealized state corresponding to the formation of a single macroscopic black hole that remains unperturbed until complete evaporation, $S \gtrsim A/4$, where A is a constant of motion characterizing the initial area of the black hole. In general, the entropy can be anything if the state has no simple continuum interpretation or relation to the situation one would like to model.

14.10 Discussion

The problem of unitarity in black hole formation and evaporation necessitates a deep understanding of the strong quantum gravity regime near the singularity beyond black hole horizons. This presents a challenge for both the *holographic* and *non-holographic* perspectives. In the holographic view, macroscopic black holes are believed to possess a finite number of internal degrees of freedom bounded by their horizon area. Consequently, they are expected to follow a Page curve, wherein the exterior entropy starts decreasing at the macroscopic stage known as the Page time. Conversely, in the non-holographic perspective, which is the framework of this work, the number of internal degrees of freedom inside black holes is unrelated to their horizon area (in agreement with the generalized second law as proven in quantum field theory). In this scenario, the exterior entropy continues to grow while the black hole is macroscopic, and the purification occurs after the semiclassical regime by the end of evaporation.

In non-holographic scenarios, where information is recovered after the complete evaporation of a black hole, it becomes essential to identify the appropriate purifying degrees of freedom. Subsequently, understanding how the correlations between the

Hawking partners are dynamically transferred to these degrees of freedom becomes crucial.

Suitable degrees of freedom are those that can persist after complete evaporation without significantly affecting the energy budget, considering that only a Planck mass energy is available for the multitude of degrees of freedom needed to purify the emitted Hawking radiation until the black hole's total evaporation. Approaches to quantum gravity, which postulate a discrete fundamental nature of matter and geometry at the Planck scale, offer a natural set of such degrees of freedom. In these formulations, the concepts of smooth geometry and continuous fields are approximate and expected to emerge only when probed with sufficiently coarse measurements. Within this framework, flat configurations such as Minkowski spacetime with the Minkowski vacuum for matter fields should exhibit high degeneracy in a coarse-grained sense, meaning that a single macroscopic configuration corresponds to a large number of microscopic states with negligible mass. Any deviations among these microscopic states can be viewed as defects in the fundamental structure that are indistinguishable at the macroscopic level. These precisely are the types of degrees of freedom suitable for purification. They are numerous (due to their local nature and increasing number with spatial extension) and, in the aforementioned sense, possess negligible weight or influence on the system's overall energy.

A general discussion about the manner this would apply to the process of black hole formation and evaporation has been discussed in the first part of this chapter while we have dedicated Sect. 14.8 to the description of simple models where the key ideas are realized concretely.

We have shown in Sect. 14.8.3 that such situation can be reproduced in a concrete model which describes a 2-dimensional quantum field theory in the continuum limit with microscopic defects have non trivial physical effects only if suitable high energy probes are used. The low energy physics remains indistinguishable from the physics of phonons while the microscopic structure of degrees of freedom admits the kind of decoherence without energy cost required by the scenario we propose.

Once suitable degrees of freedom are identified, one needs to understand the dynamical process by which unitarity is preserved as these are tracked during the black hole formation and evaporation. In particular, while the black hole is macroscopic and emits Hawking radiation, it is crucial to describe precisely the mechanism by which the initially maximally correlated Hawking pairs evolve when strong quantum gravity effects are suitably taken into account. This process should lead to one of the particles escaping to infinity as Hawking radiation, while the other entangles with the Planckian degrees of freedom near the singularity inside the black hole. As the black hole undergoes complete evaporation, these degrees of freedom should eventually emerge as purifying defects, as discussed in the previous paragraph.

We have shown that the germ of discreteness present on certain quantum gravity toy models inspired by quantum gravity allows for the precise illustration of the mechanism in a (certainly simplified) quantum gravitational context. The simplest context is the one offered by the loop quantum cosmology calculation reviewed in Sect. 14.8.4 and the better suited generalization provided by the model of quantum black holes analyzed in Sect. 14.8.5. This last model is expected to capture the dynamics of an infalling scalar excitation when it is spherically symmetric in the near-singularity approximation. It can be used to model the near-singularity dynamics of a Hawking pair of scalar particles, including the back reaction of the inside partner as it approaches the near-singularity regime.

On the one hand, the toy model used here mimics the quantization of gravity in the framework of loop quantum gravity and thus contains the type of microscopic 'weightless' degrees of freedom. More precisely, the spectrum of the mass operator is infinitely degenerate due to the existence of quantum hair associated with (what we call) the ϵ sectors, Sect. 14.8.5. On the other hand, we reviewed explicitly how correlations are transferred dynamically to the 'hidden' quantum hair as the inside excitation approaches the near-singularity regime.

Of course, such models do not explain how a real black hole evaporates and how the information remains after complete evaporation coded in the fundamental 'hair.' This would possibly require making the highly non-trivial step of allowing local degrees of freedom and treating their strong quantum dynamics near the singularity. We are not sure if one can find intermediate models between our simplistic one and the fully general situation that requires treatment without any possible approximation.

Current theoretical approaches are not yet precise enough to address the information question in a fully satisfactory manner. However, on the phenomenological side, posing the question may inspire ideas with potentially observable consequences (e.g., Sect. 14.5.4), potentially revealing new conceptual and mathematical insights necessary for progress in understanding the nature of physics at the fundamental scale.

14.11 A Wish List for a Convincing Resolution

1. **A story of how information is retrieved is given**: By this we mean that a spacetime interpretation (as story) of what happens with the information when a black hole forms and evaporates must be provided by the candidate proposal. Examples of statements where this is not the case would be for instance:
   "*According to the AdS-CFT correspondence the bulk dynamics is dual to a CFT on the AdS boundary. As the CFT is unitary, black hole formation and evaporation must be unitary.*"
   or
   "*In the canonical quantum gravity program* á la *Dirac all observables are constants of motion; therefore, black hole formation and evaporation must be unitary: nothing changes, nothing can be lost.*"
2. **Quantum gravity dynamics near the singularity must be accounted for**: Effective models of regular black holes have become very popular, but they cannot be realistic in describing what happens near the singularity. Worse yet, accounts where one insists that the information issue is not concerned with the high-curvature regime, as curvatures are low for outside observers and macro-

scopic black holes, are (in our view) akin to adopting the ostrich strategy in the face of threats: head in the sand and hindquarters dangerously exposed.

Recall that the statement of the information paradox in terms of the purification of the Hawking radiation is an idealization that aims at simplifying the discussion. There is a much more basic related problem, which corresponds to the simple question of what happens with anything (and the information it may carry) when it happens to fall inside a black hole.

In particular, black holes must have an inside: there is spacetime beyond the horizon of a macroscopic black hole. Classically, this is a consequence of the equivalence principle, implying that nothing special happens when crossing a black hole horizon. Quantum mechanically, things are more subtle, and insisting on the validity of the *central dogma* may lead to tensions and large quantum effects. For instance, some proposals evoke the emergence of a firewall at the horizon for sufficiently old black holes [95, 96], while others remove the inside completely from the system [106]. Even if one takes that possibility seriously, before the formation of the firewall (when the macroscopic black hole is young), the physics must remain approximately well described by the classical picture. Thus, the inside, as well as the interior singularity, remain accessible, and their dynamical description is necessary.

3. **Black hole formation and evaporation must be time asymmetric**: The information must be purified in a way that is compatible with the irreversibility of gravitational collapse. Accounts where information emerges from the future of the quantum gravity region (the would-be-singularity) in terms of very low-energy standard quantum field theory modes must be highly suspicious from this perspective, as this would require that a region with the expected extreme large quantum fluctuations (the singularity, as seen from the past) becomes extremely tamed and capable of emitting low-energy modes emerging from it as seen from its future. In our view, this is in contradiction with the second law in its most fundamental form.

   Another consequence of the time asymmetry of gravitational collapse is that real macroscopic black holes do not contain any Einstein-Rosen bridge connecting parallel universes or vastly remote regions of spacetime. It might be amusing to consider such situations mathematically, but they do not seem related to the problem at hand.
4. **The proposal works in the absence of asymptotic boundaries**: Asymptotic boundaries and their defining boundary conditions are useful to set up the framework for discussion. However, if a proposal works, then it should offer an explanation in the absence of boundaries, for instance, in cosmology.

Appendix 1: Decoherence Function

It is often said that the entanglement of a system with its environment tends to kill the off diagonal terms of its density matrix. The consequence is that a system initially pure becomes mixed due to this interaction. It is possible to make a more precise statement thanks to the decoherence functions. Let us consider a general subsystem $\rho(t)$ written in a basis that we will note $\{|n\rangle\}_{n\in\mathbb{N}}$. This subsystem is initially pure and interacts with its environment. The decoherence functions will be noted $\Gamma_{nm}(t)$ and are defined in the following way

$$|\langle n|\,\rho(t)\,|m\rangle\,| = \exp(\Gamma_{nm}(t)). \tag{14.115}$$

These functions describe the time evolution of the off diagonal terms of the density matrix and therefore encode the evolution of a subsystem from a pure state to a mixed one. For a system of dimension N there are $N(N-1)/2$ independent off diagonal terms in the density matrix, and as many decoherence functions. One can notice that these functions depend on the basis we chose to write the density matrix. There is however a natural basis, the 'energy basis', in which the diagonal terms are time independent . We will work in this basis here.

Appendix 2: Mutual Information as a Correlation Measure

Relative entropy is a notion defined between two states described by the density matrix ρ and σ in the following way

$$S(\rho|\sigma) := \mathrm{Tr}(\rho\log\rho - \rho\log\sigma). \tag{14.116}$$

One can show that $S(\rho|\sigma) \geq 0$ and it is equal to zero when $\rho = \sigma$. It quantifies the distinguishability of ρ from σ. An important property of the relative entropy is that for any bounded operator $\hat{O}$, the relative entropy satisfies

$$\frac{\left(\mathrm{Tr}(\hat{O}\rho) - \mathrm{Tr}(\hat{O}\sigma)\right)^2}{||\hat{O}||} \leq S(\rho|\sigma), \tag{14.117}$$

with $||\hat{O}|| = \sup\sqrt{\langle\psi|\,\hat{O}\,|\psi\rangle/\langle\psi|\,|\psi\rangle}$. The mutual information is a specific case of the relative entropy. Let us consider a system that can be decomposed into three subsystems A, B and C. We assume that the total system is in a pure state $|\psi\rangle$, the subsystem $A+B$ is described by the density matrix ρ_{AB} and the two subsystems A and B are respectively described by the density matrix ρ_A and ρ_B. The mutual information (between A and B when C is traced out) is given by

$$I_{AB|C} := S(\rho_{AB}|\rho_A \otimes \rho_B). \tag{14.118}$$

The mutual information is quantifies how much a mixed state is distinguishable from the uncorrelated state that we can get by treating the different subsystems as independent. By using (14.116), one can show that

$$I_{AB|C} = S_A + S_B - S_{AB}, \tag{14.119}$$

where S is the usual Von Neumann entropy. An important property of the mutual information comes from (14.117), which, in the case $\rho = \rho_{AB}$ and $\sigma = \rho_A \otimes \rho_B$ gives, for all $\hat{O} = \hat{O}_A \otimes \hat{O}_B$ such that $\hat{O}_A$ and $\hat{O}_B$ are bounded, the following relation

$$\frac{\left(\langle\psi|\,\hat{O}_A\hat{O}_B\,|\psi\rangle - \langle\psi|\,\hat{O}_A\,|\psi\rangle\,\langle\psi|\,\hat{O}_B\,|\psi\rangle\right)^2}{2||\hat{O}_A||||\hat{O}_B||} \leq I_{AB|C}(|\psi\rangle). \tag{14.120}$$

This relation means that the mutual information is an upper bound of the correlations between the subsystems A and B. Thus, we will use this notion to track the correlations between the different subsystems in our model because it has a clear physical meaning.

References

1. S.W. Hawking, Particle creation by black holes. Commun. Math. Phys. **43**, 199–220 (1975) [Erratum: Commun. Math. Phys. **46**, 206 (1976)]
2. R.M. Wald, On particle creation by black holes. Commun. Math. Phys. **45**, 9–34 (1975)
3. L. Parker, Probability distribution of particles created by a black hole. Phys. Rev. D **12**, 1519–1525 (1975)
4. S. Hawking, Breakdown of predictability in gravitational collapse. Phys. Rev. D **14**, 2460–2473 (1976)
5. D.N. Page, Is black hole evaporation predictable? Phys. Rev. Lett. **44**, 301 (1980)
6. J.D. Bekenstein, A universal upper bound on the entropy to energy ratio for bounded systems. Phys. Rev. D **23**, 287 (1981)
7. R. Bousso, The holographic principle. Rev. Mod. Phys. **74**, 825–874 (2002). arXiv:hep-th/0203101
8. T. Banks, L. Susskind, M.E. Peskin, Difficulties for the evolution of pure states into mixed states. Nucl. Phys. B **244**, 125 (1984)
9. D.N. Page, Information in black hole radiation. Phys. Rev. Lett. **71**, 3743–3746 (1993). arXiv:hep-th/9306083
10. C.R. Stephens, G. 't Hooft, B.F. Whiting, Black hole evaporation without information loss. Class. Quant. Grav. **11**, 621–648 (1994). https://doi.org/10.1088/0264-9381/11/3/014
11. W.G. Unruh, R.M. Wald, On evolution laws taking pure states to mixed states in quantum field theory. Phys. Rev. D **52**, 2176–2182 (1995). arXiv:hep-th/9503024
12. A. Strominger, C. Vafa, Microscopic origin of the Bekenstein-Hawking entropy. Phys. Lett. B **379**, 99–104 (1996). arXiv:hep-th/9601029
13. J.M. Maldacena, The large N limit of superconformal field theories and supergravity. Adv. Theor. Math. Phys. **2**, 231–252 (1998). arXiv:hep-th/9711200

14. W. Unruh, Decoherence without dissipation. Trans. R. Soc. Lond. **370**, 4454 (2012). arXiv:1205.6750
15. A. Perez, No firewalls in quantum gravity: the role of discreteness of quantum geometry in resolving the information loss paradox. Class. Quant. Grav. **32**(8), 084001 (2015). arXiv:1410.7062
16. M. Hotta, R. Schützhold, W.G. Unruh, Partner particles for moving mirror radiation and black hole evaporation. Phys. Rev. **D91**(12), 124060 (2015). arXiv:1503.06109
17. R.M. Wald, Particle and energy cost of entanglement of Hawking radiation with the final vacuum state. Phys. Rev. **D100**(6), 065019 (2019). arXiv:1908.06363
18. T. Thiemann, Modern canonical quantum general relativity. arXiv:gr-qc/0110034
19. C. Rovelli, *Quantum Gravity*. Cambridge monographs on mathematical physics (Cambridge University Press, 2004)
20. A. Perez, Introduction to loop quantum gravity and spin foams, gr-qc/0409061, published in *2nd International Conference on Fundamental Interactions (ICFI 2004)*, 6–12 June 2004, Domingos Martins, Espirito Santo, Brazil
21. L. Freidel, E.R. Livine, D. Pranzetti, Gravitational edge modes: from Kac–Moody charges to Poincaré networks. Class. Quant. Grav. **36**(19), 195014 (2019). arXiv:1906.07876
22. L. Bombelli, J. Lee, D. Meyer, R. Sorkin, Space-time as a causal set. Phys. Rev. Lett. **59**, 521–524 (1987)
23. R. Loll, Quantum gravity from causal dynamical triangulations: a review. Class. Quant. Grav. **37**(1), 013002 (2020). arXiv:1905.08669
24. D. Oriti, The group field theory approach to quantum gravity. arXiv:gr-qc/0607032
25. R. Wald, *General Relativity* (University of Chicago Press, Chicago, 1984)
26. A. Ashtekar, B. Krishnan, Isolated and dynamical horizons and their applications. Living Rev. Rel. **7**, 10 (2004). arXiv:gr-qc/0407042
27. R. Bousso, N. Engelhardt, Proof of a new area law in general relativity. Phys. Rev. D **92**(4), 044031 (2015). arXiv:1504.07660
28. L. Freidel, R. Oliveri, D. Pranzetti, S. Speziale, Extended corner symmetry, charge bracket and Einstein's equations. JHEP **09**, 083 (2021). arXiv:2104.12881
29. S.W. Hawking, G.F.R. Ellis, *The Large Scale Structure of Space-Time*. Cambridge monographs on mathematical physics (Cambridge University Press, 2011)
30. R. Penrose, The question of cosmic censorship. J. Astrophys. Astron. **20**, 233–248 (1999)
31. E.T. Newman, R. Penrose, Note on the Bondi-Metzner-Sachs group. J. Math. Phys. **7**, 863–870 (1966)
32. O.M. Moreschi, Supercenter of mass system at future null infinity. Class. Quant. Grav. **5**, 423–435 (1988)
33. T. Josset, A. Perez, D. Sudarsky, Dark energy as the weight of violating energy conservation. Phys. Rev. Lett. **118**(2), 021102 (2017). arXiv:1604.04183
34. A. Perez, D. Sudarsky, Dark energy from quantum gravity discreteness. Phys. Rev. Lett. **122**(22), 221302 (2019). arXiv:1711.05183
35. A. Perez, D. Sudarsky, J.D. Bjorken, A microscopic model for an emergent cosmological constant. Int. J. Mod. Phys. **D27**(14), 1846002 (2018). arXiv:1804.07162
36. A. Perez, D. Sudarsky, Black holes, Planckian granularity, and the changing cosmological "constant". Gen. Rel. Grav. **53**(4), 40 (2021). arXiv:1911.06059
37. A. Perez, D. Sudarsky, E. Wilson-Ewing, Resolving the H_0 tension with diffusion. Gen. Rel. Grav. **53**(1), 7 (2021). arXiv:2001.07536
38. S.J. Landau, M. Benetti, A. Perez, D. Sudarsky, Cosmological constraints on unimodular gravity models with diffusion. Phys. Rev. D **108**(4), 043524 (2023). arXiv:2211.07424
39. L. Amadei, A. Perez, Planckian discreteness as seeds for cosmic structure. Phys. Rev. D **106**(6), 063528 (2022). arXiv:2104.08881
40. G.R. Bengochea, G. Leon, A. Perez, Is Planckian discreteness observable in cosmology? arXiv:2405.12534
41. M. Christodoulou, A. Perez, C. Rovelli, Detecting Planck-scale dark matter with quantum interference. Phys. Rev. Lett. **133**(11), 111001 (2024). arXiv:2309.08238

42. A.C. Wall, A proof of the generalized second law for rapidly changing fields and arbitrary horizon slices. Phys. Rev. D **85**, 104049 (2012). arXiv:1105.3445 [Erratum: Phys. Rev. D **87**(6), 069904 (2013)]
43. A. Rignon-Bret, Black hole thermodynamic potentials for asymptotic observers. arXiv:2406.15843
44. J.F. Barbero, A. Perez, *Quantum Geometry and Black Holes* (WSP, 2017), pp. 241–279. arXiv:1501.02963
45. A. Perez, Black holes in loop quantum gravity. Rept. Prog. Phys. **80**(12), 126901 (2017). arXiv:1703.09149
46. L. Bombelli, R.K. Koul, J. Lee, R.D. Sorkin, A quantum source of entropy for black holes. Phys. Rev. D **34**, 373–383 (1986)
47. H. Casini, M. Huerta, Entanglement entropy in free quantum field theory. J. Phys. A **42**, 504007 (2009). arXiv:0905.2562
48. A. Perez, Statistical and entanglement entropy for black holes in quantum geometry. Phys. Rev. **D90**(8), 084015 (2014). arXiv:1405.7287
49. E. Bianchi, Horizon entanglement entropy and universality of the graviton coupling. arXiv:1211.0522
50. J.F. Barbero, A. Perez, Quantum geometry and black holes (2015). arXiv:1501.02963 (to appear in the World Scientific series)
51. T. Thiemann, *Modern canonical quantum general relativity*. Cambridge monographs on mathematical physics (Cambridge University Press, 2007)
52. J.D. Bekenstein, Black holes and entropy. Phys. Rev. D **7**, 2333–2346 (1973)
53. R. Bousso, H. Casini, Z. Fisher, J. Maldacena, Proof of a quantum Bousso bound. Phys. Rev. D **90**(4), 044002 (2014). arXiv:1404.5635
54. R. Bousso, H. Casini, Z. Fisher, J. Maldacena, Entropy on a null surface for interacting quantum field theories and the Bousso bound. Phys. Rev. D **91**(8), 084030 (2015). arXiv:1406.4545
55. M. Christodoulou, T. De Lorenzo, Volume inside old black holes. Phys. Rev. **D94**(10), 104002 (2016). arXiv:1604.07222
56. C. Rovelli, Black holes have more states than those giving the Bekenstein-Hawking entropy: a simple argument. arXiv:1710.00218
57. S. Raju, How does information emerge from a black hole? arXiv:2404.00374
58. T. Jacobson, P. Nguyen, Diffeomorphism invariance and the black hole information paradox. Phys. Rev. D **100**(4), 046002 (2019). arXiv:1904.04434
59. A. Laddha, S.G. Prabhu, S. Raju, P. Shrivastava, The holographic nature of null infinity. SciPost Phys. **10**(2), 041 (2021). arXiv:2002.02448
60. A. Perez, S. Viollet, Discreteness unravels the black hole information puzzle: insights from a quantum gravity toy model. Entropy **25**(11), 1479 (2023). arXiv:2307.10254
61. A. Ashtekar, J. Lewandowski, Quantum theory of geometry. 2. Volume operators. Adv. Theor. Math. Phys. **1**, 388–429 (1998). arXiv:gr-qc/9711031
62. A. Ashtekar, J.C. Baez, K. Krasnov, Quantum geometry of isolated horizons and black hole entropy. Adv. Theor. Math. Phys. **4**, 1–94 (2000). arXiv:gr-qc/0005126
63. W.G. Unruh, A unimodular theory of canonical quantum gravity. Phys. Rev. D **40**, 1048 (1989)
64. L. Smolin, Unimodular loop quantum gravity and the problems of time. Phys. Rev. D **84**, 044047 (2011). arXiv:1008.1759
65. L. Smolin, The quantization of unimodular gravity and the cosmological constant problems. Phys. Rev. D **80**, 084003 (2009). arXiv:0904.4841
66. D.-W. Chiou, M. Geiller, Unimodular loop quantum cosmology. Phys. Rev. D **82**, 064012 (2010). arXiv:1007.0735
67. L. Amadei, A. Perez, Hawking's information puzzle: a solution realized in loop quantum cosmology. arXiv:1911.00306
68. L. Amadei, H. Liu, A. Perez, Unitarity and information in quantum gravity: a simple example. Front. Astron. Space Sci. **8**, 46 (2021). arXiv:1912.09750
69. R. Penrose, *The Road to Reality: A Complete Guide to the Laws of the Universe* (Alfred A. Knopf, New York, 2004)

70. R. Penrose, Singularities and time-asymmetry, in *General Relativity* (1979)
71. S. Chakraborty, S. Singh, T. Padmanabhan, A quantum peek inside the black hole event horizon. JHEP **06**, 192 (2015). arXiv:1503.01774
72. J. Abedi, H. Arfaei, Obstruction of black hole singularity by quantum field theory effects. JHEP **03**, 135 (2016). arXiv:1506.05844
73. D.N. Page, Particle emission rates from a black hole: massless particles from an uncharged, nonrotating hole. Phys. Rev. D **13**, 198–206 (1976)
74. A. Fabbri, J. Navarro-Salas, *Modeling Black Hole Evaporation* (Imperial College London, London, 2005), p. 334
75. M. Bojowald, Absence of singularity in loop quantum cosmology. Phys. Rev. Lett. **86**, 5227–5230 (2001). arXiv:gr-qc/0102069
76. A. Ashtekar, T. Pawlowski, P. Singh, Quantum nature of the big bang: improved dynamics. Phys. Rev. D **74**, 084003 (2006). arXiv:gr-qc/0607039
77. A. Ashtekar, M. Bojowald, Black hole evaporation: a paradigm. Class. Quant. Grav. **22**, 3349–3362 (2005). arXiv:gr-qc/0504029
78. R.M. Wald, *Quantum Field Theory in Curved Space-Time and Black Hole Thermodynamics*. Chicago lectures in physics (University of Chicago Press, Chicago, IL, 1995)
79. R.M. Wald, *Quantum Field Theory in Curved Spacetime and Black Hole Thermodynamics* (University of Chicago Press, Chicago, 1994)
80. L.E. Parker, D. Toms, *Quantum Field Theory in Curved Spacetime*. Cambridge monographs on mathematical physics (Cambridge University Press, 2009)
81. R.M. Wald, *General Relativity* (University of Chicago Press, Chicago, 1984)
82. A. Ashtekar, Black hole evaporation: a perspective from loop quantum gravity. Universe **6**(2), 21 (2020). arXiv:2001.08833
83. E. Bianchi, M. Christodoulou, F. D'Ambrosio, H.M. Haggard, C. Rovelli, White holes as remnants: a surprising scenario for the end of a black hole. Class. Quant. Grav. **35**(22), 225003 (2018). arXiv:1802.04264
84. P. Chen, Y.C. Ong, D.-H. Yeom, Black hole remnants and the information loss paradox. Phys. Rept. **603**, 1–45 (2015). arXiv:1412.8366
85. S.B. Giddings, Comments on information loss and remnants. Phys. Rev. D **49**, 4078–4088 (1994). arXiv:hep-th/9310101
86. N.V. Prokof'ev, P.C.E. Stamp, Theory of the spin bath. Rept. Prog. Phys. **63**(4), 669 (2000)
87. L.D. Landau, Diamagnetismus der Metalle. Zeitschrift für Physik A Hadrons and Nuclei **64**(9), 629–637 (1930)
88. K. Klitzing, G. Dorda, M. Pepper, New method for high-accuracy determination of the fine-structure constant based on quantized Hall resistance. Phys. Rev. Lett. **45**(6), 494–497 (1980)
89. N. Armitage, E. Mele, A. Vishwanath, Weyl and Dirac semimetals in three-dimensional solids. Rev. Mod. Phys. **90**(1), 015001 (2018)
90. K.S. Novoselov, A.K. Geim, S.V. Morozov, D. Jiang, M. Katsnelson, I.V. Grigorieva, S. Dubonos, A. Firsov, Two-dimensional gas of massless Dirac fermions in graphene. Nature **438**(7065), 197–200 (2005)
91. L.D. Landau, E.M. Lifshitz, *Quantum Mechanics: Non-relativistic Theory, Course of Theoretical Physics*, vol. 3, 3rd edn. (Pergamon Press, Oxford, 1977)
92. A. Perez, S. Viollet, Decoherence without dissipation: phonons in a lattice with local defects (to appear) (2025)
93. A. Ashtekar, P. Singh, Loop quantum cosmology: a status report. Class. Quant. Grav. **28**, 213001 (2011). arXiv:1108.0893
94. L. Amadei, H. Liu, A. Perez, Unitarity and information in quantum gravity: a simple example (2019)
95. A. Almheiri, D. Marolf, J. Polchinski, J. Sully, Black holes: complementarity or firewalls? JHEP **1302**, 062 (2013). arXiv:1207.3123
96. S.L. Braunstein, S. Pirandola, K. Życzkowski, Better late than never: information retrieval from black holes. Phys. Rev. Lett. **110**(10), 101301 (2013). arXiv:0907.1190

97. D. Marolf, The black hole information problem: past, present, and future. Rept. Prog. Phys. **80**(9), 092001 (2017). arXiv:1703.02143
98. A. Perez, S. Ribisi, S. Viollet, Modeling quantum particles falling into a black hole: the deep interior limit. Universe **9**(2), 75 (2023). arXiv:2301.03951
99. A. Ashtekar, New variables for classical and quantum gravity. Phys. Rev. Lett. **57**, 2244–2247 (1986)
100. J.F. Barbero, Real Ashtekar variables for Lorentzian signature space times. Phys. Rev. **D51**, 5507–5510 (1995). arXiv:gr-qc/9410014
101. J. Lewandowski, A. Okolow, H. Sahlmann, T. Thiemann, Uniqueness of diffeomorphism invariant states on holonomy-flux algebras. Commun. Math. Phys. **267**, 703–733 (2006). arXiv:gr-qc/0504147
102. A. Ashtekar, M. Varadarajan, Gravitational dynamics—a novel shift in the Hamiltonian paradigm. Universe **7**(1), 13 (2021). arXiv:2012.12094
103. M. Varadarajan, Anomaly free quantum dynamics for Euclidean LQG. arXiv:2205.10779
104. M. Varadarajan, A. Perez, Public and private discussion during LOOPs 22 conference (2022)
105. S. Viollet, Study of perturbative and non perturbative aspects of quantum black holes. Thesis manuscript
106. G. 't Hooft, The firewall transformation for black holes and some of its implications. Found. Phys. **47**(12), 1503–1542 (2017). arXiv:1612.08640

Chapter 15
Information Retrieval from Euclidean Path Integral

Dong-han Yeom

Abstract In this chapter, we review the information loss paradox in the spirit of the Euclidean path integral approach. First, we argue that there is a long debate about the information loss paradox, and the non-perturbative quantum gravitational wave function must include the clue to the paradox. The Euclidean path integral approach provides the best way to describe the wave function. From this wave function, we can notice that there are not only semi-classical but also non-perturbative contributions, which are highly suppressed but preserved information. Information retrieval will be sufficiently explained if such non-perturbative contributions must be dominated by the late time. We will show that there is sufficient evidence that this scenario can be realized in generic circumstances. Finally, we compare this scenario with alternative approaches. Also, we comment on some unresolved issues that need to be clarified.

15.1 Introduction

The information loss paradox [34] is one of the fundamental questions that must be resolved by the quantum theory of gravity, or the Theory of Everything. Black holes will evaporate due to Hawking radiation in a finite time [33]. However, Hawking radiation seems to depend only on the information of its horizon, e.g., the mass M, the charge Q, and the angular momentum J. After the evaporation, can Hawking radiation carry quantum mechanical information? If not, we will eventually lose the unitarity of quantum mechanics and the fundamental predictability. On the other hand, if information must be preserved, how can Hawking radiation carry information without causing further inconsistency? This paradox drastically reveals the tension between quantum mechanics and general relativity. By struggling to reconcile this

D. Yeom (✉)
Department of Physics Education, Pusan National University, Busan, Republic of Korea
e-mail: innocent.yeom@gmail.com

Leung Center for Cosmology and Particle Astrophysics, National Taiwan University, Taipei, Taiwan

Department of Physics and Astronomy, University of Waterloo, Waterloo, ON, Canada

A. Akil and C. Bambi (eds.), *The Black Hole Information Paradox*, Springer Series in Astrophysics and Cosmology, https://doi.org/10.1007/978-981-96-6170-1_15

tension, we will learn many properties that must be included in the consistent quantum theory of gravity.

This chapter will use the Euclidean path integral approach [30] to study the information loss problem. We will summarize the information loss paradox before discussing further details of this approach. We will briefly discuss why the information loss paradox is difficult to solve and what the logically possible candidate answers are.

15.1.1 Information Loss Paradox

We define the information loss paradox as an inconsistency between well-confirmed theories. We may summarize such well-confirmed theories or assumptions as follows [14].

- 1. The entire process of black hole evaporation satisfies *unitarity*.
- 2. Spacetime satisfies equations according to *general relativity* except for a singularity.
- 3. *Local quantum field theory* is satisfied near the horizon, and Hawking radiation is emitted from the horizon.
- 4. The statistical entropy of a black hole is proportional to the area of the black hole A, i.e., $\log \Omega = A/4$, where Ω is the number of accessible states of the black hole.
- 5. There is an observer who can collect, count, and compute information or entropy from Hawking radiation.

These five assumptions cause inconsistency. Let us think that a black hole is formed due to a gravitational collapse. The collapsed matter will carry some quantum information. Due to Assumptions 1 and 3, a black hole should emit Hawking radiation, and the Hawking radiation should have information about collapsed matter after a specific time. On the other hand, according to Assumption 2 and Assumption 3, the collapsed matter will satisfy the local quantum field theory and general relativity; hence, without experiencing a dramatic event, information will be safely carried to the singularity.

Then, it seems there are two copies of information, one inside the horizon and the other outside. According to Assumptions 1 and 3, the outside should have information; according to Assumptions 2 and 3, the inside should have the original information, too. This violates the no-cloning theorem of quantum mechanics [58].

However, this naive argument might have some loopholes. When will the black hole emit information? If it is duplicated, can any witness see the copied information, violating the no-cloning theorem? To answer these questions, we need to discuss the entanglement entropy and the Page curve.

15.1.2 Entanglement Entropy and the Page Time

Let us ask how to quantify the amount of information that is emitted during the evaporation process. To answer this question, let us first define a bipartite system, where outside the black hole is A and inside the black hole is B. One can determine the total density matrix as ρ. From this, one can define the density matrix of a subsystem $\rho_A = \mathrm{tr}_B\, \rho$ and $\rho_B = \mathrm{tr}_A\, \rho$. From this, one can define the entanglement entropy between A and B:

$$S_A = -\mathrm{tr}_A \rho_A \log \rho_A, \tag{15.1}$$

where $S_A = S_B$ and $S_{AB} = 0$ for a pure state. The maximum bound of the entanglement entropy is the Boltzmann entropy of the system, i.e., if the number of states of A, B is m, n, respectively, then

$$S_A \leq \log m, \tag{15.2}$$

$$S_B \leq \log n \tag{15.3}$$

are satisfied.

There is no information if the radiation is thermal, and the entanglement entropy is the same as its Boltzmann entropy. On the other hand, if there is a deviation between the Boltzmann entropy and the entanglement entropy, it is possible to distinguish information in principle. Hence, we can define the *thermodynamic depth* [43], or information of subsystem, as follows:

$$I_A \equiv \log m - S_A. \tag{15.4}$$

Of course, I_A is positive definite and is zero only if the state is perfectly thermal (Fig. 15.1).

As the black hole evaporates, one can think about the three kinds of information: (1) information of part A, i.e., I_A, (2) information of part B, i.e., I_B, and (3) information of entanglement, i.e., the mutual information $I_{A:B}$, where

$$I_{A:B} \equiv S_A + S_B - S_{AB}. \tag{15.5}$$

(S_{AB} is the entanglement entropy of the system $A \cup B$, where this is zero for a pure state.) For an information-preserving system or a pure state, the sum of three pieces of information will be conserved [39]:

$$I_A + I_B + I_{A:B} = \log mn, \tag{15.6}$$

where $mn =$ const. in a pure state.

Now, we can ask about the relationship between entanglement entropy and Boltzmann entropy. Of course, this depends on the details of its quantum state, but assuming

Fig. 15.1 Boltzmann entropies of radiation ($\log n$), black hole ($\log m$), and the entanglement entropy ($S(A) = S(B)$). The total number of states is a constant ($\log mn$). The red and black dashed curves are I_A and I_B, respectively

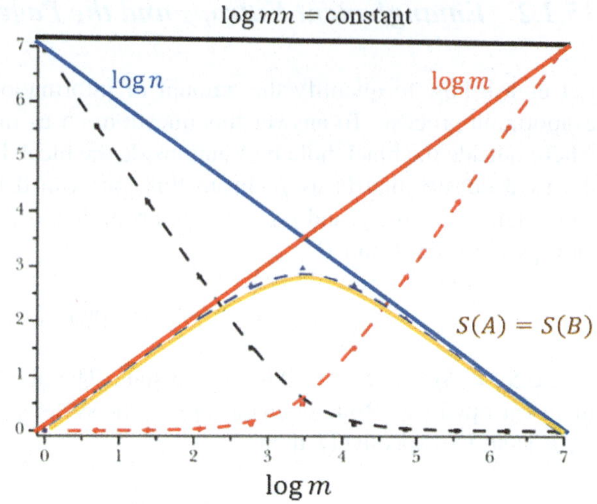

the most typical or random state, one can estimate typical relations. According to Page [49], if $m < n$, approximately,

$$S_A \simeq \log m, \tag{15.7}$$

where analytic and numerical methods can easily confirm this result. This implies that if $m < n$, then $I_A \simeq 0$; hence, Hawking radiation does not carry sufficient information. After the black hole emits more than half of the number of states, in other words, if $m > n$, then $I_B \simeq 0$ and I_A should grow linearly and eventually all information must be escaped. Hence, it is reasonable to explain that (1) in the early stage of evaporation, almost all information is inside the horizon (I_B is dominated), (2) as time goes on, inside and outside of the black hole are entangled and the mutual information is dominated ($I_{A:B}$ is dominated), and (3) eventually, as the number of states of outside dominates, all information will escape from inside to outside by Hawking radiation (I_A is dominated).

This is quite a reasonable explanation of information transfer from inside to outside the black hole. However, naturally, one can raise the following question: *when can we distinguish information from Hawking radiation*? Up to now, there is no principle to answer about this. However, if we introduce Assumption 4, i.e., if the Boltzmann entropy is proportional to its areal entropy, one can estimate the time scale, so-called the *Page time* [50]:

$$\tau_P \simeq M^3. \tag{15.8}$$

This time scale is long, but the black hole can still be semi-classical. Therefore, Hawking radiation is the only channel that carries information around the Page time.

Therefore, based on Assumptions 1–4, we can conclude that the black hole should emit information when the black hole is still in the semi-classical scale. In the following subsection, let us study why this result will cause inconsistency.

15.1.3 Trouble with Entanglements

We can see the details of inconsistency between assumptions in several ways. We clarify this issue in the following three ways. These three approaches are slightly different from each other but present the same aspect of the paradox of black hole information.

- 1. *Duplication of information* [61]: According to Assumption 5, there can be an observer who can measure information from Hawking radiation. Let us assume that an asymptotic observer measures information from Hawking radiation after the Page time. After the observer measures information, let us assume that the observer falls into the black hole. The question is whether the original information that collapsed into the black hole can be sent to the observer. If it is, in principle, allowed, the observer can witness the violation of the unitarity, which reveals the inconsistency among the assumptions.
  The possibility of this *duplication experiment* is related to the uncertainty principle [56]. To successfully send information to the observer, the collapsed matter should send a signal after the time scale Δt (Fig. 15.2), where

$$\Delta t \simeq e^{-\frac{\tau}{r}}, \tag{15.9}$$

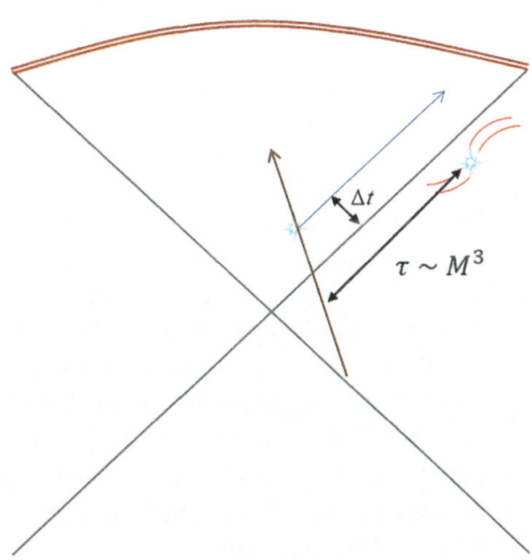

Fig. 15.2 Duplication experiment: after a bit of information falls into the horizon, it escaped after the Page time τ. An observer can measure the duplicated information if the original signal is sent to the outgoing direction inside the horizon between the time Δt

where $r \simeq M$ is the spatial size of the black hole and τ is the time-scale that emits information, e.g., $\tau_P \sim M^3$, or the temporal size of the black hole. Quantum mechanically, to send some information during the time-scale Δt, one needs energy $\Delta E \sim 1/\Delta t$ due to the uncertainty principle. If ΔE is too large, so to speak, if $\Delta E > M$, then this duplication experiment is, in principle, disallowed. However, in the semi-classical regime, this is not prevented in principle [37, 60, 61]. Consider a number N of matter fields contributing to Hawking radiation. In addition, let us scale the spatial size of the black hole such as $r \to r' = \sqrt{N}r$. Then, the temporal scale will be stretched by $\tau \sim M^3 \to \tau' = (\sqrt{N}M)^3/N = \sqrt{N}M^3 \sim \sqrt{N}\tau$. This implies that, in the large N limit, the following quantity is invariant:

$$\frac{\tau}{r} = \frac{\tau'}{r'}, \tag{15.10}$$

while the spatial and temporal size of the black hole increases in the Planck units (hence, semi-classicality is still satisfied). Therefore, $\Delta t \to \sqrt{N}\Delta t$ and $\Delta E \to 1/(\sqrt{N}\Delta t)$ is obtained. By choosing a parametrically large N, one can obtain the condition that $\Delta E < M$. If we think about the time-scale of information emission using the scrambling time [36], $\tau_S \sim M \log M$, then even with a small number of matter fields N, this energy condition $\Delta E < M$ might be easily satisfied, and hence, the duplication experiment is possible [61]. (Also, it is possible to construct such a successful duplication experiment without assuming extremely large N in lower dimensional examples, e.g., see [41].) Therefore, it is evident that the five assumptions will eventually be inconsistent.

- 2. *Inconsistency of entanglement entropy relations* [1]: Let us distinguish the three systems: Hawking radiation before the Page time $\mathscr{E}$, Hawking radiation after the Page time $\mathscr{L}$, and its counterpart $\mathscr{F}$ (Fig. 15.3). For these three systems, the strong subadditivity and subadditivity relations must be satisfied:

$$S_{\mathscr{L}} + S_{\mathscr{E}\mathscr{L}\mathscr{F}} \leq S_{\mathscr{E}\mathscr{L}} + S_{\mathscr{L}\mathscr{F}}, \tag{15.11}$$

$$|S_{\mathscr{E}} - S_{\mathscr{L}\mathscr{F}}| \leq S_{\mathscr{E}\mathscr{L}\mathscr{F}} \leq S_{\mathscr{E}} + S_{\mathscr{L}\mathscr{F}}. \tag{15.12}$$

First, due to Assumptions 1, 3, and 4, after the Page time, information should be emitted. This means that the entanglement entropy should begin to decrease. Note that $\mathscr{E} \cup \mathscr{L}$ is the entire Hawking radiation, and hence $S_{\mathscr{E}\mathscr{L}}$ is the entanglement entropy of Hawking radiation. This should decrease from its maximum value $S_{\mathscr{E}}$, i.e., the entanglement entropy at the Page time. Hence, we obtain $S_{\mathscr{E}\mathscr{L}} < S_{\mathscr{E}}$. Second, due to Assumptions 1, 2, and 3, Hawking radiation does not violate the unitarity. Hence, the only possible way to create particle and anti-particle pairs is that the particle ($\mathscr{L}$) and anti-particle ($\mathscr{F}$) pairs are separable states, i.e., the created particles are not entangled with the background system. Therefore, $S_{\mathscr{L}\mathscr{F}} = 0$ is required.
Then, by applying the strong subadditivity and subadditivity relations, one can conclude that $S_{\mathscr{L}} < 0$; of course, this result is inconsistent. According to

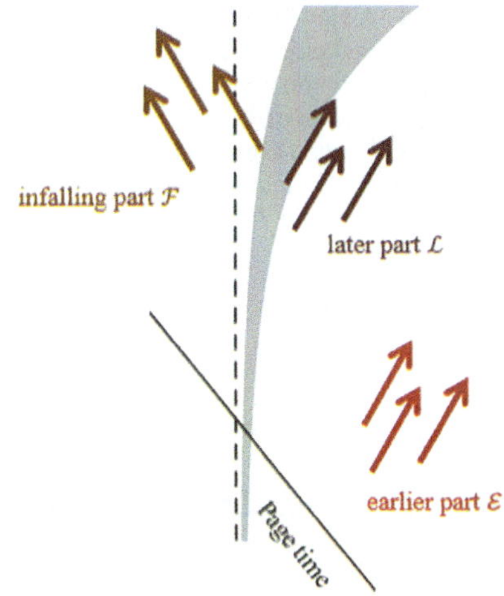

Fig. 15.3 Three parts for the entanglement entropy: radiation before the Page time $\mathscr{E}$, after the Page time $\mathscr{L}$, and its counterpart $\mathscr{F}$

Assumption 5, there is an observer who measures entanglement entropy of $\mathscr{L}$ and the observer will notice that some of the previous assumptions must be wrong because $S_{\mathscr{L}}$ cannot be negative definite. (See also an analog problem in moving mirror cases, in [13].)

- 3. *Monogamy of entanglements* [46]: According to Assumptions 1, 3, and 4, after the Page time, the entanglement entropy must decrease as time goes on. This means that, after the Page time, the number of states outside the horizon should increase monotonically (due to continuous evaporation) while the entanglement entropy of Hawking radiation decreases. The only logically possible way is for there to be entanglements between emitted particles. Therefore, after the entanglement purification protocol, one can find a maximally entangled pair (say, H and H') among Hawking particles after the Page time (Fig. 15.4).
  Assumptions 1, 2, and 3 show that a Hawking radiation pair (particle and antiparticle pair) cannot violate the unitarity. Hence, the only possible way to create them is for the pair to form a separable state. Therefore, the emitted Hawking particle (say, H) should be maximally entangled with its counterpart (say, H''). However, this causes an inconsistency. After the Page time, the emitted Hawking particle H is maximally entangled with its earlier part H'; simultaneously, the emitted Hawking particle H is maximally entangled with its counterpart H''. However, this is disallowed due to the monogamy of quantum entanglements. According to Assumption 5, if an observer detects H' and finds out H'' inside the horizon, the observer will witness the violation of assumptions. This monogamy argument is deeply related to the previous entanglement entropy argument [40].

These three arguments agree that five assumptions cause problems in terms of information or entanglements. Which assumption should be modified?

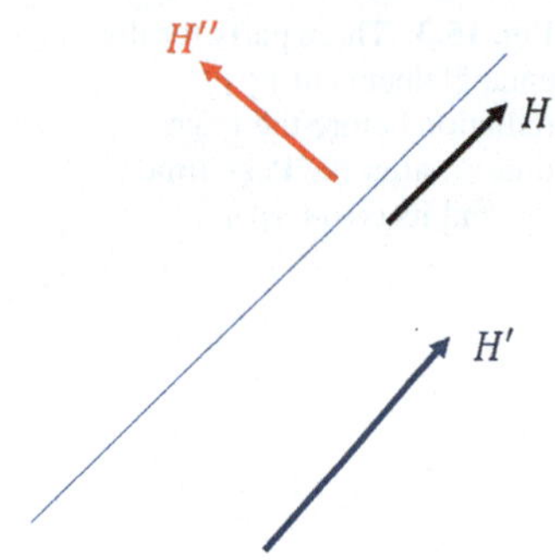

Fig. 15.4 A Hawking particle H that emitted after the Page time is maximally entangled with an earlier radiation H', and at the same time, maximally entangled with its counterpart H''

15.1.4 Candidate Resolutions

Logically, one can classify the candidate resolutions of the paradox as five possibilities [48].

- 1. *Unitarity is violated* [57]. However, this requires further explanations. In addition, the violation of the unitarity principle is tense with the AdS/CFT correspondence [44].
- 2. *General relativity is violated at a considerable length scale.* For example, a firewall [1], a fuzzball [47], a stretched horizon [55], a membrane [12], a large remnant [14], or any other strange object appears at the horizon scale. However, these objects can affect future infinity and, perhaps, can be falsified from astrophysical experiments [17, 38].
- 3. *Quantum effects can be non-local* [51]. In this case, one must explain in which circumstances such a non-local effect is triggered, even within the semi-classical regime.
- 4. *The areal entropy does not represent the Boltzmann entropy* [14]. This possibility may result in the remnant picture. However, one can ask whether the remnant can carry sufficiently large entropy; if the remnant is the Planck scale, then it might be infinitely produced; also, one can ask whether the remnant scenario can successively work in generic situations [14].
- 5. *No observer can detect information* [15, 53]. If this is the case, then one needs to ask how to realize this possibility and whether such a realization mechanism can avoid the inconsistency arguments of the previous subsection.

In this chapter, we may focus on the possibility that these five possibilities are *all true in a way*. What does it mean?

15.1.5 The Answer Is in the Wave Function

The previous inconsistency arguments reveal the trouble between *semi-classical black hole dynamics* and the *unitarity of quantum mechanics.*

Let us analogously see the particle and wave nature of quantum mechanics. Of course, two properties are inconsistent, but in reality, there is no contradiction between the two natures because there is a *quantum measurement* that distinguishes two sides; before the measurement, we only see the wave nature, while after the measurement, we only see the particle nature. The unitarity is related to the wave nature, while classical dynamics is associated with the particle nature.

Can we say that the same thing happens in the black hole physics? This requires the *quantum gravitational wave function* that explains the entire black hole dynamics [22]. In terms of the wave function, losing information is impossible; hence, the unitarity must be preserved. However, in this wave function picture, we must sum over all classical contours/histories; therefore, the semi-classical equations cannot be satisfied. Therefore, Assumption 1 is satisfied, while Assumptions 2 and 3 are no more apparent; hence, it is not surprising to expect non-local physics or dramatic effects at the horizon scale because semi-classical equations are no longer valid.

On the other hand, what 'we' will measure or observe is a semi-classical history. Of course, this semi-classical history should satisfy Assumptions 2 and 3, i.e., general relativity and local quantum field theory. However, it is not surprising that this semi-classical history may not fulfill the unitarity because we do not see the entire wave function of the universe.

If this picture is correct, then there are two points of view: the wave nature and the semi-classical nature. Hence, there are two kinds of observers: the *unitary observer* who can see the entire wave function and the *semi-classical observer* who can follow only one semi-classical history. We are semi-classical observers, and hence, we will lose information. Still, the global wave function will carry information anyway, and the unitarity must be satisfied eventually, although it is beyond our empirical scopes.

This scenario seems to provide an exciting point of view that smoothly avoids the previous tensions. A big question remains: how can we consistently compute the entire wave function? Of course, at this moment, we do not have a consistent theory of everything, and it is fair to say that we do not know how to deal with this problem. However, there is an excellent way to see the brief structure of the wave function, at least approximately. The highway to glimpse the quantum gravitational wave function is to follow the *Euclidean path integral approach.* Now, let us start to discuss what it looks like.

15.2 The Wave Function of the Universe

There are two main approaches to quantum gravity. First, one considers perturbations around a fixed classical background and computes expectation values of perturbations using Feynman diagrams [26]. Second, one directly solves the wave function by introducing the quantum Hamiltonian constraint equation [25]. The former is called by the *covariant approach* because one can use the path integral and the path integral has the manifestly covariant Lagrangian; the latter is called by the *canonical approach*, because this requires the Hamiltonian and canonical quantization relations, etc. Both methods have their problems, e.g., the non-renormalizability problem, and at the same time, some efforts are being made to avoid the issues, e.g., the string theory or loop quantum gravity.

To see the entire dynamics of black hole physics, we need the non-perturbative nature of the wave function, and hence, we should rely on the canonical approach. According to the canonical approach of quantum gravity, for a given 3-hypersurface h_{ab} and field values at the hypersurface ϕ, every information is in the wave function $\Psi[h_{ab}, \phi]$, where this should satisfy

$$\hat{\mathcal{H}}\Psi[h_{ab}, \phi] = 0, \tag{15.13}$$

where $\hat{\mathcal{H}}$ is the quantum Hamiltonian constraint operator; this equation is known as the *Wheeler-DeWitt equation* [25]. There are also quantum momentum constraint equations; sometimes, these are very important to understand the effects of quantum gravity. However, this chapter will focus on the Wheeler-DeWitt equation to cover the system's dynamics.

This Wheeler-DeWitt equation is indeed very complicated and perhaps ill-defined because this is not a simple differential equation but a *functional* differential equation. Avoiding this nature of functional differential equations is impossible except for cosmological cases [42]. Then, how can we solve this wave function? We do not know the best answer yet, but writing a *formal* solution is possible. That is to use the *path integral.*

15.2.1 Euclidean Path Integral Approach

In some sense, the Euclidean path integral [30] solves the equation of the canonical approach non-perturbatively, using the path integral that is manifestly covariant; hence, the Euclidean path integral approach is between two approaches.

One can write the propagator between an in-state $|h_{ab}^{\rm in}, \phi^{\rm in}\rangle$ to an out-state $|h_{ab}^{\rm out}, \phi^{\rm out}\rangle$

$$\left\langle h_{ab}^{\rm out}, \phi^{\rm out} | h_{ab}^{\rm in}, \phi^{\rm in}\right\rangle = \Psi\left[h_{ab}^{\rm out}, \phi^{\rm out}; h_{ab}^{\rm in}, \phi^{\rm in}\right] = \int \mathcal{D}g_{ab}\mathcal{D}\Phi e^{iS[g_{\mu\nu},\Phi]}, \tag{15.14}$$

where $S[g_{\mu\nu}, \Phi]$ is the action with the metric $g_{\mu\nu}$ and a matter field Φ, and this path integral is the sum over all histories that connect from the in-state $|h_{ab}^{\text{in}}, \phi^{\text{in}}\rangle$ to the out-state $|h_{ab}^{\text{out}}, \phi^{\text{out}}\rangle$. This is a path integral and, hence, is a formal solution of the Wheeler-DeWitt equation.

However, one problem is that this path integral is badly behaved. A good strategy to compute the wave function is to Wick-rotate the time from Lorentzian to Euclidean, $dt = -id\tau$, and to obtain

$$\left\langle h_{ab}^{\text{out}}, \phi^{\text{out}} | h_{ab}^{\text{in}}, \phi^{\text{in}} \right\rangle = \int \mathscr{D}g_{ab}\mathscr{D}\Phi e^{-S_{\text{E}}[g_{\mu\nu}, \Phi]}. \tag{15.15}$$

In many examples of quantum mechanics, this Wick-rotation explains the ground state wave function [30]. In the quantum gravitational case, it is impossible to define the ground state. However, we can analogously regard that the Euclidean path integral describes a putative ground state or a state with thermal equilibrium [28].

The exact computation of the Euclidean path integral is still complicated, but now it is easier to *approximate* the wave function. We can introduce the steepest-descent approximation by plugging on-shell histories, i.e., by summing over *instantons* [30]:

$$\left\langle h_{ab}^{\text{out}}, \phi^{\text{out}} | h_{ab}^{\text{in}}, \phi^{\text{in}} \right\rangle \simeq \sum_{\text{on-shell}} e^{-S_{\text{E}}^{\text{on-shell}}[g_{\mu\nu}, \Phi]}. \tag{15.16}$$

These instantons can show the wave function's skeleton structures, including both perturbative and non-perturbative effects.

15.2.2 Hawking Radiation as Instantons

Let us apply this Euclidean path integral approach to black hole dynamics. We should start from an initial state that describes gravitational collapses, but for simplicity, we assume that the classical collapsing process was finished; hence, let us start with the initial state of the Schwarzschild black hole with mass M:

$$|i\rangle = (\text{Schwarzschild black hole with mass } M). \tag{15.17}$$

It will evolve to an out-state $|f\rangle$ (supposed to be defined at the future infinity), where this out-state is in general a superposition of classical boundaries $|f^j\rangle$ ($j = 1, 2, ...$):

$$|f\rangle = \sum_j a_j |f^j\rangle, \tag{15.18}$$

where $p_j \equiv |a_j|^2$ is the probability to transit from $|i\rangle$ to $|f^j\rangle$. Now, from the Euclidean path integral approach, one can write the relation:

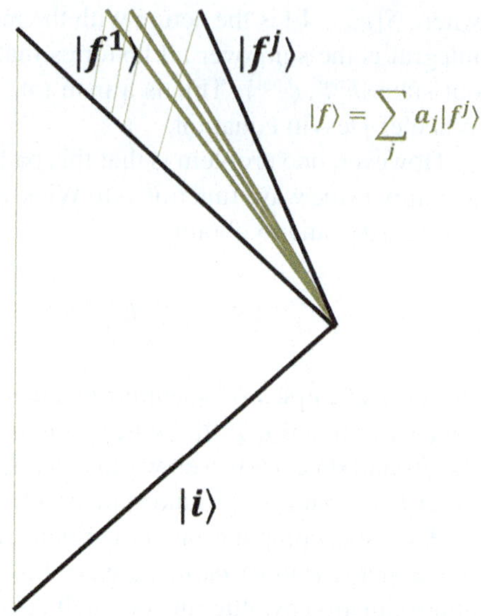

Fig. 15.5 The Euclidean path integral connects from the in-state $|i\rangle$ to the out-state $|f\rangle = \sum_j a_j |f^j\rangle$, where the out-state is a superposition of several classical boundaries

$$a_j^* = \langle f^j|i\rangle \simeq \sum_{|i\rangle \to |f^j\rangle} e^{-S_{\mathrm{E}}^{\mathrm{on-shell}}}, \tag{15.19}$$

where we sum over all instantons that connects from $|i\rangle$ to $|f^j\rangle$ (Fig. 15.5).

Now let us define $|f^j\rangle$ as follows:

$$\begin{aligned} \left|f^j\right\rangle = &\ (\text{Schwarzschild black hole with mass } M') \\ &+ (\text{matter field emission with energy } \delta M), \end{aligned} \tag{15.20}$$

where $M - M' = \delta M > 0$. Then, how can we compute a_j?

To compute the transition probability, what we need to consider are as follows [20]:

- 1. *Action*: The black hole can emit a scalar field, a photon field, a graviton field, or other massive particles, etc. To consider the simplest case, let us assume that the action contains a free scalar field:

$$S = \int_{\mathcal{M}} \sqrt{-g} dx^4 \left[\frac{R}{16\pi} - \frac{1}{2} (\nabla \Phi)^2 \right] + \int_{\partial \mathcal{M}} \frac{K - K_0}{8\pi} \sqrt{-h} dx^3, \tag{15.21}$$

  where $\mathcal{M}$ is the 4-manifold, $\partial\mathcal{M}$ is the boundary of $\mathcal{M}$, R is the Ricci scalar, and K and K_0 are the Gibbons-Hawking boundary terms of the solution and the Minkowski background (evaluated only at infinity) [28], respectively.
  If we consider a Euclidean on-shell solution with a free scalar field, the four-volume integration term automatically vanishes, and we obtain

$$S = \int_{r=\infty} \frac{K - K_0}{8\pi} \sqrt{-h} dx^3 + \text{(boundary term at horizon)}, \quad (15.22)$$

where the first term is evaluated at infinity.

- 2. *Complexification*: At the future infinity, we will consider an outgoing pulse of the matter field. If we evolve the solution backward in time, what will happen? We need to find out about Euclidean geometries. There might be many examples, but one of the trivial examples is that one Wick-rotates to the Euclidean Schwarzschild black hole with mass M'. Then, the matter field should Wick-rotate to the Euclidean time on the background of the Euclidean Schwarzschild black hole with M' [20] (Fig. 15.6).
  The problem is that the matter field usually has non-vanishing differentials for time at the Wick-rotation point. Therefore, due to the Cauchy-Riemann or analyticity conditions, the matter field must be complex-valued after the Wick-rotation; hence, the metric must also be complexified at the Euclidean domain.
  This might be strange, but this is already expected in the Euclidean path integral approach, and indeed, it is an essential property for a consistent Euclidean path integral approach [32]. In the path integral, we will observe the in-state and out-state; in the intermediate process, we do not know what happens. So, it is unsurprising that some strange things happen in the intermediating process. However, we must require the reality condition at infinity [59]; this condition is sometimes called by the *classicality condition* [32], which must be needed for the future infinity. Note that this complexification does not change the computation of Euclidean actions.
- 3. *Regularization of cusps*: Now the final task is to compute the Euclidean action from a Schwarzschild black hole with mass M to a Schwarzschild black hole with mass M' with matter emission δM. Since the Euclidean Schwarzschild black holes with mass M and mass M' are disconnected in Euclidean signatures, one can compute the transition probability as

$$p_j = \left|\langle f^j | i \rangle\right|^2 \simeq e^{-2\left(S_{\mathrm{E}}^{f^j} - S_{\mathrm{E}}^{i}\right)}, \quad (15.23)$$

  where $S_{\mathrm{E}}^{f^j}$ and S_{E}^{i} are the Euclidean action of $|f^j\rangle$ and $|i\rangle$, respectively.
  As we have mentioned, the bulk integration does not contribute to the Euclidean action, but there are contributions at the boundaries of the manifolds. For S_{E}^{i}, this is an Euclidean Schwarzschild manifold, and hence, the only contribution is at infinity (the horizon is regular). On the other hand, for $S_{\mathrm{E}}^{f^j}$, the infinity is indistinguishable from S_{E}^{i} because the asymptotic mass is the same, but at the horizon, the local mass is M'. This causes the mismatch of the Euclidean time period; asymptotically, the Euclidean time period is $8\pi M$, while the regularity condition required at the horizon is $8\pi M'$ [18].
  Due to this mismatch, we require a regularization technique at the horizon [29]. As a result, when we evaluate $S_{\mathrm{E}}^{f^j} - S_{\mathrm{E}}^{i}$, the boundary term at infinity will be

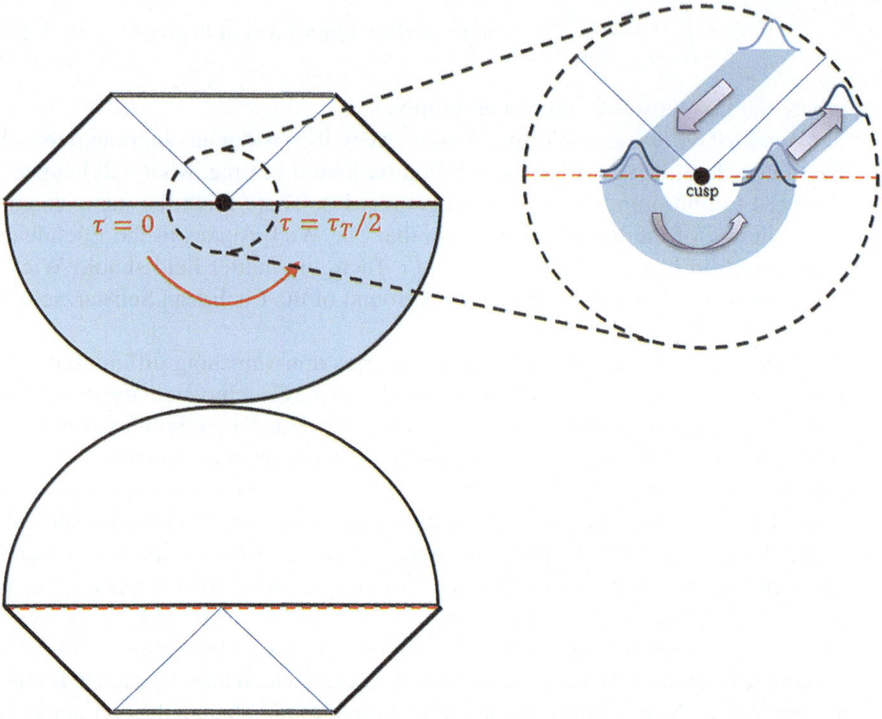

Fig. 15.6 The steepest-descent approximation that connects from a black hole with mass M to a black hole with mass M' and radiation δM. At a constant time hypersurface (red dashed line), we inserted Euclidean Schwarzschild spacetime for past and future diagrams (with the Euclidean time period τ_T). There must be a cusp at the horizon (thick black dot). The scalar field will propagate along the arrows in the black dashed circle. Imposing the classicality condition outside the horizon, the scalar field must be complexified at the Euclidean domain and inside the horizon

canceled perfectly and the boundary term at the horizon (in $S_{\rm E}^{f^j}$) will contribute only. After the regularization, we obtain that

$$2\left(S_{\rm E}^{f^j} - S_{\rm E}^{i}\right) = \frac{A - A'}{4}, \tag{15.24}$$

where A and A' are the areas of the horizon with mass M and mass M', respectively [16].

Therefore, the transition probability from a black hole with mass M to mass M' (with emission) is

$$p_j \simeq e^{-4\pi\left(M^2 - M'^2\right)}. \tag{15.25}$$

If $M - M' = \delta M \ll M$, then this probability approximately becomes the Boltzmann factor $e^{-\delta M/T}$, where $T \equiv 1/8\pi M$ recovers the correct Hawking temperature.

Therefore, in some sense, we can conclude that *Hawking radiation is a kind of instantons* [20]. The Euclidean path integral with a matter field very naturally contains the evaporation process of a black hole, although we need to carefully integrate individual Hawking radiation emission processes successfully.

15.2.3 Paths Toward Trivial Geometries

One further remark is that, in the discussion of the previous subsection, we do not need to assume small δM. Even with δM in the same order of M, the computations will be the same as long as we rely on the free scalar field. Therefore, in the extreme limit $M' = 0$, the black hole can transit to a Minkowski background, and the transition probability is

$$p_j \simeq e^{-4\pi M^2}. \tag{15.26}$$

This probability is exponentially suppressed, but it is fair to say that this is non-vanishing.

In terms of the Euclidean manifold, black holes have the topology $D^2 \times S^2$, where S^2 is due to the spherical symmetry and D^2 covers radial and (Euclidean) time directions [35]. On the other hand, a Minkowski metric has the topology $D^3 \times S$, where S is the Euclidean time direction that can be identified with an arbitrary period. This periodically identified Euclidean Minkowski is called by a *trivial geometry* because it has no horizon [35] (Fig. 15.7).

One critical remark is that, as long as we assume a free scalar field, the Euclidean path integral includes a non-vanishing transition channel toward a trivial geometry (hence, results in a topology change). Of course, we needed to assume a free scalar field, but this is not a strong assumption; for example, the same computations can be easily generalized to graviton modes. Therefore, a tunneling channel toward the trivial geometry is a generic consequence of the Euclidean path integral approach.

15.2.4 Information Survived: Importance of Trivial Geometries

Then, what is the importance of the trivial geometries? Intuitively speaking, after the Wick-rotation, for non-trivial geometries, there exists a horizon and a singularity; on the other hand, for trivial geometries, there is no horizon nor singularity. Therefore, with trivial geometries, there is no distinction between inside and outside the horizon; also, there is no way to lose information.

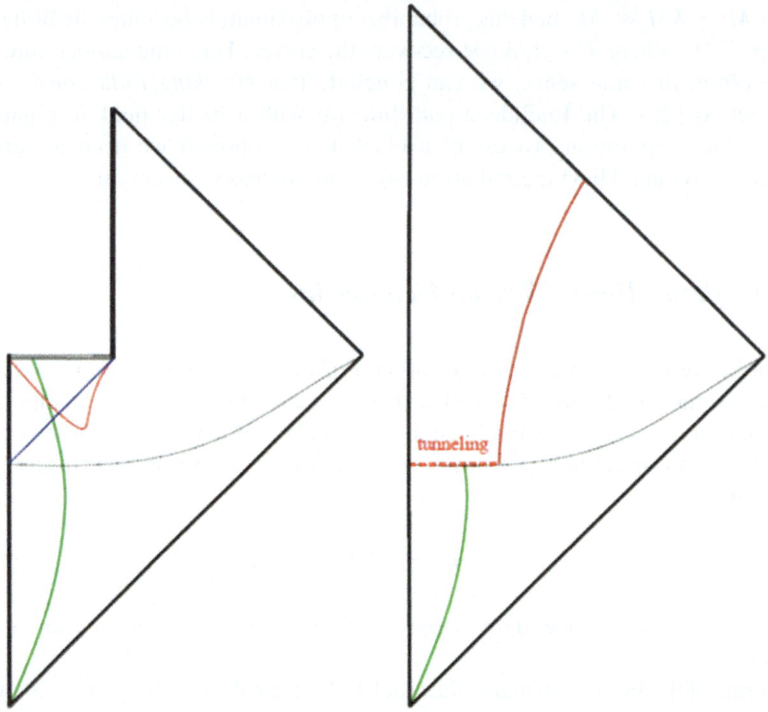

Fig. 15.7 Left: The causal structure of a semi-classical spacetime, where the blue line is the event horizon and a thin red curve is the apparent horizon. Right: The causal structure after the tunneling toward the trivial geometry, where we cut the slice (black dotted curve) of the tunneling outside the horizon. The green curve denotes the collapsing matter, and the red curve denotes the emitted matter after the tunneling

This can be demonstrated by computing correlation functions, especially relying on the AdS/CFT correspondence [45]. (All our computations can easily be extended to the anti-de Sitter background.) In an Euclidean black hole background, when we compute correlation functions (say, $\langle\phi\phi\rangle$) as a function of time, this exponentially decays to zero, which might be related to the loss of information [45]. On the other hand, in the periodically identified geometry backgrounds, this correlation function does not vanish, although the probability is exponentially suppressed [35, 45]. Conceptually,

$$\langle\phi\phi\rangle \simeq p_1\langle\phi\phi\rangle_1 + p_2\langle\phi\phi\rangle_2 + ..., \tag{15.27}$$

where p_1 is the probability of a non-trivial geometry, $\langle\phi\phi\rangle_1$ is the correlation function evaluated at the non-trivial geometry background, p_2 is the probability to a trivial geometry, and $\langle\phi\phi\rangle_2$ is the correlation function evaluated at the trivial geometry background [15, 53]. Although $p_1 \sim 1$ and $p_2 \sim e^{-M^2} \ll 1$, as time goes on,

$\langle\phi\phi\rangle_1 \to 0$, while $\langle\phi\phi\rangle_2 \to$ const., and hence, information will be preserved by the wave function through the trivial geometry.

15.2.5 'New' Wave-Particle Duality

Then, what about the geometries? In this picture, we have two observers. In terms of the unitary observer's point of view, the effective metric is obtained by summing over histories [15, 53]:

$$\langle g_{\mu\nu}\rangle \simeq p_1 g_{\mu\nu}^{(1)} + p_2 g_{\mu\nu}^{(2)} + \ldots, \tag{15.28}$$

where $g_{\mu\nu}^{(1)}$ is a non-trivial geometry and $g_{\mu\nu}^{(2)}$ is a trivial geometry, where both satisfies the semi-classical equations of motion.

Here, since p_1 is dominated, the non-trivial geometry is dominated, and hence, the semi-classical description is approximately good. However, the actual geometry $\langle g_{\mu\nu}\rangle$ of a unitary observer cannot satisfy semi-classical equations of motion. Therefore, in terms of the unitary observer, Assumption 1 is satisfied, but Assumptions 2 and 3 are not satisfied. On the other hand, for a dominant semi-classical observer, $g_{\mu\nu}^{(1)}$ satisfies Assumptions 2 and 3, while Assumption 1 is not satisfied.

In conclusion, no observer can see the inconsistency of all assumptions and hence realizes the fifth way of candidate resolutions. A semi-classical observer (like us) will lose information effectively, but the entire wave function of the universe will keep information in principle. This is in some sense a *new version of the wave-particle duality* in quantum gravity.

15.3 Recovery of the Page Curve

This picture successively resolves the conceptual tensions between assumptions, but we remain the following questions:

- 1. We can agree that information remains in some part of the wave function. However, is it *sufficient* to recover information, even if the contribution is exponentially small? Can we see the retrieval process of information?
- 2. What can we say about Assumption 4? Is the relation between the areal entropy and the Boltzmann entropy always true, or can it be violated at some moment?

In this section, we will answer these questions. To see the information recovering process clearly, we need to see the late-time behaviors of p_i.

15.3.1 Essential Conditions for the Unitary Page Curve

To see more details of the time evolution of information, it will be sufficient to see the time evolution of the entanglement entropy. Let us define A as being outside the black hole and B as inside the black hole. Initially, there is no Hawking radiation; hence, $S_A = 0$. As time passes, the entanglement entropy between the two sides should increase. Eventually, as the black hole evaporation is finished unitarily, there is no interior of the black hole, and hence $S_A = 0$, once again. If we can smoothly explain that S_A becomes zero, eventually, we can resolve the information loss problem constructively.

Now, to explain this unitary Page curve, we require the following two conditions [23] (Fig. 15.8):

- 1. *Multi-history condition*: In the computation of entanglement entropies, we need to sum over more than two kinds of contributions (histories, saddles, steepest-descents, or whatever) that dominantly contribute, say h_1 and h_2. We further assume that h_1 is an information-losing history, while h_2 is an information-preserving history.
- 2. *Late-time dominance condition*: Initially, the probability of h_1 (say, p_1) is dominated by the probability of h_2 (say, p_2). However, as time goes on, p_2 is dominated than p_1 eventually.

If these two conditions are assumed, one can easily explain the unitary Page curve because one can compute the entanglement entropy

$$\langle S \rangle \simeq p_1 S_1 + p_2 S_2 + ..., \tag{15.29}$$

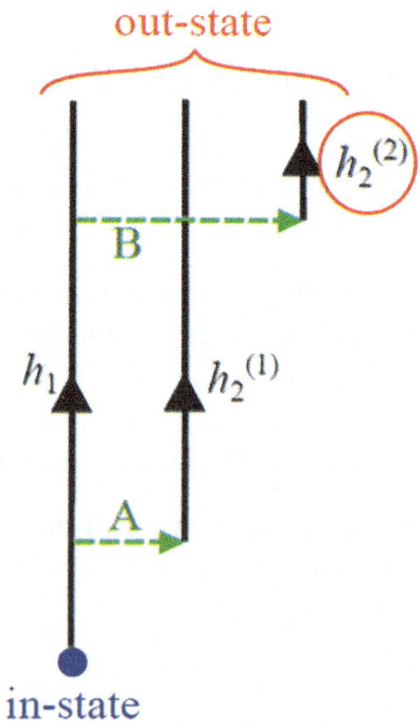

Fig. 15.8 A conceptual picture of the quantum gravitational wave function. From the in-state, histories are evolving and branching from bottom to top. h_1 is the information-losing history, while $h_2^{(1)}$ and $h_2^{(2)}$ are information-preserving histories, where the tunneling to $h_2^{(1)}$ happens in the earlier time and $h_2^{(2)}$ happens in the later time. Due to the late-time dominance condition, $h_2^{(2)}$ will be dominated eventually and save the information

where in the early time, $p_2 \ll p_1 \sim 1$, and in the late time, $p_1 \ll p_2 \sim 1$. In the early times, information seems to have disappeared, but eventually, S_2 goes to zero (because h_2 is an information-preserving history). Hence, $\langle S \rangle \to 0$ will be realized.

The critical question is, *does the Euclidean path integral approach realize these two*? The answer is yes.

First, the Euclidean path integral provides multiple histories. For trivial geometries, there is no distinction between inside and outside the horizon, and hence, the entanglement entropy is automatically zero, so to speak, $S_2 = 0$. For non-trivial geometries, we do not know how to compute entanglement entropies exactly, but let us naively assume that the entanglement entropy monotonically increases: so to speak, $S_1 = S_0 - S$, where S_0 is the initial entropy and S is the areal entropy of the black hole.

The more complicated part is checking the late-time dominance condition. Let us continue to discuss this in the following subsection.

15.3.2 Late-Time Dominance of Trivial Geometries

Now, let us compute the probabilities toward trivial geometries. Of course, this is not a trivial process and depends on the detailed tunneling mechanism. However, we can provide a reasonable guideline for computing the time-dependent probabilities even with this limitation.

First, as we have observed in the previous subsection, Hawking radiation can be regarded as a kind of instantons [20], and, in some limit, such a free scalar field instanton can describe a tunneling process toward a trivial geometry. Therefore, the existence of such a channel at any time is a generic consequence. At each time, the tunneling probability toward the trivial geometry is $\sim e^{-S}$, where S is the Bekenstein-Hawking entropy at the time.

Second, the Euclidean time period is not unique. Of course, the probability decreases as the Euclidean time period increases; hence, we can ignore it in general. However, as the black hole size decreases, the contribution from the multiple time periods may contribute more and more: the tunneling rate toward a trivial geometry is then

$$\Gamma = \sum_{n=1}^{\infty} e^{-S(2n-1)} = \frac{1}{e^{S} - e^{-S}}, \tag{15.30}$$

where n is introduced to present the multipleness of the Euclidean time period [23].

Finally, let us assume that as we average over all other non-trivial geometries, we obtain the semi-classical history, which is an information-losing history. The sum of all histories must be a unity, and the tunneling rate from a semi-classical history to a trivial geometry is Γ. If the probability of the semi-classical history is p_1 and the probability toward a trivial geometry is p_2, then $p_1 + p_2 = 1$ and $\Gamma = p_2/p_1$. From this, we obtain

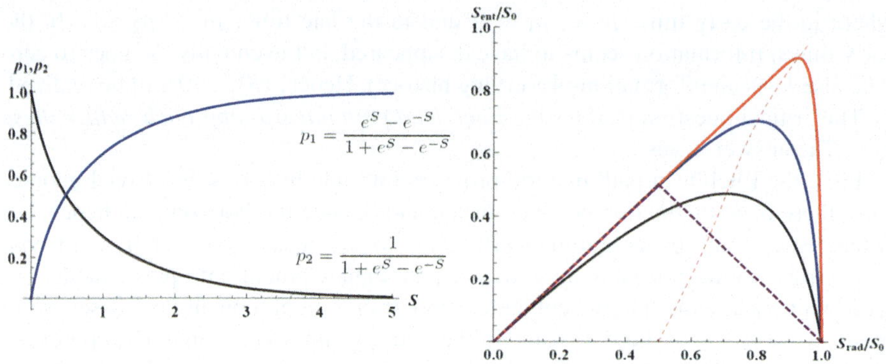

Fig. 15.9 Left: p_1 and p_2 as a function of S. The blue and black curves cross around the Planck scale. Right: $\langle S \rangle$, where black, blue, and red curves correspond larger and larger S_0. See [22, 23] for more details

$$p_1 = \frac{e^S - e^{-S}}{1 + e^S - e^{-S}}, \tag{15.31}$$

$$p_2 = \frac{1}{1 + e^S - e^{-S}}. \tag{15.32}$$

These probabilities were computed with many simplifications but still contained our expected essential features. In the beginning, i.e., if $S \gg 1$, p_1 is much more dominated, and p_2 is exponentially suppressed. However, in the late time, as S approaches to the Planck scale, the golden cross between p_1 and p_2 appears, and eventually, in the $S \to 0$ limit, $p_1 \to 0$ while $p_2 \to 1$. This realizes the *late-time dominance* from the Euclidean path integral approach (left of Fig. 15.9).

15.3.3 Recovery of the Page Curve

As a result, we can compute the entanglement entropy as follows [22, 23]:

$$\langle S \rangle = p_1 S_1 + p_2 S_2 \tag{15.33}$$

$$= (S_0 - S)\left(\frac{e^S - e^{-S}}{1 + e^S - e^{-S}}\right), \tag{15.34}$$

where we assumed that S_0 is the initial Bekenstein-Hawking entropy, $S_1 = S_0 - S$ is the radiation entropy, and hence, the entanglement entropy of the semi-classical history monotonically increases, and $S_2 = 0$, because there is no interior of the black hole in the trivial geometry. This explains the unitary Page curve (right of Fig. 15.9).

However, there are some important remarks in this final result.

- First, there exists a moment when the entanglement entropy goes beyond the entropy bound. In some sense, Assumption 4 is violated instantly because the entanglement entropy might be greater than the areal entropy. However, this does not cause the infinite production problem because the entropy accumulation is just the order of the initial entropy. Also, for a given semi-classical history, the unitarity is effectively broken, and hence, it is not surprising that there is no apparent problem even with the violation of Assumption 4.
- Second, we assumed $S_1 = S_0 - S$, but this is not necessarily true in general. This might overestimate the entanglement entropy of the semi-classical history. If a new mechanism can take entanglements from inside to outside or breaks maximum entanglements, it is not surprising that S_1 might be modified; this will be partly helpful to recover the entropy-bound relation.
- Third, if multiple channels exist toward the trivial geometry [24], the transition to tunnel to the trivial geometry becomes higher. Hence, it will be helpful to decrease the maximum value of the entanglement entropy.

In conclusion, our approach can explain the unitary Page curve within very reasonable and conservative assumptions of quantum gravity. This nicely reveals the analog relations between the wave and particle duality in the quantum gravitational version, where this is the crucial idea to overcome inconsistencies between assumptions. There exists a moment when the entropy bound is violated (hence, Assumption 4 is violated), but it seems that this has not caused any serious problems yet.

Can the Euclidean path integral *does* solve the information loss problem? Do we still have any uncertain issues? In the next section, we will finally summarize future perspectives.

15.4 Future Perspectives

15.4.1 Comparison with Alternative Ideas

First, let us compare this Euclidean path integral picture to various candidate ideas to resolve the information loss problem.

15.4.1.1 Black Hole Complementarity and Black Hole Firewalls

The black hole complementarity principle was proposed initially to overcome the inconsistency between Assumptions 1–5 [55]. If there is no witness to experience inconsistencies between assumptions, it is still consistently acceptable; this is somehow related to the complementarity principle of wave and particle natures. However, the problem is that it is possible to construct an observer who experiences inconsistency between assumptions, as discussed in the previous section [37, 60, 61].

However, the original intuition might be correct in some sense. Like the wave and particle natures, the assumptions are intuitively inconsistent. However, if a *measurement* divides wave and particle nature, one can avoid the apparent inconsistency between contradictory properties. The original black hole complementarity principle was wrong because everything was drawn in the same causal diagram. On the other hand, in our wave function picture, there are two points of view: one is a unitary observer who sees the entire wave function which resembles a wave nature, and the other is a semi-classical observer who sees only a single semi-classical history, that resembles with a particle nature [15, 53]. *Two observers are not living in the same causal diagram* [31]; hence, it is impossible to construct an inconsistency experiment, which was possible in the original black hole complementarity.

As a result, if we think about the unitary observer, who experiences the unitarity of the entire process, the observer will notice that the semi-classical equations are not satisfied; unitarity enforces the violation of general relativity even in the horizon scale. This might be related to the *black hole firewall* [1]. Initially, the effects of the firewall were regarded to be restricted inside the horizon, but this is not logically natural. In our Euclidean wave function picture, it is not surprising, even though effects from the firewall reach infinity [17, 38, 41].

15.4.1.2 Quantum Extremal Surfaces and Replica Wormholes

A proposal was made that the entanglement entropy can be computed using the replica trick [2]. Before we compute the entanglement entropy, we first compute the Renyi entropy, and a specific limit of the Renyi entropy corresponds to the entanglement entropy. It was interesting to observe that there are two main contributions to the Renyi entropy; one is the Gibbons-Hawking-like contribution that explains the increasing part of the entanglement entropy, and the other is a new type of contribution, so-called *replica wormholes* [4].

Then, which contribution is dominant when we compute the entanglement entropy? In terms of *quantum extremal surfaces*, there is only one contribution before the Page time; hence, the entanglement entropy monotonically increases. However, two quantum extremal surfaces appear after the Page time, and we need to choose only one to compute the entanglement entropy. Eventually, one can explain the Page curve that we expected [3].

However, there are still some uncertain questions about this approach. For example, the new contributions, so-called replica wormholes, are only observed in terms of the density matrix [52]; that means there is no corresponding contribution in the wave function itself, although the wave function is a more fundamental notion than the density matrix. Interestingly, for some examples with a holographic dual picture, one can see that the corresponding quantum extremal surfaces before and after the Page time are changed in its dual pictures [4]; however, there is no smooth time-dependent description between the two limits before and after the Page time. Perhaps, to describe such a smooth time evolution around the Page time, we need to know the wave function rather than the density matrix.

However, it is fair to say that this picture shares the same philosophy as the Euclidean path integral picture [22, 23]. We remark on the two essential conditions. First, we need a multi-history condition; in the replica trick, there are more than two types of contribution, one information-losing (Gibbons-Hawking contributions) and the other information-preserving (replica wormholes). Second, we need a late-time dominance condition; after the Page time, we only select the contributions from the information-preserving history. Therefore, we share the same idea to resolve the information loss paradox: information will be preserved by the non-perturbative contributions of the entire path integral. However, there is a difference in whether the path integral is of the wave function (our picture) or the density matrix (replica trick).

15.4.1.3 Bulk-to-Boundary Versus Boundary-to-Bulk

The most substantial evidence of the unitarity of black hole evolution is the AdS/CFT correspondence or the holographic principle [44]. The physics of bulk and boundary correspond, while all physical boundary descriptions are manifestly unitary. Therefore, it is evident that the bulk dynamics, including the formation and evaporation of a black hole, are also unitary, even though there is no constructive description yet.

However, it is important to note one interesting philosophical difference between the typical holography and quantum cosmology: *whether we fix the boundary first or the bulk first* (Fig. 15.10). In the usual approaches of holography, we first fix the boundary and sum over all bulk contributions that satisfy the fixed classical boundary. On the other hand, in the usual quantum cosmological approaches, diverse

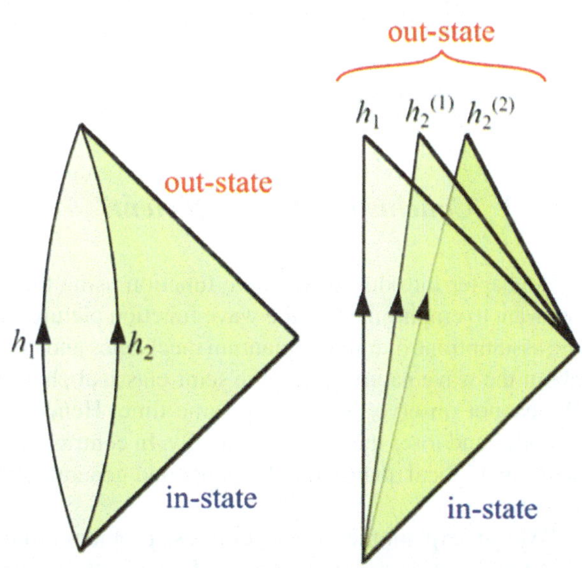

Fig. 15.10 Left: we sum over all histories by fixing the boundary. Right: without fixing the boundary, we follow the bulk history branching process

boundaries can be branched from a given quantum state [31], which is known as the *Everett branching* [27]. Hence, the final boundary is eventually a superposition of various classical boundaries [15, 53].

For example, in the Euclidean path integral approach, considering bubble nucleation from the bulk side may eventually change the boundary [19]. However, if we first fix the boundary at infinity, we must neglect such a process. So, the ER = EPR conjecture might be falsified for the former case [19, 21], while in the latter case, the ER = EPR conjecture cannot be violated [46].

Then, we can ask which approach is more suitable for unitary time evolution. Perhaps both approaches should have consistent unitary descriptions, and it will be worthwhile to study each approach independently. We leave this for an interesting future research topic.

15.4.1.4 Remnant Picture and Entropy Bounds

We noticed that our approach violates the entropy bound for a while, i.e., the statistical entropy is larger than its Bekenstein-Hawking entropy. Is this possible in principle? Interestingly, there is no fundamental proof to secure this entropy-bound relation. Our approach is not the same as the Planck-scale remnant scenario [14] but shares one point: to allow the violation of the entropy bound.

Within the semi-classical framework, it is possible to construct a thought experiment or an algorithm that explicitly violates the entropy bound [5, 11]. However, this does not seem natural from a holographic point of view. Also, several observations show that the Bekenstein-Hawking entropy correctly provides the Boltzmann entropy. However, these computations rely on a zero-temperature system [54], while we need fully dynamic situations. It is fair to say that the belief in the entropy-bound relation cannot be strongly supported in dynamic situations. Can we have a consistent point of view between the semi-classical and holographic approaches? We also leave this question for future research.

15.4.2 Conclusion: Is It Sufficient?

This chapter introduced the wave function using the Euclidean path integral. First, we want to emphasize that the wave function picture can explain the tensions among the assumptions of unitary quantum mechanics and general relativity. The unitarity is about the wave nature, while the semi-classical physics is about the particle nature. We cannot observe them at the same time. Hence, a unitary observer collects all histories and loses the semi-classicality. In contrast, a semi-classical observer like us will satisfy local quantum field theory and general relativity while effectively losing unitarity.

We can explain the information loss problem more technically. A summation of instantons can approximate the Euclidean path integral. The instantons include not only Hawking radiation but also a non-perturbative process that explains tunneling from a black hole to a Minkowski or trivial geometry. If there is a black hole, it is unsurprising to lose information, and the entanglement entropy will increase monotonically. On the other hand, if the spacetime tunnels to a trivial geometry, there is no horizon; hence, there is no way to lose information, and the entanglement entropy will be zero. We have shown that the existence of such a tunneling channel is evident within very natural assumptions.

Moreover, we computed the time dependence of each history, where one is information-losing, and the other is information-preserving. We observed that initially, the former is dominated, while finally, the latter is dominated. Therefore, we can recover a unitary Page curve, although the entropy-bound relation seems to be violated for a while.

Finally, we must mention that we relied on the Euclidean analytic continuation, although the trivial geometry preserves information; hence, we only considered the propagation of geometry outside the event horizon. Using the Lorentzian time description, can we explicitly extend a consistent description that covers both inside and outside the horizon? This may require a method to solve the Wheeler-DeWitt equation *directly* rather than using the path integral. Explicitly, we mention possible future research topics.

- Can we explain the same process by directly solving the Wheeler-DeWitt equation without relying on the Euclidean path integral?
- In this Lorentzian time consideration, can we explain the time evolution of slices that cover both inside and outside the horizon?
- Can we explain information propagation more constructively?
- Can we constructively explain the late-time dominance condition independently and consistently in this direct approach?
- In this case, what is the role or importance of the singularity? Should we provide any boundary condition at the singularity, e.g., the DeWitt boundary condition [7, 10], or does this have nothing to do with the unitary time evolution?
- What is the relationship with various ideas that resolve the singularity, e.g., regular black hole pictures [8] or loop quantum gravity pictures [6, 9]? Can we provide a consistent picture that incorporates the Euclidean path integral approach?

We will reach the ultimate and consistent answer to the information loss paradox as we answer these questions.

Acknowledgements This work was supported by the National Research Foundation of Korea (Grant No. : 2021R1C1C1008622, 2021R1A4A5031460).

References

1. A. Almheiri, D. Marolf, J. Polchinski, J. Sully, JHEP **02**, 062 (2013). arXiv:1207.3123 [hep-th]
2. A. Almheiri, T. Hartman, J. Maldacena, E. Shaghoulian, A. Tajdini, JHEP **05**, 013 (2020). arXiv:1911.12333 [hep-th]
3. A. Almheiri, R. Mahajan, J. Maldacena, Y. Zhao, JHEP **03**, 149 (2020). arXiv:1908.10996 [hep-th]
4. A. Almheiri, T. Hartman, J. Maldacena, E. Shaghoulian, A. Tajdini, Rev. Mod. Phys. **93**(3), 035002 (2021). arXiv:2006.06872 [hep-th]
5. J.M. Bae, D.J. Lee, D. Yeom, H. Zoe, Symmetry **14**(8), 1649 (2022). arXiv:2002.03543 [hep-th]
6. M. Bojowald, S. Brahma, D. Yeom, Phys. Rev. D **98**(4), 046015 (2018). arXiv:1803.01119 [gr-qc]
7. M. Bouhmadi-López, S. Brahma, C.Y. Chen, P. Chen, D. Yeom, JCAP **11**, 002 (2020). arXiv:1911.02129 [gr-qc]
8. M. Bouhmadi-López, C.Y. Chen, X.Y. Chew, Y.C. Ong, D. Yeom, Eur. Phys. J. C **81**(4), 278 (2021). arXiv:2005.13260 [gr-qc]
9. S. Brahma, C.Y. Chen, D. Yeom, Phys. Rev. Lett. **126**(18), 181301 (2021). arXiv:2012.08785 [gr-qc]
10. S. Brahma, C.Y. Chen, D. Yeom, Eur. Phys. J. C **82**(9), 772 (2022). arXiv:2108.05330 [gr-qc]
11. L. Buoninfante, F. Di Filippo, S. Mukohyama, JHEP **10**, 081 (2021). arXiv:2107.05662 [hep-th]
12. C.G. Callan, J.M. Maldacena, Nucl. Phys. B **472**, 591–610 (1996). arXiv:hep-th/9602043 [hep-th]
13. P. Chen, D. Yeom, Phys. Rev. D **96**(2), 025016 (2017). arXiv:1704.08613 [hep-th]
14. P. Chen, Y.C. Ong, D. Yeom, Phys. Rept. **603**, 1–45 (2015). arXiv:1412.8366 [gr-qc]
15. P. Chen, G. Domènech, M. Sasaki, D. Yeom, JCAP **04**, 013 (2016). arXiv:1512.00565 [hep-th]
16. P. Chen, Y.C. Hu, D. Yeom, Phys. Rev. D **94**, 024044 (2016). arXiv:1512.03914 [hep-th]
17. P. Chen, Y.C. Ong, D.N. Page, M. Sasaki, D. Yeom, Phys. Rev. Lett. **116**(16), 161304 (2016). arXiv:1511.05695 [hep-th]
18. P. Chen, G. Domènech, M. Sasaki, D. Yeom, JHEP **07**, 134 (2017). arXiv:1704.04020 [gr-qc]
19. P. Chen, C.H. Wu, D. Yeom, JCAP **06**, 040 (2017). arXiv:1608.08695 [hep-th]
20. P. Chen, M. Sasaki, D. Yeom, Eur. Phys. J. C **79**(7), 627 (2019). arXiv:1806.03766 [hep-th]
21. P. Chen, M. Sasaki, D. Yeom, Int. J. Mod. Phys. D **29**(12), 2050086 (2020). arXiv:2005.07011 [gr-qc]
22. P. Chen, M. Sasaki, D. Yeom, J. Yoon, Int. J. Mod. Phys. D **31**(14), 2242001 (2022). arXiv:2205.08320 [gr-qc]
23. P. Chen, M. Sasaki, D. Yeom, J. Yoon, Entropy **25**(12), 1663 (2023). arXiv:2206.10251 [gr-qc]
24. P. Chen, X.Y. Chew, M. Sasaki, D. Yeom, Phys. Rev. D **110**(4), 044043 (2024). arXiv:2311.00311 [gr-qc]
25. B.S. DeWitt, Phys. Rev. **160**, 1113–1148 (1967)
26. B.S. DeWitt, Phys. Rev. **162**, 1195–1239 (1967)
27. H. Everett, Rev. Mod. Phys. **29**, 454–462 (1957)
28. G.W. Gibbons, S.W. Hawking, Phys. Rev. D **15**, 2752–2756 (1977)
29. R. Gregory, I.G. Moss, B. Withers, JHEP **03**, 081 (2014). arXiv:1401.0017 [hep-th]
30. J.B. Hartle, S.W. Hawking, Phys. Rev. D **28**, 2960–2975 (1983)
31. J. Hartle, T. Hertog, Phys. Rev. D **92**(6), 063509 (2015). arXiv:1502.06770 [hep-th]
32. J.B. Hartle, S.W. Hawking, T. Hertog, Phys. Rev. D **77**, 123537 (2008). arXiv:0803.1663 [hep-th]
33. S.W. Hawking, Commun. Math. Phys. **43**, 199–220 (1975) [Erratum: Commun. Math. Phys. **46**, 206 (1976)]
34. S.W. Hawking, Phys. Rev. D **14**, 2460–2473 (1976)
35. S.W. Hawking, Phys. Rev. D **72**, 084013 (2005). arXiv:hep-th/0507171 [hep-th]
36. P. Hayden, J. Preskill, JHEP **09**, 120 (2007). arXiv:0708.4025 [hep-th]
37. S.E. Hong, D. Hwang, E.D. Stewart, D. Yeom, Class. Quant. Grav. **27**, 045014 (2010). arXiv:0808.1709 [gr-qc]

38. D. Hwang, B.H. Lee, D. Yeom, JCAP **01**, 005 (2013). arXiv:1210.6733 [gr-qc]
39. J. Hwang, D.S. Lee, D. Nho, J. Oh, H. Park, D. Yeom, H. Zoe, Class. Quant. Grav. **34**(14), 145004 (2017). arXiv:1608.03391 [hep-th]
40. J. Hwang, H. Park, D. Yeom, H. Zoe, J. Korean Phys. Soc. **73**(10), 1420–1430 (2018). arXiv:1712.00347 [hep-th]
41. W. Kim, B.H. Lee, D. Yeom, JHEP **05**, 060 (2013). arXiv:1301.5138 [gr-qc]
42. K.V. Kuchar, Phys. Rev. D **50**, 3961–3981 (1994). arXiv:gr-qc/9403003 [gr-qc]
43. S. Lloyd, H. Pagels, Ann. Phys. **188**, 186 (1988)
44. J.M. Maldacena, Adv. Theor. Math. Phys. **2**, 231–252 (1998). arXiv:hep-th/9711200 [hep-th]
45. J.M. Maldacena, JHEP **04**, 021 (2003). arXiv:hep-th/0106112 [hep-th]
46. J. Maldacena, L. Susskind, Fortsch. Phys. **61**, 781–811 (2013). arXiv:1306.0533 [hep-th]
47. S.D. Mathur, Fortsch. Phys. **53**, 793–827 (2005). arXiv:hep-th/0502050 [hep-th]
48. Y.C. Ong, D. Yeom. https://doi.org/10.1142/9789813203952_0076. arXiv:1602.06600 [hep-th]
49. D.N. Page, Phys. Rev. Lett. **71**, 1291–1294 (1993). arXiv:gr-qc/9305007 [gr-qc]
50. D.N. Page, Phys. Rev. Lett. **71**, 3743–3746 (1993). arXiv:hep-th/9306083 [hep-th]
51. D.N. Page, JCAP **06**, 051 (2014). arXiv:1306.0562 [hep-th]
52. G. Penington, S.H. Shenker, D. Stanford, Z. Yang, JHEP **03**, 205 (2022). arXiv:1911.11977 [hep-th]
53. M. Sasaki, D. Yeom, JHEP **12**, 155 (2014). arXiv:1404.1565 [hep-th]
54. A. Strominger, C. Vafa, Phys. Lett. B **379**, 99–104 (1996). arXiv:hep-th/9601029 [hep-th]
55. L. Susskind, L. Thorlacius, J. Uglum, Phys. Rev. D **48**, 3743–3761 (1993). arXiv:hep-th/9306069 [hep-th]
56. L. Susskind, L. Thorlacius, Phys. Rev. D **49**, 966–974 (1994). arXiv:hep-th/9308100 [hep-th]
57. W.G. Unruh, R.M. Wald, Phys. Rev. D **52**, 2176–2182 (1995). arXiv:hep-th/9503024 [hep-th]
58. W. Wooters, W. Zurek, Nature **299**(5886), 802–803 (1982)
59. D. Yeom, Universe **7**(10), 367 (2021). arXiv:2106.10790 [gr-qc]
60. D. Yeom, H. Zoe, Phys. Rev. D **78**, 104008 (2008). arXiv:0802.1625 [gr-qc]
61. D. Yeom, H. Zoe, Int. J. Mod. Phys. A **26**, 3287–3314 (2011). arXiv:0907.0677 [hep-th]

Chapter 16
Quantum Extremal Surfaces and the Page Curve

Raghu Mahajan

Abstract Entanglement islands arise from nontrivial saddle points in the semi-classical path integral computation of the entropy of Hawking radiation via the replica method. Their contribution to the entropy as a function of time yields results that are consistent with the black hole evaporation process being unitary. In this chapter, we discuss a brief derivation of Hawking radiation from the perspective of correlation functions, a description of entropy paradoxes in the eternal black hole, and details of the associated entanglement island computations in two-dimensional models.

16.1 Introduction

Einstein's theory of general relativity describes classical gravity [1, 2]. The equations of motion of general relativity, also called Einstein's equations, determine the geometry of spacetime from appropriate initial conditions. Needless to say, the theory is extremely well-tested experimentally [3]. It is known that gravitational collapse of stars can result in the formation of black holes [4]. Black holes are solutions to Einstein's equations which have horizons. Horizons are co-dimension one null surfaces in spacetime which have the property that light emitted from the region behind the surface cannot make it to infinity. Without any stress-energy sources, a black hole is completely characterized by its mass and angular momentum. A rotating black holes is called a Kerr black hole, and its non-rotating limit is known as the Schwarzschild black hole. The detection of gravitational waves from the merger of a pair of black holes [5] and the results from the Event Horizon Telescope [6] have established beyond doubt the existence of black holes in our universe.

It is natural to expect that classical general relativity will arise as the limit of a quantum theory of gravity. A quantum theory of gravity applicable to our universe continues to evade us, but string theory is currently the most serious candidate [7, 8].

R. Mahajan (✉)
International Center for Theoretical Sciences, Bengaluru, Karnataka, India
e-mail: raghu.mahajan@icts.res.in

Department of Physics, Stanford University, Stanford, CA, USA

A. Akil and C. Bambi (eds.), *The Black Hole Information Paradox*, Springer Series in Astrophysics and Cosmology, https://doi.org/10.1007/978-981-96-6170-1_16

The full power of quantum gravity would be needed to address physics questions that are sensitive to singularities in general relativity.[1] Given that the exact nature of a full theory of quantum gravity remains unknown, we can ask whether there are interesting phenomena that happen when the quantum effects are, in some sense, "milder" than the effects near singularities.

It is a celebrated result of Hawking [9, 10] that, when quantum effects are taken into account, black holes emit particles. This emitted gas of particles has a nonzero temperature, and has come to be referred to as Hawking radiation. For a non-rotating black hole with horizon radius r_+, this temperature T_H is given by the following expression:

$$k_B T_H = \hbar \cdot \frac{c}{r_+} \cdot \frac{1}{4\pi}. \tag{16.1}$$

Note that the temperature goes to zero in the classical limit $\hbar \to 0$. Another important fact is that this effect is intimately tied to the existence of horizons.

There is another remarkable fact about black holes. This is the fact that black holes have an entropy, in the sense of statistical physics, which is equal to the area of the horizon in Planck units [11–13]

$$S_{\mathrm{BH}} = k_B \cdot \frac{1}{4} \cdot 4\pi r_+^2 \cdot \frac{c^3}{G\hbar}. \tag{16.2}$$

In other words, the black hole spacetime should be thought of as an ensemble of a large number of microscopic states, with the number of states being $\exp(S_{\mathrm{BH}})$. The fact that this entropy is finite and devoid of additive ambiguities is another imprint of the underlying quantum theory of gravity.[2]

Let us take the initial star that undergoes gravitational collapse to form the black hole to be in a zero-entropy state. The emission of Hawking radiation by the black hole causes it to lose mass, due to which the black hole horizon starts to shrink in size, and eventually disappears completely—it "evaporates." At this point, something very perplexing has happened: The black hole has disappeared, and so the initial star that was in a zero-entropy state has been completely converted into a gas of particles with a nonzero amount of entropy (since particles with nonzero temperature have nonzero entropy). This would lead us to the disturbing conclusion that the time evolution process that takes the initial star into the final Hawking radiation (with no black hole) is not reversible, at complete odds with the rest of our knowledge about the physical world [16]. This seeming paradox has come to be known as the "black hole information paradox," or the "black hole information problem."

This result has ignited a lot of heated debate in the roughly fifty years that we have known about it. This is not the place to review the full history of how our

---

[1] Concretely, we can think about either the singularity inside black holes or the big bang singularity.

[2] It is not directly relevant to our considerations here, but it is one of the remarkable successes of string theory that it can precisely reproduce the area-entropy relation of certain special but large class of black holes from a counting of microscopic configurations of strings and membranes [14, 15].

understanding of this paradox evolved. We will take a pedagogical approach and simply state the modern perspective.

1. Time evolution processes, whether or not they involve the evaporation of black holes, are unitary, and hence, reversible. This is supported by the AdS/CFT correspondence [17], which tells us that black holes in the presence of a negative cosmological constant are completely equivalent to a plasma of quarks and gluons in a nonabelian gauge theory without gravity [18, 19].[3] Certainly, the time evolution in a nonabelian gauge theory without gravity is reversible.
2. It is not useful to think about the black hole information paradox at only the final stages of black hole evaporation. The horizon radius is then very small and we would have to think about the full theory of quantum gravity. Luckily, a paradox exists even when the black hole has shrunk by only an order one fraction (larger than one-half). This is the result of Page [20–22]. The Page version of the paradox stems from the fact that, for a quantum many body system in a generic pure state, the bipartite von Neumann entropy of a subsystem and its complement are equal, and equal to the number of degrees of freedom in the *smaller* subsystem. Let us take the two subsystems to be (a) the black hole, at a time which is some order one fraction of its lifetime, and (b) the Hawking radiation that has been emitted up to that point of time. On one hand, Hawking's result says that the entropy of the emitted radiation should be a monotonically increasing, almost linear function of time. On the other hand, the above result of Page suggests that after the half-way evaporation stage, the entropy of Hawking radiation should start going down and be equal to the area-entropy of the black hole, which is decreasing with time. This leads to a contradiction not just in the final-stages of the black hole evaporation, but also when the black hole is still large. This makes the problem amenable to a semi-classical treatment, which is a huge step forward for a deeper understanding of the problem. Given the importance of this result, the time when the black-hole has evaporated half-way is referred to as the "Page time," and the tent-shaped time-dependence of the entropy of Hawking radiation is referred to as the "Page curve."
3. Any question about quantum theory of gravity must be formulated as a path integral problem with boundary conditions that are appropriate to the question being asked. If we want to compute the von Neumann entropy of Hawking radiation from first principles, we must first find a way to formulate the question as a path integral with the correct boundary conditions. This is possible via the replica technique [23], which is a generalization of the fundamental work of Gibbons and Hawking [13]. Since this point has been the source of innumerable confusions, we cannot emphasize enough that the replica method is deployed on the boundary, and not in bulk of the gravitational spacetime. Since we are asking questions in a regime where semi-classical methods are valid, we can now compute the path integral by the saddle point method. Two fundamental questions are: What are the relevant saddle points and what are the values of their on-shell action?

[3] The nonabelian gauge theory is defined on the boundary of the black hole spacetime.

4. When one formulates the question this way using first principles, one discovers that, at early times, the standard black hole saddle point dominates. This leads to the growth of the entropy of Hawking radiation according to Hawking's prediction. However, after the Page time, a different saddle point starts to have lower action and thus dominates the answer. This saddle point is a "replica wormhole" [24, 25], which is a bulk geometry that connects the various replicas of the boundary that enter the replica technique. The on-shell action of this new saddle point leads to an entropy that tracks the shrinking area of the black hole, hence reproducing the downward slope of the Page curve after the Page time. The existence of this new saddle-point is highly nontrivial and follows a series of developments in the computation of von Neumann entropies in the context of the AdS/CFT: the Ryu-Takayanagi (RT) formula [26], the one-loop correction to the RT formula [27], the quantum extremal surface prescription [28], and a path integral understanding of these results [23]. The Page curve was derived using the quantum extremal surface prescription in the remarkable papers [29, 30]. These two papers in turn motivated the works [24, 25] which identified the replica-wormhole saddle points in the path integral setup of entropy computations.

The structure of this chapter is as follows. We will first present a derivation of the phenomenon of Hawking radiation using a lesser known technique [31] that focuses on the Kubo-Martin-Schwinger (KMS) property [32, 33] property of thermal two-point correlators as opposed to the usual method using Bogoliubov coefficients [10]. Then we will discuss some necessary formulas for von Neumann entropy in two-dimensional conformal field theories. We will then introduce our model of an eternal two-dimensional black hole in Jackiw-Teitelboim (JT) gravity [34] coupled to a two-dimensional conformal field theory. Following [35, 36], we will discuss the version of the entropy paradox and the Page curve in this eternal black-hole setup (as opposed to a black hole formed from collapse). We will then use the quantum extremal surface prescription [28] to derive the Page curve via a competition between two candidate extrema. We restrict to the case of an eternal black-hole since it is simpler compared to an evaporating black hole but contains all the necessary ingredients. We will not discuss replica-wormholes in detail, see the review [37] and also the chapter on "Replica Wormholes" by Jingru Lu, Manqian Ou and Zhenbin Yang in this book, but we hope that we have conveyed the basic idea in the introduction.

16.2 The KMS Property and the Thermality of Hawking Radiation

16.2.1 The KMS Property

Consider a quantum system in the thermal equilibrium density matrix at inverse-temperature $\beta > 0$. We define a strip region in the complex time plane

$$S := \{t \in \mathbb{C} : -\beta < \mathrm{Im}(t) < 0\}. \tag{16.3}$$

In other words, we consider a horizontal open strip lying between the lines $y = 0$ and $y = -\beta$. Let ϕ be an operator in the quantum system under consideration. For $t \in S$, we define the correlation function

$$G(t) := \frac{1}{Z}\,\mathrm{Tr}\left(e^{-\beta H}\phi(t)\phi(0)\right) \qquad (\text{for } t \in S) \tag{16.4}$$

$$= \frac{1}{Z}\,\mathrm{Tr}\left(e^{-\beta H}e^{+\mathrm{i}Ht}\phi\, e^{-\mathrm{i}Ht}\phi\right), \tag{16.5}$$

$$\text{where } Z := \mathrm{Tr}\, e^{-\beta H}. \tag{16.6}$$

Here H is the Hamiltonian, or the operator that generates translations of the time coordinate t. Note that the first exponential in (16.5) is well defined since $-\beta + \mathrm{i}t = -\beta - \mathrm{Im}(t) + \mathrm{i}\mathrm{Re}(t)$ has negative imaginary part (recall that $t \in S$ and hence $\mathrm{Im}(t) > -\beta$). Similarly, the second exponential is well defined since $-\mathrm{i}t = \mathrm{Im}(t) - \mathrm{i}\mathrm{Re}(t)$ also has negative imaginary part. In other words, $G(t)$ is an analytic function in the strip S. In most systems of interest, $G(t)$ has interesting singularities on the boundary of S.

The Kubo-Martin-Schwinger (KMS) property [32, 33] states that

$$G(t_1) = G(t_2) \quad \text{if} \quad \frac{t_1 + t_2}{2} = -\mathrm{i}\,\frac{\beta}{2}. \tag{16.7}$$

This follows from the definition of $G(t_1)$ in (16.5) since using $\beta = \mathrm{i}(t_1 + t_2)$, we can rewrite the numerator of (16.5) as $\mathrm{Tr}(e^{-\mathrm{i}Ht_2}\phi\, e^{-\mathrm{i}Ht_1}\phi)$. This latter expression is symmetric under the exchange of t_1 and t_2 because of the cyclic property of the trace. Geometrically, the KMS property (16.7) says that the function $G(t)$ is equal at two points in S such that the mid-point of the line segment joining those two points coincides with the "mid-point" $-\mathrm{i}\frac{\beta}{2}$ of the strip; see the left panel of Fig. 16.1. If we evaluate the correlator at a purely imaginary value of t, say $t = -\mathrm{i}\tau$ with $0 < \tau < \beta$, the KMS property (16.7) says that the correlator is invariant under $\tau \leftrightarrow \beta - \tau$; see the right panel of Fig. 16.1.

There are straightforward generalizations of the KMS property when the two operators are not identical, or when the operators are Grassmann odd.

The KMS property is an alternative *equivalent* characterization of thermal states [38, 39]. This is useful conceptually since quantum systems usually encountered in quantum field theory do not have a well-defined density matrix. So, it is not rigorous to define the thermal state of a general quantum system by declaring its density matrix to be $e^{-\beta H}$. It is thus conceptually cleaner to define a thermal state via the fact that correlation functions in a thermal state satisfy the KMS property.

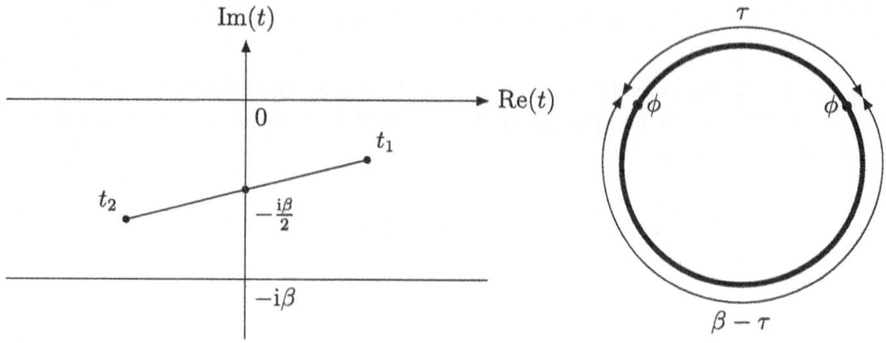

Fig. 16.1 The KMS property (for identical scalar Grassmann-even operators). The thermal correlator $\langle \phi(t)\phi(0)\rangle$ is analytic in the strip $-\beta < \mathrm{Im}(t) < 0$. The KMS property states that $G(t_1) = G(t_2)$ if $\frac{t_1+t_2}{2} = \frac{-i\beta}{2}$. If t is purely imaginary, say $t = -i\tau$, then the correlator can be represented via a purely Euclidean path integral. The KMS property is then the statement that the separation between the two operators can equally well be taken to be τ or $\beta - \tau$

16.2.2 KMS Property of Rindler Correlators in the Minkowski Vacuum

In this subsection we want to discuss the physics of the Rindler wedge, vis-a-vis the physics of Minkowski space. This is a simple example which is very useful in the understanding of Hawking radiation. Our presentation will focus on the important ideas that are needed later. We refer the reader to the extremely detailed and pedagogical review article [40] for a more information. Some original references are [41–43].

We denote D-dimensional Minkowski space by $\mathbb{R}^{1,D-1}$. We use coordinates $(x^0, \ldots, x^{D-1})$ with the line element

$$ds^2 = -(dx^0)^2 + (dx^1)^2 + \cdots + (dx^{D-1})^2. \tag{16.8}$$

The right Rindler wedge is defined as the subset

$$R^+ := \{(x^0, \ldots, x^{D-1}) \in \mathbb{R}^{1,D-1} : x^1 > |x^0|\}. \tag{16.9}$$

Note that it is defined as an open set. Let us explain the significance of R^+.

Consider a quantum field theory on $\mathbb{R}^{1,D-1}$. Full Minkowski space has a Cauchy hypersurface defined simply by $x^0 = 0$. If we restrict ourselves to half of this hypersurface defined by $x^1 > 0$, then the region R^+ is the domain of dependence of this half-Cauchy surface. If we restrict states of quantum fields in full Minkowski space to R^+, we will typically obtain mixed states since we are effectively tracing out degrees of freedom in the other half of the Cauchy hypersurface. A basic, yet deep fact is that the Minkowski vacuum state is in fact a thermal state when restricted to R^+.

Let us introduce coordinates $(\eta, \xi, x^2, \ldots, x^{D-1})$ that are suited to R^+. The coordinates η and ξ are defined via

$$x^0 =: \xi \sinh \eta, \tag{16.10}$$

$$x^1 =: \xi \cosh \eta, \tag{16.11}$$

$$\eta \in (-\infty,\infty), \quad \xi \in (0, \infty). \tag{16.12}$$

The boundary of R^+, given by $x^1 = |x^0|$, maps to $\xi = 0$. The line element (16.9) in these coordinates reads

$$\mathrm{d}s^2 = -\xi^2 \mathrm{d}\eta^2 + \mathrm{d}\xi^2 + (\mathrm{d}x^2)^2 + \cdots + (\mathrm{d}x^{D-1})^2. \tag{16.13}$$

Note that ∂_η is a Killing vector, and is in fact equal to the boost generator $x^0\partial_1 + x^1\partial_0$.

The curve with fixed $(\xi, x^2, \ldots, x^{D-1})$, that is, with only the η coordinate varying, is a timelike trajectory with constant proper acceleration, see Fig. 16.2. This

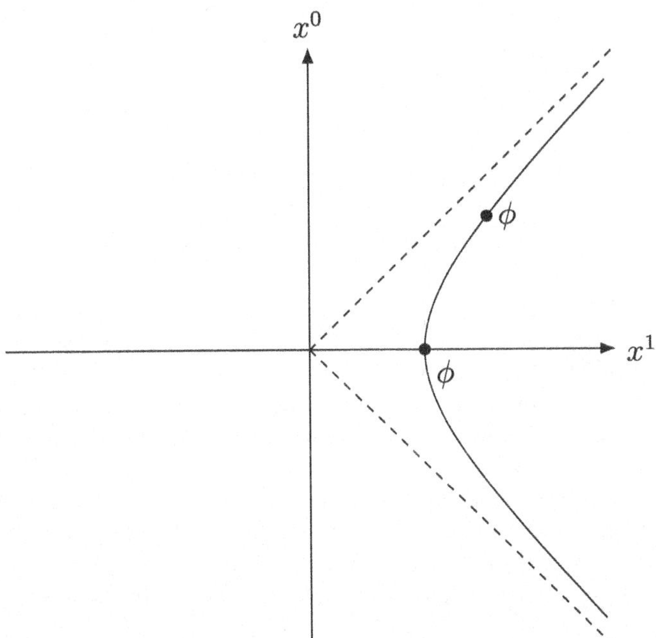

Fig. 16.2 The right Rindler wedge is the region R^+ obeying the inequality $x^1 > |x^0|$. A curve of constant ξ is depicted, it is a timelike trajectory with constant proper acceleration. We consider the two-point function of an operator in the Minkowski vacuum state, with the two-points separated purely in the η direction. This correlator obeys the KMS property with $\beta = 2\pi$ and with respect to the generator of η-translations

can be verified by representing the (x^0, x^1) part of the trajectory with proper time coordinate λ, that is, via the parameterization $\left(\xi_0 \sinh \frac{\lambda}{\xi_0}, \xi_0 \cosh \frac{\lambda}{\xi_0}\right)$. We can now see that the norm of the acceleration vector $\frac{\mathrm{d}^2 x^\mu}{\mathrm{d}\lambda^2}$ is constant and is equal to $\frac{1}{\xi_0}$. In fact, since the Rindler wedge is translationally invariant in the $(x^2, \ldots, x^{D-1})$ directions, we can say that we have an infinitely big flat slab, placed at some fixed ξ_0 that is constantly accelerating perpendicular to itself. Invoking the principle of equivalence, this is a model for an infinitely big and flat planet with a uniform gravitational acceleration [40, 44].

Consider a CFT on Minkowski space and a scalar operator ϕ in this CFT with dimension Δ. We place the CFT in the Minkowski vacuum state. For ease of notation, we also denote the Minkowski coordinates by $(t, x, \vec{x})$ where $\vec{x}$ is a $(D-2)$-dimensional vector. The two-point function of ϕ is given by[4]

$$\langle \phi(t_1, x_1, \vec{x}_1)\, \phi(t_2, x_2, \vec{x}_2) \rangle = \left[-(t_1 - t_2)^2 + (x_1 - x_2)^2 + (\vec{x}_1 - \vec{x}_2)^2 \right]^{-\Delta} . \tag{16.14}$$

We now consider the situation where $(t_1, x_1, \vec{x}_1) \in R^+$ and also $(t_2, x_2, \vec{x}_2) \in R^+$, such that the two points have the same ξ and $\vec{x}$ coordinates, see Fig. 16.2. In other words, if there was no time separation between them, the two operators would be identical.[5]

Let us now calculate the correlator in terms of the η, ξ coordinates. The squared separation between the two points appearing in the correlator (16.14) becomes

$$-(t_1 - t_2)^2 + (x_1 - x_2)^2 = -\xi^2 (\sinh \eta_1 - \sinh \eta_2)^2 + \xi^2 (\cosh \eta_1 - \cosh \eta_2)^2 \tag{16.15}$$

$$= 2\xi^2 \left(1 - \cosh(\eta_1 - \eta_2)\right) \tag{16.16}$$

Without loss of generality, we let $\eta_2 = 0$. Clearly since $\cosh(z) = \cosh(z')$ when $z + z' = -2\pi \mathrm{i}$, we conclude that the correlator[6] $\langle \phi(\eta)\, \phi(0) \rangle$ obeys the KMS property (16.7) with respect to the generator of η-translations and with $\beta = 2\pi$. It is clear from the structure of the argument that this result does not depend on us having a CFT or considering a scalar operator; we only used the expression of the Minkowski-invariant interval in Rindler coordinates.

The operator that generates the η-translations is the boost operator in Minkowski space, but its interpretation in the wedge R^+ in the coordinate system (16.13) is that of a Hamiltonian since η is the natural time coordinate in R^+. Thus we arrive at the conclusion that Minkowski vacuum state (i.e. the ground state of the operator

---

[4] We consider a Wightman function and there is an implicit $\mathrm{i}\epsilon$ prescription in the formula below. The $\mathrm{i}\epsilon$ is not relevant for our considerations here, so we omit it for the sake of clarity.

[5] We consider this case for simplicity, since from the perspective of our discussion of the KMS property, two field operators at different spatial locations are different operators.

[6] We emphasize that this correlator is computed in the Minkowski vacuum.

that generates translations in x^0) is a thermal state from the point of the view of the Rindler Hamiltonian with $\beta = 2\pi$.

Usually, this derivation is presented using a path integral in Euclidean space that prepares the Minkowski vacuum. That derivation is, of course, physically correct and provides some very useful pieces of intuition. We have chosen to present things using purely Lorentzian correlators since, for a black hole formed from the collapse of a star, the relevant state is not that of global thermal equilibrium. The strategy of [31] to derive Hawking radiation is to show that two-point functions computed at late times in the black hole spacetime resulting from the collapse of a star obey the KMS property.

The quick way to figure out the temperature is to consider the line element (16.13) with purely imaginary η. The proper framework to think about this is that the Minkowski manifold under consideration is a half-dimensional subspace of a larger manifold with complex dimension D. If the quantities under study are analytic in this larger manifold, one can take different half-dimensional sheets. In particular, let us take $\eta = -\mathrm{i}\theta$ with $\theta \in \mathbb{R}$. The line element (16.13) becomes

$$\mathrm{d}s^2 = \xi^2 \mathrm{d}\theta^2 + d\xi^2 + \mathrm{d}\vec{x}^2. \tag{16.17}$$

Now the (θ, ξ) part of this metric is just a 2D Euclidean plane in polar coordinates. If we demand the absence of a conical singularity at $\xi = 0$ in this geometry, we must identify θ with periodicity 2π. The absence of a conical singularity is just the statement that the Minkowski vacuum state on $\mathbb{R}^{D-1}$ can be prepared using a path integral on half of $\mathbb{R}^{D+1}$.

16.2.3 Temperature of an Eternal Schwarzschild Black Hole

The above discussion of Rindler space vis-a-vis Minkowski space naturally generalizes to the eternal Schwarzschild black hole. Consider, for concreteness, the line element of a 4D Schwarzschild black hole in flat space

$$\mathrm{d}s^2 = -f(r)\mathrm{d}t^2 + \frac{\mathrm{d}r^2}{f(r)} + r^2\mathrm{d}\Omega_2^2, \tag{16.18}$$

$$f(r) = 1 - \frac{r_+}{r}, \quad r_+ > 0. \tag{16.19}$$

Here r_+ is the Schwarzschild radius, or the location of the horizon, and $\mathrm{d}\Omega_2^2$ is the standard metric on a 2-sphere. Note that $f(r_+) = 0$ and $f'(r_+) \neq 0$.

We consider a quantum field propagating in this eternal Schwarzschild black hole. Since there is no global timelike Killing vector, we cannot identify a natural analog of the Minkowski vacuum state, which is the state with lowest eigenvalue of the quantum operator that generates translations along that timelike Killing vector.

However, the metric (16.18) has a natural Euclidean continuation. As in the Rindler case, let $t = -i\theta$ with $\theta \in \mathbb{R}$.

$$ds^2 = f(r)d\theta^2 + \frac{dr^2}{f(r)} + r^2 d\Omega_2^2, \tag{16.20}$$

$$ds^2 \approx f'(r_+)(r - r_+)\, d\theta^2 + \frac{dr^2}{f'(r_+)(r - r_+)} + r_+^2 d\Omega_2^2, \tag{16.21}$$

where in the second line we have written the near-r_+ behavior of the metric. If we define a new coordinate $\rho := \frac{2}{\sqrt{f'(r_+)}}\sqrt{r - r_+}$, then we can write the near-horizon metric as

$$ds^2 \approx \frac{f'(r_+)^2}{4}\rho^2\, d\theta^2 + d\rho^2 + r_+^2 d\Omega_2^2. \tag{16.22}$$

Thus, in the (θ, ρ) section of the metric, we again get a 2D plane in polar coordinates. Demanding the absence of a conical singularity leads us to demand that the θ coordinate should be periodic with period

$$\beta = 2\pi \cdot \frac{2}{f'(r_+)} = \frac{4\pi}{f'(r_+)}. \tag{16.23}$$

The Hartle-Hawking state of quantum fields in the eternal black hole geometry is the unique state such that correlators of operators in one of the exterior regions of the eternal Schwarzschild geometry are analytic in the strip $-\beta < \mathrm{Im}(t) < 0$. They also satisfy the KMS condition with respect to the generator of t-translations with $\beta = \frac{4\pi}{f'(r_+)}$. Thus, the Hartle-Hawking state is a nonzero temperature state when restricted to one of the outside regions. Of course, it is a pure state when we consider it as a state on the full Cauchy slice of the eternal Schwarzschild black hole geometry, which includes both the left and the right exterior regions.

16.2.4 A Spectral Paradox in the Eternal Black Hole Spacetime

The most important difference between the Schwarzschild and the Rindler cases is that the Schwarzschild black hole is supposed to describe a finite temperature state of some putative quantum system with a *finite* entropy. The value of this finite entropy is given by the Bekenstein-Hawking area formula (16.2). Since the area of the Rindler horizon is infinite (this is the area spanned by coordinates $(x^2, \ldots, x^{D-1})$), the value of the entropy, even if a meaningful notion existed, would be infinite. For the black hole case, the argument is most strong when the asymptotic spacetime is negatively curved anti-de Sitter space [17, 18]. There, the black hole describes a finite temperature state of quarks and gluons in some gauge theory.

Consider now the operator H that generates t-translations, where t is the Schwarzschild time coordinate in an eternal AdS Schwarzschild black hole. We will describe a paradox by first presenting an argument that the spectrum of H is discrete and then presenting an argument that the spectrum of H is continuous. The argument we present is a disguised version of [19].

The argument that the spectrum of H is discrete is simply that near the boundary of anti-de Sitter space, the t-translations are the time translations of the boundary quantum system. The Hamiltonian of the boundary gauge theory with finite N and when placed on a spatial manifold of finite volume has a discrete spectrum.

The argument that the spectrum of H is continuous can be phrased mathematically by saying that, in the bulk, H is proportional to the logarithm of the density matrix of bulk quantum fields restricted to, say, the right exterior region. This is also known as the modular Hamiltonian. The algebra of observables restricted to the right exterior region forms a so-called type-III_1 von Neumann algebra, and the associated modular Hamiltonian always has a continuous spectrum [45]. Physically, the continuous spectrum can also be seen by studying the time-separated two-point function of a simple boundary operator, and noting that the presence of the black hole horizon implies an exponential decay of the two-point function in time. A correlation function cannot decay to zero in a quantum system with finite entropy [19].

Of course, there is a loophole in this paradox. The loophole is simply that the argument that shows that the spectrum of H is continuous relies on a single classical bulk geometry, and is thus only valid in the strict $N = \infty$ limit. The boundary quantum system only has finite entropy at finite N. When N is finite, the bulk geometry undergoes fluctuations, and the Schwarzschild black hold may not be the only relevant geometry. Indeed, wormhole contributions to time-separated boundary two-point functions have been recently shown to give rise to small contributions, suppressed by exponential of the entropy, but which *grow* in time [46, 47]. This example teaches us a valuable lesson, which is that we should never directly compare results obtained using the classical black hole geometry to microscopic predictions when those microscopic results rely on the finer aspects of the structure of the boundary quantum theory.

16.2.5 Black Holes Formed from Collapse

It is harder to analyze a black hole formed from the collapse of a star as compared to an eternal black hole. The main idea is that after the transients settle down, the late time geometry is indeed described by the Schwarzschild metric, but there is no past horizon. The eternal black hole is kept in equilibrium by a balance between the radiation emitted by the black hole and the incoming radiation impinging on the black hole from past null infinity. In the absence of a past horizon, there is nothing to compensate the radiation emitted by the black hole, and hence the black hole "evaporates" [9, 10].

We will present a short account of the derivation of Hawking radiation given by Fredenhagen and Haag [31], leaving most of the details to the original reference. Say we have a free massless scalar field that is minimally coupled to the background metric of the spacetime. The main idea is to compute a time-separated two-point function of the scalar field at a fixed radial location far away from the black hole horizon, with both time points being sufficiently late compared to the moment when the star crosses its event horizon.

The main idea is that two-point function of a scalar field also obeys the Klein-Gordon wave-equation. So, given the hyperbolic nature of the wave-equation, the two-point function at late times can, very roughly, be thought as being transported along light rays. Now, light rays that arrive at a faraway location at late times all emanate from a region very close to the black hole horizon.

Let us see this in more detail. Let r_* be the tortoise coordinate defined via

$$r_* := r + r_+ \log\left(\frac{r}{r_+} - 1\right). \tag{16.24}$$

Note that $r_* \in (-\infty, \infty)$ as $r \in (r_+, \infty)$, and also that $\mathrm{d}r_* = \frac{\mathrm{d}r}{f(r)}$. Near infinity, we have $r_* \approx r$ up to a logarithmic correction, while near the horizon it is the logarithmic term that dominates $r_* \approx r_+ \log\left(\frac{r}{r_+} - 1\right)$.

Define u and v coordinates via the relations

$$v := t + r_*, \quad u := t - r_*. \tag{16.25}$$

Since the (t, r) part of the Schwarzschild metric may be written equivalently as $-f\, \mathrm{d}u\mathrm{d}v$, the curves of constant v or constant u represent light rays. The curves with constant v are light rays that go into the black hole. The curves with constant u are light rays that go out towards infinity. Now define a new time coordinate τ via

$$\tau := v - r = t + r_* - r. \tag{16.26}$$

The idea is that while the (t, r) coordinates only cover the exterior region, the coordinates (v, r) or (τ, r) also cover the region behind the horizon. Thus the slices $\tau =$ constant can serve as nice slices that cross the black hole horizon smoothly, see Fig. 16.3. Note also that $\tau \approx t$ far away from the black hole.

Imagine a light ray that arrives at a point (t, R) where t and R are both sufficiently large. We can imagine a detector placed at the location R that measures Hawking radiation arriving at that location. We know that, a long time in the past, say at some small value of τ, this light ray must have been very close to the horizon. We want to work out the precise relationship. Recalling that outgoing light rays have constant value of the u coordinate, and using $u = \tau + r - 2r_*$, we can equate the values of u at the final and the initial points on the light ray to get

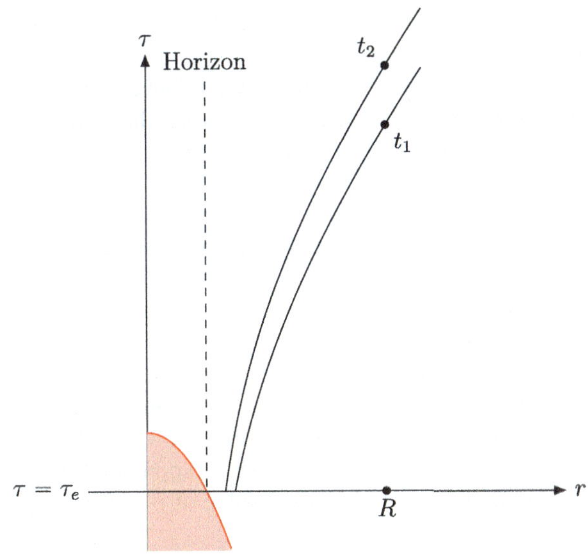

Fig. 16.3 The geometry of a black hole formed from collapse. The red shaded region represents the collapsing star and the dotted vertical line is the horizon. We are interested in computing the two-point function of a free scalar field, with the two operators inserted at radial location R and at late times t_1 and t_2. The outgoing light rays passing through these two points are shown, they emanate from locations very close to the horizon, and thus from locations that are close to each other. This two-point function is controlled by a short-distance two-point function on the $\tau = \tau_e$ slice which is universal for smooth states in a field theory with a known UV fixed point

$$t - R \approx \tau_e + r_+ - 2\left(r_+ + r_+ \log\left(\frac{r}{r_+} - 1\right)\right). \tag{16.27}$$

We have substituted $r \approx r_+$ everywhere except inside the logarithm. On the left hand side, we imagine keeping R fixed and taking larger and larger value of t. The dominant contribution from the right hand side is the logarithmic dependence of the tortoise coordinate near the horizon, and thus we get the relationship

$$r - r_+ \approx r_+ \exp\left(-\frac{t-R}{2r_+}\right) \times \text{(an order one number)}. \tag{16.28}$$

This equation tells us that for a light ray to arrive at later and later times at some fixed location R far away from the black hole, the light ray must have been emitted closer and closer to the horizon: $r - r_+ \sim r_+ \exp(-\frac{t}{2r_+})$, where r is the location from where the ray started at the early time $\tau = \tau_e$; see Fig. 16.3.

This means that if we have two points (t_1, R) and (t_2, R) at which we are trying to measure the correlation function of a free quantum field, that correlator will be controlled, via a hyperbolic wave equation, by the two-point correlator at points

(τ_e, r_1) and (τ_e, r_2) where r_1 and r_2 are both very close to the horizon, with $r_1 - r_+ \sim r_+ \exp(-\frac{t_1}{2r_+})$ and $r_2 - r_+ \sim r_+ \exp(-\frac{t_2}{2r_+})$.

Now comes another crucial piece of input. We believe that the state of quantum fields in the vicinity of the horizon must be smooth, since there are no curvature invariants getting large at the horizon. Any state that is smooth has a universal behavior of correlation functions in the short-distance limit: it is simply that the short-distance correlation function should be the same as that in the vacuum state [48–50]. Violations of this condition are possible if the state under consideration is singular, or, in other words, if it contains energy densities at arbitrarily short length scales. We assume this is not the case.[7]

Since the vacuum two-point function of a free scalar field in four spacetime dimensions is d^{-2}, where d is the distance between the two points, we conclude that the late time two-point function of interest in the black hole geometry is proportional to $(r_1 - r_2)^{-2}$. Equation (16.28) then tells us that the variables t_1 and t_2 enter via the exponentials $\exp(-\frac{t_1}{2r_+})$ and $\exp(-\frac{t_2}{2r_+})$. This exponential dependence on time implies the KMS condition with the required value of $\beta = 2\pi \cdot 2r_+ = 4\pi r_+$ (16.1). We refer the reader to [31] for further details.

It is possible to approximately calculate the expectation value of the stress tensor in a model of the background geometry and the quantum state and to compute how this stress tensor backreacts and modifies the geometry [53, 54]. While the calculations are very hard in four dimensions, it is possible to make models in two-dimensions and analytically calculate the Hawking flux [55, 56]. It is important to note a crucial difference between the eternal and the evaporating black holes: The off-diagonal component of the stress-tensor T^{0i} which measures the energy flux is zero in the eternal black hole geometry, but it is nonzero in the evaporating black hole. In the eternal black hole, the flux has positive and negative contributions that cancel, as expected in a thermal equilibrium state.

16.3 Page Curve for an Eternal Black Hole

16.3.1 Setup

Since mathematical treatment of an evaporating black hole is harder, we focus on the case of an eternal black hole. A version of the entropy paradox and the Page curve exists even in this geometry [35, 36]. We now describe this paradox.

Imagine that we have a stable eternal black hole in anti-de Sitter space. This geometry describes the thermofield double state of the boundary CFT [19]. The thermofield double state of any quantum system with energy eigenvalues E_i and corresponding eigenvectors $|E_i\rangle$ is defined as the following state on two-copies of the system:

---

[7] This assumption was challenged in the light of the Mathur-AMPS paradox [51, 52], or what has come to be knows as the "firewall problem."

$$\left|\Psi_{\beta}^{\mathrm{tfd}}\right\rangle := \frac{1}{\sqrt{Z(\beta)}} \sum_{i} e^{-\frac{\beta E_i}{2}} \left|E_i\right\rangle \left|E_i\right\rangle . \tag{16.29}$$

This state has the property that if we trace out one of the copies, the reduced density matrix on the other copy is the equilibrium thermal density matrix.

Now, a stable AdS black hole does not evaporate, but one can model the evaporation as follows. We couple the left copy of the boundary quantum system to one heat bath, and the right copy of the boundary quantum system to another heat bath. Here, by coupling we mean that the Hamiltonian contains terms that couple the boundary quantum system and the heat bath. We imagine the number of degrees of freedom in the heat bath to be much larger than the entropy of the boundary quantum system at the chosen temperature. We put the combined system into a left-right thermofield double state. Note that the left and right systems are not coupled by any terms in the Hamiltonian, but the state is left-right entangled. Now, we evolve the combined system (made of four total sub-parts) by the Hamiltonian

$$H = H_{\mathrm{left}} + H_{\mathrm{right}}. \tag{16.30}$$

It is easy to see that the state (16.29) evolves non-trivially when we apply the operator $e^{-\mathrm{i}Ht}$. This simple model of time evolution in what would otherwise have been an equilibrium state was studied in [57].

Let Q_1 and Q_2 denote the left and right copy of the boundary quantum system that is holographically dual to the theory of quantum gravity in AdS. Let R_1 and R_2 denote the two heat reservoirs. The reservoir R_1 is coupled to Q_1, and the reservoir R_2 is coupled to Q_2. Then, with the initial state and the time-evolution operators described above, we ask the question: What is the von Neumann entropy of the system $R_1 \cup R_2$ as function of time?

Since the initial state is not a stationary state with respect to the time-evolution operator, we expect that, at early times, the von Neumann entropy will increase linearly with time [57, 58], as it does in situations involving a so-called "quantum quench." However, this increase cannot continue forever. The combined system $Q_1R_1Q_2R_2$ is in a pure state and so the entropy of $R_1 \cup R_2$ is the same as the entropy of $Q_1 \cup Q_2$. However, the entropy of $Q_1 \cup Q_2$ is bounded above by $2S(\beta)$, where $S(\beta)$ is the entropy of the thermal density matrix of either of the quantum systems Q_1 and Q_2. In the gravity description, this upper bound will be $2S_{\mathrm{BH}}$, where S_{BH} is the Bekenstein-Hawking area-entropy of a black hole at inverse temperature β. Thus, we expect that the entropy will increase linearly until it reaches the upper bound, and which point, there should be a phase transition in the behavior of the entropy and the entropy curve should then become flat.

The gravity description of this system is as follows. The system $Q_1 \cup Q_2$ is dual to a two-sided AdS black hole. To model the heat reservoirs, we simply let the quantum fields in the black hole spacetime leak out from the boundary of AdS space, rather than let them reflect back in. We can imagine that the energy leaks from AdS space into (and back from) a flat space region which does not have gravity, and which represents the reservoir into which the Hawking radiation escapes. So the question about the entropy of Hawking radiation becomes a question about the entropy of these reservoir systems. In the absence of an incoming flux, this will lead to continuous loss of energy from the black hole, and the black hole will shrink [30]. The introduction of this energy loss mechanism comes at the cost of losing some rigor in applying the AdS/CFT correspondence, and also about the nature of gravity in the presence of such boundary conditions [59], but the issues that arise are strongly believed to be red herrings for the purposes of the entropy computations that we will discuss. The only purpose that the reservoirs are serving is to allow us to split the total subsystem into a black hole region and a radiation region, so we can define and compute what we mean by the entropy of Hawking radiation.

To summarize the above discussion, we have an eternal two-sided black hole geometry in AdS, and the two boundaries are coupled to flat space reservoirs with no gravity. The full system is in the thermofield double state and we evolve the system forwards on both sides (as opposed to Killing time flow, for which the time evolution of the two sides are oppositely directed). Now, the information paradox in this setting is that, in the absence of an intervening mechanism, semi-classical gravity will predict that the linear-in-time increase of the entropy of the reservoirs will continue forever. However, we know that the entropy must saturate.[8] In gravity, we will see that the quantum extremal surface (QES) prescription [28], with an allowance for disconnected entanglement wedges [29, 30, 61], reproduces the Page curve. The nontrivial QES stems from wormhole contributions to path integral replica computations of the entropy.

In this section, we describe the computations of [36] that show this in detail. However, before that, in the next two subsections, we introduce some background material. First, we present some results for the entanglement entropy in 2d CFTs, and then we introduce JT gravity (coupled to matter and with heat reservoirs), the gravity theory for in which the QES computations will be performed.

We end this prelude by briefly introducing the QES formula. We want to compute the entropy of some boundary spatial subregion A, with the full boundary theory being in a specified quantum state. The holographic computation of the entropy of A proceeds by considering candidate co-dimension one regions R in the bulk geometry that is dual to the boundary state. The only restriction is that the boundary of R should be homologous to A; this allows R to have disconnected components. Then one constructs the generalized entropy function

$$S_{\text{gen}}(R) := \frac{\text{Area}(\partial R)}{4G} + S_{\text{bulk}}(R), \tag{16.31}$$

where $S_{\text{bulk}}(R)$ is the entropy of matter fields in the bulk region R when we trace out the complement of R in the bulk. The result [28, 62] is that the entropy of A is given

[8] These is no debate that the Page curve exists for realistic quantum systems. See [60] for explicit computations which also defines a coarse-grained entropy that continues to increase after the Page time.

by S_{gen} evaluated on the region $R_\star$ which minimizes the generalized entropy.[9] The region $R_\star$ is the entanglement wedge of A: all bulk EFT operators that are localized in $R_\star$ can be represented only in the boundary region A [24, 63–65].

16.3.2 Some Entanglement Entropy Formulas in Two Dimensions

Consider a 2D CFT in its ground state on an infinite spatial line. We take the spacetime metric to be

$$\mathrm{d}s^2 = -\mathrm{d}t^2 + \mathrm{d}x^2 = -\mathrm{d}x^+\,\mathrm{d}x^-, \quad \text{where} \tag{16.32}$$

$$x^+ := t + x \quad \text{and}\ x^- := t - x. \tag{16.33}$$

Consider the reduced state on an interval PQ of (proper) length ℓ. It is well-known result that [66–68] that the von Neumann entropy of this reduced state $S(\ell)$ is given by

$$S(\ell) = \frac{c}{3}\log\frac{\ell}{\varepsilon}, \tag{16.34}$$

where c is the central charge of the theory and ε is the UV cutoff with dimensions of length. If the line segment PQ is such that P and Q do not have the same value of the time coordinate, then the entropy formula is the covariant version of the above (note that the interval PQ has to be spacelike)

$$S(PQ) = \frac{c}{6}\log(-(x^+(Q) - x^+(P))(x^-(Q) - x^-(P))), \tag{16.35}$$

$$= \frac{c}{6}\log|\Delta x^+| + \frac{c}{6}\log|\Delta x^-|, \tag{16.36}$$

where we have dropped the cutoff ε in this formula. Note that the argument of the logarithm in (16.35) is positive since the interval PQ is spacelike.

Next, we would like to have some formulas for the entanglement entropy when the background spacetime is not flat but instead has some curvature, with the metric related to the Minkowski line element by a Weyl factor:

$$\mathrm{d}s^2 := \Omega(t,x)^{-2}\,(-\mathrm{d}t^2 + \mathrm{d}x^2). \tag{16.37}$$

We consider the state in this curved space which is the Weyl transform, by the function Ω, of the ground state in the standard flat metric $-\mathrm{d}t^2 + \mathrm{d}x^2$. The Weyl factor contributes to the entropy as

---

[9] The "minimization" needs some qualifiers which are not important for our purposes.

$$S(PQ) = \frac{c}{6}\log(-(x^+(Q) - x^+(P))(x^-(Q) - x^-(P))) - \frac{c}{6}\log\Omega(P) - \frac{c}{6}\log\Omega(Q). \tag{16.38}$$

The final two terms are the extra pieces that we obtain in curved space. These contributions can be deduced by noting that the UV cutoff on the coordinates depends on Ω in such a way that the UV cutoff on the proper length is some fixed number. These contributions can also be seen as arising due to the Weyl transform of the twist operators whose correlation functions compute the Renyi entropies [67]. Note that these extra contributions depend locally on the end-points of the interval.

Let us use these results to compute the entropy of an interval in the thermal state of a 2D CFT, still on an infinite spatial line. Let $-dy^+ dy^-$ be the metric on Minkowski space M_1 where this 2D CFT in the thermal state lives. Following the discussion of the Bisognano-Wichmann theorem, we embed this thermal CFT on M_1 as the right Rindler wedge of a bigger Minkowski spacetime M_2, see Fig. 16.4. The state of the CFT on M_2 will be the vacuum state and thus we can use the entropy formulas discussed above. We let $\beta = 2\pi$ for now; we will restore β in the final formula at the end. Let

$$x^+ := \exp(y^+), \quad x^- := -\exp(y^-). \tag{16.39}$$

Note that M_1, which is covered by the range $y^+ \in (-\infty, \infty)$ and $y^- \in (-\infty, \infty)$ gets mapped to the region $x^+ \in (0, \infty)$ and $x^- \in (-\infty, 0)$ inside M_2. The full space M_2 is covered by $x^+ \in (-\infty, \infty)$ and $x^- \in (-\infty, \infty)$. The metrics are related by

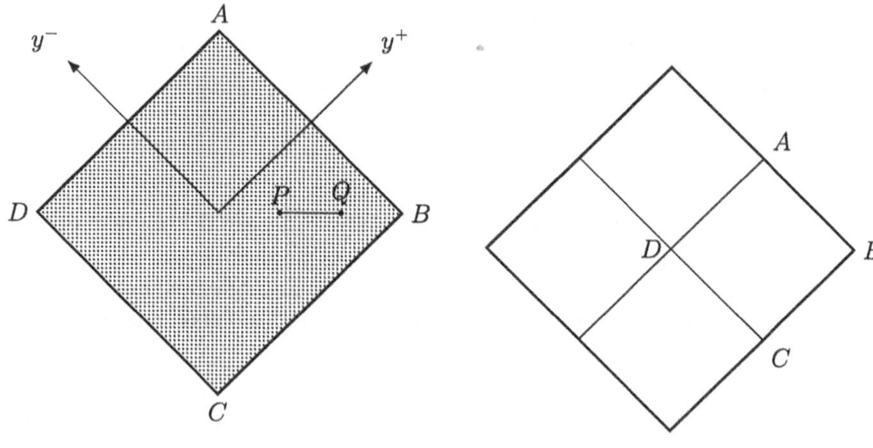

Fig. 16.4 The left figure shows the Penrose diagram $ABCD$ of a 2D Minkowski space M_1 in which a 2D CFT at nonzero temperature is placed. We have depicted this hot matter by the dots. We would like to compute the von Neumann entropy of the spatial interval PQ. In the right figure, the Minkowski space M_1 gets embedded as the right Rindler wedge of a bigger Minkowski spacetime M_2. The state in M_2 in the vacuum. The entropy of interest can be computed using the vacuum formulas in M_2 and applying the necessary Weyl transformation

$$-\,dy^+ dy^- = -\frac{dx^+ dx^-}{-x^+ x^-}, \quad \text{with } x^+ > 0 \text{ and } x^- < 0, \tag{16.40}$$

and thus comparing to (16.37), we have

$$\Omega(x^+, x^-) = (-x^+ x^-)^{\frac{1}{2}}. \tag{16.41}$$

Let us consider an interval PQ (see left panel of Fig. 16.4) with its various coordinates specified as follows

	P	Q
(t, y)	$(0, a)$	$(0, b)$
(y^+, y^-)	$(a, -a)$	$(b, -b)$
(x^+, x^-)	$(e^a, -e^a)$	$(e^b, -e^b)$

(16.42)

Here (t, y) coordinates are related to (y^+, y^-) by $y^+ = t + y$ and $y^- = t - y$. So we simply have an interval of length $b - a$ (we take $b > a$) on an equal time slice. We compute the entropy using (16.38):

$$\begin{aligned} S(PQ) &= \frac{c}{6} \log|\Delta x^+| + \frac{c}{6} \log|\Delta x^-| - \frac{c}{6} \log \Omega(P) - \frac{c}{6} \log \Omega(Q) && (16.43)\\ &= 2 \times \frac{c}{6} \log(e^b - e^a) - \frac{c}{6} \log(e^a) - \frac{c}{6} \log(e^b) && (16.44)\\ &= \frac{c}{3} \log\left(2 \sinh \frac{b-a}{2}\right). && (16.45) \end{aligned}$$

Restoring β, we find

$$S(PQ) = \frac{c}{3} \log\left(\frac{\beta}{\pi} \sinh \frac{\pi \ell}{\beta}\right). \tag{16.46}$$

This is a well-known result [68]. We can perform two sanity checks: (a) In the zero temperature limit, $\beta \to \infty$ and we recover the vacuum result (16.34), (b) When $\ell \gg \beta$, we get an entropy proportional to the length ℓ, which is simply a statement of the extensivity of thermal entropy.

16.3.3 JT Gravity

In order to perform detailed concrete computations of the entropy of Hawking radiation, one needs a simple model of gravity. One such model is JT gravity, which is a two-dimensional theory of dilaton gravity [34, 69, 70]. In certain parameter regimes, it is dual to a disordered quantum mechanical system of Majorana fermions, known as the Sachdev-Ye-Kitaev (SYK) model [71–73]. The details of the SYK model will not be relevant for us; we just need to know that it can be thought of a holographic system dual to JT gravity.

The Euclidean action for JT gravity is

$$\begin{aligned} 16\pi G\, I_{\mathrm{Euc}} = &- \phi_0 \left(\int \mathrm{d}^2x\, \sqrt{g}\, R + \int \mathrm{d}x\, \sqrt{\gamma}\, 2K \right) \\ &- \left(\int \mathrm{d}^2x\, \sqrt{g}\, \phi(R+2) + \int \mathrm{d}x\, \sqrt{\gamma}\, 2\phi\, (K-1) \right) \quad (16.47) \\ = &-\phi_0\, \chi - \left(\int \mathrm{d}^2x\, \sqrt{g}\, \phi(R+2) + \int \mathrm{d}x\, \sqrt{\gamma}\, 2\phi\, (K-1) \right). \quad (16.48) \end{aligned}$$

Here χ is the Euler characteristic of the surface on which we are studying the theory. The quantity ϕ_0 serves as the genus counting parameter of this theory, and supplies an additive constant to the entropy of black holes (so it represents an entropy that is present even at zero temperature). The quantity $\phi_0 + \phi$ is the total dilaton, and can be thought of as the area of a near-extremal higher dimensional black hole that has been dimensionally reduced to two dimensions. The -1 in $K - 1$ is a holographic counterterm.

The action needs to be supplemented with boundary conditions. They are [34]

$$g_{tt}\big|_{\mathrm{bdry}} = -\,\varepsilon^{-2}, \quad (16.49)$$

$$\phi\big|_{\mathrm{bdry}} = \phi_r\, \varepsilon^{-1}. \quad (16.50)$$

The constant ϕ_r has dimensions of length and is a dimensionful parameter of the theory. This will set the length scales of various things in the theory. Often we will encounter it in dimensionless combinations like ϕ_r/β. Also ε is a small quantity that should be understood to be the smallest length scale in the problem (UV cutoff), just like when we compute things by using cutoffs near the boundary in AdS.

We would now like to write down the equations of motion for this theory and find some simple black hole solutions. For that, let us remind ourselves how the Ricci scalar transforms when we make a variation of the metric. It is useful to vary the inverse metric:

$$\delta R = R_{ab}\, \delta g^{ab} - \nabla_a \nabla_b\, \delta g^{ab} + \nabla^c \nabla_c\, g_{ab} \delta g^{ab}. \quad (16.51)$$

Terms that look like total derivatives are important since the Ricci scalar multiplies the dilaton in the action. We can now vary the fields and get the equations of motion.

$$R + 2 = 0, \quad (16.52)$$

$$\left(\nabla_c \nabla^c \phi\right) g_{ab} - \nabla_a \nabla_b \phi - \phi\, g_{ab} = 0. \quad (16.53)$$

We will couple this gravity theory to matter to get the physics of Hawking radiation. The matter sector will be taken to be a CFT_2 with central charge c that is minimally coupled to the metric, and not at all to the dilaton.

16.3.4 QES at Zero Temperature

The equation of motion $R + 2 = 0$ says that the metric should have constant negative curvature. The zero temperature solution is the metric of AdS_2 in the Poincaré patch

$$ds^2 = \frac{1}{z^2}\left(-dt^2 + dz^2\right). \tag{16.54}$$

The boundary of AdS is at $z = 0$ but, contrary to usual convention, for later convenience, we take the range of the z coordinate in AdS space to be $(-\infty, 0)$. See Fig. 16.5.

For ϕ, we will make the ansatz that it only depends on z. Then the zz component of the dilaton equation yields the most stringent equation which is first order, namely $z\phi' + \phi = 0$. This yields

$$\phi(z) = \frac{A}{-z}. \tag{16.55}$$

Imposing the boundary conditions (16.49) and (16.50), we see that the boundary curve should be the simple locus $z = -\varepsilon$ and we also must have $A = \phi_r$.

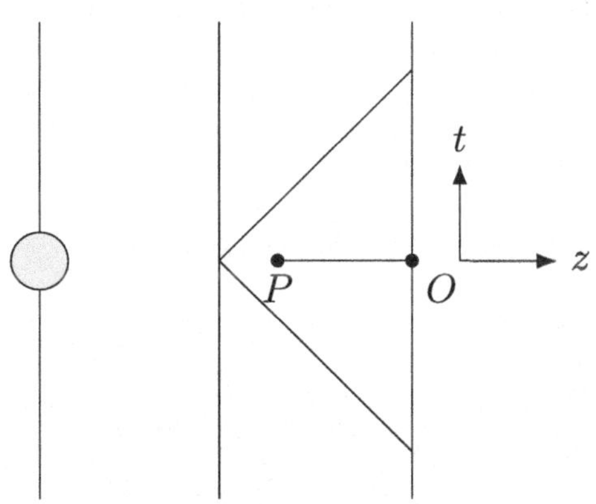

Fig. 16.5 The left figure depicts an SYK quantum mechanical system, with the time direction indicated. The system is in its ground state. The dual geometry is the Poincaré patch of AdS_2. We want to compute the entropy of the whole system. The candidate QES is shown as the point P. The entropy of bulk matter fields in the region PO contributes to the generalized entropy function

This field configuration should be thought of as describing the SYK model in its zero temperature state.

We want to apply the QES prescription [28] to compute the entropy and entanglement wedge of the full boundary system. Clearly, we know that the entanglement wedge should be the full AdS geometry. Let us see this from a computation.

Since the matter CFT has fully reflecting boundary conditions, we do not get independent contributions to the entropy from the left and right movers as in (16.36); we just get one contribution with coefficient $\frac{c}{6}$. Let the candidate surface be a point P with coordinates $(0, -a)$ with $a > 0$. The the generalized entropy function (16.31) is

$$S_{\text{gen}}(a) = \frac{\phi_r}{a} + \left(\frac{c}{6}\log(a) - \frac{c}{6}\log\Omega(P)\right) = \frac{\phi_r}{a}. \tag{16.56}$$

We have set $4G = 1$ and used the fact that the value of the dilaton plays the role of area in JT gravity.[10] The second and third term in the middle formula canceled each other. Clearly, $S_{\text{gen}}(a)$ is extremized at $a_\star = \infty$. This says that the QES is located at the Poincaré horizon and thus the entanglement wedge is all of AdS, as expected.

We now consider a different problem. Let us couple the SYK system to a reservoir and place the combined system in its ground state. The reservoir takes the form of a semi-infinite "wire" and the SYK system is coupled to it at its endpoint. We now want to compute the entropy and the entanglement wedge of the SYK model, after tracing out the reservoir. We can also keep a small contiguous part of the reservoir together with SYK and consider this as the subsystem of interest. So, we are interested in the entropy and entanglement wedge of the subsystem $[0, b]$ of this quantum mechanical system. It is very important to emphasize that the entropy that is being computed is that of a subsystem of a non-gravitational system.

In the gravity description, the coupling to the reservoir is modeled by letting the matter CFT fields have transparent boundary conditions at the boundary of AdS. After crossing the boundary of AdS, they enter a non-gravitating flat space region. This means that now the left and right movers are indeed independent and we will get both contributions from (16.36).

Let the candidate QES be located at the point P with (t, z) coordinates equal to $(0, -a)$. The generalized entropy function takes the form

$$S_{\text{gen}}(a) = \frac{\phi_r}{a} + \left(2 \times \frac{c}{6}\log(b+a) - \frac{c}{6}\log\Omega(P)\right) \tag{16.57}$$

$$= \frac{\phi_r}{a} + \frac{c}{3}\log(a+b) - \frac{c}{6}\log a. \tag{16.58}$$

The extremization condition is

$$\frac{\phi_r}{a_\star} = \frac{c}{6}\frac{a_\star - b}{a_\star + b}. \tag{16.59}$$

---

[10] We have also dropped the constant ϕ_0 since it is not important for the extremization.

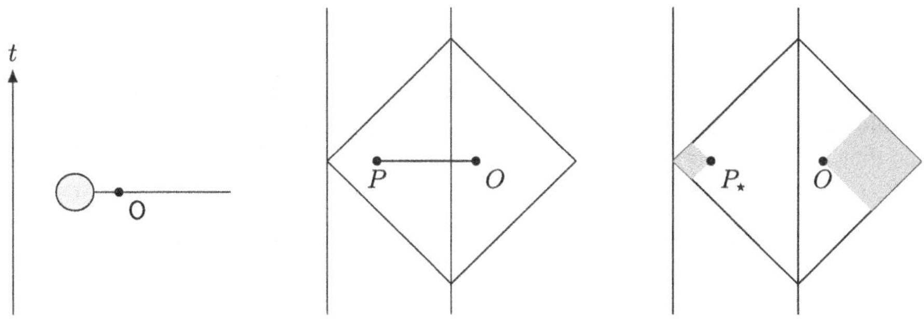

Fig. 16.6 The left figure shows an SYK model coupled to a reservoir which is a semi-infinite wire supporting a 2D CFT. The system is in its combined ground state. We want to compute the entropy and entanglement wedge of a system consisting of the SYK model plus a portion of the wire contiguous to it, up to the point labeled by O. This is modeled in the gravity description (middle figure) on the right by attaching a non-gravitating flat space region to AdS_2. The matter fields can now propagate freely across the interface. The candidate QES is depicted by the point P. The actual QES $P_\star$ for this bipartite decomposition of the system lies away from the Poincaré horizon. This means that the entanglement wedge of the subsystem to the right of O in the left figure, shown by the grey shaded region on the right, is disconnected

This is some quadratic equation that we can solve. We record the answer for $b = 0$, when we get $a_\star = \phi_r \frac{6}{c}$.

The "surprise" is that, for any value of b, even if we don't include any portion of the wire, the QES will be at some location away from the Poincaré horizon [36]. This means that the entanglement wedge of the complement system, the reservoir, contains a disconnected piece far away from it, near the horizon. See the rightmost diagram in Fig. 16.6. Such disconnected pieces of an entanglement wedge have been dubbed "islands" [61]. This is initially surprising because we might have (naively) argued that it is the SYK model which is dual to the AdS space, so how can some part of AdS space map to the reservoir under the AdS/CFT map? The point is simply that this argument is too naive, and the full geometry is dual to the coupled SYK+reservoir system. There is no paradox in having a disconnected entanglement wedge.

We have considered the case of fully reflecting and fully transparent boundary conditions for the matter fields at the AdS boundary. If we take the matter to be free fermions, it is possible to perform the matter entropy computations for partially transmitting boundary conditions [74]. As expected, the QES interpolates between the fully reflecting and the fully transmitting cases.

16.3.5 QES at Nonzero Temperature

Let us now discuss the Euclidean solution of JT gravity that corresponds to a black hole with nonzero temperature. The topology of a Euclidean 2D black hole is that of a disk, so the appropriate metric with $R = -2$ is given by

$$ds^2 = d\rho^2 + (\sinh\rho)^2\, d\theta^2. \tag{16.60}$$

Here $\rho = 0$ is the horizon and $\rho = \infty$ is the boundary. Let us see what the dilaton should be. First simplify the Eq. (16.53) by taking the trace, obtaining $\nabla^2\phi = 2\phi$, and then plugging this back into (16.53), yielding $\nabla_a\nabla_b\phi = \phi\, g_{ab}$. The left hand side is $\partial_a\partial_b\phi - \Gamma^c_{ab}\,\partial_c\phi$. When we look for a solution which is only a function of ρ, the $\theta\theta$ component gives us a first order equation. It is $\frac{\phi'}{\phi} = \frac{\sinh\rho}{\cosh\rho}$, which gives

$$\phi(\rho) = \phi_r\, \cosh\rho, \tag{16.61}$$

with the constant of proportionality being fixed by the boundary conditions.

The analytic continuation needed to get the Lorentzian solution is to make θ purely imaginary in (16.60). However, let us first try to write the disk metric in a conformally flat form. Define a coordinate y using

$$\frac{d\rho}{\sinh\rho} =: dy. \tag{16.62}$$

This implies that

$$\sinh\rho = -\frac{1}{\sinh y}, \qquad y \in (-\infty, 0) \tag{16.63}$$

$$\cosh\rho = -\frac{\cosh y}{\sinh y}. \tag{16.64}$$

So the metric (16.60) becomes

$$ds^2_{\text{Euc}} = (\sinh y)^{-2}\left(dy^2 + d\theta^2\right) \tag{16.65}$$

$$ds^2_{\text{Lor}} = (\sinh y)^{-2}\left(dy^2 - dt^2\right). \tag{16.66}$$

This is a good way to write the metric because we can obtain this via a simple reparametrization from the zero temperature solution. Let us see this. The zero temperature solution is

$$ds^2 = -\frac{dt^2 + dz^2}{z^2} = -\frac{4}{(z^+ - z^-)^2}\, dz^+\, dz^-, \tag{16.67}$$

where we have defined $z^\pm = t \pm z$. If we transform to (y^+, y^-) coordinates using $z^+ = \tanh\frac{y^+}{2}$ and $x^- = \tanh\frac{y^-}{2}$, we get

$$ds^2 = -\left(\sinh\frac{y^+ - y^-}{2}\right)^{-2} dy^+\, dy^-, \tag{16.68}$$

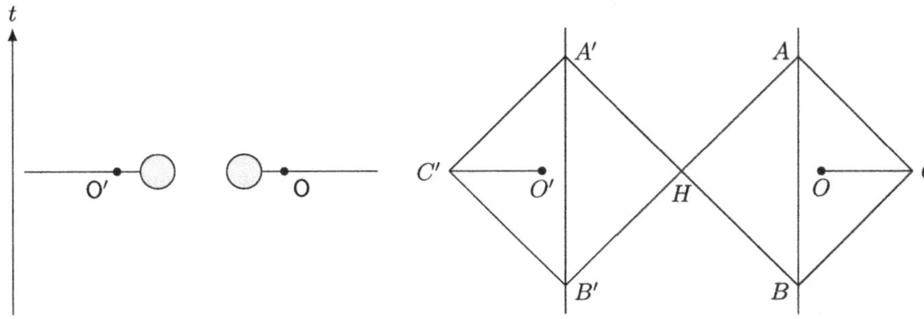

Fig. 16.7 Left: We take two copies of the SYK model, each coupled to a reservoir. The right SYK+reservoir is in a thermal state. The left SYK+reservoir acts as its canonical purification. The "radiation" region is defined to be the union of the reservoir sub-regions to the left of the point O′ and to the right of the point O. We want to compute the entropy and entanglement wedge of this region, this will be our model for the entropy of Hawking radiation in the eternal black hole setup. We evolve the system with an operator that generates forward time evolution on both copies. Right: The gravity dual of the setup. The point H is the bifurcation surface, AB and $A'B'$ are the boundaries of AdS, and the regions ABC and $A'B'C'$ are the non gravitating flat space regions

which is the same as (16.66) with $y^{\pm} = t \pm y$.[11] Pictorially, since $y < 0$ in the AdS part, we see that the (y^+, y^-) coordinates cover the triangle AHB in Fig. 16.7, with the horizon at $y = -\infty$ and the AdS boundary at $y = 0$. After we introduce the flat space bath, it lies in the $y > 0$ region and now the (y^+, y^-) coordinates cover the diamond $AHBC$.

The dilaton profile can be written using the y variable as

$$\phi = \phi_r \cosh \rho = -\phi_r \coth y. \tag{16.69}$$

The point is that the y coordinate is similar to the usual Poincare coordinate z near the boundary. The dilaton behaves like y^{-1} there.

In the previous discussion, we set $\beta = 2\pi$, but it will be useful to restore factors of β, after which we get:

$$ds^2 = -\left(\frac{\beta}{2\pi} \sinh \frac{\pi}{\beta}(y^+ - y^-)\right)^{-2} dy^+ dy^-, \tag{16.70}$$

$$\Omega = \frac{\beta}{2\pi} \sinh \frac{2\pi(-y)}{\beta}, \quad y \in (-\infty, 0), \tag{16.71}$$

$$\phi(y) = \phi_r \frac{2\pi}{\beta} \coth \frac{2\pi(-y)}{\beta}. \tag{16.72}$$

Let us record the value of the black hole entropy. It is given by

[11] This t is different than the t in (16.67). This should not cause any confusion, since from now on we will only use $t = \frac{y^+ - y^-}{2}$.

$$S_{\rm BH}(\beta) = \phi_0 + \phi_r \frac{2\pi}{\beta} = \phi_0 + 2\pi\phi_r T. \tag{16.73}$$

This linear in temperature behavior of the entropy is typical of near-extremal black holes. The role of the area is played by value of the dilaton at a particular point.

The Lorentzian eternal two-sided black hole represents two copies of the dual quantum system in the thermofield double state. As discussed in the introduction and in the beginning of this section, we would like to couple each of these two systems to a reservoir and consider the setup shown on the left in Fig. 16.7. We would like to compute the entropy of the "radiation region", which is the union of the region to the left of the point O′ and the region to the right of the point O in the left panel of Fig. 16.7. Recall that we are evolving the system with a generator that evolves both systems forwards in time. The thermofield double state is a quenched state with respect to this Hamiltonian, so we expect the entropy to initially increase linearly in time. In the next subsection, we calculate this linear increase.

16.3.5.1 Early Time Linear Growth

The early time linear growth is a purely CFT computation. We compute the entropy of the complement of the radiation system, which is the interval OO'.

We imagine that the points O and O' in the right panel of Fig. 16.7 have been evolved forward by time t. The state of the matter CFT is thermal when restricted to the diamond $AHBC$. Thus, we have to import some results from Sect. 16.3.2. In particular, we have to introduce (x^+, x^-) coordinates via Eq. (16.39).

	O	O'
(t, y)	(t, b)	
(y^+, y^-)	$(t+b, t-b)$	
(x^+, x^-)	$(e^{t+b}, -e^{-t+b})$	$(-e^{-t+b}, e^{t+b})$

(16.74)

We do not write the (t, y) or the (y^+, y^-) coordinates of the point O' since this point lies outside the region that is covered by these coordinates. The (x^+, x^-) coordinates of O' are related to those of O by reflection in the vertical line $x^+ = x^-$, which simply swaps the two coordinates.

Thus, the coordinate separation between O and O' is $\Delta x^+ = e^b \cdot 2\cosh t$. The separation along x^- is the same. For both points, $\Omega = (-x^+x^-)^{\frac{1}{2}} = e^b$. Thus, using the formula (16.38), we get

$$S_{\rm rad}(t) = 2 \cdot \frac{c}{6}\log(e^b \cdot 2\cosh t) - \frac{c}{6}\log e^b - \frac{c}{6}\log e^b \tag{16.75}$$

$$= \frac{c}{3}\log(2\cosh t) \approx \frac{c}{3}\frac{2\pi t}{\beta}, \tag{16.76}$$

where we restored β in the last expression after taking the limit $t \gg \beta$. Thus, we get a linear-in-time increase in the entropy.

Now, if this linear increase were to continue forever, we would run into an entropy paradox, which is very similar to Page's: the combined system is in a pure state and the complement system (which for us is the union of the two SYK models) has bounded entropy. We will now see how disconnected entanglement wedges save us from this entropy paradox.

16.3.5.2 Late Time Saturation and Islands

The claim is that the entanglement wedge of the radiation region contains a disconnected "island" region that spans the interior of the black hole. So, as a candidate QES configuration, we should compute the generalized entropy function of a region consisting of two intervals $O'P'$ and OP. Since two-interval entropies depend on the details of the CFT, we use a further approximation that, at late times, the answer will just be a sum of the entropies.[12]

So we consider computing the entropy of just the right SYK system plus a piece of the reservoir of length b. We construct the generalized entropy functional for the region OP, where the point P has coordinates $(t, y) = (0, -a)$, see Fig. 16.8. It takes the form

$$S_{\text{gen}}(a) = \phi_r \frac{2\pi}{\beta} \coth \frac{2\pi a}{\beta} + \frac{c}{3} \log\left(\frac{\beta}{\pi} \sinh \frac{\pi}{\beta}(b + a)\right) - \frac{c}{6} \log\left(\frac{\beta}{2\pi} \sinh \frac{2\pi a}{\beta}\right). \tag{16.77}$$

The first term in this formula is the dilaton (16.72), the second term is the thermal entropy of the interval OP, given by (16.46), and the final term is the correction due to the Weyl factor (16.71) at the point P. Setting the derivative of equal to zero, after some simplification we find the QES condition

$$\phi_r \frac{2\pi}{\beta} \frac{1}{\sinh \frac{2\pi a_\star}{\beta}} = \frac{c}{6} \frac{\sinh \frac{\pi}{\beta}(a_\star - b)}{\sinh \frac{\pi}{\beta}(a_\star + b)}. \tag{16.78}$$

Let us solve this equation in the slow emission/absorption limit, which is

$$\frac{\phi_r}{\beta} \gg c. \tag{16.79}$$

This condition says that the entropy of the black hole above extremality (16.73) is large compared to the entropy that it will lose or gain per unit thermal time due to

[12] Technically speaking, the four-point function of twist operators in this limit is controlled by an operator product expansion (OPE), and only the identity operator contributes in the intermediate channel.

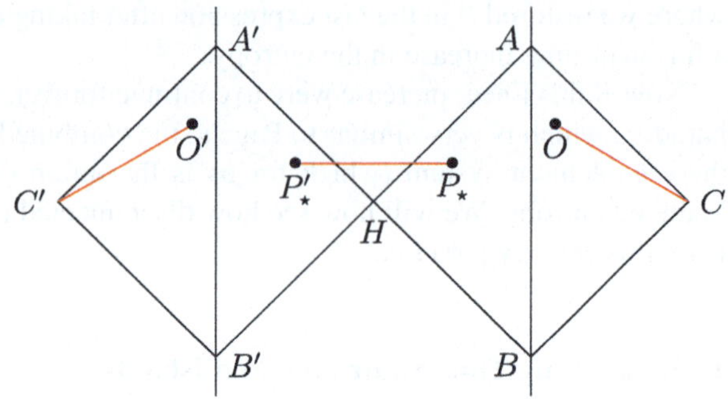

Fig. 16.8 The entanglement wedge of the radiation region contains a disconnected region $P_\star P'_\star$ at late times, which is called the island. The entropy computed using this QES does not grow with time. In fact, it almost equals the sum of the thermodynamic entropy of the two black hole, a constant

emission and absorption of radiation quanta. In this limit, we can approximate all sinh's in (16.78) with exponentials which simplifies the equation to

$$\log\left(\phi_r \frac{2\pi}{\beta}\frac{6}{c}\right) + \log 2 - \frac{2\pi a_\star}{\beta} \approx -\frac{2\pi b}{\beta}, \quad \text{or} \tag{16.80}$$

$$a_\star \approx \frac{\beta}{2\pi}\log\left(\phi_r \frac{2\pi}{\beta}\frac{12}{c}\right) + b. \tag{16.81}$$

Here b itself could be small or large, we could just set it zero. Since $\frac{\phi_r}{\beta c} \gg 1$, we have $a_\star \gg 1$, which just says that the QES is close to the horizon. The actual value of the entropy at this extremum is approximately given by the first term in (16.77) and with the coth replaced by one.[13] This is just the black hole entropy (16.73). (We dropped by constant ϕ_0 from our expression for $S_{\text{gen}}(a)$ since it does not contribute to the extremization condition, but it is really present and we should add it back.) Thus, we conclude that

$$S_{\text{rad}}(t) \approx 2S_{\text{BH}}(\beta) \quad \text{for } \frac{ct}{\beta} \gg S_{\text{BH}}. \tag{16.82}$$

The factor of two on the right hand side comes from having two intervals $OP_\star$ and $O'P'_\star$.

To summarize, the entanglement entropy of the radiation region initially grows linearly in time (16.76). At a time of order $\beta\frac{S_{\text{BH}}}{c}$, this linear growth is halted by the appearance of a disconnected portion of the entanglement wedge of the radiation. This disconnected portion lies in the gravitating region and spans the interior of the two-sided black hole, see Fig. 16.8. The entropy computed using the QES formula at late times agrees with the predictions from general principles of unitary time evolution. This early time increase, followed by the late time saturation is the analog of the Page curve for the eternal black hole.

[13] Dividing by c, the first term is order $\frac{\phi_r}{\beta c}$ and the others are of order $\log\frac{\phi_r}{\beta c}$.

16.4 Concluding Remarks

The fact that the entanglement wedge of the radiation region, which lies in a region of weak gravity, contains a part of the black hole interior often causes major discomfort. One may start worrying about "nonlocal effects" and some "physical transport of information" happening from the island region to the radiation region. Nothing of the sort is being claimed.

Let us recap the logic. We first decide which quantity Q on the boundary we want to compute. For the Page curve computations, it is the von Neumann entropy, or perhaps the Renyi entropies. Then one sets up the boundary path integral that is relevant for computing the quantity Q. Then one uses AdS/CFT to compute this path integral via gravitational saddle-points. There might be multiple competing saddle-points, and a saddle-point that is initially suppressed might become dominant when some parameter is varied. This is the full logic of the calculation, and it fits within very standard Gibbons-Hawking and AdS/CFT computations. If one is worried about "how the information is getting out", one should first formulate a precise version of this question on the boundary and identify a quantity that one wants to compute. Then one should set up the path integral on the boundary and compute the quantity using gravitational saddle points.

In this context, it seems like an interesting question to the author to ask what computation on the boundary captures the Hawking particle creation process near the black hole horizon.

We would like to stress again that the QES formula or the "island prescription" is not fundamental. It is the above path integral setup that is fundamental. In the table below, we record three entropy "prescriptions" in the left column and their corresponding derivations using the boundary path integral logic. The entropy "prescriptions" work in the original geometry when the problem under study is simple enough. The right column should be treated as more fundamental.

1	Bekenstein-Hawking area formula	Gibbons-Hawking path integral
2	Ryu-Takayanagi formula	Lewkowycz-Maldacena
3	QES formula with islands	Boundary path integral with replica wormholes

Let ρ be the density matrix of Hawking radiation. Note that $-\operatorname{Tr}(\rho \log \rho)$ is just one real number made out of an exponentially large number of matrix entries of ρ. The surprise of the papers [24, 25, 29, 30] is that this quantity happened to be computable using semi-classical gravity, albeit using some nontrivial saddle points. It is also a surprise, hailing back to the 1970s, that semi-classical gravity knows about the entropy of the black hole. It is a more recent surprise that semi-classical gravity knows something about the chaotic nature of the eigenvalues of the boundary Hamiltonian [46].

However, it is also clear that semi-classical gravity cannot know all the details of the microscopic boundary answers. For example, it does not seem to know about the large erratic fluctuations that exist on top of an averaged "ramp" feature in the spectral form factor or real-time thermal two-point functions [75]. It also does not seem to know about the individual exponentially large number of matrix elements of the full density matrix of Hawking radiation. That's a good thing. We know, after all, that Einstein gravity suffers from UV problems and, so semi-classical GR coupled to matter is definitely not the ultimate theory of quantum gravity. However, it is still an interesting question to ask: What other subtle quantities of the boundary quantum theory can be computed using semi-classical gravity, improved by wormholes?

References

1. C.W. Misner, K.S. Thorne, J.A. Wheeler, *Gravitation* (W. H. Freeman, San Francisco, 1973). ISBN 978-0-7167-0344-0, 978-0-691-17779-3
2. R.M. Wald, *General Relativity* (Chicago Univ. Press, Chicago, 1984). https://doi.org/10.7208/chicago/9780226870373.001.0001
3. C.M. Will, The confrontation between general relativity and experiment. Living Rev. Rel. **17**, 4 (2014). https://doi.org/10.12942/lrr-2014-4. arXiv:1403.7377 [gr-qc]
4. J.R. Oppenheimer, H. Snyder, On continued gravitational contraction. Phys. Rev. **56**, 455–459 (1939). https://doi.org/10.1103/PhysRev.56.455
5. B.P. Abbott, et al. (LIGO Scientific, Virgo), Observation of gravitational waves from a binary black hole merger. Phys. Rev. Lett. **116**, 061102 (2016). https://doi.org/10.1103/PhysRevLett.116.061102. arXiv:1602.03837 [gr-qc]
6. K. Akiyama, et al. (Event Horizon Telescope), First M87 event horizon telescope results. I. The shadow of the supermassive black hole. Astrophys. J. Lett. **875**, L1 (2019). https://doi.org/10.3847/2041-8213/ab0ec7. arXiv:1906.11238 [astro-ph.GA]
7. J. Polchinski, *String Theory. Vol. 1: An Introduction to the Bosonic String*. Cambridge monographs on mathematical physics (Cambridge University Press, 2007). https://doi.org/10.1017/CBO9780511816079. ISBN 978-0-511-25227-3, 978-0-521-67227-6, 978-0-521-63303-1
8. J. Polchinski, *String Theory. Vol. 2: Superstring Theory and Beyond*. Cambridge monographs on mathematical physics (Cambridge University Press, 2007). https://doi.org/10.1017/CBO9780511618123. ISBN 978-0-511-25228-0, 978-0-521-63304-8, 978-0-521-67228-3
9. S.W. Hawking, Black hole explosions. Nature **248**, 30–31 (1974). https://doi.org/10.1038/248030a0
10. S.W. Hawking, Particle creation by black holes. Commun. Math. Phys. **43**, 199–220 (1975). https://doi.org/10.1007/BF02345020 [Erratum: Commun. Math. Phys. **46**, 206 (1976)]
11. J.D. Bekenstein, Black holes and entropy. Phys. Rev. D **7**, 2333–2346 (1973). https://doi.org/10.1103/PhysRevD.7.2333
12. J.M. Bardeen, B. Carter, S.W. Hawking, The four laws of black hole mechanics. Commun. Math. Phys. **31**, 161–170 (1973). https://doi.org/10.1007/BF01645742
13. G.W. Gibbons, S.W. Hawking, Action integrals and partition functions in quantum gravity. Phys. Rev. D **15**, 2752–2756 (1977). https://doi.org/10.1103/PhysRevD.15.2752
14. A. Strominger, C. Vafa, Microscopic origin of the Bekenstein-Hawking entropy. Phys. Lett. B **379**, 99–104 (1996). https://doi.org/10.1016/0370-2693(96)00345-0. arXiv:hep-th/9601029
15. A. Sen, Black hole entropy function, attractors and precision counting of microstates. Gen. Rel. Grav. **40**, 2249–2431 (2008). https://doi.org/10.1007/s10714-008-0626-4. arXiv:0708.1270 [hep-th]
16. S.W. Hawking, Breakdown of predictability in gravitational collapse. Phys. Rev. D **14**, 2460–2473 (1976). https://doi.org/10.1103/PhysRevD.14.2460
17. J.M. Maldacena, The large N limit of superconformal field theories and supergravity. Adv. Theor. Math. Phys. **2**, 231–252 (1998). https://doi.org/10.4310/ATMP.1998.v2.n2.a1. arXiv:hep-th/9711200
18. E. Witten, Anti-de Sitter space, thermal phase transition, and confinement in gauge theories. Adv. Theor. Math. Phys. **2**, 505–532 (1998). https://doi.org/10.4310/ATMP.1998.v2.n3.a3. arXiv:hep-th/9803131

19. J.M. Maldacena, Eternal black holes in anti-de Sitter. JHEP **04**, 021 (2003). https://doi.org/10.1088/1126-6708/2003/04/021. arXiv:hep-th/0106112
20. D.N. Page, Average entropy of a subsystem. Phys. Rev. Lett. **71**, 1291–1294 (1993). https://doi.org/10.1103/PhysRevLett.71.1291. arXiv:gr-qc/9305007
21. D.N. Page, Information in black hole radiation. Phys. Rev. Lett. **71**, 3743–3746 (1993). https://doi.org/10.1103/PhysRevLett.71.3743. arXiv:hep-th/9306083
22. D.N. Page, Time dependence of Hawking radiation entropy. JCAP **09**, 028 (2013). https://doi.org/10.1088/1475-7516/2013/09/028. arXiv:1301.4995 [hep-th]
23. A. Lewkowycz, J. Maldacena, Generalized gravitational entropy. JHEP **08**, 090 (2013). https://doi.org/10.1007/JHEP08(2013)090. arXiv:1304.4926 [hep-th]
24. G. Penington, S.H. Shenker, D. Stanford, Z. Yang, Replica wormholes and the black hole interior. JHEP **03**, 205 (2022). https://doi.org/10.1007/JHEP03(2022)205. arXiv:1911.11977 [hep-th]
25. A. Almheiri, T. Hartman, J. Maldacena, E. Shaghoulian, A. Tajdini, Replica wormholes and the entropy of Hawking radiation. JHEP **05**, 013 (2020). https://doi.org/10.1007/JHEP05(2020)013. arXiv:1911.12333 [hep-th]
26. S. Ryu, T. Takayanagi, Holographic derivation of entanglement entropy from AdS/CFT. Phys. Rev. Lett. **96**, 181602 (2006). https://doi.org/10.1103/PhysRevLett.96.181602. arXiv:hep-th/0603001
27. T. Faulkner, A. Lewkowycz, J. Maldacena, Quantum corrections to holographic entanglement entropy. JHEP **11**, 074 (2013). https://doi.org/10.1007/JHEP11(2013)074. arXiv:1307.2892 [hep-th]
28. N. Engelhardt, A.C. Wall, Quantum extremal surfaces: holographic entanglement entropy beyond the classical regime. JHEP **01**, 073 (2015). https://doi.org/10.1007/JHEP01(2015)073. arXiv:1408.3203 [hep-th]
29. G. Penington, Entanglement wedge reconstruction and the information paradox. JHEP **09**, 002 (2020). https://doi.org/10.1007/JHEP09(2020)002. arXiv:1905.08255 [hep-th]
30. A. Almheiri, N. Engelhardt, D. Marolf, H. Maxfield, The entropy of bulk quantum fields and the entanglement wedge of an evaporating black hole. JHEP **12**, 063 (2019). https://doi.org/10.1007/JHEP12(2019)063. arXiv:1905.08762 [hep-th]
31. K. Fredenhagen, R. Haag, On the derivation of Hawking radiation associated with the formation of a black hole. Commun. Math. Phys. **127**, 273 (1990). https://doi.org/10.1007/BF02096757
32. R. Kubo, Statistical-mechanical theory of irreversible processes. I. General theory and simple applications to magnetic and conduction problems. J. Phys. Soc. Jpn. **12**, 570–586 (1957)
33. P.C. Martin, J. Schwinger, Theory of many-particle systems. I. Phys. Rev. **115**, 1342 (1959)
34. J. Maldacena, D. Stanford, Z. Yang, Conformal symmetry and its breaking in two dimensional Nearly Anti-de-Sitter space. PTEP **2016**, 12C104 (2016). https://doi.org/10.1093/ptep/ptw124. arXiv:1606.01857 [hep-th]
35. S.D. Mathur, What is the dual of two entangled CFTs? (2014). arXiv:1402.6378 [hep-th]
36. A. Almheiri, R. Mahajan, J. Maldacena, Islands outside the horizon (2019). arXiv:1910.11077 [hep-th]
37. A. Almheiri, T. Hartman, J. Maldacena, E. Shaghoulian, A. Tajdini, The entropy of Hawking radiation. Rev. Mod. Phys. **93**, 035002 (2021). https://doi.org/10.1103/RevModPhys.93.035002. arXiv:2006.06872 [hep-th]
38. H. Araki, G.L. Sewell, KMS conditions and local thermodynamical stability of quantum lattice systems. Commun. Math. Phys. **52**, 103–109 (1977). https://doi.org/10.1007/BF01625778
39. G.L. Sewell, KMS conditions and local thermodynamical stability of quantum lattice systems. II. Commun. Math. Phys. **55**, 53–61 (1977)
40. S. Takagi, Vacuum noise and stress induced by uniform acceleration: Hawking-Unruh effect in Rindler manifold of arbitrary dimension. Prog. Theor. Phys. Suppl. **88**, 1–142 (1986). https://doi.org/10.1143/PTP.88.1
41. W. Rindler, Kruskal space and the uniformly accelerated frame. Am. J. Phys. **34**, 1174 (1966). https://doi.org/10.1119/1.1972547

42. W.G. Unruh, Notes on black hole evaporation. Phys. Rev. D **14**, 870 (1976). https://doi.org/10.1103/PhysRevD.14.870
43. J.J Bisognano, E.H. Wichmann, On the duality condition for quantum fields. J. Math. Phys. **17**, 303–321 (1976). https://doi.org/10.1063/1.522898
44. D.W. Sciama, P. Candelas, D. Deutsch, Quantum field theory, horizons and thermodynamics. Adv. Phys. **30**, 327–366 (1981). https://doi.org/10.1080/00018738100101457
45. E. Witten, APS medal for exceptional achievement in research: invited article on entanglement properties of quantum field theory. Rev. Mod. Phys. **90**, 045003 (2018). https://doi.org/10.1103/RevModPhys.90.045003. arXiv:1803.04993 [hep-th]
46. P. Saad, S.H. Shenker, D. Stanford, A semiclassical ramp in SYK and in gravity (2018). arXiv:1806.06840 [hep-th]
47. P. Saad, Late time correlation functions, baby universes, and ETH in JT gravity (2019). arXiv:1910.10311 [hep-th]
48. R. Haag, H. Narnhofer, U. Stein, On quantum field theory in gravitational background. Commun. Math. Phys. **94**, 219 (1984). https://doi.org/10.1007/BF01209302
49. K. Fredenhagen, R. Haag, Generally covariant quantum field theory and scaling limits. Commun. Math. Phys. **108**, 91 (1987). https://doi.org/10.1007/BF01210704
50. E. Witten, Why does quantum field theory in curved spacetime make sense? And what happens to the algebra of observables in the thermodynamic limit? (2022). arXiv:2112.11614 [hep-th]
51. S.D. Mathur, The information paradox: a pedagogical introduction. Class. Quant. Grav. **26**, 224001 (2009). https://doi.org/10.1088/0264-9381/26/22/224001. arXiv:0909.1038 [hep-th]
52. A. Almheiri, D. Marolf, J. Polchinski, J. Sully, Black holes: complementarity or firewalls? JHEP **02**, 062 (2013). https://doi.org/10.1007/JHEP02(2013)062. arXiv:1207.3123 [hep-th]
53. M. Visser, Gravitational vacuum polarization. 4: Energy conditions in the Unruh vacuum. Phys. Rev. D **56**, 936–952 (1997). https://doi.org/10.1103/PhysRevD.56.936. arXiv:gr-qc/9703001
54. S. Abdolrahimi, D.N. Page, C. Tzounis, Ingoing Eddington-Finkelstein metric of an evaporating black hole. Phys. Rev. D **100**, 124038 (2019). https://doi.org/10.1103/PhysRevD.100.124038. arXiv:1607.05280 [hep-th]
55. C.G. Callan, Jr., S.B. Giddings, J.A. Harvey, A. Strominger, Evanescent black holes. Phys. Rev. D **45**, R1005 (1992). https://doi.org/10.1103/PhysRevD.45.R1005. arXiv:hep-th/9111056
56. A. Strominger, Les Houches lectures on black holes, in *NATO Advanced Study Institute: Les Houches Summer School, Session 62: Fluctuating Geometries in Statistical Mechanics and Field Theory* (1994). arXiv:hep-th/9501071
57. T. Hartman, J. Maldacena, Time evolution of entanglement entropy from black hole interiors. JHEP **05**, 014 (2013). https://doi.org/10.1007/JHEP05(2013)014. arXiv:1303.1080 [hep-th]
58. P. Calabrese, J.L. Cardy, Evolution of entanglement entropy in one-dimensional systems. J. Stat. Mech. **0504**, P04010 (2005). https://doi.org/10.1088/1742-5468/2005/04/P04010. arXiv:cond-mat/0503393
59. S. Raju, Lessons from the information paradox. Phys. Rept. **943**, 1–80 (2022). https://doi.org/10.1016/j.physrep.2021.10.001. arXiv:2012.05770 [hep-th]
60. M. Saha, M. Kulkarni, A. Dhar, Generalised hydrodynamics description of the Page curve-like dynamics of a freely expanding fermionic gas (2024). arXiv:2402.18422 [quant-ph]
61. A. Almheiri, R. Mahajan, J. Maldacena, Y. Zhao, The Page curve of Hawking radiation from semiclassical geometry. JHEP **03**, 149 (2020). https://doi.org/10.1007/JHEP03(2020)149. arXiv:1908.10996 [hep-th]
62. X. Dong, A. Lewkowycz, Entropy, extremality, Euclidean variations, and the equations of motion. JHEP **01**, 081 (2018). https://doi.org/10.1007/JHEP01(2018)081. arXiv:1705.08453 [hep-th]
63. X. Dong, D. Harlow, A.C. Wall, Reconstruction of bulk operators within the entanglement wedge in gauge-gravity duality. Phys. Rev. Lett. **117**, 021601 (2016). https://doi.org/10.1103/PhysRevLett.117.021601. arXiv:1601.05416 [hep-th]
64. J. Cotler, P. Hayden, G. Penington, G. Salton, B. Swingle, M. Walter, Entanglement wedge reconstruction via universal recovery channels. Phys. Rev. X **9**, 031011 (2019). https://doi.org/10.1103/PhysRevX.9.031011. arXiv:1704.05839 [hep-th]

65. C.-F. Chen, G. Penington, G. Salton, Entanglement wedge reconstruction using the Petz map. JHEP **01**, 168 (2020). https://doi.org/10.1007/JHEP01(2020)168. arXiv:1902.02844 [hep-th]
66. C. Holzhey, F. Larsen, F. Wilczek, Geometric and renormalized entropy in conformal field theory. Nucl. Phys. B **424**, 443–467 (1994). https://doi.org/10.1016/0550-3213(94)90402-2. arXiv:hep-th/9403108
67. P. Calabrese, J.L. Cardy, Entanglement entropy and quantum field theory. J. Stat. Mech. **0406**, P06002 (2004). https://doi.org/10.1088/1742-5468/2004/06/P06002. arXiv:hep-th/0405152
68. P. Calabrese, J. Cardy, Entanglement entropy and conformal field theory. J. Phys. A **42**, 504005 (2009). https://doi.org/10.1088/1751-8113/42/50/504005. arXiv:0905.4013 [cond-mat.stat-mech]
69. R. Jackiw, Lower dimensional gravity. Nucl. Phys. B **252**, 343–356 (1985). https://doi.org/10.1016/0550-3213(85)90448-1
70. C. Teitelboim, Gravitation and Hamiltonian structure in two space-time dimensions. Phys. Lett. B **126**, 41–45 (1983). https://doi.org/10.1016/0370-2693(83)90012-6
71. S. Sachdev, J. Ye, Gapless spin fluid ground state in a random, quantum Heisenberg magnet. Phys. Rev. Lett. **70**, 3339 (1993). https://doi.org/10.1103/PhysRevLett.70.3339. arXiv:cond-mat/9212030
72. J. Maldacena, D. Stanford, Remarks on the Sachdev-Ye-Kitaev model. Phys. Rev. D **94**, 106002 (2016). https://doi.org/10.1103/PhysRevD.94.106002. arXiv:1604.07818 [hep-th]
73. A. Kitaev, S. Josephine Suh, The soft mode in the Sachdev-Ye-Kitaev model and its gravity dual. JHEP **05**, 183 (2018). https://doi.org/10.1007/JHEP05(2018)183. arXiv:1711.08467 [hep-th]
74. J. Kruthoff, R. Mahajan, C. Murdia, Free fermion entanglement with a semitransparent interface: the effect of graybody factors on entanglement islands. SciPost Phys. **11**, 063 (2021). https://doi.org/10.21468/SciPostPhys.11.3.063. arXiv:2106.10287 [hep-th]
75. J.S. Cotler, G. Gur-Ari, M. Hanada, J. Polchinski, P. Saad, S.H. Shenker, D. Stanford, A. Streicher, M. Tezuka, Black holes and random matrices. JHEP **05**, 118 (2017). https://doi.org/10.1007/JHEP05(2017)118. arXiv:1611.04650 [hep-th] [Erratum: JHEP **09**, 002 (2018)]

Chapter 17
Final State Projection and Non-isometric Codes

Chris Akers

Abstract I review the final state proposal and non-isometric codes as possible resolutions to the information paradox. While they appear similar in their use of post-selection, their physical interpretations are different, and non-isometric codes seem to be a resolution more explicitly consistent with holography.

17.1 Introduction

17.1.1 Paradox

Hawking argued [1] that a theory cannot simultaneously have three things,

(1) A finite black hole entropy with a state-counting interpretation.
(2) A unitary black hole S-matrix.
(3) A black hole interior which is described to a good approximation by gravitational effective field theory.

This is the black hole information paradox, and the argument proceeds as follows. Consider gravity coupled to matter fields, and a classical solution in which these matter fields collapse to form a black hole, with conformal diagram

C. Akers (✉)
University of Colorado Boulder, Boulder, USA
e-mail: chris.akers@colorado.edu

A. Akil and C. Bambi (eds.), *The Black Hole Information Paradox*, Springer Series in Astrophysics and Cosmology, https://doi.org/10.1007/978-981-96-6170-1_17

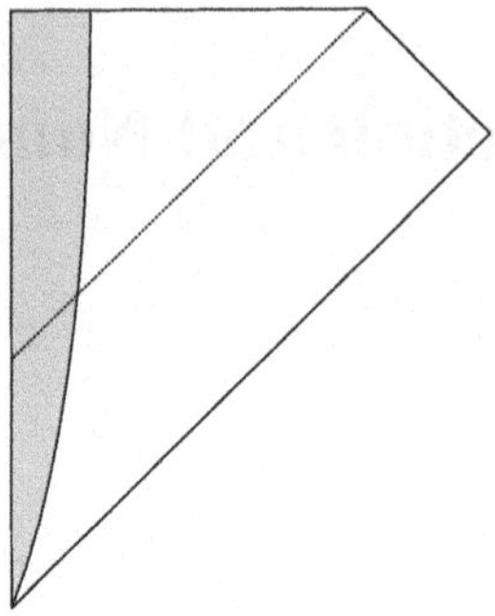

Then, following Hawking [2] and assuming (3) above, quantize the fluctuations in the matter and the metric $g_{\mu\nu}$ about this solution, with the boundary condition that these fluctuations start in the vacuum state in the far past. At late times, we can choose a timeslice that crosses the horizon, with the horizon dividing the timeslice into two parts, an inside and an outside. We can decompose the Hilbert space of the fluctuations into two parts, $\mathcal{H}_r$ and $\mathcal{H}_R$, describing the wavefunctions localized in the interior or exterior respectively. Because the metric is time dependent, so is the vacuum state, and the original vacuum state evolves to the *Unruh state* $|U\rangle \in \mathcal{H}_r \otimes \mathcal{H}_R$, which contains a flux of outgoing particles in $\mathcal{H}_R$, called Hawking radiation [3].

Of course, this evolving quantum state is coupled to the metric, and incorporating its backreaction, the black hole will slowly evaporate, losing its mass as the radiation carries away energy. Now we use assumption (1): if the black hole has finite entropy with a state-counting interpretation, then as the black hole loses mass, it loses entropy and has fewer and fewer states available to it. When it fully evaporates, the only states left are those associated to the radiation outside.

This contradicts a unitarity S-matrix, assumption (2). Imagine the initial collapsing matter fields could have been in one of many states in a Hilbert space $\mathcal{H}_M$. For simplicity, let these states all have basically the same given geometry, differing only in the details of the quantum state. This evaporation process essentially converts a state $|\psi\rangle_{\mathrm{M}}$ to a state $|\psi'\rangle_M |U\rangle_{rR}$ at late times. But if the only states left are those associated to the radiation R outside, we have a problem, because we obtain the state of R by taking $|\psi'\rangle_M |U\rangle_{rR}$ and tracing out $\mathcal{H}_M \otimes \mathcal{H}_r$. The Unruh state $|U\rangle$ is highly entangled between R and r, so tracing out M and r produces a mixed state on R, and in particular the *same* mixed state for every initial state $|\psi\rangle_M$, impossible under unitary time evolution.

17.1.2 Options

What do we make of this tension? It turns out we cannot reject (1) or (2), in general. Strominger and Vafa constructed black holes in string theory that manifestly satisfy (1) [4]. Maldacena demonstrated that certain gravitational theories in Anti de Sitter

(AdS) space are dual to conformal field theories (CFT) [5]. These CFTs manifestly have finite density of states and evolve unitarily, so any black holes on the AdS side must satisfy (1) and (2) as well. Consistent quantum theories of gravity arise from string theory and AdS/CFT, even if we do not yet know how related they are to our own universe. Therefore we must accept that there is some way to resolve the information paradox by keeping (1) and (2). This leaves us with two options.

Option A, we give up on (3) and accept that black hole interiors are very different from what gravitational effective field theory predicts. This is the option espoused by "firewall" [6] proposals, or many versions of the "fuzzball" proposals [7, 8]. Naively this seems natural; near the singularity, spacetime curvature becomes arbitrarily large. Gravitational effective theory is an expansion that only seems valid at small curvatures and energies, so like any good Wilsonian effective theory it predicts its own breakdown near the singularity. The problem, however, is that rejecting (3) involves much more than this. It includes throwing out effective theory *even far away* from the singularity, near the horizon, because it was at the horizon that Hawking assumed effective theory was valid to derive his contradiction. There's no obvious reason effective theory should break down here, in a regime of arbitrarily low curvature (for an arbitrarily large black hole). What's more, the equivalence principle—the bedrock of general relativity—tells us the horizon should be no place special. If we reject (3), throwing out the equivalence principle in a place where effective theory doesn't predict its own breakdown, how do we explain the success of general relativity everywhere else?

Option B, we find some way to reconcile (3) with (1) and (2). Is there something that Hawking missed in his argument, some hidden assumption? Is there a slight modification of (3) that mitigates the tension? In this chapter, we focus on this option, and two proposals in particular. The first is the "black hole final state" proposal, originating in 2003 by Horowtiz and Maldacena [9]. The second is "non-isometric codes", originating in 2022 by Akers, Engelhardt, Harlow, Penington, and Vardhan [10]. These are related, as we'll see, by utilizing a similar mathematical loophole in Hawking's assumptions. In spite of this, however, they are quite different physically. We focus on these because they lead us to arguably the only concrete realization of Option B known to date.

17.2 Final State

17.2.1 Proposal

Horowitz and Maldacena [9] noticed an interesting mathematical fact, which seems (at least at first) to allow us to keep all assumptions (1)–(3). The trick will be to modify quantum mechanics, but do it close to the singularity where we don't trust effective theory anyway. The modification they make is to include *final state postselection*.

To explain it, let's first review measurements in ordinary quantum mechanics, closely following the nice review [11]. Say we have a quantum system described by an initial density matrix ρ_i. We wish to make a series of measurements $A_1, A_2, ..., A_n$. The joint probability distribution for the measurement outcomes is

$$P(a_1, a_2, ..., a_n) = \mathrm{tr}(\Pi_{a_n}...\Pi_{a_1}\rho_i\Pi_{a_1}...\Pi_{a_n}), \tag{17.2.1}$$

where Π_{a_j} is the projection operator onto outcome a_j from measuring A_j. This single formula captures both the Born rule and the collapse of the wavefunction, along with their important corollaries, for example:

- If the sequential measurements do not commute, then the probability depends on the order of measurement. If they commute then it does not.
- The probability distribution for the outcomes of the first m measurements is independent of all later measurements. For example,

$$P(a_1) = \sum_{a_2} P(a_1, a_2) = \mathrm{tr}(\Pi_{a_1}\rho_i\Pi_{a_1}). \tag{17.2.2}$$

  This is crucial for causality, implying that the outcome of measurements we do today does not depend on what we choose to measure tomorrow.

To generalize this to include final state postselection, we replace (17.2.1) with

$$P(a_1, a_2, ..., a_n) = \frac{1}{\mathcal{N}}\mathrm{tr}(\rho_f\Pi_{a_n}...\Pi_{a_1}\rho_i\Pi_{a_1}...\Pi_{a_n}), \tag{17.2.3}$$

where ρ_f is a density matrix we call the final state, and

$$\mathcal{N} = \sum_{a_1',a_2',...,a_n'} \mathrm{tr}(\rho_f\Pi_{a_n'}...\Pi_{a_1'}\rho_i\Pi_{a_1'}...\Pi_{a_n'}). \tag{17.2.4}$$

This defines a normalized set of probabilities for any measurements we might do. One major caveat: it is no longer true that we can average out later measurements as in (17.2.2). Final state postselection violates causality! For the moment, let us set this concern aside and focus on the original question: can final state postselection be used to make a theory satisfying assumptions (1)–(3) (or at most a minor modification of them)?

To incorporate final states into black holes, it will be helpful to be specific about the form of the Unruh state. We can view Hawking radiation as pair creation of a positive energy particle that goes to infinity and a negative energy particle that falls into the black hole, with this pair created in a particular entangled state, such that the Unruh state can be viewed as an entangled thermal state. For our discussion, let us simplify this by considering the microcanonical ensemble, ending up with an equal combination of all states in $\mathcal{H}_R$ with fixed energy equal to the original black hole mass. We can write

$$|U\rangle = \frac{1}{\sqrt{|M|}} \sum_i |i\rangle_r |i\rangle_R , \tag{17.2.5}$$

where these radiation Hilbert spaces and the matter Hilbert space $\mathcal{H}_M$ all have dimension $|M| = e^{S_{BH}}$, where S_{BH} is the entropy of the original black hole. The total state is

$$|\psi\rangle_M |U\rangle_{rR} . \tag{17.2.6}$$

We describe the entire late time slice with the Hilbert space $\mathcal{H}_M \otimes \mathcal{H}_r \otimes \mathcal{H}_R$, where the black hole interior is given by $\mathcal{H}_M \otimes \mathcal{H}_r$, a factorization which is somewhat artificial but valid enough for our purposes for now. We will revisit this once we consider the effects of interactions between $\mathcal{H}_M$ and $\mathcal{H}_r$.

Horowitz and Maldacena introduce a final state

$$\rho_f = |\chi\rangle \langle\chi|_{Mr} \otimes \frac{\mathbb{1}_R}{|M|} \tag{17.2.7}$$

with

$$|\chi\rangle_{Mr} = \frac{1}{\sqrt{|M|}} \sum_i S^\dagger |i\rangle_M |i\rangle_r , \tag{17.2.8}$$

where S is a unitary operator acting on $\mathcal{H}_M$. The claim is that now the information in $\mathcal{H}_M$ actually gets out to R. To see this, say we want to compute the probability distribution $P(b)$ for some set of measurements on R. Represent the sequence of projectors for a set of outcomes as $C_b = \Pi_{b_1} \ldots \Pi_{b_n}$, and then we can derive that

$$\frac{\operatorname{tr}\left(\rho_f C_b^\dagger \rho_i C_b\right)}{\sum_{b'} \operatorname{tr}\left(C_{b'}^\dagger \rho_i C_{b'}\right)} = \left\langle \psi | S^\dagger C C^\dagger S | \psi \right\rangle . \tag{17.2.9}$$

In words, the outcomes of all measurements on the outside radiation are consistent with just taking it to be in the pure state $S|\psi\rangle_R$ and doing ordinary quantum mechanics. We see the unitary S has the interpretation the S-matrix, converting the in-states $|\psi\rangle$ to the out-states $S|\psi\rangle$.

17.2.2 Relation to Quantum Teleportation

It will be illuminating to compare final state postselection to a famous protocol in quantum information theory called quantum teleportation. Imagine Alice has a qubit A in state $|\psi\rangle_A$ that she wishes to give to Bob. However, she is unable to hand it to Bob directly, or send it through any kind of channel that keeps quantum information intact. Fortunately, she has another qubit A' that is maximally entangled with a qubit B that Bob has. The state of all three qubits is

$$|\psi\rangle_A \left(\frac{1}{\sqrt{2}} |00\rangle_{A'B} + \frac{1}{\sqrt{2}} |11\rangle_{A'B} \right) \tag{17.2.10}$$

Alice can achieve her goal by following the quantum teleportation protocol: first, she measures her two qubits A, A' in the Bell basis, obtaining one of four possible outcomes:

$$\begin{aligned} |\phi_+\rangle_{AA'} &= \frac{1}{\sqrt{2}} (|00\rangle_{AA'} + |11\rangle_{AA'}) \\ |\phi_-\rangle_{AA'} &= \frac{1}{\sqrt{2}} (|00\rangle_{AA'} - |11\rangle_{AA'}) \\ |\psi_+\rangle_{AA'} &= \frac{1}{\sqrt{2}} (|01\rangle_{AA'} + |10\rangle_{AA'}) \\ |\psi_-\rangle_{AA'} &= \frac{1}{\sqrt{2}} (|01\rangle_{AA'} - |10\rangle_{AA'}) \end{aligned} \tag{17.2.11}$$

Second, she communicates *classically* to Bob which of the four outcomes she measured. (This she can do by a phone call—no need to have a channel that preserves *quantum* information.) Then Bob acts on B with a unitary that depends on that outcome,

$$\begin{aligned} |\phi_+\rangle_{AA'} &\implies \mathbb{1}_B \\ |\phi_-\rangle_{AA'} &\implies Z_B \\ |\psi_+\rangle_{AA'} &\implies X_B \\ |\psi_-\rangle_{AA'} &\implies Z_B X_B \end{aligned} \tag{17.2.12}$$

and voila, Bob's qubit is now in the state $|\psi\rangle_B$. For example, say $|\psi\rangle = a\,|0\rangle + b\,|1\rangle$, and when Alice measures AA' she finds $|\psi_-\rangle$. The resulting state is obtained by acting $|\psi_-\rangle\,\langle\psi_-|_{AA'}$ on the initial state and normalizing, giving

$$|\psi_-\rangle_{AA'} (a\,|1\rangle_B - b\,|0\rangle_B)\,, \tag{17.2.13}$$

and we see indeed that acting X_B and then Z_B will put B in the correct state. This protocol trivially generalizes from sending one qubit to sending any number of qubits: for example, if Alice has n qubits she wants to send, she can teleport each qubit individually (as long as she and Bob share n Bell pairs).

Final state postselection can be understood as a kind of quantum teleportation protocol, where the information in M is quantum teleported out to the radiation. We have some late time state $|\psi\rangle_M\,|U\rangle_{rR}$, and the singularity "measures" Mr in the state $|\chi\rangle_{Mr}$. The resulting state is (up to normalization)

$$\langle\chi|_{Mr} \left(|\psi\rangle_M\,|U\rangle_{rR} \right) = S\,|\psi\rangle_R\,. \tag{17.2.14}$$

There is one key difference between this and conventional quantum teleportation: in final state postselection Mr is always acted on by $\langle\chi|$. There are no alternative outcomes. It is as though someone made a measurement and knew what they would get ahead of time. Therefore there is no need to communicate a measurement outcome to the outside—it's already known! This is the big difference between conventional teleportation (by measurement) and postselected teleportation.

17.2.3 Issues

There are many nice aspects of this proposal. At a superficial level, it seems like it might reconcile assumptions (1)–(3). We manifestly have (1), a finite dimensional Hilbert space. Unitarity (2) seems plausible from from (17.2.9), as does valid effective theory (3) because we directly modified quantum mechanics only at the singularity where we already don't trust effective theory.

However, really testing (2) and (3) turns out to be subtle, and the original Horowitz-Maldacena final state (17.2.8) leads to issues. Gottesman and Preskill point out [12] that unitarity is jeopardized under slightly more realistic physics in the black hole interior. Say degrees of freedom in the black hole interior (here M and r) can interact. We can model this by saying the state need not be of the form (17.2.6), but instead can be of the form

$$W\,|\psi\rangle_M\,|U\rangle_{rR} \tag{17.2.15}$$

where W is some unitary operator on $\mathcal{H}_M \otimes \mathcal{H}_r$. But now postselecting with (17.2.8) takes the form

$$\langle\chi|_{Mr}\,W\,|\psi\rangle_M\,|U\rangle_{rR}\,, \tag{17.2.16}$$

instead of (17.2.14), which is completely different. To quantum teleport $|\psi\rangle$ to R, it was vital that $|\chi\rangle$ was maximally entangled. (Conventional quantum teleportation won't work at all if Alice measures AA' in an unentangled basis, and it will work less than perfectly if the state is less than maximally entangled.) But we can interpret (17.2.16) as postselecting $|\psi\rangle_M\,|U\rangle_{rR}$ onto the state $W^\dagger\,|\chi\rangle_{Mr}$, which for general W need not be highly entangled at all! Therefore once interactions are allowed in the interior, we can no longer expect exact unitarity. We might hope that physically reasonable interactions don't lead to $W^\dagger\,|\chi\rangle$ that are *too* unentangled, and that maybe unitarity is generally *mostly* preserved. However, in practice this can be difficult to realize, requiring fine tuned effective theory dynamics W.

There are also subtleties with (3). To really test effective theory, we need to check if it agrees with the predictions (17.2.3) of postselected quantum theory in regions where effective theory should be valid. One check: normal effective theory should be valid at the horizon, where it predicts a 100% chance of an observer measuring the state $|U\rangle_{rR}$. Final state postselection passes this check: from (17.2.3), we see any single measurement on rR in (17.2.6) has zero probability of finding anything orthogonal to $|U\rangle$. That said, doing *more than one* measurement raises issues—we run straight into the violations of causality that final state quantum mechanics generally has. They appear here as follows. Say Alice is watching an evaporating black hole, and at some point makes a measurement on R to check if the state is consistent with $S|\psi\rangle$. Then she decides: does she jump in and measure rR, or doesn't she. (She need not have waited for evaporation to be finished to do the first measurement on R—it suffices for her to make a measurement on a little more than half of R, so the black hole can still be there afterwards! [6]) It turns out the outcome of her first measurement on R *depends on whether she later chooses to measure* rR.

The simplest model of this: let M, r, and R all be qubits, and say the state is

$$|0\rangle_M |\phi_+\rangle_{rR} . \tag{17.2.17}$$

Let the final state be

$$|\phi_+\rangle \langle\phi_+|_{Mr} \otimes \frac{\mathbb{1}_R}{2}. \tag{17.2.18}$$

Say Alice does one measurement on R, measuring in a basis $\{|0\rangle, |1\rangle\}$. By (17.2.3) her probabilities are

$$\begin{aligned} P(0) &= 1 \\ P(1) &= 0. \end{aligned} \tag{17.2.19}$$

But now say Alice first does the R measurement $\{|0\rangle, |1\rangle\}$, then does an rR measurement in the Bell basis. There are eight possible outcomes, for the two outcomes of the first measurement and the four of the second. Again by (17.2.3), the four with non-zero probability are

$$\begin{aligned} P(0, \phi_+) &= 1/4 \\ P(0, \phi_-) &= 1/4 \\ P(1, \phi_+) &= 1/4 \\ P(1, \phi_-) &= 1/4. \end{aligned} \tag{17.2.20}$$

Now the probability of getting 1 on the first measurement is non-zero! If Alice measures $|1\rangle_R$, she knows for sure she will be measuring rR later. Therefore Alice at the horizon (measuring rR) is able to signal to past Alice in the exterior (measuring R) by deciding whether or not to perform the rR measurement. This is a sharp example of acausality well in the regime where effective theory should be valid!

17.2.4 Ameliorating Some Issues: Random Unitaries

Given these breaches of unitarity *and* effective theory, we might be tempted to give up on the final state proposal. However, there's a way to ameliorate these problems somewhat, pointed out by Lloyd and Preskill [13]. Instead of (17.2.7), use a *random* final state,

$$|\chi\rangle_{Mr} = U\,|0\rangle_{Mr}\,, \tag{17.2.21}$$

where $|0\rangle$ is some fiducial state, and U is a unitary on Mr drawn *at random*.

How do we pick a unitary at random? Say we have a Hilbert space $\mathcal{H}$ with dimension d. The group of unitaries acting on this Hilbert space is $U(d)$. On this group there is a unique measure dU that is both left and right invariant,

$$dU = d(WU) = d(UW), \tag{17.2.22}$$

where $W \in U(d)$. We call this the Haar measure, will normalize it to $\int dU = 1$, and will pick unitaries uniformly from it.

The claim is that with very high probability over the random choice of U, (17.2.21) will very nearly satisfy assumptions (1), (2), and (3) in some sense. Let's see how it works to nearly give unitarity. To start, we must acknowledge more details of realistic black holes. A black hole of entropy S_{BH} has a lifetime of $O(S_{\mathrm{BH}}^{3/2})$ (in Planck units). The vast majority of ways to create black holes look like time reversed evaporation processes, taking this very long $O(S_{\mathrm{BH}}^{3/2})$ timescale. Black holes that form from fast collapse form in $O(\sqrt{S_{\mathrm{BH}}})$ time, and are much less numerous, having entropy $O(S_{\mathrm{BH}}^{3/4})$. We want to focus on these fast collapse black holes, for now. (The more generic kind of black holes are more subtle, and we will return to them later.) Then we can model the matter Hilbert space $\mathcal{H}_M$ as having two factors, $\mathcal{H}_M = \mathcal{H}_{M_1} \otimes \mathcal{H}_{M_2}$, where $\mathcal{H}_{M_1}$ are all of the degrees of freedom that can be excited in a fast collapse, and $\mathcal{H}_{M_2}$ is everything else. What's important is that the dimension of $\mathcal{H}_{M_1}$ is relatively small,

$$|M_1|/|M| = \mathrm{e}^{-O(S_{\mathrm{BH}})} \ll 1. \tag{17.2.23}$$

We want to ask whether the information in $\mathcal{H}_{M_1}$ can be decoded from the external Hawking radiation. We will do this differently from before, now in a more robust way popular in the field of quantum information theory. Consider an input state in which M_1 is maximally entangled with a reference system we'll call L. We'll check if L ends up (nearly) maximally entangled with R. If it does then we can declare success, the information in M_1 has made it out to R. The system M_2 will be in some fixed state $|\psi_0\rangle_{M_2}$.

The state after final state projection is

$$|\Psi(U)\rangle_{RL} = \langle 0|_{Mr}\, U^{\dagger}\, |\mathrm{MAX}\rangle_{M_1 L}\, |0\rangle_{M_2}\, |U\rangle_{rR}\,, \tag{17.2.24}$$

which we can illustrate as

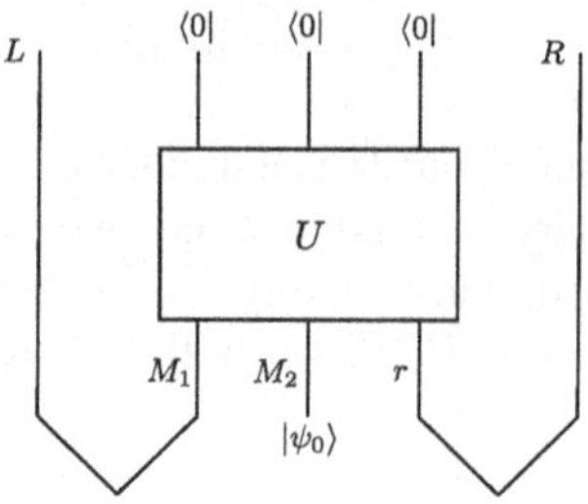

where the angled lines represent maximally entangled states. (We have drawn the postselection $\langle 0|_{Mr}$ as a separate postselection on each of three factors—this is just a convenient choice of fiducial state $|0\rangle_{Mr}$.) This is some pure state on LR, so L being maximally entangled with R is equivalent to the reduced density matrix on L being maximally mixed,

$$\text{Unitarity implies:} \ \ \rho_L(U) := \mathrm{tr}_R \, |\Psi(U)\rangle \, \langle\Psi(U)| = \frac{\mathbb{1}_L}{|L|}. \tag{17.2.25}$$

To rigorously measure how close ρ_L really is to maximally mixed, we need to introduce some notion of distance between density matrices. For this we use the L_1 *norm* $\|\rho_1 - \rho_2\|_1$, where

$$\|X\|_1 = \mathrm{tr}\sqrt{X^\dagger X}. \tag{17.2.26}$$

One can show that $\|\rho_1 - \rho_2\|_1 = 0$ if and only if the density matrices are exactly the same, growing larger the more distinguishable they are, achieving the maximum of 2 if they are orthogonal. Operationally this is a sensible notion of distance, because if $\|\rho_1 - \rho_2\|_1 < \varepsilon$, then for any projection operator Π it holds that $\mathrm{tr}(\Pi(\rho_1 - \rho_2)) < \varepsilon$, so the probability is less than ε to obtain a different measurement outcome on the two states. We will also be interested in the L_2 norm, defined as

$$\|X\|_2 = \sqrt{\mathrm{tr}\, X^\dagger X}, \tag{17.2.27}$$

which one can show obeys

$$\|X\|_2 \leq \|X\|_1 \leq \sqrt{d}\|X\|_2, \tag{17.2.28}$$

where d is the dimension of the Hilbert space on which X acts.

One can compute the following upperbound on the average distance,

$$\int dU \, \|\rho_L(U) - \frac{\mathbb{1}_L}{|L|}\|_1 \leq \sqrt{\frac{|M_1|}{|M|}}. \tag{17.2.29}$$

This is exponentially small by (17.2.23), so for almost all U, $\rho_L(U)$ is very close to maximally mixed!

To derive (17.2.29), it's helpful to use the following tricks:

$$\begin{aligned}\left(\int dU \|\rho_L(U) - \frac{\mathbb{1}_L}{|L|}\|_1\right)^2 &\leq \int dU \|\rho_L(U) - \frac{\mathbb{1}_L}{|L|}\|_1^2 \\ &\leq \sqrt{|L|} \int dU \|\rho_L(U) - \frac{\mathbb{1}_L}{|L|}\|_2^2.\end{aligned} \tag{17.2.30}$$

In the first line we have used Jensen's inequality, and in the second we have replaced the L_1 norm with the L_2 norm, which is easier to work with. Now each term in the integrand has an integer number of U and $U^\dagger$, and we can use known expressions for integrals over the Haar measure[1]

$$\begin{aligned}&\int dU U_{i_1 j_1} U^\dagger_{j_1' i_1'} = \frac{1}{d} \delta_{i_1 i_1'} \delta_{j_1 j_1'}, \\ &\int dU U_{i_1 j_1} U_{i_2 j_2} U^\dagger_{j_1' i_1'} U^\dagger_{j_2' i_2'} = \frac{1}{d^2 - 1} \Big[\delta_{i_1 i_1'} \delta_{i_2 i_2'} \delta_{j_1 j_1'} \delta_{j_2 j_2'} + \delta_{i_1 i_2'} \delta_{i_2 i_1'} \delta_{j_1 j_2'} \delta_{j_2 j_1'} \\ &\qquad - \frac{1}{d} \left(\delta_{i_1 i_1'} \delta_{i_2 i_2'} \delta_{j_1 j_2'} \delta_{j_2 j_1'} + \delta_{i_1 i_2'} \delta_{i_2 i_1'} \delta_{j_1 j_1'} \delta_{j_2 j_2'} \right) \Big],\end{aligned} \tag{17.2.31}$$

An integral over an unequal number of U and $U^\dagger$ will vanish. This completes the derivation of (17.2.29) once we recall $|L| = |M_1|$ and $|R| = |M|$.

That concludes the demonstration that a random final state (17.2.21) leads to near unitarity. What have we gained? First of all, *near* unitarity seems like a downgrade, but arguably it's not so bad. Again, if ρ_L is exponentially close to maximally mixed then it is very difficult to tell the difference, and anyway we might hope that there are unaccounted for non-perturbative gravitational effects at this scale $e^{-O(S_{BH})}$ which fix it to exact unitarity. The real gain in using (17.2.21) comes when we include interactions in the black hole interior, which led to possibly *large* deviations in unitarity in the original final state.

As in Sect. 17.2.3, consider some dynamics in the interior of the black hole modeled by some unitary operation W on Mr. The problem before was that $W^\dagger |\chi\rangle$ need not be highly entangled between M and r, leading to poor postselected quantum teleportation. However, let's now consider W to only act on a relatively small portion of r we'll call r_1, calling the rest r_2, with $\mathcal{H}_r = \mathcal{H}_{r_1} \otimes \mathcal{H}_{r_2}$. Let's consider the state we get after acting $W^\dagger$ and then $\langle 0|_{M_2}$ on (17.2.21),

$$|\Phi(U)\rangle_{M_1 r_1 r_2} := \langle 0|_{M_2} W^\dagger U |0\rangle_{Mr}, \tag{17.2.32}$$

and define the density matrix

$$\Phi_X(U) := \mathrm{tr}_{\overline{X}} |\Phi(U)\rangle \langle \Phi(U)|, \tag{17.2.33}$$

[1] These can be derived using the left and right invariance of the Haar measure to express the integral as a sum of invariant tensors with unknown coefficients, and then determining the coefficients by contracting indices and using $U^\dagger U = \mathbb{1}$.

where $\mathrm{tr}_{\overline{X}}$ denotes tracing out all systems but X. We can compute that

$$\int dU \|\Phi_{M_1 r_1}(U) - \frac{\mathbb{1}_{M_1}}{|M_1|} \otimes \Phi_{r_1}(U)\|_1 \leq \sqrt{\frac{|M_1||r_1|}{|r_2|}} \tag{17.2.34}$$

using the same techniques as before. This means that M_1 is approximately maximally entangled with r, but M_1 and r_1 are approximately unentangled for small enough $|r_1|$. That implies that M_1 and r_2 are approximately maximally entangled. Therefore, W does not change the entanglement between M and r very much, and postselected quantum teleportation will work with nearly maximal fidelity! Unitarity is approximately preserved under these types of reasonable interactions in the interior.

17.2.5 Ameliorating Other Issues: Complexity

What about the final issue, acausality? In principle it is present with this random final state (17.2.21) as much as it is with the original (17.2.7). However, there is an interesting idea to mitigate the issue, applicable to both cases, originating from Harlow and Hayden [14].[2]

The idea is to invoke tools from the theory of *quantum computational complexity*, which we will introduce carefully in a moment. With these tools, the argument is that actually performing the experiment that would see acausality would take an outrageous amount of time, $\mathrm{e}^{O(S_{\mathrm{BH}})}$, much longer than the evaporation time $O(S_{\mathrm{BH}}^{3/2})$. Therefore maybe we shouldn't worry about it!

Why should it be hard to do the acausal experiment? Recall it involved two steps: first do a measurement on the exterior radiation (part of R) that learns some information about the initial state, and second jump in to do a measurement on rR. The black hole still needs to be around for the second step to be viable, so we'll need to assume that the first measurement is at most on some fraction (at least half) of the qubits in R. Let's say $\mathcal{H}_R = \mathcal{H}_{R_1} \otimes \mathcal{H}_{R_2}$, with R_1 the fraction we measure in the first step, and

$$|R_1|/|R| = 1/|R_2| = \mathrm{e}^{-O(S_{\mathrm{BH}})}. \tag{17.2.35}$$

Now recall the Hawking radiation ends up in a state (17.2.24), in which L is highly entangled with R. Let's imagine L is just a single qubit here (if it's larger, that will only make the experiment harder). We can write this state as[3]

[2] They originally made their argument to help evade the firewall paradox [6], but it turns out to be a broadly impactful idea. We will invoke it heavily in Sect. 17.3.

[3] Keep in mind, the real state is likely not *exactly* of this form, but will be close in L_1 norm.

$$|\Psi\rangle = \frac{1}{\sqrt{2|R_2|}} \sum_h |0\rangle_L |f(0), h\rangle_{R_1} |h\rangle_{R_2} + |1\rangle_L |f(1), h\rangle_{R_1} |h\rangle_{R_2}, \qquad (17.2.36)$$

where f is some invertible function. What's important is that L is maximally entangled with R_1. R_2 is also maximally entangled with R_1, but not at all with L. When we assume that we can do a measurement on R_1 to learn information about the initial state, we are effectively assuming we have the power to act on R_1 to distill out a single qubit that is maximally entangled with L, i.e. that we can bring in some ancilla qubit $\mathcal{H}_A$ and act a unitary U_{AR_1} such that

$$\begin{aligned} U_{AR_1} |0\rangle_A |f(0), h\rangle_{R_1} &= |0\rangle_A |g, h\rangle_{R_1} \\ U_{AR_1} |0\rangle_A |f(1), h\rangle_{R_1} &= |1\rangle_A |g, h\rangle_{R_1}, \end{aligned} \qquad (17.2.37)$$

for some state $|g, h\rangle_{R_1}$. In other words, we can *distill out the purification of* L,

$$U_{AR_1} |\Psi\rangle |0\rangle_A = \frac{1}{\sqrt{2}}(|0\rangle_L |0\rangle_A + |1\rangle_L |1\rangle_A) \otimes \frac{1}{\sqrt{|R_2|}} \sum_h |g, h\rangle_{R_1} |h\rangle_{R_2}. \qquad (17.2.38)$$

The rub is that we expect actually implementing U_{AR_1} to require a very long time $e^{O(S_{\mathrm{BH}})}$ [15]! Therefore we expect the measurement on R_1 to also be very hard. To explain why, we'll need to turn to quantum complexity theory.

Quantum complexity theory teaches us that not all unitary operators are equally easy to act. Say we're in a real lab and have n qubits on a table. Our lab equipment gives us the ability to act some unitary operators on these qubits, but only from a limited set and on a few qubits at a time. The word *gate* refers to these fundamental unitaries we have access to (reminiscent of logic gates in classical computers, which act on one or two bits at a time). We might imagine, for example, that our set of gates includes all unitary operators on any two qubits. If we want to act a bigger, arbitrary unitary then we need to build it up out of many of these gates. We'll use *circuit* to denote such combinations of gates. You could wonder: given some unitary U, is there always a circuit that produces it? The answer is yes: there are *universal gate sets* that can build up any unitary [16]. One example is the one we defined above: all unitaries on any two qubits.

However, just because we *can* build any unitary out of gates doesn't mean it's *easy*. It turns out to be useful to divide unitaries acting on n qubits into two types: those which require *polynomially* many gates $O(\text{poly}(n))$, and those which require *exponentially* many gates $e^{O(n)}$. Polynomial unitaries are "easy", exponential ones are "hard". There is a simple counting argument that shows almost all unitaries are in the exponential category [16].

Ideally, we would like to prove that U_{AR_1} from (17.2.37) is exponential. Unfortunately, proving things like that rigorously are famously difficult. Instead we will do what quantum complexity theorists do: argue that if it were easy to implement U_{AR_1}, then it would also be easy to do some task that is widely expected to be difficult. This difficult task will be *inverting one-way functions*, and our argument originates from

Aaronson's improvement of the Harlow-Hayden argument [15]. We can think of a one-way function as a map f from m-bit strings to n-bit strings with $n \geq m$, such that there exists a polynomial circuit U_f such that

$$U_f \left|x, 0\right\rangle = \left|x, f(x)\right\rangle , \tag{17.2.39}$$

but no polynomial circuit $U_{f^{-1}}$ exists where

$$U_{f^{-1}} \left|0, f(x)\right\rangle = \left|x, 0\right\rangle . \tag{17.2.40}$$

Note that even though U_f is invertible, this inverse won't invert the function. If we act $U_f^\dagger$ on $|0, f(x)\rangle$, it need not do anything useful. Should such f exist? Again, it's not proven but is widely believed, and in fact almost all of modern cryptography is based on the assumption they exist!

Returning to (17.2.37), we see that if we could in general find such a U_{AR_1} then that would imply that we could in general invert injective one-way functions. We take this as evidence that U_{AR_1} is hard. Exactly how hard turns out to scale with the dimension $|R_2|$, which we recall from (17.2.35) is of order $\mathrm{e}^{O(S_{\mathrm{BH}})}$. The upshot is that the smallest circuit that implements U_{AR_1} has a number of gates exponential in the black hole entropy.

Let us summarize. With final state postselection, in principle there can be acausal signalling, but we have good reason to believe that to actually implement the acausal protocol would require applying an exponentially complicated unitary to the Hawking radiation.

17.2.6 Current Status

We have examined final state postselection as a possible solution to the information paradox. It has some nice features, including a beautiful use of postselected quantum teleportation to end up with the Hawking radiation carrying the right information. It also had some issues, including violating unitarity once interactions are included in the interior and allowing acausality. We saw these issues can be somewhat ameliorated by introducing random final states and ideas from quantum complexity theory. Now let's ask: altogether, where does final state stand as a proposal for reconciling assumptions (1), (2), and (3)? It certainly doesn't *perfectly* reconcile all three, but maybe that's too much to hope for. We can ask: are the deviations from each acceptable and consistent with known theories of quantum gravity?

In my view, the acausality is still a big enough issue to warrant discarding this proposal, because it is inconsistent with what we expect from AdS/CFT. The CFT is a completely normal quantum theory with no acausality whatsoever. If we imagine a black hole is evaporating in the AdS, we'll run into a contradiction. It's no help that implementing the acausal experiment takes a long time in the AdS—using what we know about the dual CFT, it turns out there's no limit to how little time it takes to act the exponentially long circuit in the experiment [17, 18].

Even though we will now turn away from this proposal, we won't ignore the very interesting mathematical tools it offers. In the next section, we'll turn to a different proposal that keeps the nice idea of quantum postselection but is explicitly consistent with AdS/CFT from the beginning.

17.3 Non-isometric Codes

We're going to start by setting aside the black hole information paradox and instead focus on a different puzzle regarding the emergence of spacetime, which arises in AdS/CFT. Understanding this puzzle will lead us back to the information paradox, now with a solution.

17.3.1 AdS/CFT Basics

It will help to recall more details of the AdS/CFT duality. On one side, we have a quantum theory of gravity living in anti de Sitter space. On the other side, we have a conformal field theory, which is a quantum field theory with special "conformal" symmetries, and in particular has no gravity at all. The archetypical example is the one first described by Maldacena [5], where in the AdS we have Type IIB string theory[4] and the CFT is $\mathcal{N} = 4$ supersymmetric Yang-Mills theory. While these theories appear completely different, they are in a deep sense *exactly the same*. This means in particular they are quantum theories with isomorphic Hilbert spaces.

One reason this is so useful is that we can use the duality to study one theory with the other. We don't yet understand the quantum gravity side fully. We have much better control over the CFT. So one thing we'd like to do is formulate our questions on the gravity side, map them to the CFT via this duality, and then try to answer them there.

In particular, we would like to formulate our questions in *semiclassical gravity*—the effective theory—not the full theory of quantum gravity itself. After all, our derivation of the information paradox in Sect. 17.1.1 was all semiclassical—it was a paradox within the framework of effective theory. If we knew how to formulate and study that setup satisfactorily in the language of the full quantum gravity theory, we would already be done—it would tell us the right answer! But we don't know the full

[4] Really this string theory is living on $AdS_5 \times S_5$, where the isometries of the S_5 correspond to the R-symmetry group of the CFT. We will largely ignore these extra dimensions in the rest of our discussion.

quantum gravity theory well enough. AdS/CFT allows us to go around this issue, mapping our questions directly from semiclassical gravity into the CFT, which again we do understand.

So our goal, for now, is to understand the dictionary that relates semiclassical gravity to the CFT. To describe what this means carefully, we need to explain a bit more about the basics of CFTs and semiclassical gravity in AdS.

Vacuum AdS in $d+1$ dimensions, written in global coordinates, has metric

$$\text{Vacuum AdS:}\quad ds^2 = -\left(1+\frac{r^2}{\ell_{\text{AdS}}^2}\right)dt^2 + \frac{dr^2}{1+\frac{r^2}{\ell_{\text{AdS}}^2}} + r^2 d\Omega_{d-1}^2, \tag{17.3.1}$$

where ℓ_{AdS} is the "AdS radius," which characterizes the curvature scale of the spacetime, $r \in [0,\infty)$, $t \in (-\infty,\infty)$, and Ω_{d-1} denotes the round metric on $\mathbb{S}^{d-1}$. What's important about this metric is that it solves Einstein's equations,

$$R_{\mu\nu} - \frac{1}{2}Rg_{\mu\nu} = 8\pi G T_{\mu\nu} \tag{17.3.2}$$

with energy-momentum tensor $T_{\mu\nu} = \rho_0 g_{\mu\nu}$ and $\rho_0 = d(d-1)/16\pi G\ell^2$, which is a negative cosmological constant. Indeed, (17.3.1) is the *maximally symmetric* solution of Einstein's equations with negative cosmological constant. This has conformal diagram

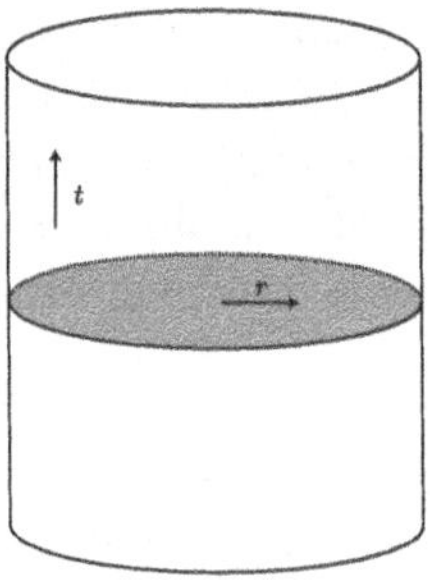

(17.3.3)

where the gray circle indicates a constant t slice, t grows upwards, and r points radially outwards. The boundary of the cylinder is at $r=\infty$.

More generally, we will be interested in metrics that approach (17.3.1) as $r \to \infty$, in particular the AdS-Schwarzschild geometry:

$$\begin{aligned} ds^2 &= -f(r)dt^2 + \frac{dr^2}{f(r)} + r^2 d\Omega_{d-1}^2 \\ f(r) &= 1 + r^2 - \frac{16\pi GM}{(d-1)\Omega_{d-1}}\frac{1}{r^{d-2}}. \end{aligned} \tag{17.3.4}$$

For $d > 2$ this is the unique family of spherically symmetric solutions to (17.3.2) with negative cosmological constant. In fact this metric describes two asymptotically AdS regions, connected by a wormhole—one at $r \to \infty$ and the other at $r \to -\infty$. This has conformal diagram

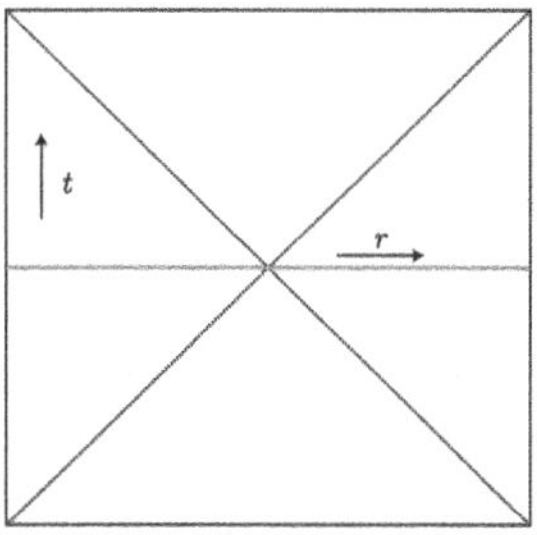

where the thick gray line indicates a $t = 0$ slice, again t grows upwards, and the right and left vertical lines represent the two asymptotic boundaries. The top edge is the black hole singularity, the bottom edge is a white hole singularity, and the diagonal dotted lines represent the event horizons.

Besides gravity, the bulk can also contain matter fields, which we will treat as quantum field theories living on this curved space, perturbatively coupled to gravity. This defines our "semiclassical gravity" in AdS, which we describe with Hilbert space $\mathcal{H}_{\text{bulk}}$.

Conformal field theories are quantum field theories with both Poincare invariance and invariance under two more transformations,

$$\text{dilations:} \qquad x^{\mu\prime} = \lambda x^{\mu} \tag{17.3.5}$$

$$\text{special conformal transformations:} \qquad x^{\mu\prime} = \frac{x^{\mu} + a^{\mu}x^2}{1 + 2a_{\nu}x^{\nu} + a^2x^2}. \tag{17.3.6}$$

Their generators are D and K_{μ}, and together with the Poincare generators P_{μ} and $J_{\mu\nu}$ generate the conformal group $SO(d, 2)$. A local operator O that transforms in a representation of this group has scaling dimension Δ defined by

$$e^{iD\alpha}O(x)e^{-iD\alpha} = e^{\alpha\Delta}O(e^{\alpha}x). \tag{17.3.7}$$

Not all CFTs have nice AdS duals, but we don't yet have a complete set of necessary and sufficient conditions to classify which do. Roughly speaking, we expect the CFT needs a large number of degrees of freedom and a large "gap" in the spectrum, meaning relatively few operators with small Δ. For more details see [19] or Sect. 2.4 of [20].

CFTs with nice AdS duals can be said to live on the conformal boundary of the AdS space, obtained from the limit $r \to \infty$ of the bulk metric with suitable rescaling to keep things finite, for example by multiplying (17.3.1) by $1/r^2$ to limit to the unit cylinder. Therefore we can regard the circle bounding the gray disk in (17.3.3) to be a timeslice of the dual CFT. For this reason we will often refer to the CFT as "the boundary theory," and even write its Hilbert space as $\mathcal{H}_{\text{bdry}}$.

What's particularly nice about CFTs is that they are UV complete. So again, while we do not fully understand quantum gravity directly in AdS, its UV complete dual is just this quantum field theory, which we understand relatively well.

17.3.2 Reconstructing the Bulk

Now we can formulate our question: which CFT operators *reconstruct* a given bulk operator? Let's state this more precisely. Our starting structure is as follows: again, the bulk has a Hilbert space $\mathcal{H}_{\text{bulk}}$ and the CFT a Hilbert space $\mathcal{H}_{\text{bdry}}$. We describe the relationship between them with a "holographic map"

$$V : \mathcal{H}_{\text{bulk}} \to \mathcal{H}_{\text{bdry}}. \tag{17.3.8}$$

Note: while the CFT had the same Hilbert space as the full theory of quantum gravity, it need not be the same as $\mathcal{H}_{\text{bulk}}$! Semiclassical gravity is an effective theory, which could in principle have a very different Hilbert space than the fundamental theory, so V might be highly non-trivial.

Now consider some operator O that acts on $\mathcal{H}_{\text{bulk}}$. Our goal is to figure out which operator (or operators) $\widetilde{O}$ acting on $\mathcal{H}_{\text{bdry}}$ satisfy

$$\widetilde{O} V |\psi\rangle_{\text{bulk}} = V O |\psi\rangle_{\text{bulk}} \tag{17.3.9}$$

for all bulk states $|\psi\rangle$, or if not for all bulk states then at least for some subspace of states we care about. The CFT subspace that is the image of this special subspace is often called the *code subspace* $\mathcal{H}_{\text{code}}$, though often we will simply refer to the bulk states themselves as the "code subspace". Also note: we won't be strict about exact equality in (17.3.9). It's generally sufficient for answering questions if those states are approximately equal, in the sense $\|\widetilde{O} V |\psi\rangle_{\text{bulk}} - V O |\psi\rangle_{\text{bulk}}\| < \varepsilon$ for some small $\varepsilon > 0$.

How do we find these $\widetilde{O}$? The starting point is the so-called *extrapolate dictionary*, fundamental to AdS/CFT and first written explicitly in [21] in 1998 (though c.f. [22, 23] for an equivalent formulation). The extrapolate dictionary guarantees that given some operator O in the bulk, there exists an operator $\widetilde{O}$ in the boundary theory such that

$$\lim_{r\to\infty} r^{n\Delta} \langle O(x_1, r) O(x_2, r) \ldots O(x_n, r)\rangle_{\text{bulk}} = \left\langle \widetilde{O}(x_1) \widetilde{O}(x_2) \ldots \widetilde{O}(x_3)\right\rangle_{\text{bdry}}, \tag{17.3.10}$$

where Δ is the scaling dimension of $\widetilde{O}$. (Part of the dictionary is also that O and $\widetilde{O}$ are promised to satisfy some additional relations; for example the mass of O is a function of Δ, and both O and $\widetilde{O}$ have the same spin.) This in principle specifies the duality, and is manifestly useful when we want to compute CFT correlation functions using the bulk, or correlation functions of near-boundary bulk operators, like

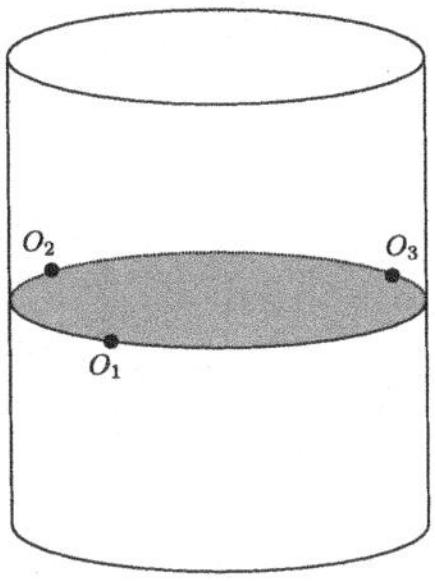

(17.3.11)

It is less clear how to use it for more general questions about the bulk, like correlation functions of operators far from the boundary, like

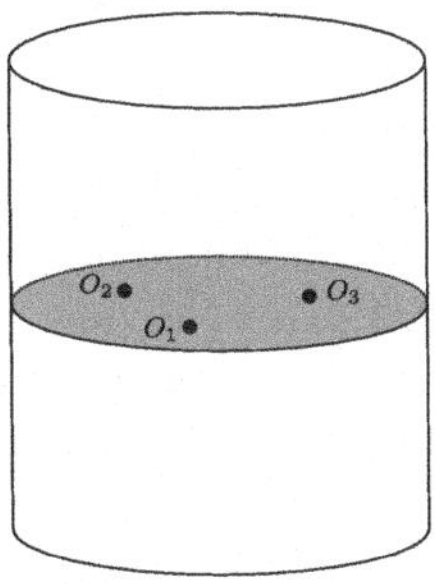

(17.3.12)

For these we will need to work harder. In fact there are now multiple explicit schemes to do this, all fundamentally rooted in the extrapolate dictionary, most working to reconstruct operators in some limited cases but not others. One of the most general schemes is the *Petz map* [24] which was originally defined in the context of abstract quantum channels and then later applied in AdS/CFT [25–27]. The exact form of this map is somewhat ugly and does not concern us. The point is that there does exist a known general purpose map $\mathcal{P}$ from bulk operators to boundary operators such that

$$\mathcal{P}(O)V\,|\psi\rangle \approx V\,O\,|\psi\rangle\,, \tag{17.3.13}$$

for some code subspace of bulk states.

Now an amazing fact. When you study these reconstruction maps, you learn that *they are not unique* [28]! Given some operator O, there are generally multiple *inequivalent* $\widetilde{O}$ that reconstruct it, satisfying (17.3.9).

Let's see how that's proven. To start, we need to introduce one of the most impactful discoveries to come out of AdS/CFT, called the *quantum extremal surface (QES) formula*. Consider a spatial subregion B of the boundary, say

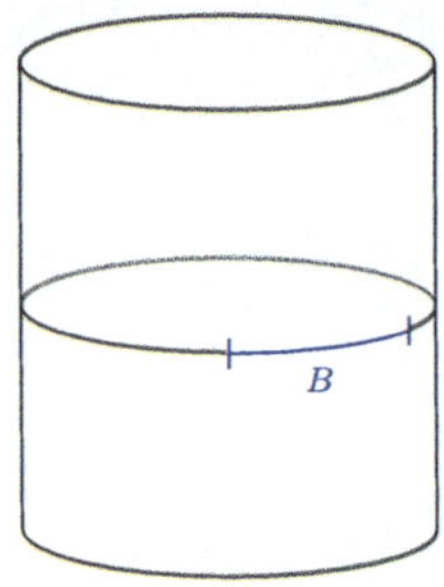

Here we are imagining the boundary Hilbert space to factorize as $\mathcal{H}_{\text{bdry}} = \mathcal{H}_B \otimes \mathcal{H}_{\overline{B}}$.[5] Now, let the boundary be in some state $|\phi\rangle$. We associate to B a *von Neumann entropy*

$$S(B)_{|\phi\rangle} := -\operatorname{tr}(\rho_B \log \rho_B), \tag{17.3.14}$$

where $\rho_B = \operatorname{tr}_{\overline{B}} |\phi\rangle \langle\phi|$ is the density matrix of B. The von Neumann entropy has many nice properties and is a convenient way to measure the entanglement between B and its complement [16], and there are a number of methods for computing it for subregions of quantum field theories (for a review see e.g. [32]). The QES formula is a tool for computing it *using the bulk*. It says that given a bulk state $|\psi\rangle$ and its corresponding boundary state $V|\psi\rangle$,

$$S(B)_{V|\psi\rangle} = \frac{A(b)}{4G} + S(b)_{|\psi\rangle} \tag{17.3.15}$$

where b is a particular subregion of the bulk, $S(b)$ is its von Neumann entropy, $A(b)$ is the area of the surface bounding b, and G is the bulk Newton's constant. The particular subregion b must have a boundary ∂b homologous to B, and is the region that extremizes the right hand side of (17.3.15), or if there are multiple extremal regions then it is the one among those that minimizes the right hand side. The *entanglement wedge of B* is defined as the domain of dependence of b satisfying (17.3.15), i.e. the region of the bulk spacetime that is fully determined by the data in b. Sometimes we will abuse this name and call b the entanglement wedge. In our example b might look something like

[5] This is a simplification—in reality the Hilbert space of a quantum field theory does not factorize. In fact, we will commit an even greater simplification for most of our discussion, working as though the Hilbert spaces involved are *finite*. We could be more careful by explicitly introducing a lattice regulator, which would lead to a factorizing Hilbert space, or by talking about the von Neumann subalgebras associated to these two regions, see e.g. [29]. Either method seems to lead to the same physics, see e.g. [30, 31].

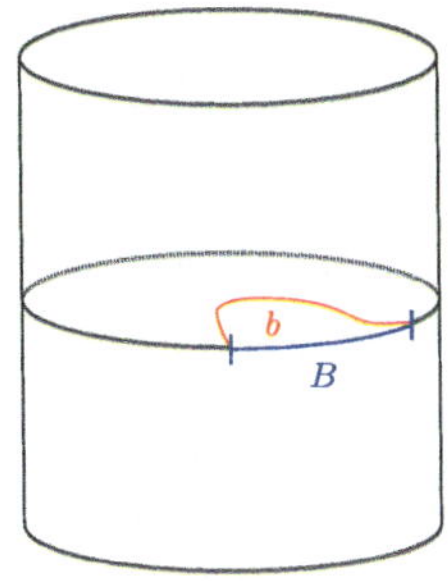

where ∂b is the red line. Note that the entanglement wedge of $\overline{B}$ will simply be the complement $\bar{b}$ of b in the bulk, if the bulk is in a pure state! (If the bulk is in a mixed state, then $S(b) \neq S(\bar{b})$ in general so the QES for each can be different.)

The history of this formula dates back to 2006, when Ryu and Takayanagi conjectured a special case of this formula, which holds only at moments of time-reflection symmetry and gives the answer to leading order in a $1/G$ expansion [33, 34]. This formula was eventually derived in [35] from the AdS/CFT dictionary, gravitational path integral, and some reasonable assumptions on the niceness of the bulk state. In 2007, [36] conjectured how that formula should generalize to settings without time-reflection symmetry, and later [37] derived this from similar assumptions to before. In 2013, [38] derived an improvement to the formula, giving the answer to next to leading order in the $1/G$ expansion. The QES formula was conjectured in its final form in [39], then derived under (once again) similar assumptions in [27, 40, 41].

How much do we trust (17.3.15)? When are the assumptions in its derivation valid? The short answer is: we expect we can trust it to all orders in a $1/G$ expansion, for most states that we care about. However, we certainly do *not* expect it ever to be exact. Its derivation clearly shows that there are small corrections of size $\mathrm{e}^{-O(1/G)}$. This will turn out to be very important. Situations in which the error is larger are fairly well understood, generally corresponding to states near phase transitions (where the error is often $O(1/\sqrt{G})$) [42, 43] or states with lots of incompressible bulk matter [44] (where the error can be as large as $O(1/G)$). We will stay away from these situations.

The most important aspect of the QES formula is what it implies about reconstruction. Indeed, there is an amazing theorem, developed over a number of papers [38, 45–47], culminating in the form below in [48].[6] Some context: recall that we are trying to understand that bulk operators can be reconstructed in multiple inequivalent ways in the boundary theory. This theorem is a key step in this argument because it will show us that general bulk operators can be reconstructed using just a *subregion* of the CFT. That is, we in general do not need to act an operator supported on the entire CFT timeslice, to implement a bulk operator. It will suffice to use just a

[6] This is a simplified version of the theorem appearing in [48], and I encourage the reader to learn the full details from the paper directly, or from Sect. 5.3 of the lectures [20]. This simplified form will be enough for us.

certain subregion. This theorem fills in the details, telling us that given some boundary subregion B, exactly which bulk operators can be reconstructed using just B. First a definition:

Definition 17.3.1 (*Isometry*) An isometry $K : \mathcal{H}_X \to \mathcal{H}_Y$ is a linear map satisfying $K^\dagger K = \mathbb{1}_X$. Note that it can only exist if $\dim \mathcal{H}_Y \geq \dim \mathcal{H}_X$.

Theorem 17.3.2 (Entanglement wedge reconstruction) *Let $\mathcal{H} = \mathcal{H}_B \otimes \mathcal{H}_{\overline{B}}$ and $\mathcal{H}_{\text{code}} = \mathcal{H}_b \otimes \mathcal{H}_{\bar{b}}$ be finite dimensional Hilbert spaces. Let $V : \mathcal{H}_{\text{code}} \to \mathcal{H}$ be an isometry. Then the following are equivalent:*

1. *For all operators O_b (respectively $O_{\bar{b}}$), there exist $\widetilde{O}_B$ (respectively $\widetilde{O}_{\overline{B}}$) such that for all $|\psi\rangle \in \mathcal{H}_{\text{code}}$,*

$$\begin{aligned} \widetilde{O}_B \widetilde{O}_{\overline{B}} V |\psi\rangle &= V O_b O_{\bar{b}} |\psi\rangle \,, \\ \widetilde{O}_B^\dagger \widetilde{O}_{\overline{B}}^\dagger V |\psi\rangle &= V O_b^\dagger O_{\bar{b}}^\dagger |\psi\rangle \,. \end{aligned} \tag{17.3.16}$$

2. *There exists an operator $\hat{A}$ on $\mathcal{H}_{\text{code}}$ such that for all states ρ on $\mathcal{H}_{\text{code}}$*

$$S(B)_{V\rho V^\dagger} = \text{tr}(\rho \hat{A}) + S(b)_\rho, \tag{17.3.17}$$

$$S(\overline{B})_{V\rho V^\dagger} = \text{tr}(\rho \hat{A}) + S(\bar{b})_\rho. \tag{17.3.18}$$

The point of this theorem is to demonstrate that the QES formula—condition 2—tells us exactly which operators can be reconstructed by a CFT subregion—condition 1. This operator $\hat{A}$ in condition 2 should be understood as the operator measuring Area$/4G$ of the quantum extremal surface, when this theorem is applied to AdS/CFT. In general it depends on V, and it can be written down explicitly. Note that it does not depend on the state ρ.

Let me emphasize the incredible lesson this theorem is teaching us. If it were just the case that condition 2 implied condition 1, then we would learn that B could reconstruct any operator in the entanglement wedge, but we wouldn't know if B could reconstruct more than that. Fortunately, we can rule that out using that condition 1 implies condition 2. Imagine B could reconstruct the operators at a point x. Then because condition 1 implies condition 2, this x will be a part of the b on the right hand side of (17.3.17)! Hence B can reconstruct exactly the operators in its entanglement wedge and no more. This fact is often referred to as "entanglement wedge reconstruction".

Entanglement wedge reconstruction leads us to a shocking feature of the AdS/CFT holographic map V [28]: bulk information is encoded in an interesting *non-local* way in the boundary state. More precisely, given a general bulk operator O, no one local operator in the CFT is required in its reconstruction! For example, consider an operator acting in the middle of vacuum AdS,

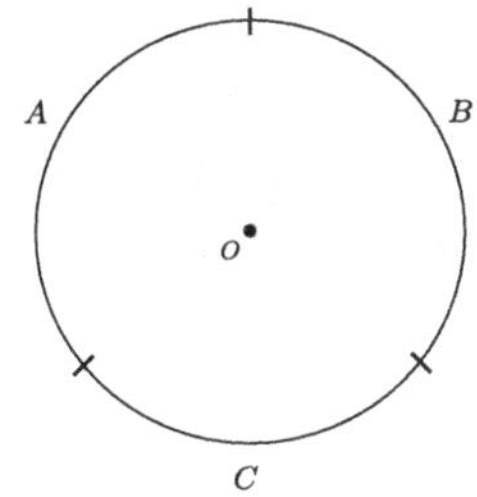

(17.3.19)

where this disk is just a single timeslice, and notice that we have divided the boundary circle into thirds A, B, and C. To proceed carefully, we should define a code subspace. Let's consider the subspace of bulk states $\mathcal{H}_{\text{code}} \subseteq \mathcal{H}_{\text{bulk}}$ below some large energy, perhaps that required to make a black hole. In this subspace there is limited gravitational backreaction, so we can straightforwardly find the entanglement wedge associated to each region A, B, C for each state, as well as that of pair regions AB, BC, AC. What you find is that the entanglement wedge of any one region *excludes* the middle where O acts, while the entanglement wedge of any pair *includes* it,

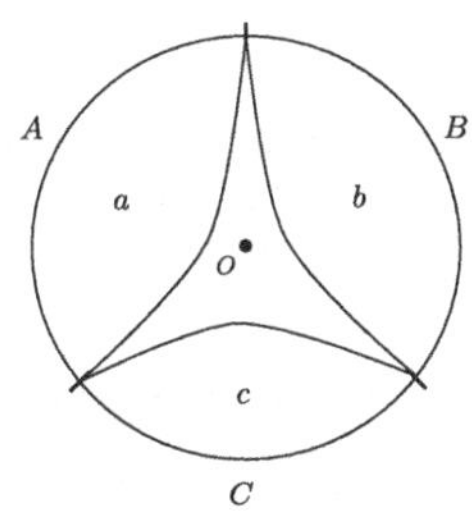

where a, b, c are the entanglement wedges of A, B, C respectively. Recall the pair regions have entanglement wedge that are the complement of the complementary single region. For example, the entanglement wedge of AB is the complement of c.

Therefore by Theorem 17.3.2, there exist reconstructions $\widetilde{O}$ that act on *any two* of A, B, C, i.e.

$$\begin{aligned}\widetilde{O}_{AB} \otimes \mathbb{1}_C V |\psi\rangle &= V O |\psi\rangle \\ \widetilde{O}_{AC} \otimes \mathbb{1}_B V |\psi\rangle &= V O |\psi\rangle \\ \widetilde{O}_{BC} \otimes \mathbb{1}_A V |\psi\rangle &= V O |\psi\rangle .\end{aligned} \tag{17.3.20}$$

What's remarkable about this is that these three $\widetilde{O}$ must be completely different CFT operators! To show this, assume for contradiction that they are the same $\widetilde{O}$. Then any operator W_C local to C must commute with $\widetilde{O}$,

$$[W_C, \widetilde{O}] = [W_C, \widetilde{O}_{AB} \otimes \mathbb{1}_C] = 0, \tag{17.3.21}$$

and similarly for any operators local to A or B. But that means $\widetilde{O}$ commutes with all local operators in the CFT, which by Schur's lemma means $\widetilde{O}$ must be proportional to the identity. This is a contradiction because it cannot be true that all bulk operators are reconstructed by something proportional to the CFT identity operator.

How can these $\widetilde{O}$ all be different CFT operators if they act like the same bulk operator? The answer is that we should understand V to map $\mathcal{H}_{\text{code}}$ into a *subspace* of the CFT, not the entire CFT Hilbert space. These $\widetilde{O}$ all act the same on this subspace, but differently on the rest of the Hilbert space.

Again, this is all important because it tells us that bulk information is stored in the boundary state in an interesting non-local way: no particular local operator in the CFT is required to reconstruct a given bulk operator. This is a deep and frankly curious fact about the holographic map V. How is it even possible? After all, the no-cloning theorem tells us that in quantum mechanics it is impossible to have the same quantum information in two different places. How is that consistent with (17.3.20)? To understand this, we will need to introduce quantum codes.

17.3.3 Quantum Codes

How do we understand the curious multiplicity of operator reconstruction, such as (17.3.20)? It turns out that this structure was already understood in the 1990s by the quantum computation community, who labelled it *quantum error correcting codes*, or for short quantum codes.

Here is a simple example [28, 49]. Let *qutrit* denote a quantum system with a three dimensional Hilbert space, and say we have a qutrit q in some state $|\psi\rangle_q \in \mathcal{H}_q$. We can embed this state into the Hilbert space of three qutrits A, B, C with the following isometry $V : \mathcal{H}_q \to \mathcal{H}_A \otimes \mathcal{H}_B \otimes \mathcal{H}_C$

$$\begin{aligned} V\,|0\rangle_q &= \frac{1}{\sqrt{3}}\left(|000\rangle + |111\rangle + |222\rangle\right), \\ V\,|1\rangle_q &= \frac{1}{\sqrt{3}}\left(|012\rangle + |120\rangle + |201\rangle\right), \\ V\,|2\rangle_q &= \frac{1}{\sqrt{3}}\left(|021\rangle + |210\rangle + |102\rangle\right). \end{aligned} \tag{17.3.22}$$

This is called the *three qutrit code*. We can show that given any one of A, B, C we can learn nothing about the encoded state $|\psi\rangle_q$, but given any *two* of A, B, C we can learn everything. For example, if we have access to only A and B, we can implement a unitary that for any $i \in \{0, 1, 2\}$ satisfies

$$U_{AB} V\,|i\rangle_q = |i\rangle_A \otimes \frac{1}{\sqrt{2}}\left(|00\rangle_{BC} + |11\rangle_{BC} + |22\rangle_{BC}\right). \tag{17.3.23}$$

By linearity, it follows that given any state $|\psi\rangle_q$,

$$U_{AB} V |\psi\rangle_q = |\psi\rangle_A \otimes \frac{1}{\sqrt{2}} (|00\rangle_{BC} + |11\rangle_{BC} + |22\rangle_{BC}) . \tag{17.3.24}$$

Explicitly, this U_{AB} acts as

$$\begin{array}{lll} |00\rangle \to |00\rangle & |11\rangle \to |01\rangle & |22\rangle \to |02\rangle \\ |01\rangle \to |12\rangle & |12\rangle \to |10\rangle & |20\rangle \to |11\rangle . \\ |02\rangle \to |21\rangle & |10\rangle \to |22\rangle & |21\rangle \to |20\rangle \end{array} \tag{17.3.25}$$

By the symmetry of (17.3.22), something similar is possible given access to only BC or AC. From this and (17.3.24) it is straightforward to show that the reduced density matrix of any one qutrit will be maximally mixed, independent of $|\psi\rangle$.

This protocol is helpful in real life. Say we have a state $|\psi\rangle_q$ and we would like to send it to our friend Bob in the mail, but we know there is a probability p it will get lost. Now say we encode it with (17.3.22), and we send each of A, B, C in separate envelopes with their letter on the front. Now each has a probability p of getting lost, but Bob will successfully get the state if any two of the three make it. The probability that at most one gets lost is $(1-p)^3 + 3p(1-p)^2 = 1 - 3p^2 + 2p^3$. This is greater than $1 - p$, our success rate just sending q, as long as p is sufficiently small!

In the previous subsection we talked about operator reconstruction, whereas so far here we have discussed state reconstruction. We can recast this discussion in terms of operators as follows. Say we have some operator O on q. We can reconstruct it on A, B with the operator

$$\widetilde{O}_{AB} = U_{AB}^\dagger O_A U_{AB}, \tag{17.3.26}$$

where O_A is O but on qutrit A. It is straightforward to show this works using (17.3.24). Again by symmetry, we can find such an operator on AC and BC, and we have exactly the structure of (17.3.20)! (It is an exercise to convince yourself that while $\widetilde{O}_{AB}$, $\widetilde{O}_{AC}$ and $\widetilde{O}_{BC}$ act the same on any state of the form $V|\psi\rangle$, they are different operators on the full Hilbert space $\mathcal{H}_A \otimes \mathcal{H}_B \otimes \mathcal{H}_C$.)

This resemblance between AdS/CFT and this three qutrit code is more than superficial. Quantum error-correcting codes can be defined in a few equivalent ways, and it turns out the structure (17.3.20) is just one of those equivalent formulations. This is embodied in the following theorem [48, 50].

Theorem 17.3.3 (Equivalence of quantum code definitions) *Let $V : \mathcal{H}_{\text{code}} = \mathcal{H}_b \otimes \mathcal{H}_{\bar{b}} \to \mathcal{H}_B \otimes \mathcal{H}_{\overline{B}}$ be an isometry between finite dimensional Hilbert spaces. Then the following three statements are equivalent:*

1. *(Schrödinger picture) There exists a "recovery isometry" $W_B : \mathcal{H}_B \to \mathcal{H}_b \otimes \mathcal{H}_E$ such that for any state ρ_{code},*

$$\mathrm{tr}_{\overline{B}E}[W_B V \rho_{\text{code}} V^\dagger W_B^\dagger] = \rho_{\text{code}} . \tag{17.3.27}$$

2. *(Heisenberg picture) There exists a "recovery isometry" $W_B : \mathcal{H}_B \to \mathcal{H}_b \otimes \mathcal{H}_{\bar{E}}$ such that for any Hermitian operator O_b, the Hermitian operator $O_B := W_B^\dagger O_b W_B$ satisfies*

$$\langle \psi | V^\dagger O_B V | \psi \rangle = \langle \psi | O_b | \psi \rangle \tag{17.3.28}$$

for all $|\psi\rangle \in \mathcal{H}_{\text{code}}$.

3. *(Operator reconstruction) For any operator O_b, there exists an operator O_B such that for all $|\psi\rangle \in \mathcal{H}_{\text{code}}$,*

$$\begin{aligned} O_B V |\psi\rangle &= V O |\psi\rangle \\ O_B^\dagger V |\psi\rangle &= V O^\dagger |\psi\rangle \end{aligned} \tag{17.3.29}$$

Note there are even more equivalent definitions of a quantum code which we are leaving out for brevity. Also note that if any of these conditions also holds for the complements, i.e. $\overline{B}$ reconstructs $\bar{b}$, then we are promised that condition 1 of (17.3.2) holds. In other words, any quantum code with *complementary recovery* has a QES-like formula! This surprising fact was first pointed out in [48] and further understood in [50].

This theorem is important because of what it tells us about the holographic map V in AdS/CFT, showing us that we can understand it as a quantum error correcting code and leverage the powerful tools developed for such systems. This has had significant implications for the study of AdS/CFT. For one thing, it led to a flurry of technical results, including the discovery of Theorem 17.3.2 and much of what we know about bulk reconstruction in AdS/CFT [10, 17, 25, 27, 47, 48, 50, 51]! Indeed our general operator reconstruction map, the Petz map, was taken as-is from the quantum error correction literature [25].

One of the most impactful implications was to resolve a longstanding puzzle in AdS/CFT [28]: *how is it even possible to have a duality between two theories living in a different number of spacetime dimensions?* Both the bulk and boundary theories are relativistic; operators at some point x commute with all operators acting at points spacelike to x. This suggests we could in principle look at the algebra of operators of the CFT, see which operators commute with which others, and figure out that this must be a theory living in, say, d dimensions. But we could also do the same in the bulk, deducing from the bulk algebra that it is a theory living in $d + 1$ dimensions. Naively this raises a puzzle: if the bulk and boundary are dual, then shouldn't the set of operators be isomorphic? But then if so, don't we have a contradiction because we should be able to infer the spacetime dimension from the algebraic structure?

This puzzle is completely solved by the above discovery. The holographic map V has the structure of a quantum error correcting code, implying that the sets of bulk and boundary operators are *not* isomorphic! There are many different CFT operators that all reconstruct the same bulk operator! To make this work, we have to imagine that the bulk and boundary Hilbert spaces are not in 1-to-1 correspondence. Instead, V maps the bulk Hilbert space into just a subspace of $\mathcal{H}_{\text{bdry}}$. That way, each of these many different CFT operators can act the same on all states that come from acting V on a bulk state, but can be different as operators in the CFT.

Reminder: it is *semiclassical* gravity whose Hilbert space is apparently embedded into just a subspace of the CFT's. The full theory of quantum gravity presumably does have a Hilbert space isomorphic to the CFT. Some states just apparently are not semiclassical, i.e. can't be described well by the theory of semiclassical gravity. This is a very sensible statement, given that semiclassical gravity is just an effective theory, and effective theories generally work only for a subset of states! What we've learned is how important this fact is mathematically for the emergence of the extra bulk spatial dimension.

All of this seems to point toward a sensible mathematical structure: $\mathcal{H}_{\text{bulk}}$ is embedded into $\mathcal{H}_{\text{bdry}}$ *via some isometry* V with quantum error correcting properties. V being an isometry seems important because that is part of the assumptions in Theorems 17.3.2 and 17.3.3, and also by Definition 17.3.1 it ensures that $\dim \mathcal{H}_{\text{bdry}} \geq \dim \mathcal{H}_{\text{bulk}}$, and that the bulk is embedded into a just a subspace of the boundary. Indeed, until the year 2021 the fruitful work on operator reconstruction worked under this assumption that V is an isometry, e.g. [17, 25, 27, 28, 47, 48, 51, 52].

However, it turns out this assumption is in general invalid. The bulk can have black holes, and as we'll see in the next section, *old black holes require* V *is not an isometry!* This will raise many puzzles, and force us to revisit everything we thought we knew about quantum codes, including Theorems 17.3.2 and 17.3.3.

17.3.4 Black Holes Require Non-isometric Codes

Consider a large black hole in AdS, which has a conformal diagram

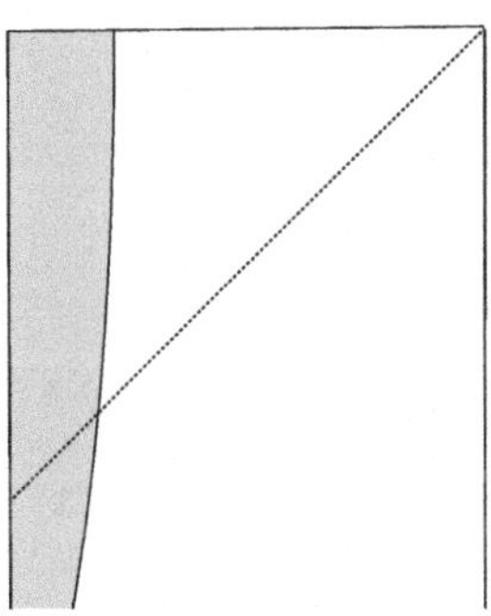

where the right vertical line is the asymptotic boundary (where the CFT lives), the left vertical line is $r = 0$, the top horizontal line is the singularity, and the dotted line is the horizon. The bottom is cutoff, and should really extend infinitely far to the past. Like any black hole, this one will radiate. To figure out in detail what happens over time, we need to specify boundary conditions at the asymptotic boundary. It is common to choose reflecting boundary conditions, in which every excitation that reaches the boundary bounces back. With this choice, a large black hole would never fully evaporate: its radiation is reflected back into it before it can finish.[7] Let us instead choose open boundary conditions, in which each excitation has some probability to pass out of the spacetime. To keep track of everything, we can imagine there is some other spacetime R whose boundary conditions are such that the radiation passes into it once it leaves the AdS. This second spacetime can really be anything, even something non-gravitational, as long as we choose the boundary conditions appropriately. For concreteness, let's say our AdS theory of semiclassical gravity consists of matter fields collectively denoted Φ, all coupled to the metric $g_{\mu\nu}$. We'll let the second spacetime R be Minkowski and include matter fields Φ but no dynamical metric. The boundary conditions will be open to the matter, such that excitations in Φ in the AdS can pass through to the Minkowski space. (R is a suitable name for this second spacetime, because it will be collecting the exterior Hawking radiation, which we have been calling R.) For further details of such constructions, see [17, 18]. We can draw this with conformal diagram

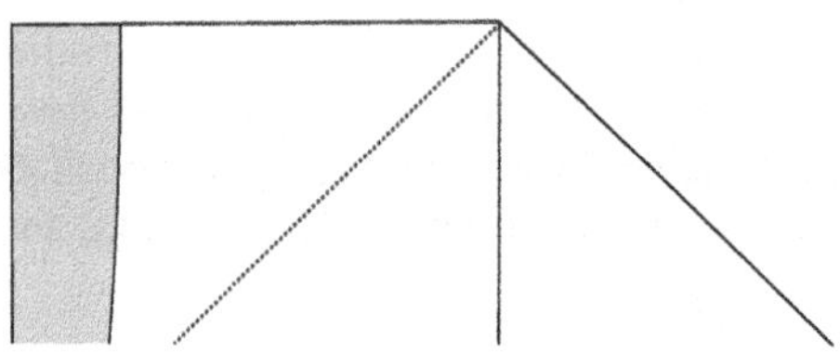

(17.3.30)

where the new diagonal line represents null infinity of R. There is a dual description of this. The standard AdS/CFT dictionary tells us that a choice of boundary conditions in the AdS corresponds to a choice of sources in the CFT [22]. Here that implies we are *coupling* the CFT to R. We can draw that as

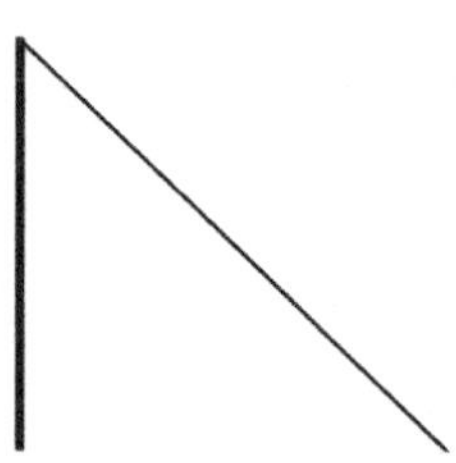

(17.3.31)

where the thick black line represents the CFT, which we'll call B for short, and we have represented its coupling to R by placing it next to null infinity.

What's important here is how the holographic map V relates these two descriptions (17.3.30) and (17.3.31). Consider some timeslice in the gravitational description,

7 How does the radiation make it the infinite distance to the boundary, and then back, before the black hole finishes evaporating? The answer lies in the metric (17.3.4). While the radiation is travelling an infinite distance, the time dilation gives it infinite time to do so. It will return to the middle of the bulk in a finite amount of time from the point of view of any observer who stayed in the middle.

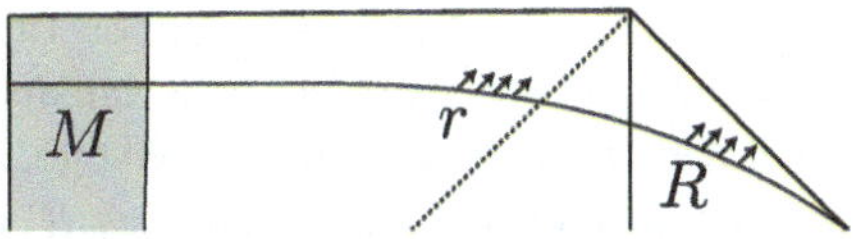

(17.3.32)

where we have labelled relevant degrees of freedom on this timeslice: the matter M, the external radiation R, and the interior partners r. The dual description is obtained from this the same way as always, by acting the holographic map $V : \mathcal{H}_{\text{bulk}} \to \mathcal{H}_{\text{bdry}}$. Now that we have R, all we do is tensor on $\mathbb{1}_R$. The correspondence between AdS states and CFT states is independent of what other systems might be hanging around. We can draw it as

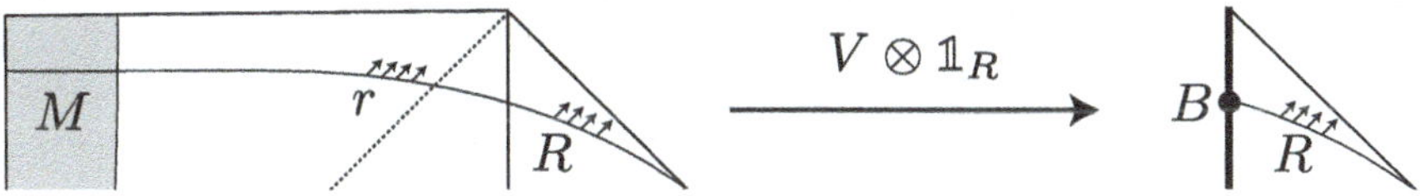

(17.3.33)

where on the right we have labelled the boundary theory as B and indicated a timeslice of it with a thick black dot. We will call the left description the *effective description*, and the right the *fundamental description*. (These terms generalize "bulk" and "boundary" to situations with a reference system R, and also foreshadows that we will eventually apply these lessons to spacetimes besides AdS.)

Now we can set up a puzzle. According to the effective description, as the black hole radiates the entanglement between r and R grows monotonically until evaporation finishes, at which point it plateaus:

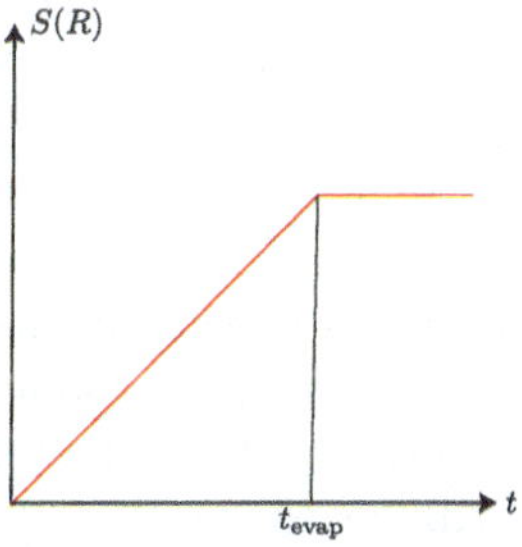

(17.3.34)

However, in the fundamental description R is entangled with a CFT B whose energy is monotonically decreasing as it radiates into R. A holographic CFT has a finite density of states, which decreases as the energy decreases. Putting numbers to this, a CFT state dual to a black hole of area A can have support on roughly $e^{A/4G}$ states.[8] The maximum von Neumann entropy available to a density matrix of dimension $e^{A/4G}$ is $S = A/4G$. But B and R together are in a pure state, and it is a basic feature of von Neumann entropy that the entropy of a system (say R) must equal the entropy of its complement (here B) [16].[9] Therefore in the fundamental description it is clear that $S(R)$ cannot follow the curve (17.3.34) but must instead stay below $A/4G$ as A decreases from its initial value to 0,

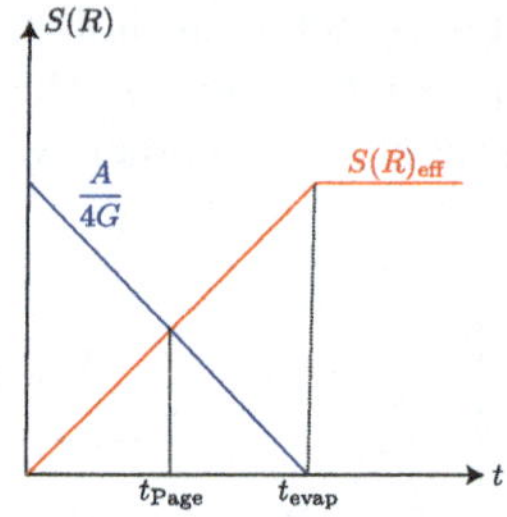

(17.3.35)

The time t_{Page} is the time at which the blue and red lines cross, which is the latest possible time at which the effective description's entropy could remain valid. So far this all just places an upperbound on $S(R)_{\text{fund}}$; what do we expect in detail? Holographic CFTs are highly chaotic systems, and it turns out one can argue on general principles that the entropy of these systems, in this setting, follows what's now known as the "Page curve" [53, 54], which follows $S(R)_{\text{eff}}$ as long as possible, until about t_{Page}, at which point it turns back down and follows $A/4G$,

$$S(R)_{\text{fund}} \approx \min\{S(R)_{\text{eff}}, A/4G\}. \tag{17.3.36}$$

This is approximate because there are generally $e^{-O(1/G)}$ corrections, which can be especially large very near t_{Page}.

We care about this difference between $S(R)_{\text{eff}}$ and $S(R)_{\text{fund}}$ because of the puzzle it raises about V. After the Page time we have $S(R)_{\text{eff}} > S(R)_{\text{fund}} \approx A/4G$, which means the effective description apparently involves a *larger* Hilbert space than the fundamental description! This is important, so let's say it more carefully: The effective description involves the Hilbert space $\mathcal{H}_M \otimes \mathcal{H}_r \otimes \mathcal{H}_R$. If we consider a pure state on this Hilbert space, then again it must be the case that $S(R) = S(Mr) \leq \log |M||r|$, where $|M|$ and $|r|$ are the respective dimensions

[8] Think of this like a thermal state: the CFT state might have a little support on infinitely many states, but the significant part of the state is supported in a fairly narrow energy window in the thermodynamic limit.

[9] A quick way to prove this is to consider an arbitrary state $|\psi\rangle \in \mathcal{H}_B \otimes \mathcal{H}_R$ and use the Schmidt decomposition to write it as $|\psi\rangle = \sum_i \lambda_i |i\rangle_B |i\rangle_R$, from which it immediately follows from (17.3.14) that $S(B) = S(R)$.

of $\mathcal{H}_M$ and $\mathcal{H}_r$, and the upperbound follows from the definition (17.3.14). This implies that after the Page time,

$$\log|M||r| \geq S(R)_{\text{eff}} > A/4G = \log|B|. \tag{17.3.37}$$

Therefore for these old black holes, $V : \mathcal{H}_M \otimes \mathcal{H}_r \to \mathcal{H}_B$ is apparently mapping states from a larger Hilbert space to states in a smaller Hilbert space. It is impossible, then, that V is an isometry. V is *non-isometric*. It must have a kernel, i.e. some states in $\mathcal{H}_M \otimes \mathcal{H}_r$ that it annihilates.

Let me emphasize that it really is low energy effective theory degrees of freedom that are outnumbering $|B|$ after the Page time. That is, you could worry that all of this counting of Hilbert space dimensions is ill-defined because the effective description really involves a quantum field theory, which has infinitely many degrees of freedom in some finite region of space, like the black hole interior. So what do we mean when we say $|M||r|$ grows with time? The answer is that we know the effective description is not valid at high energies, say if we excite Planck-scale modes. Hence we aren't counting those among the effective theory degrees of freedom in the black hole interior. We only count the sub-Planckian modes, of which there are finitely many in the interior. What's important is that the number of *these* modes grows with time, because the radial length of the black hole interior grows with time (along "nice slices" that stay away from too near the singularity). It is around the Page time where their Hilbert space dimension starts to be larger than $e^{A/4G}$.

17.3.5 Non-isometric Codes: How Can They Work?

At first, it is perplexing that apparently the holographic map V is non-isometric in general. This seems like it would destroy information! V non-isometric means we can divide the effective description Hilbert space into two parts, $\mathcal{H}_{\text{eff,pres}} \oplus \mathcal{H}_{\text{eff,kernel}}$, where $\mathcal{H}_{\text{eff,kernel}}$ is the kernel of V. Any part of the state on that subspace has zero influence on the resulting fundamental description state. What can we make of this?

This problem becomes even more severe when recall that the fundamental description should *define* the system. The effective description is a valid description only insofar as it matches the fundamental description. If the two descriptions make slightly different predictions, we should trust the fundamental one. In particular, say we have two states $|\psi\rangle$, $|\phi\rangle \in \mathcal{H}_{\text{eff}}$. We can compute their inner product $\langle\psi|\phi\rangle$ in the effective description, but we interpret their *true* inner product as the one you get by first mapping them to the fundamental description with V and then taking their inner product,

$$\text{Fundamental inner product:} \quad \langle\psi|V^\dagger V|\phi\rangle. \tag{17.3.38}$$

But that means that any state $|\psi\rangle \in \mathcal{H}_{\text{eff,kernel}}$ has zero norm, i.e. it is a *null state*! This is troubling, because it a very big correction to the inner product. We might reasonably expect these null states to imply massive, easily observable problems with the effective description.

In addition to this null states problem, we have the puzzle that much of what we knew about bulk reconstruction came from the techniques of quantum codes, in particular Theorems 17.3.2 and 17.3.3. But those theorems all assume isometric V. Our tools like the Petz map require isometric V. With all of these issues, we might be tempted to give up on even our basic idea, that in this setting there even exists a linear holographic map $V : \mathcal{H}_{\text{bulk}} \to \mathcal{H}_{\text{bdry}}$.

However, it turns out that there *is* a way to make sense of non-isometric codes [10]! To understand this, we have to ask the question: what is the regime of validity of the effective description, of semiclassical gravity? We should not expect that states outside the regime of validity match the fundamental description! A non-isometric V can work if only those states are in the kernel!

Naively we might think the regime of validity is low energies. As long as we aren't exciting Planck-energy modes, normal Wilsonian effective theory principles would tell us the effective description is okay. However, we already know this can't be right. If we accept a unitary black hole S-matrix, then we know that at late times the Hawking radiation will not be exactly thermal. There will exist some operations that probe the information that was thrown into the black hole. But this will be a *low energy* operation, because it only needs to act on the exterior Hawking modes, which all have long wavelengths. So clearly there do exist low energy operations for which the effective description gives the wrong answer. (And anyway, our argument that V is non-isometric only needed to think about low-energy states!)

But if low energy operations can see large violations of effective theory, then why don't we see violations all the time? It seems there is some quality that delineates operations that do and don't break the effective description that somehow explains why it is "hard" to see the violations. This leads us to recall the powerful lesson of Harlow-Hayden from Sect. 17.2.5: decoding the Hawking radiation is an *exponential complexity operation*. That is, they already taught us that the known low-energy violations of effective theory require high complexity.

Therefore, being conservative, we might speculate that those are the only kinds of low energy operations for which effective theory is invalid. Now we arrive at the main idea: *perhaps V is approximately isometric on all low complexity states, but not necessarily high complexity states*. That is, perhaps V satisfies

$$\langle\psi|V^{\dagger}V|\phi\rangle \approx \langle\psi|\phi\rangle \tag{17.3.39}$$

for all $|\psi\rangle$ and $|\phi\rangle$ that can be prepared by a polynomial size circuit from some reference state $|0\rangle$.[10] This does not seem obviously possible, because the set of states preparable by a short circuit are not a subspace. For example, given n qubits in the state $|00...00\rangle$, we can obtain any basis state $|00...01\rangle$, $|00...10\rangle$, ..., $|11...11\rangle$ by

---

[10] What should the reference state be? It won't matter too much for our discussion below, which will focus on whether non-isometric codes protected by complexity would work for *some* reference state. But in general it seems reasonable for it to be something simple, perhaps the vacuum state.

acting only n gates, one on each qubit. Low complexity states span the entire Hilbert space! And we know V cannot satisfy (17.3.39) for *all* $|\psi\rangle$, $|\phi\rangle \in \mathcal{H}_{\rm eff}$, because it has a kernel.

Surprisingly, it is indeed possible to construct such non-isometric V, satisfying (17.3.39) for all simple states [10]! In the next subsection we will demonstrate this. This turns out to resolve all of our issues above: even though V is non-isometric, as long as we don't do any experiments of $e^{O(S_{\rm BH})}$ complexity, we won't notice it. All null states have exponential complexity; that's why we don't see large problems with effective theory all the time. Moreover, we'll see the important Theorems 17.3.2 and 17.3.3 about quantum codes suitably generalize to apply to these kinds of non-isometric codes [10, 50]. So even though non-isometric V naively seem impossible to make work, it turns out they work perfectly once we incorporate the known relationship between effective theory and complexity.

17.3.6 Non-isometric Codes Protected by Complexity

Here we prove that there exist non-isometric maps V that are approximately isometric for all *low complexity* states, closely following [10]. We would like to define a map

$$V : \mathcal{H}_{M_1} \otimes \mathcal{H}_r \to \mathcal{H}_B. \tag{17.3.40}$$

Here $\mathcal{H}_M = \mathcal{H}_{M_1} \otimes \mathcal{H}_{M_2}$ is the Hilbert space of the matter that formed the black hole, which as in Sect. 17.2.4 we decompose into two factors where M_1 are the degrees of freedom that can be excited by standard fast collapse. We do not bother including M_2 in the input to V, because as discussed exciting those modes in general requires $e^{O(S_{\rm BH})}$ time, and we are restricting to low complexity states. For this reason, we will set M_2 to always be in some state $|\psi_0\rangle$.

Specialize for now to a situation with dimensions related by

$$|M_1||M_2||r| = |P||B| \tag{17.3.41}$$

for some positive integer P. Then we have the following two tensor decompositions of the Hilbert space:

$$\mathcal{H}_{M_1} \otimes \mathcal{H}_{M_2} \otimes \mathcal{H}_r = \mathcal{H}_B \otimes \mathcal{H}_P. \tag{17.3.42}$$

Now define the map

$$V := \sqrt{|P|}\, \langle 0|_P\, U\, |\psi_0\rangle_{M_2}\,. \tag{17.3.43}$$

Here $|0\rangle_P$ is some fixed state on $\mathcal{H}_P$ and U is a random unitary drawn from the Haar measure of unitaries acting on $\mathcal{H}_{M_1} \otimes \mathcal{H}_{M_2} \otimes \mathcal{H}_r$. The role of the $\sqrt{|P|}$ will be explained momentarily. This map can be illustrated as

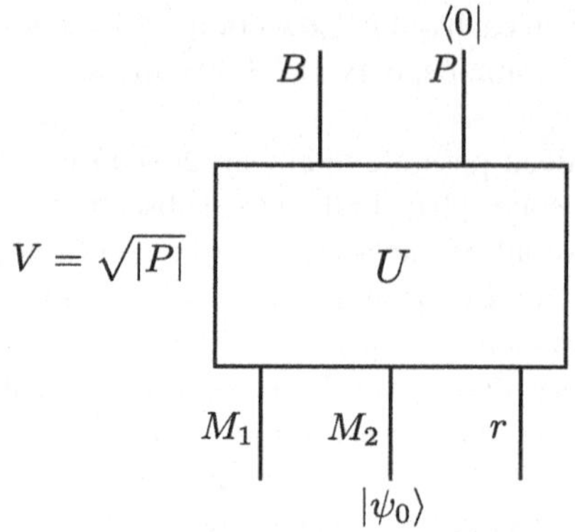

(17.3.44)

Note that with this postselection on $|0\rangle_P$, this is a non-isometric map as long as $|P| > |M_2|$.

Notice the similarity with the final state proposal (2.4). Both involve a random unitary followed by postselection. In fact, this is not a special choice in our case. All linear maps V can always be put in a form like this, as some unitary followed by postselection. Indeed, given any $V : \mathcal{H}_{M_1} \otimes \mathcal{H}_r \to \mathcal{H}_B$ of finite operator norm[11] $\|V\|_\infty$, we can define $\hat{V} := V/\|V\|_\infty$ and

$$W := |0\rangle_P \otimes \hat{V} + |1\rangle_P \otimes \sqrt{\mathbb{1} - \hat{V}^\dagger \hat{V}}, \tag{17.3.46}$$

where $\mathcal{H}_P$ is an ancillary qubit. W satisfies $W^\dagger W = \mathbb{1}$ so is an isometry, and therefore we can write it as $W = U\,|\psi_0\rangle_{M_2}$ for some unitary U and ancillary system M_2. It follows that

$$V = \|V\|_\infty \, \langle 0|_P \, U \, |\psi_0\rangle_{M_2} \, . \tag{17.3.47}$$

Therefore the form of (17.3.43) is completely general, applicable to any linear map. The only choice is the U we are using is drawn randomly.

In spite of this similarity to the final state model, there are two differences. First, here we are not necessarily postselecting on everything, just the factor we call $|P|$. This is related to the second difference: the physical interpretation is completely distinct. In the final state proposal, the postselection is happening because of novel quantum gravity dynamics. Here the postselection is happening *in the map* from effective theory to fundamental theory—it is not somehow showing up in dynamics.

[11] The operator norm for an operator X is defined as

$$\|X\|_\infty := \max_{\||\psi\rangle\|=1} \|X\,|\psi\rangle\|. \tag{17.3.45}$$

A key property of the map (17.3.43) is that is preserves the inner product on average between any two states $|\psi_1\rangle, |\psi_2\rangle \in \mathcal{H}_{M_1} \otimes \mathcal{H}_r$:

$$\int dU \left\langle\psi_2|V^\dagger V|\psi_1\right\rangle = \langle\psi_2|\psi_1\rangle . \tag{17.3.48}$$

This calculation can be done using the unitary integrals (17.2.31). More relevant for us, if we introduce a reference system L with $|L| = |M_1|$ and a reservoir R with $|R| = |r|$, and we take $|\psi_1\rangle, |\psi_2\rangle \in \mathcal{H}_L \otimes \mathcal{H}_{M_1} \otimes \mathcal{H}_r \otimes \mathcal{H}_R$ then we have

$$\int dU \left\langle\psi_2|(V^\dagger V \otimes \mathbb{1}_{LR})|\psi_1\right\rangle = \langle\psi_2|\psi_1\rangle . \tag{17.3.49}$$

It seems surprising at first that a non-isometric map can on average preserve the inner product, but we can clarify this by computing the fluctuation in the inner product:

$$\int dU \left|\left\langle\psi_2|(V^\dagger V \otimes \mathbb{1}_{LR}|\psi_1\right\rangle\right|^2 = \frac{|P|-1}{|P|^2|B|^2-1}\left[|P||B| \operatorname{tr}\left((\psi_1)_{LR}(\psi_2)_{LR}\right) - |\langle\psi_2|\psi_1\rangle|^2\right]. \tag{17.3.50}$$

Here $(\psi_1)_{LR}$ and $(\psi_2)_{LR}$ are the reduced states of $|\psi_1\rangle$ and $|\psi_2\rangle$ on LR. The right hand side vanishes when $|P| = 1$, which it must because in that case there is no postselection and V is an isometry. Otherwise it is non-zero, matching our expectation that V is not an isometry.

It is illuminating to simplify this a bit more. For any states $|\psi_1\rangle, |\psi_2\rangle$, it holds that $0 \leq \operatorname{tr}\left((\psi_1)_{LR}(\psi_2)_{LR}\right) \leq 1$, so using $\frac{1}{x-1} \leq \frac{2}{x}$ for $x \geq 2$, we have that

$$\int dU \left|\left\langle\psi_2|(V^\dagger V \otimes \mathbb{1}_{LR}|\psi_1\right\rangle\right|^2 \leq \frac{2}{|B|} \tag{17.3.51}$$

for $|B| > 1$. This is an interesting bound, because the right hand side is so small! Recall that here $|B|$ represents the CFT, which can be described with a Hilbert space of dimension $e^{A/4G}$. So even though V is not an isometry, the fluctuations in the inner product between $(V \otimes \mathbb{1}_{LR})|\psi_1\rangle$ and $(V \otimes \mathbb{1}_{LR})|\psi_2\rangle$ away from that of $|\psi_1\rangle$ and $|\psi_2\rangle$ are exponentially small in the black hole entropy. V apparently almost always preserves the inner product, even for an old black hole where

$$1 \ll |B| \ll |M_1||r|. \tag{17.3.52}$$

How is (17.3.51) compatible with the fact that for $|B| \ll |M_1||r|$ there are necessarily many states annihilated by V? One thing to note: in the derivation of (17.3.51), we implicitly assumed that the states $|\psi_1\rangle, |\psi_2\rangle$ were independent of U. If we'd allowed them to depend on U (as the set of null states does) then we would have gotten an answer different from (17.3.51). Physically, that tells us that an observer who wants to see the failure of V to be an isometry will have to construct a state that is finely tuned to the details of the map V. We might hope that for a typical U it should be exponentially difficult to construct such finely tuned states, and momentarily we will show that this is indeed the case.

First, we can gain more intuition for (17.3.51) by introducing a simplified model that we'll call the *phase model*. To motivate it, note that if $|i\rangle_{M_1}$ and $|n\rangle_r$ are orthonormal bases for $\mathcal{H}_{M_1}$ and $\mathcal{H}_r$ and $|b\rangle_B$ is an orthonormal basis for $\mathcal{H}_B$, then

$$V\,|in\rangle_{M_1 r} = \sqrt{|P|}\sum_b |b\rangle_B \langle b|_B \langle 0|_P \, U\, |\psi_0\rangle_{M_2} |in\rangle_{M_1 r}\,. \tag{17.3.53}$$

We can think of each matrix element in the sum as a random overlap between two states in $\mathcal{H}_B \otimes \mathcal{H}_P$, which is generally of order $1/\sqrt{|P||B|}$ times a random phase. The phase model simply takes this to be the holographic map:

$$V_{\text{phase}}\,|in\rangle_{M_1 r} := \frac{1}{\sqrt{B}}\sum_b \mathrm{e}^{i\theta(i,n;b)}\,|b\rangle_B\,, \tag{17.3.54}$$

where the phases $\mathrm{e}^{i\theta(i,n;b)}$ are given by a single sample from $|M_1||r||B|$ independent copies of the uniform distribution on $U(1)$. One can compute that

$$\begin{aligned}\langle in|\,V^\dagger_{\text{phase}}V_{\text{phase}}\,|jm\rangle &= \frac{1}{|B|}\sum_b \mathrm{e}^{i\theta(j,m;b)-i\theta(i,n;b)}\\ &= \begin{cases} 1 & (i,n)=(j,m)\\ O\left(\frac{1}{\sqrt{|B|}}\right) & (i,n)\neq(j,m)\end{cases}.\end{aligned} \tag{17.3.55}$$

The second line follows from a "random walk" argument: adding up $|B|$ independent phases at large $|B|$ typically equals something of order $\sqrt{|B|}$. This is nicely compatible with (17.3.51), with both holding even in the regime $1 \ll |B| \ll |M_1||r|$.[12] So while it's not possible to exactly isometrically map an orthonormal basis of a larger Hilbert space to an orthonormal basis of a smaller one, we can do it up to errors that are exponentially small.

Returning to our map (17.3.43), we have seen that $V \otimes \mathbb{1}_{LR}$ is likely to preserve the inner product between any particular pair of states in $\mathcal{H}_L \otimes \mathcal{H}_{M_1} \otimes \mathcal{H}_r \otimes \mathcal{H}_R$ up to errors of size $\mathrm{e}^{-O(S_{\text{BH}})}$. For the phase model we saw that in particular we could achieve this for a particular basis of states, as long as $|M_1||r| \ll \mathrm{e}^{|B|}$. But what we really care about is the larger set of all low complexity states, obtained from basis states by acting with a polynomial size quantum circuit. We now argue that in fact $V \otimes \mathbb{1}_{LR}$ is likely to preserve to exponential accuracy the inner product between all states on $\mathcal{H}_L \otimes \mathcal{H}_{M_1} \otimes \mathcal{H}_r \otimes \mathcal{H}_R$ whose circuit complexity is sub-exponential.[13] Our

[12] There is a limit to how big $|M_1||r|$ can be: if $|M_1||r| \gg \mathrm{e}^{|B|}$, then the random walk argument no longer gives (17.3.55), because by chance there will be occasional pairs of orthogonal basis states in $\mathcal{H}_{M_1} \otimes \mathcal{H}_r$ whose images under V have an inner product which is $O(1)$.

[13] A sub-exponential function is one that as $x \to \infty$ grows more slowly than e^{ax} for all a. An example is $\mathrm{e}^{\sqrt{x}}$. Polynomials are a subclass of sub-exponentials.

approach will be closely related to the Johnson-Lindenstrauss lemma, which says that for any set of m points in $\mathbb{R}^N$ there is a linear map to $\mathbb{R}^n$ which approximately preserves their distances to accuracy ε as long as $n > 8\frac{\log m}{\varepsilon^2}$. However, that lemma does not suffice for us because we want a statement on $\mathcal{H}_L \otimes \mathcal{H}_{M_1} \otimes \mathcal{H}_r \otimes \mathcal{H}_R$ but only want V to act on $\mathcal{H}_{M_1} \otimes \mathcal{H}_r$.

Our argument will be based on the powerful idea of *measure concentration*, which is basically the statement that for certain probability distributions on certain high-dimensional manifolds, the fluctuations of any reasonable observable will be exponentially small in the dimension of the manifold. One example of this is Levy's lemma, which says this is true for real functions on the sphere,

Lemma 17.3.4 (Levy) *Let* $F : \mathbb{S}^d \to \mathbb{R}$ *be k-Lipschitz, meaning that* $|F(x) - F(y)| \leq k|x - y|$ *for all* $x, y \in \mathbb{S}^d$, *with* $|x - y|$ *being the Euclidean distance on* $\mathbb{S}^d$. *Then in the uniform probability measure on* $\mathbb{S}^d$ *it holds that*

$$\Pr\Big[|F - \langle F\rangle| \geq \varepsilon\Big] \leq 2\exp\left(-\frac{(d-1)\varepsilon^2}{2k^2}\right) \tag{17.3.56}$$

where $\langle F\rangle$ *is the expectation value of* F *and* $\Pr[\cdot]$ *is the probability that "$\cdot$" is true.*

A pedestrian example[14]: say you are going to pick a random point on the surface of the Earth—what's the probability that point is within 1 mile of the equator? Here F is "distance north of the equator" measured in units of the radius of the Earth, with points on the southern hemisphere having negative values. This distance function is indeed k-Lipschitz for $k = 1$. Because $d = 2$, we don't get a very good bound for $\varepsilon = 1$ mile, which is small in our units, matching our intuition that it is unlikely our random point landed so close to the equator! However, for a d-dimensional Earth with d very large, the probability becomes exponentially close to 1! Our random point is almost guaranteed to land within one mile of d-dimensional Earth's equator!

This at first can seem counter-intuitive. Here is one way to understand it. First, observe that the median value F_{med} of our function naturally cuts the sphere into two equal-volume halves: one half with $F > F_{\text{med}}$ and one half with $F < F_{\text{med}}$. In a small neighborhood of this cut, it follows from the Lipschitz continuity of F that $F \approx F_{\text{med}}$. Therefore the probability of a large fluctuation in F is upperbounded by the fractional volume of the sphere that is *not* close to the cut. The volume of the sphere far from the cut is maximized when the cut is an equator—that's just the intuitive fact depicted here:

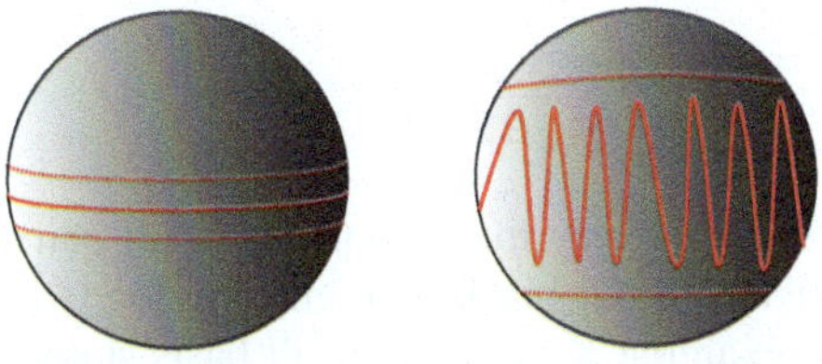

(17.3.57)

[14] This originates from Patrick Hayden.

where on the left the solid red line lies on the equator while on the right it follows a more general path. In both cases, the dashed red lines contain all points at most ε away from the solid line. We see the dotted lines include more of the sphere on the right. So again: an equator is the cut that maximizes the fraction of the sphere far from it (this fact is known as the "isoperimetric inequality"). Now we can finish arguing for Levy's lemma: the fractional volume of any ball of opening angle $\theta \in [0, \pi/2)$ on $\mathbb{S}^d$ is

$$\frac{\Omega_{d-1}}{\Omega_d} \int_0^\theta d\theta' (\sin\theta')^{d-1} \leq \frac{\Gamma((d+1)/2)}{\sqrt{\pi}\Gamma(d/2)} \theta \sin(\theta)^{d-1} \leq \frac{\sqrt{d}2\pi}{\theta} (\sin\theta)^{d-1}, \tag{17.3.58}$$

which vanishes exponentially at large d for any $\theta \in [0, \pi/2)$. Therefore almost all of the volume of $\mathbb{S}^d$ is very close to an equator, and it follows that the probability of a large fluctuation in F is exponentially small, even in the "worst-case scenario" that the $F = F_{\text{med}}$ cut lies on an equator.

For the argument we want to make, we will need a slightly different measure concentration bound:

Lemma 17.3.5 (Meckes) *Let $F : U(N) \to \mathbb{R}$ be k-Lipschitz in the sense that $|F(U_1) - F(U_2)| \leq k\|U_1 - U_2\|_2$, with $\|X\|_2 := \sqrt{\operatorname{tr} X^\dagger X}$. Then in the Haar measure on $U(N)$ it holds that*

$$\Pr\left[|F - \langle F \rangle| \geq \varepsilon\right] \leq 2\exp\left(-\frac{\varepsilon^2 N}{12k^2}\right). \tag{17.3.59}$$

For a self-contained proof of both this lemma and Levy's, see e.g. appendix B of [10]. This lemma can be used to prove the following theorem:

Theorem 17.3.6 *Let V be defined as in* (17.3.43), *and let S be a collection of N_S states in $\mathcal{H}_L \otimes \mathcal{H}_{M_1} \otimes \mathcal{H}_r \otimes \mathcal{H}_R$. Then for all $|B| \geq 16$ and $0 < \gamma < 1/2$ it holds that*

$$\Pr\left[\sup_{|\psi\rangle, |\phi\rangle \in S} \left|\left\langle \psi | V^\dagger V \otimes \mathbb{1}_{LR} | \phi \right\rangle - \langle \psi | \phi \rangle\right| \geq \sqrt{18}|B|^{-\gamma}\right] \leq 6N_s(N_s - 1)\exp\left(-\frac{|B|^{1-2\gamma}}{24}\right). \tag{17.3.60}$$

The proof can be found in Appendix C of [10]. This theorem tells us that as long as

$$\frac{\log N_S}{|B|^{1-2\gamma}} \ll 1, \tag{17.3.61}$$

then encoding by $V \otimes \mathbb{1}_{LR}$ is very likely to preserve all pairwise inner products in S to exponential precision in the black hole entropy $\log |B|$. And this can be true for quite large N_S, for example S can include all basis states of $\mathcal{H}_{M_1} \otimes \mathcal{H}_r$ as long as $|M_1||r|$ is not exponentially large in $|B|$, just as we observed in the phase model.

However, now we say even more, and ask whether S can include all states of sub-exponential complexity. We will now show that it indeed can, by literally counting how many states there are of sub-exponential complexity.[15]

To rigorously define a notion of quantum complexity for a state, we need to specify a model of computation. We will take our model to be built out of "qudits", each of which has a Hilbert space of dimension $q > 1$ which we think of as being some fixed $O(1)$ number (perhaps two). Now, we don't want to demand that any of our quantum systems M_1, r, R have to be made out of literal qudits, themselves. The point is that we can nonetheless always think of them in terms of qudits, as follows. Say we have an arbitrary quantum system with Hilbert space $\mathcal{H}_A$. To encode it into qudits we need at least

$$m = \log_q \lceil A \rceil \tag{17.3.62}$$

qudits, where $\lceil A \rceil$ is defined as the smallest integer power of q greater than or equal to $|A|$. So by definition, $|A| \leq \lceil A \rceil < q|A|$. We will furthermore assume that this encoding can be done in a "simple" way, such that the operations which are simple in terms of qudits are also simple to implement on A. For example, any local structure on A should be respected by the encoding. Now, we specify our gate set (our set of basic operations) to consist of some $O(1)$ number Γ of unitaries, which can be applied in any order to any pair of the physical qudits.[16]

When we talk about the complexity of a state, when being rigorous we mean the following. Recall a quantum circuit C is an ordered sequence of gates $g_1, g_2, \ldots$ with the number of gates $\mathcal{C}(C)$ being the *circuit complexity*. We are mostly interested in how this complexity scales asymptotically with number of qudits (e.g. polynomially or exponentially), so we generally consider a sequence of circuits $C_1, C_2, \ldots$ acting on a sequence of systems $A_1, A_2, \ldots$, and we say the circuit family C_n is *sub-exponential* if for any $\alpha > 0$ it holds that

$$\mathcal{C}(C_n) \leq (\lceil A_n \rceil)^{\alpha} \tag{17.3.63}$$

for all but maybe finitely many n. A family of states $|\psi_1\rangle \in \mathcal{H}_{A_1}, |\psi_2\rangle \in \mathcal{H}_{A_2}, \ldots$ is called sub-exponential if there exists a sub-exponential circuit family that prepares the version of each (encoded on qudits) starting from the all-zero state of the qudits.

Now we want to count how many sub-exponential states there are. Consider the more general problem: how many states of complexity $\mathcal{C}_n$ are there on $\mathcal{H}_{A_n}$? This is upper-bounded by the number of circuits of complexity $\mathcal{C}_n$ on $\mathcal{H}_{A_n}$ there are, which is simply

[15] Unfortunately, we cannot find a theorem like Theorem 17.3.6 for the phase model, because the kind of measure concentration we need does not occur in the high dimensional torus T^d on which we can imagine the random phases live—this is unlike the high dimensional sphere.

[16] In Sect. 17.2.5 we noted that a reasonable gate set could be the set of all two-qudit unitaries. It will be convenient here to be even more restrictive, and pick a *finite* set of two qudit unitaries. This is actually equivalent, because one can show that there exist universal finite gate sets, and that they can efficiently implement any unitary that can be efficiently implemented by *any* finite gate set, as long as we are okay with some tiny ε error in the unitary [16], which we always are.

$$N_{C_n} = \left(\binom{\log_q \lceil A_n \rceil}{2}\Gamma\right)^{C_n} \leq e^{\left(2\log\log_q\lceil A\rceil + \log\Gamma\right)C_n} \tag{17.3.64}$$

because for each gate we pick which gate we want and which two qudits to act on. Because for any $\alpha > 0$ a sub-exponential circuit family always eventually obeys (17.3.63), for any $\alpha > 0$ it therefore eventually holds that

$$N_{C_n} = \left(\binom{\log_q \lceil A_n \rceil}{2}\Gamma\right)^{C_n} \leq e^{\left(2\log\log_q\lceil A\rceil + \log\Gamma\right)(\lceil A_n \rceil)^{\alpha}}. \tag{17.3.65}$$

Returning to black holes, what we want is to consider circuits that are subexponential in the entropy S_{BH} of the black hole, which in the language of our model is $S_{\text{BH}} = \log|B|$. So, a "sub-exponential circuit family" is one such that for any $\alpha > 0$ we eventually have $\mathcal{C}(C_n) \leq |B_n|^{\alpha}$.[17] Let us drop the n to match our previous notation. Then, for any $\alpha > 0$, the number of sub-exponential states $N_{\text{sub-exp}}$ will always eventually obey

$$N_{\text{sub-exp}} \leq e^{\left(2\log\log_q(\lceil|M_1||r|\rceil + \log\Gamma\right)(\lceil B \rceil|)^{\alpha}} \sim e^{|B|^{\alpha}\log\log(|M_1||r|)}. \tag{17.3.66}$$

Combining this with Theorem 17.3.6, we have

$$\Pr\left[\sup_{|\psi\rangle,|\phi\rangle\in S}\left|\left\langle\psi|V^{\dagger}V\otimes\mathbb{1}_{LR}|\phi\right\rangle - \langle\psi|\phi\rangle\right| \geq \sqrt{18}|B|^{-\gamma}\right] \lesssim \exp\left(|B|^{\alpha}\log\log(|M_1||r|) - \frac{|B|^{1-2\gamma}}{24}\right). \tag{17.3.67}$$

We see that choosing $\alpha < 1 - 2\gamma$ implies that the inner products between all sub-exponential states will almost surely be preserved to exponentially small accuracy for sufficiently large black holes, provided $|M_1||r|$ is not doubly exponentially large in $|B|$). As a corollary, we see that all null states are exponentially complex.

Should this restriction on the size of $|M_1|$ and $|r|$ bother us? Not really. Recall the point of (17.3.67) is to tell us when the effective description inner products match the real inner products. (Its answer: will match at least for all simple states if S_{BH} is large enough.) If we are ever in a situation where $|M_1||r| \sim \exp(\exp(|B|))$, then either the black hole is extremely tiny or we spent an expoentially long time slowly feeding the black hole (it had to be this slow so that $|B|$ remained small by evaporating). But if the black hole is tiny, then we already know the effective description breaks for some operations, such as probing the unitarity of the Hawking radiation. And if we fed it over an exponentially long time, that counts as an exponentially complicated operation, so we again we expect the effective description to break down here. So, breaking when $|M_1||r|$ is that large is a feature, not a bug!

---

[17] Notice that while we started with one black hole of entropy $\log|B|$, we are now discussing a family of black holes with entropy $\log|B_n|$. How do we think about this family? Physically, we are considering black holes of increasingly large entropy $S_{\text{BH}} = A/4G$, which can correspond to dialing G to be smaller and smaller (at fixed area) or growing A.

This concludes our constructive proof of this subsection: there exist non-isometric maps V that are approximately isometric for all low complexity states! Again this matters because it resolves the puzzle about AdS/CFT raised in Sect. 17.3.4: how is it even possible that the effective description of black holes involves a much larger Hilbert space than the fundamental description without severe information loss? The answer: it can still be approximately isometric on all low complexity states.

This suggests interesting physics. For low complexity states, the effective description is trustworthy, giving the same answers to physical questions like the overlap of two states. The same is not generally true for high complexity operations, which can construct and probe high complexity states. For such operations, we cannot trust the effective description.

Again, this nicely matches what we know about black holes in quantum gravity. We already know that some high complexity (but low energy) experiments *should* see a failure of the effective description, such as probing the unitarity of the Hawking radiation. This model concretely explains how such a thing can be possible.

17.3.7 Unitarity, the QES Formula, and Operator Reconstruction

We have seen that there is nothing inconsistent about the holographic map V being non-isometric, as long as we only demand the bulk effective theory is valid for low complexity operations. This resolves a problem in AdS/CFT of how the holographic map can possibly work without losing information, when there are old black holes. Now it's time for us to return to the black hole information paradox, where it turns out these non-isometric codes have a significant lesson to teach.

To start, let's consider a state analogous to what we'd have at some stage of an evaporating black hole,

$$|\psi_{\text{Hawk}}\rangle = \left|\chi^{\{\text{in}\}}\right\rangle_{LM_1} \otimes \left|\chi^{\{\text{out}\}}\right\rangle_{rR}. \tag{17.3.68}$$

We might think of $\left|\chi^{\{\text{in}\}}\right\rangle_{LM_1}$ as the purification of the possibly mixed state of the infalling matter that created the black hole. L is just an arbitrary reference system. The state $\left|\chi^{\{\text{out}\}}\right\rangle_{rR}$ is the entangled state of the interior and exterior Hawking modes (like the Unruh state from Sect. 17.2.1 except $\left|\chi^{\{\text{out}\}}\right\rangle_{rR}$ could represent the state at even an early time, when $|R|$ is small). Sometimes we'll make the assumption that these are highly entangled, i.e. $S(\chi^{\{\text{in}\}}_{M_1})$ and $S(\chi^{\{\text{out}\}}_{R})$ are both near the maxima $\log|M_1|$ and $\log|R|$. Define

$$\Psi_{LBR}(U) := (V \otimes \mathbb{1}_{LR})\,|\psi_{\text{Hawk}}\rangle\,\langle\psi_{\text{Hawk}}|\,(V^{\dagger} \otimes \mathbb{1}_{LR}) \tag{17.3.69}$$

as the encoded version of the Hawking state. Note the U dependence on the right hand side is inside the V. Our goal is to compute the entropy of $\Psi_R(U) = \operatorname{tr}_{LB} \Psi_{LBR}(U)$.

We want to understand the relationship between the entropy of the system R in the effective description versus in the fundamental description. The key point will be: these entropies do *not need to be the same*! Indeed, acting $V \otimes \mathbb{1}_{LR}$ does not in general leave the state on R the same, nor its entropy, because V is not an isometry! (Recall how in the final state proposal, the state of R changed after postselection on $M_1 r$—the same math is at work here.) We will find that the entropy in the fundamental description is computed by the QES formula in the effective description, and this in turn will imply a unitary evaporation process.

First let's compute the second Renyi entropy, because it's easier and follows the same basic idea. The Renyi entropies are a generalization of the von Neumann entropy,

$$S_n(\rho) := \frac{1}{1-n} \log \operatorname{tr}(\rho^n), \tag{17.3.70}$$

defined for any real n. Note that the von Neumann entropy is the limit $S = \lim_{n \to 1} S_n$. Using our unitary integrals (17.2.31), we can compute

$$\begin{aligned}\int dU e^{-S_2(\Psi_R(U))} &= \int dU \operatorname{tr}(\Psi_R(U)^2) \\ &= \frac{|P|^2|B|^2}{|P|^2|B|^2-1}\left[\left(1-\frac{1}{|P||B|^2}\right)e^{-S_2(\chi_R^{\{\mathrm{out}\}})}+\left(1-\frac{1}{|P|}\right)\frac{e^{-S_2(\chi_{M_1}^{\mathrm{in}})}}{|B|}\right],\end{aligned} \tag{17.3.71}$$

which says when $|B|, |P| \gg 1$ that

$$S_2(\Psi_R) \approx \min\left[S_2(\chi_R^{\{\mathrm{out}\}}),\ \log|B| + S_2(\chi_{M_1}^{\{\mathrm{in}\}})\right]. \tag{17.3.72}$$

Let's unpack this remarkable answer in the language of the evaporating black hole. At early times, when $|R|$ is small and $|B|$ is big, the second Renyi entropy of R in the fundamental description is just equal to that in the effective description. Later, once the entanglement between R and r becomes large and $|B|$ is smaller, $S_2(\Psi_R)$ is very different from its effective description answer, now equaling the area of the black hole plus the Renyi entropy of whatever is in the interior. This is just minimizing the generalized entropy of the entanglement wedge—the QES formula (but for Renyi entropy)!

Of course, what we are really interested in is the von Neumann entropy. This is difficult to directly compute with our unitary integrals, but it is possible to compute it with the following trick (see Sect. 4 of [10] for more details). First, compute the S_n for general n. This is impractical to do in full, but we can make a useful approximation if we assume that either $S_2(\chi_R^{\{\mathrm{out}\}}) \gg \log|B| + S_2(\chi_{M_1}^{\{\mathrm{in}\}})$ or $S_2(\chi_R^{\{\mathrm{out}\}}) \ll \log|B| + S_2(\chi_{M_1}^{\{\mathrm{in}\}})$. Then there is a useful "large n" form of the unitary integrals, and if we assume $|\chi^{\{\mathrm{in}\}}\rangle_{LM_1}$ and $|\chi^{\{\mathrm{out}\}}\rangle_{rR}$ are highly entangled as well as $|P|, |B| \gg 1$, then we obtain

$$S_n(\Psi_R) \approx \min\left[S_n(\chi_R^{\{\mathrm{out}\}}),\ \log|B| + S_n(\chi_{M_1}^{\{\mathrm{in}\}})\right] \tag{17.3.73}$$

for the Renyi entropies and

$$S(\Psi_R) \approx \min\left[S(\chi_R^{\{\text{out}\}}),\ \log|B| + S(\chi_{M_1}^{\{\text{in}\}})\right] \tag{17.3.74}$$

for the von Neumann entropy. The last equation is exactly what we expect from the QES formula, as figured out in [17, 18].[18] Indeed, the entanglement wedge of an evaporating black hole undergoes a phase transition at the Page time. We can draw this as follows:

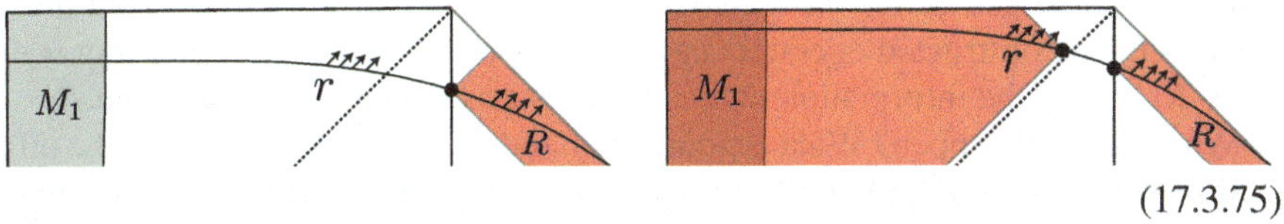

(17.3.75)

On the left we have a black hole early on, before the Page time, with the entanglement wedge of R shaded in red. We see it just includes R, so the QES formula says $S(R)_{\text{fund}} = S(R)_{\text{eff}}$. On the right we have the black hole after the Page time, with the entanglement wedge of R again shaded in red, now including both R and $M_1 r$ in the interior of the black hole. According to the QES formula, the entropy $S(R)$ now equals the area of the surface bounding the entanglement wedge—here the $A/4G$ of the black hole (the are area bounding R does not contribute because R is non-gravitational)—plus the entropy of everything in the entanglement wedge. That's just the entropy of M_1, because r and R together are in a pure state and both are in the entanglement wedge. These entropies, computed in [17, 18], exactly match (17.3.74).

This transition is significant because it implies a unitary evaporation! Indeed, as a model for real evaporation we can imagine decreasing $\log|B|$ while increasing $\log|R|$ the same amount, and plot the entropy (17.3.74) as a function of of this $|R|$:

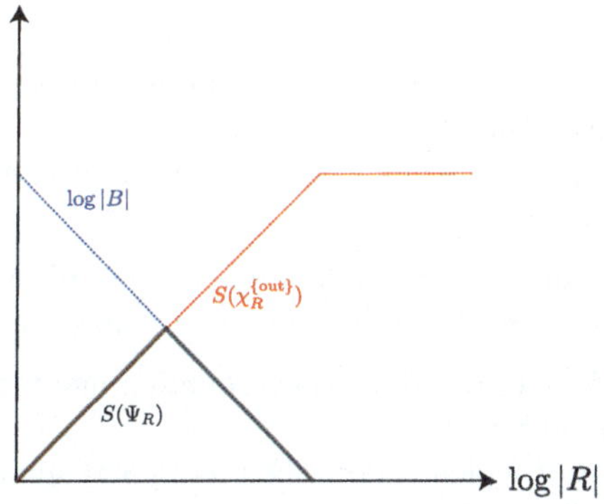

(17.3.76)

[18] The fact that the Renyi entropies satisfy a similar formula is an indicator that our model has "fixed areas" in the sense of [55, 56]. To see how this model can easily be generalized to have "non-fixed areas", see footnote 16 of [10].

The entropy $S(\Psi_R)$ follows the Page curve, which we recall from (17.3.36) is what we should expect for unitary evaporation!

We can make unitarity even clearer by remembering the connection between the QES formula and operator reconstruction. Now, the Theorem 17.3.2 that related them above assumes isometric V, so just from that we wouldn't know whether the QES formula implied any sort of operator reconstruction here. However, there is in fact a generalization of that theorem, discovered in [50], which does apply to non-isometric V. The full details of that theorem are more complicated than we need; it will suffice to say that in general the entanglement wedge is exactly the region in which operators can be reconstructed, much like as in Theorem 17.3.2, but the set of states the reconstructed operators have to work on is smaller.[19] The upshot is that once $|R| > |B|$, the information in the interior starts to be recoverable from R.

Why does that happen? What is going on in the map $V \otimes \mathbb{1}_{LR}$ such that the information in the interior starts becoming recoverable from the radiation? The answer is: postselected quantum teleportation! Just in the final state proposal, here the postselection is teleporting the interior information out to the exterior. What might be surprising is that this happens as soon as P is over half of the output—i.e. it's important that we postselect on most of the output, but we don't need to postselect on everything. The root cause is the scrambling done by U. More information can be found in [10, 51].

17.3.8 Resolving the Information Paradox: A Dynamical Model

We have seen that the non-isometric nature of the holographic map is connected to the unitarity of the evaporation process. Here we return to the information paradox in its full glory, to sharply see the resolution offered by non-isometric codes.

Recall the paradox: there is a tension between any theory having the following three things,

(1) A finite black hole entropy with a state-counting interpretation
(2) A unitary black hole S-matrix
(3) A black hole interior which is described to a good approximation by gravitational effective field theory.

What we will prove in this section is that a modest modification of assumption (3) completely erases the tension. The following three things *are* consistent:

(1) A finite black hole entropy with a state-counting interpretation
(2) A unitary black hole S-matrix
(3*) A black hole interior which is described to a good approximation by gravitational effective field theory *for low complexity operators*.

[19] In fact, the operators in general only need to work on a single state, and not an entire code subspace. This weaker sort of reconstruction is sometimes called "state-specific" reconstruction.

We will prove these are not inconsistent by explicitly constructing a system with all three properties. The key input will be the use of a non-isometric code. We regard this as a resolution of the information paradox because it reconciles these three assumptions in a maximally conservative way, preserving all of them in every regime we have experimentally tested and giving up on effective theory only in a regime we already knew it was invalid: high complexity.

So far, we have considered the holographic map at a fixed time, with dynamics present only implicitly via the dimensions of the subsystems and the choice of state. A complete model—like we construct here—should have the effective description and fundamental description each evolving under their own dynamics, and these dynamics should be compatible with the holographic map in the sense that we get the same thing if we time evolve and then encode or encode then time evolve:

$$\begin{array}{ccc} |\psi\rangle_{\mathrm{eff}} & \xmapsto{V} & |\widetilde{\psi}\rangle_{\mathrm{fund}} \\ e^{-itH_{\mathrm{eff}}}\big\downarrow & & \big\downarrow e^{-itH_{\mathrm{fund}}} \\ |\psi'\rangle_{\mathrm{eff}} & \xmapsto{V} & |\widetilde{\psi}'\rangle_{\mathrm{fund}} \end{array} \tag{17.3.77}$$

A map V with this property is said to be "equivariant". The model we now construct, closely following [10], is designed to incorporate the basic dynamics that goes into Hawking's information paradox. There will be matter in some adjustable state, which collapses to form a black hole, which then gradually evaporates into radiation, all under unitary dynamics. We will see this is compatible with existence of the interior in the effective description.

Let's start in the fundamental description. At time $t = -1$ there are m_0 qudits of collapsing matter, which are converted by isometry V_0 to a black hole at time $t = 0$ consisting of n_0 qudits. We imagine this is a "fast collapse", where the m_0 qudits make up M_1 and the other $n_0 - m_0$ qudits make up M_2. We can always write $V_0 = U_0 |\psi_0\rangle_{M_2}$, where U_0 is a unitary and $|\psi_0\rangle_{M_2}$ is some fixed state. The black hole is surrounded by a countably-infinite number of "ingoing" qudits we'll call R_{in}, as well as a countably-infinite number of "outgoing" qudits R_{out}. We'll let every qudit have Hilbert space dimension q. From $t = 0$ onwards, we evolve in discrete time steps, where in each time step the following dynamics occur:

1. One ingoing qudit falls into the black hole.
2. The black hole qudits interact with each other according to some unitary time evolution U_{t+1}.
3. Two outgoing qudits come out of the black hole.

Hence at time $t \geq 0$, the number of qudits that have entered the black hole is $m_0 + t$, the number of qudits that have come out is $2t$, and the remaining number of qudits in the black hole is $n_0 - t$. Hence the black hole completely evaporates in n_0 steps. We can draw this as

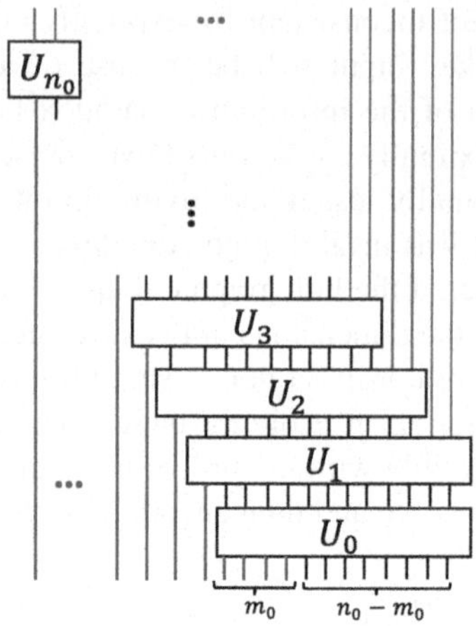

(17.3.78)

where the red lines are degrees of freedom that fall into the black hole, blue lines are degrees of freedom exiting the black hole, and black lines are those that are remaining in the black hole during that time step. This figure comes directly from [10]. The full Hilbert space of the fundamental description is therefore

$$\mathcal{H}_{\text{fund}} = \mathcal{H}_{R_{\text{in}}} \otimes \mathcal{H}_{R_{\text{out}}} \otimes \mathcal{H}_{\text{BH}} \tag{17.3.79}$$

where the black hole Hilbert space $\mathcal{H}_{\text{BH}}$ is a direct sum over different sizes of black hole

$$\mathcal{H}_{\text{BH}} \cong \oplus_t \mathcal{H}_{B^{(t)}} \tag{17.3.80}$$

where

$$\log_q |B^{(t)}| = n_0 - t. \tag{17.3.81}$$

Consider the subset of R_{out} that has already come out of the black hole after t time steps—call this $R^{(t)}$. Then we have

$$\log_q |R^{(t)}| = 2t. \tag{17.3.82}$$

Why did we choose two qudits to emerge each time step, and only one fall in? That was just a choice for the model. It allows for realistic dynamics (things can continue to fall into the black hole) while at the same time allowing the black hole to eventually evaporate away (more comes out than goes in). The one in, two out ratio is chosen just to be as simple as possible.

The final thing to specify is the particular unitaries U_i we will use at each step. Really any suitably scrambling unitary will work, but for definiteness let us consider them to each be an independently drawn random unitary.[20]

Now we turn to the dynamics in the effective description. The effective Hilbert space $\mathcal{H}_{\text{eff}}$ will consist of the ingoing and outgoing qudits as well as a direct sum over the interior degrees of freedom at different times:

$$\mathcal{H}_{\text{eff}} \cong \mathcal{H}_{R_{\text{in}}} \otimes \mathcal{H}_{R_{\text{out}}} \otimes (\oplus_t \mathcal{H}_{M_1^{(t)}} \otimes \mathcal{H}_{r^{(t)}}). \tag{17.3.83}$$

Each time step is given by the following evolution:

1. One ingoing qudit falls in, becoming part of M_1
2. Two "Hawking" pairs of qudits are added to the system, each in the maximally entangled state $|\text{MAX}\rangle := \frac{1}{\sqrt{q}} \sum_{i=0}^{q-1} |i\rangle\,|i\rangle$, with one qubit from each becoming a part of $r^{(t)}$ and the other joining R_{out}.

The initial state is m_0 matter qudits in whatever was the input state to V_0 in the fundamental description. These are part of $M_1^{(0)}$. We take $r^{(0)}$ to be empty. We can illustrate a couple steps of time evolution as

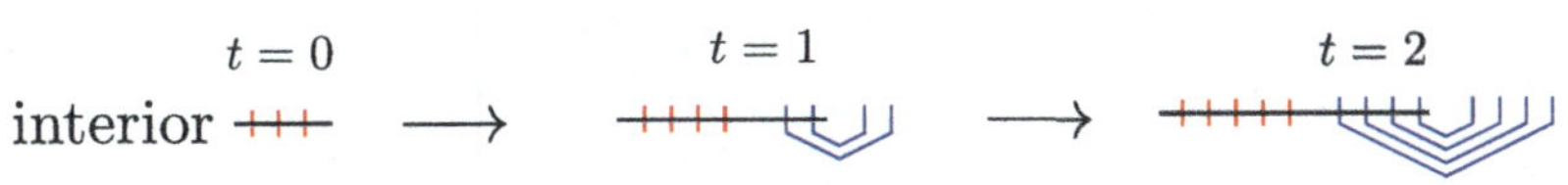

(17.3.84)

where M_1 qudits are depicted in red and grow by one each time step, and Hawking quanta are depicted in blue and grow by four, two inside (denoted by intersecting the horizontal black line) and two outside. Angled lines between two qudits denotes a maximally entangled state. We have not drawn the infinitely many qudits in R_{in}, but they should be understood to be there. We see each system grows in time as

$$\begin{aligned} \log_q |M_1^{(t)}| &= m_0 + t \\ \log_q |r^{(t)}| &= 2t. \end{aligned} \tag{17.3.85}$$

These dynamics are manifestly unitary in the sense that they preserve inner products as we advance in time. What might seem strange is the addition of new qudits in step 2, which requires some extra supply of qudits in a known state that can be brought in as needed. This is not obviously unitary. For example, say we took the state at $t = 1$

[20] This will certainly be scrambling, and calculations will be straightforward. The downside is that general random unitaries require exponentially large circuits, which is unfortunate because these are supposed to be relatively short time steps, occurring within the emission of just a couple Hawking quanta, so a proper analog would be a relatively short circuit. This more realistic physics can easily be added into the model, say by the more involved unitaries in Sect. 7 of [10] or by using *pseudorandom* unitaries [57] in place of truly random ones [58]. Again, nothing about the discussion in this section would change, because the only important feature is that the unitary is suitably scrambling.

in (17.3.84) and time evolved backwards. There is a clear rule for doing so: first undo step 2, removing the two blue qudits that are in the maximally entangled state, then move the red qudit that was just added back into R_{in}. However, say after $t = 1$ we measure the exterior qudits, changing their state, then try to evolve backwards. There is not an obvious rule that tells us how to undo step 2 now—we can't just remove the blue qudits, because then we'd wind up at the same $t = 0$ state we got when we didn't measure! This seems problematic, but actually is a feature rather than a bug: it models the fact that in effective field theory in curved spacetime, we can have a situation where modes "come down" from above the cutoff, which necessarily requires additional degrees of freedom to be introduced. If we modify the state and attempt to evolve backwards, then we typically encounter some kind of conventional breakdown of effective field theory due to large energy densities or curvatures [59, 60].

Having defined the dynamics in both effective and fundamental description, we now turn to defining the relationship between them. Consider the following holographic map, which we'll call $V_2 \otimes \mathbb{1}_R$ for reasons to become clear:

$V_2 \otimes \mathbb{1}_R =$

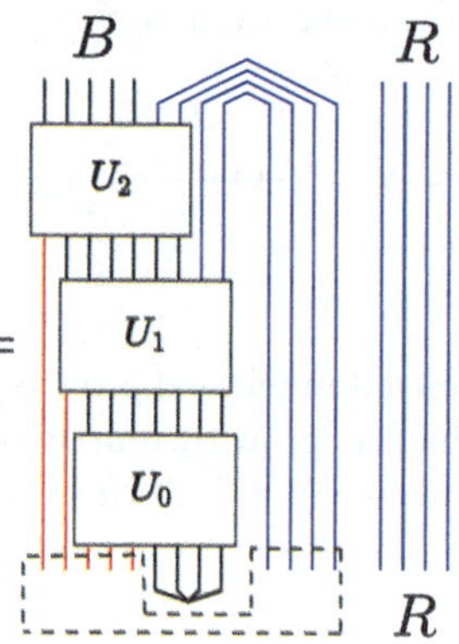

(17.3.86)

where we have indicated which inputs (at the bottom) act on degrees of freedom in the interior. The input black lines represent the state $|\psi_0\rangle_{M_2}$. The angled blue lines here, as before, represent maximally entangled states, but when upside-down like this they represent postselection by the unnormalized state $q^2 \langle \text{MAX}| = \sum_i \langle i| \langle i|$. This defines a map from the $t = 2$ state in the effective theory to the $t = 2$ state in the fundamental theory. To see this visually, simply draw this on top of the effective theory state,

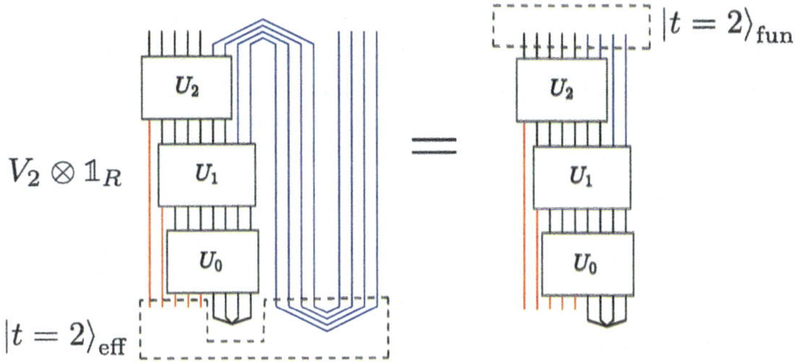

(17.3.87)

where in going from the left to right hand sides we have simply "unbent" the blue lines (which mathematically is analogous to (17.2.14)). More generally, we define the holographic map $V_t : \mathcal{H}_{M_1^{(t)}} \otimes \mathcal{H}_{r^{(t)}} \to \mathcal{H}_{B^{(t)}}$ by starting with

$$V_0 = U_0 \left|\psi_0\right\rangle_{M_2} \tag{17.3.88}$$

and then at each step acting with U_t and then postselecting on the maximally-entangled state for the two new radiation qudits:

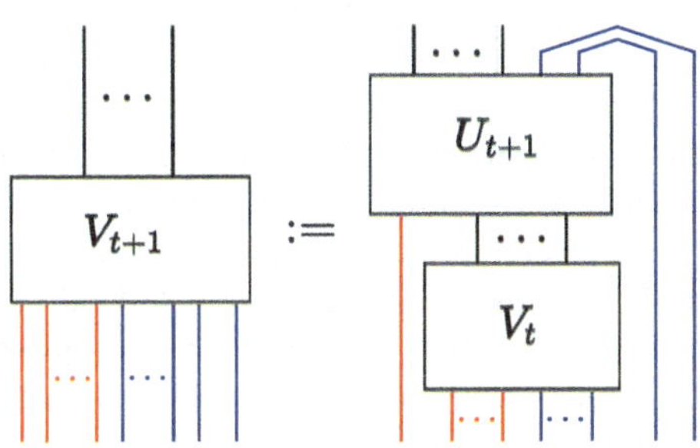

(17.3.89)

In equations,

$$V_t = \sqrt{|P^{(t)}|}\,\langle \mathrm{MAX}|_{R^{(t)}r^{(t)}}\, U_t U_{t-1} ... U_1 U_0 \left|\psi_0\right\rangle_{M_1} . \tag{17.3.90}$$

It is straightforward to argue this indeed maps the effective description state at time t to the fundamental description state at time t, using an argument similar to what we did for $t = 2$ above. The full encoding map is defined as $V : \oplus_t \mathcal{H}_{M_1^{(t)}} \otimes \mathcal{H}_{r^{(t)}} \to \mathcal{H}_{\text{BH}}$ with

$$V := \oplus_t V_t . \tag{17.3.91}$$

By the above arguments, this map V satisfies (17.3.77). Note: this is true for general inputs into V. We do not need to restrict to maximal entanglement between $r^{(t)}$ and $R^{(t)}$, even though (17.3.87) used it to motivate the definition. An important feature of this model is that, exactly as for (17.3.43), V is non-isometric (because of the postselection) but nonetheless inner products are almost exactly preserved between sub-exponential states. Therefore at any particular time we can do a sub-exponential measurement or act with a sub-exponential unitary operation and the effective description inner products will still closely match those of the fundamental description. Then we can continue to time evolve and this will continue to be true; equivariance will be maintained.

This completes the description of the model. Let us emphasize the physics it offers: the fundamental description is manifestly unitary and has a finite black hole entropy with a state counting interpretation. The effective description manifestly has a black hole interior. These are two descriptions *of the same system*, because they are related by a linear map V that is equivariant as in (17.3.77) and preserves inner products.

However, it only preserves inner products for sub-exponentially complex states! If an observer does anything too complex, the matching of the inner products breaks down, which leads to the QES formula no longer relating the entropies in the descriptions and therefore operator reconstruction becoming impossible. There stops being any sense in which the two descriptions are equivalent. In such cases, we can *only* use the fundamental description, which (as the name indicates) we take to be the defining description of the system.[21] This is a feature, and explains why we expect an observer doing a high complexity operation on the exterior radiation to see unitarity, as the fundamental description predicts, and not information loss as the effective description predicts.

We close this section by mentioning that even more sophisticated versions of this model now exist; a number of simplifying features here have been studied in later papers and are now known to be inessential. Recall the real test of the final state proposal came when we allowed interactions in the black hole interior. In this model, we did not have any kind of sophisticated interactions—the interior qubits just floated along peacefully into the future. The addition of interactions was analyzed by Kim and Preskill [61] and then also by DeWolfe and Higginbotham [62, 63], who both proposed different clever ways to add interactions without changing the equivariance of the map or the preservation of inner products. Another simplifying feature of the model here was the non-local structure of the unitaries in V; at each time step we acted with one big U_t on all of the interior degrees of freedom, which is not quite like the local scrambling dynamics we expect in real black holes. This was already ameliorated in the [10] and then even further in [64].

What of acausality? In the final state proposal, postselection led to acausality even for observers remaining outside the black hole. That does not happen here. Recall from Sect. 17.2.5 that performing the experiment that would see acausality requires an exponentially complex operation. In non-isometric codes, that means that the effective description breaks down before one can complete it—you have to use

[21] Indeed, we can *always* use the fundamental description to answer any question, even "what does an observer see who dives into a black hole." The interior is still there in the fundamental description (after all, it's equivalent to the effective description), just highly obscured.

the fundamental description to describe everything. But the fundamental description is just a normal quantum system without any postselection! Acausality never shows up.

One thing to note: while we motivated non-isometric codes by thinking about AdS/CFT, the idea is that they apply more broadly to any black hole in any spacetime. In general, we can expect a black hole to admit two descriptions: some effective description like above, which describes the black hole in terms of quantum fields living on a geometric interior, and a fundamental description with a Hilbert space of dimension $e^{A/4G}$, whose degrees of freedom are more mysterious but can be envisioned as living near the horizon of the black hole. The map V is what describes the relationship between these descriptions, and would have to be non-isometric by the same arguments as in AdS/CFT. The model in this section might as well be understood in any spacetime!

17.4 Conclusion

For almost 50 years, Hawking's information paradox has asked us to give up some core principle: either (1) a finite black hole entropy, (2) unitarity, or (3) effective theory. Since then we have learned great reasons not to give up on (1) or (2) [4, 5]. We have also learned some subtleties in (3): effective theory does breakdown for some low energy but high complexity experiments [14], such as the the experiment on the exterior Hawking radiation that tests whether the black hole is indeed evolving unitarily. Now, with the discovery of non-isometric codes, we have completed this story. We have proven that it is internally consistent to have (1) a finite black hole entropy, (2) unitarity, and (3) effective theory valid for *low complexity* operations. It is nice that there exists such a maximally conservative resolution: we don't have to give up any more than what we already knew had to go.

Of course, just because this is one way to solve the information paradox doesn't mean it's *the* way that our universe chooses to use! Maybe there are other ways to resolve the paradox that haven't been realized yet. It is important future work to build on these models, making them increasingly realistic until they fit our world and can make experimental predictions. A concrete next step is to consider solvable low-dimensional versions of quantum gravity and see whether the holographic map V can be written down for them explicitly. If so, this will answer whether those theories of quantum gravity indeed resolve the information paradox in this way. The study of non-isometric codes is just beginning—there is still much exciting work to be done.

References

1. S.W. Hawking, Breakdown of predictability in gravitational collapse. Phys. Rev. D **14**, 2460–2473 (1976)
2. S.W. Hawking, Particle creation by black holes. Commun. Math. Phys. **43**, 199–220 (1975) [Erratum: Commun. Math. Phys. **46**, 206 (1976)]
3. W.G. Unruh, Notes on black-hole evaporation. Phys. Rev. D **14**, 870–892 (1976)
4. A. Strominger, C. Vafa, Microscopic origin of the Bekenstein-Hawking entropy. Phys. Lett. B **379**, 99–104 (1996). arXiv:hep-th/9601029
5. J.M. Maldacena, The large N limit of superconformal field theories and supergravity. Adv. Theor. Math. Phys. **2**, 231–252 (1998). arXiv:hep-th/9711200
6. A. Almheiri, D. Marolf, J. Polchinski, J. Sully, Black holes: complementarity or firewalls? JHEP **02**, 062 (2013). arXiv:1207.3123
7. S.D. Mathur, The information paradox: a pedagogical introduction. Class. Quant. Grav. **26**, 224001 (2009). arXiv:0909.1038
8. B. Guo, M.R.R. Hughes, S.D. Mathur, M. Mehta, Contrasting the fuzzball and wormhole paradigms for black holes. Turk. J. Phys. **45**(6), 281–365 (2021). arXiv:2111.05295
9. G.T. Horowitz, J.M. Maldacena, The black hole final state. JHEP **02**, 008 (2004). arXiv:hep-th/0310281
10. C. Akers, N. Engelhardt, D. Harlow, G. Penington, S. Vardhan, The black hole interior from non-isometric codes and complexity. JHEP **06**, 155 (2024). arXiv:2207.06536
11. D. Harlow, Jerusalem lectures on black holes and quantum information. Rev. Mod. Phys. **88**, 015002 (2016). arXiv:1409.1231
12. D. Gottesman, J. Preskill, Comment on 'The black hole final state'. JHEP **03**, 026 (2004). arXiv:hep-th/0311269
13. S. Lloyd, J. Preskill, Unitarity of black hole evaporation in final-state projection models. JHEP **08**, 126 (2014). arXiv:1308.4209
14. D. Harlow, P. Hayden, Quantum computation vs. firewalls. JHEP **06**, 085 (2013). arXiv:1301.4504
15. S. Aaronson, *The Complexity of Quantum States and Transformations: From Quantum Money to Black Holes*, vol. 7 (2016). arXiv:1607.05256
16. M.A. Nielsen, I.L. Chuang, *Quantum Computation and Quantum Information: 10th Anniversary Edition*, 10th edn. (Cambridge University Press, 2011)
17. G. Penington, Entanglement wedge reconstruction and the information paradox. arXiv:1905.08255
18. A. Almheiri, N. Engelhardt, D. Marolf, H. Maxfield, The entropy of bulk quantum fields and the entanglement wedge of an evaporating black hole. JHEP **12**, 063 (2019). arXiv:1905.08762
19. I. Heemskerk, J. Penedones, J. Polchinski, J. Sully, Holography from conformal field theory. JHEP **10**, 079 (2009). arXiv:0907.0151
20. D. Harlow, TASI lectures on the emergence of bulk physics in AdS/CFT. PoS **TASI2017**, 002 (2018). arXiv:1802.01040
21. T. Banks, M.R. Douglas, G.T. Horowitz, E.J. Martinec, AdS dynamics from conformal field theory. arXiv:hep-th/9808016
22. E. Witten, Anti-de Sitter space and holography. Adv. Theor. Math. Phys. **2**, 253–291 (1998). arXiv:hep-th/9802150
23. S.S. Gubser, I.R. Klebanov, A.M. Polyakov, Gauge theory correlators from noncritical string theory. Phys. Lett. B **428**, 105–114 (1998). arXiv:hep-th/9802109
24. D. Petz, Sufficient subalgebras and the relative entropy of states of a von Neumann algebra. Commun. Math. Phys. **105**(1), 123–131 (1986)
25. J. Cotler, P. Hayden, G. Penington, G. Salton, B. Swingle, M. Walter, Entanglement wedge reconstruction via universal recovery channels. arXiv:1704.05839
26. C.-F. Chen, G. Penington, G. Salton, Entanglement wedge reconstruction using the Petz map. JHEP **01**, 168 (2020). arXiv:1902.02844

27. G. Penington, S.H. Shenker, D. Stanford, Z. Yang, Replica wormholes and the black hole interior. arXiv:1911.11977
28. A. Almheiri, X. Dong, D. Harlow, Bulk locality and quantum error correction in AdS/CFT. JHEP **04**, 163 (2015). arXiv:arXiv:1411.7041
29. E. Witten, APS medal for exceptional achievement in research: invited article on entanglement properties of quantum field theory. Rev. Mod. Phys. **90**(4), 045003 (2018). arXiv:1803.04993
30. M.J. Kang, D.K. Kolchmeyer, Holographic relative entropy in infinite-dimensional Hilbert spaces. Commun. Math. Phys. **1**, 31 (2023). arXiv:1811.05482
31. T. Faulkner, The holographic map as a conditional expectation. arXiv:2008.04810
32. H. Casini, M. Huerta, Entanglement entropy in free quantum field theory. J. Phys. A **42**, 504007 (2009). arXiv:0905.2562
33. S. Ryu, T. Takayanagi, Aspects of holographic entanglement entropy. JHEP **08**, 045 (2006). arXiv:hep-th/0605073
34. S. Ryu, T. Takayanagi, Holographic derivation of entanglement entropy from AdS/CFT. Phys. Rev. Lett. **96**, 181602 (2006). arXiv:hep-th/0603001
35. A. Lewkowycz, J. Maldacena, Generalized gravitational entropy. JHEP **08**, 090 (2013). arXiv:1304.4926
36. V.E. Hubeny, M. Rangamani, T. Takayanagi, A covariant holographic entanglement entropy proposal. JHEP **07**, 062 (2007). arXiv:0705.0016
37. X. Dong, A. Lewkowycz, M. Rangamani, Deriving covariant holographic entanglement. JHEP **11**, 028 (2016). arXiv:1607.07506
38. T. Faulkner, A. Lewkowycz, J. Maldacena, Quantum corrections to holographic entanglement entropy. JHEP **11**, 074 (2013). arXiv:1307.2892
39. N. Engelhardt, A.C. Wall, Quantum extremal surfaces: holographic entanglement entropy beyond the classical regime. JHEP **01**, 073 (2015). arXiv:1408.3203
40. X. Dong, A. Lewkowycz, Entropy, extremality, Euclidean variations, and the equations of motion. JHEP **01**, 081 (2018). arXiv:1705.08453
41. A. Almheiri, T. Hartman, J. Maldacena, E. Shaghoulian, A. Tajdini, Replica wormholes and the entropy of Hawking radiation. JHEP **05**, 013 (2020). arXiv:1911.12333
42. X. Dong, H. Wang, Enhanced corrections near holographic entanglement transitions: a chaotic case study. arXiv:2006.10051
43. D. Marolf, S. Wang, Z. Wang, Probing phase transitions of holographic entanglement entropy with fixed area states. arXiv:2006.10089
44. C. Akers, G. Penington, Leading order corrections to the quantum extremal surface prescription. arXiv:2008.03319
45. B. Czech, J.L. Karczmarek, F. Nogueira, M. Van Raamsdonk, The gravity dual of a density matrix. Class. Quant. Grav. **29**, 155009 (2012). arXiv:1204.1330
46. D.L. Jafferis, A. Lewkowycz, J. Maldacena, S.J. Suh, Relative entropy equals bulk relative entropy. JHEP **06**, 004 (2016). arXiv:1512.06431
47. X. Dong, D. Harlow, A.C. Wall, Reconstruction of bulk operators within the entanglement wedge in gauge-gravity duality. Phys. Rev. Lett. **117**(2), 021601 (2016). arXiv:1601.05416
48. D. Harlow, The Ryu-Takayanagi formula from quantum error correction. Commun. Math. Phys. **354**(3), 865–912 (2017). arXiv:1607.03901
49. R. Cleve, D. Gottesman, H.-K. Lo, How to share a quantum secret. Phys. Rev. Lett. **83**, 648–651 (1999). arXiv:quant-ph/9901025
50. C. Akers, G. Penington, Quantum minimal surfaces from quantum error correction. SciPost Phys. **12**(5), 157 (2022). arXiv:2109.14618
51. P. Hayden, G. Penington, Learning the alpha-bits of black holes. JHEP **12**, 007 (2019). arXiv:1807.06041
52. F. Pastawski, B. Yoshida, D. Harlow, J. Preskill, Holographic quantum error-correcting codes: toy models for the bulk/boundary correspondence. JHEP **06**, 149 (2015). arXiv:1503.06237
53. D.N. Page, Average entropy of a subsystem. Phys. Rev. Lett. **71**, 1291–1294 (1993). arXiv:gr-qc/9305007

54. D.N. Page, Information in black hole radiation. Phys. Rev. Lett. **71**, 3743–3746 (1993). arXiv:hep-th/9306083
55. C. Akers, P. Rath, Holographic Renyi entropy from quantum error correction. JHEP **05**, 052 (2019). arXiv:1811.05171
56. X. Dong, D. Harlow, D. Marolf, Flat entanglement spectra in fixed-area states of quantum gravity. JHEP **10**, 240 (2019). arXiv:1811.05382
57. F. Ma, H.-Y. Huang, How to construct random unitaries. arXiv:2410.10116
58. N. Engelhardt, R. Folkestad, A. Levine, E. Verheijden, L. Yang, Cryptographic censorship. arXiv:2402.03425
59. G. 't Hooft, On the quantum structure of a black hole. Nucl. Phys. B **256**, 727–745 (1985)
60. Y. Kiem, H.L. Verlinde, E.P. Verlinde, Black hole horizons and complementarity. Phys. Rev. D **52**, 7053–7065 (1995). arXiv:hep-th/9502074
61. I.H. Kim, J. Preskill, Complementarity and the unitarity of the black hole S-matrix. JHEP **02**, 233 (2023). arXiv:2212.00194
62. O. DeWolfe, K. Higginbotham, Non-isometric codes for the black hole interior from fundamental and effective dynamics. JHEP **09**, 068 (2023). arXiv:2304.12345
63. O. DeWolfe, K. Higginbotham, Bulk reconstruction and non-isometry in the backwards-forwards holographic black hole map. JHEP **06**, 126 (2024). arXiv:2311.12921
64. G. Bueller, O. DeWolfe, K. Higginbotham, Tensor networks for black hole interiors: non-isometries, quantum extremal surfaces, and wormholes. JHEP **10**, 012 (2024). arXiv:2407.01666

Chapter 18
The Unitarity Crisis, Nonviolent Unitarization, and Implications for Quantum Spacetime

Steven B. Giddings

Abstract This contribution overviews the information paradox, or perhaps more aptly "unitarity crisis," and a proposed resolution called nonviolent unitarization. It begins by examining the conflict of principles that yields the crisis, which can be phrased in terms of a "black hole theorem" summarizing how basic assumptions come into conflict. Proposed resolutions of the conflict, along with problems with them, are overviewed. The very important underlying question of localization of information and its role is discussed at some length, taking into account effects of perturbative gravity. The difficulty in finding a consistent scenario for black hole evolution strongly suggests new interactions on event horizon scales; a "minimal" set of assumptions about these is parameterized in nonviolent unitarization. Possible criticisms of this scenario, and some responses, are given. New interactions at event horizon scales potentially lead to observable effects, via gravitational wave or electromagnetic channels, which are briefly discussed. A possible origin of nonviolent unitarization effects from a more fundamental description of quantum spacetime, and possible implications for such a description, are also briefly discussed.

18.1 Introduction

The black hole information paradox has evolved from what to many probably seemed an esoteric curiosity to what now appears to be a full blown crisis challenging the current foundations of physics. In short, this problem appears to reveal an essential inconsistency between the basic principles of quantum mechanics, the principle of locality, and the principles of relativity, particularly the symmetries of Poincaré or general coordinate invariance. These are the principles that provide the foundation of local quantum field theory—which so far has beautifully described all natural phenomena except gravity—and general relativity. As will be discussed further, this

S. B. Giddings (✉)
Department of Physics, University of California, Santa Barbara, CA, USA
e-mail: giddings@ucsb.edu

A. Akil and C. Bambi (eds.), *The Black Hole Information Paradox*, Springer Series in Astrophysics and Cosmology, https://doi.org/10.1007/978-981-96-6170-1_18

is quite likely a crisis in the sense of Kuhn [79], pointing toward the need for fundamental revision of current principles. Since an essential question is that of unitary evolution, this is appropriately described as the *unitarity crisis*.

There have been many proposed resolutions to this crisis. Typically these involve introduction of some extra structure, and then additional questions and problems arise; none have given a convincing resolution. Frequently in the history of physics a conservative or minimalist approach has been successful, seeking a minimal revision to existing principles without introducing extraneous structure; many such revisions have proved to be profoundly radical.[1] This contribution will explore such a conservative approach, starting with some basic observations. The first is that there is increasingly good evidence that black hole-like objects are ubiquitous in the cosmos, and so far appear to be well-described by classical general relativity. Moreover, quantum mechanics has not only been tested in a wide range of contexts, but it has also been found to be very difficult to modify its basic principles without encountering extremely problematic behavior. Finally, quantum field theory has proven to be an extremely accurate description of non-gravitational physics, in contexts with weakly-curved backgrounds. Combining these with a careful formulation of the problem and assuming that evolution is ultimately unitary suggests that a quantum black hole must have interactions with its surroundings that are not well-described by quantum field theory plus perturbatively-quantized general relativity, but that can be in a certain sense a small correction to that physics, and that ultimately unitarize evolution. In particular, the failure so far to observe effects of structure near black hole horizons, and extrapolation of the statement that quantum field theory provides a highly-accurate description in weakly-curved regions, suggests that these interactions don't have violent effects on observers that might fall into a big black hole. These are basic tenets of *nonviolent unitarization*.

Necessary background for reading this piece is a basic understanding of general relativity and of standard quantum field theory including techniques used in a traditional derivation of Hawking radiation [71]. In outline, the next section will give a description of the essential conflict, formulated as a "black hole theorem." Section 18.3 will then give a more detailed description of aspects of the Hawking process, extending beyond the traditional analysis to a description of the evolving quantum state, and describing aspects of perturbative gravitational corrections, providing additional evidence that subtleties in such a description don't resolve the crisis, and providing a starting point for parameterizing corrections to such a standard description. This initial discussion is important for explaining background motivation, in the spirit of the famous A. C. Doyle quote, through the words of Sherlock Holmes: "How often have I said to you that when you have eliminated the impossible, whatever remains, however improbable, must be the truth?" Section 18.4 then elaborates on the basic tenets of nonviolent unitarization, and Sect. 18.5 describes its parametrization, and constraints on this, as a departure from the standard description. Section 18.6 discusses some of the criticisms that have appeared in the literature or discussions,

---

[1] Examples include Newtonian gravity, Special Relativity, Quantum Mechanics, General Relativity, and Quantum Field Theory.

and responds to them. Section 18.7 introduces prospects for observational signatures, given our new era of multichannel observation of black holes. Section 18.8 concludes by discussing the question of connecting to a more foundational picture of quantum gravity, and the possibly quite radical ultimate implications of such a conservative approach.

18.2 A Black Hole Theorem and the Crisis

18.2.1 The Black Hole Theorem

The unitarity crisis deals with the essential question of localization and transport of quantum information in quantum gravity. In finite quantum systems, or in locally-finite ones such as lattice approximations to field theory, localization is described in terms of a factorization of the Hilbert space, e.g. $\mathscr{H} = \mathscr{H}_A \otimes \mathscr{H}_B$. With such a structure, we can consider independent excitations of subsystems A or B, e.g. corresponding to localized information; transfer of information can then be described by a Hamiltonian that couples excitations in A to those in B.

Before including gravity, local quantum field theory (LQFT) has a similar structure. We may think of A and B as corresponding to spacelike-separated regions in spacetime. However, there are additional subtleties associated to the infinite degrees of freedom of an unregulated theory. Specifically, if we consider a region U of a spacelike slice and its complement $\bar{U}$, well-behaved states of field theory don't live in a factorized Hilbert space $\mathscr{H}_U \otimes \mathscr{H}_{\bar{U}}$.[2] Instead, we consider separately and locally exciting degrees of freedom in the complementary regions by acting with localized operators, for example involving a field operator integrated against a function with support restricted to U. Such operators commute at spacelike separations and can be used to describe information localized to either region, with the general statement being that there are commuting subalgebras of observables associated with spacelike separated regions, and that these define subsystems. This is the basic formulation of locality in an algebraic approach to the subject [70].

Now consider evolution of the quantum state in a black hole (BH) spacetime. We may describe states on spacelike slices, such as shown in Fig. 18.1, and write these as

$$|\Psi\rangle = |\psi_{\rm BH}, \psi_{\rm env}\rangle. \tag{18.1}$$

Here the labels $\psi_{\rm BH}$ and $\psi_{\rm env}$ describe the excitations localized to the black hole interior, or to the black hole environment, respectively. For many practical purposes, one may consider modeling such a state by one in a factorized Hilbert space, which

[2] For example, an attempt to write such a factorized state typically leads to problems like infinite energy density. The obstacle results from the type III property of the algebras associated to quantum field theories [7], and is associated with "infinite entanglement" between the two regions; for reviews of some of their features see [70, 107, 108].

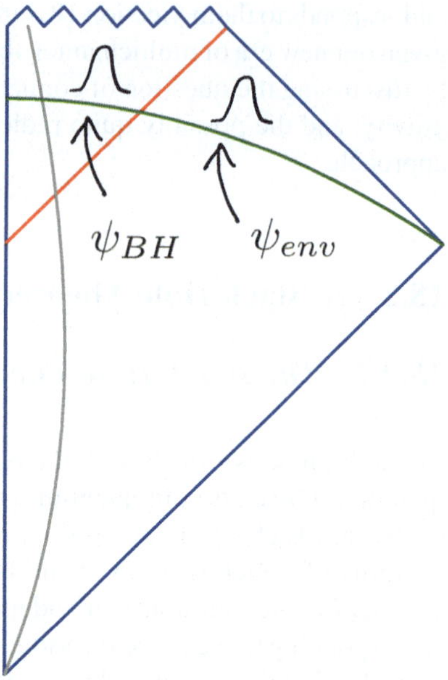

Fig. 18.1 Shown is a Penrose diagram of a black hole, with one spatial slice of a family, on which we can define the state and its evolution. The state is characterized by excitations of the BH interior, and exterior or environment, parameterized by ψ_{BH} and ψ_{env}, respectively; in a LQFT description these can be described as excited by operators localized to the respective regions

captures much of the same structure and can be written $|\Psi\rangle = |\psi_{\rm BH}\rangle \otimes |\psi_{\rm env}\rangle$. In the evolution described by Hawking [71], a state that initially looks close to vacuum, e.g as measured by observers falling into the BH, evolves to one with exterior and outgoing Hawking excitations entangled with interior excitations. A toy model for this can be written in terms of qubits,[3] with an initial vacuum evolving as

$$|\Psi_0\rangle \to \frac{1}{2^{\frac{N}{2}}}\left(|\hat{0}\rangle|0\rangle + |\hat{1}\rangle|1\rangle\right)^{\otimes N}, \tag{18.2}$$

where hatted states denote interior excitations, and unhatted states describe the excitations of N quanta of outgoing Hawking radiation.

The "black hole theorem [55]" states that the following assumptions are inconsistent:

0. The combined system of a black hole and its environment undergoes unitary quantum evolution;
1. A black hole can be treated as a subsystem of the larger system including its environment;
2. Distinct black hole states have identical exterior evolution;
3. A black hole disappears at the end of its evolution.

This theorem can be understood as follows. First, consider a collection of black hole states, labeled with index i, with the subsystem structure as described above,

[3] For further discussion of such toy models, see [56] and references therein.

$|\psi_{\mathrm{BH},i}, \psi_{\mathrm{env}}\rangle$. Postulate (2), identical exterior evolution, states that these evolve to later states

$$|\psi_{\mathrm{BH},i}, \psi_{\mathrm{env}}\rangle \to |\Psi(t)\rangle = |\psi'_{\mathrm{BH},i}, \psi'_{\mathrm{env}}\rangle \tag{18.3}$$

where the description of the exterior state, ψ'_{env} is independent of the BH state i. In the LQFT description, this follows from the locality property: an internal excitation of the BH cannot influence observables outside the BH. If the BH subsequently disappears, that then corresponds to evolution

$$|\psi'_{\mathrm{BH},i}, \psi'_{\mathrm{env}}\rangle \to |\psi''_{env}\rangle. \tag{18.4}$$

But, this is not unitary, contradicting postulate (0)—it is not even one-to-one, and "forgets" the information contained in the label i. One could for example diagnose this by entangling the initial states $|\psi_{\mathrm{BH},i}, \psi_{\mathrm{env}}\rangle$ with a set of auxiliary states $|\tilde{i}\rangle$; the information in the BH state can be described by the von Neumann (entanglement) entropy $S_{vN} = -\mathrm{Tr}(\rho \ln \rho)$ of the density matrix gotten by tracing out BH states. This entropy is preserved by (18.3) but would jump to zero under evolution (18.4); the "missing" entanglement entropy parameterizes the magnitude of the violation of unitarity. If one considers evolution from a black hole of initial mass M, the corresponding information loss is comparable to the Bekenstein-Hawking entropy,

$$S_{BH} = \frac{A}{4G} \sim \left(\frac{M}{M_p}\right)^2, \tag{18.5}$$

where G is Newton's constant and M_p is the Planck mass, and we work in units with $\hbar = 1$.

This is a crisis because violation of any of the postulates (0–4) appears to lead to a contradiction with basic physical principles. To see this, we consider possible ways to escape this contradiction.

18.2.2 Proposed Alternatives

A first proposal, originally made by Hawking [72], is to abandon the unitary postulate of quantum mechanics. However, it was later realized that this is extremely problematic. In essence, information transfer and energy transfer are closely associated. So, information loss, such as Hawking proposed, will be associated with energy nonconservation. An argument for this was given by Banks, Peskin, and Susskind in [10]. How bad is the energy nonconservation? If other principles of quantum mechanics remain intact, and locality is intact enough to allow basic effective field theory reasoning, then BH formation, evaporation, and corresponding information loss can take place as part of virtual processes. The natural cutoff scale for these would be of order the Planck mass, M_p, giving an estimated temperature for our "vacuum" that is $T \sim M_p$—a preposterous departure from observation. While it has been argued [105] that information loss can be disassociated from such energy nonconservation, models that do so [103] have similar features to models of baby universe emission. There, it was discovered [19, 68, 66] that such information loss is not "true" information loss, in that it does not persist with repeated experiments, here of black hole formation and evaporation; it is more like an initial indeterminacy in the coupling constants, that can be resolved with subsequent experiments, after which there is no more information loss. A basic message appears to be that the principles of quantum mechanics are not so easily modified.

An obvious alternative is to drop postulate (3), disappearance, as described by (18.4). For example, perhaps the state information parametrized by i transfers into the decay products of the BH, once its size is near-Planckian and a semiclassical/perturbative description is expected to fail.[4] However, this is also severely constrained by basic principles. Since the information to be accounted for is roughly of size S_{BH}, encoding that in outgoing radiation of ordinary quanta would require of order $N \sim S_{BH}$ outgoing quanta. Since the available energy is that of the near-Planckian BH, each quantum can then only carry energy $E \sim M_p/S_{BH}$. Producing such a low-energy quantum takes a time $\sim 1/E$, and so emitting N such quanta would take a time $T \sim M^4/M_p^5$. In such a scenario, BHs leave extremely long-lived remnants.

Such microscopic remnants would have masses $\sim M_p$, but would have to have an unboundedly large number of internal states, since they could form from evaporation of an arbitrarily large BH. Such a spectrum of objects—whether long-lived or stable—is also extremely problematic. Again, by basic quantum principles, these could be pair produced in ordinary physical processes, perhaps with a tiny amplitude $\mathscr{A}$ for a given remnant state. However, the total probability of production, including all the remnant states, is then $\sim N|\mathscr{A}|^2$, with N an unboundedly large number—all physical processes with sufficient total energy would be unstable to unbounded production of such remnants [93, 32, 34]. While there have been attempts to argue for dynamics escaping the basic effective field theory reasoning behind this [11, 12, 94], it does not appear that such a consistent description has been found. Simple models [33] support this basic understanding. Also, in such a remnant scenario one would also expect to have unboundedly large numbers of internal states of charged black holes, and there one can analyze a semiclassical (instantonic) production process and see how the unbounded spectrum would contribute to unbounded production [34]. Finally, there are also related arguments against such an unbounded spectrum based on effects of virtual such remnants [97].

If we have ruled out violating postulates (0) and (3), that leaves the possibility of some violation of postulates (1) and/or (2), which in either case would represent a departure from locality as described in LQFT.[5] It is certainly true that in a perturbative approach to quantizing general relativity, the LQFT description of locality in terms of

[4] For example, this scenario is now advocated in much of the loop quantum gravity community [9].

[5] We might think of postulate (1) as describing *localization* of information, and postulate (2) as associated with *local propagation* of information.

commuting subalgebras is modified. An important question, which will be discussed further in the next section, is the nature and role of that modification, and the nature and role of additional quantum gravitational effects beyond this.

Before further discussing locality, note that there is a significant number of other proposed scenarios for a resolution of the crisis. Many of these now involve some modification of conventional physics on horizon scales; one example is [98, 99]. A large number fall within the general category of "massive remnant" scenarios [31], where at some stage in its evolution a BH is replaced by an object that is not approximately described as having a vacuum-like horizon (or, trapped surface), but instead has some completely different structure replacing that region; a conventional analog of this is a neutron star, where the star's surface and interior replaces the BH interior. One could consider either classical or quantum modifications to general relativity as a possible origin for such a picture. Examples of such a general scenario are gravastars [90], fuzzballs [88], earlier versions of Planck stars [95], frozen stars [15], and presumably firewalls [5], if a physical realization of the latter scenario could be found.

The difficulties with such scenarios are that they both typically also involve some nonlocal physics, *and* they require considerable extra physical structure that potentially creates other problems and is challenging to realize in a physically consistent scenario. How nonlocality is invoked depends on when the massive remnant forms. If a massive remnant is to form in a collapse before the would be horizon forms, this appears to require nonlocal physics because the local description of the collapsing matter is in a very ordinary regime, for a big black hole. Specifically, with collapse of a gas of total mass M, the horizon forms when gas density reaches $\rho \sim M_p^6/M^2$, so e.g. less than the density of air in our atmosphere for a large enough black hole. It would appear to require a nonlocal effect on the scale of the horizon to replace such a collapsing gas by an exotic object. Alternately, such a massive remnant might form after the trapped surface of a BH had formed. But at that point, there is a large amount of information from e.g. the collapsing matter that is inside the BH, and that information would have to be transferred at spacelike separation to the exterior as part of formation of the exotic object. The combined difficulties of nonlocality and extra structure motivate seeking a more minimal scenario.

Finally, a very popular view is that the problems of quantum gravity, and of quantum black holes in particular, are resolved through an AdS/CFT map between gravitational physics in anti de Sitter space and conformal field theory in one fewer dimension [83], or analogs to it with other asymptotic structure. The problem here is understanding the detailed nature of the "holographic map" between the CFT and the bulk quantum gravitational physics, which should in turn match onto an approximate bulk description in terms of LQFT evolution and semiclassical GR. There has been significant discussion about the origin and nature of such a map. The most promising proposal seems to be that it is explained via essential properties of gravitational physics [85, 86], as opposed to be due to, e.g other effects in string theory. However, closer examination seems to show that the holographic map between "bulk" and "boundary" relies on solving the gravitational constraints, or their full quantum generalization, and in gravity this appears inseparable from having a description of the bulk quantum-gravitational evolution [51]. This suggests that in order to determine the holographic map, one needs to first understand the quantum-gravitational evolution in the bulk, whereas an original hope was that the boundary evolution plus holographic map determined the bulk evolution. This resulting circularity so far appears to preclude using the boundary theory to describe bulk evolution, and provide a resolution to the crisis that includes BH evolution that completes a bulk description. An open question is how to provide such a precise holographic map that would give a description of evolving black holes in terms of unitary boundary evolution.

18.3 Evolution and Localization in Local QFT and Perturbative Quantum General Relativity

This section turns to further description of LQFT evolution in BH spacetimes, and some of the effects of perturbative gravity. This will add detail to the preceding discussion, begin to address questions of locality in quantum gravity, and set the stage for parametrization of possible quantum gravity effects going beyond such a description.

18.3.1 LQFT Evolution

Many of the essential issues can be described in terms of evolution of a spinless scalar field on a non-spinning black hole; spin of either kind adds additional structure, but does not appear to change the basic nature of the problem. We therefore consider a scalar action

$$S_m = -\frac{1}{2}\int d^4x\sqrt{|g|}\left[(\nabla\phi)^2 + m^2\phi^2\right] \tag{18.6}$$

with a background Schwarzschild metric, initially ignoring gravitational background and fluctuations. It is useful to write the metric in ingoing Eddington-Finkelstein coordinates, (x^+, r), since those are regular across the horizon, as

$$ds^2 = -f(r)dx^{+2} + 2dx^+dr + r^2d\Omega^2, \tag{18.7}$$

with $f(r) = 1 - R/r$ and $R = 2GM$ the Schwarzschild radius. This geometry is pictured in Fig. 18.1 or equivalently by the Eddington-Finkelstein diagram of Fig. 18.2. Hawking's derivation [71] and many subsequent versions of it are based on comparing modes at future infinity to modes near the horizon or in the past, and deriving the resulting Bogoliubov transformations. Alternately, if one wants to describe an evolving quantum state like (18.1), one can introduce a time slicing and a Schrödinger picture description of the evolution; this also has the virtue of more readily extending

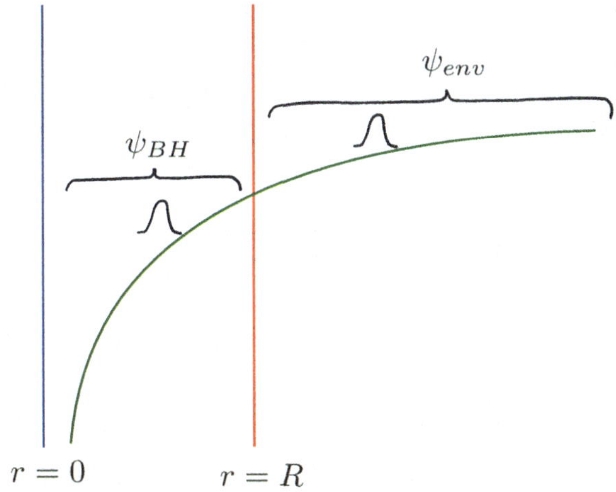

Fig. 18.2 Shown is an Eddington-Finkelstein diagram for a BH, corresponding to the Penrose diagram of Fig. 18.1. In such a diagram, evolution can be described by simply translating a spatial slice like that shown vertically up or down

to interacting theories. The choice of slicing is part of the choice of coordinates; for Schwarzschild spacetime one may choose a slice like those shown in Fig. 18.2, and then define the family of slices by translating it forward or backward using the Killing vector that translates in Schwarzschild time, with the asymptotic Schwarzschild time used to parametrize the slices.

A general slicing and metric can be described in the ADM form [8],

$$ds^2 = -N^2 dt^2 + q_{ij}(dx^i + N^i dt)(dx^j + N^j dt), \tag{18.8}$$

with lapse N, shift $N^i = q^{ij} N_j$, and spatial metric q_{ij}. If in the Schwarzschild metric (18.7) we choose a set of slices parametrized by a "slicing function" $S(r)$ as

$$x^+ = t + S(r), \tag{18.9}$$

then these ADM variables become [46]

$$N = \frac{1}{\sqrt{S'(2 - fS')}}, \quad N_r = 1 - fS', \quad q_{rr} = S'(2 - fS'). \tag{18.10}$$

The evolving state can be described by introducing the canonical variables and Hamiltonian. The canonical momentum is given in terms of the normal

$$n^\mu = \frac{1}{N}(1, -N^i) \tag{18.11}$$

to the slices as $\Pi = \sqrt{q}\partial_n \phi$, and the canonical commutators are

$$[\Pi(t, x^i), \phi(t, x^{i'})] = -i\delta^3(x - x'). \tag{18.12}$$

Comparing the Hamiltonian form of the action

$$S_m = \int d^4x \left(\Pi\dot{\phi} - \mathcal{H}\right) \tag{18.13}$$

to the matter action (18.6) gives the Hamiltonian

$$H = \int d^3x \mathcal{H} = \int d^3x \left[\frac{N}{2}\left(\frac{\Pi^2}{\sqrt{q}} + \sqrt{q}q^{ij}\partial_i\phi\partial_j\phi\right) + N^i\Pi\partial_i\phi\right] \tag{18.14}$$

and the resulting unitary evolution of the state on the slices,

$$|\Psi(t)\rangle = e^{-iH(t-t_0)}|\Psi(t_0)\rangle. \tag{18.15}$$

The structure (18.1) of the state is found by beginning with an appropriate fiducial "vacuum" state on the slice, and considering excitations of it involving combinations of the operators ϕ, Π localized either to the BH interior, or to the exterior.[6] For example, different such operators localized inside the BH create BH states with different labels i, as described in (18.3). Because of the locality property of the QFT, described above in terms of commuting subalgebras, different such operators commute with observables exterior to the BH on subsequent slices; the corresponding excitations propagate within the lightcone, and so these excitations don't affect the external part of the state.

In addition, the evolution (18.15) describes creation of correlated pairs of excitations inside and outside the BH, like the toy model evolution (18.2). This can be seen by studying the explicit evolution (18.15) written in terms of creation/annihilation operators, using the explicit form of the ADM variables (18.10), as was done in two dimensions in [52, 53] and for higher-dimensional BHs in [60]. One finds that the Hamiltonian (18.14) has a term bilinear in creation operators, which creates the correlated pairs. This leads to a refinement of the evolution law(18.3),

$$|\psi_{\mathrm{BH},i}, \psi_{\mathrm{env}}\rangle \to |\Psi(t)\rangle = \sum_\alpha |\psi'_{\mathrm{BH},i}, \psi'_{\mathrm{BH},\alpha}; \psi'_{\mathrm{env},\alpha}, \psi'_{\mathrm{env}}\rangle \tag{18.16}$$

in which the exterior is still independent of i, but has built up additional entanglement with other interior degrees of freedom, labelled by α, through subsequent evolution. The specific form of the Hamiltonian depends on the choice of slicing and coordinates on the slices, which correspond to a choice of gauge in the gravitational context.

This basic structure appears to extend to spinning BHs, with additional complications of detail. While there are additional subtleties in describing evolution for time-dependent geometries and/or slicings [75, 101, 102, 2, 59], it nonetheless appears that such unitary and local evolution can still be defined.

---

[6] A more careful statement involves using unitary operators that are localized in the respective regions.

18.3.2 *Perturbative Quantized General Relativity (QGR)*

A critical question is what gravitational corrections modify the preceding story, and how they do so.

Gravitational backreaction is an essential part of the discussion, since it describes the shrinkage of the BH. A first approach to this is by studying the average backreaction as a correction to the semiclassical metric. This has in particular been done in two-dimensional models, and specifically in dilaton gravity in [16] and many subsequent references. These provide models which are also expected to capture basic features of higher-dimensional evolution, since for large BHs this evolution is dominantly spherically symmetric, and s-wave dynamics can be described in such a two-dimensional theory. Specifically one does also explicitly find buildup of a large entanglement/missing information in such models describing a BH shrinking towards zero mass, as was shown in [58], and so at this level backreaction appears not to greatly alter the discussion of the preceding section.

The situation is more subtle at the quantum level. To see this, we can investigate the leading quantum corrections to the classical or semiclassical evolution perturbatively in Newton's constant G. The gauge symmetry of gravity now plays an important role, and in particular the basic field operators $\phi(x)$ are no longer observables, since they are not gauge invariant. When one replaces them by gauge-invariant observables e.g. by including *gravitational dressing*, these are no longer local, and the preceding formulation of locality and subsystems in terms of commuting sublagebras of observables apparently must be modified.

There isn't space in this review for a complete discussion of gravitational observables, dressing, and subsystems, and significant open questions remain. But, it is important to take some time to summarize some of the basic features of such a discussion, with further details provided in [23–26, 50, 57, 61], since this drives at the very important question of understanding localization in quantum gravity, and connecting with the discussion of Sect. 18.2.

18.3.2.1 Perturbative Gravitational Dressing

One can formally describe the evolving state by extending the Hamiltonian framework to include gravity. A starting point is the Lagrangian

$$S = \frac{1}{16\pi G}\int d^4x\sqrt{|g|}\mathscr{R} + S_\partial + S_m, \tag{18.17}$$

where $\mathscr{R}$ is the Ricci scalar, and S_∂ is a boundary term.[7] Canonical variables are defined using the ADM decomposition (18.8). Rewriting the action (18.17) in these variables, the canonical momentum becomes

[7] For further details of the following treatment, see [61].

$$P^{ij} = \frac{\delta S}{\delta \dot{q}_{ij}} = -\frac{1}{16\pi G}\sqrt{q}\left(K^{ij} - q^{ij}K\right) \tag{18.18}$$

where dot denotes $\partial/\partial t$ and K_{ij} is the extrinsic curvature of the slices,

$$K_{ij} = \frac{1}{2N}\left(-\dot{q}_{ij} + D_i N_j + D_j N_i\right), \tag{18.19}$$

with D_i the covariant derivative defined from q_{ij}. This is taken to satisfy commutation relations

$$[P^{ij}(t,x), q_{kl}(t,x')] = -\frac{i}{2}\left(\delta^i_k\delta^j_l + \delta^i_l\delta^j_k\right)\delta^3(x-x'). \tag{18.20}$$

The momenta conjugate to N, N^i vanish, in association with the presence of gauge constraints.

Comparing the Hamiltonian form of the action

$$S = \int d^4x\left(P^{ij}\dot{q}_{ij} + \Pi\dot{\phi} - \mathcal{H}\right) \tag{18.21}$$

to the action (18.17) yields the Hamiltonian in two different equivalent forms,

$$H = \int d^3x\left(N\mathcal{C}_n + N^i\mathcal{C}_i\right) + H_\partial = \int d^3x\left(\mathcal{H}_g + \mathcal{H}_m\right). \tag{18.22}$$

In the first expression, we define

$$\mathcal{C}_\mu = \sqrt{q}\left(-\frac{G_{\mu\nu}}{8\pi G} + T_{\mu\nu}\right)n^\nu, \tag{18.23}$$

and have used the index convention $\mathcal{C}_n = n^\mu\mathcal{C}_\mu$, with Einstein tensor $G_{\mu\nu}$ and stress tensor $T_{\mu\nu}$. Vanishing of the $\mathcal{C}_\mu$ expresses the constraints associated with the gauge symmetry; correspondingly, N, N^i appear as Lagrange multipliers in the expression. H_∂ is a boundary term at spatial infinity; e.g. in the case of Minkowski asymptotics it reduces to the ADM boundary Hamiltonian. Since the constraints have second derivatives of the metric variables, the latter term is needed so that after integration by parts one has a Hamiltonian of the standard form with just quadratic terms in momenta and the spatial derivatives of the fields, together with interaction terms. The second expression with $\mathcal{H} = \mathcal{H}_g + \mathcal{H}_m$ summarizes that form, with $\mathcal{H}_m$ the matter Hamiltonian, and $\mathcal{H}_g$ an effective gravitational Hamiltonian.

When the constraints $\mathcal{C}_\mu$ vanish, the Hamiltonian thus reduces to a pure boundary term. Since the $\mathcal{C}_\mu$ generate the gauge symmetries (diffeomorphisms) on shell,[8] their

[8] For an explicit discussion, see [61].

vanishing is an expression of gauge invariance. Formally the evolution of the state $|\Psi\rangle$ is then determined by the Schrödinger equation

$$i\frac{\partial}{\partial t}|\Psi\rangle = H|\Psi\rangle \tag{18.24}$$

and is nontrivial if we for example fix the time parameter t at infinity, like we have done in (18.9), (18.10). If the second expression in (18.22) is used, (18.24) takes a standard and unsurprising form, with both the matter and gravity Hamiltonians contributing to evolution. But for a state annihilated by the constraints, this evolution could alternately be expressed in terms of the action of the boundary Hamiltonian H_∂.[9]

For operators, the condition of gauge invariance is expressed as[10]

$$[\mathscr{C}_\mu, \mathscr{O}] = 0 \tag{18.25}$$

Analogous to the discussion in 18.3.1, we can then consider creating different states by acting with different such operators $\mathscr{O}_i$ on some fiducial physical state $|\Psi_0\rangle$. However, in gravity gauge invariant operators cannot be local [100]. Correspondingly, solving (18.25) will yield *nonlocal* operators, and these will generically not satisfy the local algebra of LQFT; locality in gravity must apparently be defined some other way [43, 47, 48]

The preceding discussion is somewhat formal, and there are significant challenges in solving the constraints and determining the evolution in the full nonlinear theory. However, one can begin to understand the corresponding structures by working perturbatively in G, and in particular considering the leading perturbative contributions. These are expected to dominate the behavior for weak gravitational fields, and for example should therefore capture the leading long-distance behavior, since the field falls off at long distances. This leading behavior already nontrivially illustrates some of this novel gravitational structure, as well as its modification to the discussion of the preceding sections.

For example, one may begin with an operator $\mathscr{O}$ that is a purely matter operator, and add gravitational contributions—that is, "gravitationally dress" it—to make an operator satisfying (18.25). Such gravitationally dressed operators have been constructed in [23–26, 47, 57],[11] and recently in nontrivial backgrounds such as BHs in [61]. To construct these we perturb about a background metric $g_{\mu\nu}$,

$$\tilde{g}_{\mu\nu} = g_{\mu\nu} + \kappa h_{\mu\nu}, \tag{18.26}$$

[9] The actual story is a little more complicated since typically one only wants "half" of the constraints to annihilate the state; then the remainder are expected to generate contributions that decouple. Some further discussion of this appears in [26, 61].

[10] Again, modulo the equations of motion.

[11] Earlier related work includes [74, 78]. The first derived related nontrivial commutators as arising from the constraints, but didn't give the dressed operators, and the second focused on deriving commuting bulk operators.

with $\kappa^2 = 32\pi G$, and define momenta perturbations, over background momenta P^{ij},

$$\tilde{P}^{ij} = P^{ij} + p^{ij}/\kappa \tag{18.27}$$

so that h_{ij}, p^{ij} also satisfy the commutators (18.20). To leading nontrivial order in κ, we can take a matter operator $\mathcal{O}(t, x)$ or a more general nonlocal operator $\mathcal{O}(t)$, and then its gravitationally dressed version is

$$\hat{\mathcal{O}}_t = e^{i\int d^3x\sqrt{q}V^\mu(t,x)T_{n\mu}}\mathcal{O}(t)e^{-i\int d^3x\sqrt{q}V^\mu(t,x)T_{n\mu}}, \tag{18.28}$$

where the gravitational dressing V_μ is at this order linear in h_{ij} and p^{ij}. Specific such expressions are given in the preceding references, and take the general form [61]

$$V^n(x) = \frac{\kappa}{2}\int d^3x'\left[\breve{h}_{ij}(x, x')p^{ij}(x') - \breve{p}^{ij}(x', x)h_{ij}(x')\right], \tag{18.29}$$

$$V^i(x) = \kappa\int d^3x'\left[G^{ijk}(x', x)h_{jk}(x') + H^i_{jk}(x', x)p^{jk}(x')\right] \tag{18.30}$$

where $\breve{h}_{ij}$, p^{ij}, G^{ijk}, H^i_{jk} are c-number Green functions satisfying a corresponding set of equations. For a given matter configuration, there are different dressings corresponding to different solutions of the Green function equations; a simple example is to dress a localized mass with either a Coulomb-like or a line-like gravitational dressing [23].

Since, with the example of Minkowski or AdS asymptotics, the support of the dressing generically extends to infinity [24, 57], the operators (18.28) based at different spacelike separated points generically don't commute [43, 23], concretely illustrating the preceding general comments about modification of the LQFT algebra.

18.3.2.2 Gravitational Splitting and Localization of Information

While this raises the question of how in general to define locality and subsystems of gravitational systems, there is a perturbative construction that is close to recovering the discussion of the preceding sections. This nonetheless leaves important open questions about localization of information in the full theory of quantum gravity. The question is most easily illustrated by perturbing about a flat metric, but the more general description of dressing outlined above also serves as the starting point for extending the discussion to nontrivial background metrics such as BHs. This construction was called a "gravitational splitting" in [25, 26, 50].

The starting point is the essential question in quantum gravity of whether information can, to a good approximation, be described as localized to a region—for example interior to a BH—like it is in LQFT, as described above. Or, do gravitational effects lead to important modifications to this, or, serve to transfer the information out of the region?

We will investigate this question in perturbative QGR.[12] A useful toy model to do so arises from considering two degenerate scalar fields, $\phi_a = (\phi_1, \phi_2)$, governed by a Lagrangian which is two equal-mass copies of (18.6), which has a global symmetry $\phi_a \to O_{ab}\phi_b$ relating the fields. Of course there is significant doubt about the existence of such global symmetries in a full quantum theory of gravity, but it so far appears consistent to consider the perturbative coupling of such a theory to gravity. Such a toy model then furnishes an example of what are plausibly key features of information localization in a more complete description of the states in a more complete theory.

We begin with a fiducial state $|\psi_0\rangle$—which in the Minkowski case can just be the Minkowski vacuum $|0\rangle$—and consider excitations of it that describe information localized in a region U, such as a small open set. In QFT we take these to be created by unitary operators U_I acting on $|\psi_0\rangle$, $|\psi_I\rangle = U_I|\psi_0\rangle$. For concreteness consider examples such as

$$U_I = e^{-i\int d^3x J_a^I(x)\phi_a(x)}, \tag{18.31}$$

where the functions J_a^I have compact support restricted to U; or one could consider more general localized unitary operators. These states have the property that for any local operator $\mathcal{O}(x')$, with x' spacelike separated from U,

$$\langle\psi_I|\mathcal{O}(x')|\psi_I\rangle = \langle\psi_0|U_I^\dagger\mathcal{O}(x')U_I|\psi_0\rangle = \langle\psi_0|\mathcal{O}(x')|\psi_0\rangle : \tag{18.32}$$

measurements of local observables spacelike to U are insensitive to the state created in U. A concrete example is given by the functions $J_b^a(x) = \delta_b^a J(x)$, with a common $J(x)$ of compact support—measurements outside U do not distinguish whether one acted using J^1 or J^2, that is which "color" of particles are present in the state.

This discussion must be modified in quantum gravity, since the operators U_I are not gauge invariant. We can make them gauge invariant, at least to leading order in κ, using a dressing $V_\mu(x)$ as described above,

$$\hat{U}_I = e^{i\int d^3x V^\mu(x)T_{n\mu}}U_I e^{-i\int d^3x V^\mu(x)T_{n\mu}}. \tag{18.33}$$

As in (18.29), (18.30), the dressing extends outside U, potentially affecting measurements there.

However, as noted above there are different possible dressings, and we may use this freedom to find gravitational states minimizing the information "seen" from outside U. This can be done by constructing [25, 26, 50] a "standard dressing" associated to the neighborhood U itself. Given such a dressing V_S^μ, we may write

$$V^\mu(x) = V^\mu(x, y) + V_S^\mu(y) + \frac{1}{2}(x-y)_\nu\left[\partial^\nu V_S^\mu(y) - \partial^\mu V_S^\nu(y)\right]. \tag{18.34}$$

[12] For further discussion see [54].

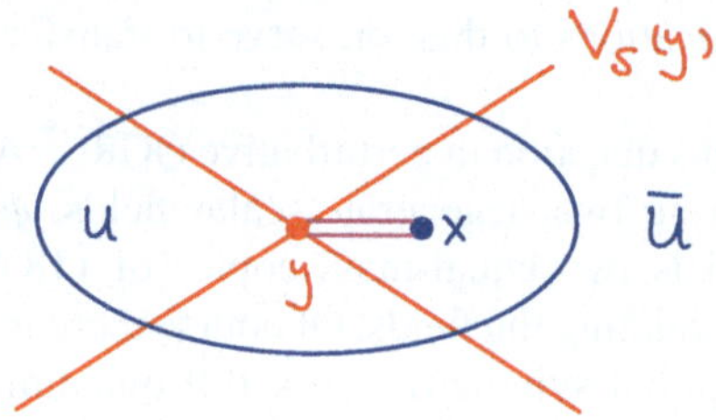

Fig. 18.3 Illustration of the standard dressing construction. A standard dressing $V_S(y)$ is chosen that only depends on the neighborhood U and e.g. a given point within it. One may choose different forms for this dressing, e.g. line-like or Coulomb-like. Dressing of operators within U is then constructed by dressing from a general point $x \in U$ to y, as illustrated; an example construction is via a line-like dressing. The result is the dressing expression (18.34)

Here y is a specified point of U, and the dressing $V^\mu(x, y)$ has the property that it has no support outside a bounded set containing both x and U—for $x \in U$, it may be taken to vanish outside U. Colloquially, we say that we dress from the point x to y—which for example may be accomplished via a line dressing like in [23, 47]—and then the dressing of the point y extends to infinity; see Fig. 18.3.

Given such a dressing, consider measuring an operator $\mathcal{O}'$ in the now quantum-gravitational state[13] $|\hat{\psi}_I\rangle = \hat{U}_I|\psi_0\rangle$. We find

$$\langle\hat{\psi}_I|\mathcal{O}'|\hat{\psi}_I\rangle = \langle\psi_0|\hat{U}_I^\dagger\mathcal{O}'\hat{U}_I|\psi_0\rangle = \langle\psi_0|\mathcal{O}'|\psi_0\rangle + \langle\psi_0|\hat{U}_I^\dagger[\mathcal{O}', \hat{U}_I]|\psi_0\rangle, \quad (18.35)$$

where, for $\mathcal{O}'$ localized in the spacelike complement to U, the leading order in κ contribution to the difference term may be evaluated by a straightforward commutator calculation, using (18.33), to yield

$$\delta\langle\hat{\psi}_I|\mathcal{O}'|\hat{\psi}_I\rangle = \langle\psi_0|\hat{U}_I^\dagger[\mathcal{O}', \hat{U}_I]|\psi_0\rangle \simeq i\int d^3x\langle\psi_0|U_I^\dagger\,[\mathcal{O}', V^\mu(x)]\,[T_{0\mu}(x), U_I]\,|\psi_0\rangle). \quad (18.36)$$

For example, consider the case where $\mathcal{O}'$ is purely constructed from the gravitational perturbation. Then the latter expression factorizes,

$$\langle\psi_0|U_I^\dagger\,[\mathcal{O}', V^\mu(x)]\,[T_{0\mu}(x), U_I]\,|\psi_0\rangle = \langle 0|\,[\mathcal{O}', V^\mu(x)]\,|0\rangle\langle\psi_0|U_I^\dagger\,[T_{0\mu}(x), U_I]\,|\psi_0\rangle\,; \quad (18.37)$$

the second factor has support only for $x \in U$, but then the first factor receives no contribution from $V^\mu(x, y)$ in (18.34) and so only has x-dependence from the standard dressing parts of (18.34). As a result, the integral over x just yields Poincaré generators, and

[13] The state $|\psi_0\rangle$ may also need dressing, which we do not make explicit. Also, more generally, $\mathcal{O}'$ may be dressed, e.g. to infinity.

$$\delta\langle\hat{\psi}_I|\mathcal{O}'|\hat{\psi}_I\rangle \simeq -i\langle\psi_0|U_I^\dagger\left[P_\mu, U_I\right]|\psi_0\rangle\,\langle 0|\left[\mathcal{O}', V_S^\mu(y)\right]|0\rangle \\ -\frac{i}{2}\langle\psi_0|U_I^\dagger\left[M_{\mu\nu}, U_I\right]|\psi_0\rangle\,\langle 0|\left[\mathcal{O}', \partial^\mu V_S^\nu(y)\right]|0\rangle. \quad (18.38)$$

This expression depends on the state created by U_I only through the wavefunction for its total Poincaré charges. For example in our model with global symmetry, this tells us that measurement of gravitational operators outside U do not distinguish the color of the state created by U_I; more generally, states with equal Poincaré moments will yield the same measurements of $\mathcal{O}'$.[14] This basic construction was called a "gravitational splitting" in [25, 26, 50]. One can understand this as a perturbative quantum generalization of certain classical constructions, where the classical gravitational field outside a given region may be taken to be of a specified form, e.g. Kerr or a cone-like distribution [17, 21], depending only on the momentum and angular momentum within the region.

An interesting modification of this story arises if $\mathcal{O}'$ is a mixed matter/gravitational operator; consider the simple example of a product

$$\mathcal{O}' = \mathcal{O}'_h\mathcal{O}'_\phi. \quad (18.39)$$

In this case, (18.36) becomes

$$\delta\langle\hat{\psi}_I|\mathcal{O}'|\hat{\psi}_I\rangle \simeq i\int d^3x\langle 0|\left[\mathcal{O}'_h, V^\mu(x)\right]|0\rangle\langle\psi_0|U_I^\dagger\mathcal{O}'_\phi\left[T_{0\mu}(x), U_I\right]|\psi_0\rangle. \quad (18.40)$$

Again the second factor is only nonvanishing for $x \in U$, and so only the standard dressing terms in the first commutator contribute. The integral over x then again yields Poincaré generators, with the result

$$\delta\langle\hat{\psi}_I|\mathcal{O}'|\hat{\psi}_I\rangle = -i\langle\psi_0|U_I^\dagger\mathcal{O}'_\phi\left[P_\mu, U_I\right]|\psi_0\rangle\,\langle 0|\left[\mathcal{O}', V_S^\mu(y)\right]|0\rangle \\ -\frac{i}{2}\langle\psi_0|U_I^\dagger\mathcal{O}'_\phi\left[M_{\mu\nu}, U_I\right]|\psi_0\rangle\,\langle 0|\left[\mathcal{O}', \partial^\mu V_S^\nu(y)\right]|0\rangle. \quad (18.41)$$

Now, the measurement *can* depend on details of the state. For example in the toy model, if one takes $\mathcal{O}'_\phi = \phi_1(x')$, the first correlator depends on which color particles J_a^I creates [54]. However, the non-vanishing result arises from a correlator $\langle\phi_a(x)\phi_b(x')\rangle$, and so is exponentially suppressed $\sim\exp\{-m|x-x'|\}$ for x' spacelike to U.

The presence of such nonvanishing correlators is certainly interesting, but due to their smallness, what is not clear is their relevance for significantly altering the story of localization and transfer of information.[15] These small corrections have only been

[14] In our model we achieve such an exact degeneracy with the global symmetry. While there are indications that a complete theory of gravity may not have such global symmetries, there may be degeneracies or very near degeneracies in Poincaré charges which are in-practice unresolvable via measurements outside U.

[15] This is despite arguments to the contrary [18].

calculated to leading order in κ, and so there are additional corrections to account for at higher orders. Moreover, one can ask how the gravitational dressing modifies the evolution of the state in a perturbative description. While in the perturbative description one does find that the state (18.15) must be dressed, there is no obvious indication that this alters the internal degrees of freedom of the BH, or the pattern of entanglement between them and the exterior degrees of freedom, as described e.g. in (18.16). So, for practical purposes it so far seems that evolution of a BH as a subsystem like in the non-gravitational LQFT description is a good approximate description.

Interestingly, at the nonperturbative level related arguments formally extend to the leading proposal for the explanation of the property of "holography," as associated with solving the gravitational constraints. Such an explanation was first given by Marolf [85, 86], who argued that an operator, say $\phi(x)$, in the middle of AdS could be related to an operator in its future lightcone on the boundary of AdS, but then since the Hamiltonian can be thought of as a boundary term, the latter operator can be translated back in time in the boundary algebra to give a boundary operator at equal time to $\phi(x)$—constructing a "holographic map" between bulk and boundary operators. Crucially, this depends on solving the gravitational constraints in AdS [51], which amounts to constructing the dressing, since that is needed to express the Hamiltonian as a pure boundary term, as seen in (18.22).

Indeed, a related but simpler formal description of a possible holographic map was given in [25]. Suppose, now, that $|\hat{\psi}_I\rangle$ are fully-nonperturbative dressed versions of the states $\hat{U}_I|\psi_0\rangle$ that were described above. From a momentum analog of the formula (18.22), one expects that for states annihilated by the constraints, the translation generators are given by their boundary, ADM, versions.[16] This means that

$$e^{iP_i^{ADM}c^i}|\hat{\psi}_I\rangle \tag{18.42}$$

should translate the state by the amount c^i; e.g. in the case of U_I given by (18.31) this is equivalent to translating the support of $J(x)$. For large c^i we can translate to "near to the boundary," and so for example the expression

$$\langle\hat{\psi}_I|\phi_i(x')e^{iP_i^{ADM}c^i}|\hat{\psi}_I\rangle, \tag{18.43}$$

for x' near to the boundary, is sensitive to details of the state. By taking a limit of infinite c^i and x', this is the starting point for an argument that bulk states can be detected—and created—by boundary operators.

However, this formal argument appears to require a full solution of the constraints, presumably with nonperturbative accuracy, since one must implement a large translation to construct the holographic map. Since solving the constraints is equivalent to finding the bulk evolution, this appears to indicate that in order to construct the holographic map one must first understand the bulk evolution—and in particular have

16 There are additional subtleties with this which we leave for other discussion.

a consistent description of black hole evolution [51]. The resulting apparent circularity in the argument thus leaves its status, particularly with somehow resolving the unitarity crisis, in question.

Returning to the question of BH evolution, the black hole theorem appears to tell us that there need to be effects that modify localization and/or transfer of information in the BH context. The upshot of the preceding discussion is that in a leading perturbative analysis, there is so far no clear indication that the relevant effects arise. One finds a description of the evolution that to leading order includes the buildup of entanglement of Hawking's picture, and at next order in G incorporates gravitational dressing of the resulting entangled states; there is not a clear indication for elimination or significant rearrangement of the underlying entanglement. This suggests that there must be effects outside such a perturbative description that modify the BH evolution.

Whether such effects arise from some nonperturbative completion of the dressing construction, or from some other fundamental picture that goes beyond quantized GR, we will now shift to a pragmatic view: we will seek to give a principled "parameterization of ignorance" of effects that can resolve the unitarity crisis. Ultimately, one would like to derive such effects as arising from a more fundamental description of quantum geometry, but for the time being we instead take such an "effective" approach, in the spirit of effective field theory descriptions that have been successful in many other contexts.

18.4 Nonviolent Unitarization: Motivation and Basic Approach

To briefly summarize the preceding motivations, we have outlined that there are significant difficulties altering quantum mechanics. In order to instead save it, specifically its principle of unitarity, while avoiding other disastrous behavior, we apparently need some modification to the QFT description of locality of information, and/or apparently more radical measures.

The latter typically involve considerable additional structure which is problematic from multiple viewpoints, and moreover black hole observations so far appear to match classical predictions for BH behavior. Since LQFT has served to give an excellent and well-tested description of all other non-gravitational phenomena, these together help motivate taking a conservative approach of seeking the "minimal" modifications to the LQFT description of BH evolution, as it was reviewed above, that are needed to restore unitary evolution. One facet of this is that we seek a consistent description of BHs that involves a minimal modification to the equivalence principle. These are key aspects of the general proposal of "nonviolent unitarization [36–39, 41, 46, 49, 64, 65]" (NVU).

A starting point for discussing NVU is therefore the LQFT evolution; it appears easiest to begin with a Hamiltonian description of it, as summarized above. We then explore the question of what corrections to this picture, that are in a certain sense "small," are needed to restore unitarity to the complete BH evolution. As also summarized above, the LQFT evolution produces increasing entanglement between internal states of the BH and the outgoing radiation. If the more complete gravitational description permits, at least to the necessary approximation, a description of the BH as a subsystem that contains information and becomes entangled with the environment as in the preceding leading order description, and if we accept postulates (0), ultimate unitarity and (3), no remnants, for the reasons described above, then the black hole theorem tells us that we need a violation of postulate (2): distinct BH states should *not* have identical exterior evolution, or, put differently, there should be some interaction between the BH states and the BH exterior. This must apparently be outside the realm of LQFT, but we will seek to parametrize it as a departure from LQFT.

In doing so, we could begin with the leading-order corrected version of the LQFT description, where states and observables have the leading-order gravitational dressing described above. However, since that has not been found to yield the needed modifications to the original uncorrected LQFT description, and produces extra complications, we will instead work with a description that neglects these effects, and focus on other corrections to the evolution that are needed.

Specifically, in such a description, we will take a "principled" approach of parameterizing our ignorance, seeking the minimal modifications to the underlying LQFT evolution that are needed to restore ultimate unitarity.

18.5 Parametrizing Nonviolent Unitarization

Given the preceding discussion, our effective description begins with the Hamiltonian formulation, e.g. with Hamiltonian (18.14) (or, more generally, (18.22) with gravitational corrections), which we now refer to as H_0; we will parameterize corrections to this that can restore ultimate unitarity. Of course this Hamiltonian depends on a choice of slicing, or gauge, but in taking this effective approach we will explore the possible interactions in such a specific gauge, and defer the question of gauge invariance to a later time when we have a more complete understanding of the fundamental underlying dynamics.

We also expect the states to be modified from their LQFT description. However, plausibly the modification to their structure is minimal in the vicinity of the BH horizon, and they are for practical purposes well-parametrized there by the LQFT description, in alignment with our approach of minimality. We do anticipate, though, that there could be quite significant modification to the description of the degrees of freedom (DOF) deep inside the BH, and also that the description there depends significantly on the choice of gauge which extends the classical slice dependence. Concretely, we could parametrize the state at a given time t as (see also Fig. 18.4)

$$|M; K; \psi, t\rangle. \tag{18.44}$$

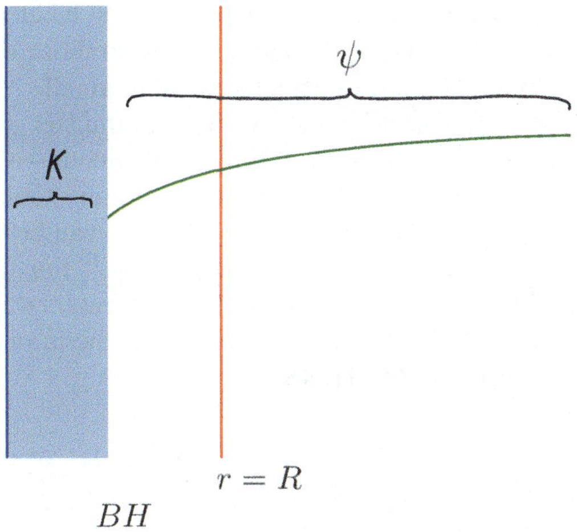

Fig. 18.4 Schematic of the state as parameterized in (18.44). At sufficiently large radius, including a region interior to the horizon, we expect that to a good approximation the LQFT description furnishes a good description of the degrees of freedom. Degrees of freedom deeper in the BH are expected to receive larger corrections to their description, and will be parameterized by a state label K

Here M is the mass of the BH; in practice we may want to describe BH states in a mass range $(M + \Delta M, M - \Delta M)$, with say $\Delta M \sim 1/R$, with R the horizon radius (spin could also be included, but will be neglected for simplicity). ψ gives the description of the state of the environment, outside the horizon, as well as the near-horizon interior,[17] say inward to some modest fraction of R, and K are labels of DOF in the deeper interior of the BH, where significant deviation from LQFT is expected. Quite possibly the internal states are finite in number, and so for the mass range $(M + \Delta M, M - \Delta M)$, we have $K = 1, \ldots, N$, where $N = \exp\{S_{bh}\}$ for a BH entropy S_{bh} that may, for example, be comparable to the Bekenstein-Hawking value $S_{BH} = A/4G$.[18]

We then parametrize the full (effective) Hamiltonian as

$$H = H_0 + \Delta H, \tag{18.45}$$

where again H_0 is given by the LQFT expression (18.14), or in a more complete treatment, its perturbative QGR completion (18.22). Once again, the LQFT expression for H_0 respects postulate (2), identical exterior evolution, and so to violate this, and restore ultimate unitarity, ΔH must incorporate interactions between the BH internal DOF and those of its environment. ΔH may also include significant corrections to the evolution of the internal DOF—say as they reach near the would-be singularity. We will separate these two contributions as

$$\Delta H = \Delta H_I + \Delta H_{bh}, \tag{18.46}$$

---

[17] To describe localized states, one may for example work in a wavepacket basis.

[18] Being more careful, one should include internal excitations from ψ in the count of internal BH states.

respectively. So, ΔH_{bh} acts on the DOF labeled by K in (18.44), and may be large, and ΔH_I couples the internal DOF to those of the BH exterior; for minimality, it should in a sense to be determined be small.

The simplest interaction that can transfer information or entanglement between two subsystems—here BH to environment—is a product of operators acting on the respective DOF. A basis of such operators acting on the states labeled by K is given by the $U(N)$ generators λ^A, and these can be coupled to different operators acting on the part of the state described by ψ.[19] For simplicity and minimality we take the latter to be superpositions of local operators $\mathcal{O}^b(x)$, although more general nonlocal operators could be considered. Then, the information-transferring interaction takes the form [38, 65, 41, 46]

$$\Delta H_I = \sum_{Ab} \int dV_x \lambda^A \mathcal{O}^b(x) G_{Ab}(x), \tag{18.47}$$

where the integral is over the timeslice of the environment and near-horizon interior, and dV_x is the volume element induced by the background metric. The functions $G_{Ab}(x)$ are parameters of our effective description, behaving somewhat like "structure functions" of the BH states, which we expect to ultimately be determined by a more complete fundamental theory but for now we regard as parametrizing our ignorance. We will aim to find constraints on these based on other, general, considerations.

Of course when described from the perspective of the semiclassical BH geometry, (18.47) is nonlocal; such a violation of the semiclassical notion of locality is again apparently *required*, as part of violating postulate (2). When described from the full quantum gravity perspective, the situation is less clear, since we currently lack a complete understanding of locality in that context; perhaps, in a sense, the interactions (18.47) are part of parameterizing the departure of quantum geometry and dynamics from its semiclassical approximation. Of course departures from locality are also *potentially* problematic. For example, in Minkowski space, a nonlocal interaction can be converted into one that acts backward in time by a boost, and then such interactions can be combined to send signals into the backward lightcone of an observer. This violates causality, and produces paradoxes such as the "grandmother paradox." However, in the present context, in which we work about the BH background, such apparently nonlocal interactions cannot necessarily be converted into interactions acting backward in time [35, 36], since the background does not have the full Lorentz symmetry and in effect chooses a frame. This suggests that such inconsistencies can be avoided.

[19] We allow for couplings between the BH deep-interior states and the environment as well as to the near-horizon interior, since we do not necessarily expect a sharp transition in behavior at the horizon. However, only the pieces coupling to the environment DOF are relevant for the information transfer to the environment that violates postulate (2).

18.5.1 Scales and Nonviolence

This discussion connects to the question of the spatial dependence of the $G_{Ab}(x)$—for example if they had support to large x, that could parametrize arbitrarily long-range nonlocality, which *is* expected to produce paradoxes. But if the new interactions are associated with the quantum structure of the BH, a natural assumption is that they extend a distance ΔR_a outside the horizon, where ΔR_a is not greatly larger than R. It has sometimes been presumed that ΔR_a is microscopically small, e.g. with a scale set by the Planck length. However, such a rapid variation of the G_{Ab} would then produce very violent behavior, for example as seen by infalling observers [30, 14, 5]; the latter reference coined the name "firewall" for the resulting object. This, for example, represents a large violation of the equivalence principle, since an infalling observer would experience large interactions in a region where small curvatures are expected classically. Such behavior may be avoided if ΔR_a instead scales with R, as

$$\Delta R_a \sim R^p \tag{18.48}$$

for $0 < p \leq 1$, that is, has its scale set by the natural scale in the problem, that of the BH radius [42, 45, 46]. Such a scaling has the virtue that the spatial variation implied by the falloff of G_{Ab} becomes weaker for larger BHs, in accord with the expectation that the near-horizon region of a very large BH is increasingly well-approximated by flat space. The simplest possibility is $p = 1$, though for example [13] have more recently argued for a new physics scale with $p = 1/2$ based on arguments involving "island" contributions to certain extensions of the integral over geometries (see further discussion in Sect. 18.8.1).

Within such a limited domain for the $G_{Ab}(x)$, they in principle could also vary more or less rapidly with distance. Again, if the scale of such variations were set to be microscopic, e.g. the Planck scale, then infalling observers would experience interactions with large momentum transfers, again yielding a large departure from the equivalence principle, and something that might be called a firewall.[20] We instead assume a variation scale [45, 46]

$$L = R^q \tag{18.49}$$

that also grows with R, with $0 < q \leq p$. The simplest possibility is again $q = 1$; notice also that if $p = q = 1$, so that both scales are set to $\sim R$, that greatly limits the number of possible functions $G_{Ab}(x)$.

These combined assumptions on p and q, which can be described in terms of minimal departure from the equivalence principle, align with our general approach of seeking minimal departures from the known physics of LQFT. This can be thought of as a principle of "nonviolence" of the interactions, for example for observers falling into large black holes. They can also be viewed [49] as part of a "correspondence

[20] Microscopic structure as with the fuzzball picture of BHs is likewise expected to generically yield hard interactions with large momentum transfers [42], and thus large departure from the equivalence principle.

principle" for gravity, that in weak gravity limits, the fundamental theory is well approximated by LQFT plus semiclassical gravity.

Another aspect of this principle of nonviolence is that the interactions in (18.47) should not significantly couple BH states with large energy differences, $\Delta E \gtrsim 1/R$; doing so would also result in a large departure from the spectrum predicted by Hawking [71]. In fact we have already implemented this in the above description, by assuming the couplings are to states in the band between $M \pm \Delta M$. A priori we can trivially extend the description to a wider range of states, with couplings to states with significantly larger or smaller energy than M, but we limit the range as part of the nonviolence assumption.

18.5.2 Universality

A next question in our principled parametrization of ignorance is that of which operators $\mathcal{O}^b$ are coupled by the interactions. Here there are several possible guides. First, the leading order Hawking process is approximately thermal, and fits well with the beautiful story of BH thermodynamics. One way to understand this is as arising from the universal coupling of gravity to all fields, through the stress tensor.[21] If, for example, the couplings (18.47) only involved certain fields, that presents problems for detailed balance, if one tries to bring a BH into equilibrium with a thermal bath. A related argument comes from the process of BH *mining* [104, 81, 29, 28]. Perhaps the simplest form of mining involves threading a BH with a cosmic string. Then, there are modes that propagate along the string, and the contribution of these modes speeds the evaporation process. If we had interactions like (18.47) that transferred information/entanglement to some modes, but not to the string modes, then the latter would lead to additional buildup of entanglement between the BH and the environment. This would return us to the situation where the BH reaches the Planck mass with a large amount of remaining entanglement with the environment, necessitating information loss (violation of unitarity) or long-lived remnants. This can be avoided by assuming a principle of universality, that the interactions (18.47) responsible for the entanglement transfer couple universally to *all* modes.

There is a question about how to implement such universality. Possibly, essentially universal behavior could be achieved by a collection of couplings such as (18.47) to different operators involving all possible fields, although it is not entirely clear how to do so in a fashion which yields universal behavior. But, we do have an operator that is universal—the stress tensor $T_{\mu\nu}$, which couples to all fields.[22] If we assume that the coupling goes through this universal channel, then (18.47) becomes [65, 41, 46]

---

[21] To include gravitons in the Hawking radiation, one may augment this with their pseudo stress tensor.

[22] Here, again, to include graviton interactions this can be extended to include a gravitational stress pseudo tensor.

$$\Delta H_I = \int dV_x \sum_A \lambda^A G_A^{\mu\nu}(x) T_{\mu\nu}(x) \tag{18.50}$$

where now the more limited set of functions $G_A^{\mu\nu}(x)$ parametrize our ignorance. This coupling has an alternate interpretation: it behaves like a metric perturbation

$$H^{\mu\nu}(x) = \sum_A \lambda^A G_A^{\mu\nu}(x) \tag{18.51}$$

which is an operator depending on the quantum state of the BH. In accord with the preceding discussion of correspondence and nonviolence, these perturbations extend over a range ΔR_a as in (18.48) from the semiclassical horizon, and have a spatial variation scale L as in (18.49). These scales limit the number of possible functions entering (18.51), particularly for $\Delta R_a \sim L \sim R$.

18.5.3 Interaction Strengths

Another important question is that of the size of the couplings, e.g. given by the typical size of the dimensionless $H^{\mu\nu}$. This is constrained by the requirement that ΔH leads to ultimate unitarity. The Hawking process builds up entanglement between the BH and environment at a characteristic rate

$$\frac{dS_{vN}}{dt} \sim \frac{1}{R}, \tag{18.52}$$

which yields a missing information $\sim S_{BH}$ by the time the BH evaporates to near the Planck mass. The interaction ΔH_I must "undo" this entanglement, and if it is to do so without a greatly different spectrum of emission, must therefore transfer entanglement from BH to environment at a rate

$$\frac{dI}{dt} \sim 1/R. \tag{18.53}$$

This raises an interesting general question: if we have two subsystems which are coupled through a bilinear interaction, like our ΔH_I of (18.47), (18.50), how fast does that transfer entanglement between the subsystems, given certain assumptions also about dynamics (such as thermalization or information transport) of the individual subsystems?

We first note that if $|\tilde{\psi}\rangle$ is a typical BH state, and if the typical size of $H_{\mu\nu}$ is given by

$$\langle\tilde{\psi}|H_{\mu\nu}(x)|\tilde{\psi}\rangle \sim \mathcal{O}(1) \tag{18.54}$$

then that seems to easily transfer information at the required rate (18.53). The reason is that this means the coupling is behaving like a classical $\mathscr{O}(1)$ metric fluctuation which is varying on spatial and temporal scales $\sim R$; this can easily produce quanta at rates $\sim 1/R$, which carry the outgoing information.

But, in the spirit of minimality, there is a question of the *necessary* size. Reference [46, 63] began a study of the general rate question, using e.g. toy models, but a simple heuristic is that if a coupling between subsystems can induce transitions at a given rate—an example would be an interaction causing an atom to decay and emit an outgoing photon—then that also gives an estimate of the rate at which the interaction can transfer information. While couplings of size (18.54) are sufficient to transfer information at the needed rate, it appears that this can also be done with weaker interactions; for this reason the size (18.54) will be referred to as corresponding to a *strong* version of nonviolent unitarization.

To estimate the interaction size of a minimal, *weak*, version of nonviolent unitarization, following the preceding heuristic, we need to estimate the rate at which ΔH_I induces transitions. A simple estimate of that rate comes from first-order perturbation theory, via Fermi's Golden Rule,

$$\frac{dI}{dt} \sim \frac{dP}{dt} = 2\pi\rho(E_f)|\Delta H_I|^2, \tag{18.55}$$

where $\rho(E_f)$ is the density of final states of the system, and $|\Delta H_I|$ is the typical size of a matrix element of ΔH_I. At this point we see that there is apparently a way to achieve the rate (18.53) with much smaller couplings: for generic couplings between states the density of final states includes the large factor $N \sim \exp\{S_{bh}\}$ of the number of BH states, and so very tiny couplings, of typical interaction size

$$|\Delta H_I| \sim \frac{1}{\sqrt{N}} = e^{-S_{bh}/2} \tag{18.56}$$

appear to suffice.[23]

18.5.4 Summary of Principles and Parametrization, and Questions

To summarize this section, adopting a conservative "minimalistic" approach, we have begun a principled parametrization of ignorance, investigating the simplest interactions that can undo the problematic entanglement or missing information generated by the Hawking process, and restore ultimate unitarity of the BH decay

[23] A similar numerology, but corresponding to a different proposed physical mechanism, was described by Mathur [89]. In Mathur's case, the small amplitude describes transitions to fuzzball states from a BH, and the large number of such fuzzball states leads to an analogous compensation.

process. In general, this involves couplings (18.47) between the BH and environment subsystems.

A correspondence principle to LQFT and semiclassical gravity implies that these couplings only extend a distance ΔR_a from the semiclassical horizon, over a region which we will refer to as a *quantum halo* of the BH, and moreover that their spatial and temporal variation is on scales that grow with R, as in (18.49). This notion of nonviolence can alternately be thought of as part of seeking a minimal violation of the equivalence principle.

The desire to at least approximately preserve BH thermodynamics, the universal nature of gravity, and Gedanken experiments involving mining all point to the hypothesis that the couplings are universal. A simple way to achieve this that is via a coupling to the stress tensor, of the form (18.50). Such couplings can be thought of as describing BH state-dependent metric perturbations.

The necessity of ultimate unitarity implies that the couplings must have a sufficient strength. While the strong scenario, with (18.54), suffices, apparently one can achieve the necessary information transfer rate with couplings that are in a sense much smaller, with typical matrix elements given by (18.56).

The resulting limitations are significantly constraining. The simplest interactions are parametrized by the finite number of functions $G_A^{\mu\nu}(x)$. The limited support of these functions, to the range ΔR_a from the horizon, combined with the limit L on their spatial variation scale, greatly limits the number of possible independent functions; for example, for $\Delta R_a \sim L \sim R$, this number is $\mathcal{O}(1)$ rather than very large. These serve as a starting point for a more explicit parameterization; see [62] for some initial work on this. Beyond giving such systematic parametrizations, and seeking further constraints on the couplings, other important questions remain.

The first is what underlying fundamental quantum gravity dynamics explains such couplings and determines their values. While comments were made on this above, and this will also be touched on in Sect. 18.8, we view this as a very important open question for future work, and stress that the parametrization given in this section is to be viewed as a parametrization of an effective description of the important effects of the more basic physics. Of course, ultimately such effective descriptions may well give very important clues regarding the structure of the more fundamental dynamics.

Another question is whether there are any inconsistencies in such an effective theory, either internal or with known aspects of physics, or if there are other compelling objections, for example as a result of general principles. We will briefly discuss some criticisms that have been raised in the next section.

Finally, corrections to the evolution that extend outside the would-be horizon imply that matter and fields near the BH can interact differently with it than with a classical BH. The revolutionary opening of observational windows on the near-horizon regime, both through very long baseline interferometry and through observation of gravitational waves, raises the very important question regarding whether such new interactions could in fact have observable effects. We will briefly discuss some work on this question in Sect. 18.7.

18.6 Responses to Some Criticisms

This section will briefly discuss some of the common prejudices, concerns, and objections regarding the basic ideas of nonviolent unitarization, and will provide some responses to them. Since it is an important question how to reconcile quantum BH evolution with general principles, it is in particular important to understand whether NVU encounters any essential conflict with known physics or general principles.

1. *Shouldn't any new physics be restricted to the immediate vicinity of the singularity?* This has been a longstanding belief in much of the quantum gravity community, but the nature of the essential conflict of the unitarity crisis appears to show this cannot be correct. For example, delaying new physics to the final stages of BH evaporation yields either information loss or microscopic remnants, whose problematic nature was reviewed above. After decades of discussion, we appear to have learned that the unitarity crisis is demanding new physics in the vicinity of the would-be BH horizon—and now most proposals on how to resolve the crisis involve some kind of new physics on these horizon scales.
2. *Shouldn't any new physics be restricted to within a Planckian distance of the would-be horizon?* While this has been a common prejudice, it is not clear that there is any principled argument in its favor. And, in fact, there are now multiple reasons to question this assumption. First, such an assumption—as discussed above—generally leads to violent behavior near the horizon [30, 14, 5], e.g. on infalling observers, aka "firewalls." Put differently, this assumption leads to a large violation of the equivalence principle. Secondly, a basic motivation for this assumption, that the Hawking radiation is somehow "produced" in a microscopic neighborhood of the horizon, appears to be a *basis dependent* statement without unambiguous foundation. Reexamination of this argument from multiple perspectives argues that instead the Hawking radiation is produced in a quantum "thermal atmosphere" of thickness $\sim R$ around the BH horizon [44, 52, 53, 60]. Third, restricting new physics or interactions to an $\mathcal{O}(l_{Pl})$ vicinity appears to require an unnatural fine tuning; there is a natural scale in the problem—given by the horizon size, or by that of the thermal atmosphere, so it seems most natural for new physics to arise on some longer scale set by that. Finally, there are now even arguments from other approaches [13] for the same kind of nonviolent scaling, with $\Delta R_a \sim R^{1/2}$ as in (18.48).
3. *New interactions, such as described in* (18.47), *clearly violate the Equivalence Principle.* The Equivalence Principle is a key principle underlying *classical* general relativity. However, without classical spacetime and classical geometry—which appear to be approximations—it is not clear how to formulate it. Indeed, one appears to see departures from it even from leading quantum effects, such as QFT effects related to Hawking radiation. Perhaps we should not be surprised if the complete quantum theory, which goes beyond classical geometry, involves some modification to the Equivalence Principle. What does seem surprising is an

extreme violation, such as would be associated with interactions with a microscopic scale ΔR_a, producing a firewall. The nonviolent scaling (18.48) yields a much more mild departure from the classical Equivalence Principle.

4. *If one considers BH mining Gedanken experiments* [104, 81, 29, 28], *the new interactions no longer lead to unitary evolution.* Indeed, [4] went so far as to say that unitarity would require "an implausible conspiracy" with the mining equipment. However, while further careful examination of mining arguments are merited, as was argued above, universal couplings to matter/fields appears to largely address the objection. Mining—e.g. using a cosmic string to increase the number of excitations in the Hawking radiation—increases the rate at which a BH shrinks, yielding a possible inconsistency. But, if the information-transferring couplings are universal, then the information transfer from the BH will also increase in a commensurate fashion.
5. *The interactions of NVU present an inconsistency with BH thermodynamics.* Again, this appears to be to a significant degree addressed if the couplings are universal. However, this question also merits further careful investigation. For example, the interactions of (18.47) are expected to increase the rate at which the BH emits radiation. However, since the BH is a localized object, if it thermally emitted the rate would be given by the Stefan-Boltzmann law. A greater emission rate may be interpreted as due to the change of the effective size of the emitting surface, rather than due to a change of the temperature-energy relationship. It is true that in the presence of such interactions the formula for BH entropy S_{bh} may not necessarily yield the exact Bekenstein-Hawking entropy S_{BH}; some discussion of this appears in [40]. However, the status of the Bekenstein-Hawking entropy as the true entropy of the BH is also something that is not well-understood from a microscopic perspective.
6. *The interactions* (18.47) *are acausal.* It is true that, in Minkowski space, nonlocal interactions can yield acausal signaling. The reason is that an interaction transferring information at spacelike separation transfers information backward in time (but still at spacelike separation) from the viewpoint of a boosted frame. Two such interactions can be combined to transfer information into the backward lightcone of an observer, yielding acausality. However, if the interactions (18.47) are associated to the structure of the quantum BH, and localized in its vicinity, it is not clear there is an analogous argument. Specifically, the BH classically specifies a frame, and the symmetries are different; if, for example, the information transfer is associated with a time delay in that frame, it is not clear how any observer experiences an inconsistency [35]. Of course, part of understanding the more fundamental description of the theory is understanding why such interactions are present in the vicinity of BHs but not elsewhere—but that is beyond the scope of our effective description.[24]
7. *The interactions* (18.47) *are slicing dependent.* In the absence of a more fundamental understanding of gauge invariance in full quantum gravity, the structure

[24] It might also be noted that such bilocal interactions have subsequently been increasingly investigated in the context of BHs in AdS/CFT, beginning with [84].

of the effective theory has been described after choosing a classical analog of a gauge, namely a slicing. The question of full gauge invariance remains one for a deeper understanding of the more complete theory.

8. *It is not clear that AdS/CFT predicts such interactions.* The question of *how* AdS/CFT solves the unitarity crisis has remained a central one since it was first proposed. A significant part of that problem is finding the "holographic map" which allows one to describe quantities in the "bulk" theory of gravity in AdS in terms of the boundary quantities of the CFT. Indeed, one of the most convincing proposals for explaining the holographic map [85, 86, 76, 77, 51], as noted above, arises from the assumption that the gravitational constraints have been solved. But, as pointed out in [51], that would mean that in order to construct the holographic map, we must first have a description of evolution in the bulk—an unsatisfying circularity. That in particular seems to mean we would have to understand BH evolution in order to construct the holographic map, rather than determining BH evolution from the boundary theory plus a holographic map constructed by other means. Since AdS/CFT is the leading approach to addressing the crisis from string theory, it is also not clear that there is anything else in string theory that will resolve the problem.

18.7 Possible Observational Signatures

Our discussion of the "black hole theorem" suggests that to avoid inconsistency we must have some interactions that make the evolution of the BH exterior dependent on the BH quantum state—these interactions extend outside the semiclassical horizon. Indeed, while there is not yet consensus in the community, many of the current proposals for a resolution of the unitarity crisis share this feature, that there are new interactions or structure outside what would be the BH horizon. Since we are now able to observe the vicinity of the BH horizon through very long baseline interferometry [3] and through gravitational wave detection [1], that suggests the possibility of *observational* effects of any such new physics.

18.7.1 Gravitational Wave Signatures

For example, new interactions or structure outside the horizon would be expected to modify the scattering of gravitational waves (GWs) from that of a classical BH. When two BHs inspiral into each other, a component of the GW signal arises from the scattering from the individual BHs; one can think of one BH as emitting GWs as it orbits, and some of those waves scatter from the other BH and contribute to the outgoing GW signal.

One can therefore adopt, as an intermediate parameterization, the amplitudes for scattering of GWs from a BH, and investigate the effect of their deviation from classical expectations. That approach has been recently taken in [27], building on earlier work by [22, 82] and others. The effect on GW observations is most easily studied, as in [27], in the case of an extremal mass ratio inspiral (EMRI), which is well-approximated as a one-body problem. One can infer the contribution of the scattering amplitude modification $\Delta\mathscr{T}$ to the outgoing signal by Green function methods. Of course, for small $\Delta\mathscr{T}$ the modification is small, and may be challenging to detect directly in the observed GW signal. But, there is also an amplifier: $\Delta\mathscr{T}$ can change the *rate* of energy emission from the inspiral, and if the latter has many orbits, then one can build up a significant net phase shift, which is potentially more easily observed. For example, initial estimates [27], extending [22], suggest possible sensitivity to a benchmark scenario for near horizon interactions, parameterized in terms of a near-horizon reflection coefficient, with sensitivity to such coefficients of size down to $\mathscr{R} \sim 10^{-6}$.

It is generally expected that more extreme scenarios, such as massive remnants (gravastars, fuzzballs, firewalls, frozen stars, ...) would have substantial $\Delta\mathscr{T}$, perhaps making them easier to rule out, but our focus is on NVU. In particular, the weak scenario for NVU is most conservative, and seems most difficult to probe. But, simple estimates suggest the possibility of doing so, despite the "smallness" of the interactions. To see this, consider scattering a graviton from a quantum BH which has an effective description including the interactions ΔH_I of (18.50), including a gravitational pseudostress tensor. The probability for the interaction to cause a transition in the graviton state can also be estimated by Fermi's Golden Rule, (18.55), with a rate

$$\frac{dP}{dt} = 2\pi\rho(E_f)|\langle\beta|\Delta H_I|\alpha\rangle|^2 \tag{18.57}$$

where now the matrix element is between an initial graviton state together with the initial BH state, together denoted $|\alpha\rangle$, and a final state with a scattered graviton and another BH state, $|\beta\rangle$. In the weak scenario where typical couplings are exponentially small as in (18.56), one can still have a significant (e.g. $\mathcal{O}(1/R)$) transition rate as a result of the exponentially-large density of final states. This suggests that the interactions can yield $\Delta\mathscr{T} = \mathcal{O}(1)$ at the quantum level.

Scattering of this size seems well within the possible sensitivity described above. One additional caveat is that the classical $\Delta\mathscr{T}$ arises from a large number of gravitons scattering from a quantum BH, and this must be related to the quantum $\Delta\mathscr{T}$ whose calculation we have just outlined. A more precise connection between these scattering amplitudes is a subject for future work.

Of course, the strong scenario of NVU is expected to directly yield a significant $\Delta\mathscr{T}$, and thus to plausibly be well within sensitivity of overall phase measurements. One can also consider other effects, including the direct contribution to the outgoing GW signal [27]. Most recently, [96] have considered additional stochastic contributions to GW signals that could originate from this strong scenario, and that possibly fall within range of measurement through future terrestrial-based GW observations.

18.7.2 Electromagnetic Signatures

Very long baseline interferometry (VLBI) has allowed us to observe light from the near-horizon region of supermassive BHs, as was first spectacularly demonstrated by the Event Horizon Telescope (EHT) [3]. As with the preceding discussion of scattering of GWs, we can investigate the effect of, e.g., the universal interactions of (18.50) on light propagation near the horizon. Once again, at the level of a single photon, the interactions yield estimated scattering probabilities given by the Fermi's Golden Rule formula (18.57), now with photon states replacing graviton states. Again, the compensation of large and small exponentials apparently can produce an $\mathcal{O}(1)$ probability. However, now there is an important difference—for nonviolent interactions, the scaling of Sect. 18.5.1 tells us that the $G_A^{\mu\nu}$ of (18.50) have typical variation on distance scales $\sim R$, in the simplest case, and therefore that the $\mathcal{O}(1)$ transition amplitudes are between states whose momentum differs by $\sim 1/R$. Since the radiation observed by EHT is in the mm to sub-mm range, and we are observing supermassive BHs with radii $\sim 10^7 - 10^{10}$ km, even such $\mathcal{O}(1)$ scattering at first sight appears to have a negligible effect on the signal.[25]

In the event a strong NVU scenario is the correct description, there can be much more significant effects on observation. The interactions of strong NVU mimic significant classical fluctuations in the metric, on spatial and temporal scales $\mathcal{O}(R)$. A preliminary investigation of the effect of such fluctuations on images seen through VLBI was performed in [62]. The effective fluctuations can produce quite dramatic time dependence of the VLBI images, and even significant distortions of individual snapshots. These are significant enough that it appears that with further analysis present observations can begin to rule out a range of parameters of these scenarios, even in the absence of continuous observation over the relevant timescales, which are of order months for M87. Needless to say, observation of any image distortion fitting these scenarios would be very important to explore.

18.8 Implications for Quantum Gravity Foundations

Our discussion of the "black hole theorem" in Sect. 18.2.1, and of the treatment of LQFT and perturbative gravity in Sect. 18.3, has attempted to lay out a succinct summary for why the crisis is such a crisis, calling for a radical departure from the known physics of LQFT and semiclassical and perturbative GR. The possible resolutions are highly constrained by general principles, and most naïve proposals either seem to run into inconsistencies—either internally or with such general principles—or to need a large amount of additional structure, raising additional questions and

[25] The same kind of argument can be made for any other short wavelength matter or radiation scattering from the BH, indicating tiny momentum transfers and thus very small effect, and also extended to scattering of matter which enters the BH—specifically to infalling observers.

problems. In short, the unitarity crisis appears to imply a failure of LQFT and semi-classical/perturbative GR to describe evolution of a BH in our quantum universe, and is strongly indicating the need for some modification of the principles underlying the foundations of known physics. The need to resolve the crisis thus seems like it will play a key role in the foundations of quantum gravity, and the nature of the problem is plausibly providing important clues towards those foundations.

One important message seems to be that quantum gravity modifications to our known physics are not just restricted to Planck scales, as has been widely assumed in much of the discussion of quantum gravity not aimed at understanding quantum BHs. A systematic outline of consequences of possible modifications that are restricted to Planck scales appears to result in the unacceptable and unphysical alternatives described in Sect. 18.2. Resolution of the crisis appears to demand new physics on scale sizes of order the size of whatever BH is being studied, which can in principle be enormously large. This is evidenced not only by general arguments, but also by the near-convergence of attempted resolutions on *some* new physics on these scales.

Needless to say, important modifications to such long-scale physics would be both radical, and are expected to be highly constrained by our body of observed physics. It is difficult to provide scenarios without significant additional structure—e.g. associated with massive remnant scenarios such as fuzzballs, gravastars, firewalls, etc., resulting in new objections and potentially problematic behavior. Instead, in NVU one takes the viewpoint that one should begin by parameterizing the effects needed for consistency, with a minimum of extra structure and minimal deviation from known physics. Difficulties for extra structure are one motivation for this. But there is another important one, arising from the fact that time and again physics has taken the path of minimal—but profound—modification of basic structure when a theory with new principles replaces a previous physical theory.

Of course ones understanding is not complete until one has a deeper understanding of the principles of a more basic explanation of the underlying physics. As an effective description, NVU is not directly addressing that—but one expects that if it captures the correct effective physics, it *is* providing very important clues toward the nature of that more complete foundation. It is certainly worth considering what that more basic physics could look like, in working toward its ultimate formulation.

The remainder of this piece will briefly discuss two possibilities for a deeper explanation of the interactions of NVU. One is based on an extrapolation of the geometric constructions of GR, involving spacetime wormholes and baby universes. A second is based on the possibility that spacetime is not fundamental, but instead emerges from a more deeper quantum reality. (These may, possibly, even overlap.)

18.8.1 Replica Wormholes and Baby Universes

The dynamical geometry of GR suggests extrapolation to dynamical topology, which has long been considered from various viewpoints. A significant fraction of the community has become excited that a set of hypothesized rules for extending the euclidean sum over geometries yields entropy formulas that match the rise of the BH entropy with the entanglement growth exemplified by (18.2) in the initial stages of BH evaporation, but then transition to a decreasing entropy as is expected for a shrinking system, as was concretely spelled out by Page in [91, 92].

These rules are based on formal prescriptions for summing over "replica geometries," which have nontrivial topology evidenced through "replica wormholes;" for a recent review see [6], and for some discussion of underlying assumptions see [66]. They do not yet give a set of quantum amplitudes between states—say described in the setting of lorentzian evolution—from which the entropy formulas could be computed from a more basic viewpoint. However, it seems that if there is a more basic explanation of the physical mechanism at work, that arises from a more complete description of spacetime topology change, and specifically from the emission and absorption of baby universes, as was emphasized by Marolf and Maxfield [87].

Such baby universe emission/absorption began to be seriously considered in the 1980s [80, 73, 67]. While it was initially believed to lead to information loss from a large-scale "parent universe" such as our own [80, 73, 67], a more systematic investigation showed that, at least in a certain approximation, this is not true [19, 68, 69]. Specifically, the emission of a baby universe (BU) from our universe can be parameterized by an interaction of the form

$$H_{BU} = a_i^\dagger \mathcal{O}_{BUi} + h.c., \tag{18.58}$$

where $a_i^\dagger$ acts on the state of BUs to create an additional BU of type i, and $\mathcal{O}_{BU,i}$ is an operator that parameterizes the effect of that emission on the fields in our Universe. This "third-quantized" Hamiltonian, which is analogous to those for particle emission from a system in traditional second quantization, may also need to be supplemented by terms describing BU interactions or other effects [69], which will be neglected in the present approximation. It has been proposed [87] that such interactions with an ensemble of BUs could explain the underlying physics behind the replica wormhole calculations, and in particular the decrease of entropy of an evaporating BH.

In examining this claim, one can go back to the older discussion of [19, 68]. There it was argued that one can find BU states that diagonalize the operators $a_i + a_i^\dagger$, with eigenvalues α_i, and that in these "alpha eigenstates" the effects of the interactions (18.58) are simply new terms in a standard Hamiltonian or Lagrangian, with coupling constants given by those eigenvalues. This relied on the operators $\mathcal{O}_{BUi}$ not carrying net energy or momentum (or other local charges), as would be expected when emitting a separate closed universe. Also part of this picture is that the operators $\mathcal{O}_{BUi}$ will be integrals of operators that are local on scales large compared to the associated BU size R_i, and so in the long distance effective theory one simply has a modification to various coupling constants of the local effective theory. Further details are explained in the preceding references, and in the large body of literature that followed.

In such a picture, one could ask how BUs could contribute to unitarization of BH evolution. One possibility is that BU interactions are important, in which case the preceding discussion needs to be modified, but let us assume that we work in the noninteracting BU approximation. Then, if we work in an alpha eigenstate, physics in our universe is described by an ordinary LQFT Hamiltonian or Lagrangian, together with extra contributions of operators $\mathcal{O}_{BU,i}$ with couplings given by the α_i's.

On distance scales large compared to the R_i's, this is once again an effective LQFT, and so one seems to be returned to the discussion of Sect. 18.2, in which the crisis remains. What is new, now, is that the replica wormholes can have scales comparable to the BHs whose evolution they modify, and one expects this to be true of the corresponding BUs, although a more concrete description of the emission processes and corresponding operators is still needed. But this suggests that in (18.58) the operators in the Hamiltonian can be nonlocal on the BH size scale $\sim R$. And, it appears that can make contact with the more general NVU discussion involving (18.45) and (18.46), and provide a mechanism (BU emission) for the interactions of NVU.

This is not yet a complete story, and leaves several questions. The first regards the actual form of the Hamiltonian (18.58) and its operators, and how this yields unitary evolution. First, the couplings depend on the alpha eigenstate of the BUs, and so in principle could be weak or strong. As was outlined in Sect. 18.5.3, one seems to need a specific size interaction strength to achieve the needed information transfer rate for ultimate unitarity. Secondly, we do not have a good understanding of which operators enter (18.58). One associated question is how such a collection of operators achieves universality, as described in Sect. 18.5.2—if that is indeed the correct physics, and if not, what is the alternative.

One other concern is that such a scenario is tied to a traditional geometrical picture of spacetime, which from other viewpoints we expect to not necessarily be fundamental. An interesting point, if this *is* the correct scenario, is that it seems to argue that topology changing processes are *required* in order for the consistency of quantum gravity with quantum unitarity.

18.8.2 An Origin in a Foundational Picture for Quantum Spacetime

There are ample reasons to believe that spacetime may not be fundamental. One is that it is very hard to make sense of a spacetime description at distances shorter than the Planck length both from the point of view of theory, or from the point of view of any experiment, Gedanken or otherwise, to probe this regime. And, if manifold structure is not part of the fundamental description at short distances, it is apparently only an approximate description of the structure at any distances.

If this is true, an important question is what *is* the mathematical starting point for a description of physics. Since quantum mechanics is both well-tested and appears difficult to modify, an attractive possibility is that the correct fundamental structure is that of quantum mechanics, and is based on Hilbert space or an appropriate generalization, with additional mathematical structure needed to describe the physics of gravity and other excitations [47, 48]. This can be motivated from the algebraic description of LQFT (see e.g. [70]), in which the basic structure is Hilbert space together with a net of local algebras that encode the structure of the underlying manifold.

Indeed, this discussion connects to another possible objection to NVU, regarding the lack of an obvious mechanical picture behind the interactions of NVU, if they do not for example arise from baby universe emission as described above. However, there are different possible perspectives on this. As an example, we commonly think of particles and their interactions as providing a mechanical picture for much of physics. But, a different perspective is that one can start with the principles of quantum mechanics, relativity, and locality and *derive* LQFT as a theory reconciling these principles, in the process discovering the origin of particles from the more basic set of principles.[26] This viewpoint also illustrates the power of effective quantum theories, since one can begin with the problem of formulating a quantum-mechanical theory respecting these further principles, and arrive at LQFT. From such a quantum-first perspective, the more fundamental question is what principles, beyond those of quantum mechanics, furnish the correct starting point.

If spacetime is not fundamental and a more basic quantum construct is, then a description of physics in terms of local quantum fields on spacetime is necessarily only an approximation, and is expected to be subject to corrections. Possibly, the interactions of NVU are part of the corrections to an erroneous description of physics in terms of fundamental geometrical spacetime, and so parameterize some of the effects of the departure of that description from the true theory. If so, then better understanding of these interactions is likely to provide us with important clues to the more fundamental quantum dynamics underlying gravity and lying at the foundations of physics.

Acknowledgements I thank my multiple collaborators on these subjects over the years, and also numerous colleagues for many discussions. I also thank R. Bousso and G. 't Hooft for comments on an earlier version. The current work is supported in part by the Heising-Simons Foundation under grants #2021-2819, #2024-5307, and by the U.S. Department of Energy, Office of Science, under Award Number DE-SC0011702.

References

1. B.P. Abbott et al., [LIGO Scientific and Virgo], Observation of gravitational waves from a binary black hole merger. Phys. Rev. Lett. **116**(6), 061102 (2016). https://doi.org/10.1103/PhysRevLett.116.061102. arXiv:1602.03837 [gr-qc]
2. I. Agullo, A. Ashtekar, Unitarity and ultraviolet regularity in cosmology. Phys. Rev. D **91**(12), 124010 (2015). https://doi.org/10.1103/PhysRevD.91.124010. arXiv:1503.03407 [gr-qc]

[26] Examples of this basic approach are provided in [20, 106] as well as in the algebraic approach to the subject.

3. K. Akiyama et al., [Event Horizon Telescope], First M87 event horizon telescope results. I. The shadow of the supermassive black hole. Astrophys. J. Lett. **875**, L1 (2019). https://doi.org/10.3847/2041-8213/ab0ec7. arXiv:1906.11238 [astro-ph.GA]
4. A. Almheiri, D. Marolf, J. Polchinski, D. Stanford, J. Sully, An apologia for firewalls. JHEP **09**, 018 (2013). https://doi.org/10.1007/JHEP09(2013)018. arXiv:1304.6483 [hep-th]
5. A. Almheiri, D. Marolf, J. Polchinski, J. Sully, Black holes: complementarity or firewalls? JHEP **02**, 062 (2013). https://doi.org/10.1007/JHEP02(2013)062. arXiv:1207.3123 [hep-th]
6. A. Almheiri, T. Hartman, J. Maldacena, E. Shaghoulian, A. Tajdini, The entropy of Hawking radiation. Rev. Mod. Phys. **93**(3), 035002 (2021). https://doi.org/10.1103/RevModPhys.93.035002. arXiv:2006.06872 [hep-th]
7. H. Araki, Type of von Neumann algebra associated with free field. Prog. Theor. Phys. **5**, 1–13 (1964)
8. R.L. Arnowitt, S. Deser, C.W. Misner, The dynamics of general relativity. Gen. Rel. Grav. **40**, 1997–2027 (2008). https://doi.org/10.1007/s10714-008-0661-1. arXiv:gr-qc/0405109 [gr-qc]
9. A. Ashtekar, Black hole evaporation in loop quantum gravity. arXiv:2502.04252 [gr-qc]
10. T. Banks, L. Susskind, M.E. Peskin, Difficulties for the evolution of pure states into mixed states. Nucl. Phys. B **244**, 125–134 (1984). https://doi.org/10.1016/0550-3213(84)90184-6
11. T. Banks, A. Dabholkar, M.R. Douglas, M. O'Loughlin, Are horned particles the climax of Hawking evaporation? Phys. Rev. D **45**, 3607–3616 (1992). https://doi.org/10.1103/PhysRevD.45.3607. arXiv:hep-th/9201061 [hep-th]
12. T. Banks, M. O'Loughlin, A. Strominger, Black hole remnants and the information puzzle. Phys. Rev. D **47**, 4476–4482 (1993). https://doi.org/10.1103/PhysRevD.47.4476. arXiv:hep-th/9211030 [hep-th]
13. R. Bousso, G. Penington, Islands far outside the horizon. arXiv:2312.03078 [hep-th]
14. S.L. Braunstein, S. Pirandola, K. Życzkowski, Better late than never: information retrieval from black holes. Phys. Rev. Lett. **110**(10), 101301 (2013). https://doi.org/10.1103/PhysRevLett.110.101301. arXiv:0907.1190 [quant-ph]
15. R. Brustein, A.J.M. Medved, T. Simhon, Black holes as frozen stars. Phys. Rev. D **105**(2), 024019 (2022). https://doi.org/10.1103/PhysRevD.105.024019. arXiv:2109.10017 [gr-qc]
16. C.G. Callan, Jr., S.B. Giddings, J.A. Harvey, A. Strominger, Evanescent black holes. Phys. Rev. D **45**(4), R1005 (1992). https://doi.org/10.1103/PhysRevD.45.R1005. arXiv:hep-th/9111056 [hep-th]
17. A. Carlotto, R.M. Schoen, Localizing solutions of the Einstein constraint equations. Invent. Math. **205**, 559 (2016)
18. C. Chowdhury, V. Godet, O. Papadoulaki, S. Raju, Holography from the Wheeler-DeWitt equation. JHEP **03**, 019 (2022). https://doi.org/10.1007/JHEP03(2022)019. arXiv:2107.14802 [hep-th]
19. S.R. Coleman, Black holes as red herrings: topological fluctuations and the loss of quantum coherence. Nucl. Phys. B **307**, 867–882 (1988). https://doi.org/10.1016/0550-3213(88)90110-1
20. S. Coleman, *Lectures of Sidney Coleman on Quantum Field Theory* (World Sci. Press, Hackensack, 2018)
21. J. Corvino, R.M. Schoen, On the asymptotics for the vacuum Einstein constraint equations. J. Diff. Geom. **73**(2), 185–217 (2006). arXiv:gr-qc/0301071 [gr-qc]
22. S. Datta, R. Brito, S.A. Hughes, T. Klinger, P. Pani, Tidal heating as a discriminator for horizons in equatorial eccentric extreme mass ratio inspirals. Phys. Rev. D **110**(2), 024048 (2024). https://doi.org/10.1103/PhysRevD.110.024048. arXiv:2404.04013 [gr-qc]
23. W. Donnelly, S.B. Giddings, Diffeomorphism-invariant observables and their nonlocal algebra. Phys. Rev. D **93**(2), 024030 (2016) [Erratum: Phys. Rev. D **94**(2), 029903 (2016)]. https://doi.org/10.1103/PhysRevD.93.024030. arXiv:1507.07921 [hep-th]
24. W. Donnelly, S.B. Giddings, Observables, gravitational dressing, and obstructions to locality and subsystems. Phys. Rev. D **94**(10), 104038 (2016). https://doi.org/10.1103/PhysRevD.94.104038. arXiv:1607.01025 [hep-th]

25. W. Donnelly, S.B. Giddings, How is quantum information localized in gravity? Phys. Rev. D **96**(8), 086013 (2017). https://doi.org/10.1103/PhysRevD.96.086013. arXiv:1706.03104 [hep-th]
26. W. Donnelly, S.B. Giddings, Gravitational splitting at first order: quantum information localization in gravity. Phys. Rev. D **98**(8), 086006 (2018). https://doi.org/10.1103/PhysRevD.98.086006. arXiv:1805.11095 [hep-th]
27. K. Fransen, S.B. Giddings, Gravitational wave signatures of departures from classical black hole scattering. Phys. Rev. D **110**(6), 064029 (2024). https://doi.org/10.1103/PhysRevD.110.064029. arXiv:2405.05970 [gr-qc]
28. V.P. Frolov, Cosmic strings and energy mining from black holes. Int. J. Mod. Phys. A **17**, 2673–2676 (2002). https://doi.org/10.1142/S0217751X0201159X
29. V.P. Frolov, D. Fursaev, Mining energy from a black hole by strings. Phys. Rev. D **63**, 124010 (2001). https://doi.org/10.1103/PhysRevD.63.124010. arXiv:hep-th/0012260 [hep-th]
30. S.B. Giddings, Quantum mechanics of black holes. arXiv:hep-th/9412138 [hep-th]
31. S.B. Giddings, Black holes and massive remnants. Phys. Rev. D **46**, 1347–1352 (1992). https://doi.org/10.1103/PhysRevD.46.1347. arXiv:hep-th/9203059 [hep-th]
32. S.B. Giddings, Comments on information loss and remnants. Phys. Rev. D **49**, 4078–4088 (1994). https://doi.org/10.1103/PhysRevD.49.4078. arXiv:hep-th/9310101 [hep-th]
33. S.B. Giddings, Constraints on black hole remnants. Phys. Rev. D **49**, 947–957 (1994). https://doi.org/10.1103/PhysRevD.49.947. arXiv:hep-th/9304027 [hep-th]
34. S.B. Giddings, Why aren't black holes infinitely produced? Phys. Rev. D **51**, 6860–6869 (1995). https://doi.org/10.1103/PhysRevD.51.6860. arXiv:hep-th/9412159 [hep-th]
35. S.B. Giddings, Nonlocality versus complementarity: a conservative approach to the information problem. Class. Quant. Grav. **28**, 025002 (2011). https://doi.org/10.1088/0264-9381/28/2/025002. arXiv:0911.3395 [hep-th]
36. S.B. Giddings, Black holes, quantum information, and unitary evolution. Phys. Rev. D **85**, 124063 (2012). https://doi.org/10.1103/PhysRevD.85.124063. arXiv:1201.1037 [hep-th]
37. S.B. Giddings, Models for unitary black hole disintegration. Phys. Rev. D **85**, 044038 (2012). https://doi.org/10.1103/PhysRevD.85.044038. arXiv:1108.2015 [hep-th]
38. S.B. Giddings, Nonviolent information transfer from black holes: a field theory parametrization. Phys. Rev. D **88**(2), 024018 (2013). https://doi.org/10.1103/PhysRevD.88.024018. arXiv:1302.2613 [hep-th]
39. S.B. Giddings, Nonviolent nonlocality. Phys. Rev. D **88**, 064023 (2013). https://doi.org/10.1103/PhysRevD.88.064023. arXiv:1211.7070 [hep-th]
40. S.B. Giddings, Statistical physics of black holes as quantum-mechanical systems. Phys. Rev. D **88**, 104013 (2013). https://doi.org/10.1103/PhysRevD.88.104013. arXiv:1308.3488 [hep-th]
41. S.B. Giddings, Modulated Hawking radiation and a nonviolent channel for information release. Phys. Lett. B **738**, 92–96 (2014). https://doi.org/10.1016/j.physletb.2014.08.070. arXiv:1401.5804 [hep-th]
42. S.B. Giddings, Possible observational windows for quantum effects from black holes. Phys. Rev. D **90**(12), 124033 (2014). https://doi.org/10.1103/PhysRevD.90.124033. arXiv:1406.7001 [hep-th]
43. S.B. Giddings, Hilbert space structure in quantum gravity: an algebraic perspective. JHEP **12**, 099 (2015). https://doi.org/10.1007/JHEP12(2015)099. arXiv:1503.08207 [hep-th]
44. S.B. Giddings, Hawking radiation, the Stefan–Boltzmann law, and unitarization. Phys. Lett. B **754**, 39–42 (2016). https://doi.org/10.1016/j.physletb.2015.12.076. arXiv:1511.08221 [hep-th]
45. S.B. Giddings, Gravitational wave tests of quantum modifications to black hole structure—with post-GW150914 update. Class. Quant. Grav. **33**(23), 235010 (2016). https://doi.org/10.1088/0264-9381/33/23/235010. arXiv:1602.03622 [gr-qc]
46. S.B. Giddings, Nonviolent unitarization: basic postulates to soft quantum structure of black holes. JHEP **12**, 047 (2017). https://doi.org/10.1007/JHEP12(2017)047. arXiv:1701.08765 [hep-th]

47. S.B. Giddings, Quantum gravity: a quantum-first approach. LHEP **1**(3), 1–3 (2018). https://doi.org/10.31526/LHEP.3.2018.01. arXiv:1805.06900 [hep-th]
48. S.B. Giddings, Quantum-first gravity. Found. Phys. **49**(3), 177–190 (2019). https://doi.org/10.1007/s10701-019-00239-1. arXiv:1803.04973 [hep-th]
49. S.B. Giddings, Black holes in the quantum universe. Phil. Trans. Roy. Soc. Lond. A **377**(2161), 20190029 (2019). https://doi.org/10.1098/rsta.2019.0029. arXiv:1905.08807 [hep-th]
50. S.B. Giddings, Gravitational dressing, soft charges, and perturbative gravitational splitting. Phys. Rev. D **100**(12), 126001 (2019). https://doi.org/10.1103/PhysRevD.100.126001. arXiv:1903.06160 [hep-th]
51. S.B. Giddings, Holography and unitarity. JHEP **11**, 056 (2020). https://doi.org/10.1007/JHEP11(2020)056. arXiv:2004.07843 [hep-th]
52. S.B. Giddings, Schrödinger evolution of the Hawking state. Phys. Rev. D **102**, 125022 (2020). https://doi.org/10.1103/PhysRevD.102.125022. arXiv:2006.10834 [hep-th]
53. S.B. Giddings, Schrödinger evolution of two-dimensional black holes. JHEP **12**, 025 (2021). https://doi.org/10.1007/JHEP12(2021)025. arXiv:2108.07824 [hep-th]
54. S.B. Giddings, On the questions of asymptotic recoverability of information and subsystems in quantum gravity. JHEP **08**, 227 (2022). https://doi.org/10.1007/JHEP08(2022)227. arXiv:2112.03207 [hep-th]
55. S.B. Giddings, A 'black hole theorem,' and its implications. Class. Quant. Grav. **40**(8), 085002 (2023). https://doi.org/10.1088/1361-6382/acbe8b. arXiv:2110.10690 [hep-th]
56. S.B. Giddings, Comparing models for a unitary black hole S matrix. Phys. Rev. D **109**(8), 084055 (2024). https://doi.org/10.1103/PhysRevD.109.084055. arXiv:2212.14551 [hep-th]
57. S.B. Giddings, A. Kinsella, Gauge-invariant observables, gravitational dressings, and holography in AdS. JHEP **11**, 074 (2018). https://doi.org/10.1007/JHEP11(2018)074. arXiv:1802.01602 [hep-th]
58. S.B. Giddings, W.M. Nelson, Quantum emission from two-dimensional black holes. Phys. Rev. D **46**, 2486–2496 (1992). https://doi.org/10.1103/PhysRevD.46.2486. arXiv:hep-th/9204072 [hep-th]
59. S.B. Giddings, J. Perkins, manuscript in preparation
60. S.B. Giddings, J. Perkins, Quantum evolution of the Hawking state for black holes. Phys. Rev. D **106**(6), 065011 (2022). https://doi.org/10.1103/PhysRevD.106.065011. arXiv:2204.13126 [hep-th]
61. S.B. Giddings, J. Perkins, Perturbative quantum evolution of the gravitational state and dressing in general backgrounds. Phys. Rev. D **110**(2), 026012 (2024). https://doi.org/10.1103/PhysRevD.110.026012. arXiv:2209.06836 [hep-th]
62. S.B. Giddings, D. Psaltis, Event horizon telescope observations as probes for quantum structure of astrophysical black holes. Phys. Rev. D **97**(8), 084035 (2018). https://doi.org/10.1103/PhysRevD.97.084035. arXiv:1606.07814 [astro-ph.HE]
63. S.B. Giddings, M. Rota, Quantum information or entanglement transfer between subsystems. Phys. Rev. A **98**(6), 062329 (2018). https://doi.org/10.1103/PhysRevA.98.062329. arXiv:1710.00005 [quant-ph]
64. S.B. Giddings, Y. Shi, Quantum information transfer and models for black hole mechanics. Phys. Rev. D **87**(6), 064031 (2013). https://doi.org/10.1103/PhysRevD.87.064031. arXiv:1205.4732 [hep-th]
65. S.B. Giddings, Y. Shi, Effective field theory models for nonviolent information transfer from black holes. Phys. Rev. D **89**(12), 124032 (2014). https://doi.org/10.1103/PhysRevD.89.124032. arXiv:1310.5700 [hep-th]
66. S.B. Giddings, G.J. Turiaci, Wormhole calculus, replicas, and entropies. JHEP **09**, 194 (2020). https://doi.org/10.1007/JHEP09(2020)194. arXiv:2004.02900 [hep-th]
67. S.B. Giddings, A. Strominger, Axion induced topology change in quantum gravity and string theory. Nucl. Phys. B **306**, 890–907 (1988). https://doi.org/10.1016/0550-3213(88)90446-4
68. S.B. Giddings, A. Strominger, Loss of incoherence and determination of coupling constants in quantum gravity. Nucl. Phys. B **307**, 854–866 (1988). https://doi.org/10.1016/0550-3213(88)90109-5

69. S.B. Giddings, A. Strominger, Baby universes, third quantization and the cosmological constant. Nucl. Phys. B **321**, 481–508 (1989). https://doi.org/10.1016/0550-3213(89)90353-2
70. R. Haag, *Local Quantum Physics* (Springer, 1992)
71. S.W. Hawking, Particle creation by black holes. Commun. Math. Phys. **43**, 199–220 (1975). https://doi.org/10.1007/BF02345020 [Erratum: Commun. Math. Phys. **46**, 206 (1976)]
72. S.W. Hawking, Breakdown of predictability in gravitational collapse. Phys. Rev. D **14**, 2460–2473 (1976). https://doi.org/10.1103/PhysRevD.14.2460
73. S.W. Hawking, Quantum coherence down the wormhole. Phys. Lett. B **195**, 337 (1987). https://doi.org/10.1016/0370-2693(87)90028-1
74. I. Heemskerk, Construction of bulk fields with gauge redundancy. JHEP **09**, 106 (2012). https://doi.org/10.1007/JHEP09(2012)106. arXiv:1201.3666 [hep-th]
75. A.D. Helfer, The Hamiltonians of linear quantum fields. 1. Existence theory. arXiv:hep-th/9908011 [hep-th]
76. T. Jacobson, Boundary unitarity and the black hole information paradox. Int. J. Mod. Phys. D **22**, 1342002 (2013). https://doi.org/10.1142/S0218271813420029. arXiv:1212.6944 [hep-th]
77. T. Jacobson, P. Nguyen, Diffeomorphism invariance and the black hole information paradox. Phys. Rev. D **100**(4), 046002 (2019). https://doi.org/10.1103/PhysRevD.100.046002. arXiv:1904.04434 [gr-qc]
78. D. Kabat, G. Lifschytz, Decoding the hologram: scalar fields interacting with gravity. Phys. Rev. D **89**(6), 066010 (2014). https://doi.org/10.1103/PhysRevD.89.066010. arXiv:1311.3020 [hep-th]
79. T.S. Kuhn, *The Structure of Scientific Revolutions*, 4th ed. (University of Chicago Press, Chicago and London, 2012)
80. G.V. Lavrelashvili, V.A. Rubakov, P.G. Tinyakov, Particle creation and destruction of quantum coherence by topological change. Nucl. Phys. B **299**, 757–796 (1988). https://doi.org/10.1016/0550-3213(88)90372-0
81. A.E. Lawrence, E.J. Martinec, Black hole evaporation along macroscopic strings. Phys. Rev. D **50**, 2680–2691 (1994). https://doi.org/10.1103/PhysRevD.50.2680. arXiv:hep-th/9312127 [hep-th]
82. E. Maggio, M. van de Meent, P. Pani, Extreme mass-ratio inspirals around a spinning horizonless compact object. Phys. Rev. D **104**(10), 104026 (2021). https://doi.org/10.1103/PhysRevD.104.104026. arXiv:2106.07195 [gr-qc]
83. J.M. Maldacena, The large N limit of superconformal field theories and supergravity. Adv. Theor. Math. Phys. **2**, 231–252 (1998). https://doi.org/10.4310/ATMP.1998.v2.n2.a1. arXiv:hep-th/9711200 [hep-th]
84. J. Maldacena, D. Stanford, Z. Yang, Diving into traversable wormholes. Fortsch. Phys. **65**(5), 1700034 (2017). https://doi.org/10.1002/prop.201700034. arXiv:1704.05333 [hep-th]
85. D. Marolf, Unitarity and holography in gravitational physics. Phys. Rev. D **79**, 044010 (2009). https://doi.org/10.1103/PhysRevD.79.044010. arXiv:0808.2842 [gr-qc]
86. D. Marolf, Holography without strings? Class. Quant. Grav. **31**, 015008 (2014). https://doi.org/10.1088/0264-9381/31/1/015008. arXiv:1308.1977 [hep-th]
87. D. Marolf, H. Maxfield, Transcending the ensemble: baby universes, spacetime wormholes, and the order and disorder of black hole information. JHEP **08**, 044 (2020). https://doi.org/10.1007/JHEP08(2020)044. arXiv:2002.08950 [hep-th]
88. S.D. Mathur, The Fuzzball proposal for black holes: An elementary review. Fortsch. Phys. **53**, 793–827 (2005). https://doi.org/10.1002/prop.200410203. arXiv:hep-th/0502050 [hep-th]
89. S.D. Mathur, Tunneling into fuzzball states. Gen. Rel. Grav. **42**, 113–118 (2010). https://doi.org/10.1007/s10714-009-0837-3. arXiv:0805.3716 [hep-th]
90. P.O. Mazur, E. Mottola, Gravitational vacuum condensate stars. Proc. Natl. Acad. Sci. **101**, 9545–9550 (2004). https://doi.org/10.1073/pnas.0402717101. arXiv:gr-qc/0407075 [gr-qc]
91. D.N. Page, Black hole information. arXiv:hep-th/9305040 [hep-th]
92. D.N. Page, Average entropy of a subsystem. Phys. Rev. Lett. **71**, 1291–1294 (1993). https://doi.org/10.1103/PhysRevLett.71.1291. arXiv:gr-qc/9305007 [gr-qc]

93. J. Preskill, Do black holes destroy information? arXiv:hep-th/9209058 [hep-th]
94. C. Rovelli, F. Vidotto, Planck stars, white holes, remnants and Planck-mass quasi-particles. The quantum gravity phase in black holes' evolution and its manifestations. arXiv:2407.09584 [gr-qc]
95. C. Rovelli, F. Vidotto, Planck stars. Int. J. Mod. Phys. D **23**(12), 1442026 (2014). https://doi.org/10.1142/S0218271814420267. arXiv:1401.6562 [gr-qc]
96. B.C. Seymour, Y. Chen, Gravitational-wave signatures of non-violent non-locality. arXiv:2411.13714 [gr-qc]
97. L. Susskind, Trouble for remnants. arXiv:hep-th/9501106 [hep-th]
98. G. 't Hooft, Alternative theory for the quantum black hole and the temperature of its quantum radiation. arXiv:2410.16891 [gr-qc] (this volume)
99. G. 't Hooft, Black hole unitarity and antipodal entanglement. Found. Phys. **46**(9), 1185–1198 (2016). https://doi.org/10.1007/s10701-016-0014-y. arXiv:1601.03447 [gr-qc]
100. C.G. Torre, Gravitational observables and local symmetries. Phys. Rev. D **48**, R2373–R2376 (1993). https://doi.org/10.1103/PhysRevD.48.R2373. arXiv:gr-qc/9306030 [gr-qc]
101. C.G. Torre, M. Varadarajan, Quantum fields at any time. Phys. Rev. D **58**, 064007 (1998). https://doi.org/10.1103/PhysRevD.58.064007. arXiv:hep-th/9707221 [hep-th]
102. C.G. Torre, M. Varadarajan, Functional evolution of free quantum fields. Class. Quant. Grav. **16**, 2651–2668 (1999). https://doi.org/10.1088/0264-9381/16/8/306. arXiv:hep-th/9811222 [hep-th]
103. W.G. Unruh, Decoherence without dissipation. Trans. Roy. Soc. Lond. **370**, 4454 (2012). https://doi.org/10.1098/rsta.2012.0163. arXiv:1205.6750 [quant-ph]
104. W.G. Unruh, R.M. Wald, Acceleration radiation and generalized second law of thermodynamics. Phys. Rev. D **25**, 942–958 (1982). https://doi.org/10.1103/PhysRevD.25.942
105. W.G. Unruh, R.M. Wald, Information loss. Rept. Prog. Phys. **80**(9), 092002 (2017). https://doi.org/10.1088/1361-6633/aa778e. arXiv:1703.02140 [hep-th]
106. S. Weinberg, *The Quantum Theory of Fields. Vol. 1: Foundations* (Cambridge University Press, 2005)
107. E. Witten, Why does quantum field theory in curved spacetime make sense? And what happens to the algebra of observables in the thermodynamic limit? https://doi.org/10.1007/978-3-031-17523-7_11. arXiv:2112.11614 [hep-th]
108. J. Yngvason, Localization and entanglement in relativistic quantum physics. Lect. Notes Phys. **899**, 325–348 (2015). https://doi.org/10.1007/978-3-662-46422-9_15. arXiv:1401.2652 [quant-ph]

Chapter 19
Replica Wormhole

Jingru Lu, Manqian Ou, and Zhenbin Yang

Abstract This chapter examines the replica wormhole approach to understanding the Page curve in the context of the black hole information problem. The problem is recast in terms of employing the gravitational path integral to calculate the von Neumann entropy of Hawking radiation. By including saddle points with non-trivial topologies in the path integral, the Page curve emerges naturally. The central result is encapsulated in the *island formula*. The chapter seeks to maintain a balance between conciseness and a self-contained exposition.

19.1 Introduction

We review the black hole information problem through the lens of gravitational path integrals. The idea involves summing over all geometries consistent with the boundary conditions, including those with non-trivial topologies.

This chapter is aimed at beginning graduate students interested in recent developments in this area of research.

Our focus is on two dimensional gravities. We begin by reviewing the Cardy-Calabrese approach to calculating entanglement entropy in two dimensional Conformal Field Theories (2d CFT). Next, we discuss the entanglement entropy in AdS/CFT,[1] including the Ryu-Takayanagi formula and its derivation using gravitational path integrals. Finally, we apply the gravitational path integral to compute

[1] A duality between Anti-de Sitter space (AdS) and CFT. Anti-de Sitter space is the maximum symmetric space with negative constant Ricci curvature.

J. Lu · M. Ou · Z. Yang (✉)
Institute for Advanced Study, Tsinghua University, Beijing, China
e-mail: yangiastu@gmail.com

J. Lu
e-mail: lujr23@mails.tsinghua.edu.cn

M. Ou
e-mail: omq24@mails.tsinghua.edu.cn

A. Akil and C. Bambi (eds.), *The Black Hole Information Paradox*, Springer Series in Astrophysics and Cosmology, https://doi.org/10.1007/978-981-96-6170-1_19

the von Neumann entropy of Hawking radiation and explore the connection between replica wormholes and the violation of global symmetries in quantum gravity.

19.2 The Problem

The density matrix ρ describes the state of a quantum system. For a pure state, ρ can be expressed as $\rho = |\psi\rangle\langle\psi|$. For a mixed state, the rank of ρ is greater than one. The degree of "mixedness" of a non-pure state ρ is characterized by the von Neumann entropy, S_{vN}, defined as:

$$S_{\text{vN}}(\rho) := -\text{Tr}(\hat{\rho}\,\log\,\hat{\rho}), \tag{19.1}$$

where $\hat{\rho} := \rho/\text{Tr}(\rho)$ is the normalized density matrix. The von Neumann entropy is also referred to as the entanglement entropy, with $S_{\text{vN}} = 0$ for pure states. For an N-dimensional density matrix ρ, the maximum value of S_{vN} is $\log N$, achieved when $\rho \propto I$, meaning the state is maximally mixed with equal probabilities.

Now consider a quantum state $|\psi\rangle$ in a bipartite system $\mathscr{H}_A \otimes \mathscr{H}_B$. The total state $\rho = |\psi\rangle\langle\psi|$ is pure, so $S_{\text{vN}}(\rho) \equiv 0$. The information about subsystem A is accessed through the reduced density matrix:

$$\rho_A = \text{Tr}_B(\rho), \tag{19.2}$$

where ρ_A may not be pure. A nonzero value of $S_{\text{vN}}(\rho_A)$ arises due to entanglement between A and B. An important consequence is:

$$S_{\text{vN}}(\rho_A) \equiv S_{\text{vN}}(\rho_B), \quad \text{if } S_{\text{vN}}(\rho) = 0. \tag{19.3}$$

In the context of the black hole information paradox, the quantum system of interest is the Hawking radiation, and $S_{\text{vN}}(\rho_{\text{rad}})$ is calculated. The combined system of radiation and the black hole is bipartite. Initially, there is no radiation, so $S_{\text{vN}}(\rho_{\text{rad}}) = 0$. As radiation is emitted, it becomes entangled with the black hole, causing $S_{\text{vN}}(\rho_{\text{rad}})$ to increase. Eventually, the black hole fully evaporates, leaving only radiation. If the black hole formed from matter in a pure state, and if the time evolution of the gravitational system is unitary (i.e., governed by quantum mechanics), the final state must also be pure. This implies:

$$\text{Unitarity:}\;\; S_{\text{vN}}(\text{final } \rho_{\text{rad}}) = 0. \tag{19.4}$$

However, this prediction contradicts Hawking's original calculation [1], which was based on quantum field theory on a curved space-time background. Hawking's calculation predicts an ever-growing entropy for the radiation. The paradox, then, lies in determining whether gravitational systems are fundamentally non-unitary or if Hawking's calculation contains subtle errors. We favor the latter, guided by the

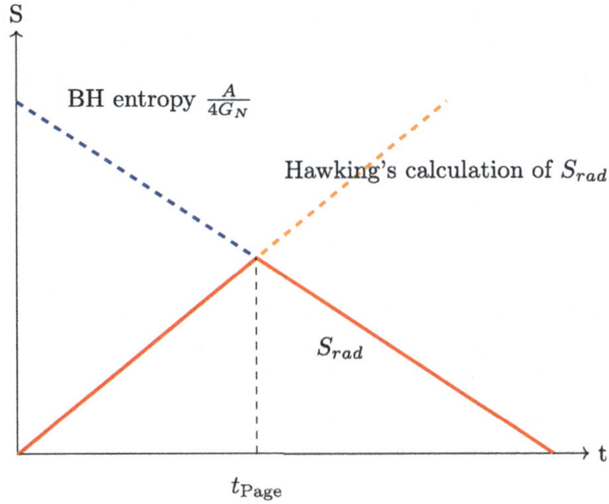

Fig. 19.1 Page curve of an evaporating black hole. The entanglement entropy of radiation is bounded by the Bekenstein-Hawking entropy (BH entropy). The entanglement entropy is fine-grained entropy, which is bounded by the coarse-grained BH entropy

holographic principle, which conjectures a correspondence between gravity and field theory [2–4]. Since the boundary theory is described by quantum field theory, we expect that bulk processes preserve unitarity.

Assuming unitarity, Don Page famously predicted the behavior of $S_{\text{vN}}(\rho_{\text{rad}})$ as a function of time t [5, 6], now known as the Page curve (Fig. 19.1). According to this prediction, the entropy first increases and then decreases, with the transition time known as the *Page time*. Before the Page time, the increase in entropy aligns with Hawking's calculation. After the Page time, the decrease in entropy follows the behavior of the Bekenstein-Hawking entropy of the black hole [7–10]. The Bekenstein-Hawking entropy relates the thermodynamic entropy of the black hole to its horizon area, measured in Planck units:

$$S_{\text{BH}} = \frac{\text{Area}}{4G_N}. \tag{19.5}$$

At the late stages of black hole evaporation, the horizon area shrinks to zero. Consequently, the von Neumann entropy of Hawking radiation should also decrease to zero at the same rate.

For an evaporating black hole, the dynamic nature of the spacetime background complicates calculations, even in classical relativity. As an alternative, consider a scenario in which pure matter is continuously fed into the black hole as it evaporates, maintaining a constant mass [11, 12].[2] In this static spacetime, the black hole radiates perpetually. Under these conditions, Hawking's calculation predicts that $S_{\text{vN}}(\rho_{\text{rad}})$ grows indefinitely, diverging over time, which contradicts the finite Bekenstein-Hawking entropy.

[2] The added matter carries no charges, such as electric charge or angular momentum, aside from stress-energy.

This chapter focuses on this rephrased model, with additional comments on the case of evaporating black holes provided at the end. But before that, let's first introduce our tool to calculate von Neumann entropies, which is called the replica trick.

19.3 Replica Trick in Field Theories

In this section, we introduce the replica trick in field theory and, following Calabrese and Cardy [13, 14], use it to derive the entanglement entropy formula for 2d CFT.

For a generic matrix ρ_A of a subsystem A, directly computing the von Neumann entropy using the logarithm formula (19.1) can be challenging. A useful technique is to compute the Rényi entropy:

$$S_A^{(n)} = -\frac{1}{n-1}\mathrm{Tr}\rho_A^n. \tag{19.6}$$

The von Neumann entropy is then obtained as the $n \to 1$ analytic continuation:

$$S_A = \lim_{n\to 1} -\frac{1}{n-1}\mathrm{Tr}\rho_A^n = -\frac{\partial}{\partial n}\mathrm{Tr}\rho_A^n\bigg|_{n=1}. \tag{19.7}$$

This approach is known as the *replica trick*. Mathematically, calculating von Neumann entropy by its definition yields the same result as using the replica trick if all entries of ρ_A are known.

Now consider a 2d CFT where the subsystem A is a single interval on a Cauchy slice Σ_t (setting $t = 0$), and the pure state $|\psi\rangle$ is the vacuum $|0\rangle$. The matrix elements of the density matrix in local basis $\phi_\pm$ are:

$$\rho_{\phi_+\phi_-} = \langle\phi_+|0\rangle\,\langle 0|\phi_-\rangle\,. \tag{19.8}$$

The wavefunction $\langle\phi_+|0\rangle$ can be obtained from the Euclidean path integral on the lower half-plane, from $t_E = -\infty$ to the given slice, with boundary condition $\Phi = \phi_+$ on the slice. Similarly, $\langle 0|\phi_-\rangle$ is derived from the Euclidean path integral on the upper half-plane.

To compute the reduced density matrix for interval A, we trace out the complement region A^c:

$$(\rho_A)_{\phi_+\phi_-} = Z_1^{-1}\int \mathcal{D}\Phi\, e^{-S_E}\prod_{x\in A}\delta(\Phi(t=0^+,x)-\phi_+(x))\delta(\Phi(t=0^-,x)-\phi_-(x)), \tag{19.9}$$

where the delta functions impose the boundary conditions and Z_1 is the normalization factor ensuring $\mathrm{Tr}(\rho) = 1$. For ρ_A^n, the path integral is replicated n times and the corresponding manifold is a n-sheeted Riemann surface:

$$
\begin{aligned}
(\rho_A^n)_{\phi_+\phi_-} =& Z_1^{-n} \int \prod_{m=1}^{n} \mathscr{D}\phi_+^m \mathscr{D}\phi_-^m \prod_{k=1}^{n} \mathscr{D}\Phi^k e^{-\sum_{k=1}^{n} S_E[\Phi^k]} \\
& \delta(\Phi_\pm^k)\delta(\phi_+^n - \phi_+)\delta(\phi_-^1 - \phi_-) \prod_{j=1}^{n-1} \delta(\phi_+^j - \phi_-^{j+1}),
\end{aligned} \tag{19.10}
$$

where

$$
\delta(\Phi_\pm^k) = \prod_{x\in A} \delta(\Phi^k(0^+, x) - \phi_+^k(x))\delta(\Phi^k(0^-, x) - \phi_-^k(x)). \tag{19.11}
$$

The delta functions impose boundary conditions for each sheet and ensure adjacent sheets are connected as per the matrix product rule. Taking the trace sets the remaining index $\phi_\pm(x \in A)$ to be equal and integrated, resulting in a path integral on an n-sheeted Riemann surface $\mathscr{R}_n$. Denoting this path integral result as Z_n, we have:

$$
\mathrm{Tr}(\rho_A^n) = \frac{Z_n}{Z_1^n}. \tag{19.12}
$$

A convenient method to compute Z_n is to introduce twist operators $\Phi_n(u)$ and $\Phi_{-n}(v)$ to impose the twist boundary conditions (see Fig. 19.2). Calculation on Riemann surface $\mathscr{R}_n$ is now understood as calculation on n independent complex planes, each plane with two twist operators Φ_n and Φ_{-n} insertions. In 2d CFT, operators are classified by their conformal dimensions. To understand the conformal dimension of the twist operators, we can examine their behavior under conformal transformation that maps $\mathscr{R}_n$ with coordinates w to the complex plane $\mathbb{C}$ parameterized by coordi-

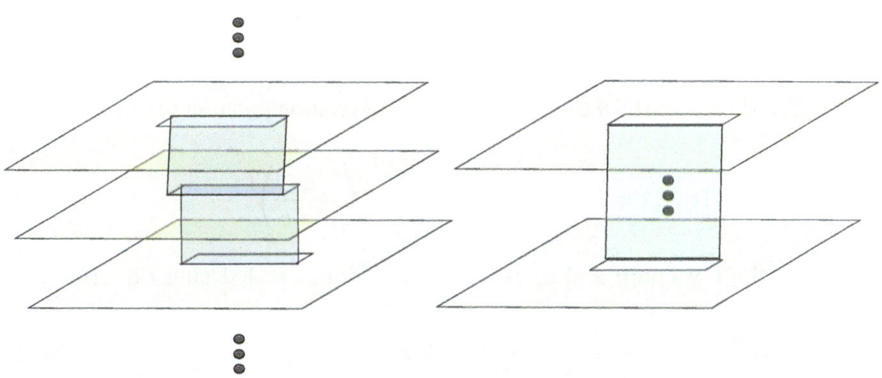

Fig. 19.2 The replica geometry used for the path integral to compute $\mathrm{Tr}(\rho^n)$. Adjacent sheets are glued according to the matrix product rule. Such boundary conditions are called twist boundary conditions

nates z. This can be done by first using the coordinate transformation $w \mapsto \zeta = \frac{w-u}{v-w}$ to map the interval $A = [u, v]$ to $[0, \infty)$, then using the conformal map $\zeta \mapsto z = \zeta^{1/n}$ to map to the complex plane. On the z-plane, the vacuum state satisfies $\langle T(z) \rangle_{\mathbb{C}} = 0$, where subscript $\mathbb{C}$ means expectation value on the complex plane, but on the w-plane, an anomaly term modifies $\langle T(w) \rangle_{\mathscr{R}_n}$ via the Schwarzian derivative[3]:

$$T(w) = \left(\frac{dz}{dw} \right)^2 T(z) + \frac{c}{12} \mathrm{Sch}(z, w), \tag{19.13}$$

where

$$\mathrm{Sch}(z, w) = \frac{\frac{d^3 z}{dw^3}}{\frac{dz}{dw}} - \frac{3}{2} \left(\frac{\frac{d^2 z}{dw^2}}{\frac{dz}{dw}} \right)^2 . \tag{19.14}$$

Thus,

$$\langle T(w) \rangle_{\mathscr{R}_n} = \frac{c(1 - (1/n)^2)}{24} \frac{(v-u)^2}{(w-v)^2 (w-u)^2}. \tag{19.15}$$

where the subscript means expectation value on the single Riemann sheet. From this result, we can read out the conformal dimensions of the twist operators $\Phi_n(u)$ and $\Phi_{-n}(v)$ with $h = \bar{h} = \frac{c}{24}(1 - (1/n)^2)$, where the above transformation can be understood from the conformal Ward identity:

$$\langle T(w) \Phi_n(u) \Phi_{-n}(v) \rangle_{\mathbb{C}} = \left(\frac{h}{(w-u)^2} + \frac{h}{(w-v)^2} + \frac{1}{w-u} \frac{\partial}{\partial u} + \frac{1}{w-v} \frac{\partial}{\partial v} \right) \langle \Phi_n(u) \Phi_{-n}(v) \rangle_{\mathbb{C}} , \tag{19.16}$$

and

$$\langle T(w) \rangle_{\mathscr{R}_n} = \frac{\langle T(w) \Phi_n(u) \Phi_{-n}(v) \rangle_{\mathbb{C}}}{\langle \Phi_n(u) \Phi_{-n}(v) \rangle_{\mathbb{C}}}. \tag{19.17}$$

Now $\mathrm{Tr}(\rho_A^n)$ is given by the n-th power of the correlation function $\langle \Phi_n(u) \Phi_{-n}(v) \rangle_{\mathbb{C}}$:

$$\mathrm{Tr}(\rho_A^n) = \langle \Phi_n(u) \Phi_{-n}(v) \rangle_{\mathbb{C}}^n = a_n \left(\frac{\epsilon}{v-u} \right)^{\frac{c}{6}(n - \frac{1}{n})} , \tag{19.18}$$

where ϵ is the UV cutoff and a_n is a constant with $a_1 = 1$. Using Eq. (19.7), we obtain [13, 14]:

$$S_A = \frac{c}{3} \log \frac{l}{\epsilon}, \tag{19.19}$$

where $l = |u - v|$ is the interval length. This result will be used in subsequent calculations.

---

[3] The c in Eq. (19.13) is an important constant in CFT called central charge.

19.4 Replica Trick in Gravity

Now we turn to the entropy calculation of Hawking radiation. The key new ingredient here is that the geometry is now dynamical, meaning that we must account for gravitational backreaction. In general, without a complete theory of quantum gravity, calculating fine-grained quantities such as the von Neumann entropy can be challenging. Fortunately, recent advancements in AdS/CFT, particularly the Ryu-Takayanagi (RT) formula for the von Neumann entropy of subregions, reveal that simple bulk observables, such as extremal surfaces, are dual to these fine-grained quantities. We shall first review the Lewkowycz-Maldacena derivation of the RT surface using gravitational path integrals [15], and then proceed to apply the same technique to study the von Neumann entropy of Hawking radiation.

The replica trick discussed in the previous section applies to subregions in field theories. Gravity shares similarities with field theories, but is the replica trick identical in the gravitational context?

Not exactly. In field theories, the geometry is fixed and we can discuss subregions that the fields live on, such as the interval $x \in (0, l)$ in $2d$ systems. However, when gravity is included, spacetime itself becomes dynamical. Subregions, and their corresponding cuts, will have gravitational fluctuations which makes it challenging to define exactly the subsystem to talk about the entropy. For instance, the gravitational fluctuation could be large enough such that the original subregion will disappear.

Fortunately, there is a way forward [10, 15]. When spacetime has spatial boundaries of codimension 1, the geometry on these *boundaries* can be **fixed**. For asymptotically AdS spacetimes, the asymptotic infinity naturally serves as the spatial boundary. For asymptotically flat spacetimes, such as a black hole in asymptotically Minkowski spacetime, spatial boundaries are manually chosen such that gravity is weak outside them, approximating the geometry as non-dynamical (Fig. 19.3). For example, the region far outside a black hole horizon is almost static. The regions enclosed by these boundaries are called the bulk.

Fixing the boundary geometry enables us to meaningfully define subregions. In other words, we cut and glue boundary subregions.

In the previous section, cutting a subregion involved calculating the reduced density matrix of the subregion. What does it mean to cut boundary subregions?[4]

According to the holographic principle, gravitational theory is conjectured to be dual to a quantum field theory (QFT) on a lower-dimensional spacetime, usually the boundary spacetime of the gravitational theory such as the AdS/CFT correspondence. Cutting boundary subregions A corresponds to cutting subregions in the boundary QFT, which are dual to **entanglement wedge** in the bulk gravitational theory. More precisely, the process involves the following steps:

1. Starting from a boundary manifold, cut a boundary subregion A of it. The bulk saddle of the original boundary is denoted as $\mathcal{M}_1$.

---

[4] Recall that all bulk geometries satisfying the prescribed boundary conditions must be summed over.

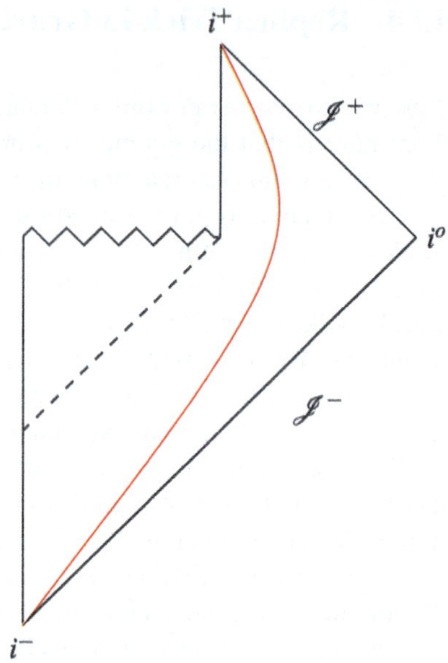

Fig. 19.3 Example of an evaporating black hole. The red curve far outside the horizon separates dynamical and non-dynamical geometries

2. Gluing n copies of the *cut* boundaries together results in a new "boundary" manifold. n is the number of replicas in replica trick.
3. "Path integrate" the bulk dual of the new boundary. In principle we need to sum over all possible bulk geometries with *the* prescribed boundary. But only the saddle bulk geometry[5] matters in saddle point approximation. Call this bulk saddle of the new saddle $\widetilde{\mathscr{M}_n}$. This procedure defines $\widetilde{\mathscr{M}_n}$ for positive integers n. Suppose the definition can be extended to $n \to 1$. The extension is discussed in Sects. 19.6 and 19.5.
4. $\widetilde{\mathscr{M}}_{n\to 1}$ resembles what we would get if we had cut some subregions $\mathscr{A}_{\text{bulk}}$ of $\mathscr{M}_1$ and had glued n such copies *directly in the bulk*. Specifically, the boundary of $\mathscr{A}_{\text{bulk}}$ is the union of A and a new codimension-2 surface in the bulk. The new surface is the RT surface (Sect. 19.5) [16–18]. This RT surface has significant implications for entropies.
5. It makes sense to say that by cutting A, we calculate the entanglement entropy of A in the boundary QFT, and equivalently, gravitational entanglement entropy of $\mathscr{A}_{\text{bulk}}$ is calculated. The causal diamond of $\mathscr{A}_{\text{bulk}}$ in the bulk is called the **entanglement wedge** of A [19, 20].
6. Notes: for a $(d+1)$-dimensional bulk, $\mathscr{M}_1$ and $\widetilde{\mathscr{M}_n}$ are $(d+1)$-dimensional; A and the RT surface are $(d-1)$-dimensional sub-manifolds on a time slice of the boundary; $\mathscr{A}_{\text{bulk}}$ is a d-dimensional sub-manifold on a time slice in the bulk.

[5] That is, the bulk geometry satisfying bulk equations of motion up to the prescribed boundary conditions.

This logic will become clearer through concrete examples. Note that sometimes there could be more than one saddle point, and in such cases, one needs to figure out the leading saddle.

In this chapter, we focus on the entanglement entropy of radiation. Radiation is collected in regions where gravity is weak, that is, outside the cutoff boundary of the gravitational bulk. The von Neumann entropy $S_{\text{vN}}(\rho_{\text{rad}})$ is calculated by directly cutting non-gravitational regions.

Why, then, are we interested in the replica trick in gravity? The answer is

- Although non-gravitational regions are directly cut, the entire spacetime includes gravitating regions, and their contribution involves the gravitational path integral.
- To test unitarity, we compare $S_{\text{vN}}(\rho_{\text{rad}})$ with $S_{\text{vN}}(\rho_{\text{BH}})$. The latter is calculated using boundary cuts.

19.5 Lewkowycz-Maldacena Derivation of Ryu-Takayanagi Formula

We begin by reviewing the derivation of RT formula using the gravitational path integral, following [15–17, 21, 22]. The RT formula states that the von Neumann entropy of a boundary region R is given by the area of RT surface, the bulk extremal surface associated with that boundary region:

$$S_{\text{vN}}(A) = \min \text{Ext}_{\mathscr{A}_{\text{bulk}}} \frac{\text{Area}(\partial \mathscr{A}_{\text{bulk}} \backslash A)}{4G_N}. \tag{19.20}$$

Here $\text{Area}(\partial \mathscr{A}_{\text{bulk}} \backslash A)$ denotes the area of the RT surface of $\mathscr{A}_{\text{bulk}}$.

To calculate the von Neumann entropy of a boundary subregion A, let's use the replica trick introduced in Eq. (19.7):

$$S = -\partial_n \left(\log \frac{\text{Tr}\rho_A^n}{(\text{Tr}\rho_A)^n} \right) \Bigg|_{n=1} = -\partial_n \left(\log \frac{Z_n}{Z_1^n} \right) \Bigg|_{n=1}, \tag{19.21}$$

Here Z_n is the unnormalized gravitational partition function on the replica manifold $\widetilde{\mathscr{M}}_n$. Using the saddle point approximation, $\log Z_n = -I_n$, we obtain (Fig. 19.4):

$$S = \partial_n I[\widetilde{\mathscr{M}}_n] \Bigg|_{n=1} - I[\mathscr{M}_1]. \tag{19.22}$$

Here, $I[\widetilde{\mathscr{M}}_n]$ represents the Einstein-Hilbert action calculated on the replica manifold. Assuming $\widetilde{\mathscr{M}}_n$ is $\mathbb{Z}_n$ invariant under the permutation of the replica index (preserves replica symmetry), we define another manifold $\mathscr{M}_n = \widetilde{\mathscr{M}}_n/\mathbb{Z}_n$, where one of

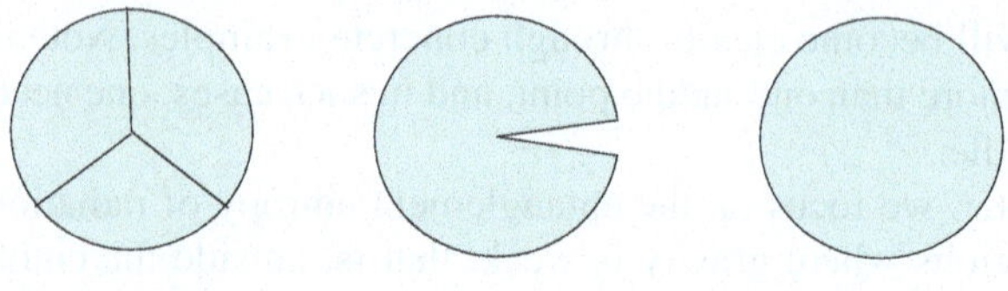

Fig. 19.4 The first figure represents path integral on $\widetilde{\mathcal{M}_n}$, and each sector represents $\mathcal{M}_n = \widetilde{\mathcal{M}_n}/\mathbb{Z}_n$. The second figure shows $\mathcal{M}_n$ when $n \to 1$. The final figure is when $n = 1$, i.e. $\mathcal{M}_1$

the n identical copies of the field theory lives on $\mathcal{M}_n$. However, $\mathcal{M}_n$ has a conical singularity with an opening angle of $2\pi/n$ at the $\mathbb{Z}_n$ fixed points, while $\widetilde{\mathcal{M}_n}$ remains smooth. Thus, the contribution from the singularities is excluded, and we have:

$$\frac{I[\widetilde{\mathcal{M}_n}]}{n} = I[\mathcal{M}_n] + T_n \int_{\Sigma_{d-2}} \sqrt{g}, \tag{19.23}$$

where the second term corresponds to the action of a cosmic brane introduced to cancel the contribution from the conical singularity at the $\mathbb{Z}_n$ fixed points in $\mathcal{M}_n$. From the conical tip contribution discussed in Sect. 19.6, we find:

$$-\frac{1}{16\pi G_N} \int_{\text{conical tip}} d^2x \sqrt{g} R = \frac{1-n}{4nG_N}. \tag{19.24}$$

Thus, the brane tension is $T_n = \frac{n-1}{4G_N n}$.

The first term, $I[\mathcal{M}_n]$, represents the on-shell action. The difference between this term and $I[\mathcal{M}_1]$ is at least $\mathcal{O}((n-1)^2)$ when $n \to 1$, so it can be replaced by $I[\mathcal{M}_1]$ in this limit. Combining everything, we find:

$$S = \partial_n \left(nI[\mathcal{M}_1]\right)\big|_{n=1} + \partial_n \left(nT_n \text{Area}\right)\big|_{n=1} - I[\mathcal{M}_1] = \frac{\text{Area}}{4G_N}. \tag{19.25}$$

This result is off-shell, as it depends on the position of the cosmic brane. By minimizing the action, we obtain the RT formula. If quantum field corrections in the bulk are considered, we must also insert twist operators on the cosmic brane. These contribute the entropy of quantum matter in the bulk, leading to the generalized entropy formula [21] (Figs. 19.5 and 19.6):

$$S_{\text{vN}}(A) = \min \text{Ext}_{\mathcal{A}_{\text{bulk}}} \left(\frac{\text{Area}(\partial \mathcal{A}_{\text{bulk}} \backslash A)}{4G_N} + S_{\text{QFT}}(\mathcal{A}_{\text{bulk}}) \right). \tag{19.26}$$

In conclusion, gravitational entropy can be interpreted as arising from the cosmic brane action at the $\mathbb{Z}_n$ fixed points. The minimal area condition follows from minimizing this action. In the following chapter, we will see that islands also arise from

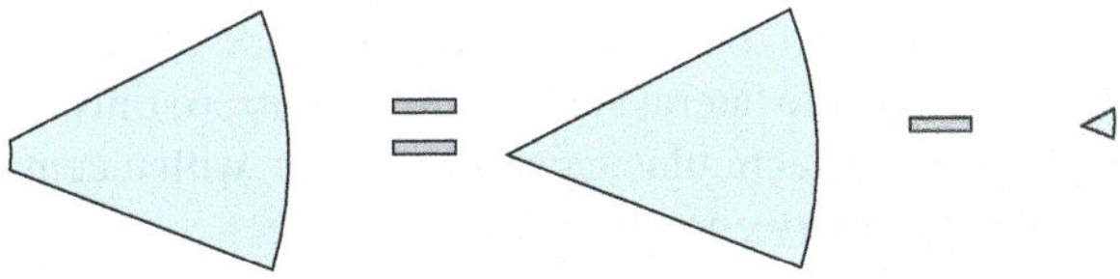

Fig. 19.5 The left hand side represents $\frac{I[\widetilde{\mathcal{M}_n}]}{n}$, excluding the contribution from the conical singularity. The first term on the right hand side represents $I[\mathcal{M}_n]$, including the singularity contribution, which vanishes when $n = 1$. The second term subtracts the conical singularity contribution

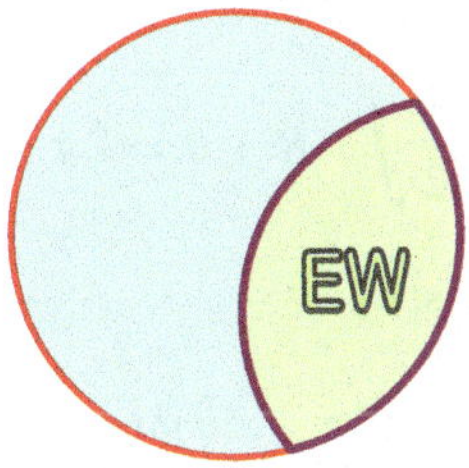

Fig. 19.6 After minimization procedure, the purple curve in the bulk is the RT surface of the purple boundary subregion. The causal development of the yellow region, which is enclosed by the RT surface and the purple boundary subregion, is the entanglment wedge of this boundary subregion

the $\mathbb{Z}_n$ fixed points of the replica manifold when computing the entropy of radiation. The connected geometry in the gravitational path integral is analyzed in a similar way.

19.6 Detour: Cone

This section derives the curvature singularity at the tip of a cone. Readers may skip this section if desired.

The lateral surfaces of cylinders and cones are extrinsically curved but intrinsically flat. What about the tip of a cone?

If the tip is smoothed out, for example, by a spherical cap, the resulting smooth surface exhibits a curvature singularity near the tip. By choosing spherical caps with progressively smaller radii, the approximation becomes more accurate, and the curvature increases. This indicates the presence of a delta-function singularity at the tip.

The calculation relies on the Gauss-Bonnet theorem for a 2d manifold $\mathcal{M}$:

$$\frac{1}{2}\int_{\mathcal{M}} R + \int_{\partial\mathcal{M}} K = 2\pi\chi = 2\pi(2 - 2g - b), \tag{19.27}$$

where R is the Ricci scalar, K is the extrinsic curvature scalar on the boundary, and χ is the Euler characteristic. This theorem is a topological statement.

A cone has the topology of a disk, so the right-hand side of Eq. (19.27) is 2π. On the left-hand side, $R = 0$ on the lateral surface and is nonzero only at the tip. A cone can be constructed by gluing a circular sector of radius r with a central angle θ. For this configuration, the boundary curvature K is:

$$K = \frac{1}{r}, \quad \int_{\partial \mathcal{M}} K = \frac{1}{r} \cdot \theta \cdot r = \theta.$$

The Gauss-Bonnet theorem then gives:

$$\frac{1}{2} \int_{\mathcal{M}} R + \theta = 2\pi. \tag{19.28}$$

Thus, the Ricci scalar at the tip is:

$$R(x) = (4\pi - 2\theta)\delta(x - x_{\text{tip}}), \tag{19.29}$$

and this singularity is referred to as a *conical singularity*.

19.7 AdS-Bath Model

We now turn to the case of Hawking radiation. In Sect. 19.2, we reformulated the discussion in terms of an eternal black hole. Here, we provide additional details [11, 12].

We restrict our analysis to spherically symmetric spacetimes, as the key feature of the black hole information problem stems from the existence of a horizon, rather than other properties such as global charges of the black hole. Consequently, only radial and temporal directions are explicitly considered. Under dimensional reduction, the area of the transverse dimensions reduces to a scalar field ϕ, called the dilaton field. Specifically, dimension reduction of a near extremal black hole, whose near horizon geometry is $AdS_2 \times \mathcal{M}_{D-2}$ ($\mathcal{M}_{D-2}$ is a compact manifold) results in Jackiw-Teitelboim (JT) gravity [23–27].

JT gravity describes nearly AdS_2 geometry with the Euclidean action:

$$I = I_{\text{top}} + I_{\text{grav}} + I_{\text{CFT}}, \tag{19.30}$$

where

$$I_{\text{top}} = -\frac{\phi_0}{16\pi G_N} \left[\int_{\mathcal{M}} \sqrt{g} R + 2 \int_{\partial \mathcal{M}} \sqrt{\gamma} K \right], \tag{19.31}$$

$$I_{\text{JT}} = -\frac{1}{16\pi G_N} \int_{\mathcal{M}} \sqrt{g}\, \phi\, (R + 2) - \frac{1}{8\pi G_N} \int_{\partial \mathcal{M}} \sqrt{\gamma}\, \phi\, K. \tag{19.32}$$

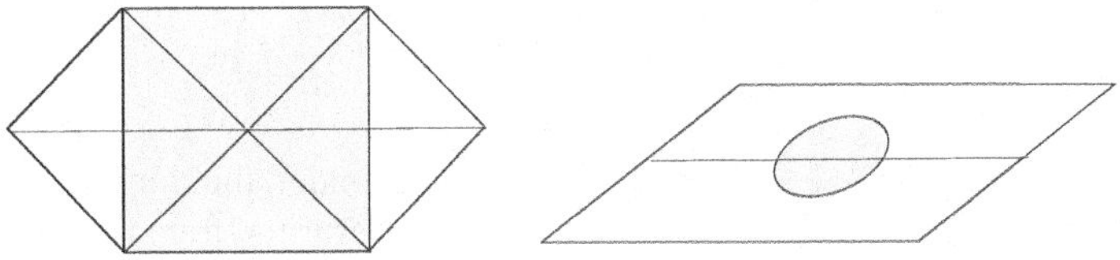

Fig. 19.7 Left: Penrose diagram of an AdS black hole coupled to Minkowski spacetime (the bath) in Lorentzian signature. Right: The corresponding Euclidean signature diagram, where the horizontal purple line represents $t_E = 0$

I_{top} is a topological term in $2d$ and simply gives $-S_0\chi$, where $S_0 \equiv \phi_0/4G_N$ and χ is the Euler number of $2d$ manifold $\mathscr{M}$. Integrating over ϕ sets the curvature to be constantly negative, and the Einstein's equation reduces to a linear equation for the dilaton field that will be solved later:

$$\nabla_\mu \nabla_\nu \phi - g_{\mu\nu}\nabla^2\phi + \phi g_{\mu\nu} = 8\pi G_N T_{\mu\nu}. \tag{19.33}$$

Studying in the semi-classical limit, we can replace $T_{\mu\nu}$ with their expectation value $\langle T_{\mu\nu}\rangle$. Now consider the Penrose diagrams of a two-sided AdS black hole and Minkowski spacetime. These diagrams can be glued together, forming a hexagon-like structure, see Fig. 19.7. The Minkowski regions are referred to as the heat bath.

In Euclidean space, the AdS region is represented as a disk with the metric:

$$ds^2 = \frac{4\pi^2}{\beta^2}\sinh^2\rho\, d\tau^2 + d\rho^2, \tag{19.34}$$

where τ is identified as $\tau \sim \tau + \beta$ to avoid a conical singularity at $\rho = 0$.

In order to glue to flat space, it's more convenient to use the black hole coordinates

$$ds^2 = \frac{4\pi^2}{\beta^2}\frac{dy\, d\bar{y}}{\sinh^2 \frac{\pi}{\beta}(y+\bar{y})}, \tag{19.35}$$

where $y = \sigma + i\tau$, $\bar{y} = \sigma - i\tau$. The flat space has the coordinates

$$ds^2 = \frac{1}{\epsilon^2} dy\, d\bar{y}. \tag{19.36}$$

For a flat Euclidean spacetime, compactifying the τ direction ($\tau \sim \tau + \beta$) transforms the spacetime into a cylinder extending infinitely in both directions. Gluing AdS to Minkowski spacetime corresponds to attaching a disk to a *half-cylinder* that extends upward only.

In complex analysis, a cylinder parameterized by $(\sigma, \tau) \sim (\sigma, \tau + \beta)$ can be conformally mapped to the complex plane using:

$$z = \exp\left(\frac{2\pi}{\beta}(\sigma + i\tau)\right) = \exp\left(\frac{2\pi}{\beta}y\right). \tag{19.37}$$

For a full cylinder, $\sigma \in (-\infty, +\infty)$. For a half-cylinder, the range is restricted to $\sigma \in [0, +\infty)$, corresponding to $|z| \geq 1$. In CFTs, conformal transformations allow us to treat Euclidean flat spacetime as a plane with a disk carved out. Gluing the hyperbolic disk ($|z| < 1$) to Euclidean flat space corresponds to filling the hole in the plane with the hyperbolic disk.

$$ds^2 = \frac{dz\, d\bar{z}}{(1 - |z|^2)^2}, \quad |z| < 1 - \frac{2\pi\epsilon}{\beta}, \tag{19.38}$$

$$ds^2 = \frac{\beta^2}{4\pi^2\epsilon^2} dz\, d\bar{z}, \quad |z| > 1 + \frac{2\pi\epsilon}{\beta}. \tag{19.39}$$

The state prepared by such Euclidean evolution is called the Hartle-Hawking state, and under analytic continuation into Lorentzian time, we obtain the time evolution of the system. In this coordinate system, the dilaton solution of (19.33) in the Hartle-Hawking state is

$$\phi = \frac{2\pi\phi_r}{\beta}\frac{1 + |z|^2}{1 - |z|^2} = -\frac{2\pi\phi_r}{\beta}\frac{1}{\tanh\frac{2\pi\sigma}{\beta}}, \quad \sigma < 0, \tag{19.40}$$

where ϕ_r is a parameter such that the dilaton at the boundary is $\phi = \phi_r/\epsilon$.

The heat bath collects radiation, and its entanglement entropy is calculated using the replica trick in field theory. In this framework, the regions $z \in (-\infty, -1 - \epsilon) \cup (1 + \epsilon, +\infty)$ are cut, where ϵ is an arbitrary positive number. The ϵ is introduced for the following reason. The boundary of the gravitational region is at $|z| = 1$, corresponding to $|z| < 1$ being the gravitating region. Cutting at $z = \pm 1$, as done in the gravitational replica trick, computes the gravitational entropy (see Sect. 19.5, where cutting at $z = 1$ calculates the entropy of a one-sided black hole). By introducing a positive ϵ, we ensure that only the entropy of the radiation is considered, while avoiding the inclusion of the gravitational region's entropy. From AdS/CFT perspective, we are coupling CFT_1 with a flat space bath, and the ϵ is introduced to separate CFT_1 from the bath.

With these preparations, we calculate the von Neumann entropy $S_{vN}(\rho_{rad})$ using the replica trick as follows:

1. Take n copies of $\mathbb{C}\backslash\{z : |z| < 1\}$. The region $\{z : |z| < 1\}$ is excluded because it is gravitating, and the geometry is not fixed.
2. On each copy, cut along $z \in (-\infty, -1 - \epsilon) \cup (1 + \epsilon, +\infty)$.
3. Glue the copies cyclically as prescribed by the replica trick.
4. Fill in the gravitational bulk, which is bounded by n copies of $|z| = 1$.

In Step 4, the gravitational bulk is bounded by n disconnected boundaries. The bulk geometry can be either connected or disconnected. In the next two sections, we consider two extreme cases: fully disconnected and fully connected bulk geometries.

Our analysis below relies on the semi-classical approximation. According to Eq. (19.21), we need to compute $\log Z_n$ for $n \approx 1$. The partition function Z_n is calculated via the gravitational path integral, utilizing the replica trick. The semi-classical ansatz is:

$$Z_n = \sum_{\gamma} A_\gamma \exp(-I_\gamma), \tag{19.41}$$

where γ represents saddle configurations, which are geometries satisfying the classical equations of motion and the given boundary conditions. Generally, there are multiple saddle solutions. I_γ is the action of saddle γ, which is typically very large.[6] A_γ is the one-loop determinant, which is smaller than the contribution from the action. The dominant contribution to Z_n comes from the saddle with the smallest I_γ. Therefore, we take the approximation:

$$-\log Z_n \approx \min_{\gamma} I_\gamma. \tag{19.42}$$

Later, we will demonstrate that for $S_{\text{vN}}(\rho_{\text{rad}})$ as a function of time, different saddles with varying bulk topologies dominate at different time scales.

19.8 Trivial Topology: Grow Forever and Ever

In this section, we consider a totally disconnected bulk replica geometry [11, 12, 28–31]. With the boundaries consisting of n circles, the simplest bulk configuration is n disks. More complicated geometries, such as handle-disks, are assumed to be subleading and are not considered here.

The contribution from the n-disk geometry to $S_{\text{vN}}(\rho_{\text{rad}})$ is equal to the field theory entanglement entropy $S_{\text{FT}}(\text{rad})$ calculated on a fixed AdS background: To calculate $S_{\text{vN}}(\rho_{\text{rad}})$, we start with n copies of $\mathbb{C} \setminus \{z : |z| < 1\}$ that describes the AdS-bath system, and then insert pairs of twist operators at the boundary of the regions $z \in (-\infty, -1-\epsilon) \cup (1+\epsilon, +\infty)$. On each copy, the geometry outside $|z| = 1$ that describes the bath is fixed, while the gravitating region bounded by $|z| = 1$ is not fixed and needs to be determined from gravitational path integral (Fig. 19.8).

When the leading geometry consists of n disks, the configuration coincides with the configuration on a fixed background, start with n copies of $\mathbb{C}$ where the geometry inside $|z| = 1$ is also fixed which then leads to the QFT entropy $S_{\text{FT}}(\text{rad})$. $S_{\text{FT}}(\text{rad})$ is what Hawking originally calculated. According to Hawking's argument, radiation

[6] The action comprises both a gravitational part, scaled by $1/G_N$, and a matter part. For CFTs, the matter part is computed in the large central charge limit.

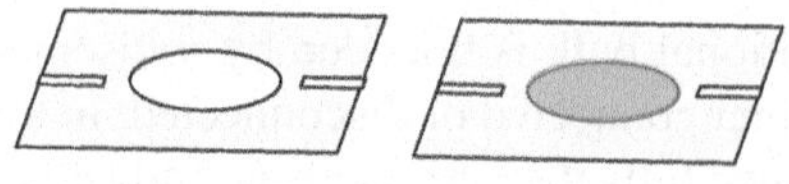

Fig. 19.8 Each figure is one of the n identical copies. In the left figure, we don't fill in the gravitational bulk region. In the right one, we fill in the bulk with disk topology. For the cut on the non-gravitational region, we impose the same boundary condition as in Fig. 19.2

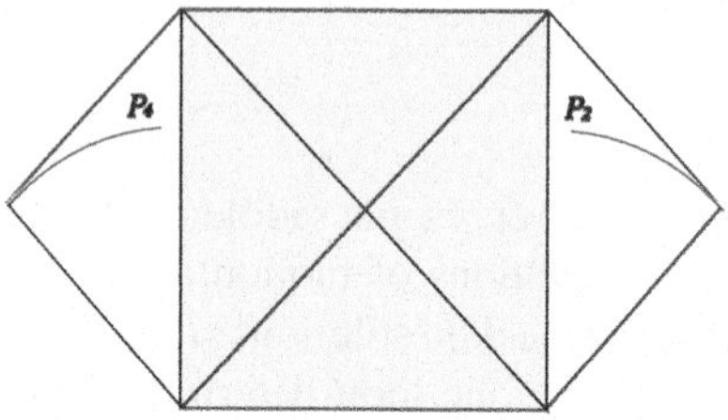

Fig. 19.9 Trivial topology in the bulk leads to calculating the entanglement entropy of the CFT

collected at different times is uncorrelated. Thus, as more radiation is collected, the effective central charge c increases, and the entropy grows.

To calculate $S_{\text{FT}}(\text{rad})$, consider the entanglement entropy of the purple intervals in Fig. 19.9. Due to the purity of the quantum state on the full Cauchy slice, this reduces to the single-interval entropy between points P_2 and P_4. As shown in Sect. 19.7, the metric in the flat region is:

$$ds^2 = \frac{\beta^2}{4\pi^2\epsilon^2} dz d\bar{z}.$$

Parametrize the points in (σ, τ) coordinates as $P_2(b, t)$ and $P_4(b, -t + i\beta/2)$. Using the entanglement entropy formula for a single interval,[7] we have:

$$S = \frac{c}{6} \log \frac{|z_2 - z_4|^2}{\Omega(z_2)\Omega(z_4)} + \text{constant}. \tag{19.43}$$

In complex plane coordinates, P_2 corresponds to e^{b+t} and P_4 to $-e^{b-t}$ (since $\tau = -it$). Substitute them in, the entanglement entropy becomes:

$$S = \frac{c}{3} \log \frac{e^{\frac{2\pi}{\beta}(b+t)} + e^{\frac{2\pi}{\beta}(b-t)}}{e^{\frac{2\pi}{\beta}b}} = \frac{c}{3} \log \left(2 \cosh \frac{2\pi}{\beta} t \right). \tag{19.44}$$

For large t, the entropy grows linearly with t:

$$S \sim \frac{c}{3} \cdot \frac{2\pi}{\beta} t. \tag{19.45}$$

[7] Here, the formula applies to a plane with the metric $ds^2 = \Omega^{-2}(z, \bar{z}) dz d\bar{z}$ rather than a flat background. The Weyl factor in the denominator accounts for the Weyl transformation properties of twist operators, as the correlation function of twist operators is computed on a non-trivial background.

This result aligns with Hawking's original calculation, where entropy grows indefinitely. However, this is impossible in the eternal black hole case, as the black hole has finite degrees of freedom, and the entanglement entropy must have an upper bound. The entropy is bounded by twice the Bekenstein-Hawking entropy of each black hole. The time at which the entropy equals twice the Bekenstein-Hawking entropy is known as the *Page time*.

The linear growth behavior reflects the growth of the black hole's interior volume. The entanglement entropy arises from the correlation between particles inside and outside the black hole. The volume of the black hole's interior, i.e., the volume of the Einstein-Rosen bridge, grows linearly with time. For large t, we find:

$$S_{\text{EE}} \sim V_{\text{interior}} \propto t. \tag{19.46}$$

In conclusion, a totally disconnected bulk fails to bound the growth of $S_{\text{vN}}(\rho_{\text{rad}})$.

19.9 Connected Topology: Island and Plateau

In this section, we explore the other extreme: a bulk geometry that connects the n $|z| = 1$ boundaries [11, 12, 28–31]. For simplicity, our discussion focuses on genus-0 (gravitating) bulk geometries.

We will first demonstrate that for a connected bulk configuration, the von Neumann entropy of radiation, $S_{\text{vN}}(\rho_{\text{rad}})$, differs from the field theory result, $S_{\text{FT}}(\text{rad})$. Specifically, additional contributions from *islands* emerge, indicating that islands can exist. Subsequently, we will show that at late times, the connected saddle dominates over the disconnected one, confirming the existence of islands in this regime.

19.9.1 Island Can Exist

The fully connected (gravitational) bulk geometry for Z_n has the topology of a sphere with n holes. Figures 19.10 and 19.11 illustrate the cases for $n = 5$ and $n = 3$, respectively.

The paradigm introduced in Sect. 19.5 can be neatly applied here, though some subtleties arise.

There are now two fixed points of $\mathbb{Z}_n$, leading to two Area$/4G_N$ terms in the entropy calculation. Additionally, in the replica-symmetric bulk picture, the bulk is divided into n identical regions by n "lines",[8] as shown by the dashed lines in Fig. 19.11. Each line extends from one fixed point to the other.

By crossing each such line, we transition from one replica to its neighbor. Focusing on replica i, the bulk contains two such lines connecting the fixed points, denoted as l_i

[8] In higher dimensions, these are codimension-1 submanifolds.

Fig. 19.10 Connected geometry of $\mathrm{Tr}\rho^5$ in the gravitational path integral. The red points represent fixed points under $\mathbb{Z}_n$ symmetry. Cuts follow the rules of Fig. 19.2

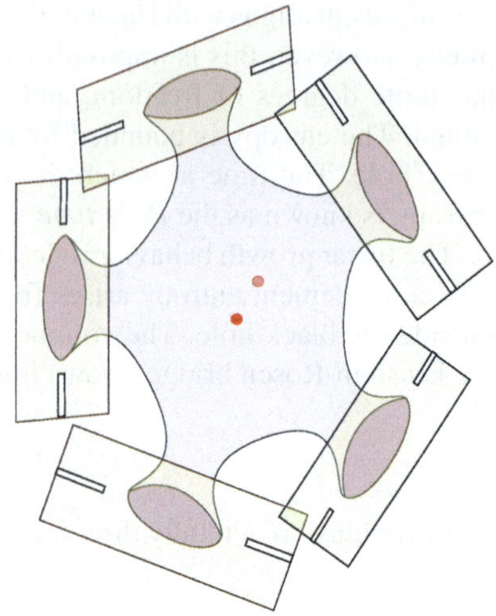

Fig. 19.11 Connected geometry in the $n = 3$ replica, showing two fixed points. Cuts follow the rules of Fig. 19.2. The two red dots are the two fixed point of $\mathbb{Z}_n$

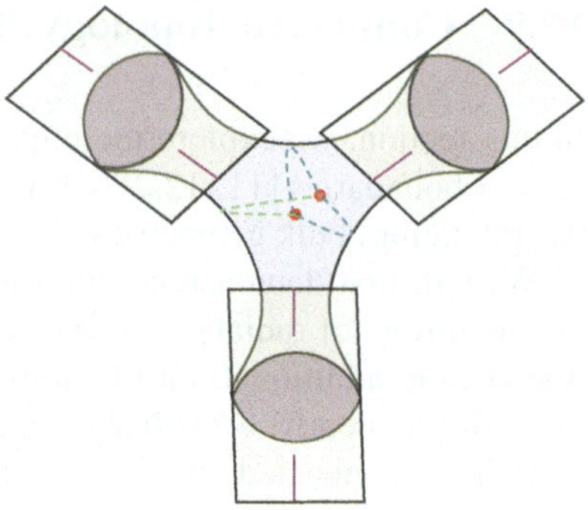

and l_i'. In the $n \to 1$ limit, l_i and l_i' nearly coincide and are separated by a branch cut. In the original (pre-quotient) $\widetilde{\mathcal{M}_n}$ bulk, crossing l_i sends us to replica $(i - 1)$, while crossing l_i' sends us to replica $(i + 1)$. These branch cuts correspond to cuts in the field theory when the replica trick is applied. Consequently, on the quotient manifold $\mathcal{M}_n$, the entanglement entropy $S_{\mathrm{FT}}(\mathrm{rad} \cup \text{branch cut})$ is calculated. The branch cut is called the *island*.

The von Neumann entropy $S_{\mathrm{vN}}(\rho_{\mathrm{rad}})$ thus comprises:

- $S_{\mathrm{FT}}(\mathrm{rad} \cup \text{branch cut})$, calculated on the classical $n = 1$ spacetime.
- Two $A/4G_N$ terms corresponding to the fixed points, also calculated on the classical $n = 1$ spacetime.

In a quantum context, all possible positions of the fixed points must be considered. Within the semi-classical approximation, following the paradigm established in Eq. (19.42), some form of minimization is required. The resulting entropy is:

$$S_{\text{vN}}(\rho_{\text{rad}}) \approx \min_{I} \text{Ext} \left\{ \sum_{\text{fixed}} \frac{\text{Area}}{4G_N} + S_{\text{FT}}(\text{rad} \cup I) \right\}, \tag{19.47}$$

where I denotes the island. This result is known as the *island formula* [28–30], a central result of this chapter. Note that the island formula gives the contribution of the connected saddle only. To determine which saddle dominates, one must compare this result with the entropy from the disconnected saddle.

Compared to the disconnected bulk configuration, the presence of an island introduces $\text{Area}/4G_N$ terms for the endpoints and an additional cut for field theory calculations. In principle, more "islands" could be included, resulting in the generalized island formula:

$$\min_{I_1,\ldots,I_m} \text{Ext} \left\{ \sum_{\text{fixed}} \frac{\text{Area}}{4G_N} + S_{\text{FT}}(\text{rad} \cup I_1 \cup \ldots \cup I_m) \right\}. \tag{19.48}$$

This formula, which accounts for bulk geometries with higher genera, is expected to yield much larger entropy contributions than Eq. (19.47), as it necessarily includes more $\text{Area}/4G_N$ terms. This is one reason why we have focused on genus-0 geometries so far. The disconnected contribution is a special case of "no island," but for convenience, we collectively refer to all cases as the "island formula."

The island formula demonstrates that islands can exist. Crucially, to determine whether an island does exist, one must compare Eq. (19.47) with the disconnected result as will be done in next section.

19.9.2 Island Does Exist

In this subsection, we analyze the eternal black hole and demonstrate that islands dominate at late times [29]. Specifically, we calculate the generalized entropy with the island contributions and show that islands resolve the issue of unbounded entropy growth.

The setup involves a black hole coupled to a heat bath, with the radiation region defined as:

$$\text{rad} = [\text{P}_4, \infty_L) \cup [\text{P}_2, \infty_R), \tag{19.49}$$

and the island as:

$$I = [\text{P}_3, \text{P}_1], \tag{19.50}$$

as shown in Fig. 19.12. The points are parameterized in (σ, τ) coordinates as:

$$\text{P}_1(-a, t_a), \quad \text{P}_2(b, t_b), \quad \text{P}_3(-a, -t_a + i\pi), \quad \text{P}_4(b, -t_b + i\pi), \tag{19.51}$$

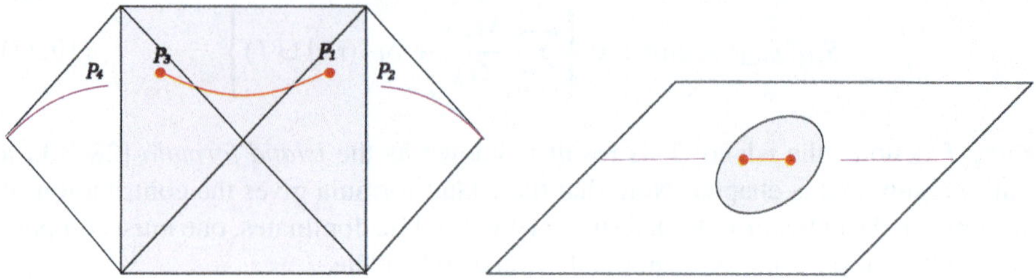

Fig. 19.12 Islands in spacetimes with Lorentzian and Euclidean signatures, respectively. The red line represents the island, with red points marking its boundaries (corresponding to the two $\mathbb{Z}_n$ fixed points). The purple lines denote the original radiation region

where we set $\beta = 2\pi$ for convenience and we have uses the parity symmetry to simplify our ansatz.

The CFT entropy formula for two intervals is not universal, so we follow [29], which considers a theory of many copies of free Dirac fermions. The resulting entropy is given by:

$$\begin{aligned} S_{\mathrm{CFT}}(I \cup \mathrm{rad}) &= \frac{c}{6}\log\left(\frac{|z_{21}z_{32}z_{43}z_{41}|^2}{|z_{31}z_{42}|^2\Omega_1\Omega_2\Omega_3\Omega_4}\right) \\ &= \frac{c}{3}\log\frac{2\cosh t_a \cosh t_b |\cosh(t_a - t_b) - \cosh(a+b)|}{\sinh a \cosh\left(\frac{a+b-t_a-t_b}{2}\right)\cosh\left(\frac{a+b+t_a+t_b}{2}\right)}. \end{aligned} \tag{19.52}$$

The generalized entropy with the island contribution is:

$$S_{\mathrm{gen}}^{\mathrm{island}} = 2S_0 + \frac{2\phi_r}{4G_N \tanh a} + S_{\mathrm{CFT}}(I \cup \mathrm{rad}), \tag{19.53}$$

where we use the dilaton profile (19.40) that gives the area term $S_0 + \frac{\phi_r}{4G_N \tanh a}$.

While without considering the island, the entropy is (19.45):

$$S_{\mathrm{gen}}^{\text{no island}} = \frac{c}{3}\log(2\cosh t_b) \sim \frac{c}{3}t_b. \tag{19.54}$$

This result grows linearly with time and cannot be correct because, as time evolves, the two boundaries will disentangle from each other and entangle with the bath, leading to:

$$S \sim 2S_{\mathrm{BH}} = 2S_0 + 2\frac{\phi_r}{4G_N}, \tag{19.55}$$

where S_{BH} is the Bekenstein-Hawking entropy and the horizon is located at $\sigma \to -\infty$ (Eq. (19.40)). Now we want to show that the island contribution correctly reproduces this result at late times.

At late times, as $t_a \to \infty$ and $t_b \to \infty$, Eq. (19.52) simplifies to:

$$S_{\text{CFT}} = \frac{c}{3} \log \frac{2|\cosh(t_a - t_b) - \cosh(a+b)|}{\sinh a}. \tag{19.56}$$

Minimizing the generalized entropy with respect to t_a ($\partial_{t_a} S_{\text{gen}} = 0$) gives $t_a = t_b$. When $\frac{\phi_r}{G_N c} \gg 1$, and at large a, the generalized entropy becomes:

$$S = 2S_0 + \frac{2\phi_r}{4G_N}(1 + e^{-2a}) - \frac{2c}{3}e^{-a} = 2S_{\text{BH}} + \left(\frac{2\phi_r}{4G_N}e^{-2a} - \frac{2c}{3}e^{-a}\right) \tag{19.57}$$

This leads to the saddle of $a \sim \log \frac{\phi_r}{G_N c}$. The CFT entropy is time-independent, yielding:

$$S_{\text{vN}}(\text{rad}) = 2S_0 + \frac{2\phi_r}{4G_N \tanh a} + \frac{c}{3} \log \frac{2\cosh(a+b)}{\sinh a}. \tag{19.58}$$

This result indeed matches twice the black hole entropy, with a small correction from the QFT. More generally, we expect the island to be close to the black hole horizon, as this region already encompasses the interior volume needed to purify the radiation. Near the horizon, the area, which is a local minimum, depends quadratically on the radius ρ: Area $\sim (\text{Area})''\rho^2$. At late times, the QFT entropy is in an OPE limit[9] and depends linearly on the radius: $S_{\text{QFT}} \sim c\rho$ [30]. This leads to a balance between the $\frac{\text{Area}}{4G_N}$ term and the S_{QFT} term, with $\rho \sim G_N c$, and the generalized entropy becomes close to $2S_{BH}$.

The dominance of this new extremal surface was initially discovered by [11, 12] and subsequently explored in [28–31].

19.10 A Simple Model: EOW Brane

Another model, where the degrees of freedom of a black hole are encoded on an End of the World (EOW) brane, has been studied. This section explores this simple model to understand the phase transition between different gravitational saddles [30, 32].

In this model, an EOW brane is introduced behind the black hole horizon (Fig. 19.13), and the gravitational action was taken to be pure Einstein-Hilbert action which is equal to the Euler characteristic in 2d. The internal degrees of freedom of the EOW brane represent the microstates of the black hole, and these states are entangled

[9] OPE (Operator Product Expansion) means that when two local operators is closed to each other, the product of them can be expanded as the sum of some single local operators. In the discussion above, it means that at late time, the four-point function of the twist operators can be factorized as two-point functions. Therefore, the total entanglement entropy of CFT is simply twice of the entanglement entropy of CFT on one side.

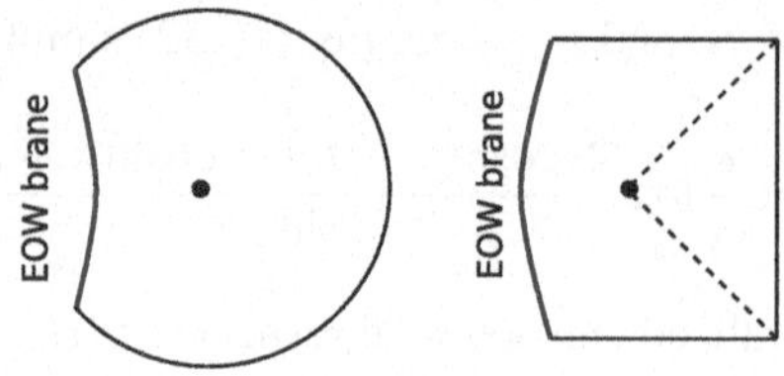

Fig. 19.13 A black hole with an EOW brane behind the horizon in Euclidean and Lorentzian signatures, respectively. Figures reproduced with permission from [30]

with an auxiliary system R, which models the radiation emitted during black hole evaporation. The entire system is described by the pure state:

$$|\Psi\rangle = \frac{1}{\sqrt{k}} \sum_{i=1}^{k} |\psi_i\rangle_B \, |i\rangle_R \,, \tag{19.59}$$

where $|\psi_i\rangle_B$ represents a black hole state, and $|i\rangle_R$ represents a radiation state. The reduced density matrix for the radiation is:

$$\rho_R = \frac{1}{k} \sum_{i,j=1}^{k} |j\rangle_R \, \langle i|_R \left\langle \psi_i | \psi_j \right\rangle_B \,, \tag{19.60}$$

where $\left\langle \psi_i | \psi_j \right\rangle_B$ is the gravitational amplitude. Naively, $\left\langle \psi_i | \psi_j \right\rangle_B$ is proportional to δ_{ij}. However, when computing $|\left\langle \psi_i | \psi_j \right\rangle_B|^2$, connected geometries also contribute, as suggested by the Euclidean gravitational path integral. This discrepancy, where $|\left\langle \psi_i | \psi_j \right\rangle_B|^2$ is not the modulus square of $\left\langle \psi_i | \psi_j \right\rangle_B$, is called the *factorization problem*. This was first pointed out by Coleman and Giddings-Strominger [33–35] and most recently by Maldacena-Maoz [36], where they showed that the inclusion of spacetime wormhole geometries will lead to hidden α parameters and the semiclassical gravitational theory requires a sort of ensemble average over the α parameters. This also raises the question of what is the bulk geometry in a fixed α state, that will be linked to the black hole microstates which is still a mystery at the current stage [36, 37]. Below, we will use the ensemble averaged interpretation and use wormhole to calculate quantities that are self-averaged such as Renyi-entropies and entanglement entropies.

To compute the von Neumann entropy, we evaluate $\mathrm{Tr}\rho^n$. For $\mathrm{Tr}\rho^2$, we have:

$$\mathrm{Tr}\rho^2 = \frac{1}{k^2} \sum_{i,j=1}^{k} |\left\langle \psi_i | \psi_j \right\rangle_B|^2, \tag{19.61}$$

where $|\left\langle \psi_i | \psi_j \right\rangle_B|^2$ includes contributions from both disconnected and connected geometries. The disconnected contribution is $\delta_{ij}\delta_{ij} Z_1^2$, while the connected contribution is $\delta_{ii}\delta_{jj}Z_2$. Here, Z_1 and Z_2 denote the gravitational path integral for each configuration. This gives:

$$\mathrm{Tr}\rho^2 = \frac{kZ_1^2 + k^2 Z_2}{(kZ_1)^2} = \frac{1}{k} + \frac{Z_2}{Z_1^2}. \tag{19.62}$$

Using the topological action, we get $Z_2/Z_1^2 = e^{-S_0}$, with S_0 being the coupling constant. This gives:

$$\mathrm{Tr}\rho^2 = \frac{1}{k} + e^{-S_0}. \tag{19.63}$$

1. At early times (k small), the first term dominates, leading to $S = \log k$, which grows indefinitely with k.
2. At late times (k large), the second term dominates, corresponding to the connected geometry. In this regime:

$$\mathrm{Tr}\rho^n = \frac{Z_n}{Z_1^n}, \tag{19.64}$$

resulting in an entropy:

$$S = S_0 = S_{\mathrm{BH}}. \tag{19.65}$$

analogous to the black hole entropy derived in Sect. 19.5 (Fig. 19.14).

Combining these results, the entropy is:

$$S = \min\{\log k,\ S_{\mathrm{BH}}\}, \tag{19.66}$$

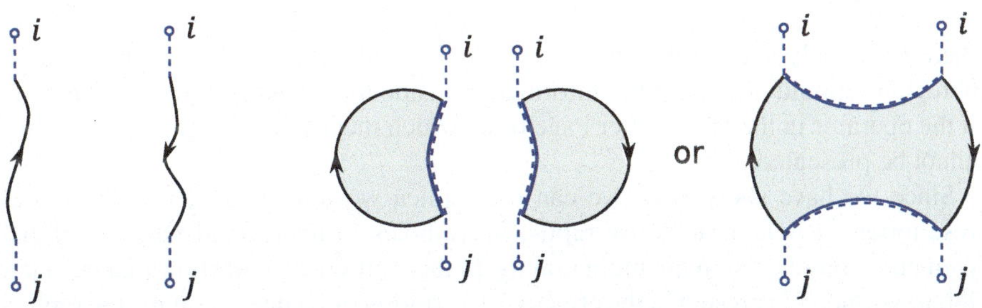

Fig. 19.14 Contributions to $|\langle\psi_i|\psi_j\rangle_B|^2$. The left figure shows the boundary conditions for the gravitational path integral, with the black line as the asymptotic boundary. Indices i and j represent EOW brane states. The middle and right figures show two possible saddles. Solid blue lines represent the EOW brane following a geodesic between endpoints, and these two figures represent Z_1^2 and Z_2 respectively. Figures reproduced with permission from [30]

which mirrors the behavior of the Page curve for an eternal black hole.

Actually, in this simple model, or a more complicated version including dynamics of JT gravity, one can relax the assumption of replica symmetry and use the technique of free probability theory to solve the full entanglement spectrum. Readers who are interested in this kind of technique can read [30].

19.11 Violation of Global Symmetry

The existence of replica wormholes also implies the violation of global symmetries in quantum gravity. Hawking argued long ago that the process of black hole evaporation can violate the superselection rule for global symmetries. This has been recently discussed by Harlow-Ooguri and Harlow-Shaghoulian, based on the concept of entanglement wedge reconstruction [38–40].

In the evaporating black hole picture, this idea can be explained as follows: We divide the system into two regions B and R. B contains the black hole and it has a holographic description. R is the radiation region where we can neglect the gravity effect. We have known that after Page time, the entanglement entropy between B and R saturates the Bekenstein-Hawking entropy and decreases to zero, which corresponds to the emergence of non-trivial quantum extremal surface in the bulk picture, and we know that the island region belongs to the entanglement wedge of R. However, if we consider a small region of R, whose thermal entropy doesn't exceed the BH entropy, it will not contain the information inside the island. Based on this, Harlow and Shaghoulian argued that if the unitary transformation generated by global symmetry can be written as the product of two parts, the unitary transformation on B and small portion R_i with $\cup R_i = R$, i.e.

$$U(g) = U(g, B) \prod_i U(g, R_i) U_{\text{edge}} \tag{19.67}$$

where U_{edge} only acts on the edges of R_i. As a result, each unitary acting on their subregions should not affect the information inside the island and therefore not act on the operator in the island after Page time, which means that the global symmetry cannot be presented.

Since we have shown that we can use replica wormholes to derive the island prescription, it is natural to use replica wormholes to understand why the global symmetry cannot exist in quantum gravity. In this section we will show that even if a global symmetry is present in the effective description of quantum gravity, the replica wormhole in the Euclidean path integral will lead to a non-perturbative violation of this global symmetry.

In a gravitational theory coupled to the matter field, which is invariant under some global transformation G, we can prepare a state via Euclidean path integral, acting the operator formed from the matter field on the Hartle-Hawking state, i.e.

$$|\psi\rangle = \mathcal{O}(\tau)\,|HH\rangle\,. \tag{19.68}$$

To see the contradiction, we consider two operators $\mathcal{O}_1$ and $\mathcal{O}_2$, transforming under representation R_1 and R_2 of G respectively. Acting them on the Hartle-Hawking state, we then have $|R_1\rangle$ and $|R_2\rangle$. If there is no singlet in $\bar{R}_1 \times R_2$, then $\langle R_1|R_2\rangle = \left\langle \mathcal{O}_1^\dagger \mathcal{O}_2 \right\rangle$ will vanish on all the configurations because of charge conservation, under the assumption that global symmetry G is not spontaneously broken. However, the physical observable-scattering amplitude $|\langle R_1|R_2\rangle|^2$ will also have contributions on connected geometry

$$|\langle R_1|R_2\rangle|^2 = \mathrm{Tr}\rho_{\text{in-in}}\rho_{\text{out-out}}, \tag{19.69}$$

where $\rho_{\text{in-in}} = |R_1\rangle\langle R_1|$ and $\rho_{\text{out-out}} = |R_2\rangle\langle R_2|$. Rewriting in density matrices, it is natural to compare with the calculation of Renyi entropy discussed above, which inspires us to consider the contribution from replica wormholes. As in Fig. 19.15, the first saddle is the disconnected part, vanishing because of charge conservation. Thus the leading contribution is from the manifold with zero genus. This contribution corresponds to $\left\langle \mathcal{O}_1^\dagger \mathcal{O}_2 \mathcal{O}_1 \mathcal{O}_2^\dagger \right\rangle$, which is non-vanishing since singlets always be contained in $\bar{R}_1 \times R_1$ and $\bar{R}_2 \times R_2$. This wormhole contribution is the non-perturbative contribution to gravitational path integral, which is $e^{-\frac{1}{G_N}}$ suppressed.

Therefore, we can conclude that the transition amplitude does not vanish, implying that the global symmetry G is violated in quantum gravity, even though there is no explicit breaking of this symmetry in the Lagrangian.

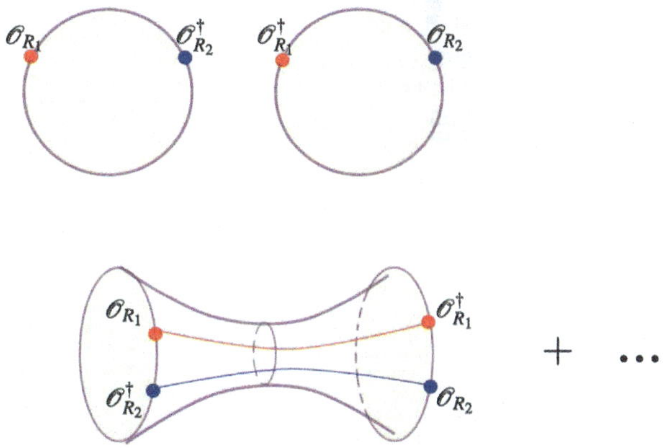

Fig. 19.15 Connected and disconnected contribution to $|\langle R_1|R_2\rangle|^2 = \mathrm{tr}\rho_{\text{in-in}}\rho_{\text{out-out}}$. The picture above represents the disconnected part, which is the vanishing contribution. The picture below is the connected wormhole contribution, which is the leading contribution now. And the ellipsis represents the subleading higher genus contribution

19.12 An Evaporating Black Hole

Discussion so far has focused on eternal black holes, where qualitative analyses are easier. It is shown that the entropy of radiation saturates to a plateau value. In this section, we will return to the original information paradox: for a black hole formed by pure matter, after evaporation, the final state consists of radiation only, and the problem is to show that the final state is pure [31]. The lessons in the previous sections will be applied boldly to black holes in asymptotic flat spacetime. As a reminder, the lessons are:

- The bulk spacetime is roughly separated into non-gravitational regions and gravitational regions. Radiation is collected in non-gravitational regions.
- Universality of island formulae (19.47) and (19.48). We need at most one island for our purpose here.
- Saddle point approximation. We will simply compare the on-shell values of different saddles and determine the dominant one.
- For the field theory entropy $S_{\rm FT}$, each Hawking radiation particle has a Hawking partner (which falls into the horizon) and the partner purifies the radiation particle. $S_{\rm FT}(\text{rad particle}) = \mathscr{O}(1)\,, \;\; S_{\rm FT}(\text{rad particle} \cup \text{partner}) \approx 0.$

Take the example of an evaporating Schwarzschild black hole (Fig. 19.16).

When one is far from the horizon, gravitation is weak and can be considered non-gravitational. One may take a codimension-2 surface enclosing the horizon to be a cut-off surface such that gravitation outside of it is negligible, for example, a

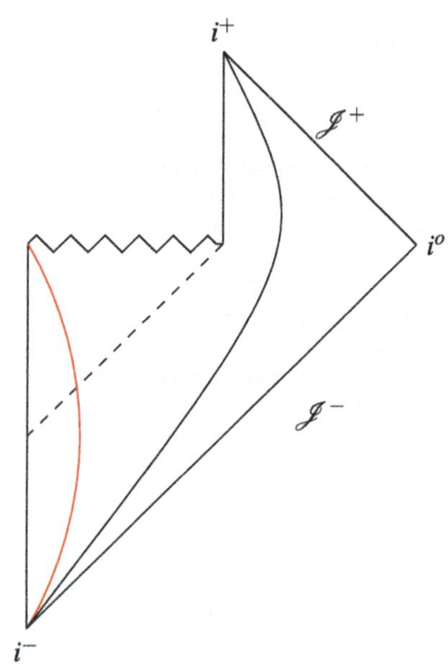

Fig. 19.16 Penrose diagram of an evaporating black hole. The red curve denotes infalling matter while the black curve denotes the cut-off surface

sphere with the radius a few times as large as the Schwarzschild radius. Radiation is collected outside the cut-off surface. Now, let us apply the "island formula" (19.48).

At an early time, if there is no island, $S_{\text{vN}} = S_{\text{FT}}(\text{rad})$. It calculates the field theory entropy of Hawking radiation, as Hawking originally did. One may wonder whether it is possible to get a smaller (more dominant) saddle by adding an island: to do that, the island should include the Hawking partners of early radiation, such that $S_{\text{FT}}(\text{rad} \cup \text{island}) \approx 0$. However, the Hawking partners are not locally encoded inside the horizon. In order to purify radiation, the hypothetical island should include the whole black hole interior, which results in a very large $\text{Area}/4G_N$, where Area is the area of the horizon. This is much larger than $S_{\text{FT}}(\text{rad})$ at an early time and thus it is not economical to include islands.

- Early time. No island. One can trust Hawking's calculation.

At a late time, $S_{\text{FT}}(\text{rad})$ becomes too large: even larger than $\text{Area}/4G_N$. Now including an island that is roughly the black hole interior reduces S_{vN} from $S_{\text{FT}}(\text{rad})$ to $\text{Area}/4G_N + S_{\text{FT}}(\text{rad} \cup \text{island}) \approx \text{Area}/4G_N$.

- Late time. One island. Hawking's calculation gets modified. $S_{\text{vN}} \sim \text{Area}/4G_N$.

One can easily dissuade herself from constructing a "more dominant" saddle by including more islands.

An evaporating black hole has a continuously decreasing horizon area. The generalized entropy formula can be trusted as long as the curvature at the horizon is not too large, which holds for most of the Page curve. Near the end of the evaporation process, the horizon area becomes small in Planck units, requiring a UV completion to fully understand the process. The physics of this final stage warrants further investigation.

19.13 Conclusion

In conclusion, we briefly reviewed the use of the gravitational path integral to derive the von Neumann entropy of Hawking radiation. The key point is the use of replica wormholes, which, in the limit $n \to 1$, lead to the quantum extremal island formula. Although this rule originates from AdS/CFT, the island formula can be extended to black holes in flat space, yielding the Page curve for evaporating black holes.

References

1. S.W. Hawking, Commun. Math. Phys. **43**, 199–220 (1975). https://doi.org/10.1007/BF02345020 [Erratum: Commun. Math. Phys. **46**, 206 (1976)]
2. L. Susskind, J. Math. Phys. **36**, 6377–6396 (1995). https://doi.org/10.1063/1.531249. arXiv:hep-th/9409089 [hep-th]

3. C.R. Stephens, G. 't Hooft, B.F. Whiting, Class. Quant. Grav. **11**, 621–648 (1994). https://doi.org/10.1088/0264-9381/11/3/014. arXiv:gr-qc/9310006 [gr-qc]
4. C.B. Thorn. arXiv:hep-th/9405069 [hep-th]
5. D.N. Page, Phys. Rev. Lett. **71**, 1291–1294 (1993). https://doi.org/10.1103/PhysRevLett.71.1291. arXiv:gr-qc/9305007 [gr-qc]
6. D.N. Page, Phys. Rev. Lett. **71**, 3743–3746 (1993). https://doi.org/10.1103/PhysRevLett.71.3743. arXiv:hep-th/9306083 [hep-th]
7. J.D. Bekenstein, Lett. Nuovo Cim. **4**, 737–740 (1972). https://doi.org/10.1007/BF02757029
8. J.D. Bekenstein, Phys. Rev. D **7**, 2333–2346 (1973). https://doi.org/10.1103/PhysRevD.7.2333
9. S.W. Hawking, Phys. Rev. D **13**, 191–197 (1976). https://doi.org/10.1103/PhysRevD.13.191
10. G.W. Gibbons, S.W. Hawking, Phys. Rev. D **15**, 2752–2756 (1977). https://doi.org/10.1103/PhysRevD.15.2752
11. A. Almheiri, N. Engelhardt, D. Marolf, H. Maxfield, JHEP **12**, 063 (2019). https://doi.org/10.1007/JHEP12(2019)063. arXiv:1905.08762 [hep-th]
12. G. Penington, JHEP **09**, 002 (2020). https://doi.org/10.1007/JHEP09(2020)002. arXiv:1905.08255 [hep-th]
13. P. Calabrese, J.L. Cardy, J. Stat. Mech. **0406**, P06002 (2004). https://doi.org/10.1088/1742-5468/2004/06/P06002. arXiv:hep-th/0405152 [hep-th]
14. P. Calabrese, J. Cardy, J. Phys. A **42**, 504005 (2009). https://doi.org/10.1088/1751-8113/42/50/504005. arXiv:0905.4013 [cond-mat.stat-mech]
15. A. Lewkowycz, J. Maldacena, JHEP **08**, 090 (2013). https://doi.org/10.1007/JHEP08(2013)090. arXiv:1304.4926 [hep-th]
16. S. Ryu, T. Takayanagi, Phys. Rev. Lett. **96**, 181602 (2006). https://doi.org/10.1103/PhysRevLett.96.181602. arXiv:hep-th/0603001 [hep-th]
17. S. Ryu, T. Takayanagi, JHEP **08**, 045 (2006). https://doi.org/10.1088/1126-6708/2006/08/045. arXiv:hep-th/0605073 [hep-th]
18. V.E. Hubeny, M. Rangamani, T. Takayanagi, JHEP **07**, 062 (2007). https://doi.org/10.1088/1126-6708/2007/07/062. arXiv:0705.0016 [hep-th]
19. X. Dong, D. Harlow, A.C. Wall, Phys. Rev. Lett. **117**(2), 021601 (2016). https://doi.org/10.1103/PhysRevLett.117.021601. arXiv:1601.05416 [hep-th]
20. D. Harlow, Commun. Math. Phys. **354**(3), 865–912 (2017). https://doi.org/10.1007/s00220-017-2904-z. arXiv:1607.03901 [hep-th]
21. T. Faulkner, A. Lewkowycz, J. Maldacena, JHEP **11**, 074 (2013). https://doi.org/10.1007/JHEP11(2013)074. arXiv:1307.2892 [hep-th]
22. X. Dong, JHEP **01**, 044 (2014). https://doi.org/10.1007/JHEP01(2014)044. arXiv:1310.5713 [hep-th]
23. R. Jackiw, Nucl. Phys. B **252**, 343–356 (1985). https://doi.org/10.1016/0550-3213(85)90448-1
24. C. Teitelboim, Phys. Lett. B **126**, 41–45 (1983). https://doi.org/10.1016/0370-2693(83)90012-6
25. A. Almheiri, J. Polchinski, JHEP **11**, 014 (2015). https://doi.org/10.1007/JHEP11(2015)014. arXiv:1402.6334 [hep-th]
26. J. Maldacena, D. Stanford, Z. Yang, PTEP **2016**(12), 12C104 (2016). https://doi.org/10.1093/ptep/ptw124. arXiv:1606.01857 [hep-th]
27. T.G. Mertens, G.J. Turiaci, Living Rev. Rel. **26**(1), 4 (2023). https://doi.org/10.1007/s41114-023-00046-1. arXiv:2210.10846 [hep-th]
28. A. Almheiri, R. Mahajan, J. Maldacena, Y. Zhao, JHEP **03**, 149 (2020). https://doi.org/10.1007/JHEP03(2020)149. arXiv:1908.10996 [hep-th]
29. A. Almheiri, T. Hartman, J. Maldacena, E. Shaghoulian, A. Tajdini, JHEP **05**, 013 (2020). https://doi.org/10.1007/JHEP05(2020)013. arXiv:1911.12333 [hep-th]
30. G. Penington, S.H. Shenker, D. Stanford, Z. Yang, JHEP **03**, 205 (2022). https://doi.org/10.1007/JHEP03(2022)205. arXiv:1911.11977 [hep-th]
31. A. Almheiri, T. Hartman, J. Maldacena, E. Shaghoulian, A. Tajdini, Rev. Mod. Phys. **93**(3), 035002 (2021). https://doi.org/10.1103/RevModPhys.93.035002. arXiv:2006.06872 [hep-th]

32. I. Kourkoulou, J. Maldacena. arXiv:1707.02325 [hep-th]
33. S.R. Coleman, Nucl. Phys. B **307**, 867–882 (1988). https://doi.org/10.1016/0550-3213(88)90110-1
34. S.B. Giddings, A. Strominger, Nucl. Phys. B **307**, 854–866 (1988). https://doi.org/10.1016/0550-3213(88)90109-5
35. S.B. Giddings, A. Strominger, Nucl. Phys. B **321**, 481–508 (1989). https://doi.org/10.1016/0550-3213(89)90353-2
36. J.M. Maldacena, L. Maoz, JHEP **02**, 053 (2004). https://doi.org/10.1088/1126-6708/2004/02/053. arXiv:hep-th/0401024 [hep-th]
37. D. Marolf, H. Maxfield, JHEP **08**, 044 (2020). https://doi.org/10.1007/JHEP08(2020)044. arXiv:2002.08950 [hep-th]
38. D. Harlow, H. Ooguri, Phys. Rev. Lett. **122**(19), 191601 (2019). https://doi.org/10.1103/PhysRevLett.122.191601. arXiv:1810.05337 [hep-th]
39. D. Harlow, H. Ooguri, Commun. Math. Phys. **383**(3), 1669–1804 (2021). https://doi.org/10.1007/s00220-021-04040-y. arXiv:1810.05338 [hep-th]
40. D. Harlow, E. Shaghoulian, JHEP **04**, 175 (2021). https://doi.org/10.1007/JHEP04(2021)175. arXiv:2010.10539 [hep-th]
41. R. Penrose, Riv. Nuovo Cim. **1**, 252–276 (1969). https://doi.org/10.1023/A:1016578408204
42. A. Almheiri, D. Marolf, J. Polchinski, J. Sully, JHEP **02**, 062 (2013). https://doi.org/10.1007/JHEP02(2013)062. arXiv:1207.3123 [hep-th]
43. P.S. Hsin, L.V. Iliesiu, Z. Yang, Class. Quant. Grav. **38**(19), 194004 (2021). https://doi.org/10.1088/1361-6382/ac2134. arXiv:2011.09444 [hep-th]

Chapter 20
How Does Information Emerge from a Black Hole?

Suvrat Raju

Abstract An examination of the constraints of quantum gravity leads to a clear physical picture for how information about the initial state is transferred to the Hawking radiation that emerges from a black hole.

20.1 Introduction

It is now generally accepted that black hole evaporation preserves unitarity. Evidence from the AdS/CFT conjecture [1–3] shows that black hole evaporation in asymptotically anti-de Sitter space (AdS) can be modelled using a dual unitary process. The key information-theoretic aspect of this duality can be generalized to asymptotically flat space [4] and asymptotically de Sitter space [5] through the principle of "holography of information" [4–13]. These results, which build on [14–18], imply that, subject to reasonable assumptions about the global state and the UV-completion of low-energy quantum gravity, information in a bounded subregion of a Cauchy slice can be accessed by observables in its complement. Since this principle applies even when the subregion is part of a black hole, an external observer can always verify the purity of the state. This provides a resolution of the information paradox [19, 20] for evaporating black holes as explained in [7, 9].

Additionally, in separate calculations involving AdS black holes coupled to non-gravitational baths [21–24], the entanglement entropy of the bath has been shown to obey a Page curve [25].

Nevertheless, the literature lacks clarity on how information from the infalling matter emerges in the final Hawking radiation. This has led to speculations that the process is mediated by exotic wormholes, or modifications of the semi-classical black hole geometry [26]. The purpose of this chapter is to explain that the transfer

S. Raju (✉)
International Centre for Theoretical Sciences, Tata Institute of Fundamental Research, Shivakote, Bengaluru, India
e-mail: suvrat@icts.res.in

A. Akil and C. Bambi (eds.), *The Black Hole Information Paradox*, Springer Series in Astrophysics and Cosmology, https://doi.org/10.1007/978-981-96-6170-1_20

of information can be understood through an analysis of the constraints of quantum gravity. This mechanism does not require us to invoke new physical phenomena.

20.2 The Classical Gauss Law

Let us start with an elementary but instructive discussion. Here, and in most of what follows, we will restrict ourselves to asymptotically flat space. Consider a star that collapses into a black hole and then evaporates. One piece of information present in the state is the total energy (or mass) of the star. Nevertheless, there has never been any dispute that an observer outside the black hole retains access to this information even after the matter that constitutes the star has disappeared behind the horizon. This is because the energy can be obtained by integrating a component of the metric at infinity using the ADM formula [27],

$$H = \frac{1}{16\pi G} \int_{S_\infty} \left(\partial_i g^i_j - \partial_j g_{ii} \right) d^2\Omega^j. \tag{20.1}$$

This is a manifestation of the Gauss law.

The precise expression for the ADM Hamiltonian involves observables at infinity. Since this chapter aims at providing physical pictures, we will somewhat loosely state that the metric "outside" the black hole has information about its energy.

Even in this context, one might ask: how does information about the energy enter the final Hawking radiation? The matter that sources this energy is causally disconnected from the region outside the black hole, where the radiation forms. But information about the energy does not "jump" from the matter to the exterior. Rather, one must recall that the matter came from outside the black hole in the past. As it collapsed, it modified the exterior metric, and a part of the metric continues to retain information about the total energy. Since the exterior metric carries information about the energy at leading order, it is not surprising that Hawking radiation from the black hole also carries information about the energy at leading order.

Moreover, information about the energy is not "cloned" in the metric at infinity. Rather the gravitational constraints instruct us to *identify* the energy in the bulk with the observable at infinity: the seemingly different ways of measuring the energy correspond to the same operator in the physical Hilbert space.

Of course, in the classical theory, observations outside the black hole are insufficient to fix the state of the interior. Classically, it is easy to construct two distinct states, for which not just the energy but all multipole moments coincide. The classical no-hair theorem suggests that as long as we treat the gravitational theory classically—as is done in textbook analyses of Hawking radiation that utilize quantum field theory in a fixed spacetime background—the Hawking radiation only carries information about the energy of the black hole and about other conserved charges that affect the classical metric.

20.3 Holography of Information

When quantum effects are included, it turns out that observations outside the black hole *are* sufficient to determine its complete state. The precise results proved in [4, 5] pertain to asymptotic algebras. But their physical interpretation is that an observer outside the black hole can completely determine the state of the interior.

These results are essentially based on the Gauss law and the uncertainty principle. The underlying physics can be understood using the following simple argument. Consider a nice slice that passes through the black hole interior and extends out to spatial infinity. Say that there exists an operator in the theory that can be used to alter some observables in the part of the slice inside the black hole while commuting with all operators outside. Such an operator must commute with the Hamiltonian, which is observable outside in a gravitational theory. However, the uncertainty principle tells us that such an operator, which has "zero" energy, must be delocalized. But then it must fail to commute with some *other* operator in the exterior. Therefore, such an operator cannot exist at all.

This is related to the usual argument that there are no gauge-invariant local operators in a theory of gravity. However, the physical consequences of this statement are not widely appreciated. For low-energy excitations about the vacuum, observers outside a bounded region can follow a well-defined perturbative protocol that completely identifies the state in the region [6]. Moreover, by adding weak assumptions about the UV-complete theory, it is possible to obtain a nonperturbative formulation of the principle of holography of information for asymptotic algebras [4]. This nonperturbative result can be applied to black holes. We describe these regimes in turn below.

20.3.1 Perturbative Example

A simple perturbative example, depicted in Fig. 20.1, illustrates this phenomenon about empty space. Let ϕ be a massless scalar field coupled to gravity. Consider the following unitary operator that acts near the "middle" of the spacetime to create an excited state from the vacuum,

$$|f\rangle = e^{-i\lambda \int f(r)\phi(u=0,r,\Omega)d^2\Omega dr}|0\rangle, \tag{20.2}$$

where $u = t - r$ is the retarded time. The information in this state is characterized by the smearing function $f(r)$ that we take to have compact support in the range $r \in (0, r_0)$ for some r_0. We also take $f(r)$ to be spherically symmetric for this example; generalizing to an angle-dependent smearing is straightforward.

Although this unitary appears to be localized, it can be detected by an observer near infinity. Consider an observer who can make observations near the past boundary of future null infinity ($\mathcal{I}^+_-$): the region obtained by first taking $r \to \infty$ and then taking

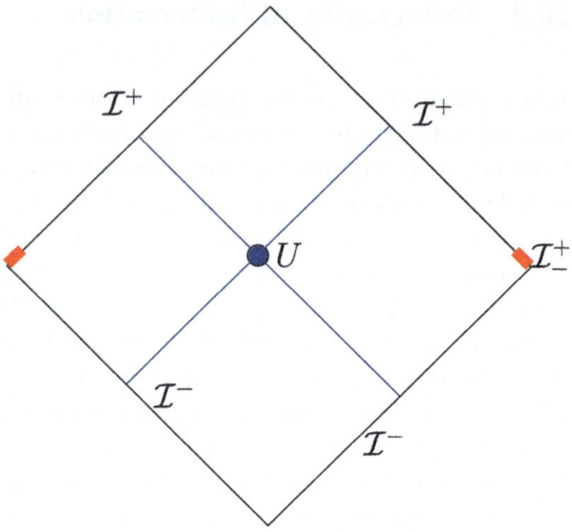

Fig. 20.1 The blue excitation can be written as the action of a "local" unitary on the vacuum but it must originate from $\mathcal{I}^-$ in a gravitational theory. This process leaves behind traces at infinity that are sufficient for an observer near $\mathcal{I}^+_-$ (red region) to determine the profile of the excitation

u to be negative and large in magnitude. Such an observer can compute the correlator of the Hamiltonian, which is available at infinity in a theory of gravity, and another insertion of the field near $\mathcal{I}^+_-$.

$$\lim_{r'\to\infty} r'\frac{\partial}{\partial\lambda}\langle f|H\phi(u', r', \Omega')|f\rangle\big|_{\lambda=0} = \frac{1}{2\pi}\int_0^{r_0} \frac{f(r)}{u'(u'-2r)}dr. \tag{20.3}$$

It is easy to see that a knowledge of the right hand side for the range $u' \in (-\infty, -\frac{1}{\epsilon})$, with any finite ϵ, is sufficient to completely reconstruct $f(r)$. For instance, series expanding the denominator in $\frac{r}{u'}$ yields all moments of $f(r)$ as coefficients of different inverse powers of u',

$$\frac{1}{2\pi}\int_0^{r_0} \frac{f(r)}{u'(u'-2r)}dr = \sum_{n=0}^{\infty} \frac{2^n}{2\pi(u')^{n+2}}\int_0^{r_0} r^n f(r)dr, \tag{20.4}$$

and these moments determine $f(r)$ to any required precision.

Therefore, we see that an observer at infinity has all information about the state. But the mechanism by means of which this happens is not mysterious. In a theory of quantum gravity, there is no such thing as a local source. Therefore, the state $|f\rangle$ must be thought of as originating on $\mathcal{I}^-$. As this excitation enters the bulk, it modifies the correlators at infinity in a manner that is consistent with the gravitational constraints. These modified correlators retain information about the state even after the excitation has seemingly fallen in, and before it emerges out to $\mathcal{I}^+$.

In ordinary nongravitational theories (including nongravitational gauge theories), it is possible to prepare a state by sending in information into a bounded region, while erasing all traces of this information outside the region. In a theory of gravity, the principle of holography of information tells us that this is impossible.

The example can be embellished to display some additional instructive features. Consider the case where the theory has a second field $\widetilde{\phi}$ that couples to gravity in precisely the same manner as ϕ. We can take ϕ and $\widetilde{\phi}$ to be related by a global symmetry and set the $\phi - \widetilde{\phi}$ two-point function to vanish. (Theories of quantum gravity might not display such global symmetries but this issue is irrelevant for the present discussion.)

Now consider the state

$$|\widetilde{f}\rangle = e^{-i\lambda \int f(r)\widetilde{\phi}(u=0,r,\Omega)d^2\Omega dr}|0\rangle, \tag{20.5}$$

which is similar to the state $|f\rangle$ defined above, but contains an excitation of $\widetilde{\phi}$ rather than ϕ.

It might naively appear that the observer at infinity cannot distinguish $|f\rangle$ and $|\widetilde{f}\rangle$. However, one finds that

$$\lim_{r'\to\infty} r' \frac{\partial}{\partial\lambda}\langle\widetilde{f}|H\widetilde{\phi}(u',r',\Omega')|\widetilde{f}\rangle\big|_{\lambda=0} = \frac{1}{2\pi}\int_0^{r_0} \frac{f(r)}{u'(u'-2r)}dr. \tag{20.6}$$

This is similar to the result above except that it is $\widetilde{\phi}$ that has been inserted in the state $|\widetilde{f}\rangle$. Moreover,

$$\lim_{r'\to\infty} r' \frac{\partial}{\partial\lambda}\langle\widetilde{f}|H\phi(u',r',\Omega')|\widetilde{f}\rangle\big|_{\lambda=0} = \lim_{r'\to\infty} r' \frac{\partial}{\partial\lambda}\langle f|H\widetilde{\phi}(u',r',\Omega')|f\rangle\big|_{\lambda=0} = 0 \tag{20.7}$$

Therefore, an observer at infinity can distinguish the two states by determining whether the two point function of H and ϕ is nonzero or whether the two point function of H and $\widetilde{\phi}$ is nonzero.

20.3.2 Nonperturbative Results

The perturbative examples provided above should not be thought of as a proof of the holography of information. A nonperturbative result for asymptotic algebras is proved in [4]. (See [7] for a simplified discussion.) In the case of asymptotically flat space, the result is that any observable on $\mathcal{I}^+$ can be approximated arbitrarily well by an observable near its past boundary, $\mathcal{I}^+_-$. Correspondingly, any observable on $\mathcal{I}^-$ can be approximated arbitrarily well by an observable near its future boundary $\mathcal{I}^-_+$.

Since we do not know the full UV-complete theory of quantum gravity in asymptotically flat space, this result requires some assumptions about that theory. The assumptions required are as follows.

1. The formalism of quantum field theory is applicable in the asymptotic region and the asymptotic algebra continues to make sense in the full UV-complete theory.
2. The Hamiltonian is bounded below.
3. The algebra at infinity (more precisely, the algebra at $\mathcal{I}^+_-$ or $\mathcal{I}^-_+$) contains projectors onto the vacua of the theory.

The first assumption is reasonable if one adopts the perspective that, even in quantum gravity, the asymptotic geometry should be held fixed. Therefore, while quantum field theory might not provide a useful description in the bulk, one still expects it to provide a useful description asymptotically. The second assumption is difficult to prove but is reasonably expected in any quantum-mechanical theory. (For convenience, we set the lower-bound of energy to be zero.) The third assumption is related to the Gauss law. In the low-energy theory, the energy can be measured at infinity and the projector onto the vacuum is simply the probability that one obtains zero in such a measurement. The vacuum might be degenerate, as in four-dimensional flat space but, in that case, the supertranslation charges that identify the different vacua are also elements of the algebra at infinity in the low-energy theory. The third assumption states that this property extends to the UV-complete theory.

20.3.3 Intermediate Regime

The principle of holography of information is important in some situations and not in others. An analogy with our understanding of unitarity in ordinary quantum field theories might help to illustrate this point. In the canonical formalism, it is possible to formally prove that the S-matrix is unitary. Unitarity can be verified through explicit calculations when one studies the scattering of simple excitations of the vacuum. However, in the presence of a heavy nonperturbative background with e^S microstates, effective field theory appears to be dissipative unless one keeps track of effects of size $\mathrm{O}\left(e^{-S}\right)$.

In precisely the same spirit, discussions of holography in gravity should be separated into three regimes. The nonperturbative regime alluded to in Sect. 20.3.2 provides the correct setting to ask fine-grained questions, such as those about the unitarity of black hole evaporation. The second regime is where one studies simple states in perturbation theory, without requiring any assumptions about the UV-complete theory; this is described in Sect. 20.3.1. A third, intermediate, regime is where one introduces a heavy background with S microstates and coarse-grains the algebra of operators [28, 29] by discarding observables suppressed by S. In this regime, holography is obscured and gravity behaves like other quantum field theories. This is consistent with our mundane experience.

The constructions of [28, 29] that build on [30, 31] effectively "dress" bulk excitations to the background state so as to obtain local operators that perturbatively commute with operators at infinity. This extends the standard procedure that is used to obtain quasilocal operators in heavy classical backgrounds [32].

However, a study of the third regime should not lead one to the erroneous conclusion that holography of information is a property of "special states" [28]. This would be analogous to studying coarse-grained models that are dissipative and using them to conclude that unitarity is only a property of "special states".

20.4 A Physical Picture for Information Transfer

The effect described above gives us a clear picture of how information emerges from the black hole. At leading order, one-point functions of the metric and other fields outside the black hole have information about the mass, charge and angular momentum. So the Hawking radiation formed in this background carries this information at leading order. Quantum effects in this system are naturally controlled by the parameter $GE^2 \propto \frac{1}{S}$, where E is the energy of a Hawking quanta and S is the entropy of the black hole. Additional information about the black hole is encoded in $\frac{1}{S}$-suppressed observables outside the black hole and this affects subleading features of the Hawking radiation. For instance, the Hawking radiation from a black hole with "canonical spread" in energy differs, at $\mathrm{O}\left(\frac{1}{S}\right)$, from the radiation that emanates from a black hole with "microcanonical spread" in energy.

The purity of the state can be verified at all times outside the black hole, provided that one keeps track of effects of size $\mathrm{O}\left(e^{-S}\right)$. This is perfectly consistent with the usual statistical arguments which suggest that it is necessary to probe the final radiation to nonperturbative order to verify that its form is consistent with unitarity [7].

To summarize: the mechanism by which information is transferred to the Hawking radiation is that the constraints of quantum gravity force it to always remain available outside the black hole horizon. Because fine-grained details of fluctuations of the metric outside the horizon and the correlations of these fluctuations with other observables are sensitive to the state of the infalling matter, it is natural that the Hawking radiation contains this information after the black hole has evaporated. Figure 20.2 summarizes the main point of our discussion.

20.5 Comparison with Other Pictures

The mechanism described in this chapter has some features in common with other proposed resolutions of the information paradox. In this section, we present a brief comparison.

Fuzzballs. The flavour of the "fuzzball resolution" is broadly similar to ours since it also suggests that the wavefunctional outside the horizon provides a complete description of the state. However, the framework presented above does not require us to modify the metric at the semi-classical level. Instead, we expect that the expectation

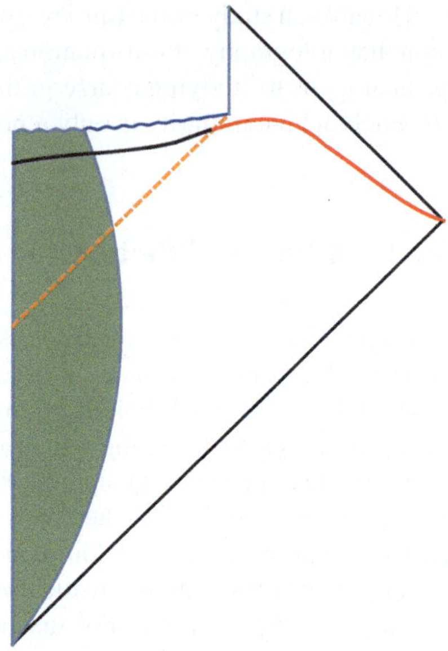

Fig. 20.2 In the classical theory, observables on the red-part of the Cauchy slice (outside the horizon) have information about the total energy. In the quantum theory, observables in this region suffice to completely identify the state of the black hole. In both cases, the mechanism is that the process of black hole formation modifies observables in the black hole exterior, and the gravitational constraints force them to retain information about the infalling matter

value of the metric will coincide up to exponential accuracy with the usual black hole metric. It is only small quantum features of observables outside the black hole that retain information about the initial state. The fuzzball proposal was motivated by a search for loopholes in the classical no-hair theorem. But the no-hair theorem fails as soon as gravity is treated quantum mechanically and one need not seek any modifications at the classical level.

Soft hair. The quantum observables outside the black hole that suffice to distinctively characterize its state can be thought of as "quantum-mechanical hair." This hair arises automatically when one correctly imposes the constraints of gravity in quantum mechanics. In [33], it was proposed that soft hair on black holes might help preserve information about the initial state. But the principle of holography of information does not rely on infrared effects; in fact, the presence of soft hair was a complication that had to be surmounted in [4]. For this reason, our discussion generalizes nicely to AdS where the cosmological constant regulates infrared effects.

Measurements of energy. An old proposal to resolve the information paradox [34] suggests that if the energy eigenstates are discrete, and energy can be measured at infinity with exponential precision then this can be used to obtain information about the microstate.

While this proposal was prescient, correlators of H, by themselves, are insufficient to determine the state. For instance, correlators of H cannot even distinguish between the pure state,

$$|\Psi\rangle = \sum_E c_E |E\rangle, \tag{20.8}$$

written as a linear combination of energy eigenstates, and the mixed density matrix,

$$\rho = \sum_E |c_E|^2 |E\rangle\langle E|. \tag{20.9}$$

As even the perturbative example shows, it is necessary for the distant observer to study observables that do not commute with H in order to identify the state.

Quantum hair. The authors of [35, 36] arrived at the same conclusion as earlier work [7, 9] that is summarized here: information about the black hole interior is always present in the exterior. However, the techniques used in [35, 36] are different from those used to prove the holography of information and, as such, they are insufficient to establish this conclusion.

The authors of [36] use an "effective action" for gravity which suggests that subleading terms in gravitational field at infinity are sensitive to the details of the density distribution of a spherical source. Regardless of the validity of this effective action, presumably, the metric obtained by this procedure should be thought of as the quantum-corrected expectation value of the metric operator. However, in general, one requires quantum correlations of the metric and other observables at infinity to obtain complete information about the state. It is insufficient to examine one-point functions of the metric as is already clear from the perturbative examples presented above.

Second, the matter fields in [36] are also not treated quantum mechanically. In [36], it is suggested that the matter be placed in an "energy eigenstate". However, the uncertainty principle implies that the wavefunctions of an energy eigenstate are always delocalized, even in the absence of gravity. So energy eigenstates should not be used as models of localized excitations.

Third, the constraints of gravity correlate the metric to the state of the matter field. Therefore, in the gravitational theory, a state corresponding to a superposition of matter excitations should be described as a sum of "dressed states" where, in each term, the state of the constrained components of the metric is related to the state of the dynamical components of the metric and other fields. But the starting point of [35] is a simple superposition of geometries, which is incorrect.

For these reasons, we direct the reader to [8], where the metric and matter fields are treated quantum mechanically and the constraints are solved to obtain the full wavefunctional. This shows that observables at infinity (see Eq. 6.6 in [8] for a precise list of observables) uniquely fix the bulk state. However, we caution the reader that while the techniques of [8] are sufficient and instructive at the perturbative level, they cannot easily be extended to the nonperturbative setting, which is relevant for black holes. In that setting, the arguments of [4] are required.

Page curve. Page [37] argued that information is initially lost in the black hole and then starts to emerge gradually after about "half" the black hole has evaporated. In contrast, the result utilized here is that the observer outside can always determine that the state is pure. This result is not in contradiction with recent derivations of the Page curve [21–24]. An essential feature [38–40] of these Page curves is that they are obtained by adding a nongravitational bath to an AdS black hole. They do not measure how information emerges from the black hole but merely how it flows from one part of the nongravitational bath to the other.

There has been some debate on whether these Page curves are relevant for realistic black holes where there is no bath. The following observations are uncontroversial.

1. Page [25] originally studied the algebra of all observables in the black hole exterior. This can be defined precisely by starting with an HKLL-type [41] representation of elementary field operators in terms of asymptotic operators and taking all possible products and linear combinations to obtain an algebra. The entropy of this algebra is always zero, if the black hole is in a pure state, and does not follow a Page curve in time.
2. It is possible to define other algebras whose entropy does obey a Page curve as a function of some parameter. Some examples are given in [4]. However, these algebras are somewhat contrived and the Page curve has no interpretation in terms of information "emerging" from the black hole.
3. Some researchers believe that it might be possible to define a "natural algebra" whose entropy follows a Page curve that can be interpreted in terms of information emerging from the black hole. However, as of this writing, such an algebra has not been specified precisely.

Returning to the case with a nongravitational bath, the discussion above helps to address a common criticism of these calculations [42], which is that they do not explain how information enters the bath. Consider a two-step process. In the first step, one prepares a black hole in AdS. The principle of holography of information tells us that information about the state of the black hole is already present in observables near the boundary of the gravitational theory. In the second step, one couples the system to a bath by turning on a coupling at the boundary of AdS. This allows the information to flow locally from the boundary into the bath, and then from one part of the bath to the other. We note that information does not have to "jump" from the black hole interior into the bath, and nor does this mechanism require any novel effects to mediate this process.

References

1. J.M. Maldacena, The large N limit of superconformal field theories and supergravity. Int. J. Theor. Phys. **38**, 1113–1133 (1999). https://doi.org/10.1023/A:1026654312961. arXiv:hep-th/9711200
2. E. Witten, Anti-de Sitter space and holography. Adv. Theor. Math. Phys. **2**, 253–291 (1998). https://doi.org/10.4310/ATMP.1998.v2.n2.a2. arXiv:hep-th/9802150
3. S. Gubser, I.R. Klebanov, A.M. Polyakov, Gauge theory correlators from noncritical string theory. Phys. Lett. B **428**, 105–114 (1998). https://doi.org/10.1016/S0370-2693(98)00377-3. arXiv:hep-th/9802109
4. A. Laddha, S.G. Prabhu, S. Raju, P. Shrivastava, The holographic nature of null infinity. SciPost Phys. **10**(2), 041 (2021). https://doi.org/10.21468/SciPostPhys.10.2.041. arXiv:2002.02448 [hep-th]

5. T. Chakraborty, J. Chakravarty, V. Godet, P. Paul, S. Raju, Holography of information in de Sitter space. JHEP **12**, 120 (2023). https://doi.org/10.1007/JHEP12(2023)120. arXiv:2303.16316 [hep-th]
6. C. Chowdhury, O. Papadoulaki, S. Raju, A physical protocol for observers near the boundary to obtain bulk information in quantum gravity. SciPost Phys. **10**(5), 106 (2021). https://doi.org/10.21468/SciPostPhys.10.5.106. arXiv:2008.01740 [hep-th]
7. S. Raju, Lessons from the information paradox. Phys. Rept. **943**, 1–80 (2022). https://doi.org/10.1016/j.physrep.2021.10.001. arXiv:2012.05770 [hep-th]
8. C. Chowdhury, V. Godet, O. Papadoulaki, S. Raju, Holography from the Wheeler-DeWitt equation. JHEP **03**, 019 (2022). https://doi.org/10.1007/JHEP03(2022)019. arXiv:2107.14802 [hep-th]
9. S. Raju, Failure of the split property in gravity and the information paradox. Class. Quant. Grav. **39**(6), 064002 (2022). https://doi.org/10.1088/1361-6382/ac482b. arXiv:2110.05470 [hep-th]
10. J. Chakravarty, D. Jain, A. Sivakumar, Holography of information in massive gravity using Dirac brackets. JHEP **06**, 109 (2023). https://doi.org/10.1007/JHEP06(2023)109. arXiv:2301.01075 [hep-th]
11. R. de Mello Koch, G. Kemp, Holography of information in AdS/CFT. JHEP **12**, 095 (2022). https://doi.org/10.1007/JHEP12(2022)095. arXiv:2210.11066 [hep-th]
12. C. Chowdhury, O. Papadoulaki, Recovering information in an asymptotically flat spacetime in quantum gravity. Class. Quant. Grav. **39**(24), 245012 (2022). https://doi.org/10.1088/1361-6382/aca192. arXiv:2203.07449 [hep-th]
13. N. Gaddam, A.H, Holography of information in a ball of finite radius. arXiv:2410.17316 [hep-th]
14. D. Marolf, Unitarity and holography in gravitational physics. Phys. Rev. D **79**, 044010 (2009). https://doi.org/10.1103/PhysRevD.79.044010. arXiv:0808.2842 [gr-qc]
15. D. Marolf, Holography without strings? Class. Quant. Grav. **31**, 015008 (2014). https://doi.org/10.1088/0264-9381/31/1/015008. arXiv:1308.1977 [hep-th]
16. S. Banerjee, J.-W. Bryan, K. Papadodimas, S. Raju, A toy model of black hole complementarity. JHEP **05**, 004 (2016). https://doi.org/10.1007/JHEP05(2016)004. arXiv:1603.02812 [hep-th]
17. T. Jacobson, Boundary unitarity and the black hole information paradox. Int. J. Mod. Phys. **D22**, 1342002 (2013). https://doi.org/10.1142/S0218271813420029. arXiv:1212.6944 [hep-th]
18. T. Jacobson, P. Nguyen, Diffeomorphism invariance and the black hole information paradox. Phys. Rev. D **100**(4), 046002 (2019). https://doi.org/10.1103/PhysRevD.100.046002. arXiv:1904.04434 [gr-qc]
19. S. Hawking, Particle creation by black holes. Commun. Math. Phys. **43**, 199–220 (1975). https://doi.org/10.1007/BF02345020 [Erratum: Commun. Math. Phys. **46**, 206 (1976)]
20. S. Hawking, Breakdown of predictability in gravitational collapse. Phys. Rev. D **14**, 2460–2473 (1976). https://doi.org/10.1103/PhysRevD.14.2460
21. G. Penington, Entanglement wedge reconstruction and the information paradox. JHEP **09**, 002 (2020). https://doi.org/10.1007/JHEP09(2020)002. arXiv:1905.08255 [hep-th]
22. A. Almheiri, T. Hartman, J. Maldacena, E. Shaghoulian, A. Tajdini, Replica wormholes and the entropy of hawking radiation. JHEP **05**, 013 (2020). https://doi.org/10.1007/JHEP05(2020)013. arXiv:1911.12333 [hep-th]
23. A. Almheiri, T. Hartman, J. Maldacena, E. Shaghoulian, A. Tajdini, The entropy of Hawking radiation. Rev. Mod. Phys. **93**(3), 035002 (2021). https://doi.org/10.1103/RevModPhys.93.035002. arXiv:2006.06872 [hep-th]
24. A. Almheiri, N. Engelhardt, D. Marolf, H. Maxfield, The entropy of bulk quantum fields and the entanglement wedge of an evaporating black hole. JHEP **12**, 063 (2019). https://doi.org/10.1007/JHEP12(2019)063. arXiv:1905.08762 [hep-th]
25. D.N. Page, Average entropy of a subsystem. Phys. Rev. Lett. **71**, 1291–1294 (1993). https://doi.org/10.1103/PhysRevLett.71.1291. arXiv:gr-qc/9305007
26. S.D. Mathur, The fuzzball proposal for black holes: an elementary review. Fortsch. Phys. **53**, 793–827 (2005). https://doi.org/10.1002/prop.200410203. arXiv:hep-th/0502050

27. R.L. Arnowitt, S. Deser, C.W. Misner, The dynamics of general relativity. Gen. Rel. Grav. **40**, 1997–2027 (2008). https://doi.org/10.1007/s10714-008-0661-1. arXiv:gr-qc/0405109
28. E. Bahiru, A. Belin, K. Papadodimas, G. Sarosi, N. Vardian, State-dressed local operators in the AdS/CFT correspondence. Phys. Rev. D **108**(8), 086035 (2023). https://doi.org/10.1103/PhysRevD.108.086035. arXiv:2209.06845 [hep-th]
29. K. Jensen, S. Raju, A.J. Speranza, Holographic observers for time-band algebras. arXiv:2412.21185 [hep-th]
30. K. Papadodimas, S. Raju, Local operators in the eternal black hole. Phys. Rev. Lett. **115**(21), 211601 (2015). https://doi.org/10.1103/PhysRevLett.115.211601. arXiv:1502.06692 [hep-th]
31. K. Papadodimas, S. Raju, State-dependent bulk-boundary maps and black hole complementarity. Phys. Rev. D **89**(8), 086010 (2014). https://doi.org/10.1103/PhysRevD.89.086010. arXiv:1310.6335 [hep-th]
32. B.S. DeWitt, Quantum theory of gravity. 1. The canonical theory. Phys. Rev. **160**, 1113–1148 (1967). https://doi.org/10.1103/PhysRev.160.1113
33. S.W. Hawking, M.J. Perry, A. Strominger, Soft hair on black holes. Phys. Rev. Lett. **116**(23), 231301 (2016). https://doi.org/10.1103/PhysRevLett.116.231301. arXiv:1601.00921 [hep-th]
34. V. Balasubramanian, D. Marolf, M. Rozali, Information recovery from black holes. Gen. Rel. Grav. **38**, 1529–1536 (2006). https://doi.org/10.1007/s10714-006-0344-8. arXiv:hep-th/0604045
35. X. Calmet, S.D.H. Hsu, Quantum hair and black hole information. Phys. Lett. B **827**, 136995 (2022). https://doi.org/10.1016/j.physletb.2022.136995. arXiv:2112.05171 [hep-th]
36. X. Calmet, R. Casadio, S.D.H. Hsu, F. Kuipers, Quantum hair from gravity. Phys. Rev. Lett. **128**(11), 111301 (2022). https://doi.org/10.1103/PhysRevLett.128.111301. arXiv:2110.09386 [hep-th]
37. D.N. Page, Information in black hole radiation. Phys. Rev. Lett. **71**, 3743–3746 (1993). https://doi.org/10.1103/PhysRevLett.71.3743. arXiv:hep-th/9306083
38. H. Geng, A. Karch, Massive islands. JHEP **09**, 121 (2020). https://doi.org/10.1007/JHEP09(2020)121. arXiv:2006.02438 [hep-th]
39. H. Geng, A. Karch, C. Perez-Pardavila, S. Raju, L. Randall, M. Riojas, S. Shashi, Information transfer with a gravitating bath. arXiv:2012.04671 [hep-th]
40. H. Geng, A. Karch, C. Perez-Pardavila, S. Raju, L. Randall, M. Riojas, S. Shashi, Inconsistency of islands in theories with long-range gravity. arXiv:2107.03390 [hep-th]
41. A. Hamilton, D.N. Kabat, G. Lifschytz, D.A. Lowe, Holographic representation of local bulk operators. Phys. Rev. D **74**, 066009 (2006). https://doi.org/10.1103/PhysRevD.74.066009. arXiv:hep-th/0606141
42. E.J. Martinec, Trouble in paradox. arXiv:2203.04947 [hep-th]

Chapter 21
Information Scrambling at Quantum Hall Interfaces and Their Analog to Black Hole Event Horizon

Ken K. W. Ma and Kun Yang

Abstract The black hole information paradox has been hotly debated for the last few decades without a full resolution. This makes it desirable to find analogues of this paradox in simple and experimentally accessible systems, whose resolutions may shed light on this longstanding and fundamental problem. Here, we review and resolve the apparent "information paradox" in two different interfaces separating Abelian and non-Abelian quantum Hall states. In both cases, the information carried by the pseudospin degree of freedom of the Abelian anyons gets scrambled when they cross the interface and enter the non-Abelian quantum Hall liquid. Nevertheless, it is found that the scrambling mechanism depends on the nature of the interface. The corresponding analogues of different concepts in black hole physics such as event horizon, black hole interior, Hawking radiation, and Page curve will also be discussed.

21.1 Introduction

The existence of black holes has received strong support from recent observations [1–3]. Instead of being a region which nothing can escape from, Hawking predicted that a black hole emits radiation and evaporates slowly [4, 5]. He also concluded that the radiation carries no information except mass, angular momentum, and charge of the black hole [6, 7]. This result points to possible loss of information in black holes. On one hand, it is consistent with the no-hair theorem [8–10]. On the other hand, quantum mechanics forbids information loss due to the unitarity of time evolution. This apparent contradiction leads to the black hole information paradox [11–15]. It

K. K. W. Ma (✉)
Orlando, FL, USA
e-mail: kwokwai_ma@alumni.brown.edu

K. Yang (✉)
Department of Physics, Florida State University and the National High Magnetic Field Laboratory, Tallahassee, FL, USA
e-mail: kunyang@magnet.fsu.edu

A. Akil and C. Bambi (eds.), *The Black Hole Information Paradox*, Springer Series in Astrophysics and Cosmology, https://doi.org/10.1007/978-981-96-6170-1_21

is believed that the resolution of this paradox may provide important clues on how to combine quantum mechanics and general relativity.

Various approaches have been proposed to resolve this black hole information paradox. Among them, the holographic principle [16–19] supports the preservation of unitarity and information. In particular, information can be encoded holographically on surfaces, such as the event horizon. This belief is substantiated by the discovery of the anti-de Sitter/conformal field theory (AdS/CFT) correspondence [20]. Recently, the firewall scenario [21] was proposed to resolve the conflict between black hole complementarity [22, 23] and monogamy of entanglement [24]. If this conjecture is correct, the firewall at the event horizon (or black hole's boundary) may also break the entanglement between the outgoing and the in-falling particles. Thus, the boundary can be as important as, or even more important than, the interior of a black hole.

Suppose black hole evaporation is a unitary process. Then how is information hidden in the black hole released from Hawking radiation? Page argued that the release of information starts slowly at the beginning, but becomes faster in the later stage of the evaporation [25–27]. If the system was initially in a pure state, entropy of the radiation (coming from its entanglement with the remainder of black hole) would first increase from zero but eventually decrease back to zero when the black hole evaporates completely, thus recovering the pure state nature of the system and all the (quantum) information it carries. This feature is now known as the Page curve. Based on quantum information theory, the thought experiment by Hayden and Preskill (Hayden-Preskill protocol) has provided further insight on retrieving information from Hawking radiation [28]. Suppose the black hole has already passed its Page time and become maximally entangled with its previously emitted Hawking radiation. If the internal dynamics of black hole can be described by an instantaneous random unitary transformation, then any additional information entering the black hole can be recovered from Hawking radiation almost immediately (a very short time compared to the lifetime of the black hole) [28, 29]. The protocol has postulated the existence of information scrambling [28, 30–34], which describes the dispersal of local information into entanglement and correlation throughout the entire system. Thus, the original information is stored nonlocally and cannot be recovered via local measurement. This concept has attracted considerable attention in the context of many-body dynamics [35–39] and quantum neural network [40–42], which has been verified in recent experiments [43, 44]. In addition, recent studies have recovered the Page curve for AdS black holes [45–50]. However, a full resolution of the paradox remains an open problem [51]. It is thus desirable to mimic the information paradox in simple and experimentally accessible systems, that allows for a complete understanding of this process.

Somewhat similar to the holographic principle, the bulk-edge correspondence relates the topologically protected edge modes and bulk topological orders in fractional quantum Hall (FQH) systems [52]. This allows us to learn about the bulk by probing the edge of the system [53]. Comparatively speaking, interfaces between a pair of FQH states are explored much less [54–73]. In fact, the physics and dynamics of interfaces are much richer than simple edges [72, 73]. For example, as we demonstrate below, certain interfaces allow quasiparticle tunneling between two different

FQH states, even if the quasiparticles are of very different nature. If they have different internal degrees of freedom, (local) information carried by them needs to be transmuted (or scrambled in a specific way) to prevent information loss. This motivates us to explore analogues of black hole information paradox in quantum Hall (QH) interfaces.

In this chapter, we review and resolve a condensed matter analogue of information paradox in two different QH interfaces [74, 75]. We first consider the interface between Halperin-331 [76] and Pfaffian (Moore-Read) [77] QH states. The latter is a famous non-Abelian state that hosts non-Abelian anyons that may be useful in topological quantum computation [78]. Meanwhile, both 331 and Pfaffian states may be realized in bilayer systems or wide quantum wells at filling factor $\nu = 1/2$ [79, 80]. Due to the competition between interlayer tunneling and intralayer Coulomb interaction, a phase transition between the 331 and Pfaffian states was predicted [81–83]. This suggests the possibility of creating a 331-Pfaffian interface by controlling the tunneling strengths in different regions of the bilayer system [57]. Interestingly, the Abelian 331 state has a pseudospin (layer) degree of freedom, which is absent in the Pfaffian state. If the original information carried by the pseudospin degree of freedom becomes irrecoverable after quasiparticles cross the interface and enter the Pfaffian liquid, it leads to an "information paradox". We demonstrate that the information is scrambled and stored nonlocally in the Pfaffian liquid and the interface. We also mimic black hole evaporation in the same system, and find it satisfies the Page curve naturally. In other words, the original pseudospin information is recovered and the "information paradox" in our model is resolved. Following the similar logic, we consider the "information paradox" in another QH interface formed by Halperin 330 state and Read Rezayi state at level four (abbreviated as RR_4 state) [84]. On one hand, the setting closely resembles the 331-Pfaffian interface by having a pseudospin degree of freedom in the 330 state that is absent in the RR_4 state. Both states may be realized in bilayer systems at $\nu = 2/3$ [85–88]. On the other hand, we will show that the nature of 330-RR_4 interface is different from the 331-Pfaffian interface. This leads to different mechanisms for scrambling the original pseudospin degree of freedom carried by the Abelian anyons in the two QH interfaces. Here, we must emphasize that we are not aiming at a resolution of the original information paradox in astrophysical black holes. This is clearly unachievable by proposing a simple analogy. Instead, we want to simulate some important concepts in resolving the original paradox in a simple and accessible manner.

21.2 Basics of Quantum Hall Effect

To set the stage for later discussion, we first review briefly some basic concepts in QH physics [89]. Electrons moving in two dimensions ($x - y$ plane) and a perpendicular magnetic field (in the z direction) have their energy levels being quantized in Landau levels. A very important parameter characterizing the system is the Landau level filling factor ν, which is the ratio between the number of electrons and the number of magnetic flux quanta enclosed by the system. Of particular interest is the case $\nu < 1$, where only the lowest Landau level is partially filled by electrons at low temperature, known as the FQH regime. Since the kinetic energy of these electrons is quenched due to Landau quantization, the interaction between them dominates the properties of the system. Various FQH states, which possess numerous fascinating properties, are realized in this strongly correlated electronic system. The exotic properties of FQH states are associated with the topological order they possess [52]. Most prominent among them is the existence of low-energy excitations (quasiparticles) that have fractional charges and obey fractional statistics (between bosonic and fermionic statistics) [90]. A famous example is the Laughlin state at $\nu = 1/3$ [91], in which quasiparticle with a fractional charge $e/3$ [92] and a fractional statistics $2\pi/3$ can exist [93, 94]. Note that both fractional charge [95, 96] and fractional statistics were observed experimentally [97, 98]. Such exotic quasiparticles are called anyons, and the possible types of anyons are associated with the specific topological order.

21.2.1 Bulk-Edge Correspondence and Conformal Field Theory

The bulk-edge correspondence, another consequence of the topological order, relates the edge structure and the bulk topological order in FQH systems. In particular, it predicts the existence of gapless edge modes described by conformal field theories (CFTs), and there is a one-to-one correspondences between the bulk topological order and edge CFT [99]. In our previous example, the edge of the Laughlin state at $\nu = 1/3$ has a single chiral bosonic edge mode ϕ, which can be described by the Lagrangian density,

$$\mathscr{L}_{1/3} = -\frac{3}{4\pi}\partial_x\phi(\partial_t - u\partial_x)\phi. \tag{21.1}$$

Here, u is the speed of the edge mode, and ϕ is a (chiral) bosonic field. In general, the edge of a FQH liquid can have more than one edge mode. For Abelian FQH states, the corresponding edge theory is described by Wen [52]

$$\mathscr{L}_{\text{edge}} = -\frac{1}{4\pi}\sum_{i,j} K_{ij}\partial_t\phi_i\partial_x\phi_j - \frac{1}{4\pi}\sum_{i,j} V_{ij}\partial_x\phi_i\partial_x\phi_j. \tag{21.2}$$

Importantly, the K matrix encodes all information of the topological order. For a FQH state in a bilayer system, it may (but not always) be described by a two-component topological order which has a 2×2 K matrix. In this situation, two different edge modes exist. Furthermore, the possible type of edge modes is not limited to bosonic mode. Other types of modes such as Majorana fermion modes exist when the topological orders are non-Abelian [100–103].

With the knowledge of the edge structure in hand, different low-energy excitations in the FQH system can be described or created by suitable CFT operators [99]. For example, a charge-$e/3$ quasiparticle and an electron in the $\nu = 1/3$ Laughlin state are created by the operators $: \exp(i\phi) :$ and $: \exp(3i\phi) :$, respectively. Here, $: \mathscr{V} :$ denotes the normal ordering of the vertex operator $\mathscr{V}$. When there is no confusion, this normal ordering notation will be dropped in the later discussion. For a FQH state being described by a multicomponent topological order, there are multiple types of anyons (described by different CFT operators) that have the same electric charge. In other words, the anyons have an additional degree of freedom. This point will become clear when we discuss our setup.

21.3 Information Paradox in 331-Pfaffian Interface

Both Halperin-331 and Pfaffian QH liquids have the same Landau level filling factor $\nu = 1/2$, which can be realized in a bilayer system. For the 331 liquid, it is described by a two-component topological order with the K matrix [52, 76],

$$K = \begin{pmatrix} 3 & 1 \\ 1 & 3 \end{pmatrix}. \tag{21.3}$$

The two different edge modes are denoted as $\phi_\uparrow$ and $\phi_\downarrow$. The two most relevant operators creating an electron are $\exp(3i\phi_\uparrow + i\phi_\downarrow)$ and $\exp(i\phi_\uparrow + 3i\phi_\downarrow)$. On the other hand, the Pfaffian liquid is described by a single-component non-Abelian order with $K = 2$. Its edge has a bosonic mode ϕ and a Majorana fermion mode ψ [100]. The corresponding electron operator is $\psi \exp(2i\phi)$. Since the Halperin-331 and Pfaffian edges have opposite chiralities at the interface, the interface is described by the Lagrangian density [57],

$$\begin{aligned}\mathscr{L} = &-\frac{1}{4\pi}\sum_{i,j} K_{ij}\partial_t\phi_i\partial_x\phi_j + \frac{2}{4\pi}\partial_t\phi_l\partial_x\phi_l - i\psi_l\partial_t\psi_l \\ &- \mathscr{H}(\phi, \psi).\end{aligned} \tag{21.4}$$

Here, the indices $i, j = \uparrow, \downarrow$ denote the layer or the pseudospin. All $\phi_\uparrow$, $\phi_\downarrow$, and ϕ_l are charge modes. The first two are right-moving along the edge of the 331 liquid, whereas the last one is left-moving along the edge of the Pfaffian liquid. The Pfaffian liquid also has a left-moving neutral Majorana fermion mode ψ_l along the edge. The edge structures of both QH states are illustrated in Fig. 21.1a. As shown by one of us, a relevant random electron tunneling between the Pfaffian and 331 edges can lead to a phase transition at the interface [57].

Now, we follow Ref. [57] and briefly summarize how different modes get localized at the interface. This also allows us to introduce useful notations for later discussion. One can define a charge mode $\phi_r = \phi_\uparrow + \phi_\downarrow$ and a neutral spin mode $\phi_n = \phi_\uparrow - \phi_\downarrow$

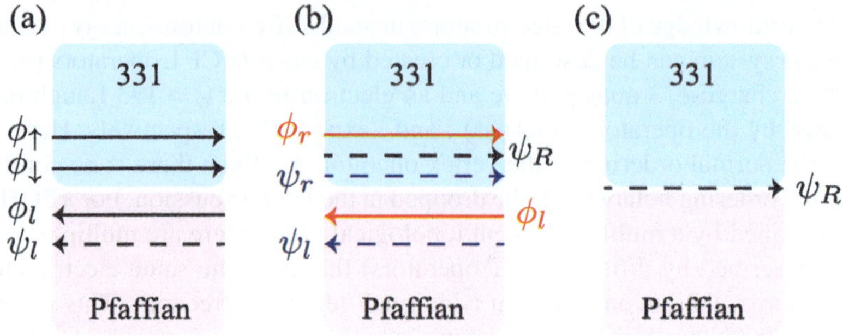

Fig. 21.1 Localization of edge modes at the interface due to random electron tunneling between 331 edge and Pfaffian edge. Solid lines denote charge modes, whereas dashed lines correspond to neutral modes. **a** The original edge modes in the 331 liquid and Pfaffian liquid. **b** Counterpropagating edge modes with the same color are gapped or localized. **c** Only a single chiral Majorana fermion mode remains gapless and propagating along the interface. Figure adopted from Ref. [74]

in the 331 liquid. Using this new set of modes, the topological term of the 331 edge becomes

$$\mathcal{L}_{331} = -\frac{2}{4\pi}\partial_t\phi_r\partial_x\phi_r - \frac{1}{4\pi}\partial_t\phi_n\partial_x\phi_n. \tag{21.5}$$

The overall charge density at the interface is given by

$$\rho(x) = \frac{1}{2\pi}\partial_x(\phi_\uparrow + \phi_\downarrow + \phi_l) = \frac{1}{2\pi}\partial_x\phi_c. \tag{21.6}$$

The random electron tunneling between the Pfaffian and 331 edges is described by

$$\begin{aligned} H_{\mathrm{T}} &= \int \xi(x)\psi_l\left(e^{3i\phi_\uparrow+i\phi_\downarrow+2i\phi_l} + e^{i\phi_\uparrow+3i\phi_\downarrow+2i\phi_l}\right) dx + \text{H.c.} \\ &= \int |\xi(x)|\psi_l(x)\psi_r(x)\cos\left[2\phi_c(x) + \varphi(x)\right] dx. \end{aligned} \tag{21.7}$$

Here, $\xi(x)$ denotes the random tunneling amplitude. In the second line, $|\xi(x)|$ and $\varphi(x)$ are the magnitude and the phase of $\xi(x)$, respectively. We have also fermionized $\exp[i\phi_n(x)] = \psi_r(x) + i\psi_R(x)$. The resulting edge modes are shown in Fig. 21.1b. If H_{T} is relevant in the renormalization group sense, then both charge modes, ψ_l, and ψ_r are localized at the interface. After the localization, only a single right-moving Majorana fermion mode remains gapless and propagates freely at long distance, as shown in Fig. 21.1c. Notice that this gapless mode is neither an original edge mode of the 331 nor the Pfaffian state.

Since Pfaffian and 331 states are QH states formed by superconducting pairing between composite fermions [104], both of these two states have quasiparticles with

Table 21.1 Primary fields in the chiral Ising CFT with a central charge $c = 1/2$

Primary field	Conformal spin	Quantum dimension
I	0	1
ψ	1/2	1
σ	1/16	$\sqrt{2}$

the smallest possible charge of $e/4$ [77, 105]. However, there is a fundamental difference between these quasiparticles. For the 331 state, there are two different types of Abelian $e/4$ quasiparticles created by the vertex operators, $e^{i\phi_\uparrow}$ and $e^{i\phi_\downarrow}$. One may view them as quasiparticles with different pseudospins. When we formulate the information paradox in the following discussion, this pseudospin will be regarded as the *degree of freedom* of the Abelian quasiparticles. For the Pfaffian state, there is only one type of $e/4$ quasiparticle created by $\sigma e^{-i\phi_l/2}$ [77]. Here, σ is the twist field with a scaling dimension $1/16$ in the chiral Ising CFT [106]. We summarize the three primary fields and their properties in Table 21.1. In particular, σ satisfies the fusion rule $\sigma \times \sigma = \psi + I$. Note that we have omitted the subscript l for the Majorana field to make the discussion of Ising CFT general. Its proper meaning should be clear from context. The fusion rule indicates that the quasiparticle is non-Abelian. An interesting question is what happens if a quasiparticle is dragged from the 331 liquid in to the Pfaffian liquid? It seems that the pseudospin information would be lost. In this sense, we can define the interface between the two different QH liquids as the "*event horizon with a firewall*" in our setup. This definition or analogy makes sense since the interface plays the role of a one-way surface of information in our setup, and the "destruction" of pseudospin information at the interface resembles a firewall conjectured in Ref. [21]. In addition, the Pfaffian liquid can be viewed as the interior of a "*black hole*". Suppose this analogous black hole can evaporate (discussed in Sect. 21.5) and the original pseudospin information cannot be recovered at the end of the evaporation. Then, the lost of information contradicts to the fact that quasiparticle tunneling is a unitary process. Thus, we have identified an apparent "information paradox".

21.3.1 331-Pfaffian Interface from Anyon Condensation

Before resolving the paradox, we reformulate the above discussion in the framework of anyon condensation [55, 107–110]. This technique has been commonly applied to study possible transitions between topologically ordered phases. In the context of QH physics, it was used to study the interface between Pfaffian and non-Abelian spin-singlet (NASS) QH states [54, 55]. In this section, we first use anyon condensation to deduce the CFT description of the 331-Pfaffian interface. In the next section, we apply the same technique to resolve the paradox. Along the way, we adopt a pedagogical approach and aim at relating the rather abstract technique to the more physical picture in Sect. 21.3. It will allow us to highlight the advantages of applying anyon condensation in studying QH interfaces.

From Sect. 21.3, we know that the edges of the Halperin-331 and Pfaffian liquids are described by CFTs with central charges 2 and 3/2, respectively. These two edges are *counterpropagating* at the interface. Hence, we expect the resulting CFT describing the 331-Pfaffian interface has a net central charge of $2 - 3/2 = 1/2$. To deduce exactly what the CFT is, it is first necessary to separate the charge and neutral sectors for both Halperin-331 and Pfaffian liquids. It is because a charge mode cannot be gapped out by coupling to a neutral mode in a usual situation. Equivalently, we do not consider the possibility of condensing charge bosons, which will break the U(1) gauge symmetry. The separation was already achieved in Sect. 21.3. In particular, the combination of charge modes $\phi_c = \phi_r + \phi_l$ was shown to be gapped out (more precisely, localized) by H_{T} [111]. Therefore, we can focus our discussion on the neutral sectors.

As stated previously, the neutral sector of the Pfaffian state is described by a chiral Ising CFT. For the Halperin-331 state, its neutral sector is governed by the spin mode ϕ_n, which is described by the $\mathrm{U}(1)_4$ CFT. Different primary fields in this Abelian CFT are summarized in Table 21.2. Note that any two vertex operators in the form $e^{i\alpha\phi_n/2}$ and $e^{i(\alpha+4\mathbb{Z})\phi_n/2}$ are identified.

The structure (remaining gapless modes) of the 331-Pfaffian interface is solely determined by anyon condensation in the neutral sectors. This condensation occurs in the $\mathrm{U}(1)_4 \times \overline{\mathrm{Ising}}$ CFT. We emphasize again that the bar denotes *conjugation* of the Ising CFT due to the opposite chiralities between the 331 and Pfaffian edges at the interface. Compared to the original CFT, anyons in the conjugate CFT have the same fusion rules, but complex conjugated topological spins and braiding phases. Alternatively, one may interpret the condensation as a coset construction [107]. We label a generic anyon as $(e^{i\alpha\phi_n/2}, \bar{t})$. Here, the parameter $\alpha = 0, 1, 2, 3$ determines the corresponding primary fields in the $\mathrm{U}(1)_4$ CFT. Meanwhile, $\bar{t} = \{\bar{I}, \bar{\psi}, \bar{\sigma}\}$ denotes the primary fields in the conjugate Ising CFT. In the present case, there is only one condensable boson, $B = (e^{i\phi_n}, \bar{\psi})$. The condensation of B leads to confinement of some of the anyons in the condensed phase. An anyon remains unconfined if and only if it has a trivial mutual statistics with B. This condition ensures that an unconfined anyon has a consistent topological spin in the condensed phase. Furthermore, two

Table 21.2 Primary fields in the $\mathrm{U}(1)_4$ CFT. Here, the normal ordering in the vertex operators are not shown explicitly. Note that $\mathcal{V}_3 = \exp(3i\phi_n/2) \simeq \exp(-i\phi_n/2)$

Symbol	Vertex operator	Conformal spin	Type
$\mathcal{V}_0$	1	0	Boson
$\mathcal{V}_1$	$\exp(i\phi_n/2)$	1/8	Anyon
$\mathcal{V}_2$	$\exp(i\phi_n)$	1/2	Fermion
$\mathcal{V}_3$	$\exp(3i\phi_n/2)$	1/8	Anyon

Table 21.3 Unconfined anyons in the $\mathrm{U}(1)_4 \times \overline{\text{Ising}}$ CFT after condensing the boson $B = (e^{i\phi_n}, \bar{\psi})$. Vertex operators $\mathcal{V}_i$ are defined in Table 21.2. Here, the symbol $\simeq$ denotes identification of anyons modulo B. The conformal spins are deduced from $s = h_1 - h_2 \pmod 1$, where h_1 and h_2 denote the conformal dimensions of primary fields in the $\mathrm{U}(1)_4$ and Ising CFTs, respectively

Sector	Unconfined anyons	Conformal spin
$\tilde{I}$	$(\mathcal{V}_0, \bar{I}) \simeq (\mathcal{V}_2, \bar{\psi})$	0
$\tilde{\psi}$	$(\mathcal{V}_0, \bar{\psi}) \simeq (\mathcal{V}_2, \bar{I})$	1/2
$\tilde{\sigma}$	$(\mathcal{V}_1, \bar{\sigma}) \simeq (\mathcal{V}_3, \bar{\sigma})$	1/16

anyons are identified when they differ from each other by a multiple of B. Using operator product expansion, it is straightforward to deduce the six (or three after identification) deconfined anyons in the condensed phase. They are listed in Table 21.3. Their corresponding topological sectors, namely $\tilde{I}$, $\tilde{\psi}$, and $\tilde{\sigma}$ are defined according to their conformal spins. From the table, we conclude that the 331-Pfaffian interface is described by a chiral Ising CFT. Note that this Ising CFT has an opposite chirality to the one describing the Pfaffian edge at the interface.

Now, one may wonder why going through such abstract and seemingly redundant procedures to find out the CFT describing the interface. Doesn't the net central charge $c = 1/2$ directly indicate that it should be an Ising CFT? There are two reasons for analyzing this simple system by anyon condensation. First of all, it is fortunate that for the 331-Pfaffian interface, the mechanism and consequences of anyon condensation can be visualized in a very transparent and physical manner, but this is a very special case. In Eq. (21.7), the electron tunneling between counterpropagating edges at the interface couples ψ_l and $e^{i\phi_n}$. This leads to a mass term and eventually gaps out the counterpropagating ψ_l and ψ_r. Only ψ_R remains gapless at the interface. This was demonstrated by fermionizing $e^{i\phi_n} = \psi_r + i\psi_R$. This type of arguments does not always generalize to more complicated interfaces. On the other hand, the condensation of B systematically captures the gaping process and leads to a correct CFT description of the 331-Pfaffian interface. More importantly, anyon condensation relates every primary field in the original and condensed phases. These relations cannot be obtained from the argument in Sect. 21.3.

21.3.2 Transmutation of Pseudospin Information

We now discuss the transmutation of pseudospin information when Abelian quasiparticles cross the interface. Let us first comment on the charge sectors. As one will see, they basically play no role in the resolution of the paradox. Since charge-$e/4$ quasiparticles are allowed in both Halperin-331 and Pfaffian liquids, dragging quasiparticles across the interface does not require any absorption of net charge at the interface. Furthermore, the gapping of ϕ_c indicates that the dragging will not create any low-energy charge excitation at the interface [112].

Thus, we focus on the neutral sectors. An Abelian quasiparticle in the Halperin-331 liquid has its pseudospin degree of freedom carried solely by the neutral mode ϕ_n. This is observed by writing

$$e^{i\phi_\uparrow} = e^{i\phi_r/2}e^{i\phi_n/2}, \tag{21.8}$$

$$e^{i\phi_\downarrow} = e^{i\phi_r/2}e^{-i\phi_n/2}. \tag{21.9}$$

Hence, the vertex operators $\mathscr{V}_1$ and $\mathscr{V}_3$ encode the spin-up and spin-down states of an Abelian charge-$e/4$ quasiparticle, respectively. These two operators are not defined in the Pfaffian liquid. To understand the transmutation of quasiparticles when they cross the interface, we need to represent the four primary fields in the U(1)$_4$ CFT as different products between two Ising CFTs. One of them describes the interface, whereas the other describes the Pfaffian order. Both Ising CFTs now have the same chirality to match the central charges, $1 = 1/2 + 1/2$. From Table 21.3, one can obtain the inverted expressions:

$$\mathscr{V}_0 \equiv I_1 = I_{1/2} \times \tilde{I} + \psi \times \tilde{\psi}, \tag{21.10}$$

$$\mathscr{V}_1 \equiv e^{i\phi_n/2} = \sigma \times \tilde{\sigma}, \tag{21.11}$$

$$\mathscr{V}_2 \equiv e^{i\phi_n} = \psi \times \tilde{I} + I_{1/2} \times \tilde{\psi}, \tag{21.12}$$

$$\mathscr{V}_3 \equiv e^{-i\phi_n/2} = \sigma \times \tilde{\sigma}. \tag{21.13}$$

The tilded and untilded fields are in the CFTs describing the interface and the Pfaffian liquid, respectively. Also, the subscripts 1 and 1/2 in the identity fields denote the central charges of the corresponding CFTs. When there is no confusion, these subscripts will be skipped.

Equation (21.10) suggests that the U(1)$_4$ CFT is obtained from condensing the boson $b = (\psi, \tilde{\psi})$ [113, 114]. This result is consistent with the orbifold construction [115]. After the condensation, one of the unconfined particles is $(\sigma, \tilde{\sigma})$. We should state clearly that these two twist fields describe excitations (anyons) at *different* regions of the system, so it is meaningless to consider their fusion. In other words, the present situation is different from the case of a Pfaffian liquid, in which σ in the bulk and σ at the edge created from vacuum must fuse to I for conserving fermion parity. Importantly,

$$(\sigma, \tilde{\sigma}) \times (\sigma, \tilde{\sigma}) = (\psi, \tilde{\psi}) + (I, \tilde{I}) + (\psi, \tilde{I}) + (I, \tilde{\psi}). \tag{21.14}$$

The first two terms on the right hand side show that two orthogonal copies of vacuum exist, so $(\sigma, \tilde{\sigma})$ needs to split into two *inequivalent* types of anyons in the resulting U(1)$_4$ CFT [113]. We denote them as $(\sigma, \tilde{\sigma})_1$ and $(\sigma, \tilde{\sigma})_2$. Both of them have conformal spins 1/8, which are identified as the vertex operators $\mathscr{V}_1$ and $\mathscr{V}_3$ in the U(1)$_4$ CFT (see Table 21.2). The fusion rules are consistent by imposing the conditions $(\sigma, \tilde{\sigma})_1 \times (\sigma, \tilde{\sigma})_1 = (\sigma, \tilde{\sigma})_2 \times (\sigma, \tilde{\sigma})_2 = \mathscr{V}_2$ and $(\sigma, \tilde{\sigma})_1 \times (\sigma, \tilde{\sigma})_2 = \mathscr{V}_0$. Following

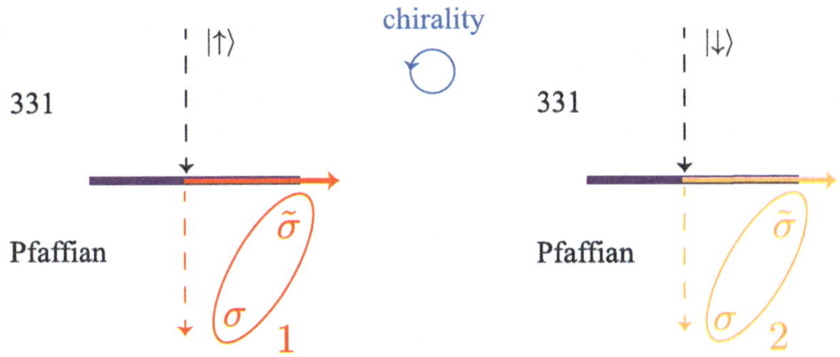

Fig. 21.2 Transmutation of an Abelian 331 quasiparticle when it crosses the interface and enters the Pfaffian liquid. Here, only the neutral sector is considered (see the main text for more details). The symbols $|\uparrow\rangle$ and $|\downarrow\rangle$ denote quasiparticles with pseudospin up and down, respectively. Their corresponding vertex operators are $\mathscr{V}_1$ and $\mathscr{V}_3$ in the U(1)$_4$ CFT. Figure adopted from Ref. [74]

Ref. [55], we interpret the above result as an incoming pseudospin-up quasiparticle transmutes into a neutral anyon $\tilde{\sigma}$ at the interface, and another anyon σ free to move in the Pfaffian liquid. To be more precise, the last anyon actually carries charge $e/4$, but we skip displaying its charge sector $e^{-i\phi_l/2}$ explicitly. The same conclusion holds for an incoming quasiparticle with pseudospin down. We illustrate the results in Fig. 21.2.

21.3.2.1 Matching of Hilbert Spaces and Analogue of Information Scrambling

It is obvious that the total quantum dimension of $(\sigma, \tilde{\sigma})_1$ and $(\sigma, \tilde{\sigma})_2$ is two. It matches the two-dimensional Hilbert space spanned by the pseudospin degree of freedom of an Abelian charge-$e/4$ quasiparticle. This matching is guaranteed mathematically by the commutativity between fusion and restriction in anyon condensation [107]. Interestingly, the information of pseudospin is being stored *nonlocally* at the interface *and* in the interior of Pfaffian liquid. There is no local measurement that can distinguish between $(\sigma, \tilde{\sigma})_1$ and $(\sigma, \tilde{\sigma})_2$. Hence, it is impossible to recover the original information from any local measurement. This feature resembles a quantum information scrambling, which can be defined as the spreading of local information into many-body entanglement and correlation in the whole system [43].

The situation becomes more interesting when we keep dragging more Abelian quasiparticles across the interface. Suppose $N-1$ charge-$e/4$ quasiparticles were already dragged. We assume the corresponding $N-1$ anyons $\tilde{\sigma}$ created at the interface are well separated from each other, so that no fusion occurs between them. We also pose the same assumption for the $N-1$ anyons created in the Pfaffian liquid. Consider dragging an additional charge-$e/4$ quasiparticle across the interface. This process increases both numbers of $\tilde{\sigma}$ and σ by one. As a result, there are N neutral anyons $\tilde{\sigma}$ at the interface, and N non-Abelian anyons in the interior of the Pfaffian liquid. The dimension of the corresponding topological Hilbert space is then increased by a factor of two, which is consistent with the one bit of information carried by the additional Abelian quasiparticle from the Halperin-331 liquid. We illustrate the example of $N = 6$ in Fig. 21.5a, b. Now, we relax the confining potential, and allow the anyons to move and braid [116]. The braiding can further scramble the original information [117]. The N anyons at the interface are indistinguishable, so are those N anyons in the Pfaffian liquid. Meanwhile, the information carried by pseudospins of the original N Abelian quasiparticles is still preserved. Both Hilbert spaces for indistinguishable anyons at the interface and indistinguishable anyons in the Pfaffian liquid have dimensions $2^{N/2}$. It is intriguing that the interface and the Pfaffian liquid store the same amount of information. This does not hold in the Pfaffian-NASS interface [54, 55, 118].

The above discussion also suggests another important feature. In addition to being stored nonlocally, the original pseudospin information is actually "hidden" in the fusion channels of the non-Abelian anyons. Hence, the scrambled information is protected topologically and will not be destroyed by any local perturbation. This property is essential in topological quantum computation (TQC) [78, 119–122].

21.3.3 *Upper Bound of Information Storage and Holographic Principle*

Our previous discussion assumed that local anyons in the system can be well separated to prevent fusion. This assumption leads to a natural question. How much information can be stored nonlocally with the topological protection that has been described?

Recall that the minimum separation between two anyons is in the order of the magnetic length ℓ_B, so that they are well defined individually and do not fuse. From this, one may naively think that the maximum amount of information can be stored is $N_A \sim A/\pi\ell_B^2$, where A denotes the area of the Pfaffian liquid. This argument is valid if the information is carried solely by anyons in the Pfaffian liquid. However, this is not true in the present case. We have assumed both Pfaffian liquid and 331-Pfaffian interface were initially in the ground state with no excitations. As we discussed, the nonlocal storage of pseudospin information of the Abelian quasiparticles from the Halperin-331 liquid requires both anyons at the interface and in the Pfaffian liquid.

For an interface with a length (perimeter) L, it can only accommodate $N_L \sim L/\ell_B$ neutral anyons $\tilde{\sigma}$. Since the radius R of a circular QH droplet satisfies $R \gg \ell_B$, one has $N_L \ll N_A$. When the number of $\tilde{\sigma}$ gets close to or exceeds N_L, different $\tilde{\sigma}$ anyons start to fuse. The resulting particles will be either a fermion or a boson that can propagate back to the Halperin-331 liquid. More explicitly, one has

$$(I, \tilde{\sigma}) \times (I, \tilde{\sigma}) = (I, \tilde{I}) + (I, \tilde{\psi}). \tag{21.15}$$

For the first fusion outcome, the two $\tilde{\sigma}$ anyons at the interface can fuse to a neutral boson with its spin part described by $\mathscr{V}_0$ [see Eq. (21.10)]. This neutral boson can then split to a pair of quasihole and quasiparticle with opposite charges but the same pseudospin, and propagate in the 331 liquid. For the second fusion outcome, the neutral fermion can split to a pair of quasihole and quasiparticle with opposite pseudospins propagating in the 331 liquid. Consequently, some of the hidden information is released and being accessible by local probes. Therefore, the "black hole" is no longer completely black. Note that the released information is not protected topologically and can suffer from quantum decoherence. The discussion shows that the length of the interface sets an upper bound of storing information nonlocally and topologically via $(\sigma, \tilde{\sigma})$ pairs. Furthermore, the magnetic length $\ell_B = \sqrt{1/eB}$ (in the unit of $h/2\pi = c = 1$) plays the role of Planck length in the present system. Here, B denotes the magnetic field.

The above observation actually resembles the argument from holographic principle in black holes. Based on this principle, the maximum amount of information can be stored in a black hole is not determined by its volume, but bounded by its area [18]. This is because the Bekenstein entropy of the black hole is proportional to its area [4, 123–125], which limits the number of degree of freedoms the black hole can have. In contrast to the QH interface, the black hole can always store and "hide" more information by increasing its area. Since the length of the QH interface is assumed to be fixed, the analogous (a weaker version of) holographic principle there implies that Hawking radiation in the form of Abelian quasiparticles and quasiholes will be released when the bound $N_L \sim L/\ell_B$ is reached. Roughly speaking, any additional incoming information is thus reflected by the interface (event horizon).

21.4 Information Paradox in 330-RR$_4$ Interface

In the previous section, we demonstrated that an original Abelian anyon from the Halperin-331 QH liquid would transmute into a pair of anyons, when it crosses the interface and enters the non-Abelian Pfaffian QH liquid. More specifically, one of the resulting anyons is neutral and being created at the interface, and the remaining one is created in the Pfaffian liquid. By proximitizing different pairs of QH liquids, a wide variety of interfaces can be formed. The corresponding quasiparticle transmutation (if any) will depend on the properties of the interface. Hence, different ways of scrambling the original information may be realized in different QH interfaces. Recall that the 331-Pfaffian interface has a chiral central charge of $c = 2 - 3/2 = 1/2$. This indicates that there must be a gapless chiral Majorana fermion mode on the interface [57]. As a result, some of the scrambled information can be stored at the interface. Meanwhile, an interface with a zero chiral central charge can be created by proximitizing a pair of different QH states with identical central charges for their edges. Interfaces of this kind have been studied extensively, and importantly, can be gapped or gapless. When both QH phases are Abelian, they are described by the K-matrix formalism [52, 126, 127]. This enables one to study the gapness of the interface through the concepts of Lagrangian subgroups [128–133] or the null vector criteria [134]. For an interface involving non-Abelian QH state(s), its gapness and the corresponding gapped phases can be explored by the anyon condensation approach [107–110, 135–139]. A rigorous mathematical formalism was developed in [140, 141]. We note that gapped interfaces between different non-Abelian topologically ordered states is still an ongoing research topic.

Motivated by the above discussion, we consider in this section the information paradox and the corresponding quasiparticle transmutation in the interface between the Abelian Halperin-330 state [76] and the non-Abelian Read-Rezayi state at level four [84]. Both states or phases may describe the FQH state in a bilayer system at the total Landau-level filling factor $\nu = 2/3$. Whether these phases are favorable or not in a realistic sample depends on the actual microscopic details of the system. Here, we assume both states and the interface between them can be realized, and study the possible consequences. Previous theoretical work suggested that a continuous phase transition between the Halperin-330 state and the RR_4 state might be triggered by tuning the interlayer tunneling strength in a bilayer FQH system [85–88]. This transition and similar phase transitions in bilayer systems at other filling factors can be studied systematically through the anyon condensation approach [88, 142]. In particular, the condensation of a specific type of anyon in the RR_4 state leads to the Halperin-330 state. Based on the "folding trick" [137–139], it is expected that a gapped boundary between these two FQH states can form. Nevertheless, the precise gapping mechanism and the possibility of having different phases for the gapped interface have not been addressed. Suppose the interface can be gapped, the scrambled pseudospin information may be entirely stored by anyons in the non-Abelian RR_4 FQH liquid with no information stored at the interface. This feature is completely different from the one in the aforementioned 331-Pfaffian interface [74].

21.4.1 Review of Halperin-330 and Read-Rezayi States

The Halperin-330 state is a two-component Abelian topological order, which is characterized by the matrix [76],

$$K = \begin{pmatrix} 3 & 0 \\ 0 & 3 \end{pmatrix}, \tag{21.16}$$

and the associated $t = (1, 1)^T$. Clearly, one has $\nu = t^T K^{-1} t = 2/3$. In a more explicit form, the corresponding topological term for the Halperin-330 edge is

$$L_0 = -\frac{3}{4\pi}\left(\partial_t \phi_\uparrow \partial_x \phi_\uparrow + \partial_t \phi_\downarrow \partial_x \phi_\downarrow\right). \tag{21.17}$$

Here, the subscripts $\uparrow$, $\downarrow$ denote the layer or the pseudospin index. Roughly speaking, each layer of the Halperin-330 state is a simple Laughlin state at $\nu = 1/3$ [143].

Since $\phi_\uparrow$ and $\phi_\downarrow$ have the same chirality, the edge is maximally chiral and has a central charge $c = 2$.

The two most relevant vertex operators for electrons are $\exp(3i\phi_\uparrow)$ and $\exp(3i\phi_\downarrow)$. These two different operators create respectively an electron in the upper and the lower layer. We call them as electrons with pseudospin up and pseudospin down. Both electrons operators have scaling dimensions $3/2$, indicating that they are indeed fermionic as required. The most fundamental anyon that the Halperin-330 state can host has charge $e/3$. Depending on its pseudospin, it is created by the operator $\exp(i\phi_\uparrow)$ or $\exp(i\phi_\downarrow)$. The operator product expansion (OPE) between any one of them and each of the electron operators is singlevalued. For example, one has the following OPEs:

$$\lim_{z\to w} e^{3i\phi_\uparrow(z)} \times e^{i\phi_\uparrow(w)} \sim (z-w)e^{4i\phi_\uparrow(w)}, \tag{21.18}$$

$$\lim_{z\to w} e^{3i\phi_\downarrow(z)} \times e^{i\phi_\uparrow(w)} \sim e^{i[3\phi_\downarrow(w)+\phi_\uparrow(w)]}. \tag{21.19}$$

Notice that $|\alpha| = 1$ is the smallest possible nonzero value for the generic operator $\exp(i\alpha\phi_\uparrow)$ to have singlevalued OPEs with both electron operators. This fact verifies that the $e/3$ anyon is the smallest-charge anyon that can be host by the Halperin-330 state.

21.4.1.1 Parafermion Conformal Field Theory and Read-Rezayi States

Different from the Halperin-330 state, the Read-Rezayi states are a series of non-Abelian topological orders introduced from the CFT approach [84]. Each RR state consists of two different types of CFTs. First, it has a compactified U(1) holomorphic boson ϕ. On the edge of the system, ϕ corresponds to the gapless charge mode described by

$$L_\phi = -\frac{1}{4\pi\nu}\partial_x\phi(\partial_t + v\partial_x)\phi. \tag{21.20}$$

This mode fixes the charge density $\rho(x) = \partial_x\phi/2\pi$ and the quantized electrical Hall conductance $\sigma_{xy} = \nu e^2/h$. The possible values of the filling factor ν depend on the second type of CFT in the RR state, which is the chiral $\mathbb{Z}_k$ parafermion CFT with the central charge [106, 144],

$$c = \frac{2(k-1)}{k+2}. \tag{21.21}$$

The above parafermion CFT can be obtained from the $\mathrm{SU}(2)_k/\mathrm{U}(1)$ coset construction [145]. Here, $k \in \mathbb{N}$ is the level of the corresponding Wess-Zumino-Witten model. Following the notations in Ref. [146] (which are more convenient for the general discussion), each Virasoro primary field in the parafermion CFT is labeled as Φ_m^ℓ with

$\ell + m \equiv 0 \pmod 2$. The index $\ell = 0, 1, \ldots, k$. By imposing the field identification, $\Phi^{\ell}_{m} = \Phi^{\ell}_{m+2k} = \Phi^{k-\ell}_{m-k}$, the index m can be restricted to the range $-\ell < m \leq \ell$, and $\ell > 0$. Hence, there are in total $k(k+1)/2$ Virasoro primary fields. Their scaling dimensions are given by

$$h^{\ell}_{m} = \frac{\ell(\ell+2)}{4(k+2)} - \frac{m^2}{4k}. \tag{21.22}$$

Since the CFT is chiral or holomorphic, the scaling dimension of the field is also its conformal spin. The fusion rule between different fields is

$$\Phi^{\ell_1}_{m_1} \times \Phi^{\ell_2}_{m_2} = \sum_{\ell=|\ell_1-\ell_2|}^{\min(\ell_1+\ell_2, 2k-\ell_1-\ell_2)} \Phi^{\ell}_{m_1+m_2}. \tag{21.23}$$

When $k \geq 2$, the $\mathbb{Z}_k$ parafermion CFT is non-Abelian. A famous example is the Ising CFT at $k = 2$, which has a non-Abelian anyon σ corresponding to the field Φ^1_1. The Ising CFT is relevant in the description of FQH state at $\nu = 5/2$ [147]. The parafermion CFTs and their related FQH states have attracted considerable amount of attention as the non-Abelian anyons there are useful in topological quantum computation [78, 122, 148–154].

An anyon (including the electron) in the RR state is created by the operator, $\eta \exp(i\omega\phi)$. Here, η is a Virasoro primary field in the parafermion CFT. It is customary to choose $\eta = \Phi^{k}_{k-2}$ for constructing the electron operator. The conformal spin for $\Phi^{k}_{k-2} e^{i\phi/\nu}$ is

$$h = \frac{k-1}{k} + \frac{1}{2\nu}. \tag{21.24}$$

By requiring $\nu > 0$ and h being an half-integer, the possible filling factor is fixed at $\nu = k/(Mk+2)$, where M is a positive odd integer. For $k = 4$ and $M = 1$, the Read-Rezayi state may describe a FQH state of electrons at $\nu = 2/3$. This matches the filling factor of the Halperin-330 state. In particular, the Lagrangian density describing the chiral Bose mode ϕ_l on the RR_4 edge is

$$L_{\phi_l} = -\frac{3}{8\pi} \partial_x \phi_l (-\partial_t + v\partial_x) \phi_l. \tag{21.25}$$

Since we will eventually study the interface between two FQH liquids, we define the chirality of ϕ_l as opposite to the chiralities of the two Bose modes on the Halperin-330 edge. The subscript l indicates that ϕ_l is a left-moving mode. As a reminder, the complete edge theory for the RR_4 state also involves the neutral parafermionic sector (see the particular discussion in Ref. [155]).

For later reference, the ten different Virasoro primary fields in the $\mathbb{Z}_4$ CFT and their fusion rules are listed in Table 21.4. Note that we have switched to another set of

Table 21.4 Fusion rules for the 10 primary fields in the $\mathbb{Z}_4$ parafermion CFT

$\times$	I	ψ_1	ψ_2	ψ_3	σ_+	σ_-	ε	ρ	χ_+	χ_-
I	I	ψ_1	ψ_2	ψ_3	σ_+	σ_-	ε	ρ	χ_+	χ_-
ψ_1	ψ_1	ψ_2	ψ_3	I	χ_-	σ_+	ρ	ε	σ_-	χ_+
ψ_2	ψ_2	ψ_3	I	ψ_1	χ_+	χ_-	ε	ρ	σ_+	σ_-
ψ_3	ψ_3	I	ψ_1	ψ_2	σ_-	χ_+	ρ	ε	χ_-	σ_+
σ_+	σ_+	χ_-	χ_+	σ_-	$\psi_1+\rho$	$I+\varepsilon$	$\sigma_+ + \chi_+$	$\sigma_- + \chi_-$	$\psi_3+\rho$	$\psi_2+\varepsilon$
σ_-	σ_-	σ_+	χ_-	χ_+	$I+\varepsilon$	$\psi_3+\rho$	$\sigma_- + \chi_-$	$\sigma_+ + \chi_+$	$\psi_2+\varepsilon$	$\psi_1+\rho$
ε	ε	ρ	ε	ρ	$\sigma_+ + \chi_+$	$\sigma_- + \chi_-$	$I+\psi_2+\varepsilon$	$\psi_1+\psi_3+\rho$	$\sigma_+ + \chi_+$	$\sigma_- + \chi_-$
ρ	ρ	ε	ρ	ε	$\sigma_- + \chi_-$	$\sigma_+ + \chi_+$	$\psi_1+\psi_3+\rho$	$I+\psi_2+\varepsilon$	$\sigma_- + \chi_-$	$\sigma_+ + \chi_+$
χ_+	χ_+	σ_-	σ_+	χ_-	$\psi_3+\rho$	$\psi_2+\varepsilon$	$\sigma_+ + \chi_+$	$\sigma_- + \chi_-$	$\psi_1+\rho$	$I+\varepsilon$
χ_-	χ_-	χ_+	σ_-	σ_+	$\psi_2+\varepsilon$	$\psi_1+\rho$	$\sigma_- + \chi_-$	$\sigma_+ + \chi_+$	$I+\varepsilon$	$\psi_3+\rho$

notations (somehow more convenient for this particular discussion) for the primary fields. The identification with the notations in the previous discussion:

$$\begin{aligned} I &= \Phi_4^4, \psi_1 = \Phi_2^4, \psi_2 = \Phi_0^4, \psi_3 = \Phi_{-2}^4, \sigma_+ = \Phi_3^3, \\ \sigma_- &= \Phi_1^1, \varepsilon = \Phi_0^2, \rho = \Phi_2^2, \chi_- = \Phi_1^3, \chi_+ = \Phi_{-1}^3. \end{aligned} \tag{21.26}$$

After determining the electron operator, all other operators for quasiparticles can be deduced in a "brute-force" manner by requiring the OPE between $\eta \exp(i\omega\phi)$ and the electron operator is single-valued. This restricts the possible values of ω in the vertex operator for each separate η as listed in Table 21.5. From the result, the charge of the corresponding quasiparticles $2\omega/3$ (in unit of e) and the whole

Table 21.5 The 10 primary fields in the $\mathbb{Z}_4$ parafermion CFT with their conformal dimensions (also conformal spins) and quantum dimensions. The possible values for ω in the vertex operator is determined by having a singlevalued OPE between the CFT operators for the anyon and the electron. Here, $m \in \mathbb{Z}$. The corresponding anyon has charge $2\omega/3$ (in the unit of e)

Primary field (η)	I	ψ_1	ψ_2	ψ_3	σ_+	σ_-	ε	ρ	χ_+	χ_-
Conformal dimension (h_η)	0	3/4	1	3/4	1/16	1/16	1/3	1/12	9/16	9/16
Quantum dimension (d_η)	1	1	1	1	$\sqrt{3}$	$\sqrt{3}$	2	2	$\sqrt{3}$	$\sqrt{3}$
Possible values for ω_η	m	$m+\frac{1}{2}$	m	$m+$ 1/2	$m+$ 1/4	$m+$ 3/4	m	$m+$ 1/2	$m+$ 1/4	$m+$ 3/4

spectrum of anyons that the Read-Rezayi state can host is determined. Notice that the quantum dimension of the anyon a is determined by the largest eigenvalue of the fusion matrix $\boldsymbol{N}_a$ with matrix elements $(\boldsymbol{N}_a)_{bc} = N_{ab}^c$. Here, N_{ab}^c is defined from the fusion between the anyons a and b, and the resulting anyon c:

$$a \times b = \sum_c N_{ab}^c c. \tag{21.27}$$

The maximum eigenvalue is positive and nondegenerate, as guaranteed by the Perron-Frobenius theorem. For an Abelian anyon, the quantum dimension is one. Otherwise, the anyon is non-Abelian.

21.4.2 Mathematical Details of Anyon Condensation

Here, we discuss the interface between the Halperin-330 RR_4 states (abbreviated as 330-RR_4 interface) from the anyon condensation perspective. Readers who are uninterested in the mathematical details may skip this subsection. The corresponding physical picture for the fully gapped interface and mechanism of scrambling the pseudospin degree of freedom in the system are reviewed in Sects. 21.4.3 and 21.4.4.

Since both edges of the QH states have the same central charge but with opposite chiralities at the interface, the interface has a zero chiral central charge. In principle, this kind of interface can be (but not guaranteed to be) fully gapped. For example, the edge of the $\mathbb{Z}_2$ toric code (i.e., the boundary between it and the vacuum) remains gapless if the chosen boundary does not break the translational symmetry [156, 157]. However, the edge in a generic situation (with the so-called smooth or rough boundary) is gapped due to the condensation of either one of the self-bosons, e or m on the edge [158, 159]. The discussion suggests that there are two different possible gapped boundaries. When the gapped boundary is obtained by condensing e, then m is confined and becomes a boundary excitation, or vice versa. Can the 330-RR_4 interface be fully gapped? If the answer is affirmative, there will be no gapless excitations on the interface. Then, it may be possible to transmute the original information carried by certain types of the Abelian anyons completely into topological information stored by anyons in the non-Abelian RR_4 liquid.

In the following discussion, we first employ the anyon condensation approach to explore the gapness of the 330-RR_4 interface. We use $\mathscr{A}$ and $\mathscr{B}$ to denote the sets of anyons for the Halperin-330 state and RR_4 state, respectively. The possible phases of the interface are determined by the resulting phases from condensing different possible anyons (if it occurs) in $\mathscr{A} \times \bar{\mathscr{B}}$. Here, $\bar{\mathscr{B}}$ indicates the conjugation of $\mathscr{B}$. In the present case, we have

$$\mathscr{A} = \left\{ e^{i\alpha\phi_\uparrow} e^{i\beta\phi_\downarrow} \right\}, \tag{21.28}$$

$$\bar{\mathscr{B}} = \{ \eta e^{i\omega_\eta \phi_l} \}. \tag{21.29}$$

Here, α, β are integers, and η denotes the Virasoro primary field in the $\mathbb{Z}_4$ parafermion CFT. For each η, the possible values of ω_η are given by the last row in Table 21.5 with $m \in \mathbb{Z}$. Since both electrons in the Halperin-330 state and $\overline{\text{RR}}_4$ state are fermions that have trivial mutual statistics with every anyon in $\mathscr{A} \times \bar{\mathscr{B}}$, $\mathscr{A} \times \bar{\mathscr{B}}$ is a fermionic topological order. Specifically, we introduce the symbol $\Psi_e = e^{3i\phi_\uparrow}$ to denote the electron with pseudospin up for the Halperin-330 state. It is a trivial fermion in $\mathscr{A} \times \bar{\mathscr{B}}$.

21.4.2.1 Determination of Lagrangian Subset

The interface can be fully gapped only if there exists a Lagrangian subset $\mathscr{L} \subset \mathscr{A} \times \bar{\mathscr{B}}$, in which the condensed anyons are the only deconfined anyons in the condensed phase. To identify some possible $\mathscr{L}$, we follow the strategy in Ref. [68] by first condensing Abelian anyons. After that, we will analyze the resulting phase and check for the necessity of condensing more anyons in the system to obtain a Lagrangian subset. Finally, we discuss the underlying reason that gives rise to the anyon condensation. It is vital to clarify that one can actually condense non-Abelian anyons, but the strategy below greatly simplifies the analysis. When the Lagrangian subset $\mathscr{L}$ only consists of *Abelian bosons*, then the explicit conditions for having a fully gapped interface are:

1. trivial monodromy for all $a \in \mathscr{L}$: $M_I^{a\bar{a}} = 1$;
2. trivial mutual statistics for all $a, b \in \mathscr{L}$: $M_{a \times b}^{ab} = 1$;
3. confinement for all $a \notin \mathscr{L}$: there exists at least one $b \in \mathscr{L}$ that has a non-trivial braiding phase with a.

The condition (i) ensures that a is a boson. If a is its own antiparticle (i.e., $a = \bar{a}$ and $a \times \bar{a} = I$), then a should have an integer conformal spin. When a is not its own antiparticle, then it further requires that $b = a \times a$ also has an integer conformal spin [107–109]. For condition (ii), it implies that the anyon $c = a \times b$ also needs to be a boson, otherwise it needs to be condensed as well [109, 110]. Note that one can also condense fermions (by fusing it with the trivial fermion) in a fermionic topological order [139], but this turns out unnecessary in our discussion.

21.4.2.2 Separation of Charge and Neutral Sectors

From Table 21.4, it is clear that the four Abelian anyons in the $\mathbb{Z}_4$ parafermion CFT are I, ψ_1, ψ_2, and ψ_3. To deduce the set of condensable Abelian bosons, one can in principle analyze all anyons with $\eta = \{I, \psi_1, \psi_2, \psi_3\}$ and generic values for α, β, and ω_η in the vertex operators. However, this general treatment turns out to be unphysical. As we will discuss in Sect. 21.4.3, the gapped interface originates from the gapping or localization of counterpropagating edge modes due to electron and quasiparticle tunneling process. Therefore, the charge modes and neutral modes should obtain expectation values independently. This suggests us to separate the charge and neutral sectors in the problem, which is achieved by introducing a new set of modes for the Halperin-330 edge:

$$\phi_r = \phi_\uparrow + \phi_\downarrow, \tag{21.30}$$

$$\phi_n = \phi_\uparrow - \phi_\downarrow. \tag{21.31}$$

They correspond to the overall charge mode and neutral spin mode for the Halperin-330 edge, respectively. Both modes are right-moving. Using the new set of modes, the topological term in Eq. (21.17) becomes

$$L_0 = -\frac{3}{8\pi}\left(\partial_t\phi_r\partial_x\phi_r + \partial_t\phi_n\partial_x\phi_n\right). \tag{21.32}$$

The t vector becomes $t = (1, 0)^T$, which verifies that ϕ_n is a neutral mode. Furthermore, we have

$$\mathscr{A} \times \bar{\mathscr{B}} = \left\{e^{iQ\phi_r}e^{iS\phi_n}\right\} \times \{\eta e^{i\omega_\eta\phi_l}\}. \tag{21.33}$$

Here, $Q = (\alpha + \beta)/2$ and $S = (\alpha - \beta)/2$. Since $\alpha \in \mathbb{Z}$ and $\beta \in \mathbb{Z}$, both Q and S can only take half-integer or integer values. The corresponding anyon in $\mathscr{A}$ has charge $2Q/3$ (in the unit of e). Furthermore, the conformal spins for the two vertex operators $\exp(iQ\phi_r)$ and $\exp(iS\phi_n)$ are

$$s_Q = h_Q = \frac{Q^2}{3}, \quad s_S = h_S = \frac{S^2}{3}. \tag{21.34}$$

Each of them will change by an integer value whenever Q or S is changed by $3\mathbb{Z}$. Physically, this corresponds to the fusion between the anyon with a trivial boson in the Halperin-330 state. Thus, we can identify a pair of anyons with (Q, S) and $(Q + 3\mathbb{Z}, S + 3\mathbb{Z})$. More explicitly, one can rescale the fields $\varphi_r = \phi_r/2$ and $\varphi_n = \phi_n/2$ and rewrite Eq. (21.32) as

$$L_0 = -\frac{6}{4\pi}\left(\partial_t\varphi_r\partial_x\varphi_r + \partial_t\varphi_n\partial_x\varphi_n\right). \tag{21.35}$$

The possible vertex operators take the form $\exp(ip\varphi_r)$ and $\exp(iq\varphi_n)$, where both p and q can only take integer values now. Thus, both φ_r and φ_n (and hence the original fields, ϕ_r and ϕ_n) are compactified bosons in the $U(1)_6$ CFT which has $\exp(6i\varphi)$ as the trivial boson. For later discussion, the six primary fields involving the neutral spin mode in the $U(1)_6$ CFT are summarized in Table 21.6. Moreover, a pair of anyons with (Q, S) and $[Q + 3(\mathbb{Z} + 1/2), S + 3(\mathbb{Z} + 1/2)]$ differ by the fusion of an odd multiples of electrons in the Halperin-330 state.

Table 21.6 The six different primary fields in the U(1)$_6$ CFT for the neutral spin mode ϕ_n. Here, all vertex operators are normal ordered. Note that two vertex operators are identified if they differ by a bosonic operator. For example, we have $\mathscr{V}_4 \sim \mathscr{V}_{-2}$ and $\mathscr{V}_5 \sim \mathscr{V}_{-1}$

Symbol	Vertex operator	Conformal spin
$\mathscr{V}_0$	1	0
$\mathscr{V}_1$	$\exp(i\phi_n/2)$	1/12
$\mathscr{V}_2$	$\exp(i\phi_n)$	1/3
$\mathscr{V}_3$	$\exp(3i\phi_n/2)$	3/4
$\mathscr{V}_4$	$\exp(2i\phi_n)$	1/3
$\mathscr{V}_5$	$\exp(5i\phi_n/2)$	1/12

21.4.2.3 Condensation of Abelian Bosons

Following the above discussion, we look for possible values of Q, ω_η, and S such that both anyons with operators $\exp(iQ\phi_r)\exp(i\omega_\eta\phi_l)$ and $\eta\exp(iS\phi_n)$ are bosonic. The conformal spin for $\exp(iQ\phi_r)\exp(i\omega_\eta\phi_l)$ is

$$s(Q,\omega_\eta) = \frac{Q^2-\omega_\eta^2}{3}. \tag{21.36}$$

Therefore, a possible solution for $s(Q,\omega_\eta)=0$ (or alternatively, the null vector for the K matrix describing the charge sector) is $(Q,\omega_\eta)=(1,1)$. Hence, we first condense the following four bosons:

$$b_0 = e^{im_0(\phi_r+\phi_l)}, \tag{21.37}$$

$$b_1 = \psi_1 e^{3i\phi_n/2} e^{i(m_1+1/2)(\phi_r+\phi_l)}, \tag{21.38}$$

$$b_2 = \psi_2 e^{im_2(\phi_r+\phi_l)}, \tag{21.39}$$

$$b_3 = \psi_3 e^{3i\phi_n/2} e^{i(m_3+1/2)(\phi_r+\phi_l)}. \tag{21.40}$$

Here, all $m_i \in \mathbb{Z}$. It is easy to check that all of the above bosons satisfy the conditions (i) and (ii) in the previous discussion. This verifies that all of them can be condensed simultaneously [108]. It is worthwhile for clarifying that being a boson does not mean that it must be condensable. For example, the condensation of non-Abelian bosons can be obstructed by a no-go theorem [160]. Also, a self-dual boson having Frobenius-Schur indicator $\varkappa = -1$ is not condensable [109, 161]. Nevertheless, none of these limitations applies here.

Since a deconfined anyon must have trivial mutual statistics with all the condensed bosons, we first put $m_2 = 0$ to eliminate a large set of possibilities. The corresponding monodromy is given by

$$M^{\eta\psi_2}_{\eta\times\psi_2} = \exp\left[-(2\pi i)\left(h_{\eta\times\psi_2} - h_\eta - h_{\psi_2}\right)\right]. \tag{21.41}$$

Hence, all anyons with $\eta = \{\sigma_+, \sigma_-, \chi_+, \chi_-\}$ are confined. This result was used by Barkeshli and Wen in showing that the phase transition from the RR_4 state to the Halperin-330 state could be described by the anyon condensation of ψ_2 [88]. By setting $m_2 \neq 0$, the monodromy between $Y = \eta \exp(iS\phi_n)\exp(iQ\phi_r)\exp(i\omega_\eta\phi_l)$ and b_2 is given by

$$M_{Y\times b_2}^{Yb_2} = \exp\left[\frac{4m_2\pi i}{3}\left(Q - \omega_\eta\right)\right], \; m_2 \in \mathbb{Z}. \tag{21.42}$$

Here, $\eta = \{I, \psi_1, \psi_2, \psi_3, \varepsilon, \rho\}$. Thus, the set of possible deconfined anyons is further reduced to $e^{i\omega_\eta(\phi_l+\phi_r)}e^{3i\gamma\phi_r/2}e^{iS\phi_n} \times \{I, \psi_1, \psi_2, \psi_3, \varepsilon, \rho\}$, with $\gamma \in \mathbb{Z}$. This set of anyons have trivial mutual statistics with b_0 also. By requiring them to have trivial mutual statistics with b_1 and b_3, it further reduces the possible set of deconfined anyons to four apparently different classes. For the first class, we have

$$\begin{aligned}
\mathcal{T}_1 &= e^{im(\phi_l+\phi_r)}e^{3i(p+1/2)\phi_r}e^{i(q+1/2)\phi_n} \times \{I, \psi_2, \varepsilon\} \\
&\sim e^{im(\phi_l+\phi_r)}\left[e^{3i(\phi_r+\phi_n)/2}\right]\{\mathcal{V}_0, \mathcal{V}_2, \mathcal{V}_4\} \times \{I, \psi_2, \varepsilon\} \\
&\sim e^{im(\phi_l+\phi_r)}\Psi_e\{\mathcal{V}_0, \mathcal{V}_2, \mathcal{V}_4\} \times \{I, \psi_2, \varepsilon\} \\
&\sim \Psi_e\{\mathcal{V}_0, \mathcal{V}_2, \mathcal{V}_4\} \times \{I, \varepsilon\}.
\end{aligned} \tag{21.43}$$

All m, p, and q in the above calculation are integers. It is recalled that $\Psi_e = \exp(3i\phi_\uparrow) = \exp[3i(\phi_r + \phi_n)/2]$ is the electron in the Halperin-330 state. In the second line, we used the fact that ϕ_n is a compactified boson in the $U(1)_6$ CFT, and labeled the corresponding vertex operators by the symbols in Table 21.6. Moreover, we used the fact that $\exp(3ip\phi_r)$ is a trivial boson to make the identification denoted by $\sim$. In the last line, the identification is made by fusing the set of anyons in the third line with the condensed boson b_2 with $m_2 = -m$.

Using similar procedures, we determine the second class of deconfined anyons as

$$\begin{aligned}
\mathcal{T}_2 &= e^{im(\phi_l+\phi_r)}e^{3ip\phi_r}e^{iq\phi_n} \times \{I, \psi_2, \varepsilon\} \\
&\sim \{\mathcal{V}_0, \mathcal{V}_2, \mathcal{V}_4\} \times \{I, \varepsilon\}.
\end{aligned} \tag{21.44}$$

By setting $\eta = \{\psi_1, \psi_3, \rho\}$, we have the remaining two classes of deconfined anyons,

$$\begin{aligned}
\mathcal{T}_3 &= e^{i(m+1/2)(\phi_l+\phi_r)}e^{3ip\phi_r}e^{i(q+1/2)\phi_n} \times \{\psi_1, \psi_3, \rho\} \\
&\sim e^{i(\phi_l+\phi_r)/2}\{\mathcal{V}_1, \mathcal{V}_3, \mathcal{V}_5\} \times \{\psi_1, \psi_3, \rho\},
\end{aligned} \tag{21.45}$$

and

$$\begin{aligned}
\mathcal{T}_4 &= e^{i(m+1/2)(\phi_l+\phi_r)}e^{3i(p+1/2)\phi_r}e^{iq\phi_n} \times \{\psi_1, \psi_3, \rho\} \\
&\sim \Psi_e\, e^{i(\phi_l+\phi_r)/2}\{\mathcal{V}_1, \mathcal{V}_3, \mathcal{V}_5\} \times \{\psi_1, \psi_3, \rho\}.
\end{aligned} \tag{21.46}$$

Meanwhile, they are actually equivalent to $\mathscr{T}_2$ and $\mathscr{T}_1$, respectively. This is observed by fusing $\mathscr{T}_3$ and $\mathscr{T}_4$ with either b_1 or b_3 to obtain

$$\mathscr{T}_3 \sim \{\mathscr{V}_0, \mathscr{V}_2, \mathscr{V}_4\} \times \{I, \varepsilon\} \sim \mathscr{T}_2, \tag{21.47}$$
$$\mathscr{T}_4 \sim \Psi_e \{\mathscr{V}_0, \mathscr{V}_2, \mathscr{V}_4\} \times \{I, \varepsilon\} \sim \mathscr{T}_1. \tag{21.48}$$

Therefore, the above four classes of deconfined anyons can be combined into

$$\begin{aligned} \mathscr{T} &= \{1, \Psi_e\} \times \{\mathscr{V}_0, \mathscr{V}_2, \mathscr{V}_4\} \times \{I, \varepsilon\} \\ &= \{1, \Psi_e\} \times \mathscr{T}_B. \end{aligned} \tag{21.49}$$

Since $\mathscr{V}_2$ and $\mathscr{V}_4$ cannot be obtained from any fusion between the bosons in the set $\{b_0, b_1, b_2, b_3\}$, this set is *not* a Lagrangian subset. In the second line of Eq. (21.49), $\mathscr{T}_B$ is a bosonic order with $\mathscr{V}_0 I$ being the trivial boson. The separable form in Eq. (21.49) is guaranteed for any *Abelian* fermionic order [162, 163]. The Abelianity of $\mathscr{T}$ becomes transparent after splitting the non-Abelian anyons in $\mathscr{T}_B$. Notice that $(\Psi_e)^2 = \exp(3i\phi_r)\exp(3i\phi_n) \sim 1$.

21.4.2.4 Splitting of Non-Abelian Anyons and the $\mathbb{Z}_3$ Toric Code

Since some of the anyons in the original phase have been identified as the trivial vacuum after the anyon condensation, the unconfined non-Abelian anyons in $\mathscr{T}$ may need to split. Consider the fusion:

$$\varepsilon \times \varepsilon = I + \psi_2 + \varepsilon \sim 2I + \varepsilon. \tag{21.50}$$

Due to the identification $\psi_2 \sim I$, the vacuum appears twice. Consequently, ε needs to split into two Abelian anyons, $\varepsilon = \varepsilon_1 + \varepsilon_2$. This matches the quantum dimension as $2 = 1 + 1$. Furthermore, the splitting and the fusion rules are consistent only if

$$\varepsilon_1 \times \varepsilon_1 = \varepsilon_2, \tag{21.51}$$
$$\varepsilon_2 \times \varepsilon_2 = \varepsilon_1, \tag{21.52}$$
$$\varepsilon_1 \times \varepsilon_2 = \varepsilon_2 \times \varepsilon_1 = I. \tag{21.53}$$

Then, the fusion rule $\rho \times \rho = I + \psi_2 + \varepsilon$ implies that the non-Abelian anyon ρ also needs to split, $\rho = \rho_1 + \rho_2$. After the splitting, a possible (but not unique) consistent set of fusion rules are

$$\rho_1 \times \rho_1 = \varepsilon_1, \;\; \rho_2 \times \rho_2 = \varepsilon_2, \;\; \rho_1 \times \rho_2 = I, \tag{21.54}$$
$$\varepsilon_1 \times \psi_1 = \rho_2, \;\; \varepsilon_2 \times \psi_1 = \rho_1, \tag{21.55}$$

$$\rho_1 \times \psi_1 = \varepsilon_2,\ \rho_2 \times \psi_1 = \varepsilon_1, \tag{21.56}$$

$$\rho_1 \times \varepsilon_1 = \psi_1,\ \rho_1 \times \varepsilon_2 = \rho_2, \tag{21.57}$$

$$\rho_2 \times \varepsilon_1 = \rho_1,\ \rho_2 \times \varepsilon_2 = \psi_3. \tag{21.58}$$

Note that $\psi_3 \sim \psi_1$ in the condensed phase since they differ by a fusion with ψ_2 (i.e, b_2 with $m_2 = 0$). Therefore, the fusion rules involving ψ_3 are the same as those involving ψ_1. By making the change(s), $\varepsilon_1 \leftrightarrow \varepsilon_2$, or/and $\rho_1 \leftrightarrow \rho_2$, one can still obtain a consistent set of fusion rules. This is reasonable as it is impossible to uniquely define or fix the anyons resulting from the splitting. On the other hand, they must be inequivalent. Without loss of generality, we will stick with the set of fusion rules in Eqs. (21.51)–(21.58) in the following discussion.

The above discussion shows that $\mathscr{T}_B$ actually consists of nine different *Abelian* anyons, so $\mathscr{T}$ is an Abelian fermionic topological order. Then, what is the topological order $\mathscr{T}$? By labeling the nine anyons in $\mathscr{T}_B$ as shown in Table 21.7, it is observed that these anyons take the form of $e^p m^q$, where $p, q \in \mathbb{Z}$. Moreover, one has $e^3 = m^3 = \mathbb{I}$. Hence, $\mathscr{T}_B$ is the $\mathbb{Z}_3$ toric code. This topological order is a generalization of the more famous $\mathbb{Z}_2$ toric code that only has four anyons, $\{1, e, m, f \equiv em\}$ [122]. Since it is the topological order built from (mathematically, the modular tensor category constructed over) the quantum double model with the finite group $\mathbb{Z}_3$ [122, 164, 165], the $\mathbb{Z}_3$ toric code is also denoted as $\mathfrak{D}(\mathbb{Z}_3)$ [166–169]. It was suggested that this special topological order can be realized by proximitizing a bilayer system of electrons and holes in separate Laughlin states with respective filling factors $\pm 1/3$ to a superconductor [170]. It was also pointed out that the more general $\mathbb{Z}_p$ toric code appears as the symmetry-enriched neutral sector of non-diagonal QH states [171]. Here, we have shown that the interface between the Halperin-330 and RR_4 states with suitable anyon condensation also leads to the $\mathbb{Z}_3$ toric code. The physical origin

Table 21.7 The identification between the nine different anyons in the $\mathbb{Z}_3$ toric code (denoted as $\mathfrak{D}(\mathbb{Z}_3)$ in the previous literature and this work) and the corresponding anyons in the bosonic phase $\mathscr{T}_B$. This identification is made such that the topological twist for $e^p m^q$ is $\theta = \exp(2pq\pi i/3)$. Also notice that we can always define the conformal spin as positive, since it is defined only up to modulo one

Symbol	Anyon in $\mathscr{T}_B$	Conformal spin ($s = h - \bar{h}$)
$\mathbb{I}$	$\gamma_0 I$	0
e	$\gamma_2 \varepsilon_1$	0
e^2	$\gamma_4 \varepsilon_2$	0
m	$\gamma_4 \varepsilon_1$	0
m^2	$\gamma_2 \varepsilon_2$	0
$e^2 m$	$\gamma_2 I$	1/3
em^2	$\gamma_4 I$	1/3
em	$\gamma_0 \varepsilon_2$	2/3
$e^2 m^2$	$\gamma_0 \varepsilon_1$	2/3

that triggers such an anyon condensation will be discussed in Sect. 21.4.3. On the application side, it was suggested that the $\mathbb{Z}_3$ toric code could be used in implementing universal topological quantum computation [167]. To summarize, the condensation of b_0, b_1, b_2, and b_3 leads to the condensed phase with deconfined anyons,

$$\mathscr{T} = \{1, \Psi_e\} \times \mathfrak{D}(\mathbb{Z}_3). \tag{21.59}$$

Note that $\mathscr{T}$ does not describe a fully gapped interface. This is because $\mathfrak{D}(\mathbb{Z}_3)$ is not a topologically trivial order. Alternatively, the edge of $\mathfrak{D}(\mathbb{Z}_3)$ is gappable but the edge remains gapless unless a further anyon condensation occurs, which we are going to discuss below.

21.4.2.5 Two Different Phases of Gapped Interfaces

As we have shown, the set $\{b_0, b_1, b_2, b_3\}$ is not a Lagrangian subset, and the condensation of this set of bosons does not lead to a fully gapped interface. On the other hand, a fully gapped interface can be achieved by further condensing some bosonic particles in $\mathfrak{D}(\mathbb{Z}_3)$. By recycling the results in Ref. [167], one can immediately conclude that there are two different types of fully gapped interfaces between the Halperin-330 FQH state and the RR_4 FQH state.

The first possible kind of fully gapped interface is obtained by a further condensation of e and e^2 in $\mathfrak{D}(\mathbb{Z}_3)$. It is straightforward to verify that both of them satisfy conditions (i) and (ii) in the previous discussion. This implies that they can be condensed simultaneously. Moreover, their condensation leads to the confinement of the remaining anyons (except the identity $\mathbb{I}$) in $\mathfrak{D}(\mathbb{Z}_3)$. Therefore, we determine the first Lagrangian subgroup [172] for $\mathscr{A} \times \bar{\mathscr{B}}$ as

$$\mathscr{L}_e = \{b_0, b_1, b_2, b_3\} \times \left\{\mathbb{I}, e, e^2\right\}. \tag{21.60}$$

The corresponding fully gapped interface is usually known as the e-boundary. Notice that the symbol $\times$ actually means picking an element from each set on the right hand side of Eq. (21.60) and then fuse them (see Eq. (21.68) for example). The product structure of $\mathscr{L}_e$ ensures that it is a maximal set of condensable bosons.

Another possible type of fully gapped interface is obtained by condensing m and m^2 in $\mathfrak{D}(\mathbb{Z}_3)$ instead. This leads to the second Lagrangian subgroup,

$$\mathscr{L}_m = \{b_0, b_1, b_2, b_3\} \times \left\{\mathbb{I}, m, m^2\right\}. \tag{21.61}$$

The corresponding fully gapped interface is known as the m-boundary. For both e- and m-boundaries, the remaining deconfined anyons outside the Lagrangian subgroups are $\mathscr{F}_0 = \{1, \Psi_e\}$. This is the trivial fermionic topological order, which indicates that the condensation of anyons in $\mathscr{L}_e$ or $\mathscr{L}_m$ leads to a fully gapped interface [139].

21.4.3 Physical Picture of the Gapped Interface

While anyon condensation has provided a systematic and mathematical approach in studying the gapness of the interface, it will be also desirable to understand the gapping of the interface in a more physical picture. We claim that the anyon condensation may originate from the electron and quasiparticle tunneling processes at the interface in two different steps.

At the beginning, the Halperin-330 and Read Rezayi states are two topologically distinct phases. Therefore, only electrons can tunnel across the two different FQH liquids. In general, the electron tunneling process is described by the following Lagrangian density,

$$\begin{aligned} L_{\text{el, tun}} &= \xi_{1,a}(x)\left(\psi_1 e^{3i\phi_l/2} e^{3i\phi_\uparrow}\right)^a \\ &\quad + \xi_{2,a}(x)\left(\psi_1 e^{3i\phi_l/2} e^{3i\phi_\downarrow}\right)^a + \text{H.c.} \\ &= \xi_{1,a}(x)\left[\psi_1 e^{3i\phi_n/2} e^{3i(\phi_l+\phi_r)/2}\right]^a \\ &\quad + \xi_{2,a}(x)\left[\psi_1 e^{-3i\phi_n/2} e^{3i(\phi_l+\phi_r)/2}\right]^a + \text{H.c.} \end{aligned} \tag{21.62}$$

Note that H.c. stands for the Hermitian conjugation. We include the exponent $a > 1$ to describe multi-electron tunneling processes. As the electron operators $\psi_1 e^{3i\phi_l/2}$ and $e^{3i\phi_\sigma}$ (where $\sigma = \uparrow, \downarrow$) enter the same number of times, $L_{\text{el, tun}}$ conserves the total electric charge. Furthermore, the tunneling amplitudes $\xi_{1,a}(x)$ and $\xi_{2,a}(x)$ are random functions in x. This is because the electron tunneling generally does not conserve momentum, and disorder needs to be involved. Due to the random nature of the tunneling, some of the processes are actually irrelevant in the renormalization group sense [173]. In this situation, we will need to assume the tunneling strength W_a, defined as $\overline{\xi_a(x)\xi_a(x')} = W_a\delta(x - x')$, is sufficiently large so that the charge modes $3a[\phi_l(x) + \phi_r(x)]/2$ can still be pinned. Then, this combination of counterpropagating charge modes obtains a nonzero expectation value, and indicates the localization (analogous to mode gapping in nonrandom tunneling) of charge modes. For the neutral sector, it is expected that the combination $\psi_1 e^{\pm 3i\phi_n/2}$ will also be gapped by $\mathscr{L}_{\text{el, tun}}$ in the strong coupling regime. By defining $(\psi_1)^2 \sim \psi_2$, $(\psi_1)^3 \sim \psi_3$, and $(\psi_1)^4 \sim I$, the tunneling processes described by $\mathscr{L}_{\text{el, tun}}$ lead to the condensation of b_0, b_1, b_2, and b_3 in Eqs. (21.37)–(21.40).

After the above condensation or charge-mode localization, the interface remains gapless. Meanwhile, quasiparticles with charges $e/3$ and $2e/3$ can tunnel across this "new" gapless interface. Let us specifically consider the tunneling of charge $e/3$ anyon with the topological sector ρ in the Read-Rezayi liquid. This process can be described by the following Lagrangian density,

$$\begin{aligned} &L_{e/3,\text{ tun}} \\ &= \zeta_1(x)\left(\rho e^{i\phi_l/2} e^{i\phi_\uparrow}\right) + \zeta_2(x)\left(\rho e^{i\phi_l/2} e^{i\phi_\downarrow}\right) + \text{H.c.} \\ &= \zeta_1(x)\left[\rho \mathscr{V}_1 e^{i(\phi_l+\phi_r)/2}\right] + \zeta_2(x)\left[\rho \mathscr{V}_5 e^{i(\phi_l+\phi_r)/2}\right] + \text{H.c.} \end{aligned} \tag{21.63}$$

Since the charge modes have been localized, we focus on the neutral sector. Now, both combinations $\rho\mathcal{V}_1$ and $\rho\mathcal{V}_5$ have zero conformal spins (see Tables 21.5 and 21.6 for reference), which indicates that they are bosonic. In principle, they may be condensed or gapped as well. Consider the set of anyons generated from any fusion between $\rho\mathcal{V}_1$ and $\rho\mathcal{V}_5$, we have

$$\{\rho\mathcal{V}_1, \rho\mathcal{V}_5\} \times \{\rho\mathcal{V}_1, \rho\mathcal{V}_5\} = \{I, \psi_2, \varepsilon\} \times \{\mathcal{V}_0, \mathcal{V}_2, \mathcal{V}_4\} \tag{21.64}$$

A further fusion with $\{\rho\mathcal{V}_1, \rho\mathcal{V}_5\}$ gives

$$\{\psi_1, \psi_3, \rho\} \times \{\mathcal{V}_1, \mathcal{V}_3, \mathcal{V}_5\}. \tag{21.65}$$

Hence, the fusion between $\rho\mathcal{V}_1$ and $\rho\mathcal{V}_5$ generates the list of anyons (more precisely, their neutral sectors) in $\mathcal{L}_e$ and $\mathcal{L}_m$. In this sense, the condensation described by the Lagrangian subgroup $\mathcal{L}_e$ or $\mathcal{L}_m$ may be understood as originating from the gapping of modes due to $e/3$ quasiparticle tunneling at the interface. However, this is only possible if we can treat the whole system as a single topological phase. Otherwise, only electrons can tunnel across two topologically distinct phases (for example, between a FQH liquid and a normal metal) [52].

21.4.4 *Transmutation of Pseudospin Information*

Previously, we found that it is possible (at least in principle) to form a gapped interface between the Halperin-330 state and the Read-Rezayi state at level four. Although both states may describe the FQH state in a bilayer system at total filling factor 2/3, they host different sets of anyons. Specifically, the Halperin-330 state is a two-component state, in which the anyons possess the layer or pseudospin degree of freedom that is absent in the RR_4 state. Meanwhile, the RR_4 state supports non-Abelian anyons that have more complicated fusion rules than the usual Abelian quasiparticles. This observation leads to a natural question: What happens when an Abelian quasiparticle from the Halperin-330 liquid crosses the gapped interface and enters the non-Abelian RR_4 FQH liquid? In the opposite direction, what happens when we drag a non-Abelian quasiparticle from the RR_4 liquid to the Halperin-330 liquid? The answers to these questions lead to the ideas of topological quantum information scrambling and Andreev-like reflection of non-Abelian anyons.

An anyon a originally in the topological phase $\mathcal{A}$ can pass through the interface (described by the condensed phase of $\mathcal{A} \times \bar{\mathcal{B}}$) and transmutes into an anyon b in $\mathcal{B}$ if and only if ab is a deconfined anyon in the condensed phase. This idea was introduced to study the transmutation between anyons in the interface between the Pfaffian and non-Abelian spin-singlet FQH states [54, 55]. Now, we employ the same kind of argument to study the transmutation of Abelian anyons in the Halperin-330 FQH liquid when they cross the interface and enter the RR_4 liquid.

21.4.4.1 Transmutation of Abelian Charge $e/3$ Anyon

First, an Abelian charge $e/3$ anyon in the Halperin-330 state can have pseudospin up or pseudospin down. We denote them as $(e/3, \uparrow)$ and $(e/3, \downarrow)$, respectively. Their associated CFT operators are

$$(e/3, \uparrow) \equiv \exp(i\phi_\uparrow) = \mathscr{V}_1 \exp(i\phi_r/2), \tag{21.66}$$

$$(e/3, \downarrow) \equiv \exp(i\phi_\downarrow) = \mathscr{V}_5 \exp(i\phi_r/2). \tag{21.67}$$

When they cross the fully gapped 330-RR_4 interface, their transmutation depends on the type of the gapped interface (i.e., whether it is an e- or m-boundary).

Suppose the fully gapped interface is described by the e-boundary. Then, the deconfined anyons are those in $\mathscr{L}_e$ in Eq. (21.60). In particular, we have the following pair of deconfined anyons:

$$e^2 \times b_1 = \left[\rho_1 \exp(i\phi_l/2)\right]\left[\mathscr{V}_1 \exp(i\phi_r/2)\right], \tag{21.68}$$

$$e \times b_1 = \left[\rho_2 \exp(i\phi_l/2)\right]\left[\mathscr{V}_5 \exp(i\phi_r/2)\right]. \tag{21.69}$$

These are the only two deconfined anyons that involve $(e/3, \uparrow)$ and $(e/3, \downarrow)$. Since $e^2 \times b_1$ and $e \times b_1$ are condensed bosons in $\mathscr{L}_e$, they are identified as the vacuum sector in the condensed phase. Based on the interpretation in Ref. [54], $(e/3, \uparrow)$ and $(e/3, \downarrow)$ will transmute into $\rho_1 \exp(i\phi_l/2)$ and $\rho_2 \exp(i\phi_l/2)$ in the RR_4 FQH state, respectively. It is important to clarify that these transmuted anyons are now in $\mathscr{B}$, but not $\bar{\mathscr{B}}$. Since $\rho_1 \neq \rho_2$, the resulting Abelian anyons are different.

Consider the case when the fully gapped interface is described by the m-boundary that has the Lagrangian subset $\mathscr{L}_m$ in Eq. (21.61). Now, we have the following pair of deconfined anyons:

$$m \times b_1 = \left[\rho_2 \exp(i\phi_l/2)\right]\left[\mathscr{V}_1 \exp(i\phi_r/2)\right], \tag{21.70}$$

$$m^2 \times b_1 = \left[\rho_1 \exp(i\phi_l/2)\right]\left[\mathscr{V}_5 \exp(i\phi_r/2)\right]. \tag{21.71}$$

Thus, $(e/3, \uparrow)$ and $(e/3, \downarrow)$ will transmute respectively into $\rho_2 \exp(i\phi_l/2)$ and $\rho_1 \exp(i\phi_l/2)$. Notice that the result is different from the one in the e-boundary. We summarize the above results in Fig. 21.3.

21.4.4.2 Transmutation of Abelian Charge $2e/3$ Anyon

Similarly, there will be a transmutation of the pseudospin information carried by a charge $2e/3$ anyon when it crosses the interface and enters the RR_4 FQH liquid. There are three different pseudospin states for the charge $2e/3$ anyon in the Halperin-330 FQH liquid. They are spin up, spin zero, and spin down. Physically, they can be viewed as the combination of two spin-up, one spin-up and one spin-down, and two spin-down $e/3$ anyons. The corresponding CFT operators are

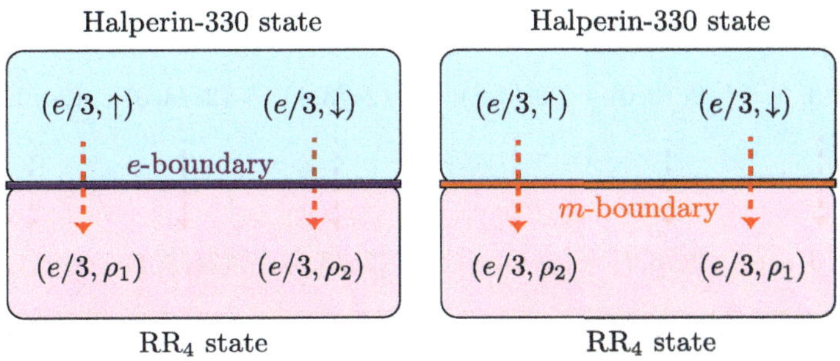

Fig. 21.3 The transmutation of an Abelian charge $e/3$ anyon from the Halperin-330 liquid when it crosses the fully gapped interface. The left (right) panel illustrates the case when the interface is described by the e-boundary (m-boundary). Due to the transmutation, the original local pseudospin information (marked by ↑ and ↓) is scrambled, and being stored by the split anyons ρ_1 and ρ_2. They have different fusion rules with other anyons in the system

$$(2e/3, \uparrow) \equiv \exp(2i\phi_\uparrow) = \mathcal{V}_2 \exp(i\phi_r), \tag{21.72}$$
$$(2e/3, 0) \equiv \exp[i(\phi_\uparrow + \phi_\downarrow)] = \mathcal{V}_0 \exp(i\phi_r). \tag{21.73}$$
$$(2e/3, \downarrow) \equiv \exp(2i\phi_\downarrow) = \mathcal{V}_4 \exp(i\phi_r). \tag{21.74}$$

Following the previous discussion, we first discuss their transmutation when the fully gapped interface is described by the e-boundary. In this case, one has the following three deconfined anyons,

$$e \times b_0 = \left[\varepsilon_1 \exp(i\phi_l)\right]\left[\mathcal{V}_2 \exp(i\phi_r)\right], \tag{21.75}$$
$$\mathbb{I} \times b_0 = \left[I \exp(i\phi_l)\right]\left[\mathcal{V}_0 \exp(i\phi_r)\right], \tag{21.76}$$
$$e^2 \times b_0 = \left[\varepsilon_2 \exp(i\phi_l)\right]\left[\mathcal{V}_4 \exp(i\phi_r)\right], \tag{21.77}$$

Thus, $(2e/3, \uparrow)$ transmutes into $\varepsilon_1 \exp(i\phi_l)$; $(2e/3, 0)$ becomes $I \exp(i\phi_l)$; $(2e/3, \downarrow$ $)$ becomes $\varepsilon_2 \exp(i\phi_l)$. For the case of having the m-boundary, one has

$$m^2 \times b_0 = \left[\varepsilon_2 \exp(i\phi_l)\right]\left[\mathcal{V}_2 \exp(i\phi_r)\right], \tag{21.78}$$
$$m \times b_0 = \left[\varepsilon_1 \exp(i\phi_l)\right]\left[\mathcal{V}_4 \exp(i\phi_r)\right]. \tag{21.79}$$

In this scenario, $(2e/3, \uparrow)$ transmutes into $\varepsilon_2 \exp(i\phi_l)$, whereas $(2e/3, \downarrow)$ becomes $\varepsilon_1 \exp(i\phi_l)$. The results are illustrated in Fig. 21.4.

The original pseudospin information of a charge $e/3$ or a charge $2e/3$ anyon from the Halperin-330 state is entirely encoded in the neutral spin mode ϕ_n. Clearly, this is a local information which can be accessed via local measurement. More specifically, we know in which layer the Abelian anyon was created. Depending on *both* the original pseudospin state and the boundary type of the fully gapped interface, the pseudospin information will be transmuted into ρ_1 or ρ_2 (for the charge $e/3$ anyon), or ε_1, ε_2, or I (for the charge $2e/3$ anyon). Although one knows that the split anyons,

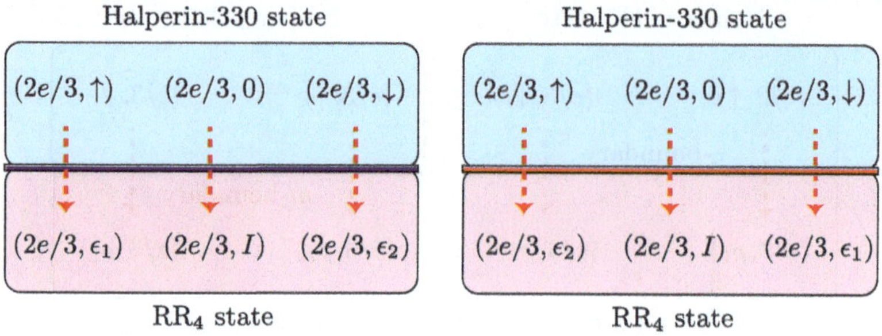

Fig. 21.4 The transmutation of an Abelian charge $2e/3$ anyon from the Halperin-330 liquid when it crosses the fully gapped interface. The left (right) panel illustrates the case when the interface is described by the e-boundary (m-boundary). The original local pseudospin state can be ↑, 0, or ↓. This information is scrambled and being stored by the split anyons ε_1, ε_2, or the trivial vacuum I

say ρ_1 and ρ_2, are different, it is impossible to distinguish between them via any local measurement. In fact, their inequivalence is only manifested topologically in their braiding and fusion rules with other anyons (including themselves) as summarized in Eqs. (21.51)–(21.58). In this sense, we claim that the original local pseudospin information is completely scrambled into a form of nonlocal information being stored and protected topologically by the anyons in the RR_4 FQH liquid. No information is stored on the interface. It is reasonable as the interface is fully gapped, and does not support any low-energy gapless excitations there. This feature is different from the situation of having a gapless QH interface. As a concrete example, our recent work [74] showed that the pseudospin information for an Abelian charge $e/4$ anyon from the Halperin-331 state will be scrambled and stored nonlocally by a pair of vortices (one on the interface, and another one in the Pfaffian FQH liquid) when the Abelian anyon crosses the gapless 331-Pfaffian interface. The comparison here demonstrates the dependence of the scrambling mechanism on the gapness of the QH interface.

21.4.5 Some Remarks

One may realize that the transmuted particle from the charge $2e/3$ anyon can be obtained directly by fusing a pair of transmuted particles from the charge $e/3$ anyons. This is not a coincidence. It is guaranteed from the commutativity between restriction and fusion in anyon condensation [107]. This important property also ensures that the total quantum dimension of the split anyons is identical to the quantum dimension of the original non-Abelian anyon. Moreover, the total conformal spin for the transmuted anyon(s) matches the one for the original anyon. In the present case, we have a fully gapepd interface, and the vertex operators describing the charge sector of the deconfined anyons always have zero conformal spins. Thus, the conformal spin for the parafermionic field η from the $\mathbb{Z}_4$ CFT must match with the conformal spin for the vertex operator describing the neutral spin mode ϕ_n. This holds in all our results, and further highlights the advantage of separating the charge and neutral sectors as we did in our analysis.

In principle, one can drag more quasiparticles from the Halperin-330 liquid into the RR_4 liquid. Suppose the resulting anyons are well separated from each other (i.e., with a separation much larger than the magnetic length). Then, the braiding between the transmuted anyons will lead to a further scrambling of the original pseudospin information [117]. In order to recover the original pseudospin information, one basically needs to "pull" all the anyons back to the Halperin-330 liquid. However, the recovered information will be in a highly entangled form, which again resembles the definition of quantum information scrambling.

21.5 Simulation of Black Hole Evaporation

In order to resolve the information paradox in our model, it is necessary to mimic a black hole evaporation in the QH interface. To simplify our discussion, we only consider the 331-Pfaffian interface here. As we will show, the process recovers the original information carried by the pseudospin degree of freedom naturally. To simplify the discussion, we assume only charge-$e/4$ Abelian quasiparticles were dragged across the interface before the "evaporation". In general, one can also drag quasiholes and quasiparticles with other charges. Under the above assumption, we argued in Sect. 21.3.2.1 that neutral anyons $\tilde{\sigma}$ and non-Abelian charge-$e/4$ quasiparticles carrying σ are created. This is illustrated in Fig. 21.5a, b.

The "black hole evaporation" is simulated by shrinking the Pfaffian region. Experimentally, it may be achieved by reducing the interlayer tunneling in the bilayer system [57, 81–83]. When a non-Abelian quasiparticle reaches the shrinking interface, it is released back to the Halperin-331 liquid. This process plays the role of Hawking radiation in the present setup. There are two different mechanisms for the conversion from a non-Abelian quasiparticle into an Abelian quasiparticle. First, the former may encounter an existing $\tilde{\sigma}$ at the interface. In this case, they recombine and transmute back to an Abelian quasiparticle [see Eqs. (21.11) and (21.13)]. This special scenario is shown in Fig. 21.5c. In order for this recombination to occur, it requires a highly delicate control of the shrinking process. Thus, it is unlikely to recombine all the existing $\tilde{\sigma}$ and σ (with charge sector skipped) in this way. On the other hand, it is likely that a non-Abelian anyon reaches the interface at a position with no neutral anyon $\tilde{\sigma}$. Being already outside the Pfaffian liquid, the non-Abelian anyon still needs to transmute into an Abelian quasiparticle. In this case, a pair of $\tilde{\sigma}$ needs to be created at the interface. One of them combines with the non-Abelian anyon to covert into an Abelian quasiparticle in the Halperin-331 liquid. The remaining one is left at the interface. This mechanism is shown in Fig. 21.5d. Since the additional pair of $\tilde{\sigma}$ are

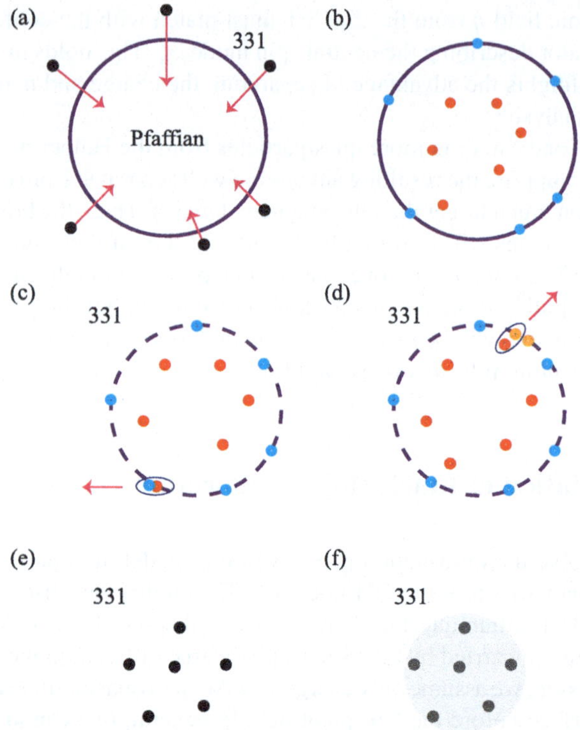

Fig. 21.5 Illustration of dragging quasiparticles in to the black hole (Pfaffian region) and simulating black hole evaporation in the 331-Pfaffian interface. Before the evaporation: **a** dragging Abelian charge-$e/4$ quasiparticles (black dots) in to the Pfaffian liquid; **b** creating neutral anyons $\tilde{\sigma}$ (blue dots) and non-Abelian charge-$e/4$ quasiparticles (red dots). Different mechanisms of releasing quasiparticles back to the 331 liquid during the evaporation: **c** combining a non-Abelian anyon with an existing $\tilde{\sigma}$ at the interface; **d** creating an additional pair of $\tilde{\sigma}$ (orange dots) and combining one of them with a non-Abelian anyon. After the evaporation: **e** the most idealistic scenario with the same number of quasiparticles as the initial configuration in (**a**); and **f** the generic situation with a superposition of different number of anyons in the Halperin-331 liquid. See the main text for more details. Figure adopted from Ref. [74]

created from the vacuum, they must have their fusion channel in the trivial topological sector. Hence, they do not carry additional information. All above processes are unitary, so information should be preserved.

21.5.1 Recovery of Page Curve

Now, we show that the above "black hole evaporation" recovers all original information and satisfies the Page curve. Here, the first subsystem consists of anyons remaining in the Pfaffian liquid and the interface. Another subsystem consists of Abelian quasiparticles in the Halperin-331 liquid. For simplicity, we call these two subsystems as (I) and (II), respectively. Since we treat the Pfaffian liquid as the black hole and the interface as an event horizon, reduced density matrices at different stages are obtained by partial tracing out (I). At the beginning of the evaporation, (II) is in a vacuum state with no quasiparticles, so its entropy is zero. When the Pfaffian liquid starts to shrink, the entropy of Abelian quasiparticles originating from their entanglement with (I) increases. However, the increase in entanglement entropy will not continue forever. By keep shrinking the Pfaffian region, the number of non-Abelian anyons and the dimension of the corresponding Hilbert space decrease. Hence, the dimensions of Hilbert spaces of (I) and (II) will become comparable and eventually equal to each other. The entanglement entropy reaches its maximum at this moment [25], which is known as the Page time. The Page time depends on the actual shrinking process. After passing the Page time, the entanglement entropy starts to decrease.

In the most idealistic (yet most unlikely) situation in which one can recombine all N non-Abelian anyons with the originally existing $\tilde{\sigma}$ (exist before the evaporation) at the interface, the Page time occurs when $N/2$ quasiparticles are released to the 331 region. This feature does not hold in a generic situation. One can actually deduce the average entanglement entropy of (II) in the most idealistic case. We assume the initial state of the total system (before shrinking the Pfaffian region) is a random pure state $|\Psi\rangle$ in the 2^N-dimensional Hilbert space. The entanglement entropy is averaged with respect to the unitary invariant Haar measure on the space of unitary vectors $|\Psi\rangle$ in the 2^N-dimensional Hilbert space [25]. Suppose j non-Abelian anyons have been dragged out from the Pfaffian region and transmuted back to Abelian quasiparticles in the Halperin-331 liquid. The corresponding Hilbert space dimensions of (I) and (II) are given by $n = 2^{N-j}$ and $m = 2^j$, respectively. When $m \leq n$, the conjecture by Page (later proved by Sen [174]) suggests that the average entanglement entropy of (II) takes the form [25],

$$\langle S_{(\mathrm{II})}\rangle \equiv S_{m,n} = \left(\sum_{k=n+1}^{mn} \frac{1}{k}\right) - \frac{m-1}{2n}. \tag{21.80}$$

For $m > n$, one obtains $\langle S_{(\mathrm{II})}\rangle$ by interchanging m and n in Eq. (21.80). By plotting $\langle S_{(\mathrm{II})}\rangle$ versus $\ln m$, one concludes that it is identical to the one in Fig. 1 of Ref. [26].

From our previous discussion, it is very likely that the number of $\tilde{\sigma}$ anyons increases during the "black hole evaporation". This leads to two consequences. First, it is possible that all non-Abelian anyons in the Pfaffian liquid have been released to the 331 liquid, but some $\tilde{\sigma}$ anyons still remain at the interface. These anyons are entangled with the Abelian anyons in the Halperin-331 region, so the entanglement entropy is still nonzero. Second, the bound $N_L \sim L/\ell_B$ can be satisfied easily during the shrinking process. The discussion in Sect. 21.3.3 showed that pairs of Abelian quasiparticles and quasiholes with opposite charges will be released to the 331 region. After eliminating the Pfaffian region completely, the total number of anyons in the

331 region needs not be equal to the number in the original configuration (before sending the anyons in the Pfaffian liquid). Only in the most idealistic situation that we mentioned previously, these two numbers are equal as shown in Fig. 21.5a, e. In general, the final state of the system will have a superposition of different total numbers of anyons in the 331 liquid. It is reasonable since the total charge in the system is still conserved. This idea is illustrated in Fig. 21.5f. In particular, Fig. 21.5f denotes a superposition state of $6 + M$ charge-$e/4$ quasiparticles and M charge-$-e/4$ quasi-holes, where M is a non-negative integer. This kind of superposition state is actually a closer analogue of the actual Hawking radiation emitted from a black hole, which consists of different types of particles or excitations.

Independent of the actual shrinking process, the system must return to a pure state when the Pfaffian region is eliminated completely. Then, the entanglement entropy goes back to zero and resembles the Page curve. The original pseudospin information is recovered but in a highly entangled form. Thus, the paradox in our model is resolved. Our above discussion suggests that the Page curve in the present system should be more complicated than the one in Fig. 1 of Ref. [26].

21.6 Summary and Discussion

To conclude, we have identified and resolved an "information paradox" in the 331-Pfaffian and the 330-RR_4 QH interface respectively [74, 75]. In both cases, the paradox originates from an apparent inability to recover the original pseudospin information carried by Abelian quasiparticles after they cross the interface and enter the non-Abelian single-component QH liquid.

Employing the technique of anyon condensation, we found that each incoming Abelian anyon needs to be transmuted into new types of anyons when the former crosses the interface. Moreover, the mechanism of transmutation and its corresponding scrambling of pseudospin information depend on the nature of QH interfaces. For the 331-Pfaffian interface, an incoming Abelian anyon is transmuted into a pair of non-Abelian anyons. One of them is created in the Pfaffian liquid, whereas the other is created on the interface. This is expected because the interface remains gapless and being described by a conformal field theory with central charge $c = 1/2$. On the other hand, the 330-RR_4 interface can be fully gapped. In this case, the incoming Abelian anyon is completely transmuted into a non-Abelian anyon in the Read-Rezayi QH liquid. Abelian anyons with opposite pseudospins would transmute into non-Abelian anyons satisfying different fusion channels.

Regardless of the transmutation mechanism, the original pseudospin information is stored nonlocally in the system. As a result it cannot be recovered by any local measurement. We believe this is a fair analogy to an object falling into a real black hole, in the sense that while the information it carries is not lost, it does become inaccessible to an (outside) observer. This resembles the idea of quantum information scrambling, which is consistent with the modern viewpoint that black holes are fast (perhaps the fastest) information scramblers [28, 175–178]. The matching between the dimensions of Hilbert spaces for the Abelian quasiparticle and non-Abelian anyons further verifies the preservation of information.

In addition, we considered the case when more quasiparticles are dragged across the interface. Focusing on the 331-Pfaffian interface, we argued that the maximum amount of information the system can store in a topologically protected way is bounded by the length of the interface. This feature is reminiscent of a similar bound in black hole set by its area due to the holographic principle and the Bekenstein entropy. Finally, we discussed the simulation of black hole evaporation by shrinking the Pfaffian region which releases quasiparticles back to the 331 liquid. We demonstrate explicitly that the corresponding entanglement entropy would follow the Page curve. Hence, the original pseudospin information is recovered and the "information paradox" in our model is resolved. Note that remnants may be left at the end of evaporation in actual astrophysical black holes [179–181]. This may provide an alternative resolution of the information paradox, which is not addressed in the present work.

It is quite surprising that the seemingly simple QH interface has a rich analogy with black hole physics. Nevertheless, we need to point out some potential differences between our model and real astrophysical black holes. For a (semi)classical black hole, the horizon is not expected to have an effect on an infalling object (the so-called "no drama scenario"), including the information carried by it. Whether this remains to be the case or not for a fully quantum-mechanical black hole is unclear. A firewall at the horizon is a possible scenario that is currently under investigation and debate [21]. In our model, the Abelian quasiparticles must be transmuted when they cross the interface. This is inevitable as the Abelian and non-Abelian QH liquids support different degrees of freedom. Thus, the interface in our model behaves like a firewall. In our opinion, this interface may be a very simple and accessible "black hole firewall", which deserves more attention. In future work, it will be tempting to examine possible analogy of black hole thermodynamics in QH interfaces. It is also interesting to examine whether QH interfaces can provide an easy simulation of (a topological version of) the Hayden-Preskill protocol.

The black-hole information paradox is arguably one of the most fundamental problems in physics, which involves gravitation, quantum field theory, and in particular, quantum information science. This long-standing problem is currently being actively studied by physicists in many different areas, and from very different perspectives (but so far only theoretically). Its resolution may well pave the way for the quantum theory of gravity, the holy grail of theoretical physics. While there is a lack of complete parallel between our model and certain "believed" processes in actual black holes (especially in the description of black hole evaporation which should be spontaneous), the analogy presented here provides a simple and accessible platform to simulate (i) apparent information loss, (ii) information scrambling, and (iii) information recovery. We believe these are arguably the most important and central concepts in understanding and resolving the original information paradox. Furthermore, our work may open a new research direction of studying how local information can be transmuted and stored nonlocally in an actual black hole. Since the concept of firewall and many other aspects in the paradox are still under intense debate, it is worthwhile to have simple analogies that capture some of the relevant concepts (but not necessarily all details precisely) in the original problem. In addition, our results have established a connection between quantum information, black hole physics and QH physics, and may bring experimentalists into this exciting research area.

Last but not least, it is worthwhile to mention that a deep connection between QH effect and gravitational physics has been revealed in previous work [182–191], leading to the recent experimental discovery of chiral graviton [192, 193]. In particular, Refs. [189–191] have suggested a possible simulation of Hawking-Unruh effect by scattering quasiparticles in QH systems. We are optimistic that more connections between black hole physics and QH physics may be discovered in the future.

Acknowledgements We thank N. E. Bonesteel for a useful discussion in our original work. This work was supported by the National Science Foundation Grants No. DMR-1932796 and DMR-2315954, and performed at the National High Magnetic Field Laboratory which is supported by National Science Foundation Cooperative Agreement No. DMR-2128556, and the State of Florida.

References

1. B.P. Abbott, et al. (LIGO Scientific Collaboration and Virgo Collaboration), Observation of gravitational waves from a binary black hole merger. Phys. Rev. Lett. **116**, 061102 (2016). https://journals.aps.org/prl/abstract/10.1103/PhysRevLett.116.061102
2. The Event Horizon Telescope Collaboration et al., First M87 event horizon telescope results. I. The shadow of the supermassive black hole. Astrophys. J. Lett. **875**, L1 (2019). https://iopscience.iop.org/article/10.3847/2041-8213/ab0ec7
3. J. Paynter, R. Webster, E. Thrane, Evidence for an intermediate-mass black hole from a gravitationally lensed gamma-ray burst. Nat. Astron. **5**, 560 (2021). https://www.nature.com/articles/s41550-021-01307-1
4. S.W. Hawking, Black hole explosions? Nature **248**, 30 (1974). https://www.nature.com/articles/248030a0
5. S.W. Hawking, Particle creation by black holes. Commun. Math. Phys **43**, 199 (1975). https://link.springer.com/article/10.1007%2FBF02345020
6. S.W. Hawking, Breakdown of predictability in gravitational collapse. Phys. Rev. D **14**, 2460 (1976). https://journals.aps.org/prd/abstract/10.1103/PhysRevD.14.2460
7. Notice that Hawking and his collaborators have recently introduced the idea of soft hair [194]
8. W. Israel, Event horizons in static vacuum space-times. Phys. Rev. **164**, 1776 (1967). https://journals.aps.org/pr/abstract/10.1103/PhysRev.164.1776
9. W. Israel, Event horizons in static electrovac space-times. Commun. Math. Phys. **8**, 245 (1968). https://link.springer.com/article/10.1007/BF01645859
10. B. Carter, Axisymmetric black hole has only two degrees of freedom. Phys. Rev. Lett. **26**, 331 (1971). https://journals.aps.org/prl/abstract/10.1103/PhysRevLett.26.331
11. S.B. Giddings, The black hole information paradox. https://arxiv.org/abs/hep-th/9508151. arXiv: hep-th/9508151
12. S.D. Mathur, The information paradox: a pedagogical introduction. Class. Quantum Gravity **26**, 224001 (2009). https://iopscience.iop.org/article/10.1088/0264-9381/26/22/224001
13. K. Lochan, S. Chakraborty, T. Padmanabhan, Information retrieval from black holes. Phys. Rev. D **94**, 044056 (2016). https://journals.aps.org/prd/abstract/10.1103/PhysRevD.94.044056

14. S. Chakraborty, K. Lochan, Black holes: eliminating information or illuminating new physics? Universe **3**, 55 (2017). https://www.mdpi.com/2218-1997/3/3/55
15. D. Marolf, The black hole information problem: past, present, and future. Rep. Prog. Phys. **80**, 092001 (2017). https://iopscience.iop.org/article/10.1088/1361-6633/aa77cc
16. G. 't Hooft, Dimensional reduction in quantum gravity. https://arxiv.org/abs/gr-qc/9310026. arXiv:gr-qc/9310026
17. L. Susskind, The world as a hologram. J. Math. Phys. **36**, 6377 (1995). https://aip.scitation.org/doi/10.1063/1.531249
18. R. Bousso, The holographic principle. Rev. Mod. Phys. **74**, 825 (2002). https://journals.aps.org/rmp/abstract/10.1103/RevModPhys.74.825
19. J.L.F. Barbón, Black holes, information and holography. J. Phys.: Conf. Ser. **171**, 012009 (2009). https://iopscience.iop.org/article/10.1088/1742-6596/171/1/012009
20. J.M. Maldacena, The large-N limit of superconformal field theories and supergravity. Int. J. Theor. Phys. **38**, 1113 (1999). https://link.springer.com/article/10.1023%2FA%3A1026654312961
21. A. Almheiri, D. Marolf, J. Polchinski, J. Sully, Black holes: complementarity or firewalls? J. High Energy Phys. **2013**, 62 (2013). https://link.springer.com/article/10.1007%2FJHEP02%282013%29062
22. L. Susskind, L. Thorlacius, J. Uglum, The stretched horizon and black hole complementarity. Phys. Rev. D **48**, 3743 (1993). https://journals.aps.org/prd/abstract/10.1103/PhysRevD.48.3743
23. C.R. Stephens, G. 't Hooft, B.F. Whiting, Black hole evaporation without information loss. Class. Quantum Gravity **11**, 621 (1994). https://iopscience.iop.org/article/10.1088/0264-9381/11/3/014
24. V. Coffman, J. Kundu, W.K. Wootters, Distributed entanglement. Phys. Rev. A **61**, 052306 (2000). https://journals.aps.org/pra/abstract/10.1103/PhysRevA.61.052306
25. D.N. Page, Average entropy of a subsystem. Phys. Rev. Lett. **71**, 1291 (1993). https://journals.aps.org/prl/abstract/10.1103/PhysRevLett.71.1291
26. D.N. Page, Information in black hole radiation. Phys. Rev. Lett. **71**, 3743 (1993). https://journals.aps.org/prl/abstract/10.1103/PhysRevLett.71.3743
27. D.N. Page, Black hole information. https://arxiv.org/abs/hep-th/9305040. arXiv:hep-th/9305040
28. P. Hayden, J. Preskill, Black holes as mirrors: quantum information in random subsystems. J. High Energy Phys. **2007**, 120 (2007). https://iopscience.iop.org/article/10.1088/1126-6708/2007/09/120
29. B. Yoshida, A. Kitaev, Efficient decoding for the Hayden-Preskill protocol. https://arxiv.org/abs/1710.03363. arXiv:1710.03363
30. Y. Sekino, L. Susskind, Fast scramblers. J. High Energy Phys. **2008**, 065 (2008). https://iopscience.iop.org/article/10.1088/1126-6708/2008/10/065
31. L. Susskind, Addendum to fast scramblers. https://arxiv.org/abs/1101.6048. arXiv:1101.6048
32. N. Lashkari, D. Stanford, M. Hastings, T. Osborne, P. Hayden, Towards the fast scrambling conjecture. J. High Energy Phys. **2013**, 22 (2013). https://link.springer.com/article/10.1007/JHEP04(2013)022
33. S.H. Shenker, D. Stanford, Black holes and the butterfly effect. J. High Energy Phys. **2014**, 67 (2014). https://link.springer.com/article/10.1007/JHEP03(2014)067
34. S.H. Shenker, D. Stanford, Stringy effects in scrambling. J. High Energy Phys. **2015**, 132 (2015). https://link.springer.com/article/10.1007/JHEP05(2015)132
35. J. Maldacena, S.H. Shenker, D. Stanford, A bound on chaos. J. High Energy Phys. **2016**, 106 (2016). https://link.springer.com/article/10.1007/JHEP08(2016)106
36. P. Hosur, X.-L. Qi, D.A. Roberts, B. Yoshida, Chaos in quantum channels. J. High Energy Phys. **2016**, 4 (2016). https://link.springer.com/article/10.1007/JHEP02(2016)004
37. D.A. Roberts, B. Yoshida, Chaos and complexity by design. J. High Energy Phys. **2017**, 121 (2017). https://link.springer.com/article/10.1007/JHEP04(2017)121

38. E. Iyoda, T. Sagawa, Scrambling of quantum information in quantum many-body systems. Phys. Rev. A **97**, 042330 (2018). https://journals.aps.org/pra/abstract/10.1103/PhysRevA.97.042330
39. B. Yoshida, N.Y. Yao, Disentangling scrambling and decoherence via quantum teleportation. Phys. Rev. X **9**, 011006 (2019). https://journals.aps.org/prx/abstract/10.1103/PhysRevX.9.011006
40. H. Shen, P. Zhang, Y.-Z. You, H. Zhai, Information scrambling in quantum neural networks. Phys. Rev. Lett. **124**, 200504 (2020). https://journals.aps.org/prl/abstract/10.1103/PhysRevLett.124.200504
41. Y. Wu, P. Zhang, H. Zhai, Scrambling ability of quantum neural network architectures. Phys. Rev. Res. **3**, L032057 (2021). https://journals.aps.org/prresearch/abstract/10.1103/PhysRevResearch.3.L032057
42. R.J. Garcia, K. Bu, A. Jaffe, Quantifying scrambling in quantum neural networks. J. High Energy Phys. **2022**, 27 (2022). https://link.springer.com/article/10.1007/JHEP03(2022)027
43. K.A. Landsman, C. Figgatt, T. Schuster, N.M. Linke, B. Yoshida, N.Y. Yao, C. Monroe, Verified quantum information scrambling. Nature **567**, 61 (2019). https://www.nature.com/articles/s41586-019-0952-6
44. Q. Zhu, Z.-H. Sun, M. Gong, F. Chen, Y.-R. Zhang, Y. Wu, Y. Ye, C. Zha, S. Li, S. Guo, H. Qian, H.-L. Huang, J. Yu, H. Deng, H. Rong, J. Lin, Y. Xu, L. Sun, C. Guo, N. Li, F. Liang, C.-Z. Peng, H. Fan, X. Zhu, J.-W. Pan, Observation of thermalization and information scrambling in a superconducting quantum processor. Phys. Rev. Lett. **128**, 160502 (2022). https://journals.aps.org/prl/abstract/10.1103/PhysRevLett.128.160502
45. G. Penington, Entanglement wedge reconstruction and the information paradox. https://arxiv.org/abs/1905.08255. arXiv:1905.08255
46. A. Almheiri, N. Engelhardt, D. Marolf, H. Maxfield, The entropy of bulk quantum fields and the entanglement wedge of an evaporating black hole. J. High Energy Phys. **2019**, 63 (2019). https://link.springer.com/article/10.1007/JHEP12(2019)063
47. G. Penington, S.H. Shenker, D. Stanford, Z. Yang, Replica wormholes and the black hole interior. https://arxiv.org/abs/1911.11977. arXiv:1911.11977
48. C. Akers, N. Engelhardt, D. Harlow, Simple holographic models of black hole evaporation. J. High Energy Phys. **2020**, 32 (2020). https://link.springer.com/article/10.1007%2FJHEP08%282020%29032
49. A. Almheiri, T. Hartman, J. Maldacena, E. Shaghoulian, A. Tajdini, Replica wormholes and the entropy of Hawking radiation J. High Energy Phys. **2020**, 13 (2020). https://link.springer.com/article/10.1007%2FJHEP05%282020%29013
50. A. Almheiri, T. Hartman, J. Maldacena, E. Shaghoulian, A. Tajdini, The entropy of Hawking radiation. Rev. Mod. Phys. **93**, 35002 (2021). https://journals.aps.org/rmp/abstract/10.1103/RevModPhys.93.035002
51. S. Raju, Lessons from the information paradox. https://arxiv.org/abs/2012.05770. arXiv:2012:05770
52. X.G. Wen, *Quantum Field Theory of Many-Body Systems: From the Origin of Sound to an Origin of Light and Electrons* (Oxford University Press, Oxford, 2004)
53. M. Heiblum, D.E. Feldman, *Fractional Quantum Hall Effects: New Developments*, ed. by B.I. Halperin, J.K. Jain (World Scientific, Singapore, 2020). https://www.worldscientific.com/doi/abs/10.1142/S0217751X20300094; Int. J. Mod. Phys. A **35**, 2030009 (2020)
54. E. Grosfeld, K. Schoutens, Non-Abelian anyons: when Ising meets Fibonacci. Phys. Rev. Lett. **103**, 076803 (2009). https://journals.aps.org/prl/abstract/10.1103/PhysRevLett.103.076803
55. F.A. Bais, J.K. Slingerland, S.M. Haaker, Theory of topological edges and domain walls. Phys. Rev. Lett. **102**, 220403 (2009). https://journals.aps.org/prl/abstract/10.1103/PhysRevLett.102.220403
56. X. Wan, K. Yang, Striped quantum Hall state in a half-filled Landau level. Phys. Rev. B **93**, 201303(R) (2016). https://journals.aps.org/prb/abstract/10.1103/PhysRevB.93.201303
57. K. Yang, Interface and phase transition between Moore-Read and Halperin 331 fractional quantum Hall states: realization of chiral Majorana fermion. Phys. Rev. B **96**, 241305(R) (2017). https://journals.aps.org/prb/abstract/10.1103/PhysRevB.96.241305

58. D.F. Mross, Y. Oreg, A. Stern, G. Margalit, M. Heiblum, Theory of disorder-induced half-integer thermal Hall conductance. Phys. Rev. Lett. **121**, 026801 (2018). https://journals.aps.org/prl/abstract/10.1103/PhysRevLett.121.026801
59. C. Wang, A. Vishwanath, B.I. Halperin, Topological order from disorder and the quantized Hall thermal metal: possible applications to the $\nu = 5/2$ state. Phys. Rev. B **98**, 045112 (2018). https://journals.aps.org/prb/abstract/10.1103/PhysRevB.98.045112
60. B. Lian, J. Wang, Theory of the disordered $\nu = 5/2$ quantum thermal Hall state: emergent symmetry and phase diagram. Phys. Rev. B **97**, 165124 (2018). https://journals.aps.org/prb/abstract/10.1103/PhysRevB.97.165124
61. S.H. Simon, M. Ippoliti, M.P. Zaletel, E.H. Rezayi, Energetics of Pfaffian-anti-Pfaffian domains. Phys. Rev. B **101**, 041302(R) (2020). https://journals.aps.org/prb/abstract/10.1103/PhysRevB.101.041302
62. W. Zhu, D.N. Sheng, K. Yang, Topological interface between Pfaffian and anti-Pfaffian order in $\nu = 5/2$ quantum Hall effect. Phys. Rev. Lett. **125**, 146802 (2020). https://journals.aps.org/prl/abstract/10.1103/PhysRevLett.125.146802
63. J. May-Mann, T.L. Hughes, Families of gapped interfaces between fractional quantum Hall states. Phys. Rev. B **99**, 155134 (2019). https://journals.aps.org/prb/abstract/10.1103/PhysRevB.99.155134
64. V. Crépel, N. Claussen, N. Regnault, B. Estienne, Microscopic study of the Halperin-Laughlin interface through matrix product states. Nat. Commun. **10**, 1860 (2019). https://www.nature.com/articles/s41467-019-09169-y
65. V. Crépel, N. Claussen, B. Estienne, N. Regnault, Model states for a class of chiral topological order interfaces. Nat. Commun. **10**, 1861 (2019). https://www.nature.com/articles/s41467-019-09168-z
66. B. Jaworowski, A.E.B. Nielsen, Model wave functions for interfaces between lattice Laughlin states. Phys. Rev. B **101**, 245164 (2020). https://journals.aps.org/prb/abstract/10.1103/PhysRevB.101.245164
67. B. Jaworowski, A.E.B. Nielsen, Model wave functions for an interface between lattice Laughlin and Moore-Read States. Phys. Rev. B **103**, 205149 (2021). https://journals.aps.org/prb/abstract/10.1103/PhysRevB.103.205149
68. R. Sohal, B. Han, L.H. Santos, J.C.Y. Teo, Entanglement entropy of generalized moore-read fractional quantum Hall state interfaces. Phys. Rev. B **102**, 045102 (2020). https://journals.aps.org/prb/abstract/10.1103/PhysRevB.102.045102
69. B. Dutta, W. Yang, R.A. Melcer, H.K. Kundu, M. Heiblum, V. Umansky, Y. Oreg, A. Stern, D. Mross, Distinguishing between non-Abelian topological orders in a quantum Hall system. Science **375**, 193–197 (2022). https://www.science.org/doi/10.1126/science.abg6116
70. M. Yutushui, A. Stern, D.F. Mross, Identifying the $\nu = 5/2$ topological order through charge transport measurements. Phys. Rev. Lett. **128**, 016401 (2022). https://journals.aps.org/prl/abstract/10.1103/PhysRevLett.128.016401
71. S. Manna, A. Das, M. Goldstein, Y. Gefen, Full classification of transport on an equilibrated 5/2 edge via shot noise. Phys. Rev. Lett. **132**, 136502 (2024). https://journals.aps.org/prl/abstract/10.1103/PhysRevLett.132.136502
72. Q. Li, K.K.W. Ma, R. Wang, Z.-X. Hu, H. Wang, K. Yang, Dynamics of quantum Hall interfaces. Phys. Rev. B **104**, 125303 (2021). https://journals.aps.org/prb/abstract/10.1103/PhysRevB.104.125303
73. O. Türker, K. Yang, String-like theory of quantum Hall interfaces. Phys. Rev. B **106**, 245138 (2022). https://journals.aps.org/prb/abstract/10.1103/PhysRevB.106.245138
74. K.K.W. Ma, K. Yang, Simple analogue of the black-hole information paradox in quantum Hall interfaces. Phys. Rev. B **105**, 045306 (2022). https://journals.aps.org/prb/abstract/10.1103/PhysRevB.105.045306
75. K.K.W. Ma, Anyon condensation, topological quantum information scrambling, and Andreev-like reflection of non-Abelian anyons in quantum Hall interfaces. https://arxiv.org/abs/2209.11119 arXiv:2209.11119
76. B.I. Halperin, Theory of the quantized Hall conductance. Helv. Phys. Acta **56**, 75 (1983)

77. G. Moore, N. Read, Nonabelions in the fractional quantum Hall effect. Nucl. Phys. B **360**, 362 (1991). https://www.sciencedirect.com/science/article/abs/pii/055032139190407O
78. C. Nayak, S.H. Simon, A. Stern, M. Freedman, S. Das Sarma, Non-Abelian anyons and topological quantum computation. Rev. Mod. Phys. **80**, 1083 (2008). https://journals.aps.org/rmp/abstract/10.1103/RevModPhys.80.1083
79. Y.W. Suen, L.W. Engel, M.B. Santos, M. Shayegan, D.C. Tsui, Observation of a $\nu = 1/2$ fractional quantum Hall state in a double-layer electron system. Phys. Rev. Lett. **68**, 1379 (1992). https://journals.aps.org/prl/abstract/10.1103/PhysRevLett.68.1379
80. J.P. Eisenstein, G.S. Boebinger, L.N. Pfeiffer, K.W. West, S. He, New fractional quantum Hall state in double-layer two-dimensional electron systems. Phys. Rev. Lett. **68**, 1383 (1992). https://journals.aps.org/prl/abstract/10.1103/PhysRevLett.68.1383
81. Z. Papić, M.O. Goerbig, N. Regnault, M.V. Milovanović, Tunneling-driven breakdown of the 331 state and the emergent Pfaffian and composite Fermi liquid phases. Phys. Rev. B **82**, 075302 (2010). https://journals.aps.org/prb/abstract/10.1103/PhysRevB.82.075302
82. W. Zhu, Zhao Liu, F.D.M. Haldane, D.N. Sheng, Fractional quantum Hall bilayers at half filling: tunneling-driven non-Abelian phase. Phys. Rev. B **94**, 245147 (2016). https://journals.aps.org/prb/abstract/10.1103/PhysRevB.94.245147
83. Y.W. Suen, H.C. Manoharan, X. Ying, M.B. Santos, M. Shayegan, Origin of the $\nu = 1/2$ fractional quantum Hall state in wide single quantum wells. Phys. Rev. Lett. **72**, 3405 (1994). https://journals.aps.org/prl/abstract/10.1103/PhysRevLett.72.3405
84. N. Read, E. Rezayi, Beyond paired quantum Hall states: parafermions and incompressible states in the first excited Landau level. Phys. Rev. B **59**, 8084 (1999). https://journals.aps.org/prb/abstract/10.1103/PhysRevB.59.8084
85. E. Rezayi, X.-G. Wen, N. Read, Condensation of factional excitons, non-Abelian states in double-layer quantum Hall systems and Z_4 parafermions https://arxiv.org/abs/1007.2022. arXiv:1007.2022
86. M. Barkeshli, X.-G. Wen, $U(1) \times U(1) \rtimes Z_2$ Chern-Simons theory and Z_4 parafermion fractional quantum Hall states. Phys. Rev. B **81**, 045323 (2010). https://journals.aps.org/prb/abstract/10.1103/PhysRevB.81.045323
87. M. Barkeshli, X.-G. Wen, Bilayer quantum Hall phase transitions and the orbifold non-Abelian fractional quantum Hall states. Phys. Rev. B **84**, 115121 (2011). https://journals.aps.org/prb/abstract/10.1103/PhysRevB.84.115121
88. M. Barkeshli, X.-G. Wen, Anyon condensation and continuous topological phase transitions in non-Abelian fractional quantum Hall states. Phys. Rev. Lett. **105**, 216804 (2010). https://journals.aps.org/prl/abstract/10.1103/PhysRevLett.105.216804
89. For a general introduction to quantum Hall effect, readers may refer to S.M. Girvin, K. Yang, *Modern Condensed Matter Physics* (Cambridge University Press, Cambridge, 2019). In particular, Chaps. 12 and 16
90. D.E. Feldman, B.I. Halperin, Fractional charge and fractional statistics in the quantum Hall effects. Rep. Prog. Phys. **84**, 076501 (2021). https://iopscience.iop.org/article/10.1088/1361-6633/ac03aa
91. D.C. Tsui, H.L. Stormer, A.C. Gossard, Two-dimensional magnetotransport in the extreme quantum limit. Phys. Rev. Lett. **48**, 1559 (1982). https://journals.aps.org/prl/abstract/10.1103/PhysRevLett.48.1559
92. R.B. Laughlin, Anomalous quantum Hall effect: an incompressible quantum fluid with fractionally charged excitations. Phys. Rev. Lett. **50**, 1395 (1983). https://journals.aps.org/prl/abstract/10.1103/PhysRevLett.50.1395
93. B.I. Halperin, Statistics of quasiparticles and the hierarchy of fractional quantized Hall states. Phys. Rev. Lett. **52**, 1583 (1984). https://journals.aps.org/prl/abstract/10.1103/PhysRevLett.52.1583
94. D. Arovas, J.R. Schrieffer, F. Wilczek, Fractional statistics and the quantum Hall effect. Phys. Rev. Lett. **53**, 722 (1984). https://journals.aps.org/prl/abstract/10.1103/PhysRevLett.53.722
95. R. de Picciotto, M. Reznikov, M. Heiblum, V. Umansky, G. Bunin, D. Mahalu, Direct observation of a fractional charge. Nature **389**, 162, (1997). https://www.nature.com/articles/38241

96. L. Saminadayar, D.C. Glattli, Y. Jin, B. Etienne, Observation of the $e/3$ fractionally charged Laughlin quasiparticle. Phys. Rev. Lett. **79**, 2526 (1997). https://journals.aps.org/prl/abstract/10.1103/PhysRevLett.79.2526
97. J. Nakamura, S. Liang, G.C. Gardner, M.J. Manfra, Direct observation of anyonic braiding statistics. Nat. Phys. **16**, 931 (2020). https://www.nature.com/articles/s41567-020-1019-1
98. H. Bartolomei, M. Kumar, R. Bisognin, A. Marguerite, J.-M. Berroir, E. Bocquillon, B. Plaçais, A. Cavanna, Q. Dong, U. Gennser, Y. Jin, Fractional statistics in anyon collisions. Science **368**, 6487 (2020). https://science.sciencemag.org/content/368/6487/173
99. T.H. Hansson, M. Hermanns, S.H. Simon, S.F. Viefers, Quantum Hall physics: hierarchies and conformal field theory techniques. Rev. Mod. Phys. **89**, 025005 (2017). https://journals.aps.org/rmp/abstract/10.1103/RevModPhys.89.025005
100. M. Milovanović, N. Read, Edge excitations of paired fractional quantum Hall states. Phys. Rev. B **53**, 13559 (1996). https://journals.aps.org/prb/abstract/10.1103/PhysRevB.53.13559
101. M. Levin, B.I. Halperin, B. Rosenow, Particle-hole symmetry and the Pfaffian state. Phys. Rev. Lett. **99**, 236806 (2007). https://journals.aps.org/prl/abstract/10.1103/PhysRevLett.99.236806
102. S.-S. Lee, S. Ryu, C. Nayak, M.P.A. Fisher, Particle-hole symmetry and the $\nu = 5/2$ quantum Hall state. Phys. Rev. Lett. **99**, 236807 (2007). https://journals.aps.org/prl/abstract/10.1103/PhysRevLett.99.236807
103. P.T. Zucker, D.E. Feldman, Stabilization of the particle-hole Pfaffian order by Landau-level mixing and impurities that break particle-hole symmetry. Phys. Rev. Lett. **117**, 096802 (2016). https://journals.aps.org/prl/abstract/10.1103/PhysRevLett.117.096802
104. N. Read, D. Green, Paired states of fermions in two dimensions with breaking of parity and time-reversal symmetries and the fractional quantum Hall effect. Phys. Rev. B **61**, 10267 (2000). https://journals.aps.org/prb/abstract/10.1103/PhysRevB.61.10267
105. C. Nayak, F. Wilzcek, $2n$-quasihole states realize 2^{n-1}-dimensional spinor braiding statistics in paired quantum Hall states. Nucl. Phy. B **479**, 529 (1996). https://www.sciencedirect.com/science/article/pii/0550321396004300
106. P. Di Francesco, P. Mathieu, D. Senechal, *Conformal Field Theory* (Springer, New York, 1997)
107. F.A. Bais, J.K. Slingerland, Condensate-induced transitions between topologically ordered phases. Phys. Rev. B **79**, 045316 (2009). https://journals.aps.org/prb/abstract/10.1103/PhysRevB.79.045316
108. F.J. Burnell, Anyon condensation and its applications. Annu. Rev. Condens. Matter Phys. **9**, 307 (2017). https://www.annualreviews.org/doi/10.1146/annurev-conmatphys-033117-054154
109. I.S. Eliëns, J.C. Romers, F.A. Bais, Diagrammatics for Bose condensation in anyon theories. Phys. Rev. B **90**, 195130 (2014). https://journals.aps.org/prb/abstract/10.1103/PhysRevB.90.195130
110. T. Neupert, H. He, C. von Keyserlingk, G. Sierra, B.A. Bernevig, Boson condensation in topologically ordered quantum liquids. Phys. Rev. B **93**, 115103 (2016). https://journals.aps.org/prb/abstract/10.1103/PhysRevB.93.115103
111. The gapping of modes is realized when the tunneling amplitude in H_T is a constant instead of a random variable in Eq. (21.7). Strictly speaking, this kind of nonrandom tunneling is unphysical as there is a momentum mismatch between different modes. However, gapped modes and localized modes discussed in Sect. 21.3 have the same physical consequences in the present discussion
112. In our opinion, this is a big advantage over the Pfaffian-NASS interface [54, 55]. In that case, the NASS quantum Hall liquid at $\nu = 4/7$ has quasiparticle with the smallest charge of $e/7$. Hence, a net charge of $e/4 - e/7 = 3e/28$ is absorbed, which leads to an accumulation of charges at the interface. Furthermore, the difference between filling factors of the two quantum Hall liquids may complicate the gapping of counterpropagating charge modes at the interface. This difference also introduces additional challenges in realizing the interface experimentally

113. T. Månsson, V. Lahtinen, J. Suorsa, E. Ardonne, Condensate-induced transitions and critical spin chains. Phys. Rev. B **88**, 041403(R) (2013). https://journals.aps.org/prb/abstract/10.1103/PhysRevB.88.041403
114. V. Lahtinen, T. Månsson, E. Ardonne, Quantum criticality in many-body parafermion chains. SciPost Phys. Core **4**, 014 (2021). https://scipost.org/SciPostPhysCore.4.2.014
115. R. Dijkgraaf, C. Vafa, E. Verlinde, H. Verlinde, The operator algebra of orbifold models. Comm. Math. Phys. **123**, 485 (1989). https://link.springer.com/article/10.1007/BF01238812
116. Although Fig. 21.1 may imply the interface is a one-dimensional object, it is not. The interface actually penetrates in both quantum Hall liquids and has a finite width (but much shorter than its length), similar to the interface between Pfaffian and anti-Pfaffian liquids [62]. Therefore, braiding of anyons can occur at the interface
117. Z.-C. Yang, K. Meichanetzidis, S. Kourtis, C. Chamon, Scrambling via braiding of non-Abelions. Phys. Rev. B **99**, 045132 (2019). https://journals.aps.org/prb/abstract/10.1103/PhysRevB.99.045132
118. In the Pfaffian-NASS interface, one may consider a similar "information paradox" by dragging a charge-$e/7$ non-Abelian quasiparticle from the NASS liquid to the Pfaffian liquid. Depending on the species, the neutral sector is given by $\sigma_\uparrow$ or $\sigma_\downarrow$ in the $c = 6/5$ CFT. Both transmute into $(\bar{\sigma}, \sigma)$, where $\bar{\sigma}$ is a primary field in the $\mathcal{M}(5, 4)$ minimal model describing the interface [54, 55]. Since $\bar{\sigma}$ has a quantum dimension $d = (1 + \sqrt{5})/\sqrt{2} > \sqrt{2}$, the Hilbert space of anyons at the interface has a larger dimension
119. J. Preskill, *Lecture Notes for Physics 219: Quantum Computation—Chapter 9. Topological Quantum Computation*
120. M.H. Freedman, A. Kitaev, M.J. Larsen, Z. Wang, Topological quantum computation. Bull. Am. Math. Soc. **40**, 31 (2003). https://www.ams.org/journals/bull/2003-40-01/S0273-0979-02-00964-3/home.html
121. It is reminded that one cannot perform universal TQC (or reproduce any unitary operation to arbitrary accuracy by braiding of anyons) by using anyons in the Ising CFT [151, 195]. More complicated anyons such as Fibonacci anyons are required for achieving the universality [196]
122. A.Y. Kitaev, Fault-tolerant quantum computation by anyons. Ann. Phys. **303**, 2 (2003). https://www.sciencedirect.com/science/article/abs/pii/S0003491602000180
123. J.D. Bekenstein, Black holes and the second law. Lett. Nuovo Cimento **4**, 737 (1972). https://link.springer.com/article/10.1007%2FBF02757029
124. J.D. Bekenstein, Black holes and entropy. Phys. Rev. D **7**, 2333 (1973). https://journals.aps.org/prd/abstract/10.1103/PhysRevD.7.2333
125. J.D. Bekenstein, Generalized second law of thermodynamics in black-hole physics. Phys. Rev. D **9**, 3292 (1974). https://journals.aps.org/prd/abstract/10.1103/PhysRevD.9.3292
126. X.G. Wen, A. Zee, Classification of Abelian quantum Hall states and matrix formulation of topological fluids. Phys. Rev. B **46**, 2290 (1992). https://journals.aps.org/prb/abstract/10.1103/PhysRevB.46.2290
127. A possible generalization of the K-matrix formalism to non-Abelian orders was discussed in T. Lan, Matrix formulation for non-Abelian families. Phys. Rev. B **100**, 241102(R) (2019). https://journals.aps.org/prb/abstract/10.1103/PhysRevB.100.241102
128. A. Kapustin, N. Saulina, Topological boundary conditions in Abelian Chern-Simons theory. Nucl. Phys. B **845**, 393 (2011). https://www.sciencedirect.com/science/article/pii/S0550321310006723
129. M. Levin, Protected edge modes without symmetry. Phys. Rev. X **3**, 021009 (2013). https://journals.aps.org/prx/abstract/10.1103/PhysRevX.3.021009
130. M. Barkeshli, C.-M. Jian, X.-L. Qi, Theory of defects in Abelian topological states. Phys. Rev. B **88**, 235103 (2013). https://journals.aps.org/prb/abstract/10.1103/PhysRevB.88.235103
131. M. Barkeshli, C.-M. Jian, X.-L. Qi, Classification of topological defects in Abelian topological states. Phys. Rev. B **88**, 241103(R) (2013). https://journals.aps.org/prb/abstract/10.1103/PhysRevB.88.241103
132. A. Kapustin, Ground-state degeneracy for Abelian anyons in the presence of gapped boundaries. Phys. Rev. B **89**, 125307 (2014). https://journals.aps.org/prb/abstract/10.1103/PhysRevB.89.125307

133. J.C. Wang, X.-G. Wen, Boundary degeneracy of topological order. Phys. Rev. B **91**, 125124 (2015). https://journals.aps.org/prb/abstract/10.1103/PhysRevB.91.125124
134. F.D.M. Haldane, Stability of Chiral Luttinger liquids and Abelian quantum Hall states. Phys. Rev. Lett. **74**, 2090 (1995). https://journals.aps.org/prl/abstract/10.1103/PhysRevLett.74.2090
135. J. Fuchs, C. Schweigert, A. Valentino, Bicategories for boundary conditions and for surface defects in 3-d TFT. Commun. Math. Phys. **321**, 543 (2013). https://link.springer.com/article/10.1007/s00220-013-1723-0
136. L.-Y. Hung, Y. Wan, Symmetry-enriched phases obtained via pseudo anyon condensation. Int. J. Mod. Phys. B **28**, 1450172 (2014). https://www.worldscientific.com/doi/abs/10.1142/S0217979214501720
137. L. Kong, Anyon condensation and tensor categories. Nucl. Phys. B **886**, 436 (2014). https://www.sciencedirect.com/science/article/pii/S0550321314002223
138. L.-Y. Hung, Y. Wan, Ground-state degeneracy of topological phases on open surfaces. Phys. Rev. Lett. **114**, 076401 (2015). https://journals.aps.org/prl/abstract/10.1103/PhysRevLett.114.076401
139. Y. Wan, C. Wang, Fermion condensation and gapped domain walls in topological orders. J. High Energ. Phys. **2017**, 172 (2017). https://doi.org/10.1007/JHEP03(2017)172
140. T. Lan, J.C. Wang, X.-G. Wen, Gapped domain walls, gapped boundaries, and topological degeneracy. Phys. Rev. Lett. **114**, 076402 (2015). https://journals.aps.org/prl/abstract/10.1103/PhysRevLett.114.076402
141. T. Lan, X. Wen, L. Kong, X.-G. Wen, Gapped domain walls between 2+1D topologically ordered states. Phys. Rev. Res. **2**, 023331 (2020). https://journals.aps.org/prresearch/abstract/10.1103/PhysRevResearch.2.023331
142. M. Barkeshli, X.-G. Wen, Phase transitions in Z_N gauge theory and twisted Z_N topological phases. Phys. Rev. B **86**, 085114 (2012). https://journals.aps.org/prb/abstract/10.1103/PhysRevB.86.085114
143. R.B. Laughlin, Anomalous quantum Hall effect: an incompressible quantum fluid with fractionally charged excitations. Phys. Rev. Lett. **50**, 1395 (1983). https://journals.aps.org/prl/abstract/10.1103/PhysRevLett.50.1395
144. A.B. Zamolodchikov, V.A. Fateev, Nonlocal (parafermion) currents in two-dimensional conformal quantum field theory and self-dual critical points in Z_N-symmetric statistical systems. Zh. Eksp. Teor. Fiz. **89**, 380 (1985)
145. D. Gepner, Z. Qiu, Modular invariant partition functions for parafermionic field theories. Nucl. Phys. B **285**, 423 (1987). https://www.sciencedirect.com/science/article/pii/0550321387903488
146. J.K. Slingerland, F.A. Bais, Quantum groups and non-Abelian braiding in quantum Hall systems. Nucl. Phys. B **612**, 229 (2001). https://www.sciencedirect.com/science/article/abs/pii/S055032130100308X?
147. For a recent review, see K.K.W. Ma, M.R. Peterson, V.W. Scarola, K. Yang, Fractional quantum Hall effect at the filling factor $\nu = 5/2$, in *Encyclopedia of Condensed Matter Physics*, 2nd edn. (2023). https://www.sciencedirect.com/science/article/abs/pii/B9780323908009001359
148. M.H. Freedman, M.J. Larsen, Z. Wang, A modular functor which is universal for quantum computation. Commun. Math. Phys. **227**, 605 (2002). https://link.springer.com/article/10.1007/s002200200645
149. N.E. Bonesteel, L. Hormozi, G. Zikos, S.H. Simon, Braid topologies for quantum computation. Phys. Rev. Lett. **95**, 140503 (2005). https://journals.aps.org/prl/abstract/10.1103/PhysRevLett.95.140503
150. L. Hormozi, G. Zikos, N.E. Bonesteel, S.H. Simon, Topological quantum compiling. Phys. Rev. B **75**, 165310 (2007). https://journals.aps.org/prb/abstract/10.1103/PhysRevB.75.165310
151. A. Ahlbrecht, L.S. Georgiev, R.F. Werner, Implementation of Clifford gates in the Ising-anyon topological quantum computer. Phys. Rev. A **79**, 032311 (2009). https://journals.aps.org/pra/abstract/10.1103/PhysRevA.79.032311

152. J.K. Pachos, *Introduction to Topological Quantum Computation* (Cambridge University Press, Cambridge, 2012)
153. A. Hutter, D. Loss, Quantum computing with parafermions. Phys. Rev. B **93**, 125105 (2016). https://journals.aps.org/prb/abstract/10.1103/PhysRevB.93.125105
154. V. Lahtinen, J.K. Pachos, A short introduction to topological quantum computation. SciPost Phys. **3**, 021 (2017). https://scipost.org/10.21468/SciPostPhys.3.3.021
155. W. Bishara, G.A. Fiete, C. Nayak, Quantum Hall states at $\nu = 2/(k+2)$: analysis of the particle-hole conjugates of the general level-k Read-Rezayi states. Phys. Rev. B **77**, 241306(R) (2008). https://link.aps.org/doi/10.1103/PhysRevB.77.241306
156. J. Yu, X.-H. Zhang, S.-P. Kou, Majorana edge states for $\mathbb{Z}_2$ topological orders of the Wen plaquette and toric code models. Phys. Rev. B **87**, 184402 (2013). https://journals.aps.org/prb/abstract/10.1103/PhysRevB.87.184402
157. P.M. Tam, J.W.F. Venderbos, C.L. Kane, Toric-code insulator enriched by translation symmetry. Phys. Rev. B **105**, 045106 (2022). https://journals.aps.org/prb/abstract/10.1103/PhysRevB.105.045106
158. S.B. Bravyi, A.Yu. Kitaev, Quantum codes on a lattice with boundary. https://arxiv.org/abs/quant-ph/9811052. arXiv:quant-ph/9811052
159. A. Kitaev, L. Kong, Models for gapped boundaries and domain walls. Commun. Math. Phys. **313**, 351 (2012). https://doi.org/10.1007/s00220-012-1500-5
160. T. Neupert, H. He, C. von Keyserlingk, G. Sierra, B.A. Bernevig, No-go theorem for Boson condensation in topologically ordered quantum liquids. New J. Phys. **18**, 123009 (2016). https://iopscience.iop.org/article/10.1088/1367-2630/18/12/123009
161. S.H. Simon, J.K. Slingerland, Straightening out the Frobenius-Schur indicator. https://arxiv.org/abs/2208.14500. arXiv:2208.14500
162. J. Cano, M. Cheng, M. Mulligan, C. Nayak, E. Plamadeala, J. Yard, Bulk-edge correspondence in $(2+1)$-dimensional Abelian topological phases. Phys. Rev. B **89**, 115116 (2014). https://journals.aps.org/prb/abstract/10.1103/PhysRevB.89.115116
163. M. Cheng, Fermionic Lieb-Schultz-Mattis theorems and weak symmetry-protected phases. Phys. Rev. B **99**, 075143 (2019). https://journals.aps.org/prb/abstract/10.1103/PhysRevB.99.075143
164. V. Drinfeld, Quasi-Hopf algebras. Algebra i Analiz **1**, 6 (1989); English translation in Leningrad Math. J. **1**, 1419 (1989)
165. M.A. Levin, X.-G. Wen, String-net condensation: a physical mechanism for topological phases. Phys. Rev. B **71**, 045110 (2005). https://journals.aps.org/prb/abstract/10.1103/PhysRevB.71.045110
166. I. Cong, M. Cheng, Z. Wang, Topological quantum computation with gapped boundaries. https://arxiv.org/abs/1609.02037. arXiv:1609.02037
167. I. Cong, M. Cheng, Z. Wang, Universal quantum computation with gapped boundaries. Phys. Rev. Lett. **119**, 170504 (2017). https://journals.aps.org/prl/abstract/10.1103/PhysRevLett.119.170504
168. I. Cong, M. Cheng, Z. Wang, Hamiltonian and algebraic theories of gapped boundaries in topological phases of matter. Commun. Math. Phys. **355**, 645 (2017). https://doi.org/10.1007/s00220-017-2960-4
169. I. Cong, M. Cheng, Z. Wang, Defects between gapped boundaries in two-dimensional topological phases of matter. Phys. Rev. B **96**, 195129 (2017). https://journals.aps.org/prb/abstract/10.1103/PhysRevB.96.195129
170. M. Barkeshli, Charge $2e/3$ superconductivity and topological degeneracies without localized zero modes in bilayer fractional quantum Hall states. Phys. Rev. Lett. **117**, 096803 (2016). https://journals.aps.org/prl/abstract/10.1103/PhysRevLett.117.096803
171. P.M. Tam, C.L. Kane, Nondiagonal anisotropic quantum Hall states. Phys. Rev. B **103**, 035142 (2021). https://journals.aps.org/prb/abstract/10.1103/PhysRevB.103.035142
172. In the general case, the maximal set of condensable bosons may not be closed under fusion due to the splitting of non-Abelian anyons. In this case, a more appropriate terminology should be Lagrangian subset [68]. Meanwhile, $\mathscr{L}_e$ and $\mathscr{L}_m$ are indeed closed under fusion in our present case

173. Since the tunneling process is random, the renormalization group (RG) equation reads $dW/d(\ln s) = (3 - 2\Delta)W$. Here, s is the scaling factor of time and space, and $\overline{\xi(x)\xi(x')} = W\delta(x - x')$. For the single electron tunneling process (i.e., $a = 1$), the scaling dimension for the operator involving the neutral modes $\psi_1 \exp(i\phi_n)$ is already $h_n = 3/4 + 3/4 = 3/2$. The scaling dimension of the operator involving charge modes $\exp[i(\phi_l + \phi_r)/2]$ is non-universal but must be positive. Therefore, the process is irrelevant in the RG sense. On the other hand, the corresponding RG equation for nonrandom tunneling is $dW'/d(\ln s) = (2 - \Delta)W'$, where W' stands for the tunneling strength. Depending on the interaction between $\phi_r(x)$ and $\phi_l(x)$, the single electron tunneling process can be relevant in the RG sense
174. S. Sen, Average entropy of a quantum subsystem. Phys. Rev. Lett. **77**, 1 (1996). https://journals.aps.org/prl/abstract/10.1103/PhysRevLett.77.1
175. Y. Sekino, L. Susskind, Fast scramblers. J. High Energy Phys. **2008**, 65 (2008). https://iopscience.iop.org/article/10.1088/1126-6708/2008/10/065
176. L. Susskind, Addendum to fast scramblers. https://arxiv.org/abs/1101.6048. arXiv:1101.6048
177. J.L.F. Barbón, J.M. Magán, Chaotic fast scrambling at black holes. Phys. Rev. D **84**, 106012 (2011). https://journals.aps.org/prd/abstract/10.1103/PhysRevD.84.106012
178. N. Lashkari, D. Stanford, M. Hastings, T. Osborne, P. Hayden, Towards the fast scrambling conjecture. J. High Energy Phys. **2013**, 22 (2013). https://link.springer.com/article/10.1007%2FJHEP04%282013%29022
179. R.J. Adler, P. Chen, D.I. Santiago, The generalized uncertainty principle and black hole remnants. Gen. Rel. Grav. **33**, 2101 (2001). https://link.springer.com/article/10.1023/A:1015281430411
180. Y. Aharonov, A. Casher, S. Nussinov, The unitarity puzzle and Planck mass stable particles. Phys. Lett. B **191**, 51 (1987). https://www.sciencedirect.com/science/article/pii/0370269387913207
181. P. Chen, Y.C. Ong, D. Yeom, Black hole remnants and the information loss paradox. Phys. Rep. **603**, 1 (2015). https://www.sciencedirect.com/science/article/pii/S0370157315004391
182. B. Yang, Z.-X. Hu, Z. Papić, F.D.M. Haldane, Model wave functions for the collective modes and the magnetoroton theory of the fractional quantum Hall effect. Phys. Rev. Lett. **108**, 256807 (2012). https://journals.aps.org/prl/abstract/10.1103/PhysRevLett.108.256807
183. S. Golkar, D.X. Nguyen, D.T. Son, Spectral sum rules and magneto-roton as emergent graviton in fractional quantum Hall effect. JHEP **01**, 21 (2016). https://link.springer.com/article/10.1007/JHEP01(2016)021
184. K. Yang, Acoustic wave absorption as a probe of dynamical geometrical response of fractional quantum Hall liquids. Phys. Rev. B **93**, 161302(R) (2016). https://journals.aps.org/prb/abstract/10.1103/PhysRevB.93.161302
185. S.-F. Liou, F.D.M. Haldane, K. Yang, E.H. Rezayi, Chiral gravitons in fractional quantum Hall liquids. Phys. Rev. Lett. **123**, 146801 (2019). https://journals.aps.org/prl/abstract/10.1103/PhysRevLett.123.146801
186. D.X. Nguyen, D.T. Son, Probing the spin structure of the fractional quantum Hall magnetoroton with polarized Raman scattering. Phys. Rev. Res. **3**, 023040 (2021). https://journals.aps.org/prresearch/abstract/10.1103/PhysRevResearch.3.023040
187. Z. Liu, A.C. Balram, Z. Papić, A. Gromov, Quench dynamics of collective modes in fractional quantum Hall bilayers. Phys. Rev. Lett. **126**, 076604 (2021). https://journals.aps.org/prl/abstract/10.1103/PhysRevLett.126.076604
188. F.D.M. Haldane, E.H. Rezayi, K. Yang, Graviton chirality and topological order in the half-filled Landau level. Phys. Rev. B **104**, L121106 (2021). https://journals.aps.org/prb/abstract/10.1103/PhysRevB.104.L121106
189. M. Stone, An analogue of Hawking radiation in the quantum Hall effect. Class. Quantum Gravity **30**, 085003 (2013). https://iopscience.iop.org/article/10.1088/0264-9381/30/8/085003
190. S.S. Hegde, V. Subramanyan, B. Bradlyn, S. Vishveshwara, Quasinormal modes and the Hawking-Unruh effect in quantum Hall systems: lessons from black hole phenomena. Phys. Rev. Lett. **123**, 156802 (2019). https://journals.aps.org/prl/abstract/10.1103/PhysRevLett.123.156802

191. V. Subramanyan, S.S. Hegde, S. Vishveshwara, B. Bradlyn, Physics of the inverted harmonic oscillator: from the lowest Landau level to event horizons. Ann. Phys. **435**, 168470 (2021). https://www.sciencedirect.com/science/article/abs/pii/S0003491621000762
192. J. Liang, Z. Liu, Z. Yang, Y. Huang, U. Wurstbauer, C.R. Dean, K.W. West, L.N. Pfeiffer, L. Du, A. Pinczuk, Evidence of chiral graviton modes in fractional quantum Hall liquids. Nature **628**, 78 (2024). https://www.nature.com/articles/s41586-024-07201-w
193. K. Yang, Graviton-like excitation observed with predicted chirality in fractional quantum Hall liquids. The Innovation **5**, 100641 (2024). https://www.cell.com/the-innovation/fulltext/S2666-6758(24)00079-1?rss=yes
194. S.W. Hawking, M.J. Perry, A. Strominger, Soft hair on black holes. Phys. Rev. Lett. **116**, 231301 (2016)
195. V. Lahtinen, J.K. Pachos, A short introduction to topological quantum computation. SciPost Phys. **3**, 021 (2017). https://scipost.org/10.21468/SciPostPhys.3.3.021
196. N.E. Bonesteel, L. Hormozi, G. Zikos, S.H. Simon, Braid topologies for quantum computation. Phys. Rev. Lett. **95**, 140503 (2005). https://journals.aps.org/prl/abstract/10.1103/PhysRevLett.95.140503
197. In principle, the spin vector s is another required topological quantum number in characterizing the topological order [198]. We will ignore s and focus on the K matrix and t vector, which is sufficient for the present discussion
198. X.G. Wen, A. Zee, Shift and spin vector: new topological quantum numbers for the Hall fluids. Phys. Rev. Lett. **69**, 953 (1992). https://journals.aps.org/prl/abstract/10.1103/PhysRevLett.69.953

The manufacturer's authorised representative in the EU is Springer Nature Customer Service Centre GmbH, Europaplatz 3, 69115 Heidelberg, Germany. If you have any concerns regarding our products, please contact ProductSafety@springernature.com

Printed and bound by CPI Group (UK) Ltd, Croydon, CR0 4YY
30/09/2026
02225300-0019